COST – the acronym for European COoperation in Science and Technology – is the oldest and widest European intergovernmental network for cooperation in research. Established by the Ministerial Conference in November 1971, COST is presently used by the scientific communities of 36 European countries to cooperate in common research projects supported by national funds.

The funds provided by COST – less than 1% of the total value of the projects – support the COST cooperation networks (COST Actions) through which, with EUR 30 million per year, more than 30.000 European scientists are involved in research having a total value which exceeds EUR 2 billion per year. This is the financial worth of the European added value which COST achieves.

A "bottom up approach" (the initiative of launching a COST Action comes from the European scientists themselves), "à la carte participation" (only countries interested in the Action participate), "equality of access" (participation is open also to the scientific communities of countries not belonging to the European Union) and "flexible structure" (easy implementation and light management of the research initiatives) are the main characteristics of COST.

As precursor of advanced multidisciplinary research COST has a very important role for the realisation of the European Research Area (ERA) anticipating and complementing the activities of the Framework Programmes, constituting a "bridge" towards the scientific communities of emerging countries, increasing the mobility of researchers across Europe and fostering the establishment of "Networks of Excellence" in many key scientific domains such as: Biomedicine and Molecular Biosciences; Food and Agriculture; Forests, their Products and Services; Materials, Physical and Nanosciences; Chemistry and Molecular Sciences and Technologies; Earth System Science and Environmental Management; Information and Communication Technologies; Transport and Urban Development; Individuals, Societies, Cultures and Health. It covers basic and more applied research and also addresses issues of pre-normative nature or of societal importance.

Web: `http://www.cost.eu`

ESF Provides the COST Office through an EC contract

COST is supported by the EU RTD Framework programme

Thomas Hermann, Andy Hunt, John G. Neuhoff (Eds.)

The Sonification Handbook

λογος

Thomas Hermann
Ambient Intelligence Group
CITEC, Bielefeld University
Universitätsstraße 21-23
33615 Bielefeld, Germany

> Thomas Hermann, Andy Hunt, John G. Neuhoff (Eds.). *The Sonification Handbook.* Logos Verlag, Berlin, Germany, 2011.

The SID logo Sonic Interaction Design SI has been designed by Frauke Behrendt.

The book cover, including the cover artwork has been designed by Thomas Hermann.

The word cloud on the back cover was rendered with Wordle (http://www.wordle.net)

The Deutsche Nationalbibliothek lists this publication in the Deutsche Nationalbibliografie; detailed bibliographic data are available in the Internet at http://dnb.d-nb.de.

ISBN 978-3-8325-2819-5

Logos Verlag Berlin GmbH
Comeniushof, Gubener Str. 47,
10243 Berlin, Germany
Tel.: +49 (0)30 42 85 10 90
Fax: +49 (0)30 42 85 10 92
INTERNET: http://www.logos-verlag.de

Preface

This book offers a comprehensive introduction to the field of Sonification and Auditory Display. Sonification is so inherently interdisciplinary that it is easy to become disoriented and overwhelmed when confronted with its many different facets, ranging from computer science to psychology, from sound design to data mining. In addition, each discipline uses its own jargon, and–because the research comes from such diverse areas–there are few agreed upon definitions for the complex concepts within the research area.

With *The Sonification Handbook* we have organized topics roughly along the following progression: perception - data - sound synthesis - sonification techniques - central application areas. While the chapters are written in the spirit of reviewing, organizing and teaching relevant material, they will hopefully also surprise, encourage, and inspire to new uses of sound. We hope that this book will support all sorts of readers, from students to experts, from HCI practitioners to domain-experts, those that seek to dive quickly or more thoroughly into Sonification, to see whether it may be useful for their application area. Due to their thematic richness the chapters can best be seen as providing mutually complementary views on a multi-disciplinary and broad emerging field. We hope that together they will help readers to better understand the whole field by looking at it from different disciplinary angles.

We decided to publish this book as an OpenAccess book because auditory display is still a small but growing community, and the easy access and sharing of information and ideas is of high importance. Free availability of publication and material lowers the barrier to enter the field and also matches the spirit of the ICAD community.

An online portal at `http://sonification.de/handbook` provides digital versions, supplementary material such as sound examples, videos and further descriptions.

The publication has been made possible and supported by the EU COST Action IC0601 "Sonic Interaction Design" (SID). In addition to providing publication costs, the COST Action SID supported the book with author involvement and expertise, in the reviewing of chapters, sharing forces with the strong involvement in authoring and reviewing from ICAD. We take this opportunity to thank all authors and reviewers and all who contributed to make this book possible.

There are few books available that introduce these topics. A well established and respected source is *Auditory Display*, edited by Gregory Kramer in 1994. This book hopes to set the next stepping stone, and we are happy that Greg relates these two books together in a Foreword to “The Sonification Handbook”.

Bielefeld, York, Wooster

Thomas Hermann, Andy Hunt, John G. Neuhoff
September, 2011

Foreword

The book you're holding, or perhaps reading on a screen, represents a sea change: the maturation of the field of Auditory Display (AD). It represents the aggregate work of a global community of inquiry as well as the labors of its individual editors and authors. Nineteen years ago–in 1992 and 1993–I was editing another book, one that would be published in 1994 as part of the Santa Fe Institute's Studies in the Sciences of Complexity, *Auditory Display: Sonification, Audification, and Auditory Interfaces*. Although there had certainly been research papers that pre-dated it, this 1994 publication seemed to have the effect of catalyzing the field of auditory display research.

Up until the seminal 1992 conference–little more than a workshop with an outsized title, International Conference on Auditory Display–only scattered attention had been given to auditory interfaces generally, and nearly none to using sound as a means of conveying data. When I edited the conference proceedings into a book (with the feature, unusual for its time, of being sold with an audio CD included), and wrote an introduction that I hoped would provide some context and orienting theory for the field, the threshold of significance was modest. The vision, the fact of these unique papers, and a little weaving of them into a coherent whole was enough.

That is no longer the case. Nearly twenty years have passed since ICAD 92. A new generation of researchers has earned Ph.D.'s: researchers whose dissertation research has been in this field, advised by longtime participants in the global ICAD community. Technologies that support AD have matured. AD has been integrated into significant (read "funded" and "respectable") research initiatives. Some forward thinking universities and research centers have established ongoing AD programs. And the great need to involve the entire human perceptual system in understanding complex data, monitoring processes, and providing effective interfaces has persisted and increased. The book that was needed twenty years ago is not the book needed now.

The Sonification Handbook fills the need for a new reference and workbook for the field, and does so with strength and elegance. I've watched as Thomas, Andy, and John have shepherded this project for several years. The job they had is very different from the one I had, but by no means easier. Finding strong contributions in 1990 often meant hunting, then cajoling, then arduous editing to make the individual papers clear and the whole project effective and coherent. Now, the field has many good people in it, and they can find each other easily (at the beginning of the 1990's, the Web was still a "wow, look at that" experiment). With the bar so much higher, these editors have set high standards of quality and have helped authors who face the same time famine as everyone else to bring their chapters to fruition. Some of the papers included in the 1994 book were excellent; some were essentially conference papers, sketches of some possibility, because that's what was available at the time. That book was both a reference source and documentation of the rise of a new field. Now there is a foundation of solid work to draw from and a body of literature to cite. In consequence, the present book is more fully and truly a reference handbook.

Just as compelling, there is a clear need for this book. When a field is first being defined, who's to say that there is any need for that field–let alone for a book proffering both a body of work and the theoretical underpinnings for it. The current need includes the obvious demand for an updated, central reference source for the field. There is also a need for a book from which to teach, as well as a book to help one enter a field that is still fabulously interdisciplinary. And there is need for a volume that states the case for some of the pioneering work such as sonification and audification of complex data, advanced alarms, and non-traditional auditory interfaces. That we still call this work "pioneering" after twenty or thirty years of effort remains a notion worth investigating.

At ICAD conferences, and in many of the labs where AD research is undertaken, you'll still find a community in process of defining itself. Is this human interface design, broadly speaking? Is it computer science? Psychology? Engineering? Even music? Old questions, but this multi-disciplinary field still faces them. And there are other now-classic challenges: when it comes to understanding data, vision still reigns as king. That the ears have vast advantages in contributing to understanding much temporally demanding or highly multi-dimension data has not yet turned the tide of funding in a significant way. There are commercial margins, too, with efforts progressing more in interfaces for the blind and less in the fields of medicine, financial data monitoring or analysis, and process control, long targets of experimental auditory displays. The cultural bias to view visually displayed data as more objective and trustworthy than what can be heard remains firmly established. Techniques to share and navigate data using sound will only become accepted gradually.

Perhaps the community of researchers that finds commonality and support at the ICAD conferences, as well as at other meetings involved with sound, such as ISon, Audio Mostly, and HAID, will have some contributions to make to understanding the human experience that are just now ready to blossom. New research shows that music activates a broad array of systems in the brain–a fact which, perhaps, contributes to its ubiquity and compelling force in all the world's cultures. Might this hold a key to what is possible in well designed auditory displays? Likewise, advances in neuroscience point to complex interactions among auditory, visual, and haptic-tactile processing, suggesting that the omission from a design process of any sensory system will mean that the information and meanings derived, and the affective engagement invoked, will be decreased; everything from realism to user satisfaction, from dimensionality to ease of use, will suffer unacceptably.

I've been asked many times, "Where are things going in this field?" I have no idea! And that's the beauty of it. Yes, AD suffers the curse of engaging so many other research areas that it struggles to find research funding, a departmental home in academia, and a clear sense of its own boundaries. The breadth that challenges also enriches. Every advance in auditory perception, sound and music computing, media technology, human interface design, and cognition opens up new possibilities in AD research. Where is it all leading? In this endeavor, we all must trust the emergent process.

When I began to put together the first ICAD conference in 1990, it took me a couple of years of following leads to find people currently doing, or recently involved in, any work in the field whatsoever. From the meager list I'd assembled, I then had to virtually beg people to attend the gathering, as if coming to Santa Fe, New Mexico, in the sunny, bright November of 1992 was insufficient motivation. In the end, thirty-six of us were there. Now, about 20 years later, a vibrant young field has emerged, with a global community of inquiry. *The Sonification Handbook* is a major step in this field's maturation and will serve to unify, advance, and challenge the scientific community in important ways. It is impressive that its authors and editors have sacrificed the "brownie point" path of publishing for maximum academic career leverage, electing instead to publish this book as OpenAccess, freely available to anybody. It is an acknowledgement of this research community's commitment to freely share information, enthusiasm, and ideas, while maintaining innovation, clarity, and scientific value. I trust that this book will be useful for students and newcomers to the field, and will serve those of us who have been deeply immersed in auditory displays all these years. It is certainly a rich resource. And yet–it's always just beginning. *The Sonification Handbook* contributes needed traction for this journey.

Orcas Island, Washington

Gregory Kramer
August, 2011

List of authors

Stephen Barrass	University of Canberra, Canberra, Australia
Jonathan Berger	Stanford University, Stanford, California, United States
Terri L. Bonebright	DePauw University, Greencastle, Indiana, United States
Till Bovermann	Aalto University, Helsinki, Finland
Eoin Brazil	Irish Centre for High-End Computing, Dublin, Ireland
Stephen Brewster	University of Glasgow, Glasgow, United Kingdom
Densil Cabrera	The University Of Sydney, Sydney, Australia
Simon Carlile	The University Of Sydney, Sydney, Australia
Perry Cook	Princeton University (Emeritus), Princeton, United States
Alberto de Campo	University for the Arts Berlin, Berlin, Germany
Florian Dombois	Zurich University of the Arts, Zurich, Switzerland
Gerhard Eckel	University of Music and Performing Arts Graz, Graz, Austria
Alistair D. N. Edwards	University of York, York, United Kingdom
Alfred Effenberg	Leibniz University Hannover, Hannover, Germany
Mikael Fernström	University of Limerick, Limerick, Ireland
Sam Ferguson	University Of New South Wales, Sydney, Australia
John Flowers	University of Nebsaska, Lincoln, Nebraska, United States
Karmen Franinovic	Zurich University of the Arts, Zurich, Switzerland
Florian Grond	Bielefeld University, Bielefeld, Germany
Anne Guillaume	Laboratoire d'accidentologie et de biomécanique, Nanterre, France
Thomas Hermann	Bielefeld University, Bielefeld, Germany
Oliver Höner	University of Tübingen, Tübingen, Germany
Andy Hunt	University of York, York, United Kingdom
Gregory Kramer	Metta Foundation, Orcas, Washington, United States
Guillaume Lemaitre	Carnegie Mellon University, Pittsburgh, Pennsylvania, United States
David McGookin	University of Glasgow, Glasgow, United Kingdom
Michael Nees	Lafayette College, Easton, Pennsylvania, United States
John G. Neuhoff	The College of Wooster, Wooster, Ohio, United States
Sandra Pauletto	University of York, York, United Kingdom
Michal Rinott	Holon Institute of Technology, Holon, Israel
Niklas Röber	University of Magdeburg, Magdeburg, Germany
Davide Rocchesso	IUAV University Venice, Venice, Italy
Julian Rohrhuber	Robert Schumann Hochschule, Düsseldorf, Germany
Stefania Serafin	Aalborg University Copenhagen, Aalborg, Denmark
Paul Vickers	Northumbria University, Newcastle-upon-Tyne, United Kingdom
Bruce N. Walker	Georgia Institute of Technology, Atlanta, Georgia, United States

Contents

1 Introduction **1**
Thomas Hermann, Andy Hunt, John G. Neuhoff
1.1 Auditory Display and Sonification 1
1.2 The Potential of Sonification and Auditory Display 3
1.3 Structure of the book 4
1.4 How to Read 6

I Fundamentals of Sonification, Sound and Perception **7**

2 Theory of Sonification **9**
Bruce N. Walker and Michael A. Nees
2.1 Chapter Overview 9
2.2 Sonification and Auditory Displays 10
2.3 Towards a Taxonomy of Auditory Display & Sonification 11
2.4 Data Properties and Task Dependency 17
2.5 Representation and Mappings 22
2.6 Limiting Factors for Sonification: Aesthetics, Individual Differences, and Training 27
2.7 Conclusions: Toward a Cohesive Theoretical Account of Sonification . . . 31

3 Psychoacoustics **41**
Simon Carlile
3.1 Introduction 41
3.2 The transduction of mechanical sound energy into biological signals in the auditory nervous system 42
3.3 The perception of loudness 46
3.4 The perception of pitch 48
3.5 The perception of temporal variation 49
3.6 Grouping spectral components into auditory objects and streams 51
3.7 The perception of space 52
3.8 Summary 59
3.9 Further reading 59

4 Perception, Cognition and Action in Auditory Displays **63**
John G. Neuhoff
4.1 Introduction 63
4.2 Perceiving Auditory Dimensions 64
4.3 Auditory-Visual Interaction 71
4.4 Auditory Space and Virtual Environments 71
4.5 Space as a Dimension for Data Representation 73

4.6 Rhythm and Time as Dimensions for Auditory Display 73
4.7 Auditory Scene Analysis 75
4.8 Auditory Cognition 77
4.9 Summary 81

5 Sonic Interaction Design **87**
Stefania Serafin, Karmen Franinović, Thomas Hermann,
Guillaume Lemaitre, Michal Rinott, Davide Rocchesso
5.1 Introduction 87
5.2 A psychological perspective on sonic interaction 88
5.3 Product sound design 94
5.4 Interactive art and music 99
5.5 Sonification and Sonic Interaction Design 103
5.6 Open challenges in SID 106

6 Evaluation of Auditory Display **111**
Terri L. Bonebright and John H. Flowers
6.1 Chapter Overview 111
6.2 General Experimental Procedures 112
6.3 Data Collection Methods for Evaluating Perceptual Qualities and Relationships among Auditory Stimuli 120
6.4 Analysis of Data Obtained from Identification, Attribute Rating, Discrimination, and Dissimilarity Rating Tasks 126
6.5 Using "Distance" Data Obtained by Dissimilarity Ratings, Sorting, and Other Tasks 130
6.6 Usability Testing Issues and Active Use Experimental Procedures 137
6.7 Conclusion 141

7 Sonification Design and Aesthetics **145**
Stephen Barrass and Paul Vickers
7.1 Background 146
7.2 Design 148
7.3 Aesthetics: sensuous perception 154
7.4 Towards an aesthetic of sonification 161
7.5 Where do we go from here? 164

II Sonification Technology **173**

8 Statistical Sonification for Exploratory Data Analysis **175**
Sam Ferguson, William Martens and Densil Cabrera
8.1 Introduction 175
8.2 Datasets and Data Analysis Methods 178
8.3 Sonifications of Iris Dataset 186
8.4 Discussion 192
8.5 Conclusion and Caveat 193

9 Sound Synthesis for Auditory Display 197

Perry R. Cook

9.1 Introduction and Chapter Overview . . . 197
9.2 Parametric vs. Non-Parametric Models . . . 197
9.3 Digital Audio: The Basics of PCM . . . 198
9.4 Fourier (Sinusoidal) “Synthesis” . . . 204
9.5 Modal (Damped Sinusoidal) Synthesis . . . 209
9.6 Subtractive (Source-Filter) synthesis . . . 213
9.7 Time Domain Formant Synthesis . . . 218
9.8 Waveshaping and FM Synthesis . . . 219
9.9 Granular and PhISEM Synthesis . . . 221
9.10 Physical Modeling Synthesis . . . 223
9.11 Non-Linear Physical Models . . . 229
9.12 Synthesis for Auditory Display, Conclusion . . . 232

10 Laboratory Methods for Experimental Sonification 237

Till Bovermann, Julian Rohrhuber and Alberto de Campo

10.1 Programming as an interface between theory and laboratory practice . . . 238
10.2 Overview of languages and systems . . . 240
10.3 SuperCollider: Building blocks for a sonification laboratory . . . 243
10.4 Example laboratory workflows and guidelines for working on sonification designs . . . 251
10.5 Coda: back to the drawing board . . . 270

11 Interactive Sonification 273

Andy Hunt and Thomas Hermann

11.1 Chapter Overview . . . 273
11.2 What is Interactive Sonification? . . . 273
11.3 Principles of Human Interaction . . . 276
11.4 Musical instruments – a 100,000 year case study . . . 280
11.5 A brief History of Human Computer Interaction . . . 283
11.6 Interacting with Sonification . . . 286
11.7 Guidelines & Research Agenda for Interactive Sonification . . . 293
11.8 Conclusions . . . 296

III Sonification Techniques 299

12 Audification 301

Florian Dombois and Gerhard Eckel

12.1 Introduction . . . 301
12.2 Brief Historical Overview (before ICAD, 1800-1991) . . . 303
12.3 Methods of Audification . . . 307
12.4 Audification now (1992-today) . . . 316
12.5 Conclusion: What audification should be used for . . . 319
12.6 Towards Better Audification Tools . . . 320

13 Auditory Icons **325**
Eoin Brazil and Mikael Fernström
13.1 Auditory icons and the ecological approach . . . 325
13.2 Auditory icons and events . . . 326
13.3 Applications using auditory icons . . . 327
13.4 Designing auditory icons . . . 331
13.5 Conclusion . . . 335

14 Earcons **339**
David McGookin and Stephen Brewster
14.1 Introduction . . . 339
14.2 Initial Earcon Research . . . 340
14.3 Creating Earcons . . . 343
14.4 Earcons and Auditory Icons . . . 349
14.5 Using Earcons . . . 352
14.6 Future Directions . . . 357
14.7 Conclusions . . . 358

15 Parameter Mapping Sonification **363**
Florian Grond, Jonathan Berger
15.1 Introduction . . . 363
15.2 Data Features . . . 365
15.3 Connecting Data and Sound . . . 367
15.4 Mapping Topology . . . 369
15.5 Signal and Sound . . . 371
15.6 Listening, Thinking, Tuning . . . 373
15.7 Integrating Perception in PMSon . . . 374
15.8 Auditory graphs . . . 376
15.9 Vowel / Formant based PMSon . . . 378
15.10 Features of PMSon . . . 380
15.11 Design Challenges of PMSon . . . 385
15.12 Synthesis and signal processing methods used in PMSon . . . 388
15.13 Artistic applications of PMSon . . . 390
15.14 Conclusion . . . 392

16 Model-Based Sonification **399**
Thomas Hermann
16.1 Introduction . . . 399
16.2 Definition of Model-Based Sonification . . . 403
16.3 Sonification Models . . . 408
16.4 MBS Use and Design Guidelines . . . 415
16.5 Interaction in Model-Based Sonification . . . 418
16.6 Applications . . . 419
16.7 Discussion . . . 421
16.8 Conclusion . . . 425

IV Applications 429

17 Auditory Display in Assistive Technology 431
Alistair D. N. Edwards

17.1 Introduction . . . 431
17.2 The Power of Sound . . . 432
17.3 Visually Disabled People . . . 433
17.4 Computer Access . . . 434
17.5 Electronic Travel Aids . . . 437
17.6 Other Systems . . . 446
17.7 Discussion . . . 449
17.8 Conclusion . . . 450

18 Sonification for Process Monitoring 455
Paul Vickers

18.1 Types of monitoring — basic categories . . . 455
18.2 Modes of Listening . . . 457
18.3 Environmental awareness (workspaces and living spaces) . . . 459
18.4 Monitoring program execution . . . 462
18.5 Monitoring interface tasks . . . 469
18.6 Potential pitfalls . . . 473
18.7 The road ahead . . . 479

19 Intelligent auditory alarms 493
Anne Guillaume

19.1 Introduction . . . 493
19.2 The concept of auditory alarms . . . 494
19.3 Problems linked to non-speech auditory alarm design . . . 495
19.4 Acoustic properties of non-speech sound alarms . . . 496
19.5 A cognitive approach to the problem . . . 498
19.6 Spatialization of alarms . . . 500
19.7 Contribution of learning . . . 501
19.8 Ergonomic approach to the problem . . . 503
19.9 Intelligent alarm systems . . . 504
19.10 Conclusion . . . 505

20 Navigation of Data 509
Eoin Brazil and Mikael Fernström

20.1 Navigation Control Loop . . . 510
20.2 Wayfinding . . . 510
20.3 Methods For Navigating Through Data . . . 511
20.4 Using Auditory Displays For Navigation Of Data . . . 515
20.5 Considerations for the Design of Auditory Displays for the Navigation of Data 521

21 Aiding Movement with Sonification in "Exercise, Play and Sport" 525
Edited by Oliver Höner

21.1 Multidisciplinary Applications of Sonification in the Field of "Exercise, Play and Sport" . . . 525

21.2 Use of Sound for Physiotherapy Analysis and Feedback 528
21.3 Interaction with Sound in auditory computer games 532
21.4 Sonification-based Sport games and Performance Tests in Adapted Physical Activity 538
21.5 Enhancing Motor Control and Learning by Additional Movement Sonification 547
21.6 Concluding Remarks 551

Index **555**

Chapter 1

Introduction

Thomas Hermann, Andy Hunt, John G. Neuhoff

1.1 Auditory Display and Sonification

Imagine listening to changes in global temperature over the last thousand years. What does a brain wave sound like? How can sound be used to facilitate the performance of a pilot in the cockpit? These questions and many more are the domain of Auditory Display and Sonification. Auditory Display researchers examine how the human auditory system can be used as the primary interface channel for communicating and transmitting information. The goal of Auditory Display is to enable a better understanding, or an appreciation, of changes and structures in the data that underlie the display. *Auditory Display* encompasses all aspects of a human-machine interaction system, including the setup, speakers or headphones, modes of interaction with the display system, and any technical solution for the gathering, processing, and computing necessary to obtain sound in response to the data. In contrast, *Sonification* is a core component of an auditory display: the technique of rendering sound in response to data and interactions.

Different from speech interfaces and music or sound art, Auditory Displays have gained increasing attention in recent years and are becoming a standard technique on par with visualization for presenting data in a variety of contexts. International research efforts to understand all aspects of Auditory Display began with the foundation of the International Community for Auditory Display (ICAD) in 1992. It is fascinating to see how Sonification techniques and Auditory Displays have evolved in the relatively few years since the time of their definition, and the pace of development in 2011 continues to grow.

Auditory Displays and Sonification are currently used in a wide variety of fields. Applications range from topics such as chaos theory, bio-medicine, and interfaces for visually disabled people, to data mining, seismology, desktop computer interaction, and mobile devices, to name just a few. Equally varied is the list of research disciplines that are required to comprehend and carry out successful sonification: Physics, Acoustics, Psychoacoustics,

Perceptual Research, Sound Engineering, Computer Science are certainly core disciplines that contribute to the research process. Yet Psychology, Musicology, Cognitive Science, Linguistics, Pedagogies, Social Sciences and Philosophy are also needed for a fully faceted view of the description, technical implementation, use, training, understanding, acceptance, evaluation and ergonomics of Auditory Displays and Sonification in particular. Figure 1.1 depicts an interdisciplinarity map for the research field.

It is clear that in such an interdisciplinary field, too narrow a focus on any of the above isolated disciplines could quickly lead to "seeing the trees instead of understanding the forest". As with all interdisciplinary research efforts, there are significant hurdles to interdisciplinary research in Auditory Display and Sonification. Difficulties range from differences in theoretical orientations among disciplines to even the very words we use to describe our work. Interdisciplinary dialogue is crucial to the advancement of Auditory Display and Sonification. However, the field faces the challenge of developing and using a common language in order to integrate many divergent "disciplinary" ways of talking, thinking and tackling problems. On the other hand this obstacle often offers great potential for discovery because these divergent ways of thinking and talking can trigger creative potential and new ideas.

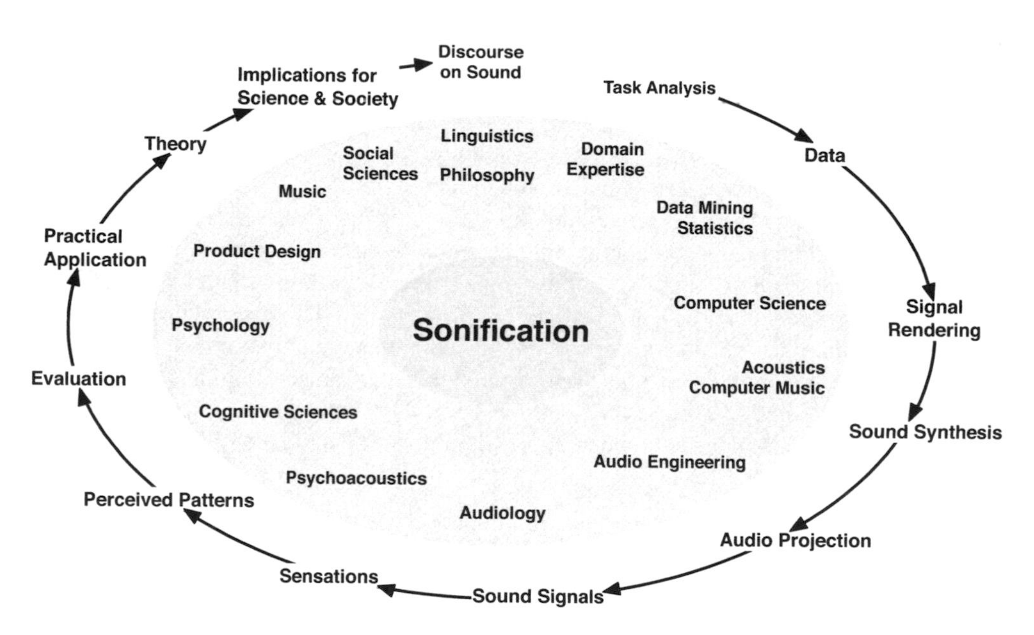

Figure 1.1: The interdisciplinary circle of sonification and auditory display: the outer perimeter depicts the transformations of information during the use cycle, the inner circle lists associated scientific disciplines. This diagram is surely incomplete and merely illustrates the enormous interdisciplinarity of the field.

1.2 The Potential of Sonification and Auditory Display

The motivation to use sound to understand the world (or some data under analysis) comes from many different perspectives. First and foremost, humans are equipped with a complex and powerful listening system. The act of identifying sound sources, spoken words, and melodies, even under noisy conditions, is a supreme pattern recognition task that most modern computers are incapable of reproducing. The fact that it appears to work so effortlessly is perhaps the main reason that we are not aware of the incredible performance that our auditory system demonstrates every moment of the day, even when we are asleep! Thus, the benefits of using the auditory system as a primary interface for data transmission are derived from its complexity, power, and flexibility.

We are, for instance, able to interpret sounds using multiple layers of understanding. For example, from spoken words we extract the word meaning, but also the emotional/health state of the speaker, and their gender, etc. We can also perceive and identify "auditory objects" within a particular auditory scene. For example, in a concert hall we can hear a symphony orchestra as a whole. We can also tune in our focus and attend to individual musical instruments or even the couple who is whispering in the next row. The ability to selectively attend to simultaneously sounding "auditory objects" is an ability that is not yet completely understood. Nonetheless it provides fertile ground for use by designers of auditory displays. Another fascinating feature is the ability to learn and to improve discrimination of auditory stimuli. For example, an untrained listener may notice that "something is wrong" with their car engine, just from its sound, whereas a professional car mechanic can draw quite precise information about the detailed error source from the same sound cue. The physician's stethoscope is a similarly convincing example. Expertise in a particular domain or context can dramatically affect how meaning is constructed from sound. This suggest that – given some opportunity to train, and some standardized and informative techniques to hear data – our brain has the potential to come up with novel and helpful characterizations of the data.

Nowadays we have access to enough computing power to generate and modify sonifications in real-time, and this flexibility may appear, at first glance, to be a strong argument for rapid development of the research field of sonification. However, this flexibility to change an auditory display often and quickly can sometimes be counter-productive in the light of the human listening system's need of time to adapt and become familiar with an auditory display. In the real world, physical laws grant us universality of sound rendering, so that listeners can adapt to real-world sounds. Likewise, some stability in the way that data are sonified may be necessary to ensure that users can become familiar with the display and learn to interpret it correctly.

Sonification sets a clear focus on the use of sound to convey information, something which has been quite neglected in the brief history of computer interfaces. Looking to the future, however, it is not only sound that we should be concerned with. When we consider how information can be understood and interpreted by humans, sound is but one single modality amongst our wealth of perceptual capabilities. Visual, auditory, and tactile information channels deliver complementary information, often tightly coupled to our own actions. In consequence we envision, as attractive roadmap for future interfaces, a better balanced use of all the available modalities in order to make sense of data. Such a generalized discipline may be coined *Perceptualization*.

Sonification in 50 years – A vision

Where might sonification be 50 years from now? Given the current pace of development we might expect that sonification will be a standard method for data display and analysis. We envision established and standardized sonification techniques, optimized for certain analysis tasks, being available as naturally as today's mouse and keyboard interface. We expect sound in human computer interfaces to be much better designed, much more informative, and much better connected to human action than today. Perhaps sonification will play the role of enhancing the appreciation and understanding of the data in a way that is so subtle and intuitive that its very existence will not be specifically appreciated yet it will be clearly missed if absent (rather like the best film music, which enhances the emotion and depth of characterization in a movie without being noticed). There is a long way to go towards such a future, and we hope that this book may be informative, acting as an inspiration to identify where, how and when sound could be better used in everyday life.

1.3 Structure of the book

The book is organized into four parts which bracket chapters together under a larger idea. Part I introduces the fundamentals of sonification, sound and perception. This serves as a presentation of theoretical foundations in chapter 2 and basic material from the different scientific disciplines involved, such as psychoacoustics (chapter 3), perception research (chapter 4), psychology and evaluation (chapter 6) and design (chapter 7), all concerned with Auditory Display, and puts together basic concepts that are important for understanding, designing and evaluating Auditory Display systems. A chapter on Sonic Interaction Design (chapter 5) broadens the scope to relate auditory display to the more general use of sounds in artifacts, ranging from interactive art and music to product sound design.

Part II moves towards the procedural aspects of sonification technology. Sonification, being a scientific approach to representing data using sound, demands clearly defined techniques, e.g., in the form of algorithms. The representation of data and statistical aspects of data are discussed in chapter 8. Since sonifications are usually rendered in computer programs, this part addresses the issues of how sound is represented, generated or synthesized (chapter 9), and what computer languages and programming systems are suitable as laboratory methods for defining and implementing sonifications (chapter 10). The chapter includes also a brief introduction to operator-based sonification and sonification variables, a formalism that serves a precise description of methods and algorithms. Furthermore, interaction plays an important role in the control and exploration of data using sound, which is addressed in chapter 11.

The different Sonification Techniques are presented in Part III. Audification, Auditory Icons, Earcons, Parameter Mapping Sonification and Model-Based Sonification represent conceptually different approaches to how data is related to the resulting sonification, and each of these is examined in detail.

Audification (chapter 12) is the oldest technique for rendering sound from data from areas such as seismology or electrocardiograms, which produce time-ordered sequential data streams. Conceptually, canonically ordered data values are used directly to define the samples of a digital audio signal. This resembles a gramophone where the data values actually determine the structure of the trace. However, such techniques cannot be used when the data sets are arbitrarily large or small, or which do not possess a suitable ordering criterion.

Earcons (chapter 14) communicate messages in sound by the systematic variation of simple sonic 'atoms'. Their underlying structure, mechanism and philosophy is quite different from the approach of Auditory Icons (chapter 13), where acoustic symbols are used to trigger associations from the acoustic 'sign' (the sonification) to that which is 'signified'. Semiotics is here one of the conceptual roots of this display technique. Both of these techniques, however, are more concerned with creating acoustic communication for discrete messages or events, and are not suited for continuous large data streams.

Parameter Mapping Sonification (chapter 15) is widely used and is perhaps the most established technique for sonifying such data. Conceptually, acoustic attributes of events are obtained by a 'mapping' from data attribute values. The rendering and playback of all data items yields the sonification. Parameter Mapping Sonifications were so ubiquitous during the last decade that many researchers frequently referred to them as 'sonification' when they actually meant this specific technique.

A more recent technique for sonification is Model-Based Sonification (chapter 16), where the data are turned into dynamic models (or processes) rather than directly into sound. It remains for the user to excite these models in order to explore data structures via the acoustic feedback, thus putting interaction into a particular focus.

Each of these techniques has its favored application domain, specific theory and logic of implementation, interaction, and use. Each obtains its justification by the heterogeneity of problems and tasks that can be solved with them. One may argue that the borders are dilute – we can for instance interpret audifications as a sort of parameter mapping – yet even if this is possible, it is a very special case, and such an interpretation fails to emphasize the peculiarities of the specific technique. None of the techniques is superior per se, and in many application fields, actually a mix of sonification techniques, sometimes called hybrid sonification, needs to be used in cooperation to solve an Auditory Display problem. Development of all of the techniques relies on the interdisciplinary research discussed above. These 'basis vectors' of techniques span a sonification space, and may be useful as mindset to discover orthogonal conceptual approaches that complement the space of possible sonification types.

Currently there is no single coherent theory of sonification, which clearly explains all sonification types under a unified framework. It is unclear whether this is still a drawback, or perhaps a positive property, since all techniques thus occupy such different locations on the landscape of possible sonification techniques. The highly dynamic evolution of the research field of auditory display may even lead to novel and conceptually complementary approaches to sonification. It is a fascinating evolution that we are allowed to observe (or hear) in the previous and following decades.

Finally, in Part IV of this book the chapters focus on specific application fields for Sonification and Auditory Display. Although most real Auditory Displays will in fact address different functions (e.g., to give an overview of a large data set and to enable the detection of hidden features), these chapters focus on specific tasks. Assistive Technology (chapter 17) is a promising and important application field, and actually aligns to specific disabilities, such as visual impairments limiting the use of classical visual-only computer interfaces. Sonification can help to improve solutions here, and we can all profit from any experience gained in this field. Process Monitoring (chapter 18) focuses on the use of sound to represent (mainly online) data in order to assist the awareness and to accelerate the detection of changing states. Intelligent Auditory Alarms (chapter 19), in contrast cope with symbolic auditory displays, which are most ubiquitous in our current everyday life, and how these can be structured to be more informative and specifically alerting. The use of sonification to assist the navigation (chapter 20) of activities is an application field becoming more visible (or should we say: audible), such as in sports science, gestural controlled audio interactions, interactive sonification etc. Finally, more and more applications deal with the interactive representation of body movements by sonification, driven by the idea that sound can support skill learning and performance without the need to attend a located visual display. This application area is presented in chapter 21).

Each chapter sets a domain-, field-, or application-specific focus and certain things may appear from different viewpoints in multiple chapters. This should prove useful in catalyzing increased insight, and be inspiring for the next generation of Auditory Displays.

1.4 How to Read

The Sonification Handbook is intended to be a resource for lectures, a textbook, a reference, and an inspiring book. One important objective was to enable a highly vivid experience for the reader, by interleaving as many sound examples and interaction videos as possible. We strongly recommend making use of these media. A text on auditory display without listening to the sounds would resemble a book on visualization without any pictures. When reading the pdf on screen, the sound example names link directly to the corresponding website at `http://sonification.de/handbook`. The margin symbol is also an active link to the chapter's main page with supplementary material. Readers of the printed book are asked to check this website manually.

Although the chapters are arranged in this order for certain reasons, we see no problem in reading them in an arbitrary order, according to interest. There are references throughout the book to connect to prerequisites and sidelines, which are covered in other chapters. The book is, however, far from being complete in the sense that it is impossible to report all applications and experiments in exhaustive detail. Thus we recommend checking citations, particularly those that refer to ICAD proceedings, since the complete collection of these papers is available online, and is an excellent resource for further reading.

Part I

Fundamentals of Sonification, Sound and Perception

Chapter 2

Theory of Sonification

Bruce N. Walker and Michael A. Nees

2.1 Chapter Overview

An auditory display can be broadly defined as any display that uses sound to communicate information. Sonification has been defined as a subtype of auditory displays that use non-speech audio to represent information. Kramer et al. (1999) further elaborated that "sonification is the transformation of data relations into perceived relations in an acoustic signal for the purposes of facilitating communication or interpretation", and this definition has persevered since its publication. More recently, a revised definition of sonification was proposed to both expand and constrain the definition of sonification to "..the data-dependent generation of sound, if the transformation is systematic, objective and reproducible..." (also see Hermann, 2008; Hermann, 2011). Sonification, then, seeks to translate relationships in data or information into sound(s) that exploit the auditory perceptual abilities of human beings such that the data relationships are comprehensible.

Theories offer empirically-substantiated, explanatory statements about relationships between variables. Hooker (2004) writes, "Theory represents our best efforts to make the world intelligible. It must not only tell us how things are, but why things are as they are" (pp. 74). Sonification involves elements of both science, which must be driven by theory, and design, which is not always scientific or theory-driven.

The theoretical underpinnings of research and design that can apply to (and drive) sonification come from such diverse fields as audio engineering, audiology, computer science, informatics, linguistics, mathematics, music, psychology, and telecommunications, to name but a few, and are as yet not characterized by a single grand or unifying set of sonification principles or rules (see Edworthy, 1998). Rather, the guiding principles of sonification in research and practice can be best characterized as an amalgam of important insights drawn from the convergence of these many diverse fields. While there have certainly been plenty of generalized contributions toward the sonification theory base (e.g., Barrass, 1997; Brazil,

2010; de Campo, 2007; Frauenberger & Stockman, 2009; Hermann, 2008; Nees & Walker, 2007; Neuhoff & Heller, 2005; Walker, 2002, 2007), to date, researchers and practitioners in sonification have yet to articulate a complete theoretical paradigm to guide research and design. Renewed interest and vigorous conversations on the topic have been reignited in recent years (see, e.g., Brazil & Fernstrom, 2009; de Campo, 2007; Frauenberger, Stockman, & Bourguet, 2007b; Nees & Walker, 2007).

The 1999 collaborative *Sonification Report* (Kramer et al., 1999) offered a starting point for a meaningful discussion of the theory of sonification by identifying four issues that should be addressed in a theoretical description of sonification. These included:

1. taxonomic descriptions of sonification techniques based on psychological principles or display applications;
2. descriptions of the types of data and user tasks amenable to sonification;
3. a treatment of the mapping of data to acoustic signals; and
4. a discussion of the factors limiting the use of sonification.

By addressing the current status of these four topics, the current chapter seeks to provide a broad introduction to sonification, as well as an account of the guiding theoretical considerations for sonification researchers and designers. It attempts to draw upon the insights of relevant domains of research, and where necessary, offers areas where future researchers could answer unresolved questions or make fruitful clarifications or qualifications to the current state of the field. In many cases, the interested reader is pointed to another more detailed chapter in this book, or to other external sources for more extensive coverage.

2.2 Sonification and Auditory Displays

Sonifications are a relatively recent subset of auditory displays. As in any information system (see Figure 2.1), an auditory display offers a relay between the information source and the information receiver (see Kramer, 1994). In the case of an auditory display, the data of interest are conveyed to the human listener through sound.

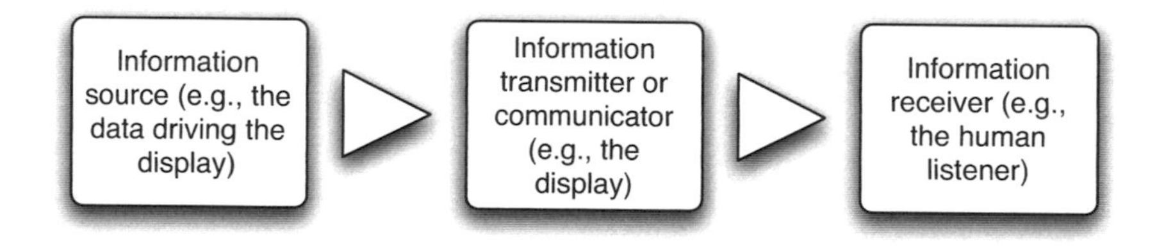

Figure 2.1: General description of a communication system.

Although investigations of audio as an information display date back over 50 years (see Frysinger, 2005), digital computing technology has more recently meant that auditory displays of information have become ubiquitous. Edworthy (1998) argued that the advent of auditory displays and audio interfaces was inevitable given the ease and cost efficiency with which electronic devices can now produce sound. Devices ranging from cars to computers to cell phones to microwaves pervade our environments, and all of these devices now use *intentional* sound[1] to deliver messages to the user. Despite these advances, there remains lingering doubt for some about the usefulness of sound in systems and ongoing confusion for many about how to implement sound in user interfaces (Frauenberger, Stockman, & Bourguet, 2007a).

The rationales and motivations for displaying information using sound (rather than a visual presentation, etc.) have been discussed extensively in the literature (e.g., Buxton et al., 1985; Hereford & Winn, 1994; Kramer, 1994; Nees & Walker, 2009; Peres et al., 2008; Sanderson, 2006). Briefly, though, auditory displays exploit the superior ability of the human auditory system to recognize temporal changes and patterns (Bregman, 1990; Flowers, Buhman, & Turnage, 1997; Flowers & Hauer, 1995; Garner & Gottwald, 1968; Kramer et al., 1999; McAdams & Bigand, 1993; Moore, 1997). As a result, auditory displays may be the most appropriate modality when the information being displayed has complex patterns, changes in time, includes warnings, or calls for immediate action.

In practical work environments the operator is often unable to *look* at, or unable to *see*, a visual display. The visual system might be busy with another task (Fitch & Kramer, 1994; Wickens & Liu, 1988), or the perceiver might be visually impaired, either physically or as a result of environmental factors such as smoke or line of sight (Fitch & Kramer, 1994; Kramer et al., 1999; Walker, 2002; Walker & Kramer, 2004; Wickens, Gordon, & Liu, 1998), or the visual system may be overtaxed with information (see Brewster, 1997; M. L. Brown, Newsome, & Glinert, 1989).

Third, auditory and voice modalities have been shown to be most compatible when systems require the processing or input of verbal-categorical information (Salvendy, 1997; Wickens & Liu, 1988; Wickens, Sandry, & Vidulich, 1983). Other features of auditory perception that suggest sound as an effective data representation technique include our ability to monitor and process multiple auditory data sets (parallel listening) (Fitch & Kramer, 1994), and our ability for rapid auditory detection, especially in high stress environments (Kramer et al., 1999; Moore, 1997).

Finally, with mobile devices decreasing in size, sound may be a compelling display mode as visual displays shrink (Brewster & Murray, 2000). For a more complete discussion of the benefits of (and potential problems with) auditory displays, see Kramer (1994), Kramer et al., 1999), Sanders and McCormick (1993), Johannsen (2004), and Stokes (1990).

2.3 Towards a Taxonomy of Auditory Display & Sonification

A taxonomic description of auditory displays in general, and sonification in particular, could be organized in any number of ways. Categories often emerge from either the function of the display or the technique of sonification, and either could serve as the logical foundation for a taxonomy. In this chapter we offer a discussion of ways of classifying auditory displays

[1] *Intentional* sounds are purposely engineered to perform as an information display (see Walker & Kramer, 1996), and stand in contrast to *incidental* sounds, which are non-engineered sounds that occur as a consequence of the normal operation of a system (e.g., a car engine running). Incidental sounds may be quite informative (e.g., the sound of wind rushing past can indicate a car's speed), though this characteristic of incidental sounds is serendipitous rather than designed. The current chapter is confined to a discussion of intentional sounds.

and sonifications according to both function and technique, although, as our discussion will elaborate, they are very much inter-related.

Sonification is clearly a subset of auditory display, but it is not clear, in the end, where the exact boundaries should be drawn. Recent work by Hermann (2008) identified data-dependency, objectivity, systematicness, and reproducibility as the necessary and sufficient conditions for a sound to be called "sonification". Categorical definitions within the sonification field, however, tend to be loosely enumerated and are somewhat flexible. For example, auditory representations of box-and-whisker plots, diagrammatic information, and equal-interval time series data have all been called sonification, and, in particular, "auditory graphs", but all of these displays are clearly different from each other in both form and function. Recent work on auditory displays that use speech-like sounds (Jeon & Walker, 2011; Walker, Nance, & Lindsay, 2006b) has even called into question the viability of excluding speech sounds from taxonomies of sonification (for a discussion, also see Worrall, 2009a).

Despite the difficulties with describing categories of auditory displays, such catalogs of auditory interfaces can be helpful to the extent that they standardize terminology and give the reader an idea of the options available for using sound in interfaces. In the interest of presenting a basic overview, this chapter provides a description, with definitions where appropriate, of the types of sounds that typically have been used in auditory interfaces. Other taxonomies and descriptions of auditory displays are available elsewhere (Buxton, 1989; de Campo, 2007; Hermann, 2008; Kramer, 1994; Nees & Walker, 2009), and a very extensive set of definitions for auditory displays (Letowski et al., 2001) has been published. Ultimately, the name assigned to a sonification is much less important than its ability to communicate the intended information. Thus, the taxonomic description that follows is intended to parallel conventional naming schemes found in the literature and the auditory display community. However, these descriptions should not be taken to imply that clear-cut boundaries and distinctions are always possible to draw or agree upon, nor are they crucial to the creation of a successful display.

2.3.1 Functions of sonification

Given that sound has some inherent properties that should prove beneficial as a medium for information display, we can begin by considering some of the functions that auditory displays might perform. Buxton (1989) and others (e.g., Edworthy, 1998; Kramer, 1994; Walker & Kramer, 2004) have described the function of auditory displays in terms of three broad categories:

1. alarms, alerts, and warnings;
2. status, process, and monitoring messages; and
3. data exploration.

To this we would add:

4. art, entertainment, sports, and exercise.

The following sections expand each of the above categories.

Alerting functions

Alerts and notifications refer to sounds used to indicate that something has occurred, or is about to occur, or that the listener should immediately attend to something in the environment (see Buxton, 1989; Sanders & McCormick, 1993; Sorkin, 1987). Alerts and notifications tend to be simple and particularly overt. The message conveyed is information-poor. For example, a beep is often used to indicate that the cooking time on a microwave oven has expired. There is generally little information as to the details of the event— the microwave beep merely indicates that the time has expired, not necessarily that the food is fully cooked. Another commonly heard alert is a doorbell— the basic ring does not indicate who is at the door, or why.

Alarms and warnings are alert or notification sounds that are intended to convey the occurrence of a constrained class of events, usually adverse, that carry particular urgency in that they require immediate response or attention (see Haas & Edworthy, 2006 and chapter 19 in this volume). Warning signals presented in the auditory modality capture spatial attention better than visual warning signals (Spence & Driver, 1997). A well-chosen alarm or warning should, by definition, carry slightly more information than a simple alert (i.e., the user knows that an alarm indicates an adverse event that requires an immediate action); however, the specificity of the information about the adverse event generally remains limited. Fire alarms, for example, signal an adverse event (a fire) that requires immediate action (evacuation), but the alarm does not carry information about the location of the fire or its severity.

More complex (and modern) kinds of alarms attempt to encode more information into the auditory signal. Examples range from families of categorical warning sounds in healthcare situations (e.g., Sanderson, Liu, & Jenkins, 2009) to helicopter telemetry and avionics data being used to modify a given warning sound (e.g., "trendsons", Edworthy, Hellier, Aldrich, & Loxley, 2004). These sounds, discussed at length by Edworthy and Hellier (2006), blur the line between alarms and status indicators, discussed next. Many (ten or more) alarms might be used in a single environment (Edworthy & Hellier, 2000), and Edworthy (2005) has critiqued the overabundance of alarms as a potential obstacle to the success of auditory alarms. Recent work (Edworthy & Hellier, 2006; Sanderson, 2006; Sanderson et al., 2009) has examined issues surrounding false alarms and suggested potential emerging solutions to reduce false alarms, including the design of intelligent systems that use multivariate input to look for multiple cues and redundant evidence of a real critical event. Sanderson et al. argued that the continuous nature of many sonifications effectively eliminates the problem of choosing a threshold for triggering a single discrete auditory warning. While it is clear that the interruptive and preemptive nature of sound is especially problematic for false alarms, more research is needed to understand whether sonifications or continuous auditory displays will alleviate this problem.

Status and progress indicating functions

Although in some cases sound performs a basic alerting function, other scenarios require a display that offers more detail about the information being represented with sound. The current or ongoing status of a system or process often needs to be presented to the human listener, and auditory displays have been applied as dynamic *status and progress indicators* (also see chapter 18 in this volume). In these instances, sound takes advantage of "the listener's ability to detect small changes in auditory events or the user's need to have their eyes free for other tasks" (Kramer et al., 1999 p. 3). Auditory displays have been developed for uses ranging from monitoring models of factory process states (see Gaver, Smith, & O'Shea, 1991; Walker & Kramer, 2005), to patient data in an anesthesiologist's workstation (Fitch & Kramer, 1994), blood pressure in a hospital environment (M. Watson, 2006), and telephone hold time (Kortum, Peres, Knott, & Bushey, 2005). Recent work (e.g., Jeon, Davison, Nees, Wilson, & Walker, 2009; Jeon & Walker, 2011; Walker, Nance, & Lindsay, 2006b) has begun to examine speech-like sounds for indicating a user's progress while scrolling auditory representations of common menu structures in devices (see sound examples **S2.1** and **S2.2**).

Data exploration functions

The third functional class of auditory displays contains those designed to permit *data exploration* (also see chapter 8 and 20 in this volume). These are what is generally meant by the term "sonification", and are usually intended to encode and convey information about an entire data set or relevant aspects of the data set. Sonifications designed for data exploration differ from status or process indicators in that they use sound to offer a more holistic portrait of the data in the system rather than condensing information to capture a momentary state such as with alerts and process indicators, though some auditory displays, such as soundscapes (Mauney & Walker, 2004), blend status indicator and data exploration functions. *Auditory graphs* (for representative work, see Brown & Brewster, 2003; Flowers & Hauer, 1992, 1993, 1995; Smith & Walker, 2005) and model-based sonifications (see Chapter 11 in this volume and Hermann & Hunt, 2005) are typical exemplars of sonifications designed for data exploration purposes.

Entertainment, sports, and leisure

Auditory interfaces have been prototyped and researched in the service of exhibitions as well as leisure and fitness activities. Audio-only versions have appeared for simple, traditional games such as the Towers of Hanoi (Winberg & Hellstrom, 2001) and Tic-Tac-Toe (Targett & Fernstrom, 2003), and more complex game genres such as arcade games (e.g., space invaders, see McCrindle & Symons, 2000) and role-playing games (Liljedahl, Papworth, & Lindberg, 2007) have begun to appear in auditory-only formats.

Auditory displays also have been used to facilitate the participation of visually-impaired children and adults in team sports. Stockman (2007) designed an audio-only computer soccer game that may facilitate live action collaborative play between blind and sighted players. Sonifications have recently shown benefits as real-time biofeedback displays for competitive sports such as rowing (Schaffert, Mattes, Barrass, & Effenberg, 2009) and speed skating (Godbout & Boyd, 2010). While research in this domain has barely scratched the surface of potential uses of sonification for exercise, there is clearly a potential for auditory displays to give useful feedback and perhaps even offer corrective measures for technique (e.g., Godbout) in a variety of recreational and competitive sports and exercises (also see chapter 21 in this volume).

Auditory displays have recently been explored as a means of bringing some of the experience and excitement of dynamic exhibits to the visually impaired. A system for using sonified soundscapes to convey dynamic movement of fish in an "accessible aquarium" has been developed (Walker, Godfrey, Orlosky, Bruce, & Sanford, 2006a; Walker, Kim, & Pendse, 2007). Computer vision and other sensing technologies track the movements of entities within the exhibit, and these movements are translated, in real time, to musical representations. For example, different fish might be represented by different instruments. The location of an individual fish might be represented with spatialization of the sound while speed of movement is displayed with tempo changes. Soundscapes in dynamic exhibits may not only make such experiences accessible for the visually impaired, but may also enhance the experience for sighted viewers. Research (Storms & Zyda, 2000) has shown, for example, that high quality audio increases the perceived quality of concurrent visual displays in virtual environments. More research is needed to determine whether high quality auditory displays in dynamic exhibits enhance the perceived quality as compared to the visual experience alone.

Art

As the sound-producing capabilities of computing systems have evolved, so too has the field of computer music. In addition to yielding warnings and sonifications, events and data sets can be used as the basis for musical compositions. Often the resulting performances include a combination of the types of sounds discussed to this point, in addition to more traditional musical elements. While the composers often attempt to convey something to the listener through these sonifications, it is not for the pure purpose of information delivery. As one example, Quinn (2001, 2003) has used data sonifications to drive ambitious musical works, and he has produced entire albums of compositions. Of note, the mapping of data to sound must be systematic in compositions, and the potentially subtle distinction between sonification and music as a conveyor of information is debatable (see Worrall, 2009a). Vickers and Hogg (2006) offered a seminal discussion of the similarities between sonification and music.

2.3.2 Sonification techniques and approaches

Another way to organize and define sonifications is to describe them according to their sonification technique or approach. de Campo (2007) offered a sonification design map (see Figure 10.1 on page 252) that featured three broad categorizations of sonification approaches:

1. event-based;
2. model-based; and
3. continuous.

de Campo's (2007) approach is useful in that it places most non-speech auditory displays within a design framework. The appeal of de Campo's approach is its placement of different types of auditory interfaces along continua that allow for blurry boundaries between categories, and the framework also offers some guidance for choosing a sonification technique. Again, the definitional boundaries to taxonomic descriptions of sonifications are indistinct and often overlapping. Next, a brief overview of approaches and techniques employed in sonification is provided; but for a more detailed treatment, see Part III of this volume.

Modes of interaction

A prerequisite to a discussion of sonification approaches is a basic understanding of the nature of the interaction that may be available to a user of an auditory display. Interactivity can be considered as a dimension along which different displays can be classified, ranging from completely non-interactive to completely user-initiated (also see chapter 11 in this volume). For example, in some instances the listener may passively take in a display without being given the option to actively manipulate the display (by controlling the speed of presentation, pausing, fast-forwarding, or rewinding the presentation, etc.). The display is simply triggered and plays in its entirety while the user listens. Sonifications at this non-interactive end of the dimension have been called "concert mode" (Walker & Kramer, 1996) or "tour based" (Franklin & Roberts, 2004).

Alternatively, the listener may be able to actively control the presentation of the sonification. In some instances, the user might be actively choosing and changing presentation parameters of the display (see Brown, Brewster, & Riedel, 2002). Sonifications more toward this interactive end of the spectrum have been called "conversation mode" (Walker & Kramer, 1996) or "query based" (Franklin & Roberts, 2004) sonification. In other cases, user input and interaction may be the required catalyst that drives the presentation of sounds (see Hermann & Hunt, 2005). Walker has pointed out that for most sonifications to be useful (and certainly those intended to support learning and discovery), there needs to be at least some kind of interaction capability, even if it is just the ability to pause or replay a particular part of the sound (e.g., Walker & Cothran, 2003; Walker & Lowey, 2004).

Parameter mapping sonification

Parameter mapping represents changes in some data dimension with changes in an acoustic dimension to produce a sonification (see chapter 15 in this volume). Sound, however, has a multitude of changeable dimensions (see Kramer, 1994; Levitin, 1999) that allow for a large design space when mapping data to audio. In order for parameter mapping to be used in a sonification, the dimensionality of the data must be constrained such that a perceivable display is feasible. Thus parameter mapping tends to result in a lower dimension display than the model-based approaches discussed below. The data changes may be more qualitative or discrete, such as a thresholded on or off response that triggers a discrete alarm, or parameter mapping may be used with a series of discrete data points to produce a display that seems more continuous. These approaches to sonification have typically employed a somewhat passive mode of interaction. Indeed, some event-based sonifications (e.g., alerts and notifications, etc.) are designed to be brief and would offer little opportunity for user interaction. Other event-based approaches that employ parameter mapping for purposes of data exploration (e.g., auditory graphs) could likely benefit from adopting some combination of passive listening and active listener interaction.

Model-based sonification

Model-based approaches to sonification (Hermann, 2002, chapter 16 in this volume; Hermann & Ritter, 1999) differ from event-based approaches in that instead of mapping data parameters to sound parameters, the display designer builds a virtual model whose sonic responses to user input are derived from data. A model, then, is a virtual object or instrument with which the user can interact, and the user's input drives the sonification such that "the sonification is the reaction of the data-driven model to the actions of the user" (Hermann, 2002 p. 40). The user comes to understand the structure of the data based on the acoustic responses of the model during interactive probing of the virtual object. Model-based approaches rely upon (and the sounds produced are contingent upon) the active manipulation of the sonification by the user. These types of sonifications tend to involve high data dimensionality and large numbers of data points.

Audification

Audification is the most prototypical method of direct sonification, whereby waveforms of periodic data are directly translated into sound (Kramer, 1994, chapter 12 in this volume). For example, seismic data have been audified in order to facilitate the categorization of seismic events with accuracies of over 90% (see Dombois, 2002; Speeth, 1961). This approach may require that the waveforms be frequency- or time-shifted into the range of audible waveforms for humans.

The convergence of taxonomies of function and technique

Although accounts to date have generally classified sonifications in terms of function or technique, the categorical boundaries of functions and techniques are vague. Furthermore, the function of the display in a system may constrain the sonification technique, and the choice of technique may limit the functions a display can perform. Event-based approaches are the only ones used for alerts, notifications, alarms, and even status and process monitors, as these functions are all triggered by events in the system being monitored. Data exploration may employ event-based approaches, model-based sonification, or continuous sonification depending upon the specific task of the user (Barrass, 1997).

2.4 Data Properties and Task Dependency

The nature of the data to be presented and the task of the human listener are important factors for a system that employs sonification for information display. The display designer must consider, among other things:

- what the user needs to accomplish (i.e., the task(s));
- what parts of the information source (i.e., the data[2]) are relevant to the user's task;

[2]The terms "data" and "information" are used more or less interchangeably here in a manner consistent with Hermann's (2008) definition of sonification. For other perspectives, see Barrass (1997) or Worrall (2009b, Chapter 3)

- how much information the user needs to accomplish the task;
- what kind of display to deploy (simple alert, status indicator, or full sonification, for example); and
- how to manipulate the data (e.g., filtering, transforming, or data reduction).

These issues come together to present major challenges in sonification design, since the nature of the data and the task will necessarily constrain the data-to-display mapping design space. Mapping data to sound requires a consideration of perceptual or "bottom up" processes, in that some dimensions of sound are perceived as categorical (e.g., timbre), whereas other attributes of sound are perceived along a perceptual continuum (e.g., frequency, intensity). Another challenge comes from the more cognitive or conceptual "top down" components of perceiving sonifications. For example, Walker (2002) has shown that conceptual dimensions (like size, temperature, price, etc.) influence how a listener will interpret and scale the data-to-display relationship.

2.4.1 Data types

Information can be broadly classified as quantitative (numerical) or qualitative (verbal). The design of an auditory display to accommodate quantitative data may be quite different from the design of a display that presents qualitative information. Data can also be described in terms of the scale upon which measurements were made. Nominal data classify or categorize; no meaning beyond group membership is attached to the magnitude of numerical values for nominal data. Ordinal data take on a meaningful order with regards to some quantity, but the distance between points on ordinal scales may vary. Interval and ratio scales have the characteristic of both meaningful order and meaningful distances between points on the scale (see Stevens, 1946). Data can also be discussed in terms of its existence as discrete pieces of information (e.g., events or samples) versus a continuous flow of information.

Barrass (1997; 2005) is one of the few researchers to consider the role of different types of data in auditory display and make suggestions about how information type can influence mappings. As one example, nominal/categorical data types (e.g., different cities) should be represented by categorically changing acoustic variables, such as timbre. Interval data may be represented by more continuous acoustic variables, such as pitch or loudness (but see Stevens, 1975; Walker, 2007 for more discussion on this issue).

Nevertheless, there remains a paucity of research aimed at studying the factors within a data set that can affect perception or comprehension. For example, data that are generally slow-changing, with relatively few inflection points (e.g., rainfall or temperature) might be best represented with a different type of display than data that are rapidly-changing with many direction changes (e.g., EEG or stock market activity). Presumably, though, research will show that data set characteristics such as density and volatility will affect the best choices of mapping from data to display. This is beginning to be evident in the work of Hermann, Dombois, and others who are using very large and rapidly changing data sets, and are finding that audification and model-based sonification are more suited to handle them. Even with sophisticated sonification methods, data sets often need to be pre-processed, reduced in dimensionality, or sampled to decrease volatility before a suitable sonification can be created. On the other hand, smaller and simpler data sets such as might be found in a high-school science class may be suitable for direct creation of auditory graphs and auditory histograms.

2.4.2 Task types

Task refers to the functions that are performed by the human listener within a system like that depicted in Figure 2.1. Although the most general description of the listener's role involves simply receiving the information presented in a sonification, the person's goals and the functions allocated to the human being in the system will likely require further action by the user upon receiving the information. Furthermore, the auditory display may exist within a larger acoustic context in which attending to the sound display is only one of many functions concurrently performed by the listener. Effective sonification, then, requires an understanding of the listener's function and goals within a system. What does the human listener need to accomplish? Given that sound represents an appropriate means of information display, how can sonification best help the listener successfully perform her or his role in the system? Task, therefore, is a crucial consideration for the success or failure of a sonification, and a display designer's knowledge of the task will necessarily inform and constrain the design of a sonification[3]. A discussion of the types of tasks that users might undertake with sonifications, therefore, closely parallels the taxonomies of auditory displays described above.

Monitoring

Monitoring requires the listener to attend to a sonification over a course of time and to detect events (represented by sounds) and identify the meaning of the event in the context of the system's operation. These events are generally discrete and occur as the result of crossing some threshold in the system. Sonifications for monitoring tasks communicate the crossing of a threshold to the user, and they often require further (sometimes immediate) action in order for the system to operate properly (see the treatment of alerts and notifications above).

Kramer (1994) described monitoring tasks as "template matching", in that the listener has a priori knowledge and expectations of a particular sound and its meaning. The acoustic pattern is already known, and the listener's task is to detect and identify the sound from a catalogue of known sounds. Consider a worker in an office environment that is saturated with intentional sounds from common devices, including telephones, fax machines, and computer interface sounds (e.g., email or instant messaging alerts). Part of the listener's task within such an environment is to monitor these devices. The alerting and notification sounds emitted from these devices facilitate that task in that they produce known acoustic patterns; the listener must hear and then match the pattern against the catalogue of known signals.

Awareness of a process or situation

Sonifications may sometimes be employed to promote the awareness of task-related processes or situations (also see chapter 18 in this volume). Awareness-related task goals are different

[3] Human factors scientists have developed systematic methodologies for describing and understanding the tasks of humans in a man-machine system. Although an in-depth treatment of these issues is beyond the scope of this chapter, see Luczak (1997) or Barrass (1996) for thorough coverage of task analysis purposes and methods.

from monitoring tasks in that the sound coincides with, or embellishes, the occurrence of a process rather than simply indicating the crossing of a threshold that requires alerting. Whereas monitoring tasks may require action upon receipt of the message (e.g., answering a ringing phone or evacuating a building upon hearing a fire alarm), the sound signals that provide information regarding awareness may be less action-oriented and more akin to ongoing feedback regarding task-related processes.

Non-speech sounds such as earcons and auditory icons have been used to enhance human-computer interfaces (see Brewster, 1997; Gaver, 1989). Typically, sounds are mapped to correspond to task-related processes in the interface, such as scrolling, clicking, and dragging with the mouse, or deleting files, etc. Whereas the task that follows from monitoring an auditory display cannot occur in the absence of the sound signal (e.g., one can't answer a phone until it rings), the task-related processes in a computer interface can occur with or without the audio. The sounds are employed to promote awareness of the processes rather than to solely trigger some required response.

Similarly, soundscapes—ongoing ambient sonifications—have been employed to promote awareness of dynamic situations (a bottling plant, Gaver et al., 1991; financial data, Mauney & Walker, 2004; a crystal factory, Walker & Kramer, 2005). Although the soundscape may not require a particular response at any given time, it provides ongoing information about a situation to the listener.

Data exploration

Data exploration can entail any number of different subtasks ranging in purpose from holistic accounts of the entire data set to analytic tasks involving a single datum. Theoretical and applied accounts of visual graph and diagram comprehension have described a number of common tasks that are undertaken with quantitative data (see, for example, Cleveland & McGill, 1984; Friel, Curcio, & Bright, 2001; Meyer, 2000; Meyer, Shinar, & Leiser, 1997), and one can reasonably expect that the same basic categories of tasks will be required to explore data with auditory representations. The types of data exploration tasks described below are representative (but not necessarily comprehensive), and the chosen sonification approach may constrain the types of tasks that can be accomplished with the display and vice versa.

Point estimation and point comparison Point estimation is an analytic listening task that involves extracting information regarding a single piece of information within a data set. Point estimation is fairly easily accomplished with data presented visually in a tabular format (Meyer, 2000), but data are quite likely to appear in a graphical format in scientific and popular publications (Zacks, Levy, Tversky, & Schiano, 2002). The extraction of information regarding a single datum, therefore, is a task that may need to be accomplished with an abstract (i.e., graphical) representation of the data rather than a table. Accordingly, researchers have begun to examine the extent to which point estimation is feasible with auditory representations of quantitative data such as auditory graphs. Smith and Walker (2005) performed a task analysis for point estimation with auditory graphs and determined that five steps were required to accomplish a point estimation task with sound. The listener must: 1. listen to the sonification; 2. determine in time when the datum of interest occurs;

3. upon identifying the datum of interest, estimate the magnitude of the quantity represented by the pitch of the tone; 4. compare this magnitude to a baseline or reference tone (i.e., determine the scaling factor); and 5. report the value.

Point comparison, then, is simply comparing more than one datum; thus, point comparison involves performing point estimation twice (or more) and then using basic arithmetic operations to compare the two points. In theory, point comparison should be more difficult for listeners to perform accurately than point estimation, as listeners have twice as much opportunity to make errors, and there is the added memory component of the comparison task. Empirical investigations to date, however, have not examined point comparison tasks with sonifications.

Trend identification Trend identification is a more holistic listening task whereby a user attempts to identify the overall pattern of increases and decreases in quantitative data. Trend in a sonification closely parallels the notion of melodic contour in a piece of music. The listener may be concerned with global (overall) trend identification for data, or she/he may wish to determine local trends over a narrower, specific time course within the sonification. Trend identification has been posited as a task for which the auditory system is particularly well-suited, and sound may be a medium wherein otherwise unnoticed patterns in data emerge for the listener.

Identification of data structure While the aforementioned tasks are primarily applicable to event-based sonification approaches, the goals of a model-based sonification user may be quite different. With model-based sonifications, the listener's task may involve identification of the overall structure of the data and complex relationships among multiple variables. Through interactions with the virtual object, the listener hopes to extract information about the relationships within, and structure of, the data represented.

Exploratory inspection Occasionally, a user's task may be entirely exploratory requiring the inspection or examination of data with no a priori questions in mind. Kramer (1994) described exploratory tasks with sound as a less tractable endeavor than monitoring, because data exploration by its nature does not allow for an a priori, known catalogue of indicators. Still, the excellent temporal resolution of the auditory system and its pattern detection acuity make it a viable mode of data exploration, and the inspection of data with sound may reveal patterns and anomalies that were not perceptible in visual representations of the data.

Dual task performance and multimodal tasking scenarios

In many applications of sonification, it is reasonable to assume that the human listener will likely have other auditory and/or visual tasks to perform in addition to working with the sonification. Surprisingly few studies to date have considered how the addition of a secondary task affects performance with sonifications. The few available studies are encouraging. Janata and Childs (2004) showed that sonifications aided a monitoring task with stock data, and the helpfulness of sound was even more pronounced when a secondary number-matching task was added. Peres and Lane (2005) found that while the addition of a visual monitoring task to an auditory monitoring task initially harmed performance of the auditory task, performance soon (i.e., after around 25 dual task trials) returned to pre-dual task levels. Brewster (1997) showed that the addition of sound to basic, traditionally visual interface operations enhanced performance of the tasks. Bonebright and Nees (2009) presented sounds that required a manual response approximately every 6 seconds while participants listened to a passage for verbal comprehension read aloud. The sound used included five types of earcons and also brief speech sounds, and the researchers predicted that speech sounds would interfere most with spoken passage comprehension. Surprisingly, however, only one condition—featuring particularly poorly designed earcons that used a continuous pitch-change mapping—significantly interfered with passage comprehension compared to a control condition involving listening only without the concurrent sound task. Although speech sounds and the spoken passage presumably taxed the same verbal working memory resources, and all stimuli were concurrently delivered to the ears, there was little dual-task effect, presumably because the sound task was not especially hard for participants.

Despite these encouraging results, a wealth of questions abounds regarding the ability of listeners to use sonifications during concurrent visual and auditory tasks. Research to date has shed little light on the degree to which non-speech audio interferes with concurrent processing of other sounds, including speech. The successful deployment of sonifications in real-world settings will require a more solid base of knowledge regarding these issues.

2.5 Representation and Mappings

Once the nature of the data and the task are determined, building a sonification involves mapping the data source(s) onto representational acoustic variables. This is especially true for parameter mapping techniques, but also applies, in a more general sense, to all sonifications. The mappings chosen by the display designer are an attempt to communicate information in each of the acoustic dimensions in use. It is important to consider how much of the intended "message" is received by the listener, and how close the perceived information matches the intended message.

2.5.1 Semiotics: How acoustic perception takes on conceptual representation

Semiotics is "the science of signs (and signals)" (Cuddon, 1991 p. 853). Clearly sonification aims to use sound to signify data or other information (Barrass, 1997), and Pirhonen, Murphy, McAllister, and Yu (2006) have encouraged a semiotic perspective in sound design. Empirical approaches, they argued, have been largely dominated by atheoretical, arbitrary sound design choices. Indeed the design space for sonifications is such that no study or series of studies could possibly make empirical comparisons of all combinations of sound manipulations. Pirhonen et al. argued for a semiotic approach to sound design that requires detailed use scenarios (describing a user and task) and is presented to a design panel of experts or representative users. Such an approach seeks input regarding the most appropriate way to use sounds as signs for particular users in a particular setting or context.

Kramer (1994) has described a representation continuum for sounds that ranges from analogic to symbolic (see Figure 2.2). At the extreme analogic end of the spectrum, the sound has the most direct and intrinsic relationship to its referent. Researchers have, for example, attempted to determine the extent to which the geometric shape of an object can be discerned by listening to the vibrations of physical objects that have been struck by mallets (Lakatos, McAdams, & Causse, 1997). At the symbolic end of the continuum, the referent may have an arbitrary or even random association with the sound employed by the display.

Keller and Stevens (2004) described the signal-referent relationships of environmental sounds with three categories: direct, indirect ecological, and indirect metaphorical. Direct relationships are those in which the sound is ecologically attributable to the referent. Indirect ecological relationships are those in which a sound that is ecologically associated with, but not directly attributable to, the referent is employed (e.g., the sound of branches snapping to represent a tornado). Finally, indirect metaphorical relationships are those in which the sound signal is related to its referent only in some emblematic way (e.g., the sound of a mosquito buzzing to represent a helicopter).

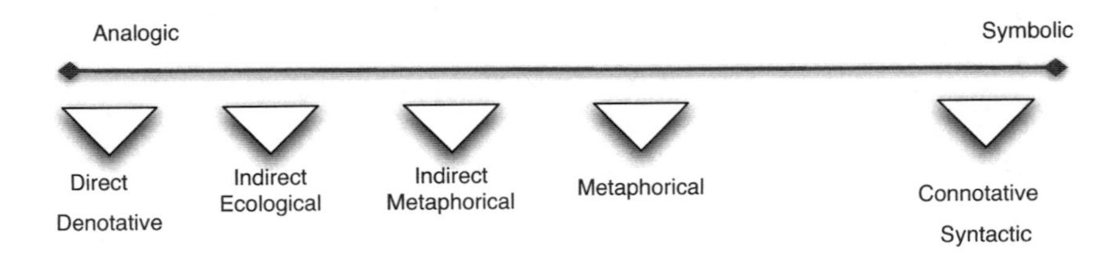

Figure 2.2: The analogic-symbolic representation continuum.

2.5.2 Semantic/iconic approach

Auditory icons, mentioned earlier, are brief communicative sounds in an interface that bear an analogic relationship with the process they represent (see chapter 13 in this volume). In other words, the sound bears some ecological (i.e., naturally-associated) resemblance to the action or process (see Gaver, 1994; Kramer, 1994). This approach has also been called *nomic mapping* (Coward & Stevens, 2004). Auditory icons are appealing in that the association between the sound and its intended meaning is more direct and should require little or no learning, but many of the actions and processes in a human-computer interface have no inherent auditory representation. For example, what should accompany a "save" action in a word processor? How can that sound be made distinct from a similar command, such as "save as"? *Earcons*, on the other hand, use sounds as symbolic representations of actions or processes; the sounds have no ecological relationship to their referent (see Blattner, Sumikawa, & Greenberg, 1989; Kramer, 1994 and chapter 14 in this volume). Earcons are made by systematically manipulating the pitch, timbre, and rhythmic properties of sounds to create a structured set of non-speech sounds that can be used to represent any object or concept through an arbitrary mapping of sound to meaning. Repetitive or related sequences or motifs may be employed to create "families" of sounds that map to related actions or processes. While earcons can represent virtually anything, making them more flexible than auditory icons, a tradeoff exists in that the abstract nature of earcons may require longer learning time or even formal training in their use. Walker and colleagues (Palladino & Walker, 2007; Walker & Kogan, 2009; Walker et al., 2006b) have discussed a new type of interface sound, the *spearcon*, which is intended to overcome the shortcomings of both auditory icons and earcons. Spearcons (see sound examples **S2.1** and **S2.3**) are created by speeding up a spoken phrase even to the point where it is no longer recognizable as speech, and as such can represent anything (like earcons can), but are non-arbitrarily mapped to their concept (like auditory icons). The main point here is that there are tradeoffs when choosing how to represent a concept with a sound, and the designer needs to make explicit choices with the tradeoffs in mind.

2.5.3 Choice of display dimension

When creating a more typical parameter-mapped sonification, such as representing rainfall and average daily temperature over the past year, the issues of mapping, polarity, and scaling are crucial (Walker, 2002, 2007; Walker & Kramer, 2004).

Data-to-display Mapping

In sonification it matters which specific sound dimension is chosen to represent a given data dimension. This is partly because there seems to be some agreement among listeners about what sound attributes are good (or poor) at representing particular data dimensions. For example, pitch is generally good for representing temperature, whereas tempo is not as effective (Walker, 2002). It is also partly because some sound dimensions (e.g., loudness) are simply not very effective in auditory displays for practical design reasons (Neuhoff, Kramer, & Wayand, 2002). Walker has evaluated mappings between ten conceptual data dimensions (e.g., temperature, pressure, danger) and three perceptual/acoustic dimensions (pitch, tempo, and spectral brightness), in an effort to determine which sounds should be used to represent a given type of data (see also Walker, 2002, 2007). This type of research will need to be extended to provide designers with guidance about mapping choices. In turn, sonification designers need to be aware that not all mappings are created equal, and must use a combination of empirically-derived guidelines and usability testing to ensure the message they are intending to communicate is being received by the listener. In addition to those already discussed, guidelines for mappings from a variety of sources should be consulted (e.g., Bonebright, Nees, Connerley, & McCain, 2001; Brown, Brewster, Ramloll, Burton, & Riedel, 2003; Flowers & Hauer, 1995; Neuhoff & Heller, 2005; Smith & Walker, 2005; Walker, 2002, 2007).

Mapping Polarity

Sonification success also requires an appropriate polarity for the data-to-display mappings. For example, listeners might agree that pitch should increase in order to represent increasing temperature (a positive mapping polarity, Walker, 2002), while at the same time feel that pitch should decrease in order to represent increasing size (a *negative* polarity). The issue of polarity is not typically an issue for visual displays, but it can be very important in auditory representations ranging from helicopter warning sounds (Edworthy et al., 2004) to interfaces for the visually impaired (Mauney & Walker, 2010; Walker & Lane, 2001). Walker (2002, 2007) lists the preferred polarities for many mappings, and points out that performance is actually impacted with polarities that do not match listener expectancies. Again, a mixture of guidelines and testing are important to ensure that a sonification is in line with what listeners anticipate.

Scaling

Once an effective mapping and polarity has been chosen, it is important to determine how much change in, say, the pitch of a sound is used to convey a given change in, for example, temperature. Matching the data-to-display scaling function to the listener's internal conceptual scaling function between pitch and temperature is critical if the sonification is to be used to make accurate comparisons and absolute or exact judgments of data values, as opposed to simple trend estimations (for early work on scaling a perceptual sound space, see Barrass, 1994/2005). This is a key distinction between sonifications and warnings or trend monitoring sounds. Again, Walker (2002, 2007) has empirically determined scaling factors for several mappings, in both positive and negative polarities. Such values begin to provide guidance about how different data sets would be represented most effectively. However, it is important not to over-interpret the exact exponent values reported in any single study, to the point where they are considered "*the*" correct values for use in all cases. As with any performance data that are used to drive interface guidelines, care must always be taken to avoid treating the numbers as components of a design recipe. Rather, they should be treated as guidance, at least until repeated measurements and continued application experience converge toward a clear value or range.

Beyond the somewhat specific scaling factors discussed to this point, there are some practical considerations that relate to scaling issues. Consider, for example, using frequency changes to represent average daily temperature data that ranges from 0-30° Celsius. The temperature data *could* be scaled to fill the entire hearing range (best case, about 20 Hz to 20,000 Hz); but a much more successful approach might be to scale the data to the range where hearing is most sensitive, say between 1000-5000 Hz. Another approach would be to base the scaling on a musical model, where the perceptually equal steps of the notes on a piano provide a convenient scale. For this reason, computer music approaches to sonification, including mapping data onto MIDI notes, have often been employed. Limiting the range of notes has often been recommended (e.g., using only MIDI notes 35-100, Brown et al., 2003). Even in that case, the designer has only 65 display points to use to represent whatever data they may have. Thus, the granularity of the scale is limited. For the daily temperature data that may be sufficient, but other data sets may require more precision. A designer may be forced to "round" the data values to fit the scale, or alternatively employ "pitch bending" to play a note at the exact pitch required by the data. This tends to take away from the intended musicality of the approach. Again, this is a tradeoff that the designer needs to consider. Some software (e.g., the Sonification Sandbox, Walker & Cothran, 2003; Walker & Lowey, 2004) provides both rounding and exact scaling options, so the one that is most appropriate can be used, given the data and the tasks of the listener.

Concurrent presentation of multiple data streams/series

Many data analysis tasks require the comparison of values from more than one data source presented concurrently. This could be daily temperatures from different cities, or stock prices from different stocks. The general theory invoked in this situation is auditory streaming (Bregman, 1990). In some cases (for some tasks), it is important to be able to perceptually separate or segregate the different city data, whereas in other cases it is preferable for the two streams of data to fuse into a perceptual whole. Bregman (1990) discusses what acoustic properties support or inhibit stream segregation. Briefly, differences in timbre (often achieved by changing the musical instrument, see Cusack & Roberts, 2000) and spatial location (or stereo panning) are parameters that sonification designers can often use simply and effectively (see also Bonebright et al., 2001; Brown et al., 2003). McGookin and Brewster (2004) have shown that, while increasing the number of concurrently presented earcons decreases their identifiability, such problems can be somewhat overcome by introducing timbre and onset differences. Pitch is another attribute that can be used to segregate streams, but in sonification pitch is often dynamic (being used to represent changing data values), so it is a less controllable and less reliable attribute for manipulating segregation.

Context

Context refers to the purposeful addition of non-signal information to a display (Smith & Walker, 2005; Walker & Nees, 2005a). In visual displays, additional information such as axes and tick marks can increase readability and aid perception by enabling more effective top-down processing (Bertin, 1983; Tufte, 1990). A visual graph without context cues (e.g., no axes or tick marks) provides no way to estimate the value at any point. The contour of the line provides some *incidental* context, which might allow an observer to perform a trend analysis (rising versus falling), but the accurate extraction of a specific value (i.e., a point estimation task) is impossible without context cues.

Even sonifications that make optimal use of mappings, polarities, and scaling factors need to include contextual cues equivalent to axes, tick marks and labels, so the listener can perform the interpretation tasks. Recent work (Smith & Walker, 2005) has shown that even for simple sonifications, the addition of some kinds of context cues can provide useful information to users of the display. For example, simply adding a series of clicks to the display can help the listener keep track of the time better, which keeps their interpretation of the graph values more "in phase" (see also Bonebright et al., 2001; Flowers et al., 1997; Gardner, Lundquist, & Sahyun, 1996). Smith and Walker (2005) showed that when the clicks played at twice the rate of the sounds representing the data, the two sources of information combined like the major and minor tick marks on the x-axis of a visual graph. The addition of a repeating reference tone that signified the maximum value of the data set provided dramatic improvements in the attempts by listeners to estimate exact data values, whereas a reference tone that signified the starting value of the data did not improve performance. Thus, it is clear that adding context cues to auditory graphs can play the role that x- and y-axes play in visual graphs, but not all implementations are equally successful. Researchers have only scratched the surface of possible context cues and their configurations, and we need to implement and validate other, perhaps more effective, methods (see, e.g., Nees & Walker, 2006).

2.6 Limiting Factors for Sonification: Aesthetics, Individual Differences, and Training

Although future research should shed light on the extent to which particular tasks and data sets are amenable to representation with sound, the major limiting factors in the deployment of sonifications have been, and will continue to be, the perceptual and information processing capabilities of the human listener.

2.6.1 Aesthetics and musicality

Edworthy (1998) aptly pointed out the independence of display performance and aesthetics. While sound may aesthetically enhance a listener's interaction with a system, performance may not necessarily be impacted by the presence or absence of sound. Questions of aesthetics and musicality remain open in the field of sonification. The use of musical sounds (as opposed to pure sine wave tones, etc.) has been recommended because of the ease with which musical sounds are perceived (Brown et al., 2003), but it remains to be seen whether the use of musical sounds such as those available in MIDI instrument banks affords performance improvements over less musical, and presumably less aesthetically desirable, sounds. Although the resolution of issues regarding aesthetics and musicality is clearly relevant, it nevertheless remains advisable to design aesthetically pleasing (e.g., musical, etc.) sonifications to the extent possible while still conveying the intended message. Vickers and Hogg (2006) made a pointed statement about aesthetics in sonification. In particular, they argued that more careful attention to aesthetics would facilitate ease of listening with sonifications, which would in turn promote comprehension of the intended message of the displays.

2.6.2 Individual differences and training

The capabilities, limitations, and experiences of listeners, as well as transient states (such as mood and level of fatigue) will all impact performance outcomes with auditory displays. Surprisingly little is known about the impact of between- and within-individual differences on auditory display outcomes. Understanding individual differences in perceptual, cognitive, and musical abilities of listeners will inform the design of sonifications in several important ways. First, by understanding ranges in individual difference variables, a designer can, where required, build a display that accommodates most users in a given context (e.g., universal design, see Iwarsson & Stahl, 2003). Furthermore, in situations where only optimal display users are desirable, understanding the relevance and impact of individual difference variables will allow for the selection of display operators whose capabilities will maximize the likelihood of success with the display. Finally, the extent to which differences in training and experience with sonifications affects performance with the displays is a topic deserving further investigation.

Perceptual capabilities of the listener

A treatment of theoretical issues relevant to sonification would be remiss not to mention those characteristics of the human listener that impact comprehension of auditory displays. The fields of psychoacoustics and basic auditory perception (see chapter 3 and 4 in this volume) have offered critical insights for the design and application of sonifications. As Walker and Kramer (2004) pointed out, these fields have contributed a widely accepted vocabulary and methodology to the study of sound perception, as well as a foundation of knowledge that is indispensable to the study of sonification.

Detection is of course a crucial first consideration for auditory display design. The listener must be able to hear the sound(s) in the environment in which the display is deployed. Psychoacoustic research has offered insights into minimum thresholds (e.g., see Hartmann, 1997; Licklider, 1951), and masking theories offer useful predictions regarding the detectability of a given acoustic signal in noise (for a discussion, see Mulligan, McBride, & Goodman, 1984; Watson & Kidd, 1994). Empirical data for threshold and masking studies, however, are usually gathered in carefully controlled settings with minimal stimulus uncertainty. As Watson and Kidd (1994) and others (e.g., Mulligan et al., 1984; Walker & Kramer, 2004) point out, such data may provide apt descriptions of auditory capabilities but poor guidelines for auditory display design. The characteristics of the environment in which a display operates may differ drastically from the ideal testing conditions and pure tone stimuli of psychophysical experiments. As a result, Watson and Kidd suggested that ecologically valid testing conditions for auditory displays should be employed to establish real-world guidelines for auditory capabilities (also see Neuhoff, 2004). Furthermore, recent work has drawn attention to the phenomenon of informational masking, whereby sounds that theoretically should *not* be masked in the peripheral hearing mechanism (i.e., the cochlea) are indeed masked, presumably at higher levels in the auditory system (see Durlach et al., 2003). Clearly, the seemingly straightforward requirement of detectability for auditory displays warrants a careful consideration of the display's user as well as the environments and apparatus (headphones, speakers, etc.) with which the display will be implemented.

Beyond basic knowledge of the detectability of sound, auditory display designers should be aware of the psychophysical limitations on judgments of discrimination (e.g., just-noticeable differences, etc.) and identification of sounds. Again, however, the data regarding discrimination or identification performance in controlled conditions may offer misleading design heuristics for less controlled, non-laboratory environments. Sonification researchers can and should, however, actively borrow from and adapt the knowledge and methods of psychoacousticians. For example, Bregman's (1990) theory of auditory scene analysis (ASA) has considerable explanatory value with respect to the pre-attentive emergence of auditory objects and gestalts, and this perspective can offer auditory display design heuristics (see, e.g., Barrass & Best, 2008). Similarly, Sandor and Lane (2003) introduced the term *mappable difference* to describe the absolute error in response accuracy one must allow for in order to achieve a given proportion of accurate responses for a point estimation sonification task. Such a metric also allowed them to identify the number of distinct values that could be represented with a given proportion of accuracy for their chosen scales. Such innovative approaches that combine the methods and tools of psychoacoustics and perception with the real-world stimuli and applications of auditory display designers may be the best approach to understanding how to maximize information transmission with auditory displays by playing to the strengths of the human perceiver.

Cognitive abilities of the listener

Researchers have posited roles for a number of cognitive abilities in the comprehension of visual displays, including spatial abilities (Trickett & Trafton, 2006), domain or content knowledge and graph-reading skill (Shah, 2002), and working memory (Toth & Lewis, 2002). The role of such cognitive abilities in the comprehension of sonifications and auditory stimuli in general, however, remains relatively unexplored. The few studies that have examined relationships between cognitive abilities and auditory perception have found results that suggest cognitive individual differences will impact auditory display performance. Walker and Mauney (2004) found that spatial reasoning ability predicts some variance in performance with auditory graphs. More research is needed to determine the full array of cognitive factors contributing to auditory display performance, and the extent to which such cognitive abilities can be accurately assessed and used to predict performance.

Additionally, questions regarding the cognitive representations formed and used by auditory display listeners remain virtually untouched. For example, if, as Kramer (1994) argued, sonification monitoring tasks employ template matching processes, then what are the properties of the stored templates and how are they formed? In the case of auditory graphs, do people attempt to translate the auditory stimulus into a more familiar visual mental representation? Anecdotal evidence reported by Flowers (1995) suggested that listeners were indeed inclined to draw visual representations of auditory graphs on scrap paper during testing. A recent qualitative study (Nees & Walker, 2008) and a series of experiments (Nees, 2009; Nees & Walker, in press) have both suggested that non-speech sound can be rehearsed in working memory as words, visual images, or as quasi-isomorphic sounds per se. Though sonification research tends to shy away from basic and theoretical science in favor of more applied lines of research, studies leading to better accounts of the cognitive representations of sonifications would favorably inform display design.

Musical abilities of the listener

For many years, researchers predicted and anticipated that musicians would outperform non-musicians on tasks involving auditory displays. Musical experience and ability, then, have been suggested as individual level predictors of performance with auditory displays, but research has generally found weak to non-existent correlations between musical experience and performance with auditory displays. One plausible explanation for the lack of relationship between musicianship and auditory display performance is the crude nature of self-report metrics of musical experience, which are often the yardstick for describing the degree to which a person has musical training. A person could have had many years of musical experience as child, yet that person could be many years removed from their musical training and exhibit no more musical *ability* than someone who received no formal training. A more fruitful approach to the measurement of musicianship in the future may be to develop brief, reliable, and valid measure of musical ability for diagnostic purposes in research (e.g., Edwards, Challis, Hankinson, & Pirie, 2000), along the lines of research in musical abilities by Seashore and others (e.g., Brown, 1928; Cary, 1923; Seashore, Lewis, & Saetveit, 1960).

Although the predictive value of individual differences in musical ability is worthy of further study and differences between musicians and non-musicians have been reported (e.g., Lacherez, Seah, & Sanderson, 2007; Neuhoff & Wayand, 2002; Sandor & Lane, 2003), the ultimate contribution of musical ability to performance with auditory displays may be minor. Watson and Kidd (1994) suggested that the auditory perceptual abilities of the worst musicians are likely better than the abilities of the worst non-musicians, but the best non-musicians are likely have auditory perceptual abilities on par with the best musicians.

Visually-impaired versus sighted listeners

Though sonification research is most often accomplished with samples of sighted students in academic settings, auditory displays may provide enhanced accessibility to information for visually-impaired listeners. Visual impairment represents an individual difference that has been shown to have a potentially profound impact on the perception of sounds in some scenarios. Walker and Lane (2001), for example, showed that blind and sighted listeners actually had opposing intuitions about the polarity of the pairing of some acoustic dimensions with conceptual data dimensions. Specifically, blind listeners expected that increasing frequency represented a decreasing "number of dollars" (a negative polarity) whereas sighted listeners expected that increasing frequency conveyed that wealth was accumulating (a positive polarity). This finding was extended upon and further confirmed in a recent study (Mauney & Walker, 2010). These data also suggested that, despite generally similar patterns of magnitude estimation for conceptual data dimensions, sighted participants were more likely to intuit split polarities than blind participants. Individual differences between visually-impaired and sighted listeners require more research and a careful testing of auditory displays with the intended user population. Potential differences between these user groups are not necessarily predictable from available design heuristics.

Training

Sonification offers a novel approach to information representation, and this novelty stands as a potential barrier to the success of the display unless the user can be thoroughly and efficiently acclimated to the meaning of the sounds being presented. Visual information displays owe much of their success to their pervasiveness as well as to users' formal education and informal experience at deciphering their meanings. Graphs, a basic form of visual display, are incredibly pervasive in print media (see Zacks et al., 2002), and virtually all children are taught how to read graphs from a very young age in formal education settings. Complex auditory displays currently are not pervasive, and users are not taught how to comprehend auditory displays as part of a standard education. This problem can be partially addressed by exploiting the natural analytic prowess and intuitive, natural meaning-making processes of the auditory system (see Gaver, 1993), but training will likely be necessary even when ecological approaches to sound design are pursued.To date, little attention has been paid to the issue of training sonification users. Empirical findings suggesting that sonifications can be effective are particularly encouraging considering that the majority of these studies sampled naïve users who had presumably never listened to sonifications before entering the lab. For the most part, information regarding performance ceilings for sonifications remains speculative, as few or no studies have examined the role of extended training in performance.

As Watson and Kidd (1994) suggested, many populations of users may be unwilling to undergo more than nominally time-consuming training programs, but research suggests that even brief training for sonification users offers benefits. Smith and Walker (2005) showed that brief training for a point estimation task (i.e., naming the Y axis value for a given X axis value in an auditory graph) resulted in better performance than no training, while Walker and Nees (2005b) further demonstrated that a brief training period (around 20 min) can reduce performance error by 50% on a point estimation sonification task. Recent and ongoing work is examining exactly what kinds of training methods are most effective for different classes of sonifications.

2.7 Conclusions: Toward a Cohesive Theoretical Account of Sonification

Current research is taking the field of sonification in many exciting directions, and researchers and practitioners have only just begun to harness the potential for sound to enhance and improve existing interfaces or be developed into purely auditory interfaces. The literature on auditory displays has grown tremendously. These successes notwithstanding, sonification research and design faces many obstacles and challenges in the pursuit of ubiquitous, usable, and aesthetically pleasing sounds for human-machine interactions, and perhaps the most pressing obstacle is the need for a cohesive theoretical paradigm in which research and design can continue to develop.

Although the field of auditory display has benefited tremendously from multidisciplinary approaches in research and practice, this same diversity has likely been an obstacle to the formation of a unified account of sound as an information display medium. To date, few theories or models of human interaction with auditory displays exist. It seems inevitable that the field of sonification will need to develop fuller explanatory models in order to realize the full potential of the field. As Edwards (1989) pointed out, the development of new models or the expansion of existing models of human interaction with information systems to include auditory displays will benefit twofold: 1) In research, models of human interaction with auditory displays will provide testable hypotheses that will guide a systematic, programmatic approach to auditory display research, and 2) In practice, auditory display designers will be able to turn to models for basic guidelines. These benefits notwithstanding, the development of theory remains difficult, especially in pragmatic and somewhat design-oriented fields like sonification (for a discussion, see Hooker, 2004).

A distinction has been drawn, however, between "theorizing" as a growing process within a field, and "theory" as a product of that process (Weick, 1995). Despite the absence of a grand theory of sonification, recent developments reflect the field's active march toward meaningful theory. Important evidence of progress toward meeting some of the conditions of a cohesive theory of sonification is emerging. Theory in sonification will depend upon a shared language, and Hermann (2008) recently initiated a much-needed discussion about definitional boundaries and fundamental terminology in the field. Theory requires a meaningful organization of extant knowledge, and de Campo's (2007) recent work offered an important step toward describing the diverse array of sonification designs within a common space. Theory will bridge the gap between research and practice, and Brazil (Brazil, 2010;

Brazil & Fernstrom, 2009) has begun to offer insights for integrating sonification design and empirical methods of evaluation (also see chapter 6 in this volume). Theory specifies the important variables that contribute to performance of the data-display-human system. Nees and Walker (2007) recently described a conceptual model of the variables relevant to auditory graph comprehension, whereas Bruce and Walker (2009) took a similar conceptual model approach toward understanding the role of audio in dynamic exhibits. Theory will result in reusable knowledge rather than idiosyncratic, ad hoc designs, and Frauenberger and Stockman (2009) have developed a framework to assist in the capture and dissemination of effective designs for auditory displays. As such, there is reason for optimism about the future of theoretical work in the field of sonification, and a shared based of organized knowledge that guides new research and best practice implementation of sonifications should be one of the foremost aspirations of the field in the immediate future.

Bibliography

[1] Barrass, S. (1994/2005). A perceptual framework for the auditory display of scientific data. *ACM Transactions on Applied Perception*, 2(4), 389–402.

[2] Barrass, S. (1996). TaDa! Demonstrations of auditory information design. *Proceedings of the 3rd International Conference on Auditory Display*, Palo Alto, CA.

[3] Barrass, S. (1997). Auditory information design. Unpublished Dissertation, Australian National University.

[4] Barrass, S. (2005). A perceptual framework for the auditory display of scientific data. *ACM Transactions on Applied Perception*, 2(4), 389–492.

[5] Barrass, S., & Best, V. (2008). Stream-based sonification diagrams. *Proceedings of the 14th International Conference on Auditory Display*, Paris, France.

[6] Bertin, J. (1983). *Semiology of Graphics* (W. J. Berg, Trans.). Madison, Wisconsin: The University of Wisconsin Press.

[7] Blattner, M. M., Sumikawa, D. A., & Greenberg, R. M. (1989). Earcons and icons: Their structure and common design principles. *Human-Computer Interaction*, 4, 11–44.

[8] Bonebright, T. L., & Nees, M. A. (2009). Most earcons do not interfere with spoken passage comprehension. *Applied Cognitive Psychology*, 23(3), 431–445.

[9] Bonebright, T. L., Nees, M. A., Connerley, T. T., & McCain, G. R. (2001). Testing the effectiveness of sonified graphs for education: A programmatic research project. *Proceedings of the International Conference on Auditory Display (ICAD2001)* (pp. 62–66), Espoo, Finland.

[10] Brazil, E. (2010). A review of methods and frameworks for sonic interaction design: Exploring existing approaches. *Lecture Notes in Computer Science*, 5954, 41–67.

[11] Brazil, E., & Fernstrom, M. (2009). Empirically based auditory display design. *Proceedings of the SMC 2009 – 6th Sound and Computing Conference* (pp. 7–12), Porto, Portugal.

[12] Bregman, A. S. (1990). *Auditory Scene Analysis: The Perceptual Organization of Sound.* Cambridge, MA: MIT Press.

[13] Brewster, S. (1997). Using non-speech sound to overcome information overload. *Displays*, 17, 179–189.

[14] Brewster, S., & Murray, R. (2000). Presenting dynamic information on mobile computers. *Personal Technologies*, 4(4), 209–212.

[15] Brown, A. W. (1928). The reliability and validity of the Seashore Tests of Musical Talent. *Journal of Applied Psychology*, 12, 468–476.

[16] Brown, L. M., Brewster, S., & Riedel, B. (2002). Browsing modes for exploring sonified line graphs. *Proceedings of the 16th British HCI Conference*, London, UK.

[17] Brown, L. M., & Brewster, S. A. (2003). Drawing by ear: Interpreting sonified line graphs. *Proceedings of*

the International Conference on Auditory Display (ICAD2003) (pp. 152–156), Boston, MA.

[18] Brown, L. M., Brewster, S. A., Ramloll, R., Burton, M., & Riedel, B. (2003). Design guidelines for audio presentation of graphs and tables. *Proceedings of the International Conference on Auditory Display (ICAD2003)* (pp. 284–287), Boston, MA.

[19] Brown, M. L., Newsome, S. L., & Glinert, E. P. (1989). An experiment into the use of auditory cues to reduce visual workload. *Proceedings of the ACM CHI 89 Human Factors in Computing Systems Conference (CHI 89)* (pp. 339–346).

[20] Bruce, C., & Walker, B. N. (2009). Modeling visitor-exhibit interaction at dynamic zoo and aquarium exhibits for developing real-time interpretation. *Proceedings of the Association for the Advancement of Assistive Technology in Europe Conference* (pp. 682–687).

[21] Buxton, W. (1989). Introduction to this special issue on nonspeech audio. *Human-Computer Interaction*, 4, 1–9.

[22] Buxton, W., Bly, S. A., Frysinger, S. P., Lunney, D., Mansur, D. L., Mezrich, J. J., et al. (1985). Communicating with sound. *Proceedings of the CHI '85* (pp. 115–119).

[23] Cary, H. (1923). Are you a musician? Professor Seashore's specific psychological tests for specific musical abilities. *Scientific American*, 326–327.

[24] Cleveland, W. S., & McGill, R. (1984). Graphical perception: Theory, experimentation, and application to the development of graphical methods. *Journal of the American Statistical Association*, 79(387), 531–554.

[25] Coward, S. W., & Stevens, C. J. (2004). Extracting meaning from sound: Nomic mappings, everyday listening, and perceiving object size from frequency. *The Psychological Record*, 54, 349–364.

[26] Cuddon, J. A. (1991). *Dictionary of Literary Terms and Literary Theory* (3rd ed.). New York: Penguin Books.

[27] Cusack, R., & Roberts, B. (2000). Effects of differences in timbre on sequential grouping. *Perception & Psychophysics*, 62(5), 1112–1120.

[28] de Campo, A. (2007). Toward a data sonification design space map. *Proceedings of the International Conference on Auditory Display (ICAD2007)* (pp. 342–347), Montreal, Canada.

[29] Dombois, F. (2002). Auditory seismology – On free oscillations, focal mechanisms, explosions, and synthetic seismograms. *Proceedings of the 8th International Conference on Auditory Display* (pp. 27–30), Kyoto, Japan.

[30] Durlach, N. I., Mason, C. R., Kidd, G., Arbogast, T. L., Colburn, H. S., & Shinn-Cunningham, B. (2003). Note on informational masking. *Journal of the Acoustical Society of America*, 113(6), 2984–2987.

[31] Edwards, A. D. N., Challis, B. P., Hankinson, J. C. K., & Pirie, F. L. (2000). Development of a standard test of musical ability for participants in auditory interface testing. *Proceedings of the International Conference on Auditory Display (ICAD 2000)*, Atlanta, GA.

[32] Edworthy, J. (1998). Does sound help us to work better with machines? A commentary on Rautenberg's paper 'About the importance of auditory alarms during the operation of a plant simulator'. *Interacting with Computers*, 10, 401–409.

[33] Edworthy, J., & Hellier, E. (2000). Auditory warnings in noisy environments. *Noise & Health*, 2(6), 27–39.

[34] Edworthy, J., & Hellier, E. (2005). Fewer but better auditory alarms will improve patient safety. *British Medical Journal*, 14(3), 212–215.

[35] Edworthy, J., & Hellier, E. (2006). Complex nonverbal auditory signals and speech warnings. In M. S. Wogalter (Ed.), *Handbook of Warnings* (pp. 199–220). Mahwah, NJ: Lawrence Erlbaum.

[36] Edworthy, J., Hellier, E. J., Aldrich, K., & Loxley, S. (2004). Designing trend-monitoring sounds for helicopters: Methodological issues and an application. *Journal of Experimental Psychology: Applied*, 10(4), 203–218.

[37] Fitch, W. T., & Kramer, G. (1994). Sonifying the body electric: Superiority of an auditory over a visual display in a complex, multivariate system. In G. Kramer (Ed.), *Auditory Display: Sonification, Audification, and Auditory Interfaces* (pp. 307–326). Reading, MA: Addison-Wesley.

[38] Flowers, J. H., Buhman, D. C., & Turnage, K. D. (1997). Cross-modal equivalence of visual and auditory scatterplots for exploring bivariate data samples. *Human Factors*, 39(3), 341–351.

[39] Flowers, J. H., & Hauer, T. A. (1992). The ear's versus the eye's potential to assess characteristics of numeric data: Are we too visuocentric? *Behavior Research Methods, Instruments & Computers*, 24(2), 258–264.

[40] Flowers, J. H., & Hauer, T. A. (1993). "Sound" alternatives to visual graphics for exploratory data analysis. *Behavior Research Methods, Instruments & Computers*, 25(2), 242–249.

[41] Flowers, J. H., & Hauer, T. A. (1995). Musical versus visual graphs: Cross-modal equivalence in perception of time series data. *Human Factors*, 37(3), 553–569.

[42] Franklin, K. M., & Roberts, J. C. (2004). A path based model for sonification. *Proceedings of the Eighth International Conference on Information Visualization (IV '04)* (pp. 865–870).

[43] Frauenberger, C., & Stockman, T. (2009). Auditory display design: An investigation of a design pattern approach. *International Journal of Human-Computer Studies*, 67, 907–922.

[44] Frauenberger, C., Stockman, T., & Bourguet, M.-L. (2007). A Survey on Common Practice in Designing Audio User Interface. *21st British HCI Group Annual Conference (HCI 2007).* Lancaster, UK.

[45] Frauenberger, C., Stockman, T., & Bourguet, M.-L. (2007). Pattern design in the context space: A methodological framework for auditory display design. *Proceedings of the International Conference on Auditory Display (ICAD2007)* (pp. 513–518), Montreal, Canada.

[46] Friel, S. N., Curcio, F. R., & Bright, G. W. (2001). Making sense of graphs: Critical factors influencing comprehension and instructional applications [Electronic version]. *Journal for Research in Mathematics*, 32(2), 124–159.

[47] Frysinger, S. P. (2005). A brief history of auditory data representation to the 1980s. *Proceedings of the International Conference on Auditory Display (ICAD 2005)*, Limerick, Ireland.

[48] Gardner, J. A., Lundquist, R., & Sahyun, S. (1996). TRIANGLE: A practical application of non-speech audio for imparting information. *Proceedings of the International Conference on Auditory Display* (pp. 59–60), San Francisco, CA.

[49] Garner, W. R., & Gottwald, R. L. (1968). The perception and learning of temporal patterns. *The Quarterly Journal of Experimental Psychology*, 20(2).

[50] Gaver, W. W. (1989). The SonicFinder: An interface that uses auditory icons. *Human-Computer Interaction*, 4(1), 67–94.

[51] Gaver, W. W. (1993). What in the world do we hear? An ecological approach to auditory event perception. *Ecological Psychoogy*, 5(1), 1–29.

[52] Gaver, W. W. (1994). Using and creating auditory icons. In G. Kramer (Ed.), *Auditory Display: Sonification, Audification, and Auditory Interfaces* (pp. 417–446). Reading, MA: Addison-Wesley.

[53] Gaver, W. W., Smith, R. B., & O'Shea, T. (1991). Effective sounds in complex systems: The ARKola simulation. *Proceedings of the ACM Conference on Human Factors in Computing Systems CHI'91*, New Orleans.

[54] Godbout, A., & Boyd, J. E. (2010). Corrective sonic feedback for speed skating: A case study. *Proceedings of the 16th International Conference on Auditory Display (ICAD2010)* (pp. 23–30), Washington, DC.

[55] Haas, E., & Edworthy, J. (2006). An introduction to auditory warnings and alarms. In M. S. Wogalter (Ed.), *Handbook of Warnings* (pp. 189–198). Mahwah, NJ: Lawrence Erlbaum

[56] Hereford, J., & Winn, W. (1994). Non-speech sound in human-computer interaction: A review and design guidelines. *Journal of Educational Computer Research*, 11, 211–233.

[57] Hermann, T. (2002). Sonification for exploratory data analysis. Ph.D. thesis, Faculty of Technology, Bielefeld University, `http://sonification.de/publications/Hermann2002-SFE`

[58] Hermann, T. (2008). Taxonomy and definitions for sonification and auditory display. *Proceedings of the 14th International Conference on Auditory Display*, Paris, France.

[59] Hermann, T. (2011). Sonification – A Definition. Retrieved January 23, 2011, from `http://sonification.de/son/definition`

[60] Hermann, T., & Hunt, A. (2005). An introduction to interactive sonification. *IEEE Multimedia*, 12(2), 20–24.

[61] Hermann, T., & Ritter, H. (1999). Listen to your data: Model-based sonification for data analysis. In G. D. Lasker (Ed.), *Advances in Intelligent Computing and Multimedia Systems* (pp. 189–194). Baden-Baden, Germany: IIASSRC.

[62] Hooker, J. N. (2004). Is design theory possible? *Journal of Information Technology Theory and Application*, 6(2), 73–82.

[63] Iwarsson, S., & Stahl, A. (2003). Accessibility, usability, and universal design–positioning and definition of concepts describing person-environment relationships. *Disability and Rehabilitation*, 25(2), 57–66.

[64] Janata, P., & Childs, E. (2004). Marketbuzz: Sonification of real-time financial data. *Proceedings of the Tenth Meeting of the International Conference on Auditory Display (ICAD04)*, Sydney, Australia.

[65] Jeon, M., Davison, B., Nees, M. A., Wilson, J., & Walker, B. N. (2009). Enhanced auditory menu cues improve dual task performance and are preferred with in-vehicle technologies. *Proceedings of the First International Conference on Automotive User Interfaces and Interactive Vehicular Applications (Automotive UI 2009)*, Essen, Germany.

[66] Jeon, M., & Walker, B. N. (2011). Spindex (speech index) improves auditory menu acceptance and navigation performance. *ACM Transactions on Accessible Computing, 3,* Article 10.

[67] Johannsen, G. (2004). Auditory displays in human-machine interfaces. *Proceedings of the IEEE*, 92(4), 742–758.

[68] Keller, P., & Stevens, C. (2004). Meaning from environmental sounds: Types of signal-referent relations and their effect on recognizing auditory icons. *Journal of Experimental Psychology: Applied*, 10(1), 3–12.

[69] Kortum, P., Peres, S. C., Knott, B., & Bushey, R. (2005). The effect of auditory progress bars on consumer's estimation of telephone wait time. *Proceedings of the Human Factors and Ergonomics Society 49th Annual Meeting* (pp. 628–632), Orlando, FL.

[70] Kramer, G. (1994). An introduction to auditory display. In G. Kramer (Ed.), *Auditory Display: Sonification, Audification, and Auditory Interfaces* (pp. 1–78). Reading, MA: Addison Wesley.

[71] Kramer, G., Walker, B. N., Bonebright, T., Cook, P., Flowers, J., Miner, N., et al. (1999). The Sonification Report: Status of the Field and Research Agenda. Report prepared for the National Science Foundation by members of the International Community for Auditory Display. Santa Fe, NM: International Community for Auditory Display (ICAD).

[72] Lacherez, P., Seah, E. L., & Sanderson, P. M. (2007). Overlapping melodic alarms are almost indiscriminable. *Human Factors*, 49(4), 637–645.

[73] Lakatos, S., McAdams, S., & Causse, R. (1997). The representation of auditory source characteristics: Simple geometric form. *Perception & Psychophysics*, 59(8), 1180–1190.

[74] Letowski, T., Karsh, R., Vause, N., Shilling, R. D., Ballas, J., Brungart, D., et al. (2001). Human factors military lexicon: Auditory displays. Army Research Laboratory Technical Report No. ARL-TR-2526

[75] Levitin, D. J. (1999). Memory for musical attributes. In P. Cook (Ed.), *Music, Cognition, and Computerized Sound: An Introduction to Psychoacoustics.* (pp. 209–227). Cambridge, MA: MIT Press.

[76] Liljedahl, M., Papworth, N., & Lindberg, S. (2007). Beowulf: A game experience built on sound effects. *Proceedings of the International Conference on Auditory Display (ICAD2007)* (pp. 102–106), Montreal, Canada.

[77] Luczak, H. (1997). Task analysis. In G. Salvendy (Ed.), *Handbook of Human Factors and Ergonomics* (2nd ed., pp. 340–416). New York: Wiley.

[78] Mauney, B. S., & Walker, B. N. (2004). Creating functional and livable soundscapes for peripheral monitoring of dynamic data. *Proceedings of the 10th International Conference on Auditory Display (ICAD04)*, Sydney, Australia.

[79] Mauney, L. M., & Walker, B. N. (2010). Universal design of auditory graphs: A comparison of sonification mappings for visually impaired and sighted listeners. *ACM Transactions on Accessible Computing*, 2(3), Article 12.

[80] McAdams, S., & Bigand, E. (1993). *Thinking in Sound: The Cognitive Psychology of Human Audition.*

Oxford: Oxford University Press.

[81] McCrindle, R. J., & Symons, D. (2000). Audio space invaders. *Proceedings of the 3rd International Conference on Disability, Virtual Reality, & Associated Technologies* (pp. 59–65), Alghero, Italy.

[82] McGookin, D. K., & Brewster, S. A. (2004). Understanding concurrent earcons: Applying auditory scene analysis principles to concurrent earcon recognition. *ACM Transactions on Applied Perception*, 1, 130–150.

[83] Meyer, J. (2000). Performance with tables and graphs: Effects of training and a visual search model. *Ergonomics*, 43(11), 1840–1865.

[84] Meyer, J., Shinar, D., & Leiser, D. (1997). Multiple factors that determine performance with tables and graphs. *Human Factors*, 39(2), 268–286.

[85] Moore, B. C. J. (1997). *An Introduction to the Psychology of Hearing* (4th ed.). San Diego, Calif.: Academic Press.

[86] Mulligan, B. E., McBride, D. K., & Goodman, L. S. (1984). A design guide for nonspeech auditory displays: Naval Aerospace Medical Research Laboratory Technical Report, Special Report No. 84–1.

[87] Nees, M. A. (2009). Internal representations of auditory frequency: Behavioral studies of format and malleability by instructions. Unpublished Ph.D. Dissertation.: Georgia Institute of Technology. Atlanta, GA.

[88] Nees, M. A., & Walker, B. N. (2006). Relative intensity of auditory context for auditory graph design. *Proceedings of the Twelfth International Conference on Auditory Display (ICAD06)* (pp. 95–98), London, UK.

[89] Nees, M. A., & Walker, B. N. (2007). Listener, task, and auditory graph: Toward a conceptual model of auditory graph comprehension. *Proceedings of the International Conference on Auditory Display (ICAD2007)* (pp. 266–273), Montreal, Canada.

[90] Nees, M. A., & Walker, B. N. (2008). Encoding and representation of information in auditory graphs: Descriptive reports of listener strategies for understanding data. *Proceedings of the International Conference on Auditory Display (ICAD 08)*, Paris, FR (24–27 June).

[91] Nees, M. A., & Walker, B. N. (2009). Auditory interfaces and sonification. In C. Stephanidis (Ed.), *The Universal Access Handbook* (pp. TBD). New York: CRC Press.

[92] Nees, M. A., & Walker, B. N. (in press). Mental scanning of sonifications reveals flexible encoding of nonspeech sounds and a universal per-item scanning cost. *Acta Psychologica*.

[93] Neuhoff, J. G. (Ed.). (2004). *Ecological Psychoacoustics*. New York: Academic Press.

[94] Neuhoff, J. G., & Heller, L. M. (2005). One small step: Sound sources and events as the basis for auditory graphs. *Proceedings of the Eleventh Meeting of the International Conference on Auditory Display*, Limerick, Ireland.

[95] Neuhoff, J. G., Kramer, G., & Wayand, J. (2002). Pitch and loudness interact in auditory displays: Can the data get lost in the map? *Journal of Experimental Psychology: Applied*, 8(1), 17–25.

[96] Neuhoff, J. G., & Wayand, J. (2002). Pitch change, sonification, and musical expertise: Which way is up? *Proceedings of the International Conference on Auditory Display* (pp. 351–356), Kyoto, Japan.

[97] Palladino, D., & Walker, B. N. (2007). Learning rates for auditory menus enhanced with spearcons versus earcons. *Proceedings of the International Conference on Auditory Display (ICAD2007)* (pp. 274–279), Montreal, Canada.

[98] Peres, S. C., Best, V., Brock, D., Shinn-Cunningham, B., Frauenberger, C., Hermann, T., et al. (2008). Auditory Interfaces. In P. Kortum (Ed.), *HCI Beyond the GUI: Design for Haptic, Speech, Olfactory and Other Nontraditional Interfaces* (pp. 147–196). Burlington, MA: Morgan Kaufmann.

[99] Peres, S. C., & Lane, D. M. (2005). Auditory graphs: The effects of redundant dimensions and divided attention. *Proceedings of the International Conference on Auditory Display (ICAD 2005)* (pp. 169–174), Limerick, Ireland.

[100] Pirhonen, A., Murphy, E., McAllister, g., & Yu, W. (2006). Non-speech sounds as elements of a use scenario: A semiotic perspective. *Proceedings of the 12th International Conference on Auditory Display (ICAD06)*, London, UK.

[101] Quinn, M. (2001). Research set to music: The climate symphony and other sonifications of ice core, radar, DNA, seismic, and solar wind data. *Proceedings of the 7th International Conference on Auditory Display (ICAD01)*, Espoo, Finland.

[102] Quinn, M. (2003). For those who died: A 9/11 tribute. *Proceedings of the 9th International Conference on Auditory Display*, Boston, MA.

[103] Salvendy, G. (1997). *Handbook of Human Factors and Ergonomics* (2nd ed.). New York: Wiley.

[104] Sanders, M. S., & McCormick, E. J. (1993). *Human Factors in Engineering and Design* (7th ed.). New York: McGraw-Hill.

[105] Sanderson, P. M. (2006). The multimodal world of medical monitoring displays. *Applied Ergonomics*, 37, 501–512.

[106] Sanderson, P. M., Liu, D., & Jenkins, D. A. (2009). Auditory displays in anesthesiology. *Current Opinion in Anesthesiology*, 22, 788–795.

[107] Sandor, A., & Lane, D. M. (2003). Sonification of absolute values with single and multiple dimensions. *Proceedings of the 2003 International Conference on Auditory Display (ICAD03)* (pp. 243–246), Boston, MA.

[108] Schaffert, N., Mattes, K., Barrass, S., & Effenberg, A. O. (2009). Exploring function and aesthetics in sonifications for elite sports. *Proceedings of the Second International Conference on Music Communication Science*, Sydney, Australia.

[109] Seashore, C. E., Lewis, D., & Saetveit, J. G. (1960). *Seashore Measures of Musical Talents* (Revised 1960 ed.). New York: The Psychological Corp.

[110] Shah, P. (2002). Graph comprehension: The role of format, content, and individual differences. In M. Anderson, B. Meyer & P. Olivier (Eds.), *Diagrammatic Representation and Reasoning* (pp. 173–185). New York: Springer.

[111] Smith, D. R., & Walker, B. N. (2005). Effects of auditory context cues and training on performance of a point estimation sonification task. *Applied Cognitive Psychology*, 19(8), 1065–1087.

[112] Sorkin, R. D. (1987). Design of auditory and tactile displays. In G. Salvendy (Ed.), *Handbook of Human Factors* (pp. 549–576). New York: Wiley & Sons.

[113] Speeth, S. D. (1961). Seismometer sounds. *Journal of the Acoustical Society of America*, 33, 909–916.

[114] Spence, C., & Driver, J. (1997). Audiovisual links in attention: Implications for interface design. In D. Harris (Ed.), *Engineering Psychology and Cognitive Ergonomics Vol. 2: Job Design and Product Design* (pp. 185–192). Hampshire: Ashgate Publishing.

[115] Stevens, S. S. (1946). On the theory of scales of measurement. *Science*, 13(2684), 677–680.

[116] Stevens, S. S. (1975). *Psychophysics: Introduction to its Perceptual, Neural, and Social Prospects*. New York: Wiley.

[117] Stockman, T., Rajgor, N., Metatla, O., & Harrar, L. (2007). The design of interactive audio soccer. *Proceedings of the 13th International Conference on Auditory Display* (pp. 526–529), Montreal, Canada.

[118] Stokes, A., Wickens, C. D., & Kite, K. (1990). *Display Technology: Human Factors Concepts*. Warrendale, PA: Society of Automotive Engineers.

[119] Storms, R. L., & Zyda, M. J. (2000). Interactions in perceived quality of auditory-visual displays. *Presence: Teleoperators & Virtual Environments*, 9(6), 557–580.

[120] Targett, S., & Fernstrom, M. (2003). Audio games: Fun for all? All for fun? *Proceedings of the International Conference on Auditory Display (ICAD2003)* (pp. 216–219), Boston, MA.

[121] Toth, J. A., & Lewis, C. M. (2002). The role of representation and working memory in diagrammatic reasoning and decision making. In M. Anderson, B. Meyer & P. Olivier (Eds.), *Diagrammatic Representation and Reasoning* (pp. 207–221). New York: Springer.

[122] Trickett, S. B., & Trafton, J. G. (2006). Toward a comprehensive model of graph comprehension: Making the case for spatial cognition. *Proceedings of the Fourth International Conference on the Theory and Application of Diagrams (DIAGRAMS 2006)*, Stanford University, USA.

[123] Tufte, E. R. (1990). *Envisioning Information*. Cheshire, Connecticut: Graphics Press.

[124] Vickers, P., & Hogg, B. (2006). Sonification abstraite/sonification concrete: An 'aesthetic perspective space' for classifying auditory displays in the ars musica domain. *Proceedings of the International Conference on Auditory Display (ICAD2006)* (pp. 210–216), London, UK.

[125] Walker, B. N. (2002). Magnitude estimation of conceptual data dimensions for use in sonification. *Journal of Experimental Psychology: Applied*, 8, 211–221.

[126] Walker, B. N. (2007). Consistency of magnitude estimations with conceptual data dimensions used for sonification. *Applied Cognitive Psychology*, 21, 579–599.

[127] Walker, B. N., & Cothran, J. T. (2003). Sonification Sandbox: A graphical toolkit for auditory graphs. *Proceedings of the International Conference on Auditory Display (ICAD2003)* (pp. 161–163), Boston, MA.

[128] Walker, B. N., Godfrey, M. T., Orlosky, J. E., Bruce, C., & Sanford, J. (2006a). Aquarium sonification: Soundscapes for accessible dynamic informal learning environments. *Proceedings of the International Conference on Auditory Display (ICAD 2006)* (pp. 238–241), London, UK.

[129] Walker, B. N., Kim, J., & Pendse, A. (2007). Musical soundscapes for an accessible aquarium: Bringing dynamic exhibits to the visually impaired. *Proceedings of the International Computer Music Conference (ICMC 2007)* (pp. TBD), Copenhagen, Denmark.

[130] Walker, B. N., & Kogan, A. (2009). Spearcons enhance performance and preference for auditory menus on a mobile phone. *Proceedings of the 5th international conference on universal access in Human-Computer Interaction (UAHCI) at HCI International 2009*, San Diego, CA, USA.

[131] Walker, B. N., & Kramer, G. (1996). Human factors and the acoustic ecology: Considerations for multimedia audio design. *Proceedings of the Audio Engineering Society 101st Convention*, Los Angeles.

[132] Walker, B. N., & Kramer, G. (2004). Ecological psychoacoustics and auditory displays: Hearing, grouping, and meaning making. In J. Neuhoff (Ed.), *Ecological psychoacoustics* (pp. 150–175). New York: Academic Press.

[133] Walker, B. N., & Kramer, G. (2005). Mappings and metaphors in auditory displays: An experimental assessment. *ACM Transactions on Applied Perception*, 2(4), 407–412.

[134] Walker, B. N., & Lane, D. M. (2001). Psychophysical scaling of sonification mappings: A comparision of visually impaired and sighted listeners. *Proceedings of the 7th International Conference on Auditory Display* (pp. 90–94), Espoo, Finland.

[135] Walker, B. N., & Lowey, M. (2004). Sonification Sandbox: A graphical toolkit for auditory graphs. *Proceedings of the Rehabilitation Engineering & Assistive Technology Society of America (RESNA) 27th International Conference*, Orlando, FL.

[136] Walker, B. N., & Mauney, L. M. (2004). Individual differences, cognitive abilities, and the interpretation of auditory graphs. *Proceedings of the International Conference on Auditory Display (ICAD2004)*, Sydney, Australia.

[137] Walker, B. N., Nance, A., & Lindsay, J. (2006b). Spearcons: Speech-based earcons improve navigation performance in auditory menus. *Proceedings of the International Conference on Auditory Display (ICAD06)* (pp. 63–68), London, UK.

[138] Walker, B. N., & Nees, M. A. (2005). An agenda for research and development of multimodal graphs. *Proceedings of the International Conference on Auditory Display (ICAD2005)* (pp. 428–432), Limerick, Ireland.

[139] Walker, B. N., & Nees, M. A. (2005). Brief training for performance of a point estimation task sonification task. *Proceedings of the International Conference on Auditory Display (ICAD2005)*, Limerick, Ireland.

[140] Watson, C. S., & Kidd, G. R. (1994). Factors in the design of effective auditory displays. *Proceedings of the International Conference on Auditory Display (ICAD1994)*, Sante Fe, NM.

[141] Watson, M. (2006). Scalable earcons: Bridging the gap between intermittent and continuous auditory displays. *Proceedings of the 12th International Conference on Auditory Display (ICAD06)*, London, UK.

[142] Weick, K. E. (1995). What theory is not, theorizing is. *Administrative Science Quarterly*, 40(3), 385–390.

[143] Wickens, C. D., Gordon, S. E., & Liu, Y. (1998). *An Introduction to Human Factors Engineering*. New York:

Longman.

[144] Wickens, C. D., & Liu, Y. (1988). Codes and modalities in multiple resources: A success and a qualification. *Human Factors*, 30(5), 599–616.

[145] Wickens, C. D., Sandry, D. L., & Vidulich, M. (1983). Compatibility and resource competition between modalities of input, central processing, and output. *Human Factors*, 25(2), 227–248.

[146] Winberg, F., & Hellstrom, S. O. (2001). Qualitative aspects of auditory direct manipulation: A case study of the Towers of Hanoi. *Proceedings of the International Conference on Auditory Display (ICAD 2001)* (pp. 16–20), Espoo, Finland.

[147] Worrall, D. R. (2009). An introduction to data sonification. In R. T. Dean (Ed.), *The Oxford Handbook of Computer Music and Digital Sound Culture* (pp. 312–333). Oxford: Oxford University Press.

[148] Worrall, D. R. (2009). Information sonification: Concepts, instruments, techniques. Unpublished Ph.D. Thesis: University of Canberra.

[149] Zacks, J., Levy, E., Tversky, B., & Schiano, D. (2002). Graphs in print. In M. Anderson, B. Meyer & P. Olivier (Eds.), *Diagrammatic Representation and Reasoning* (pp. 187–206). London: Springer.

Chapter 3

Psychoacoustics

Simon Carlile

3.1 Introduction

Listening in the real world is generally a very complex task since sounds of interest typically occur on a background of other sounds that overlap in frequency and time. Some of these sounds can represent threats or opportunities while others are simply distracters or maskers. One approach to understanding how the auditory system makes sense of this complex acoustic world is to consider the nature of the sounds that convey high levels of information and how the auditory system has evolved to extract that information. From this evolutionary perspective, humans have largely inherited this biological system so it makes sense to consider how our auditory systems use these mechanisms to extract information that is meaningful to us and how that knowledge can be applied to best sonify various data.

One biologically important feature of a sound is its *identity*; that is, the spectro-temporal characteristics of the sound that allow us to extract the relevant information represented by the sound. Another biologically important feature is the location of the source. In many scenarios an appropriate response to the information contained in the sound is determined by its relative location to the listener – for instance to approach an opportunity or retreat from a threat.

All sounds arrive at the ear drum as a combined stream of pressure changes that jointly excite the inner ear. What is most remarkable is that the auditory system is able to disentangle sort out the many different streams of sound and provides the capacity to selectively focus our attention on one or another of these streams [1, 2]. This has been referred to as the "cocktail party problem" and represents a very significant signal processing challenge. Our perception of this multi-source, complex auditory environment is based on a range of acoustic cues that occur at each ear. Auditory perception relies firstly on how this information is broken down and encoded at the level of the auditory nerve and secondly how this information is then recombined in the brain to compute the identity and location of the different sources. Our capacity to focus attention on one sound of interest and to ignore distracting sounds is also dependent, at least in part, on the differences in the locations of the different sound sources [3, 4] (sound example **S3.1**). This capacity to segregate the different sounds is essential to the extraction of meaningful information from our complex acoustic world.

In the context of auditory displays it is important ensure that the fidelity of a display is well matched to the encoding capability of the human auditory system. The capacity of the auditory system to encode physical changes in a sound is an important input criterion in the design of an auditory display. For instance, if a designer encodes information using changes in the frequency or amplitude of a sound it is important to account for the fundamental sensitivity of the auditory system to these physical properties to ensure that these physical variations can be perceived. In the complexity of real world listening, many factors will contribute to the perception of individual sound sources. Perception is not necessarily a simple linear combination of different frequency components. Therefore, another critical issue is understanding how the perception of multiple elements in a sound field are combined to give rise to specific perceptual objects and how variations in the physical properties of the sound will affect different perceptual objects. For instance, when designing a 3D audio interface, a key attribute of the system is the sense of the acoustic space that is generated. However, other less obvious attributes of the display may play a key role in the performance of users. For example, the addition of reverberation to a display can substantially enhance the sense of 'presence' or the feeling of 'being in' a virtual soundscape [5]. However, reverberation can also degrade user performance on tasks such as the localization of brief sounds (e.g., see [6, 7]).

This chapter looks at how sound is encoded physiologically by the auditory system and the perceptual dimensions of pitch, timbre, and loudness. It considers how the auditory system decomposes complex sounds into their different frequency components and also the rules by which these are recombined to form the perception of different, individual sounds. This leads to the identification of the acoustic cues that the auditory system employs to compute the location of a source and the impact of reverberation on those cues. Many of the more complex aspects of our perception of sound sources will be covered in later chapters

3.2 The transduction of mechanical sound energy into biological signals in the auditory nervous system

The first link in this perceptual chain is the conversion of physical acoustic energy into biological signals within the inner ear. Every sound that we perceive in the physical world is bound by the encoding and transmission characteristics of this system. Therefore, sound is not simply encoded but various aspects of the sound may be filtered out. Sound enters the auditory system by passing through the outer and middle ears to be transduced into biological signals in the inner ear. As it passes through these structures the sound is transformed in a number of ways.

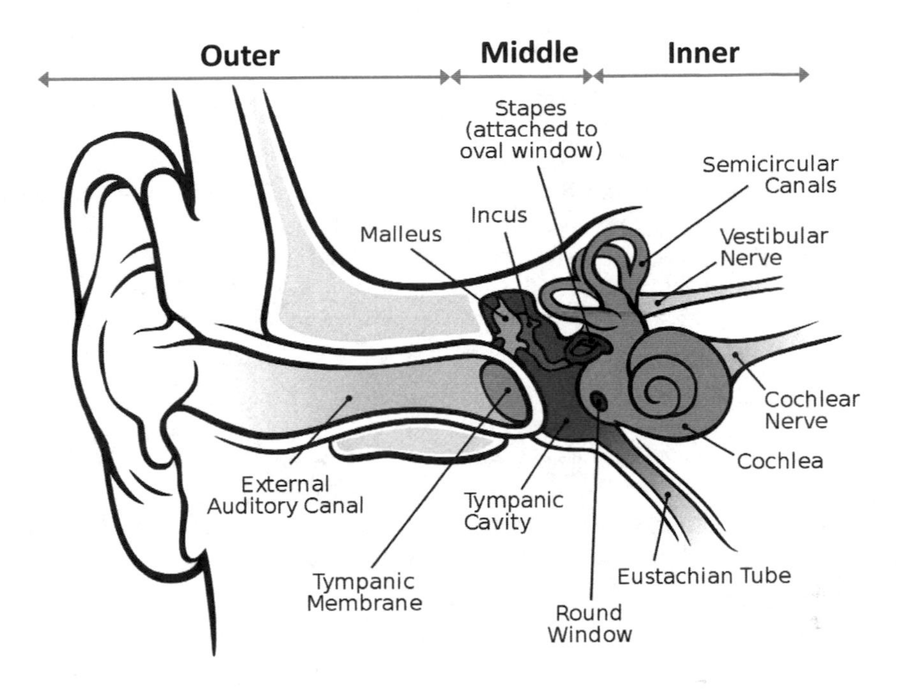

Figure 3.1: The human ear has three main groups of structures, namely the outer, middle and inner ear. The pinna and concha of the outer ear collects and filters sound and delivers this to the middle ear via the external auditory canal. The middle ear effectively transmits the sounds from the gas medium of the outer ear to the fluid medium of the inner ear. The inner ear transduces the physical sound energy to biological signals that are transmitted into the brain via the auditory nerve. Adapted from `http://en.wikipedia.org/wiki/File:Anatomy_of_the_Human_Ear.svg`.

3.2.1 The outer ear

The first step in the process is the transmission of sound through the outer ear to the middle ear. The outer ear extends from the pinna and concha on the side of the head to the end of the auditory canal at the ear drum (Figure 3.1). The relatively large aperture of the pinna of the outer ear collects sound energy and funnels it into the smaller aperture of the external auditory canal. This results in an overall gain in the amount of sound energy entering the middle ear. In common with many animals, the pinna and concha of the human outer ears are also quite convoluted and asymmetrical structures. This results in complex interactions between the incoming sounds and reflections within the ear that producing spectral filtering of the sound [8]. Most importantly, the precise nature of the filtering is dependent on the relative direction of the incoming sounds [9, 10]. There are two important consequences of this filtering.

Firstly, the auditory system uses these direction-dependent changes in the filtering as cues to the relative locations of different sound sources. This will be considered in greater detail later. This filtering also gives rise to the perception of a sound outside the head. This is best illustrated when we consider the experience generated by listening to music over headphones compared to listening over loudspeakers. Over headphones, the sound is introduced directly into the ear canal and the percept is of a source or sources located within the head and lateralized to one side of the head or the other. By contrast, when listening to sounds through loudspeakers, the sounds are first filtered by the outer ears and it is this cue that the auditory system uses to generate the perception of sources outside the head and away from the body. Consequently, if we filter the sounds presented over headphones in the same way as they would have been filtered had the sounds actually come from external sources, then the percept generated in the listener is of sounds located away from the head. This is the basis of so called virtual auditory space (VAS [9]).

Secondly, the details of the filtering are related to the precise shape of the outer ear. The fact that everybody's ears are slightly different in shape means that filtering by the outer ear is quite individualized. The consequence of this is that if a sound, presented using headphones, is filtered using the filtering characteristics of one person's ears, it will not necessarily generate the perception of an externalized source in a different listener – particularly if the listener's outer ear filter properties are quite different to those used to filter the headphone presented sounds.

3.2.2 The middle ear

The second stage in the transmission chain is to convey the sound from the air filled spaces of the outer ear to the fluid filled space of the inner ear. The middle ear plays this role and is comprised of (i) the ear drum, which is attached to the first of the middle bones - the malleus; (ii) the three middle ear bones (malleus, incus and stapes) and (iii) the stapes footplate which induces fluid movement in the cochlea of the inner ear. Through a combination of different mechanical mechanisms sound energy is efficiently transmitted from the air (gas) medium of the outer ear to the fluid filled cochlea in the inner ear.

3.2.3 The inner ear

The final step in the process is the conversion of sound energy into biological signals and ultimately neural impulses in the auditory nerve. On the way, sound is also analyzed into its different frequency components. The encoding process is a marvel of transduction as it preserves both a high level of frequency resolution as well as a high level of temporal resolution. All this represents an amazing feat of signal processing by the cochlea, a coiled structure in the inner ear no larger than the size of a garden pea!

The coiled structure of the cochlea contains the sensory transduction cells which are arranged along the basilar membrane (highlighted in red in Figure 3.2). The basilar membrane is moved up and down by the pressure changes in the cochlea induced by the movement of the stapes footplate on the oval window. Critically the stiffness and mass of the basilar membrane varies along its length so that the basal end (closest to the oval window and the middle ear) resonates at high frequencies and at the apical end resonates at low frequencies. A complex sound containing many frequencies will differentially activate the basilar membrane at the locations corresponding to the local frequency of resonance. This produces a place code

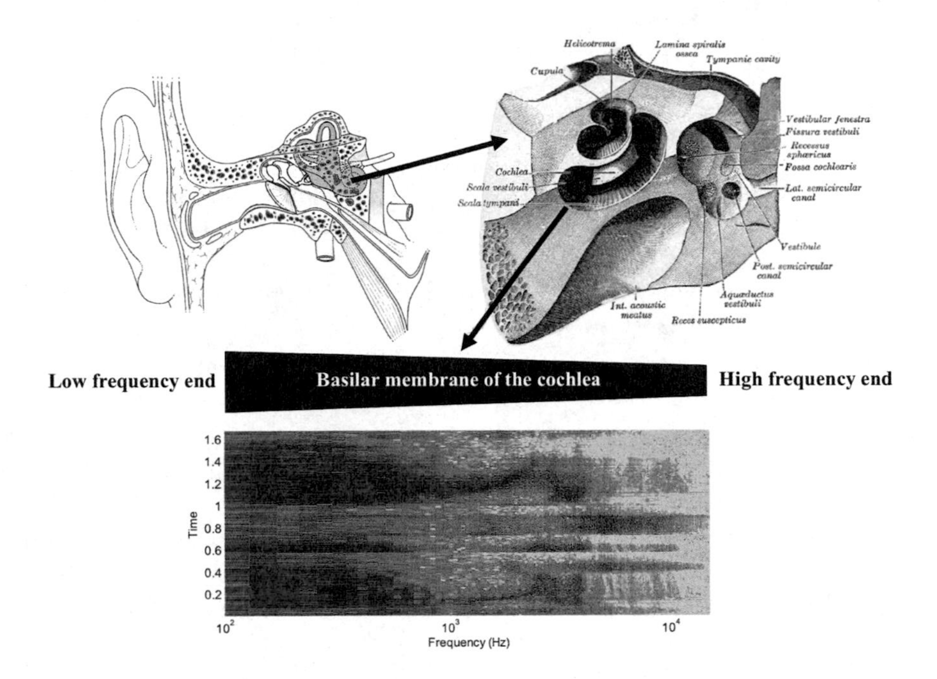

Figure 3.2: This figure shows the parts of the outer, middle and inner ear (top left), as well as an enlarged view of the inner ear with the basilar membrane in the cochlea highlighted in red (top right). The variation in frequency tuning along the length of the basilar membrane is illustrated in the middle panel and a sonogram of the words "please explain" is shown in the lower panel. The sonogram indicates how the pattern of sound energy changes over time (y-axis) and over the range of sound frequencies to which we are sensitive (x-axis). The sonogram also gives us an idea as to how the stimulation of the basilar membrane in the cochlea changes over time.

of frequency of the spectral content of the sound along the basilar membrane and provides the basis of what is called the tonotopic representation of frequency in the auditory nervous system and the so-called place theory of pitch perception (see also below).

The place of activation along the basilar membrane is indicated by the excitation of small sensory cells that are arranged along its structure. The sensory cells are called hair cells and cause electrical excitation of specific axons in the auditory nerve in response to movement of the part of the basilar membrane to which they are attached. As each axon is connected to just one inner hair cell it consequently demonstrates a relatively narrow range of frequency sensitivity. The frequency to which it is most sensitive is called its characteristic frequency (CF). The response bandwidth increases with increasing sound level but the frequency tuning remains quite narrow up to 30 dB to 40 dB above the threshold of hearing. The axons in the auditory nerve project into the nervous system in an ordered and systematic way so that this tonotopic representation of frequency is largely preserved in the ascending nervous system up to the auditory cortex. A second set of hair cells, the outer hair cells, provide a form of positive feedback and act as mechanical amplifiers that vastly improves the sensitivity and frequency selectivity. The outer hair cells are particularly susceptible to damage induced by overly loud sounds.

An important aspect of this encoding strategy is that for relatively narrow band sounds, small differences in frequency can be detected. The psychophysical aspects of this processing are considered below but it is important to point out that for broader bandwidth sounds at a moderate sound level, each individual axon will be activated by a range of frequencies both higher and lower than its characteristic frequency. For a sound with a complex spectral shape this will lead to a smoothing of the spectral profile and a loss of some detail in the encoding stage (see [15] for a more extensive discussion of this important topic).

In addition to the place code of frequency discussed above, for sound frequencies below about 4 kHz the timing of the action potentials in the auditory nerve fibres are in phase with the phase of the stimulating sound. This temporal code is called "phase locking" and allows the auditory system to very accurately code the frequency of low frequency sounds – certainly to a greater level of accuracy than that predicted by the place code for low frequencies.

The stream of action potentials ascending from each ear form the basis of the biological code from which our perception of the different auditory qualities are derived. The following sections consider the dimensions of loudness, pitch and timbre, temporal modulation and spatial location.

3.3 The perception of loudness

The auditory system is sensitive to a very large range of sound levels. Comparing the softest with the loudest discriminable sounds demonstrates a range of 1 to 10^{12} in intensity. Loudness is the percept that is generated by variations in the intensity of the sound. For broadband sounds containing many frequencies, the auditory system obeys Weber's law over most of the range of sensitivity. That is, the smallest detectable change in the intensity is related to the overall intensity. Consequently, the wide range of intensities to which we are sensitive is described using a logarithmic scale of sound pressure level (SPL) , the decibel (dB).

$$\mathrm{SPL(dB)} = 20 \log_{10} \left(\frac{\text{measured Pressure}}{\text{reference Pressure}} \right) \tag{1}$$

The reference pressure corresponds to the lowest intensity sound that we are able to discriminate which is generally taken as 20 μP.

Importantly, the threshold sensitivity of hearing varies as a function of frequency and the auditory system is most sensitive to frequencies around 4 kHz. In Figure 3.3, the variation in sensitivity as a function of frequency is shown by the lower dashed curve corresponding to the minimum audible field (MAF) . In this measurement the sound is presented to the listener in a very quiet environment from a sound source located directly in front of the listener [13]. The sound pressure corresponding to the threshold at each frequency is then measured in the absence of the listener using a microphone placed at the location corresponding to the middle of the listener's head. The shape of the minimum audible field curve is determined in part by the transmission characteristics of the middle ear and the external auditory canal and by the direction dependent filtering of the outer ears (sound example **S3.2** and **S3.3**).

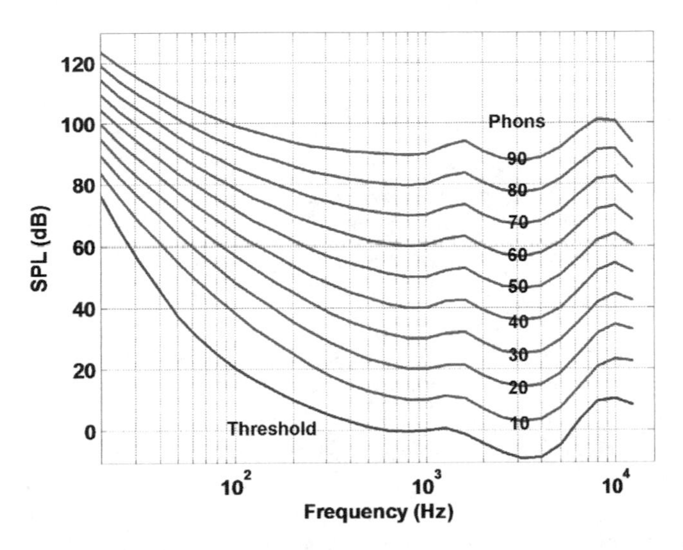

Figure 3.3: The minimum auditory field (MAF) or threshold for an externally placed sound source is illustrated by the lower (blue) line. The equal loudness contours are above this (shown in red) for 9 different loudness levels (measured in phons). A phon is the perceived loudness at any frequency that is judged to be equivalent to a reference sound pressure level at 1 kHz. For example, at 20 phons the reference level at 1 kHz is 20 dB SPL (by definition) but at 100 Hz the sound level has to be nearly 50 dB to be perceived as having the same loudness. Note that the loudness contours become progressively flatter at higher sound levels. These are also referred to as the Fletcher-Munson curves.

The equal loudness contours (Figure 3.3, red lines) are determined by asking listeners to adjust the intensity at different frequencies so that the loudness matches the loudness of a reference stimulus set at 1 kHz . The equal loudness contours become increasingly flat at high sound levels. This has important implications for the tonal quality of music and speech when mixing at different sound levels. A tonally balanced mix at low to moderate sound levels will have too much bottom end when played at high sound levels. Conversely, a mix intended for high sound levels will appear to have too much middle when played at low to moderate sound levels.

The threshold at any particular frequency is also dependent on the duration of the stimulus [14]. For shorter duration sounds the perception of loudness increases with increasing duration with an upper limit of between 100 ms to 200 ms suggesting that loudness is related to the total energy in the sound. The sounds used for measuring the absolute threshold curves and the equal loudness contours in Figure 3.3 are usually a few hundred milliseconds long. By contrast, exposure to prolonged sounds can produce a reduction in the perceived loudness, which is referred to as adaptation or fatigue (see [15]). Temporary threshold shift results from exposure to prolonged, moderate to high sound levels and the period of recovery can vary from minutes to tens of hours depending on the sound level and duration of the exposure

. Sound levels above 110 to 120 dB SPL can produce permanent threshold shift, particularly if exposure is for a prolonged period, due partly from damage to the hair cells on the basilar membrane of the inner ear.

3.4 The perception of pitch

The frequency of the sound is determined by the periodic rate at which a pressure wave fluctuates at the ear drum. This gives rise to the perception of pitch which can be defined as the sensation by which sounds can be ordered on a musical scale. The ability to discriminate differences in pitch has been measured by presenting two tones sequentially that differ slightly in frequency: the *just detectible differences* are called the frequency difference limen (FDL). The FDL in Hz is less that 1 Hz at 100 Hz and increases as an increasing function of frequency so that at 1 kHz the FDL is 2 Hz to 3 Hz (see in [15] Chapter 6, sound example **S3.4**). This is a most remarkable level of resolution, particularly when considered in terms of the extent of the basilar membrane that would be excited by a tone at a moderate sound level. A number of models have been developed that attempt to explain this phenomenon and are covered in more detail in the extended reading for this chapter. There is also a small effect of sound level on pitch perception: for high sound levels at low frequencies (< 2 kHz) pitch tends to decrease with intensity and for higher frequencies (> 4 kHz) tends to increase slightly.

The perception of musical pitch for pure tone stimuli also varies differently for high and low frequency tones. For frequencies below 2.5 kHz listeners are able to adjust a second sound quite accurately so that it is an octave above the test stimulus (that is, at roughly double the frequency). However, the ability to do this deteriorates quite quickly if the adjusted frequency needs to be above 5 kHz. In addition, melodic sense is also lost for sequences of tone above 5 kHz although frequency differences per se are clearly perceived (sound example **S3.5**). This suggests that different mechanisms are responsible for frequency discrimination and pitch perception and that the latter operates over low to middle frequency range of human hearing where temporal coding mechanisms (phase locking) are presumed to be operating.

The pitch of more complex sounds containing a number of frequency components generally does not simply correspond to the frequency with the greatest energy. For instance, a series of harmonically related frequency components, say 1800, 2000, and 2200 Hz, will be perceived to have a fundamental frequency related to their frequency spacing, in our example at 200 Hz. This perception occurs even in the presence of low pass noise that should mask any activity on the basilar membrane at the 200 Hz region. With the masking noise present, this perception cannot be dependent on the place code of frequency but must rely on the analysis different spectral components or the temporal pattern of action potentials arising from the stimulation (or a combination of both). This perceptual phenomenon is referred to as ‘residue’ pitch, ‘periodicity pitch’ or the problem of the missing fundamental (sound example **S3.6**) . Interestingly, when the fundamental is present (200 Hz in the above example) the pitch of the note is the same but timbre is discernibly different. Whatever the exact mechanism, it appears that the pitch of complex sounds like that made from most musical instruments is computed from the afferent (inflowing) information rather than resulting from a simple place code of spectral energy in the cochlea.

Another important attribute of the different spectral components in a complex sound is the perception of timbre. While a flute and a trumpet can play a note that clearly has the same fundamental, the overall sounds are strikingly different. This is due to the differences in the number, level and arrangement of the other spectral components in the two sounds. It is these differences that produce the various timbres associated with each instrument.

3.5 The perception of temporal variation

As discussed above, biologically interesting information is conveyed by sounds because the evolution of the auditory system has given rise to mechanisms for the detection and decoding of information that has significant survival advantages. One of the most salient features of biologically interesting sounds is the rapid variation in spectral content over time. The rate of this variation will depend on the nature of the sound generators. For instance, vocalization sounds are produced by the physical structures of the vocal cords, larynx, mouth etc. The variation in the resonances of voiced vocalizations and the characteristics of the transient or broadband components of unvoiced speech will depend on the rate at which the animal can change the physical characteristics of these vocal structures – for instance, their size or length and how they are coupled together.

The rate of these changes will represent the range of temporal variation over which much biologically interesting information can be generated in the form of vocalizations. On the receiver side, the processes of biologically encoding the sounds will also place limitations on the rate of change that can be detected and neurally encoded. The generation of receptor potentials in the hair cells and the initiation of action potentials in the auditory nerve all have biologically constrained time constants. Within this temporal bandwidth however, the important thing to remember is that the information in a sound is largely conveyed by its variations over time.

Mathematically, any sound can be decomposed into two different temporally varying components: a slowly varying envelope and a rapidly varying fine structure (Figure 3.4). Present data indicates that both of these characteristics of the sound are encoded by the auditory system and play important roles in the perception of speech and other sounds (e.g., [29] and below).

When sound is broken down into a number of frequency bands (as happens along the basilar membrane of the inner ear), the envelopes in as few as 3–4 bands have been shown to be sufficient for conveying intelligible speech [16] (sound example **S3.7**). Although less is known about the role of the fine structure in speech, it is known that this is encoded in the auditory nerve for the relevant frequencies and there is some evidence that this can be used to support speech processing .

Auditory sensitivity to temporal change in a sound has been examined in a number of ways. The auditory system is able to detect gaps in broadband noise stimuli as short as 2 - 3 ms [17]. This temporal threshold is relatively constant over moderate to high stimulus levels; however, longer gaps are needed when the sound levels are close to the auditory threshold.

In terms of the envelope of the sounds, the sensitivity of the auditory system to modulation of a sound varies as a function of the rate of modulation. This is termed the *temporal modulation transfer function* (TMTF). For amplitude modulation of a broadband sound, the greatest sensitivity is demonstrated for modulation rates below about 50–60 Hz. Above this range, sensitivity falls off fairly rapidly and modulation is undetectable for rates above 1000 Hz. This sensitivity pattern is fairly constant over a broad range of sound levels. The modulation sensitivity using a wide range of narrowband carries such as sinusoids (1–10 kHz) show a greater range of maximum sensitivity (100–200 Hz) before sensitivity begins to roll off (see [15], Chapter 5 for discussion).

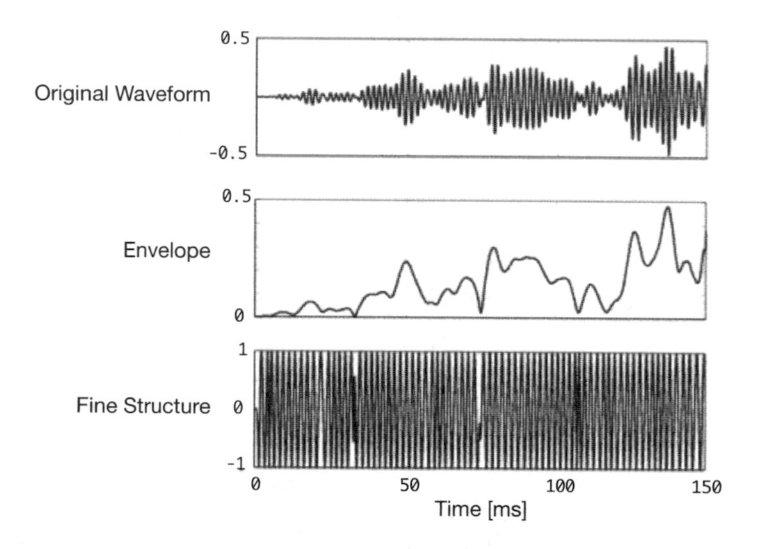

Figure 3.4: A complex sound (top panel) can be broken down into an envelope (middle panel) and a carrier (bottom panel). The top panel shows the amplitude changes of a sound wave over 150 ms as would be seen by looking at the output of a microphone. What is easily discernible is that the sound is made up primarily of a high frequency oscillation that is varying in amplitude. The high frequency oscillation is called the carrier or fine structure and has been extracted and illustrated in the lower panel. The extent of the amplitude modulation of the carrier is shown in the middle panel and is referred to as the envelope. Taken from `https://research.meei.harvard.edu/Chimera/motivation.html`

The auditory system is also sensitive to differences in the duration of a sound. In general, for sounds longer than 10 ms the smallest detectible change (the just noticeable difference, JND) increases with the duration of the sound ($T/\Delta T$: 10/4 ms, 100/15 ms, 1000/60 ms). The spectrum or the level of the sound appears to play no role in this sensitivity. However, the sensitivity to the duration of silent gaps is poorer at lower sound levels compared to moderate and higher levels and when the spectra of the two sounds defining the silent gap are different.

3.6 Grouping spectral components into auditory objects and streams

In the early 1990s, Albert Bregman's influential book Auditory Scene Analysis [1] was published which summarized the research from his and other laboratories examining the sorts of mechanisms that allow us to solve the 'cocktail party problem'. As mentioned above, our ability to segregate a sound of interest from a complex background of other sounds play a critical role in our ability to communicate in everyday listening environments. Bregman argued that the jumble of spectral components that reach the ear at any instant in time can be either **integrated** and heard as a single sound (e.g., a full orchestra playing a chord spanning several octaves) or **segregated** into a number of different sounds (the brass and woodwind playing the middle register notes versus the basses and strings playing the lower and higher register components, respectively).

Bregman argued that there are a number of innate processes as well as learned strategies which are utilized in segregating concurrent sounds. These processes rely on the so-called *grouping cues*. Some of these cues reflect some basic rules of perceptual organization (first discovered by the Gestalt psychologists in the 19th century) as well as the physical characteristics of sounds themselves.

The rules used by the auditory system in carrying out this difficult signal processing task also reflect in part, the physics of sounding objects. For instance, it is not very often that two different sounds will turn on at precisely the same time. The auditory system uses this fact to group together the spectral components that either start or stop at the same time (i.e. are synchronous, sound example **S3.8**). Likewise, many sounding objects will resonate with a particular fundamental frequency. Similarly, when two concurrent sounds have different fundamental frequencies, the brain can use the fact that the harmonics that comprise each sound will be a whole number multiple of the fundamental. By analyzing the frequency of each component, the energy at the different harmonic frequencies can be associated with their respective fundamental frequency. Each collection of spectra is then integrated to produce the perception of separate sounds, each with their own specific characteristics, timbre or tonal color (sound example **S3.9**).

If a sounding object modulates the amplitude of the sound (AM) then all of the spectral components of the sound are likely to increase and decrease in level at the same time. Using this as another plausible assumption, the brain uses synchrony in the changes in level to group together different spectral components and to fuse them as a separate sound. Opera singers have known this for years: by placing some vibrato on their voice there is a synchronous frequency and amplitude modulation of the sound. This allows the listener to perceptually segregate the singer's voice from the veritable wall of sound that is provided by the accompanying orchestra.

Once a sound has been grouped over a 'chunk' of time, these sound-chunks need to be linked sequentially over time – a process referred to as streaming. The sorts of rules that govern this process are similar to those that govern grouping, and are based in part on physical plausibility. Similarity between chunks is an important determinant of steaming. Such similarities can include the same or substantially similar fundamental frequency, similar timbre, or sounds that appear to be repeated in quick succession (sound example **S3.10**) or part of a progressive sequence of small changes to the sound (a *portamento* or *glissando*).

We then perceive these auditory streams as cohesive auditory events, such as a particular person talking, or a car driving by, or a dog barking.

Of course, like any perceptual process, grouping and streaming are not perfect and at times there can be interesting perceptual effects when these processes fail. For instance, if concurrent grouping fails, then two or more sounds may be blended perceptually, giving rise to perceptual qualities that are not present in any of the segregated sounds. Failure in streaming can often happen with speech where two syllables–or even different words–from different talkers might be incorrectly streamed together, which can give rise to misheard words and sentences (a phenomenon called *information masking*). Such confusions can happen quite frequently if the voices of the concurrent talkers are quite similar, as voice quality provides a very powerful streaming cue (sound example **S3.11**).

In the context of sound design and sonification, the auditory cues for grouping and streaming tell us a lot about how we can design sounds that either stand out (are salient) or blend into the background. By designing sounds that obey the grouping cues, the auditory system is better able to link together the spectral components of each sound when it is played against a background of other spectrally overlapping sounds. While onset synchrony is a fairly obvious rule to follow, other rules such as the harmonicity of spectral components and common frequency and amplitude modulation are not as obvious, particularly for non-musical sounds. Likewise purposely avoiding the grouping rules in design will create sounds that contribute to an overall 'wash' of sound and blend into the background. The perceptual phenomena of integration would result in such sounds subtly changing the timbral color of the background as new components are added.

3.7 The perception of space

In addition to the biologically interesting information contained within a particular sound stream, the location of the source is also an important feature. The ability to appropriately act on information derived from the sound will, in many cases, be dependent on the location of the source. Predator species are often capable of very fine discrimination of location [18] – an owl for instance can strike, with very great accuracy, a mouse scuttling across the forest floor in complete darkness. Encoding of space in the auditory domain is quite different to the representation of space in the visual or somatosensory domains. In the latter sensory domains the spatial location of the stimuli are mapped directly onto the receptor cells (the retina or the surface of the skin). The receptor cells send axons into the central nervous system that maintain their orderly arrangement so that information from adjacent receptor cells is preserved as a place code within the nervous system. For instance, the visual field stimulating the retina is mapped out in the visual nervous system like a 2D map – this is referred to as a place code of space. By contrast, as is discussed above, it is sound frequency that is represented in the ordered arrangement of sensory cells in inner ear. This gives rise to a tonotopic rather than a spatiotopic representation in the auditory system. Any representation of auditory space within the central nervous systems, and therefore our perception of auditory space, is derived computationally from acoustic cues to a sound source location occurring at each ear.

Auditory space represents a very important domain for sonification. The relative locations of sounds in everyday life plays a key role in helping us remain orientated and aware of what is happening around us – particularly in the very large region of space that is outside our visual field! There are many elements of the spatial dimension that map intuitively onto data – high, low, close, far, small, large, enclosed, open etc. In addition, Virtual Reality research has demonstrated that the sense of auditory spaciousness has been found to play an important role in generating the sense of 'presence' – that feeling of actually being in the virtual world generated by the display. This section looks at the range of acoustic cues available to the auditory system and human sensitivity to the direction, distance and movement of sound sources in our auditory world.

3.7.1 Dimensions of auditory space

The three principal dimensions of auditory spatial perception are direction and distance of sources and the spaciousness of the environment. A sound source can be located along some horizontal direction (azimuth), at a particular height above or below the audio-visual horizon (elevation) and a specific distance from the head. Another dimension of auditory spatial perception is referred to as the spatial impression, which includes the sense of spaciousness, the size of an enclosed space and the reverberance of the space (see [27]). These are important in architectural acoustics and the design of listening rooms and auditoria – particularly for music listening.

3.7.2 Cues for spatial listening

Our perception of auditory space is based on acoustic cues that arise at each ear. These cues result from an interaction of the sound with the two ears, the head and torso as well as with the reflecting surfaces in the immediate environment. The auditory system simultaneously samples the sound field from two different locations – i.e. at the two ears which are separated by the acoustically dense head. For a sound source located off the midline, the path length difference from the source to each ear results in an interaural difference in the arrival times of the sound (Interaural Time Difference (ITD) , Figure 3.5, sound example **S3.12**). With a source located on the midline, this difference will be zero. The difference will be at a maximum when the sound is opposite one or other of the ears.

As the phase of low frequency sounds can be encoded by the 'phase locked' action potentials in the auditory nerve, the ongoing phase difference of the sound in each ear can also be used as a cue to the location of a source. As well as extracting the ITD from the onset of the sound, the auditory system can also use timing differences in the amplitude modulation envelopes of more complex sounds. Psychophysical studies using headphone-presented stimuli have demonstrated sensitivity to interaural time differences as small as $13\mu s$ for tones from 500 to 1000 Hz.

As the wavelengths of the mid to high frequency sounds are relatively small compared to the head, these sounds will be reflected and refracted by the head so that the ear furthest from the source will be acoustically shadowed. This gives rise to a difference in the sound level at each ear and is known as the interaural level (or intensity) difference (ILD) cue (sound example **S3.13**). Sensitivity to interaural level differences of pure tone stimuli of as small as 1dB have been demonstrated for pure tone stimuli presented over headphones. The ITD cues are believed to contribute principally at the low frequencies and the ILD cues at the mid to high frequencies; this is sometimes referred to as the duplex theory of localisation [9].

The binaural cues alone provide an ambiguous cue to the spatial location of a source because any particular interaural interval specifies the surface of a cone centred on the interaural axis - the so called 'cone of confusion' (Figure 3.6: top left) . As discussed above, the outer ear filters sound in a directionally dependent manner which gives rise to the spectral (or monaural) cues to location. The variations in the filter functions of the outer ear, as a function of the location of the source, provide the basis for resolving the cone of confusion (Figure 3.6: top right panel). Where these cues are absent or degraded, or where the sound has a relatively narrow bandwidth, front-back confusions can occur in the perception of sound source location. That is, a sound in the frontal field could be perceived to be located in the posterior field and vice versa. Together with the head shadow, the spectral cues also explain how people who are deaf in one ear can still localize sound.

3.7.3 Determining the direction of a sound source

Accurate determination of the direction of a sound source is dependent on the integration of the binaural and spectral cues to its location [19]. The spectral cues from each ear are weighted according to the horizontal location of the source, with the cues from the closer ear dominating [20]. In general there are two classes of localisation errors: (i) Large 'front-back' or 'cone of confusion' errors where the perceived location is in a quadrant different from the source but roughly on the same cone of confusion; (ii) Local errors where the location is perceived to be in the vicinity of the actual target. Average localisation errors are generally only a few degrees for targets directly in front of the subject (SD $\pm$ 6°–7°) [21]. Absolute errors, and the response variability around the mean, gradually increase for locations towards the posterior midline and for elevations away from the audio-visual horizon. For broadband noise stimuli the front-back error rates range from 3 % to 6 % of the trials. However, localisation performance is also strongly related to the characteristics of the stimulus. Narrowband stimuli, particularly high or low sound levels or reverberant listening conditions, can significantly degrade performance.

A different approach to understanding auditory spatial performance is to examine the resolution or acuity of auditory perception where subjects are required to detect a change in the location of a single source. This is referred to as a minimum audible angle (MAA: [22]). This approach provides insight into the just noticeable differences in the acoustic cues to spatial location. The smallest MAA (1–2°) is found for broadband sounds located around the anterior midline and the MAA increases significantly for locations away from the anterior median plane. The MAA is also much higher for narrowband stimuli such as tones. By contrast, the ability of subjects to discriminate the relative locations of concurrent sounds with the same spectral characteristics is dependent on interaural differences rather than the spectral cues [23]. By contrast, in everyday listening situations it is likely that the different spectral components are grouped together as is described in section 3.6 above and the locations then computed from the interaural and spectral cues available in the grouped spectra.

The majority of localisation performance studies have been carried out in anechoic environments. Localisation in many real world environments will of course include some level of reverberation. Localisation in rooms is not as good as in anechoic space but it does appear to be better than what might be expected based on how reverberation degrades the acoustic cues to location. For instance, reverberation will tend to de-correlate the waveforms at each ear because of the differences in the patterns of reverberation that combine with the direct wavefront at each ear. This will tend to disrupt the extraction of ongoing ITD although the auditory system may be able to obtain a reasonably reliable estimate of the ITD by integrating across a much longer time window [24]. Likewise, the addition of delayed copies of the direct sound will lead to comb filtering of the sound that will tend to fill in the notches and flatten out the peaks in the monaural spectral cues and decrease the overall ILD cue. These changes will also be highly dependent on the relative locations of the sound sources, the reflecting surfaces and the listener.

3.7.4 Determining the distance of a sound source

While it is the interactions of the sound with the outer ears that provides the cues to source direction, it is the interactions between the sound and the listening environment that provide the four principal cues to source distance [25]. First, the intensity of a sound decreases with distance according to the inverse square law: this produces a 6 dB decrease in level with a doubling of distance. Second, as a result of the transmission characteristics of the air, high frequencies (> 4 kHz) are absorbed to a greater degree than low frequencies which produces a relative reduction of the high frequencies of around 1.6 dB per doubling of distance. However, with these cues the source characteristics (intensity and spectrum) are confounded with distance so they can only act as reliable cues for familiar sounds (such as speech sounds). In other words, it is necessary to know what the level and spectral content of the source is likely to be for these cues to be useful.

A third cue to distance is the ratio of the direct to reverberant energy. This cue is not confounded like the first two cues but is dependent on the reverberant characteristics of an enclosed space. It is the characteristics of the room that determine the level of reverberation which is then basically constant throughout the room. On the other hand, the direct energy is subject to the inverse square law of distance so that this will vary with the distance of the source to the listener. Recent work exploring distance perception for sound locations within arm's reach (i.e. in the near field) has demonstrated that substantial changes in the interaural level differences can occur with variation in distance [26] over this range. There are also distance related changes to the complex filtering of the outer ear when the sources are in the near field because of the parallax change in the relative angle between the source and each ear (sound example **S3.14**).

The nature of an enclosed space also influences the spatial impression produced. In particular, spaciousness has been characterized by 'apparent source width' which is related to the extent of early lateral reflections in a listening space and the relative sound level of the low frequencies. A second aspect of spaciousness is 'listener envelopment' which is related more to the overall reverberant sound field and is particularly salient with relatively high levels arriving later than 80 ms after the direct sound [27].

3.7.5 Perception of moving sounds

A sound source moving through space will produce a dynamic change in the binaural and spectral cues to location and the overall level and spectral cues to distance. In addition, if the source is approaching or receding from the listener then there will be a progressive increase or decrease respectively in the apparent frequency of the sound due to the Doppler shift.

While the visual system is very sensitive to the motion of visual objects, the auditory system appears to be less sensitive to the motion of sound sources. The minimum audible movement angle (MAMA) is defined as the minimum distance a source must travel before it is perceived as moving. The MAMA is generally reported to be somewhat larger than the MAA discussed above [28]. However, MAMA has also been shown to increase with the velocity of the moving sound source, which has been taken to indicate a minimum integration time for the perception of a moving source. On the other hand this also demonstrates that the parameters of velocity, time and displacement co-vary with a moving stimulus. Measuring sensitivity to a moving sound is also beset with a number of technical difficulties – not least the fact that mechanically moving a source will generally involve making other noises which can complicate interpretation. More recently, researchers have been using moving stimuli exploiting virtual auditory space presented over headphones to overcome some of these problems.

When displacement is controlled for it has been shown that the just noticeable difference in velocity is also related to the velocity of sound source moving about the midline [28]. For sounds moving at 15°, 30° and 60° per second the velocity thresholds were 5.5°, 9.1° and 14.8° per second respectively. However, velocity threshold decreased by around half if displacement cues were also added to these stimuli. Thus, while the auditory system is moderately sensitive to velocity changes per se, comparisons between stimuli are greatly aided if displacement cues are present as well. In these experiments all the stimuli were hundreds of milliseconds to 3 seconds long to ensure that they lasted longer than any putative integration time required for the generation of the perception of motion.

Another form of auditory motion is spectral motion where there is a smooth change in the frequency content of a sound. A trombone sliding up or down the scale or a singer sliding up to a note (glissando) are two common examples of a continuous variation in the fundamental frequency of a complex sound.

Both forms of auditory motion (spatial and spectral) demonstrate after effects. In the visual system relatively prolonged exposure to motion in one direction results in the perception of motion in the opposite direction when the gaze is subsequently directed towards a stationary visual field. This is known as the "waterfall effect". The same effect has been demonstrated in the auditory system for sounds that move either in auditory space or have cyclic changes in spectral content. For example, broadband noise will appear to have a spectral peak moving down in frequency following prolonged exposure to a sound that has a spectral peak moving up in frequency.

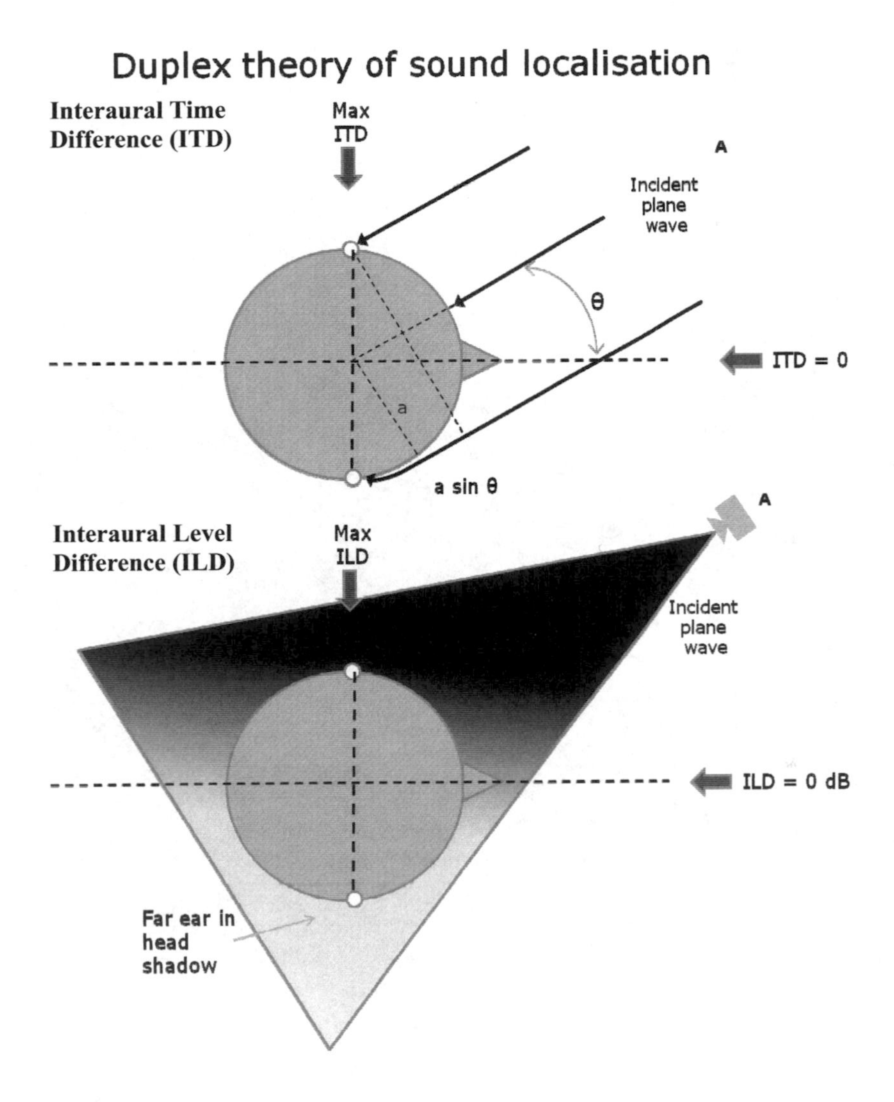

Figure 3.5: Having two ears, one on each side of an acoustically dense head, means that for sounds off the midline there is a difference in the time of arrival and the amplitude of the sound at each ear. These provide the so-called binaural cues to the location of a sound source.

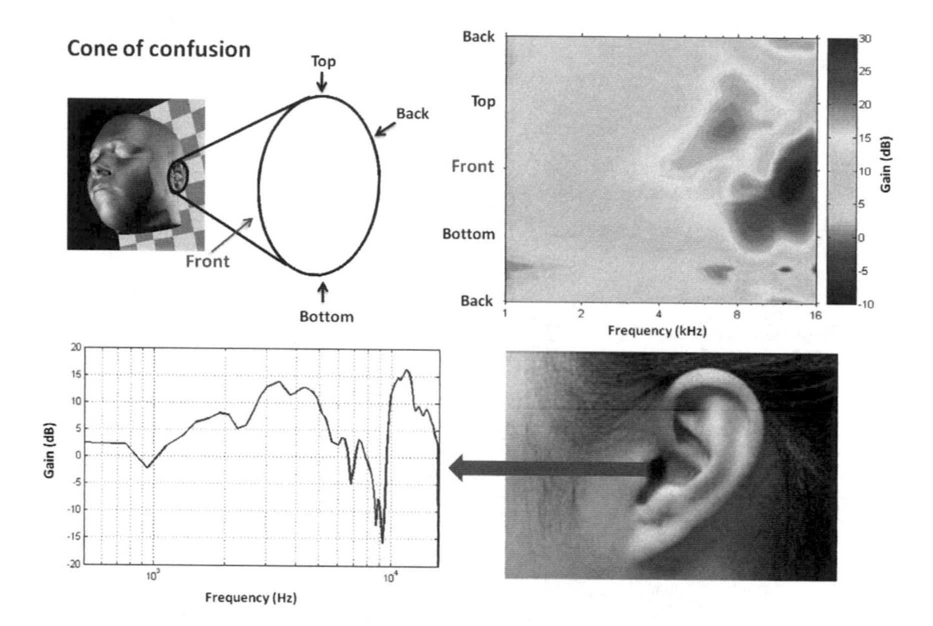

Figure 3.6: The ambiguity of the binaural cues is illustrated by the 'cone of confusion' for a particular ITD/ILD interval (top left). The complex spectral filtering of the outer ear (bottom panels) varies around the cone of confusion (top right panel), and provides an additional monaural (single ear) cue to a sound's precise location. This allows the brain to resolve the spatial ambiguity inherent in the binaural cues.

3.8 Summary

This chapter has looked at how multiple sound sources can contribute to the pattern of sound waves that occur at our ears. Objects in the sonic world are characterized by perceptual qualities such as pitch, timbre, loudness, spatial location and extent. Biologically interesting information is conveyed by temporal changes in these qualities. The outer and middle ears transmit and filter the sound from the air space around the head to the fluid spaces of the inner ear. The cochlea of the inner ear transduces the sound into biological signals. The spectral content of the sound is broken down into a spatio-topic or tonotopic code on the basilar membrane which then projects in an ordered topographical manner into the auditory nervous system and up to the auditory cortex. Temporal coding of the low to mid frequencies also plays a role in maintaining very high sensitivity to frequency differences in this range.

From the stream of biological action potentials generated in the auditory nerve, the auditory system derives the loudness, pitch, timbre and spatial location of the sound source. Different spectral components in this signal are grouped together to form auditory objects and streams which provide the basis for our recognition of different sound sources. As frequency is what is encoded topographically in the auditory system, spatial location needs to be computed from acoustic cues occurring at each ear. These cues include the interaural differences in level and time of arrival of the sound and the location dependent filtering of the sound by the outer ear (the monaural or spectral cues to location). From these cues the auditory system is able to compute the direction and distance of the sound source with respect to the head. In addition, motion of the sound source in space or continuous changes in spectral content, gives rise to motion after effects.

3.9 Further reading

General texts

- B.C.J. Moore, An introduction to the psychology of hearing (4th ed, London: Academic Press 1997)
- E. Kandel, J. Schwartz, and T. Jessel, eds. Principals of neural science. (4th Ed, McGraw-Hill, 2000). Chapters 30 and 31 in particular. 3.9.2 Acoustical and psychophysical basis of spatial perception
- S. Carlile, "Auditory space", in Virtual auditory space: Generation and applications (S. Carlile, Editor. Landes: Austin) p. Ch 1 (1996)
- S. Carlile, "The physical and psychophysical basis of sound localization", in Virtual auditory space: Generation and applications (S. Carlile, Editor. Landes: Austin) p. Ch 2 (1996)

Distance perception

- P. Zahorik, D.S. Brungart, and A.W. Bronkhorst, "Auditory distance perception in humans: A summary of past and present research", Acta Acustica United with Acustica, 91 (3), 409-420 (2005).

Bibliography

[1] A.S. Bregman, Auditory scene analysis: The perceptual organization of sound, Cambridge, Mass: MIT Press 1990)

[2] M. Cooke and D.P.W. Ellis, "The auditory organization of speech and other sources in listeners and computational models", Speech Communication, 35, 141–177 (2001).

[3] R.L. Freyman, et al., "The role of perceived spatial separation in the unmasking of speech", J Acoust Soc Am, 106 (6), 3578–3588 (1999).

[4] R.L. Freyman, U. Balakrishnan, and K.S. Helfer, "Spatial release from informational masking in speech recognition", J Acoust Soc Am, 109 (5 Pt 1), 2112–2122 (2001).

[5] N.I. Durlach, et al., "On the externalization of auditory images", Presence, 1, 251–257 (1992).

[6] C. Giguere and S. Abel, "Sound localization: Effects of reverberation time, speaker array, stimulus frequency and stimulus rise/decay", J Acoust Soc Am, 94 (2), 769–776 (1993).

[7] J. Braasch and K. Hartung, "Localisation in the presence of a distracter and reverberation in the frontal horizontal plane. I psychoacoustic data", Acta Acustica, 88, 942–955 (2002).

[8] E.A.G. Shaw, "The external ear", in Handbook of sensory physiology (W.D. Keidel and W.D. Neff, Editors. Springer-Verlag: Berlin) p. 455–490 (1974)

[9] S. Carlile, "The physical and psychophysical basis of sound localization", in Virtual auditory space: Generation and applications (S. Carlile, Editor. Landes: Austin) p. Ch 2 (1996)

[10] S. Carlile and D. Pralong, "The location-dependent nature of perceptually salient features of the human head-related transfer function", J Acoust Soc Am, 95 (6), 3445–3459 (1994).

[11] J.O. Pickles, An introduction to the physiology of hearing (Second ed, London: Academic Press 1992)

[12] R. Fettiplace and C.M. Hackney, "The sensory and motor roles of auditory hair cells", Nature Reviews Neuroscience, 7 (1), 19–29 (2006).

[13] L. Sivian and S. White, "On minimum audible sound fields", J Acoust Soc Am, 4 (1933).

[14] S. Buus, M. Florentine, and T. Poulson, "Temporal integration of loudness, loudness discrimination and the form of the loudness function", J Acoust Soc Am, 101, 669–680 (1997).

[15] B.C.J. Moore, An introduction to the psychology of hearing (5th ed, London: Academic Press) (2003).

[16] M.F. Dorman, P.C. Loizou, and D. Rainey, "Speech intelligibility as a function of the number of channels of stimulation for signal processors using sine-wave and noise-band outputs", J Acoust l Soc Am, 102 (4), 2403–2411 (1997).

[17] R. Plomp, "The rate of decay of auditory sensation", J Acoust Soc Am, 36, 277-282 (1964).

[18] S.D. Erulkar, "Comparitive aspects of spatial localization of sound", Physiological Review, 52, 238–360 (1972).

[19] J.C. Middlebrooks, "Narrow-band sound localization related to external ear acoustics", J Acoust Soc Am, 92 (5), 2607–2624 (1992).

[20] P.M. Hofman and A.J. Van Opstal, "Binaural weighting of pinna cues in human sound localization", Experimental Brain Research, 148 (4), 458–470 (2003).

[21] S. Carlile, P. Leong, and S. Hyams, "The nature and distribution of errors in the localization of sounds by humans." Hearing Res, 114, 179–196 (1997).

[22] A.W. Mills, "On the minimum audible angle", J Acoust Soc Am, 30 (4), 237–246 (1958).

[23] V. Best, A.v. Schaik, and S. Carlile, "Separation of concurrent broadband sound sources by human listeners", J Acoust Soc Am, 115, 324–336 (2004).

[24] B.G. Shinn-Cunningham, N. Kopco, and T.J. Martin, "Localizing nearby sound sources in a classroom: Binaural room impulse responses", J Acoust Soc Am, 117, 3100–3115 (2005).

[25] P. Zahorik, D.S. Brungart, and A.W. Bronkhorst, "Auditory distance perception in humans: A summary of past and present research", Acta Acustica United with Acustica, 91 (3), 409–420 (2005).

[26] B.G. Shinn-Cunningham. "Distance cues for virtual auditory space", in IEEE 2000 International Symposium on Multimedia Information Processing. Sydney, Australia, (2000).

[27] T. Okano, L.L. Beranek, and T. Hidaka, "Relations among interaural cross-correlation coefficient (iacc(e)), lateral fraction (lfe), and apparent source width (asw) in concert halls", J Acoust Soc Am, 104 (1), 255–265 (1998).

[28] S. Carlile and V. Best, "Discrimination of sound source velocity by human listeners", J Acoust Soc Am, 111 (2), 1026–1035 (2002).

[29] Gilbert, G. and C. Lorenzi, "Role of spectral and temporal cues in restoring missing speech information." J Acoust Soc Am, 128(5): EL294–EL299 (2010).

Chapter 4

Perception, Cognition and Action in Auditory Displays

John G. Neuhoff

4.1 Introduction

Perception is almost always an automatic and effortless process. Light and sound in the environment seem to be almost magically transformed into a complex array of neural impulses that are interpreted by the brain as the subjective experience of the auditory and visual scenes that surround us. This transformation of physical energy into "meaning" is completed within a fraction of a second. However, the ease and speed with which the perceptual system accomplishes this Herculean task greatly masks the complexity of the underlying processes and often times leads us to greatly underestimate the importance of considering the study of perception and cognition, particularly in applied environments such as auditory display.

The role of perception in sonification has historically been of some debate. In 1997 when the International Community for Auditory Display (ICAD) held a workshop on sonification, sponsored by the National Science Foundation, that resulted in a report entitled "*Sonification Report: Status of the Field and Research Agenda*" (Kramer, et al., 1999). One of the most important tasks of this working group was to develop a working definition of the word "sonification". The underestimation of the importance of perception was underscored by the good deal of discussion and initial disagreement over including anything having to *do* with "perception" in the definition of sonification. However, after some debate the group finally arrived at the following definition:

> "*...sonification is the transformation of data relations into perceived relations in an acoustic signal for the purposes of facilitating communication or interpretation.*"

The inclusion of the terms "*perceived relations*" and "*communication or interpretation*" in this definition highlights the importance of perceptual and cognitive processes in the development of effective auditory displays. Although the act of perceiving is often an effortless and automatic process it is by no means simple or trivial. If the goal of auditory display is to convey meaning with sound, then knowledge of the perceptual processes that turn sound into meaning is crucial.

No less important are the cognitive factors involved in extracting meaning from an auditory display and the actions of the user and interactions that the user has with the display interface. There is ample research that shows that interaction, or intended interaction with a stimulus (such as an auditory display) can influence perception and cognition.

Clearly then, an understanding of the perceptual abilities, cognitive processes, and behaviors of the user are critical in designing effective auditory displays. The remainder of this chapter will selectively introduce some of what is currently known about auditory perception, cognition, and action and will describe how these processes are germane to auditory display.

Thus, the chapter begins with an examination of "low level" auditory dimensions such as pitch, loudness and timbre and how they can best be leveraged in creating effective auditory displays. It then moves to a discussion of the perception of auditory space and time. It concludes with an overview of more complex issues in auditory scene analysis, auditory cognition, and perception action relationships and how these phenomena can be used (and misused) in auditory display.

4.2 Perceiving Auditory Dimensions

There are many ways to describe a sound. One might describe the sound of an oboe by its timbre, the rate of note production, or by its location in space. All of these characteristics can be referred to as "auditory dimensions". An auditory dimension is typically defined as the subjective perceptual experience of a particular physical characteristic of an auditory stimulus. So, for example, a primary physical characteristic of a tone is its fundamental frequency (usually measured in cycles per second or Hz). The perceptual dimension that corresponds principally to the physical dimension of frequency is "pitch", or the apparent "highness" or "lowness" of a tone. Likewise the physical intensity of a sound (or its amplitude) is the primary determinant of the auditory dimension "loudness".

A common technique for designers of auditory displays is to use these various dimensions as "channels" for the presentation of multidimensional data. So, for example, in a sonification of real-time financial data Janata and Childs (2004) used rising and falling pitch to represent the change in price of a stock and loudness to indicate when the stock price was approaching a pre-determined target (such as its thirty day average price). However, as is made clear in the previous chapter on psychoacoustics, this task is much more complex than it first appears because there is not a one-to-one correspondence between the physical characteristics of a stimulus and its perceptual correlates. Moreover, (as will be shown in subsequent sections) the auditory dimensions "interact" such that the pitch of a stimulus can influence its loudness, loudness can influence pitch, and other dimensions such as timbre and duration can all influence each other. This point becomes particularly important in auditory display, where various auditory dimensions are often used to represent different variables in a data set. The complexities of these auditory interactions have yet to be fully addressed by the research community. Their effects in applied tasks such as those encountered in auditory display are even less well illuminated. However, before discussing how the various auditory dimensions interact, the discussion turns toward three of the auditory dimensions that are most commonly used in auditory display: pitch, loudness, and timbre.

4.2.1 Pitch

Pitch is perhaps the auditory dimension most frequently used to represent data and present information in auditory displays. In fact, it is rare that one hears an auditory display that does not employ changes in pitch. Some of the advantages of using pitch are that it is easily manipulated and mapped to changes in data. The human auditory system is capable of detecting changes in pitch of less than 1Hz at a frequency of 100Hz (See Chapter 3 section 3.4 of this volume). Moreover, with larger changes in pitch, musical scales can provide a pre-existing cognitive structure that can be leveraged in presenting information. This would occur for example in cases where an auditory display uses discrete notes in a musical scale to represent different data values.

However, there are a few disadvantages in using pitch. Some work suggests that there may be individual differences in musical ability that can affect how a display that uses pitch change is perceived (Neuhoff, Kramer, & Wayand, 2002). Even early psychophysicists acknowledged that musical context can affect pitch perception. The revered psychophysicist S.S. Stevens, for example, viewed the intrusion of musical context into the psychophysical study of pitch as an extraneous variable. He tried to use subjects that were musically naive and implemented control conditions designed to prevent subjects from establishing a musical context. For example instead of using frequency intervals that corresponded to those that followed a musical scale (e.g., the notes on a piano), he used intervals that avoided any correspondence with musical scales. In commenting about the difficulty of the method involved in developing the mel scale (a perceptual scale in which pitches are judged to be equal in distance from one another), Stevens remarked "*The judgment is apparently easier than one might suppose, especially if one does not become confused by the recognition of musical intervals when he sets the variable tone.*" (Stevens & Davis, 1938, p. 81). It was apparent even to Stevens and his colleagues then that there are privileged relationships between musical intervals that influence pitch perception. In other words, frequency intervals that correspond to those that are used in music are more salient and have greater "meaning' than those that do not, particularly for listeners with any degree of musical training.

If pitch change is to be used by a display designer, the changes in pitch must be mapped in some logical way to particular changes in the data. The question of mapping the direction of pitch change used in a display (rising or falling) to increasing or decreasing data value is one of "polarity". Intuitively, increases in the value of a data dimension might seem as though they should be represented by increases in the pitch of the acoustic signal. Indeed many sonification examples have taken this approach. For example, in the sonification of historical weather data, daily temperature has been mapped to pitch using this "positive polarity", where high frequencies represent high temperatures and low frequencies represent low temperatures (Flowers, Whitwer, Grafel, & Kotan, 2001). However, the relationship between changes in the data value and frequency is not universal and in some respects depends on the data dimension being represented and the nature of the user. For example, a "negative polarity" works best when sonifying size, whereby decreasing size is best represented by *increasing* frequency (Walker, 2002). The cognitive mechanisms that underly polarity relationships between data and sound have yet to be investigated.

Walker and colleagues (Walker 2002; Walker 2007; Smith & Walker, 2002; Walker & Kramer, 2004) have done considerable work exploring the most appropriate polarity and conceptual mappings between data and sound dimensions. This work demonstrates the complexity of the problem of mapping pitch to data dimensions with respect to polarity. Not only do different data dimensions (e.g., temperature, size, and pressure) have different effective polarities, but there are also considerable individual differences in the choice of preferred polarities. Some users even show very little consistency in applying a preferred polarity (Walker, 2002). In other cases distinct individual differences predict preferred polarities. For example, users with visual impairment sometimes choose a polarity that is different from those without visual impairment (Walker & Lane, 2001).

In any case, what may seem like a fairly simple auditory dimension to use in a display has some perhaps unanticipated complexity. The influence of musical context can vary from user to user. Polarity and scaling can vary across the data dimensions being represented. Mapping data to pitch change should be done carefully with these considerations in the forefront of the design process.

4.2.2 Loudness

Loudness is a perceptual dimension that is correlated with the amplitude of an acoustic signal. Along with pitch, it is easily one of the auditory dimensions most studied by psychologists and psychoacousticians. The use of loudness change in auditory displays, although perhaps not as common as the use of pitch change, is nonetheless ubiquitous. The primary advantages of using loudness change in an auditory display are that it is quite easy to manipulate, and is readily understood by most users of auditory displays. However, despite its frequent use, loudness is generally considered a poor auditory dimension for purposes of representing continuous data sets. There are several important drawbacks to using loudness change to represent changes in data in sonification and auditory display.

- First, the ability to discriminate sounds of different intensities, while clearly present, lacks the resolution that is apparent in the ability to discriminate sounds of different frequencies.
- Second, memory for loudness is extremely poor, especially when compared to memory for pitch.
- Third, background noise and the sound reproduction equipment employed in any given auditory display will generally vary considerably depending on the user's environment. Thus, reliable sonification of continuous variables using loudness change becomes difficult (Flowers, 2005).
- Finally, there are no pre-existing cognitive structures for loudness that can be leveraged in the way that musical scales can be utilized when using pitch. Loudness, like most other perceptual dimensions, is also subject to interacting with other perceptual dimensions such as pitch and timbre.

Nonetheless, loudness change is often used in auditory display and if used correctly in the appropriate contexts, it can be effective. The most effective use of loudness change usually occurs when changes in loudness are constrained to two or three discrete levels that are mapped to two or three discrete states of the data being sonified. In this manner, discrete changes in loudness can be used to identify categorical changes in the state of a variable or to indicate when a variable has reached some criterion value. Continuous changes in loudness can be used to sonify trends in data. However, the efficacy of this technique leaves much to be desired. Absolute data values are particularly difficult to perceive by listening to loudness change alone. On the other hand, continuous loudness change can be mapped *redundantly* with changes in pitch to enhance the salience of particularly important data changes or auditory warnings. This point will be expanded below when discussing the advantageous effects of dimensional interaction.

4.2.3 Timbre

Timbre *(pronounced TAM-bur)* is easily the perceptual dimension about which we have the least psychophysical knowledge. Even *defining* timbre has been quite a challenge. The most often cited definition of timbre (that of the American National Standards Institute or ANSI) simply identifies what timbre is *not* and that whatever is left after excluding these characteristics– is timbre. ANSI's "negative definition" of timbre reads like this: "*...that attribute of auditory sensation in terms of which a listener can judge that two sounds, similarly presented and having the same loudness and pitch, are different*". In other words, timbre is what allows us to tell the difference between a trumpet and a clarinet when both are playing the same pitch at the same loudness. Part of the difficulty in defining timbre stems from the lack of a clear physical stimulus characteristic that is ultimately responsible for the perception of timbre. Unlike the physical-perceptual relationships of amplitude-loudness and frequency-pitch, there is no single dominant physical characteristic that correlates well with timbre. The spectral profile of the sound is most often identified as creating the percept of timbre, and spectrum does indeed influence timbre. However, the time varying characteristics of the amplitude envelope (or attack, sustain and decay time of the sound) has also been shown to have a significant influence on the perception of timbre.

Timbre can be an effective auditory dimension for sonification and has been used both as a continuous and a categorical dimension. Continuous changes in timbre have been proposed for example, in the auditory guidance of surgical instruments during brain surgery (Wegner, 1998). In this example, a change in spectrum is used to represent changes in the surface function over which a surgical instrument is passed. A homogeneous spectrum is used when the instrument passes over a homogeneous surface, and the homogeneity of the spectrum changes abruptly with similar changes in the surface area. Alternatively, discrete timbre changes, in the form of different musical instrument sounds can be used effectively to represent different variables or states of data. For example, discrete timbre differences have been used to represent the degree of confirmed gene knowledge in a sonification of human chromosome 21 (Won, 2005). Gene sequence maps are typically made in six colors that represent the degree of confirmed knowledge about the genetic data. Won (2005) employed six different musical instruments to represent the various levels of knowledge. When using different timbres it is critical to choose timbres that are easily discriminable. Sonification using similar timbres can lead to confusion due to undesirable perceptual grouping (Flowers, 2005).

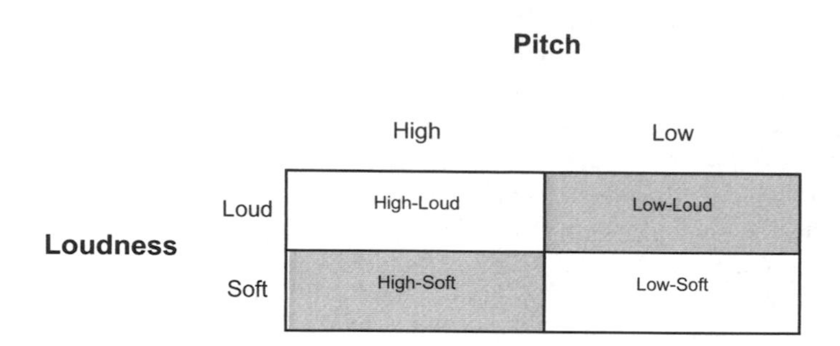

Figure 4.1: Schematic diagram of the four types of stimuli used in a speeded sorting task to test dimensional interaction. Grey and white boxes indicate "incongruent" and "congruent" stimuli respectively.

4.2.4 Interacting Perceptual Dimensions

At first glance, it would be easy to believe that distinct changes in individual acoustic characteristics of a stimulus such as frequency, intensity and spectrum would be perceived as perceptually distinct characteristics of a sound. However, there is growing evidence to the contrary. Changes in acoustic dimensions affect not only the percept of the corresponding perceptual dimension, but also specific perceptual characteristics of *other* perceptual dimensions. In other words, changes in one dimension (such as pitch) can affect perceived changes in the others (such as loudness). Given that the auditory system has evolved in an environment where stimuli constantly undergo simultaneous dynamic change of multiple acoustic parameters, perhaps this should come as no surprise. However, the implications of this kind of dimensional interaction for sonification and auditory display are important.

Perception researchers have devised a set of "converging operations" that are used to examine interacting perceptual dimensions (Garner, 1974). Listeners are typically presented with stimuli that vary along two dimensions such as pitch and loudness. They are instructed to attend to one dimension (e.g., pitch) and ignore changes in the other. In a speeded sorting task, for example, listeners would be presented with four types of sounds, with pitch and loudness each having two values (See Figure 4.1). The 2 (pitch) $\times$ 2 (loudness) matrix yields four sounds that are 1.) "high-loud", 2.) "high-soft", 3.) "low-loud", and 4.) "low soft". Listeners might be asked to perform a two-alternative forced-choice task in which they are asked to ignore loudness and simply press one of two buttons to indicate whether the pitch of the sound is "high" or "low". The researcher measures the amount of time required to make each response and the number of errors in each condition. Results typically show that responses are faster and more accurate in the "congruent" conditions of "high-loud" and "low-soft" than in the incongruent conditions. Because performance in the attended dimension is affected by variation in the *unattended* dimension, the two dimensions are said to interact. Pitch, timbre, loudness, and a number of other perceptual dimensions commonly used by display designers have all been shown to interact perceptually.

Thus, simply mapping orthogonal variables to different parts of the acoustic signal does not guarantee that they will remain orthogonal perceptually (Anderson & Sanderson, 2009;

Melara & Marks, 1990; Neuhoff, Kramer & Wayand, 2002, Walker & Ehrenstein, 2000), and therein lies a potential problem for designers of auditory displays. On the other hand, the "problem" of interacting perceptual dimensions has also been capitalized upon by redundantly mapping multiple perceptual dimensions (e.g., pitch and loudness) to a single data variable. This technique makes changes in the data more salient and is particularly effective for important signals. Depending on the context of the display, dimensional interaction can have both detrimental and advantageous effects in the context of auditory display. Each of these effects will now be explored.

4.2.5 Detrimental Effects of Dimensional Interaction

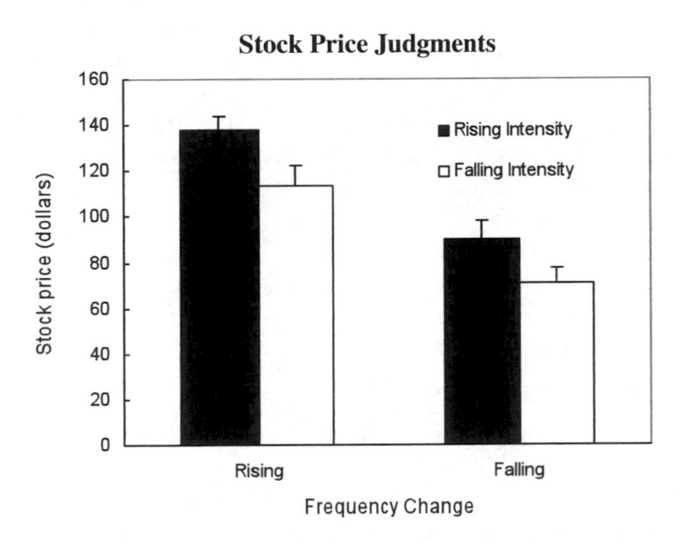

Figure 4.2: Perceptual interaction in auditory display. In a stock market sonification where terminal stock price was mapped to frequency change and the number of shares traded was mapped to intensity change, listeners gave different terminal price estimates for the same amount of frequency change depending on whether the concurrent intensity rose or fell. When frequency and intensity both rose, prices were judged to be higher than when frequency rose the same amount, but intensity fell. *(Adapted from Neuhoff, Kramer & Wayand, 2002).*

It is not at all uncommon in the context of sonification and auditory display for display designers to use various auditory dimensions to represent distinct variables in a data set. For example, when sonifying historical weather patterns, a display designer might use pitch change to represent the change in temperature, loudness change to represent the changes in the amount of precipitation, and timbre change to represent the relative change in humidity. Changes in the three weather variables can be easily represented by changes in three separate physical characteristics of the acoustic signal (frequency, amplitude, and spectrum). However, the perceptual interaction that occurs can be problematic. Although loudness is *principally* determined by the amplitude of a sound, there is good evidence that loudness is also more subtly influenced by the frequency (or pitch) of a sound. Fletcher & Munson (1933) were the first to show that the loudness of pure tones of equal intensity varied as a function of frequency. Their "equal-loudness contours" showed that listeners are most sensitive to sounds between 2 k kHz and 5 kHz. Similarly, Stevens (1935) showed that intensity can influence pitch. His "equal-pitch contours" showed that tones that differ in intensity can differ in frequency by up to 3% and still be perceived as equal in pitch. Timbre can interact with both pitch and loudness in similar ways.

Neuhoff, Kramer, & Wayand (2002) showed that pitch and loudness interact in auditory displays (see Figure 4.2). In a sonification of fictional stock market data, changes in stock price were mapped to changes in pitch, and changes in the number of shares traded were mapped to loudness. Rising pitch represented an increase in the price of a stock, and rising loudness represented an increase in the number of shares traded. In two contrasting conditions, listeners judged the terminal price of a stock. In one condition, the stock price rose while the number of shares also rose. This was represented with a sound that increased in both pitch and loudness. In the other condition, the stock price also rose (by the same amount as in the first condition). However, as the pitch increased to represent the rising price of the stock, the number of shares traded fell, thus loudness *decreased*. Despite the fact that the terminal pitch in each condition was the same and the stock price should be perceived as the same in each condition, listeners judged the price to be higher when both pitch and loudness rose than when pitch rose and loudness fell. Similarly listeners rated the price as *lower* when pitch and loudness both fell than when pitch fell and loudness rose. In other words, when the two dimensions changed in the same direction, the amount of change in one dimension was perceived as greater than when they changed in opposite directions.

4.2.6 Advantageous Effects of Dimensional Interaction

Auditory dimensions can be detrimental when separate variables are mapped to different auditory dimensions. However, there are cases when the interaction of auditory dimensions can be advantageously used in an auditory display. Mapping a single variable to multiple auditory dimensions has been shown to make the changes in that variable more salient than mapping it to single dimensions alone. For example, in sonifying changes in the volume of internet traffic on a particular site, one might use changes in loudness to denote changes in the amount of traffic, with higher loudness representing a higher volume of traffic. However, the change in traffic would be more perceptually salient if it were redundantly mapped to more than one dimension. Hansen and Ruben (2001) represented an increase in traffic by mapping it loudness, timbre, and repetition rate of a tone. So, an increase in traffic would yield a tone that gets brighter in timber, repeated faster, and also gets louder. This kind of "redundancy mapping" is effective in situations where absolute values in data are of secondary importance to changes and trends.

Redundancy mapping is also useful in auditory process monitoring tasks, particularly during "eyes busy" situations. Peres and Lane (2005) showed that redundant pitch and loudness mapping improved performance in a situation where listeners had to monitor auditory box plots while simultaneously performing a visual task. Importantly, the gains in performance due to redundancy mapping only occurred for auditory dimensions that have been shown to interact or are considered "integral" (such as pitch and loudness). When "separable" auditory dimensions (e.g., pitch and tempo) were mapped redundantly performance was not improved over the case in which only a single auditory dimension was used.

4.3 Auditory-Visual Interaction

There is a history in perceptual research of greater research efforts toward vision than audition, and a concentration on a single modality rather than on how vision and audition interact. We have relatively detailed accounts of the function of structures in the visual pathways when compared with those in audition. We know even less about the physiological interaction of the two systems. However, there are some clear examples of auditory and visual interaction at both the neurological and behavioral levels that have important implications for auditory display.

Perhaps the most famous example of auditory-visual interaction comes in the area of speech perception. The "McGurk Effect" occurs when visual and auditory speech tokens are mismatched, but presented simultaneously. For example, subjects may be presented with a video of a talker saying the syllable /ba/ with an accompanying audio track that says /ga/. In this case, listeners overwhelmingly report hearing the syllable */da/* (see video example **S4.1**). The work provides strong evidence for multimodal speech perception as does work showing that speech intelligibility increases when subjects can both hear and see the talker (Munhall, Jones, Callan, Kuratate, & Vatikiotis-Bateson, 2004; Sumby & Pollack, 1954) Thus, although the nature of auditory displays are such that they are most useful in "eyes busy" or low vision conditions, auditory displays that incorporate speech might benefit from the use of video to increase to reliability of the display if the conditions warrant.

Localization is another area in which strong auditory-visual interaction has been found. Visual performance in localization tasks is generally better than auditory performance. However, when subjects can use both their eyes and ears to localize an object, performance outpaces that which occurs in the visual only condition (Spence, 2007). The interdependence of vision and audition are particularly important in displays that require spatial estimates of a target that is both auditory and visual. In some contexts if the auditory and visual signals emanate from different locations, "visual capture" (or the "ventriloquist effect") will occur and users can perceive the audio signal as emanating from the location of the visual signal. However, in other contexts, the target can be perceived as somewhere in between the two signals (Alais, & Burr, 2004; Pick, Warren & Hay, 1969). This suggests a cognitive representation of external space that may be invariant across perceptual modalities. This arrangement allows, for example, that an auditory object that is heard but not seen can be spatially referenced with an object that is seen but not heard.

4.4 Auditory Space and Virtual Environments

The details of "how" we are able to perceive auditory space and motion are covered in the previous chapter. This section examines how this ability can be leveraged for use in auditory displays in both real and virtual environments and how the perception of auditory space interacts with vision.

Despite remarkable human ability to localize sound sources, the spatial resolution of the auditory system pales in comparison to what we can resolve visually. This, along with other visual advantages may contribute to the notion that humans are primarily "visual" beings. However, our strong reliance on vision may actually overshadow the degree to which we do rely on our ears for spatial localization and navigation in the real world. For example, there are some particular advantages that are obtained when localizing objects with our ears that cannot be obtained when localizing objects with our eyes. Obviously, the perception of auditory space does not require light. So, darkness and other poor viewing conditions do not present any great difficulty for auditory localization tasks. We can also hear objects that are hidden or occluded by other objects. Finally, while the field of vision is limited to approximately 120 degrees in front of the viewer, listeners can detect sounding objects 360 degrees around the head. Chronicling the advantages and disadvantages of the two systems might lead one to think that they are somehow in competition. However, the two systems work seamlessly together, each with strengths compensating for deficits in the other's repertoire of localization abilities. The result is an integrated multi-modal localization system that has evolved to help us localize objects and navigate a complex environment. As virtual environments become more common, our knowledge of both auditory and visual spatial perception will be crucial in creating environments that maintain a sense of presence.

In some environments where spatial auditory display is employed, the interaction of vision and audition can be of critical concern to display designers. For example, the spatial coincidence of auditory and visual representations of an object in a display will increase the sense of presence as well as the overall localization accuracy of the user. In other cases (e.g., auditory displays for the visually impaired) the focus on auditory-visual spatial interaction is decidedly less important. Nonetheless, the use of spatial information in auditory display is increasing. Advances in technology and our knowledge of how the auditory system processes spatial information has led to the emergence of virtual environments that realistically recreate 3-dimensional spatial auditory perception.

Many of these virtual auditory displays use Head Related Transfer Functions (HRTFs) to present binaural acoustic signals over headphones that mimic how the sounds would be received in a natural environment. The acoustic cues to spatial location (see chapter 3, this volume) can be manipulated as the user moves through a virtual environment. For example, a sound presented to the right of a listener will be perceived as louder in the right ear than in the left. However, when the listener's head turns to face the sound, a head-tracking device detects the movement of the head. The system detects the change in head position and the rendering system adjusts the level of the sound to be equal in the two ears, now equidistant from the source. All of the other cues to localization (e.g., interaural time differences and pinnae cues) are adjusted in a similar manner. Other systems use an array of loudspeakers that surround the listener. There are advantages and disadvantages to both approaches. However, the goal of both types of systems is to render the acoustic properties of the sound source and the environment such that the listener experiences the sound as though they were listening in the environment in which the sound would normally occur (Lokki, Savioja, Vaanaanaen, Huopaniemi, & Takala, 2002). When done well, the result is a spatial auditory display that in almost every way is more realistic and has better resolution than current visual virtual environments. Virtual auditory environments have applications in many domains. In addition to providing highly controlled environments in which researchers can study the psychology and physiology of auditory spatial perception (e.g., Nager, Dethlefsen, Münte, 2008), virtual auditory displays are used in psychiatry, aviation, entertainment, the military, as aids for the visually impaired, and in many other areas. The addition of spatialized sound in virtual environments does not only add to the auditory experience. It also increases the overall sense of presence and immersion in the environment (Hendrix & Barfield, 1996; Viaud-Delmon,

Warusfel, Seguelas, Rio, & Jouvent, 2006). Spatial coherence between visual and auditory objects in such environments are crucial to maintaining this sense of presence for the user.

Navigation performance in virtual environments is significantly better when spatial auditory information is present than when it is not (Grohn, Lokki, & Takala, 2003), and researchers have also shown that auditory localization performance for some listeners is comparable in real and virtual auditory environments (Loomis, Hebert, & Cicinelli, 1990). These latter findings are particularly encouraging to those involved in developing displays that are used for navigation (e.g., Seki & Sato, 2011).

4.5 Space as a Dimension for Data Representation

Virtual auditory environments provide a host of new possibilities for auditory display. The ability to use space as another dimension creates interesting possibilities for sonification and auditory display designers. Recent advances in technology have caused a growth in the use of spatialised sound in the areas of sonification and auditory display.

In one interesting example, Brungart and colleagues (2008) designed an auditory display for pilots that represented the attitude of the aircraft relative to the horizon (i.e. the plane's pitch and roll). Changes in roll were represented spatially by moving an audio signal back and forth as necessary between the left and right headphones. When the plane banked to the left, the signal moved to the right ear and vice versa. Additionally, changes in the plane's pitch (relative "nose-up" or "nose-down" position) were represented by a spectral filtering process. When the plane was nose-up, a spatially diffuse and low pitched characteristic was present in the stimulus, indicating that the nose of the aircraft should brought down to a more level flight position. When the aircraft was "nose-down" the signal was changed to a high pitched characteristic indicating that the nose of the plane should be pulled up. Straight and level flight was indicated by a spectrally unchanged signal that was equally centered between the right and left headphones. Importantly, the audio signal that was fed into the system could be anything, including music selected by the pilots. This technique has the advantage of greatly reduced annoyance and listener fatigue as well as higher compliance (i.e. the willingness of pilots to use the system).

Auditory spatial cueing has also been shown to be effective in automotive applications. Ho and Spence (2005) designed a display in which spatial auditory warnings facilitated visual attention in the direction of the auditory warning. Moreover, performance of emergency driving maneuvers such as braking or acceleration was improved by the use of spatial auditory displays.

4.6 Rhythm and Time as Dimensions for Auditory Display

One indication of the dominance of vision over other senses in humans is the tremendous disparity in the amount of cortex devoted to visual processing when compared to the other sensory modalities. Thus, as one might expect, the visual system tends to show better performance on many types of perceptual tasks when compared to audition (e.g., spatial localization). However, when it comes to rhythmic perception and temporal resolution, the auditory system tends to perform significantly better than the visual system. Thus, auditory display is particularly well suited for domains in which rhythmic perception and temporal discrimination are critical and domains in which the underlying data lend themselves to rhythmic and temporal variation, particularly when the rate of presentation is within the optimal sensitivity range of tempi for the user (Jones, 1976; Jones & Boltz, 1989).

Sound is inherently temporal, and differences in the timing and tempo of acoustic information has been studied extensively. Differences in tempo between displays and changes in tempo within a single display can be used effectively to convey relevant information. For example, urgency is generally perceived as higher when acoustic stimuli are presented at faster rates (Edworthy, Loxley, & Dennis, 1991; Langlois, Suied, Lageat & Charbonneau, 2008). Changes in tempo can also be used to indicate directional information (e.g., "up", "down") although tempo as an indicator of direction may be a relatively weak acoustic cue when compared to similar changes in pitch and loudness (Pirhonen & Palomäki, 2008). Semantic meaning of fast and slow rhythmic tempi has even been examined in the context of earcons (Palomäki, 2006).

Sensitivity to rhythm can also be exploited to indicate processes or data anomalies. For example, Baier, Hermann, and Stephani (2007; Baier & Herman, 2004) used variation in rhythm to indicate differences between epileptic and non-epileptic activity in human EEG data. Changes in rhythm and tempo have been also used in biofeedback systems designed for stroke rehabilitation (Wallis, et al. 2007).

Changes in rhythm and tempo can be used to indicate changes in the state or value of sonified data. However, simply speeding up the auditory display can also yield display benefits, particularly in displays that use speech. Listeners can retain a good deal of intelligibility even when speech is presented up to three times its normal rate (Janse, Nooteboom & Quené, 2003). The ability to perceive speech at faster rates than it is normally produced has been explored in a wide array of applications that range from screen readers for the visually impaired, to complex communication systems, to mobile phones. For example, many complex workstations feature simultaneous voice communication systems. Intelligibility in multiple talker systems generally decreases as the number of talkers goes up. Successful efforts to increase multi-talker intelligibility have generally focused on spatializing the talkers such that each voice emanates from a different position in space relative to the listeners (e.g., Brungart & Simpson, 2002). However, recent methods have also employed the dimension of time.

Brock, et al (2008) showed that in a four talker situation, speech that was artificially sped up by 75% and presented serially was understood significantly better than speech presented at normal speeds concurrently. Walker, Nance, and Lindsay (2006) showed that extremely fast speech can improve navigation through auditory menus in a cell phone application. As opposed to earcons or auditory iconsthe fast speech or "spearcons" yielded faster and more accurate user performance. Spearcons are produced by speeding up text-to-speech audio output until it is no longer perceived as speech. However, the spearcon still retains some similarity to the original speech signal from which it was derived. The time required to learn an auditory menu also appears to be reduced when the menu is presented with spearcons rather than earcons (Palladino & Walker, 2007). The majority of the research on spearcons has been conducted from the perspective of improving human performance in auditory display settings. Thus, little is known about the underlying cognitive mechanisms that afford the enhanced performance.

4.7 Auditory Scene Analysis

Sounds generally do not occur in isolation. At any given instant numerous sound sources create separate acoustic waves that can reach the ear simultaneously. In fact, in most environments it is unusual to hear a single sound in isolation. The hum of a computer fan is heard simultaneously with the ticking of a clock, a conversation in the next room, and the muffled sound of passing cars on the roadway outside. When the acoustic waves from these various sound sources reach the tympanic membrane (eardrum) they result in a single highly complex mechanical signal that, were it examined visually via sonograph, might appear to be almost random. Yet, perceptually listeners have little difficulty extracting the source information from this complex signal and can easily hear distinct sound sources. In other words, it is rarely difficult to tell where one sound stops and another begins. The process of segregating these auditory sources is referred to as *auditory scene analysis*, and the ease with which we accomplish the task belies its tremendous complexity.

Consider for example, the many technologies that now respond to voice commands. Cell phones, computers, automobiles, and many other devices can decipher simple voice commands and produce a requested action, provided there are no significant competing background sounds. Have just two or three people speak to a voice activated device simultaneously and the device fails to detect where one voice ends and another begins, a task that human listeners can do with relative ease. Thus, despite our tremendous technological advances, we have yet to develop voice-activated technology that might work well, for example, at a cocktail party (Cherry, 1953).

Al Bregman pioneered the study of auditory scene analysis by asking questions about audition that at the time were considered "non-traditional". His Ph.D. in cognitive psychology was completed at Yale in a laboratory that primarily studied vision. He subsequently pursued his interest in auditory perception undertaking research that was a strong departure from the traditional work being performed in psychoacoustics at the time. He applied many of the same techniques and questions that were being asked about visual phenomena to auditory phenomena. Because of his work, people now speak regularly of "auditory objects" and "auditory streams".

The perceptual work on auditory scene analysis has important implications for auditory display. The fact that listeners can attend to separate sources or auditory streams allows sonification designers to exploit our auditory perceptual organization abilities and simultaneously present distinct aspects of a multidimensional data set in distinct auditory streams.

Acoustic characteristics and attentional factors can both influence how the auditory system perceptually organizes the auditory scene. At the acoustic level, individual sound sources have certain acoustic regularities in the sounds that they produce that can be used by the auditory system to parse an auditory stream. For example, sounds that are similar in frequency are more likely to be allocated to the same source. Thus, in sonifying a multidimensional data set it is common to separate different variables by differences in frequency range. One variable might be represented by low frequency sounds and another by high frequency sounds. The separation in frequency of the two "streams" makes it more likely that the two variables being sonified will also remain independent. Similarly, sounds that are similar in timbre are more likely to be perceptually grouped together. Thus, a common technique is to use different musical instruments to represent separate aspects of the underlying data.

Pitch and timbre differences can be manipulated independently to affect how auditory grouping occurs. For example, the grouping effect that occurs by making sounds similar in pitch can be counteracted by making them dissimilar in timbre and vice versa (Singh, 1987). Conversely, the grouping effect can be made stronger by using redundant segregation cues in each stream. For example, the differences in both pitch and timbre between a bass guitar and a piccolo would provide better stream segregation than only the timbre differences of say, a saxophone and trumpet played in the same frequency range.

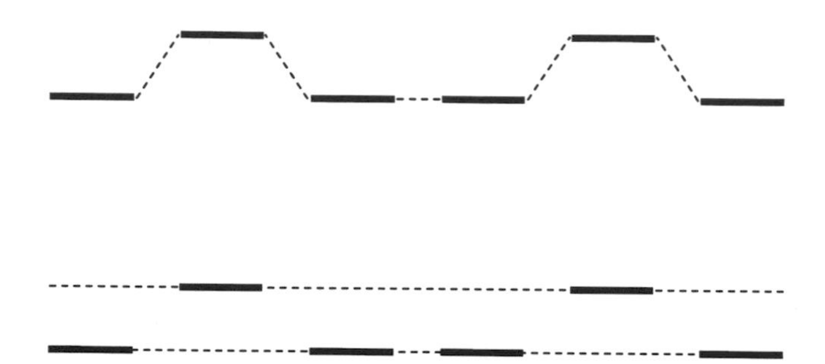

Figure 4.3: Alternating high and low pitched tones can either be perceived as one or two auditory streams depending on presentation rate and the distance in pitch between the tones.

Differences in other acoustic characteristics such as loudness and spatial location can also be used to parse sound sources. Although loudness level may not be as strong a cue to grouping as other acoustic characteristics, sounds presented at similar levels nonetheless tend to group together (Hartmann & Johnson, 1991; Van Noorden, 1975). Spatial location is a strong cue to auditory stream segregation. Sounds that come from the same location tend to be grouped together. A lack of spatial coherence often prevents sounds from being perceptually grouped together. For example, a sequence of tones presented to alternating ears tends not to form a single auditory stream (Van Noorden, 1975). Spatial separation of sources using binaural cues to localization is a particularly effective means for segregating real world sources such as multiple talkers (Hawley, Litovsky, & Culling, 2004).

Tempo and rhythm also interact with auditory stream segregation. However, rather than parsing simultaneous sounds into distinct auditory objects, tempo and rhythm effects are more likely to occur with sequentially presented stimuli. Van Noorden (1975) presented listeners with alternating high and low pitched notes that could either be perceived as one or two separate streams (see Figure 4.3). When perceived as a single stream the notes are heard as a galloping rhythm that goes up and down in pitch. When perceived as two streams the notes are heard as two repeating patterns each with a regular isochronous rhythm. The tempo at which the stimuli are presented can influence whether the notes are perceived as one stream or two with faster tempi being more likely to induce the perception of two streams. Moreover, cues which aid in simultaneous stream segregation can also influence sequential segregation (Micheyl, Hunter & Oxenham, 2010). For example, the amount of separation in frequency between the notes can influence how the streams are perceived. Greater frequency separation makes it more likely that the notes will be perceived as two streams (Bregman, 1990).

In addition to the lower level acoustic characteristics of sound sources, attention and higher order cognitive processes can affect how the auditory scene is parsed (Bregman, 1990; Carlyon, Cusack, Foxton, & Robertson, 2001; Snyder & Alain, 2007; Sussman & Steinschneider, 2009). Prior knowledge, expectations, selective attention, and expertise can all influence the landscape of the auditory scene. These cognitive processes work in concert with the acoustic characteristics when listeners parse auditory objects (Alain, Arnott, & Picton, 2001).

4.8 Auditory Cognition

There is a rich history of psychoacoustic research on the "sensory" aspects of audition. Conversely, "auditory cognition" has received comparatively little attention. Incoming acoustic information is transformed into a neural signal at the level of specialized cells in the inner ear. With the exception of speech and music, this is where the study of audition often stopped. However, in addition to the incoming acoustic signal that arrives at the eardrum, the listener's prior knowledge, experience, expertise, and expectations can all influence how acoustic information is perceived. Cognitive psychologists have come to call these kinds of effects "top-down" processing to distinguish them from the "bottom-up" processing that occurs when acoustic information is received, transformed into a sensory signal, and passed "up" to higher cortical areas. The effects of top-down processing are widespread (though perhaps not well known) in auditory display environments. Any type of effect in user performance due to the expertise of the user, training, or the expectations of the user comes under the umbrella of top-down effects (Strait, Kraus, Parbery-Clark, & Ashley, 2010; Sussman, Winkler, & Schröger, 2003).

An example of top-down cognitive processing occurs in a phenomenon called the "phonemic restoration effect". In natural listening environments speech sounds are often briefly interrupted or masked by other environmental sounds. Yet, this rarely interferes with the listener's comprehension of the message. Warren (1970) showed that if a phoneme (i.e, the smallest segment of a word that still imparts meaning) is removed from a word and replaced with noise or a cough, listeners still hear the missing phoneme. Moreover, they have great difficulty even indicating where the cough or noise occurred in the utterance. The effect has been rigorously researched and is the result of top-down perceptual processing (Samuel, 2001).

The simple act of recognizing a friend's familiar voice also requires top-down processing. Subsequent to the transformation of the acoustic signal into a neural impulse, the stimulus must be identified as a voice, likely engaging many of the mechanisms that process the various aspects speech, including syntax, semantics, and even emotion. Memory must be activated, and the incoming signal matched to a cognitive representation of your friend's voice. All of this occurs in an instant, and you can then recognize that your friend is talking to you, he wants to get something to eat, and he sounds a little sad. The prior experience, memory, and expectations of the listener can shape the perception of sound. Similar processes must occur for non-speech sounds. Recognizing and responding appropriately to the sound of a car horn, a baby's cry, or gunfire can have life or death implications.

Although researchers are beginning to make progress in understanding some of the complex processes that occur in "auditory meaning making" for speech, they are not yet completely understood. When it comes to understanding the cognitive processes of the non-speech sounds typically used in auditory display, we know even less. Thus, in order to understand sound and derive real world meaning from these neural signals, a more thorough investigation is required. Cognition and action in response to auditory stimuli are crucial not only in auditory display environments, but in almost all real world situations.

4.8.1 Cognitive Auditory Representations

Cognitive or "mental" representations of stimuli have a rich history in cognitive psychology. They are also a potentially fruitful area for designers of auditory displays. The idea that a cognitive representation of an external stimulus could even exist was at one time quite controversial, and the specifics of such representations are still debated among psychologists and cognitive scientists. There is clearly subjective or anecdotal evidence of cognitive representations. When asked, for example, to imagine their kitchen, most people can bring a visual image of their kitchen to mind and describe it in some detail. From an experimental perspective, behavioral and neuroimaging studies have provided rather convincing evidence that the brain does store some kind of representation of stimuli from the external world.

In the auditory domain, there is also evidence for cognitive representations of acoustic stimuli. As in the visual domain, there is abundant subjective and anecdotal evidence. Almost anyone will admit to being able to imagine the sound of a car horn, a bird chirping, or of eggs frying in a pan. There is also abundant experimental evidence for "auditory imagery". In one ingenious study by Halpern and Zatorre (1999), subjects listened to simple melodies while connected to a Positron Emission Tomography (PET) scanner. The PET scanner allows researchers to identify areas of brain activation during various activities or when various stimuli are presented. In one condition the subjects were simply asked to listen to the song. In another condition subjects were played only the first half of the song and asked imagine the rest by "singing it in their head". The surprising finding was that the same areas of the brain were active during the silent "imagined" portion of the song as were active when the song was actually heard. This work suggests that auditory "cognitive representations" may in fact simply be the occurrence of a pattern of neural firing in the absence of a stimulus that would occur if the stimuli were actually present.

Surprisingly, cognitive representations of real world sounds have not been widely used by sonification designers as a means of representing variable data sets. The majority use simple changes in pitch, loudness or timbre to represent changes in the variables of interest. The result is often a changing auditory signal that has no direct cognitive representation of the underlying data for the listener. This is certainly not to say that associations between the changing acoustic characteristics and the data set cannot be learned; only that it is a secondary process to understand, for example, that a change in timbre represents a change in temperature. Moreover, when multivariate datasets are sonified, simultaneous changes in pitch, loudness, and timbre are commonly used in a single signal to represent various changes in data. However, the underlying data in this example are subject to distortions from the perceptual interaction effects outlined above.

An alternative to this sonification technique has been proposed that involves mapping changes in real world auditory events to changes in the underlying data set. Gaver (1993) suggested that listeners attend to "auditory events" in a way that makes the physical characteristics of the sound source an important factor in auditory perception of non-speech sounds. So, rather than hearing "... *a quasi-harmonic tone lasting approximately three seconds with smooth variations in the fundamental frequency and the overall amplitude...*", listeners will report instead that they heard "*A single-engine propeller plane flying past*", (Gaver, 1993, p. 285–286). The upshot is that listeners consciously process events, not acoustics.

Neuhoff and Heller (2005) suggested that this "event based" representation might be effectively used in sonification. For example, rather than mapping increasing pitch to an increase in the data, a designer might instead map changes in the data to the pace of a real world auditory event that listeners are highly skilled at perceiving, such as footsteps (Li, Logan, & Pastore, 1991; Visell, et al., 2009). The advantage to this approach is twofold. First, the changes in these complex stimulus dimensions tend to be more familiar and easier to identify than changes in simple acoustic dimensions. Music novices, for example, often have difficulty describing pitch change as going "up" or "down" because they have not been had the necessary exposure to know that increases in frequency are related to "higher" pitch (Neuhoff, Knight & Wayand, 2002). However, most listeners can easily distinguish between fast and slow footsteps. Second, the problem of unwanted interacting perceptual dimensions can be avoided by using real world auditory events to represent changes in data. For example, if walking speed were used to represent one variable in a multivariate data set, the hardness of the surface might be used to represent another variable. Most listeners can identify specific properties of walking surfaces in addition to characteristics of the walker such as gender and height (Visell, Fontana, Giordano, Nordahl, Serafin& Bresin, 2009). The complexity of such an acoustic representation would yield large benefits in the simplicity of the perceptual interpretation of the data (Neuhoff & Heller, 2005).

4.8.2 Music and Data Representation

Perhaps some of the most structured auditory cognitive representations that exist are musical systems. Musical scales provide a formal structure or framework that can be leveraged in the design of effective auditory displays (Krumhansl, 1982; Jordan & Shepard, 1987; Shepard, 1982). Thus, given that one of the main goals of auditory display is to communicate information, auditory display can be informed by music theory. Rather than mapping data to arbitrary changes in frequency, many auditory displays map changes in data to changes in pitch that are constrained to standard culturally specific musical scales. For example, Vickers and Alty (1997; 2002; 2003) have employed melodic motifs to aid computer programmers in debugging code and to provide other programming feedback. Valenzuela (1998) used melodic information to provide users with integrity evaluation information about concrete and masonry structures. Melodic information in auditory display has even been used as a tool for mathematics instruction with middle school and high school students (Upson, 2002).

An advantage of using musical scales in sonification is that they may be perceived as more pleasant and less annoying than frequency change that is not constrained to musical scales. Although there has been ample work to show that differing levels of musical expertise can influence perceptual performance in a musical setting (e.g., Bailes. 2010), these differences can be minimized when the stimuli are interpreted in units that reflect the underlying data dimensions (Neuhoff, Knight, & Wayand, 2002). The effects of musical expertise on the perception of auditory displays have not been thoroughly investigated. Part of the difficulty in this area has been the lack of a well designed system for measuring musical expertise (Edwards, Challis, Hankinson & Pirie, 2000). Although there are tests of musical ability among musicians, there are few validated ways of examining musical ability among those who have no formal training in music (however, for one promising method see Ollen, 2006).

4.8.3 Perception and Action

The idea that our actions and the motor system are involved in perceiving the external world dates back to at least the late 1960s. Liberman and colleagues (1967) proposed that the speech signal is decoded in part by referring incoming speech sounds to the neuro-muscular processes that are used to produce them. In essence, we understand speech through the motor commands that are employed when we ourselves speak. The details of the "Motor Theory" of speech perception have been sharply debated over the years, but there are few who would doubt that perception and action are closely linked in many domains.

Advances in neuroimaging have yielded numerous investigations which show that regions of the brain that are responsible for motor activity are recruited to process incoming auditory stimuli, even when those stimuli are non-speech sounds. For example, Chen and colleagues (2008) showed that motor areas were active when subjects listened to a rhythmic pattern in anticipation of tapping along with the rhythm later. Even when subjects were simply asked to listen to the rhythms with no knowledge that they would be asked to tap along later, the same motor regions were active. Similarly, pianists show activation in motor areas when simply listening to a piano performance (Haueisen, Knösche, 2001; Bangert, et al, 2006). The perception-action link is further evidenced by the finding that non-pianists (who presumably would not have the motor plans for a piano performance) do not show activation in motor areas when presented with the same music.

In another study, subjects were presented with "action sounds" that were consistent with human motor behavior (e.g., crunching, opening a zipper, crushing an aluminum can) and "non-action" sounds that did not require any motor behavior (e.g., waves on a beach, a passing train, or wind). The motor areas of the brain activated when the "action sounds" were presented were the same ones activated when the subjects actually performed the actions depicted in the sounds. However, motor areas were not recruited when listeners were presented with the non-action sounds. In addition to processing incoming stimuli, these so called auditory "mirror neurons" may be involved in facilitating communication and simulation of action (Kohler, Simpson l., 2002).

An important point taken from these studies is that the articulatory gestures that are used to produce "action sounds" may be as important as the acoustic structure of the sounds themselves. In other words, the link between the auditory and motor system appears to capitalize on the knowledge of the actions used to produce the sounds as much as the specific acoustic attributes per se. Thus, the use of real world sounds in auditory display discussed previously may tap into perceptual and "meaning making" processes that cannot be accessed with sounds that are more artificial. An additional distinction among real world sounds has been made by Giordano, McDonnel, and McAdams (2010). They used a sound sorting task with "living sounds" and "non-living sounds" and found that listeners differentiate non-living action and non-action sounds with an iconic strategy that does indeed focus on acoustic characteristics of the sound. The evaluation of living sounds, on the other hand, relied much more on a symbolic cognitive representation of the sound referent.

From the perspective of designing auditory displays, these findings suggest that the judicial use of environmental sounds rather than simpler artificial sounds might provide a better means of communicating the information to be displayed. Millions of years of evolution have produced neural and cognitive architecture that is highly sensitive to meaningful real world environmental sounds. Perceptual processing of these sounds appears to happen in a way that is fundamentally different from that which occurs with simple arbitrary beeps and buzzes. We know that simply mapping a sound that has a clear environmental referent (i.e. auditory icons see chapter 13) to a particular display dimension increases user response time and accuracy in the display over more arbitrary mappings (McKeown & Isherwood, 2007). Future research may demonstrate even greater gains with environmental sounds have a clear *behavioral* referent which maps to a specific motor action.

4.9 Summary

The ease with which we perceive the auditory world masks the complexity of the process of transforming acoustic waves into meaning and responsive behavior. Basic acoustic dimensions such as pitch, loudness and timbre can be used to represent various aspects of multidimensional data. However, extreme care and an intentional approach should be taken in understanding the perceptual interactions that occur with these kinds of dimensions. Auditory perception acts in concert with other sensory modalities, and cross modal influences with vision and other senses can influence perception and performance in an auditory display. Higher order acoustic characteristics, including time, and space, are also common vehicles through which acoustic information is used to represent data. These factors interact with the cognitive processes involved in auditory scene analysis, music and speech, and perception-action relationships to form a complex foundation upon which effective auditory displays can be designed.

Bibliography

[1] Alain C., Arnott S. R., Picton T. W. (2001). Bottom-up and top-down influences on auditory scene analysis: evidence from event-related brain potentials. J Exp Psychol Hum Percept Perform. 27(5):1072–89.

[2] Alais, D. & Burr, D. (2004). The ventriloquist effect results from near-optimal bimodal integration. Current Biology, 14, 257–262.

[3] Anderson, J., & Sanderson, P. (2009). Sonification design for complex work domains: Dimensions and distractors. Journal of Experimental Psychology: Applied. 15(3), 183–198.

[4] Baier, G. & Hermann, T. (2004). The sonification of rhythms in human electroencephalogram. Proceedings of the International Conference on Auditory Display.

[5] Baier, G., Hermann T., & Stephani U. (2007). Event-based sonification of EEG rhythms in real time. Clinical Neurophysiology. 1377–1386.

[6] Bailes, F. (2010). Dynamic melody recognition: Distinctiveness and the role of musical expertise. Memory & Cognition 38(5), 641–650.

[7] Bangert, M., Peschel, T., Schlaug, G., Rotte, M., Drescher, D., Hinrichs, H, Heinze, H.J., & Altenmüllera, E. (2006). Shared networks for auditory and motor processing in professional pianists: Evidence from fMRI conjunction. NeuroImage, 30, 917–926.

[8] Bregman, A. S. 1990. Auditory Scene Analysis. MIT Press, Cambridge, MA.

[9] Brock, D., McClimens, B., Wasylyshyn, C., Trafton, J. G., & McCurry, M. (2008). Evaluating listeners'

attention to and comprehension of spatialized concurrent and serial talkers at normal and a synthetically faster rate of speech. Proceedings of the International Conference on Auditory Display.

[10] Brungart D. S. & Simpson, B. D. (2002). The effects of spatial separation in distance on the informational and energetic masking of a nearby speech signal. J. Acoust. Soc. 112, 664–676.

[11] Brungart D. S., & Simpson, B. D., (2008). Design, validation, and in-flight evaluation of an auditory attitude ondicator based on pilot-selected music. Proceedings of the International Conference on Auditory Display.

[12] Carlyon, R. P., Cusack, R., Foxton, J. M., & Robertson, I. H. (2001). Effects of attention and unilateral neglect on auditory stream segregation. Journal of Experimental Psychology: Human Perception and Performance, 27, 115–127.

[13] Chen, J. L., Penhune, V. B., & Zatorre, R. J. (2008) Moving on time: brain network for auditory-motor synchronization is modulated by rhythm complexity and musical training. Journal of Cognitive Neuroscience, 20(2):226–239.

[14] Cherry, E. C. (1953). Some Experiments on the Recognition of Speech, with One and with Two Ears. Journal of Acoustic Society of America 25 (5): 975–979.

[15] Edwards, A. D. N., Challis, B. P., Hankinson, J. C. K., & Pirie, F. L. (2000). Development of a standard test of musical ability for participants in auditory interface testing. Proceedings of the International Conference on Auditory Display.

[16] Edworthy, J., Loxley, S., & Dennis, I. (1991). Improving auditory warning design – relationship between warning sound parameters and perceived urgency. Human Factors, 33(2), 205–231.

[17] Fletcher, H. & Munson, W. A. (1933). Loudness, its definition, measurement and calculation. Journal of the Acoustical Society of America, 5:82–108.

[18] Flowers, H. (2005). Thirteen years of reflection on auditory graphing: Promises, pitfalls, and potential new directions, Proceedings of the International Conference on Auditory Display, 406–409.

[19] Flowers, J. H.,Whitwer, L. E., Grafel, D. C., & Kotan, C. A. (2001). Sonification of daily weather records: Issues of perception, attention and memory in design choices. Proceedings of the International Conference on Auditory Display.

[20] Garner, W. R. (1974). The Processing of Information and Structure. Potomac, MD: Erlbaum.

[21] Gaver, W. W. (1993). How do we hear in the world? Explorations of ecological acoustics. Ecological Psychology, 5(4): 285–313.

[22] Giordano, B. L., McDonnell, J., & McAdams, S. (2010). Hearing living symbols and nonliving icons: Category-specificities in the cognitive processing of environmental sounds. Brain & Cognition, 73, 7–19.

[23] Gröhn, M., Lokki, T., & Takala, T. (2003). Comparison of auditory, visual, and audio-visual navigation in a 3D space. Proceedings of the International Conference on Auditory Display, 200–203.

[24] Halpern, A. R. & R. J. Zatorre. 1999. When that tune runs through your head: an PET investigation of auditory imagery for familiar melodies. Cerebral Cortex 9: 697–704.

[25] Hansen, M. H. & Rubin, B (2001). Babble online: applying statistics and design to sonify the internet. Proceedings of the International Conference on Auditory Display.

[26] Hartmann, W. M., and Johnson, D. (1991). Stream segregation and peripheral channeling. Music Perception 9, 155–183.

[27] Haueisen, J., & Knösche, T. R. (2001). Involuntary motor activity in pianists evoked by music perception. Journal of Cognitive Neuroscience, 13(6), 786–79.

[28] Hawley, M. L., Litovsky, R. Y., & Culling, J. F. 2004. The benefit of binaural hearing in a cocktail party: effect of location and type of interferer. Journal of the Acoustical Society of America, 115, 833–843.

[29] Hendrix, C. & Barfield, W. (1996). The sense of presence within auditory virtual environments. Presence: Teleoperators and Virtual Environments. 5, (3), 290–301.

[30] Ho, C., & Spence, C. (2005). Assessing the effectiveness of various auditory cues in capturing a driver's visual attention. Journal of Experimental Psychology: Applied, 11, 157–174.

[31] Janata, P., & Childs, E. (2004). MarketBuzz: Sonification of real-time financial data. Proceedings of The 10th

Meeting of the International Conference on Auditory Display.

[32] Janse, E., Nooteboom, S. & Quené, H. (2003). Word-level intelligibility of time-compressed speech: Prosodic and segmental factors. Speech Communication 41, 287–301.

[33] Jones, M. R. (1976). Time, our lost dimension: Toward a new theory of perception, attention, and memory. Psychological Review 83: 323–355.

[34] Jones, M. R., & Boltz, M. (1989). Dynamic attending and responses to time. Psychological Review 96: 459–491.

[35] Jordan, D. S. & Shepard, R. N. (1987). Tonal schemas: Evidence obtained by probing distorted musical scales. Special Issue: The understanding of melody and rhythm, 41, 489–504.

[36] Kohler, E., Keysers, C., Umiltà, M.A., Fogassi, L., Gallese, V. and Rizzolatti, G. (2002) Hearing sounds, understanding actions: action representation in mirror neurons. Science, 297: 846–848.

[37] Kramer, G., Walker, B., Bonebright, T., Cook, P., Flowers, J., Miner, N.; Neuhoff, J., Bargar, R., Barrass, S., Berger, J., Evreinov, G., Fitch, W., Gröhn, M., Handel, S., Kaper, H., Levkowitz, H., Lodha, S., Shinn-Cunningham, B., Simoni, M., Tipei, S. (1999). The Sonification Report: Status of the Field and Research Agenda. Report prepared for the National Science Foundation by members of the International Community for Auditory Display. Santa Fe, NM: ICAD.

[38] Krumhansl, C. L. (1983). Perceptual structures for tonal music.," Music Perception, 1, 28–62.

[39] Langlois, S., Suied, C., Lageat, T. & Charbonneau, A. (2008). Cross cultural study of auditory warnings. Proceedings of the International Conference on Auditory Display.

[40] Li, X.-F., Logan, R. J., & Pastore, R. E. (1991). Perception of acoustic source characteristics: Walking sounds. Journal of the Acoustical Society of America, 90, 3036–3049.

[41] Liberman, A. M., Cooper, F. S., Shankweiler, D. P., & Studdert-Kennedy, M. (1967). Perception of the speech code. Psychological Review 74 (6): 431–461.

[42] Lokki, T., Savioja, L., Väänänen, R., Huopaniemi, J., Takala, T. (2002). Creating interactive virtual auditory environments. IEEE Computer Graphics and Applications, special issue "Virtual Worlds, Real Sounds", 22(4), 49–57.

[43] Loomis, J. M., Hebert, C., & Cicinelli, J. G. (1990). Active localization of virtual sounds. Journal of Acoustical Society of America, 88, 1757–1764.

[44] McKeown, D., & Isherwood, S. (2007). Mapping candidate within-vehicle auditory displays to their referents. Human Factors, 49(3), 417–428.

[45] Melara, R. D., & Marks, L. E. (1990). Interaction among auditory dimensions: Timbre, pitch, and loudness. Perception and Psychophysics, 48, 169–178.

[46] Micheyl, C., Hunter, C., Oxenham, A. J. (2010). Auditory stream segregation and the perception of across-frequency synchrony. Journal of Experimental Psychology: Human Perception and Performance. 36(4), 1029–1039.

[47] Munhall, K. G., Jones, J. A., Callan, D. E., Kuratate, T., & Vatikiotis-Bateson, E. (2004). Visual prosody and speech intelligibility: Head mov

[48] Nager W., Dethlefsen C., Münte T. F. (2008). Attention to human speakers in a virtual auditory environment: brain potential evidence. Brain Research, 1220, 164–170.

[49] Neuhoff, J. G. & Heller, L. M. (2005). One small step: Sound sources and events as the basis for auditory graphs. Proceedings of the 11th International Conference on Auditory Display.

[50] Neuhoff, J. G., Knight, R. & Wayand, J. (2002). Pitch change, sonification, and musical expertise: Which way is up? Proceedings of the International Conference on Auditory Display.

[51] Neuhoff, J. G., Kramer, G., & Wayand, J. (2002). Pitch and loudness interact in auditory displays: Can the data get lost in the map? Journal of Experimental Psychology: Applied. 8 (1), 17–25.

[52] Ollen, J. E. (2006). A criterion-related validity test of selected indicators of musical sophistication using expert ratings. Dissertation, Ohio State University.

[53] Palladino, D., & Walker, B. N. (2007). Learning rates for auditory menus enhanced with spearcons versus

earcons. Proceedings of the International Conference on Auditory Display, 274–279.

[54] Palomäki, H. (2006). Meanings conveyed by simple auditory rhythms. Proceedings of theInternational Conference on Auditory Display, 99-104.

[55] Peres, S. C., & Lane, D. M. (2005) Auditory Graphs: The effects of redundant dimensions and divided attention, Proceedings of the International Conference on Auditory Display, 169–174.

[56] Pick, H. L., Warren, D. H., , & Hay, J. C. (1969). Sensory conflict in judgments of spatial direction. Perception & Psychophysics, 6(4), 203–205.

[57] Pirhonen, A., Palomäki, H. (2008). Sonification of Directional and Emotional Content: Description of Design Challenges. Proceedings of the International Conference on Auditory Display.

[58] Samuel, A. G. (2001). Knowing a word affects the fundamental perception of the sounds within it. Psychological Science, 12, 348–351.

[59] Seki Y., Sato T. A. (2011) Training system of orientation and mobility for blind people using acoustic virtual reality. IEEE Trans Neural Syst Rehabil Eng. 1, 95–104.

[60] Shepard, R. N. (1982). Geometrical approximations to the structure of musical pitch. Psychological Review, 89, 305–333.

[61] Singh, P. G. (1987). Perceptual organization of complex-tone sequences: a tradeoff between pitch and timbre? J Acoust Soc Am 82: 886–899.

[62] Smith, D. R. & Walker, B. N. (2002). Tick-marks, axes, and labels: The effects of adding context to auditory graphs. Proceedings of the International Conference on Auditory Display, 362–36

[63] Snyder, J. S., & Alain, C. (2007). Sequential auditory scene analysis is preserved in normal aging adults. Cerebral Cortex, 17, 501–5.

[64] Spence, C. (2007). Audiovisual multisensory integration. Acoustical Science and Technology, 28, 61–70.

[65] Stevens, S. S. (1935). The relation of pitch to intensity. Journal of the Acoustical Society of America, 6, 150–154.

[66] Stevens, S. S. & Davis, H. (1938). Hearing: Its psychology and physiology. Oxford, England: Wiley.

[67] Strait D. L., Kraus, N. Parbery-Clark, A., & Ashley, R. (2010). Musical experience shapes top-down auditory mechanisms: Evidence from masking and auditory attention performance. Hearing Research 261, 22–29.

[68] Sumby, W. H., & Pollack, I. (1954). Visual Contribution to Speech Intelligibility in Noise. Journal of the Acoustical Society of America, 26 (2) 212–215.

[69] Sussman E., Steinschneider, M. (2009). Attention effects on auditory scene analysis in children. Neuropsychologia 47(3):771–785.

[70] Sussman, E., Winkler, I., & Schröger, E. (2003). Top-down control over involuntary attention switching in the auditory modality. Psychonomic Bulletin & Review, 10(3), 630–637.

[71] Upson, R. (2002). Educational sonification exercises: Pathways for mathematics and musical achievement. International Conference on Auditory Display, Kyoto, Japan.

[72] Valenzuela, M. L. (1998). Use of synthesized sound for the auditory display of impact-echo signals: Design issues and psychological investigations. Dissertation Abstracts International: Section B: The Sciences and Engineering. Vol 58(12-B).

[73] van Noorden, (1975) Temporal Coherence in the Perception of Tone Sequences. Dissertation, Technical University Eindhoven.

[74] Viaud-Delmon I, Warusfel O, Seguelas A, Rio E, Jouvent R (2006). High sensitivity to multisensory conflicts in agoraphobia exhibited by virtual reality, European Psychiatry, 21(7), 501–508.

[75] Vickers, P. and Alty, J. L. (1996) CAITLIN: A Musical Program Auralisation Tool to Assist Novice Programmers with Debugging. Proceedings of the International Conference on Auditory Display.

[76] Vickers, P. and Alty, J. L. (2002) Using Music to Communicate Computing Information. Interacting with Computers, 14 (5). 435–456.

[77] Vickers, P. and Alty, J. L. (2003) Siren Songs and Swan Songs: Debugging with Music. Communications of

the ACM, 46 (7). 86–92.

[78] Visell, Y., Fontana, F., Giordano, B.L., Nordahl, R., Serafin, S., & Bresin, R. (2009). Sound design and perception in walking interactions. International Journal of Human-Computer Studies, 67 (11), 947–959.

[79] Walker, B. N. (2002). Magnitude estimation of conceptual data dimensions for use in sonification. Journal of Experimental Psychology: Applied, 8, 4, 211–221.

[80] Walker, B. N. (2007). Consistency of magnitude estimations with conceptual data dimensions used for sonification. Applied Cognitive Psychology, 21(5), 579–599.

[81] Walker, B. N., & Ehrenstein, A. (2000). Pitch and pitch change interact in auditory displays. Journal of Experimental Psychology: Applied, 6, 15–30.

[82] Walker, B. N., & Kramer, G. (2004). Ecological psychoacoustics and auditory displays: Hearing, grouping, and meaning making. In J. Neuhoff (Ed.), Ecological Psychoacoustics (pp.150–175). New York: Academic Press.

[83] Walker, B. N., & Lane, D. M. (2001). Psychophysical scaling of sonification mappings: A comparison of visually impaired and sighted listeners. Proceedings of the International Conference on Auditory Display, 90–94.

[84] Walker, B. N., Nance, A., & Lindsay, J. (2006). Spearcons: Speech-based Earcons Improve Navigation Performance in Auditory Menus. Proceedings of the International Conference on Auditory Display, 63–68.

[85] Wallis, I., Ingalls, T., Rikakis, T., Olsen, L., Chen, Y., Xu, W., & Sundaram, H. (2007). Real-time sonification of movement for an immersive stroke rehabilitation environment. Proceedings of the International Conference on Auditory Display, 497–503.

[86] Warren, R. M. (1970). Perceptual restoration of missing speech sounds. 167, 392–393.

[87] Wegner, K. (1998). Surgical navigation system and method using audio feedback. Proceedings of the International Conference on Auditory Display.

[88] Won, S. Y. (2005). Auditory display of genome data: Human chromosome 21. Proceedings of the International Conference on Auditory Display, 280–282.

Chapter 5

Sonic Interaction Design

Stefania Serafin, Karmen Franinović, Thomas Hermann, Guillaume Lemaitre, Michal Rinott, Davide Rocchesso

5.1 Introduction

Sonic Interaction Design (SID) is an interdisciplinary field which has recently emerged as a combined effort of researchers and practitioners working at the intersection of sound and music computing, interaction design, human-computer interaction, novel interfaces for musical expression, product design, music psychology and cognition, music composition, performance and interactive arts.

SID explores ways in which sound can be used to convey information, meaning, aesthetic and emotional qualities in interactive contexts. One of the ultimate goals of SID is the ability to provide design and evaluation guidelines for interactive products with a salient sonic behavior. SID addresses the challenges of creating interactive, adaptive sonic interactions, which continuously respond to the gestures of one or more users. At the same time, SID investigates how the designed gestures and sonic feedback is able to convey emotions and engage expressive and creative experiences.

SID also aims at identifying new roles that sound may play in the interaction between users and artifacts, services, or environments. By exploring topics such as multisensory experience with sounding artifacts, perceptual illusions, sound as a means for communication in an action-perception loop and sensorimotor learning through sound, SID researchers are opening up new domains of research and practice for sound designers and engineers, interaction and interface designers, media artists and product designers, among others[1].

SID emerges from different established disciplines where sound has played an important role. Within the field of human-computer studies, the subtopics of auditory display and sonification have been of interest for a couple of decades, as extensively described in this

[1] When talking about designers, we use the definition proposed by [66]

handbook.

In sound and music computing, researchers have moved away from the mere engineering reproduction of existing musical instruments and everyday sounds in a passive context, towards investigating principles and methods to aid in the design and evaluation of sonic interactive systems. This is considered to be one of the most promising areas for research and experimentation [61]. Moreover, the design and implementation of novel interfaces to control such sounds, together with the ability to augment existing musical instruments and everyday objects with sensors and auditory feedback, is currently an active area of exploration in the New Interfaces for Musical Expression (NIME) community [13].

Among scholars in perception and cognition, there has been a shift in attention, from the human as a receiver of auditory stimuli, to the perception-action loops that are mediated by acoustic signals [43]. Such loops have become an important topic of research also in the sonification domain, where the topic of interactive sonification has emerged. This topic is described in Section 5.5, as well as in chapter 11 of this handbook.

Several efforts in these research areas were unified under the Sonic Interaction Design umbrella thanks to a European COST (CoOperation in Science and Technology) action which started in 2006 [1][2]. The different areas of exploration of SID, which are reflected in this action, are described in the following.

5.2 A psychological perspective on sonic interaction

Before addressing sonic interaction design from the perspective of product design, interactive arts and sonification in the next sections, the next paragraphs will consider some basic psychological phenomena involved in sonic interactions. To do so, they will examine a specific type of sonic interaction: closed-loop interactions. During such interactions, the users manipulate an interface that produces sound, and the sonic feedback affects in turn the users' manipulation (see Chapter 11). Such interactions have been used in applied [57, 19] and experimental settings [41, 50][3]. In fact, the design of these interactions brings under a magnifying glass a phenomenon that has recently received a great deal of attention on the part of psychologists interested in perception: the tight coupling between auditory perception and action [3].

Let us first consider a recent example of such an interaction: the real-time sonification of a rowing boat aiming to improve the athletes' performance [57]. In this design, the athletes' movements modulated the auditory feedback in real time. In turn, the sound helped the athletes to adapt their movements. Sounds had a great advantage in this case, because auditory perception and action are naturally and tightly coupled. Therefore, the intention was that the rowers would not be expected to consciously "decode" the information conveyed by the sounds, nor to think about how modifying their action would modify the sound. The sound-action loop was supposed to be intuitive. After all, this is what happens in "natural" interactions through sound. A user filling a vessel with water does not need to understand the relationship between pitch and volume to fill a recipient without overflowing [9]. Nor does

[2]http://sid.soundobject.org

[3]The ISon conferences provide a useful repository of such approaches http://www.interactive-sonification.org

a beginner violinist need to be aware of the physics of the bow-string interaction to avoid squeaky sounds (at least after a bit of practice).

In a designed sonic interaction, the richness of the added auditory feedback has the potential to let the users explore the complex patterns, and discover how their actions can modulate the sound. In turn, the auditory feedback guides the actions. As such, sonic interactions have a great potential to help a user become more proficient at the fine movements required in sports, as illustrated by the rowing example, but also in music, dance, surgery, and the complicated manipulation of tools [7]. As discussed later in this chapter, there are also other aspects of sounds to consider. The next section shows how recent research in psychology sheds light on the phenomenon of action-sound coupling.

5.2.1 The auditory perception-action loop

This section covers the importance of action, perception and multimodal feedback when designing interactive sounds.

The brain specifically processes the sounds of actions

Recent neuropsychological research has consistently suggested that the brain processes the sounds of actions made by an agent differently from other sounds. This line of research was initiated by the identification of audio-visual mirror neurons in monkeys' brains [36]. These are neurons that react both when the monkey subject does, sees, or hears the action.

Some recent experiments on human subjects led scientists to hypothesize the existence of two different brain mechanisms processing sounds caused by a human action (e.g., the sound of someone walking) and non-action sounds (e.g., thunder) [48]. They suggested that, on one hand, action-related sounds activate the mirror system, together with a specific motor action program. This system represents "how the sound was made". On the other hand, non-action sounds rely solely on the acoustic and perceptual properties of the sound itself, without the possibility of activating any action-related representation. This is for instance illustrated by the results of Lahav and co-workers [37] who showed that non-musician subjects had their brain premotor areas activated while they were listening to a piano piece they just had learned to play. When they listened to pieces that they had not learned, the motor area was not activated: for these latter sounds, they had no motor representation available.

Listening to sounds might not only activate a representation of how the sound was made: it might also prepare the listener to react to the sound [14]. Cognitive representations of sounds might be associated with action-planning schemas, and sounds can also unconsciously cue a further reaction on the part of the listener. This is exactly the principle of a closed-loop sonic interaction. Since the mirror system is also activated when the subject is seeing the action, some scientists introduced the idea of an abstract representation of the meaning of the actions, parallel to the activation of the motor plans [23]. And it might be that this abstract representation integrates multimodal inputs, and particularly audition and vision [4].

Multimodality and naturalness

During any interaction, users receive visual, haptic, and proprioceptive information in addition to sound. Even in the case of "passive" auditory displays, sounds influence the identification and interpretation of visual images [10]. With regard to the perceived quality of products, there are many cases (e.g., potato chips, electric toothbrushes) where the sound of a product affects the perception of its quality [63]. In the example of the iPod clickwheel described in section 5.3.1, a sound feedback may create pseudo-haptic sensations. Such a phenomenon has also been used to create pseudo-haptic interfaces [20].

Sonically augmented interfaces offer the psychologists the possibility of exploring the relationships between different modalities (audition, vision and touch). Important issues are those of the temporal synchrony between stimulations of different sensory modalities, and the related perception of causality[4] [30]. For example, whether two moving discs with crossing trajectories are perceived as bouncing or overlapping is heavily affected by the presence, timing and nature of a sound occurring at the contact instant [26].

Synchrony between sounds and gestures is important for sonic interactions because it influences the perception of causality. And the perception of causality is important for sonic interaction, because designers often choose to use a causal or iconic representation, rather than an arbitrary one, based on the hypothesis that sonic interactions should not require excessive cognitive effort on the part of users. In other words, by using the sounds that users could commonly expect as a result of their gestures, the designer assumes that users will intuitively understand how their gestures influence the sonic feedback. Such commonly expected sounds which result from gestures (e.g., the sound of an impact arising from the striking of an object) are here referred to as "natural". The natural relationships between a sound and a gesture are those driven by the laws of physics.[5]

The use of causal sonic feedback was explored in two recent studies. In the first study, an arbitrary (e.g., a bicycle bell) or causal (the sound of keystroke) feedback sound was added to a numerical keypad of an ATM cash machine [64]. Subjects judged the causal sounds as natural, and the arbitrary sounds as being less natural, and found that using the keypad with arbitrary sounds was more unpleasant and less efficient than with the causal sounds

(for an example of different kinds of sonic feedback, see video **S5.1**). In another study [41], the researchers designed a tangible interface (the Spinotron, see Figure 5.1) based on the metaphor of a child's spinning top. When the users pumped the Spinotron, they drove a physical model of a ratcheted wheel that produced a characteristic clickety-clack sound. The participants were required to pump the interface and to reach and maintain a precise and constant pace. By using sonic feedback which modeled the dynamic behavior of a spinning top the users' performance was improved significantly compared to more arbitrary feedback.

The design of sonic interactions based on the physical modeling of natural interaction seems to have two advantages. Firstly, the listeners find the interaction more pleasant, natural and engaging. Secondly, it seems that the interfaces are easier to use because the subjects already know, from their previous experience with everyday objects, how sound and gesture

[4]As discussed later, the sense of agency - the perception that one is causing the sound - is a particular and very important case of causality.

[5]Note that using a natural or causal relationship may have its own drawbacks - e.g., users having an overly deterministic vision of the feedback model based on prior expectations from the "natural" situation.

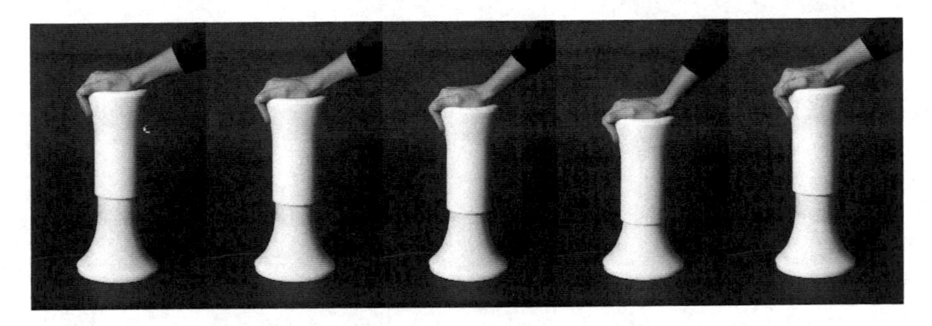

Figure 5.1: When a user pumps the Spinotron, a physical model of a ratcheted wheel produces a characteristic clickety-clack sound.

are related. It is unclear if interactions based on the modeling of natural interaction work well because they use a sound-action relationship pre-learned by the users, or because they provide rich, complex and redundant information that users just have to pick up. Maybe interactive interfaces based on natural interaction are easier to learn and master. However, natural sounds are in most of the cases preferred by users over artificial ones.

The evaluation of performance in sonic interactions

The evaluation of sonic interactions has a lot in common with what is done in product design. Laboratory studies enable the designer to evaluate the effectiveness of the interaction. As illustrated by the example of the Spinotron, the evaluation of the success of a sonically augmented interactive interface requires the designer to measure how the sound influences the user's interaction with the interface. This paradigm is therefore different from that of the sonification of passive auditory displays, where the evaluation consists in assessing whether the user is capable of consciously decoding the information conveyed by the sounds. In the case of closed-loop sonic interactions, what is important is not that users are consciously aware of the information, but that they can successfully adapt their movements and gestures.

The study of human-computer interaction offers an interesting point of comparison. Many of the methods that have been developed in this discipline measure reaction times, movement times or other chronometrical measurements. But what is probably more important is how well and fast users can learn to manipulate an interface, or successfully modify their actions. The quality of the design becomes indexed by the users' performance, and by the speed of their learning.

In the case of the Spinotron, the participants were required to pump an interface and to reach and maintain a precise and constant pace (indicated by a visual target). Half of the participants were provided with a continuous auditory feedback (the sounds of a virtual spinning top set into motion by their pumping gesture), half with a discrete visual feedback only. Only the participants who were provided with the auditory feedback were able to improve their performance across trials. The speed of learning was the actual measure used to quantify the success of the auditory feedback. However, when asked to describe their appraisal of the sonic feedback, the subjects reported two interesting comments. First, they were not aware that the sound actually helped them improving their performance. Second, they found the sound very irritating.

Therefore, evaluating the functional aspect of a sonic interaction is only one side of the coin. Designers should not forget that sounds create strong aesthetical and emotional reactions in users.

5.2.2 Affective and emotional reactions to sonic interactions

In fact, the sounds of interactive interfaces have the power to influence the users' emotions, as it is the case with any artificially added sound. The "pleasantness", "aesthetic", and "annoyance" of the sonic interaction are an important part of their appraisal by the users, and require investigation.

What are emotions?

The study of emotions is the subject of intense debate. Most modern emotion theorists agree that an emotion episode is a dynamic process consisting of coordinated changes in several cognitive, neurophysiological, and motor components [55, 59]. Among these components, feelings have a particular status: they serve as a monitoring function, and are consciously accessible. Feelings thus represent the component of an emotion episode that a subject can report. And, importantly, it is the component that the researcher can observe. Physiological measures (heart rate, skin conductance, facial EMG, startle reflex, etc.) can indicate neurophysiological activities, action tendencies and motor expressions. Self-reports can provide insights into the feelings of the subjects. The results of many studies have very often suggested that the feelings observed in, or reported by subjects can be accounted by a few principal dimensions. Furthermore, these dimensions can be related to different types of appraisals [58]:

- *Valence* results from the appraisal of intrinsic pleasantness (a feature of the stimulus) and goal conduciveness (the positive evaluation of a stimulus that helps reaching goals or satisfying needs).
- *Arousal* results from the appraisal of the stimulus' novelty and unexpectedness (when action is needed unexpectedly).
- *Dominance* results from the appraisal of the subject's coping potential.

Therefore, concerning the sounds of interactive interfaces, the appraisal of the features of a sound may have an influence on the valence (appraisal of pleasantness) and arousal (appraisal of novelty) dimensions of the feelings. Possibly, if the sound has a function in the interaction, it may also have an influence on the appraisal of the goal conduciveness (imagine an alarm clock that does not sound loud enough to wake you up).

Emotions and auditory feedback

Sound quality studies[6] provide, indirectly, some insights into the relationships between acoustic features of product sounds and emotions. For example, it has been reported that attractive products are perceived as easier to use [46]. Emotional reactions to the sounds of everyday products have been studied in terms of pleasantness or annoyance [29] or preference [65].

Sounds are also used in many forms of human-computer interfaces. And, because computer interfaces (and more particularly computer games) have the potential to induce emotions through different types of appraisal, they can also be used as an experimental technique to elicit emotions in subjects in a laboratory setting, and to enable the study of emotion processes [51].

In a recent study, the emotions felt by users manipulating a computationally and acoustically augmented artifact were assessed [40] (see interaction video **S5.2**). The artifact consisted of an interface similar to a glass (the Flops, see Figure 5.2), that the users tilted to pour a number of virtual items, that they could only hear. The task was to pour exactly a predetermined

Figure 5.2: When a user tilts the Flops, a number of virtual items, that can only be heard, are poured out.

number of items. Both the sound design (making more or less pleasant sounds), and the dynamics of the interaction (making the manipulation more or less difficult) were manipulated, and users had to report their feelings. The difficulty of the task, obstructing or

[6]We refer here to academic studies that explore the quality of everyday sounds: e.g., air-conditioning noises, transportation noises, vacuum cleaners, car horns, etc. - see [42] for an overview.

facilitating the users' goal conduciveness, modulated the valence and dominance dimensions of their feelings. However, the acoustic qualities of the sounds also influenced the feelings reported by participants. The quality of the sounds (indexed by their sharpness and their naturalness) systematically influenced the valence of the users' feelings, independently from the difficulty of the task. These results demonstrate that sonic interactions have the potential to influence the users' emotions: the quality of the sounds has a clear influence on the pleasantness of the interaction, and the difficulty of the manipulation (which, in some cases, results directly from the quality of the sound design) influences whether the user feels in control or not.

5.2.3 Summary of the psychological perspective

Closed-loop sonic interactions are different from *passive* auditory displays in that they involve users in actively manipulating an interface (or performing some action). The action modulates the sound, and the sound informs the users on how to modify their actions.

From the design perspective, the main question is how to create a multimodal interface that engages users in active manipulation, that provides them with auditory feedback complex enough to discover new patterns, and intuitive enough to successfully modulate their actions and gestures.

However, as with other forms of auditory interfaces, sonic interaction also affects the users' emotions. This is true partly because sounds can be more or less pleasant, but also, in the case of sonic interaction, it is the sound that can make the interaction successful or not.

The next section describes how sonic interactions have already been designed and implemented in real products, and discusses the issues that these examples highlight.

5.3 Product sound design

When we interact with physical objects in the world, these interactions often create sound. The nature of this sound is a combined product of our actions and of the physical attributes of the objects with which we interact – their form, materials and dynamics, as well as the surrounding environment. People possess a natural capacity for deriving information from sound: we can infer, from the sound arriving at our ears, rich information about its source [24].

Today more and more sounds for products are being designed. This includes both sounds that are produced through physical phenomena, and sounds that are digitally created. As an example of both types, the physical manipulation of materials and fine-tuning of internal components have been used to create the distinct sound of the Harley Davidson engine, a sound that the company tried to protect as a trademark[7]. With the recent advent of electric cars that create very little noise [53], digitally produced sounds have been introduced into cars both for pedestrian safety and for driver experience [38]. The long-awaited Fisker Karma, the first hybrid sports car, is said to have external speakers that generate "a sound somewhere between a Formula One car and a starship", but can be configured by the owner[8].

[7] http://articles.latimes.com/2000/jun/21/business/fi-43145

[8] http://www.popsci.com/cars/article/2010-04/price-karma

Obviously, these corporations realize the impact of sound on the perception of the product quality.

The field of sound design for products – specifically the design of non-speech, non-musical sounds – is quite young. A main source of knowledge on which it builds is the domain of film, where sound has been used extensively and in complex ways to affect the viewer's experience. Michel Chion, a researcher of film sound, has referred to two types of added value of sound in film: informative and expressive [11]. These are useful in thinking about sound for products as well: sound can add information in the use of a product, and can enhance its perceived quality and character. The development of the field of sound design is such that sound designers today use their skills to create auditory logos and signals (such as the attention-getting tone – or attenson [32] – that precedes an announcement in a train station), sound effects for website navigation and for computer games, and more.

Interactive physical products bring a new level of potential and challenge into this field. The lack of an inherent relation between form and functionality, as found in many consumer-electronics products, makes feedback a prominent factor. The complexity of functions makes the dialog between user and system more critical. Fortunately, these products are embedded with technological components and can be equipped with micro-controllers and sound producing elements Thus there is great potential for rich responsive sound in interactive products.

When we think of the sounds of products, we may still think about the beeps and bleeps of our household appliances, or the "ding" of the PC error. However, things are changing. Our input methods for digital products are no longer limited to pressing or pointing, and continuous interactions such as finger gestures and body movements are those for which sonic feedback may be the most beneficial [52]. Knowledge from the realm of interaction design, sound design and software development is needed to tackle continuous interactive sound projects.

The next section reviews a few examples of existing products and prototypes with informative and expressive sound, with an emphasis on the continuous nature of the interaction.

5.3.1 Key issues in designing interactive sonic products

Not surprisingly, some of the best examples of continuous sound for interaction come from the world of mobile devices. The reasons are twofold: the price and positioning of these products make the embedding of high quality audio components most feasible, and also the fact that these devices are used "on the move" motivates the provision of information in a non-visual way.

The iPod Clickwheel

The first iPod "Classic" model (see Figure 5.3) used a mechanical scroll wheel as an input device: a wheel that turned to allow scrolling between menu items. Consequent iPod versions replaced the mechanical wheel with the click wheel: a round, touch sensitive surface on which users slide their finger clockwise and counterclockwise, as if on a moving wheel.

One element that was introduced to the click wheel is the clicker: a clicking sound that

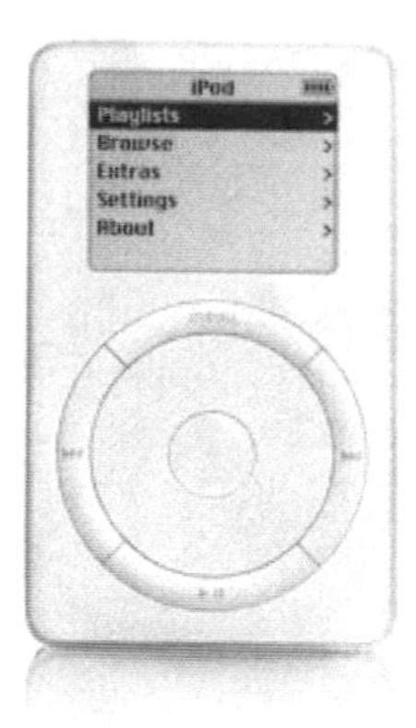

Figure 5.3: The first iPod “classic" with its mechanical scroll wheel.

Figure 5.4: The Apple Mighty Mouse, the Apple Magic Mouse, and the Microsoft Arc Touch Mouse, all viewed from top.

provides feedback for the movement between menu items. This feature gives a tactile feel to the click wheel (a pseudo-haptic illusion), somewhat similarly to the rotary dial on old phones, making the scrolling more expressive and more informative. Since the scrolling reacts to acceleration – the more you scroll the faster menu items move per rotation – the clicker provides information that is not evident from the scrolling action per se. The click sound is the only sound made by the iPod outside of the headphones, and is generated via a small, piezoelectric speaker inside the device.

Sonic, silent, and purring mice

The Apple Mighty Mouse (see Figure 5.4), introduced in 2005, contained an embedded speaker that gave sonic feedback to scrolling gestures. Apple seemed to abandon this line completely in 2009, when the Magic Mouse was introduced. This symmetric, uniformly smooth, and perfectly silent object supported multi-touch gestures and contained no apparent usability clues. Interestingly, despite the success of the Magic Mouse, Microsoft decided to go the other way and in 2010 unveiled the Arc Touch Mouse, that includes both haptic and sonic feedback to scrolling gestures over a central capacitive scroll strip.

Nintendo Wii Controller feedback

The Wii remote is the primary controller for Nintendo Wii game console, introduced in 2006. A main feature of the Wii Remote is its motion sensing capability, which allows the user to interact with and manipulate items on screen via gesture recognition and pointing through the use of accelerometer and optical sensor technology. The Wii Remote has basic audio functionality, via its own independent speaker on the face of the unit. This audio is used in different games to enhance the experience of the gestures through tightly coupled sound. Sonic and vibro-tactile feedback can be experienced, for example, in the Wii Tennis (a swish sound when swinging the racket), or in The Legend of Zelda: Twilight Princess (the sound is altered as the bow is shot to give the impression of the arrow traveling away from the player).

The sonified moka

The moka coffee maker is an Italian household accessory, composed of a bottom water chamber, a middle filter and a top container. To make coffee, the water chamber needs to be filled with water and the filter with ground coffee; the three parts then need to be connected by means of a screw connection. In a prototype [52], the screwing action was sonified to inform the user of the right degree of tightness. Sound dynamically changes its timbral quality as the coupling becomes tighter, starting from the sound of glass harmonica for loose coupling, assuming a rubber quality for the right tightness, and resembling the sound of a squeaking hinge when the coupling becomes too tight. This example shows a possible future direction of designed sonic feedback in consumer products, a direction that goes against an otherwise increasing clutter of beeps and bleeps[9].

5.3.2 Key issues in designing interactive products with sound

In the following we examine the different elements which relate to the design of interactive products with a salient sonic behavior.

Sounds and behaviors

One of the main challenges in creating sound for products is finding the design language – the selection of sound type and sound character to fit the product and the interaction. Now that we are no longer limited by piezoelectric buzzers in our products, the wealth of possible sound is great; which sounds should we choose? From which category? Musical sounds, speech sounds and everyday sounds all hold benefits. If our microwave wants to tell us that

[9]In the same category of coffee makers, the Bialetti Moka Sound incorporates a musical auditory alert that, given its poor sound quality, gives a significant contribution to lowering the quality of domestic soundscapes.

the meat is defrosted, should it moo? Play a tune? Emit clicks? Call out to us in words? And how should simple objects sound, as compared to complex products such as robots?

Thinking and sketching

Creating sounds for continuous interaction, where the sonic behavior changes rapidly and dynamically, is a challenging task. To the designer, thinking and sketching in sound is not as readily accessible as pen and paper, whiteboards and Post-its.

A number of methods have been proposed to help designers think and sketch sound. Different ways to increase designers' sensitivity to the auditory domain include, for example, sound walks [67, 2]. Vocal sketching [18] is simply the practice of describing sounds using the voice while operating a prop; the idea being that with the right setting, designers can easily and intuitively communicate sonic ideas through non-verbal vocal sound. It has been shown that people spontaneously use vocal imitations in everyday conversations, and that imitating a sound allows a listener to recover what has been imitated [5, 39]. Methods from interaction design, mostly focused on the visual domain, have been adapted to the sonic domain. Sonic Overlay refers to video prototypes in which sound is designed and overlaid over the video footage at a later time, to create a "fake" sonic interaction for the viewer. The "Wizard of Oz" technique[10] [27] has been useful for sound behaviors, and methods of developing narrative through sound, inspired by film sound, have been used to develop narrative interactive objects [35].

Creating functional prototypes, which enable the direct experience of interaction firsthand, is of great value in iterating and improving designs. Microcontroller kits such as Arduino[11] and Phidgets[12], which enable the easy connection of sensors to sound-producing software such as Max/MSP[13] and PureData[14], together create a way to embed (at least part of) the electronics inside objects and to prototype sound behaviors. Parameter-based sound models such as the Sound Design Toolkit [15] help to link between sensor input and dynamic output.

Challenges of evaluation

There is much work to be done in assessing the value that sound brings to interactive products. Evaluation can be performed through laboratory experimentation, or via analysis of products in the market. Both paths have their own challenges, since products have complex behaviors and usage patterns, and discerning the role of sound is not obvious. Some initial work shows promise, and can draw knowledge from existing research in interaction design [34, 60].

The laboratory experimentation with the Spinotron, for example, has shown that sonic feedback may aid users in learning to control the object [41]. In particular, as stated in section 5.2 the controllability of the interface and pleasantness of the sonic feedback are two important factors which need to be taken into consideration when evaluating interactive

[10] This techniques refers to a computer system which is apparently autonomous, but where infact a human is operating it.
[11] `http://www.arduino.cc/`
[12] `http://www.phidgets.com/`
[13] `http://cycling74.com`
[14] `http://puredata.info/`

products with a salient sonic behavior.

As an additional challenge, sound does not exist in isolation. Sound has the potential to intrude and annoy when wrongfully designed. Designers of sonic artifacts need to scrutinize closely the context in which their product will be used, considering both the direct user and the indirect, unintended users around. The existing soundscape also needs to be considered since it will determine whether the added sounds will be heard and how they will be perceived.

5.3.3 Summary of Product Sound Design

Digital technologies and scale economies have enabled new possibilities in using sound in interactive products. Interaction can be coupled with feedback in the auditory domain, potentially benefiting objects and use-situations in which the auditory channel is superior to the visual one, such as with users who are mobile. The degree to which this potential will be achieved depends on the value sound will have for the users. This is to some extent cyclical, since this value will depend on good sound quality and good interaction design, which, especially in small objects, is still a technological challenge and a costly endeavor. Good processes for working with sound, and research directed at showing the value of sonic interaction, will help designers to push forward sonic interactions. Most importantly, designers must create interactions that, through sound, enhance the beauty and utility of experiences.

An important source of inspiration and knowledge comes from the worlds of art and music, as described in the next section.

5.4 Interactive art and music

Visionary inspiration and aesthetic experimentation in art and music have always been valuable for design. Artistic projects working with interactive sound expand the notions of interactivity, performance and participation which have become an integral part of our everyday life. Artists question our own sonic agency in everyday life [6], involve non-expert users in sound creation [45], deal with mobile music making [25], explore collaboration through sound [21], experiment with interactive metaphors [31] and overall enable novel sonic expressions. These projects not only exemplify novel approaches to designing interactive sound, but also situate and probe possible social and phenomenological sonic experience within everyday contexts.

5.4.1 Listening and Doing with Sound

"Impression is only half of perception. The other half is expression", wrote the father of soundscape research Murray Schafer, reminding us that sonic acting is as important as listening [56]. In sonic interaction design, the involvement of art and music researchers focuses mainly on "exploiting the role of enactive engagement with sound-augmented interactive objects."[15]. The enactive approach challenges the dominant models of sound

[15]Memorandum of Understanding of the COST Action on Sonic Interaction Design, 2007: `http://w3.cost.esf.org/index.php?id=110&action_number=IC0601`

reception in which users' activity is limited to listening only. Rather, working with sound is an active multisensory experience which bridges the gap between perception and action. Sound making is considered to be a meaningful aesthetic experience not only for musicians but also for users who do not posses expert musical skills. This shift from reception-based to performance-based experience brings new challenges to sound design and sonification practices. Although "doing with sound" has been sparsely researched outside of the realm of professional music performance, examples of audience involvement in sound manipulation have been present since the 1960s, for example in certain experiments with audiotape.

In the Random Access Music installation by Nam Jun Paik (1963), visitors could generate sounds by moving the audio recorder head over the audiotapes arranged in abstract shapes on the wall. By changing the control of the head from an automatic mechanism to the human hand, a functional piece of technology was converted into an expressive instrument. The rearrangement of a technological device offered the visitors a rich sonic experience through their direct engagement with sound material. The unpredictability of visitors' gestures created sounds that the artist could not compose or predict. Abandoning the traditional listening role of the audience meant that the artist was giving up control by making his artifact accessible to all. Today, audience engagement is an integral part of many sound installations as well as social and participatory media projects.

5.4.2 Molding Sound: Ease or Virtuosity?

Sonic interaction has been challenged and shaped by the tension between the ease of interaction and virtuosity of musical expression.

Although highly expressive, many interfaces demand musical virtuosity and are not suitable for non-expert users (e.g., The Hands by M. Waisvisz, 1984). However, molding sound may be an experience as natural as pouring water [22] or bending a flexible tube [62]. Intuitive interaction can be facilitated through everyday objects such as the kitchenware used in the Crackle Family (Waisvisz, 1976) and the Gamelunch [49]. In the AudioShaker project, for example, [31] an ordinary cocktail shaker is used to mix sounds rather than liquids. Users can open the object, speak into it to record sounds, shake it to mix them and then literally pour out the sound mix. The sounds keep the reference to the recorded sound but are transformed according to intensity and repetition of shaking gestures. The project shows that the close coupling of body movement and sonic responses of an object plays an important role in increasing the malleability of sound. The design affordances of the AudioShaker invite familiar manipulation, letting the sonic material be molded under the force of users' physical gestures.

The use of everyday, rather than expert musical movements creates the potential for intuitive interaction without the need for instruction and learning. However, the balance between expression and effortless interaction remains to be explored beyond the triggering of habitual movements. Understanding the learning processes that underlie familiarization and exploration is a key issue in opening new possibilities for sound design [17].

5.4.3 Embodying Emotions

The emotional power of sound is often harnessed in artistic projects. When embodied in an object, interactive sound may be associated with the object's behavior and identity. For example, Blendie [16] is a blender that a user can control by vocally imitating its motor sounds. Such conversation based on the interplay between the artifact's machine sounds and the user's vocal expressions creates an emerging identity of the object which appears to respond emotionally. Blendie shows that objects can acquire an emotional character not simply by using the semantic qualities of sound, but rather by activating its relational potential.

The vibrotactile sensations caused by being in contact with a sounding object can also amplify its emotional power. While researchers are working with vibratory feedback to explore audio-haptic and sensorymotor interplay [47], artists are imagining worlds in which such responses could gain new meanings. For example, the ScreamBody (Dobson 1998-2004) is a wearable object which silences, stores and reproduces its user's screams. The user wears it on the chest and can replay his or her recorded screams by a strong and sudden squeeze of the object. This gesture and the vibrational feedback on the user's body help the user to re-enact the actual screaming movements, hopefully relieving the user of associated and unexpressed emotions. The ScreamBody excites the users' auditory, tactile and kinesthetic senses in multiple ways, allowing them to play, express and share emotional states, both in an intimate (when offering the scream to another person) and social (when performed in front of others) manner.

5.4.4 Contextualizing

A range of artistic projects are challenging and criticizing our sonic behaviors in everyday contexts, as well as probing our possible sonic futures. The SoMo5 phone by Ideo and Crispin Jones challenges the annoying uses of mobile phones in public spaces by allowing the user to virtually hurt a person who is talking too loudly on the phone. The user pulls a catapult-like device mounted on their phone, aiming and releasing it towards the offending person in order to activate an abrupt warning sound emitted from the other person's phone. The catapulting gesture's spatial directness and sonic consequences create the feeling that something physical has been thrown at the annoying person. The physical release of anger is thus expressed and enacted through a sonic gesture that exploits a new malleability of sound material.

Other artists explore collaborative composition and play as a means of encountering strangers in public space. For example, projects by the Zero-Th group aim to bring the transient sonic information floating in urban locations into the hands of passers-by [21]. In the Recycled Soundscapes project (see Figure 5.4.4), the sculptural interfaces enable citizens to intuitively capture, transform and compose soundscapes, thus bringing awareness to their own sonic actions and contributing to the ever-evolving urban compositions. Sound is once again treated as material which can be caught within public objects as well as liberated and transformed through physical action. Such experiments in phenomenology and sociality reveal existing social behaviors, question sonic privacy in public space, challenge the composition strategies and engage the playful relations among strangers in urban locations through sound.

Figure 5.5: The Recycled soundscape installation.

5.4.5 Sonic Awareness

Designing sound for action requires a shift of perspective from unconscious hearing or even ignoring one's sonic agency to becoming aware that one can shape one's sonic contributions in the world.

As Murray Schafer suggested, the awareness of our sonic contributions may be the key to re-shaping the quality of our everyday surroundings [56]. The problem is that during ergoaudition, the term that Michel Chion uses to describe the experience of hearing the self-produced sound, we are often less conscious of the sounds we make than of those that others produce [12].

In digitally-augmented artefacts, our agency is often "schizophonically"[16] displaced from the sound that is produced, not allowing us to be aware of the sonic effects we generate. In such context, our interpretation of the cause of the sound event is challenged, and, due to the blurred relationship between action and sound, this may decrease the responsibility for the sound we produce. However, in our cacophonic world, taking responsibility for self-produced sound is an ethical issue and the transparency between our actions and their sonic effects must be considered within sonic interaction design.

Learning from artistic and musical creations may help sonic interaction designers to raise awareness of human agency in everyday life. However, many questions and challenges remain. Artworks are often temporary experiments or imaginary narratives that cannot probe the evolution of interactive sonic systems on a long term scale. Although artists borrow from ethnography and psychology to bring insights to design and technology, the transfer of

[16] Schafer coined the term "schizophonia" to describe this phenomenon of separating sound from its source through technological means [56].

knowledge often remains hidden as tacit knowledge or may be reduced to dry facts using scientific methods. This challenge of abstracting and sharing knowledge has begun to be addressed by the community of sonic interaction design through the development of tools, methods and strategies accessible to designers and artists.

5.5 Sonification and Sonic Interaction Design

The previous sections in this chapter have provided an overview of the emerging field of sonic interaction design, which is situated at the intersection of interaction design and sound computing. This section addresses more specifically the relation between this field and sonification, discusses some examples and proposes a research agenda of relevant scientific questions.

Sonification, as defined in [33] and in chapter 1 and 2 in this volume, provides *information* in an auditory, typically non-speech, form. When looking at interaction with objects in everyday contexts we can pose questions about (a) what information the sound conveys, (b) how exactly sonic interaction depends on relevant variables and (c) when and how the sounds occur and structure the overall interaction. This analysis may give us inspiration as to how new technical devices, or normally silent artifacts or interfaces, can better profit from auditory display.

5.5.1 Examples of sonic information in everyday contexts

Let us consider two everyday examples where we probably underestimate the information value of sound: (a) walking along a corridor, and (b) filling a kettle with water.

When walking along a corridor, we generate a contact sound with each footstep. This sound not only provides us with the information that we have touched the floor as acknowledgement to proceed to the next step, but also gives detailed information about the material of the shoe sole and the floor, the impact energy and velocity, etc. [44]. In the sequence of these sounds we can attend to the walking speed, walking style, eventually even gender, emotion or gait problems to some extent. Beyond that we also obtain a sonic response from the reflections of these sounds from the walls and other objects, even allowing visually impaired pedestrians to stay in the middle of the corridor without other cues [54]. Normally we are not aware of this information since our sensory-motor system integrates them so seamlessly into our overall behavior programmes.

The second example shows that we may also profit more explicitly from interactive sounds to direct our actions. When filling a kettle with water, we typically attend to the accompanying water sounds which systematically change with fill level. The pitch rises during filling the kettle and thereby suggests a time until task completion [9]. Also, the sound depends on the water speed, kettle material, jet shape, etc., conveying even more detail beyond our primary interest. Often people explicitly make use of the resonance sound and only look to the fill level when the pitch starts to rise quickly.

These two examples make clear that there is much information in sound, and particularly in interaction sound, and we often exploit it effortlessly, and even without being aware of it. Only when a problem or a change occurs, for instance if electrical car indicators are installed where the usual "tick-tack" sound from the relays is missing, do we become conscious of the missing information.

How can we explicitly profit from sound and establish interaction sounds so that they support and enhance the interaction with task-relevant information? How can objects sound even without interaction so that we can keep peripheral awareness of relevant information without interference with verbal communication? Sonification provides the answer and the following sections shed light on the functions that are supported by information-carrying sound.

5.5.2 Functions of informative components in object sounds

The following functions of information-loaded everyday interaction sounds, and also of sonification-based additional interaction sounds, can be identified:

- Sound provides an *acknowledgement* of the completion of an action step, supporting us to structure more complex actions. The information is basically binary and conveyed by the mere occurrence of the sound. An everyday example is that of closing a door until you hear the "click" sound of the latch which indicates that it is now firmly closed. A sonification example is the "file deleted" sound when dragging and releasing a file icon onto the trashcan icon on a computer desktop (see **S5.3** for an example using parameterized auditory icons).
- Feedback sounds allow users to refine their actions. An everyday example has already been given above with "filling a kettle with water". A good sonification example is the sonification-enhanced drilling machine [28] which indicates by pulsing sounds how far the actual orientation of the drilling axis deviates from intended vertical and horizontal angles to the surface, in other words: a parking aid for the drilling machine (see interaction video **S5.4**).
- Sound can lead to characteristic sonic interaction *gestalts* which allow us to compare repeated instances of interactions. For instance, the sound of a gait becomes a pattern from which a person can be identified. For sonification of body movements, a complex movement such as a pirouette in dance or a racket serve in tennis may be turned into a sonic contour which can be compared to an ideal movement execution in timing and expression (see interaction video **S5.5**, which shows movement sonification in a sensor augmented German wheel).
- Sound can enhance awareness of certain information of interest: traffic sounds or environmental sounds (birds, cafeteria noises) are "passive sound" examples where we are not interacting. An interactive everyday example is the reverberant response following any sound (e.g., contact sound, footstep, verbal utterance) by which we become aware of the size, depth, wall/surface materials in a room or place. This latter principle inspired auditory augmentation, a sonification type where the real physical structure-born sound of real-world objects such as a keyboard or table is recorded and modified in real-time. This enables us to perceive - on top and tightly coupled to the original sound - the sonification which keeps us in touch with any information of interest. In [8] this is demonstrated with a modification of keystroke sounds by weather data (as shown in example video **S5.6**).

For SID, the inclusion of sound for the normally unhearable bears the potential to enable novel functions currently unavailable. For instance, a cooking oil bottle could sonically communicate how many millilitres have been poured out, making it easier for the chef to follow the recipe without using spoons or scales.

5.5.3 Interaction design consequences for sonification design

Sound in interaction is certainly a multi-faceted phenomenon which can be understood on various levels including the aesthetic, emotional, affective, coordination, information and even social and cultural level. In everyday interaction with objects, sound is mainly the result of the object properties and the interaction details, so sound design mostly operates on the level of the design of object properties. There are basic bindings between the interaction and sonic response which are fully determined by the laws of physics: the more energy is put into a system, the louder is typically the sound signal, the higher the tension, the higher the pitch, etc..

For sonifications, however, more freedom exists on how exactly to connect information with sound. Mapping data variables to sound parameters is a common approach for that. The designer here needs to take many decisions which influence the effectiveness of the system. If, for instance, the energy during interaction is a critical variable, it may seem sensible to map it to pitch, a sonic variable where we have a much higher sensitivity to perceive changes compared to sound level. However, such a mapping would be highly counterintuitive in the light of natural bindings, and this could increase learning time and even cause misunderstandings.

Therefore the designer needs to balance various factors and adjust designs to find an optimal working point. Learnability versus effectiveness is just one example. There may be sound categories with very salient sonic parameters which are perhaps very intuitive, yet the sound would be less pleasant for long-term use, or even irritating or provoking an unwanted emotional reaction.

A possible procedure would be (a) to sort all factors according to their importance for the given application context, (b) to optimize the sonification in light of the most important factor, (c) to refine the sound design within limits in light of the secondary factors, and (d) to iterate this until no further improvement can be made. Ideally this procedure needs to be followed with different seed designs, and user studies and questionnaires are the only way to compare their acceptance, utility and effectiveness.

Sonification within SID brings into the focus of attention that sound, and particularly sound in interaction contexts, can carry a large amount of information, which designers can shape and refine. This information-carrying aspect should not be underestimated only because we obviously do not pay so much conscious attention to it in everyday situations. For sonic interaction design, sonification can offer powerful tools and know-how about how to shape sounds according to measured or available information to generate additional benefits. The experiences in interactive sonification can furthermore inspire "classical" sound design where the information level has not yet been developed. What if car horn sound level and direction depended on the car's velocity? Or if the urgency level of the alarm clock depended on the time until the first appointment in the user's calendar? The sounds of technical products could possibly be enhanced in most cases if an information-based view would be taken to the sound.

5.5.4 Research topics in sonification for sonic interaction design

There are many open research questions on how best to integrate sonification in sonic interaction design, which are brought together in this section as a research agenda. Starting backwards from the perspective of the application, perhaps the most difficult question is how to evaluate the characteristics of complex sound in interaction. What questionnaires are to be used to gather information about the relevant factors? Are questionnaires at all a valid tool for evaluating sonic interactions? Can we investigate an interaction at all in experimental settings where an ecological acoustic context is missing? How can we make general statements about the utility of mappings from observations or studies with specific data-to-sound mappings, given the fact that users are so highly adaptive to accept and learn even inconvenient mappings? How to extrapolate the interaction data in light of the users' adaptivity to learn even inconvenient mappings?

From the other side there are questions such as: How can designers weigh the factors (perceptability, pleasantness, intuitiveness, long-term acceptability, etc.) for a specific application?

From the side of the sonification itself, the most important question is how to create metaphors that are convincing to the user, need little explanation, are in unison with the user's expectation and create sounds so rich in complexity that users are not bored or annoyed by them. A promising way is to adopt ideas from physical modelling, or directly to use Model-Based Sonification (see chapter 16) and trust that with learning the user will discover the relevant bindings between data variables and sonic characteristics.

5.5.5 Summary of Sonification in sound design

Sonification addresses the information level in sound, how information can be conveyed with sound. Thereby sonification provides a distinct perspective on the design process in sonic interaction design, which complements other perspectives such as aesthetic or emotional qualities of sound or branding/identification aspects. Sonification and its techniques are extensively introduced, described and characterized throughout the whole of this volume. A particular recommendation to the reader is to observe interaction in everyday contexts with a fresh and unconditioned mind, attending to how sound reflects and conveys a fantastic richness of information in real-time. Since our human sensory-motor systems are so well optimized to effortlessly make sense of this information, these observations can offer much inspiration on how to shape technology, and technical interaction sounds in particular, to be useful from a functional perspective. While starting from such a functional and information-oriented perspective will hopefully lead to interesting interaction design ideas, later these need to be refined to be in balance with the other relevant design criteria.

5.6 Open challenges in SID

This chapter has introduced the novel discipline of SID, outlining different applications. The importance of multimodality in SID has been underlined by presenting different examples of commercial products, artistic applications and research projects where the tight connection between sound and touch has been exploited. The different examples presented all have in common the presence of an action-perception loop mediated by sound, together with the need of creating aesthetically pleasurable sonic experiences, which might be of an exploratory and artistic nature, or possibly providing some new information.

The development of SID follows the trends of the so-called third wave of human-computer interaction, where culture, emotion and experience, rather than solely function and efficiency, are included in the interaction between humans and machines [46].

From a methodological point of view, this requires novel perspectives that move away from the rigid guidelines and techniques which have been traditionally adopted in the auditory research community. Strict engineering guidelines and formal listening tests are not valid as such in SID, but need to be replaced by design and evaluation principles which are more exploratory in nature. These include participatory workshops and active listening experiences, which support the importance of an ecological approach to SID, together with the need to investigate sound in an action-perception loop. This distinguishes SID from most previous efforts in auditory perception and cognition research, where the role of sound has merely been connected to the investigation of basic psychophysical phenomena. It also represents one of the biggest challenges in SID, i.e., how to evaluate the characteristics of a complex sound in interaction. Different possibilities have been proposed, ranging from using questionnaires, to measurement of user behavior to informal observations of users.

Together with the issue of evaluation, another open question is how to design the sound themselves, balancing between pleasantness versus annoyance, artistic expression or ability to understand the message conveyed by sounds as in the case of interactive sonification. The design challenges proposed by SID are no longer predominantly of a technical nature. The wide availability of sound design, synthesis and processing tools, together with physical computing resources, allows practitioners who are not technically trained to easily produce sonic interactive artifacts. Instead, the challenges are mostly focused on the ways in which designers may successfully create meaningful, engaging and aesthetically pleasing sonic interactions. To come closer to reaching the ambitious goal of becoming an established discipline, the field of SID will benefit from advances in knowledge in many related areas, including the perceptual, cognitive, and emotional study of sonic interactions, improved sound synthesis and design methods and tools, a better understanding of the role of sound while performing actions, and finally design and evaluation methods addressing the objective and subjective qualities of sounding objects, especially in active settings. For a new generation of sound designers to be capable of addressing the interdisciplinary problems the field raises, a more solid foundation of methodologies in those related disciplines needs to be developed.

Bibliography

[1] D. Rocchesso. *Explorations in Sonic Interaction Design*. Logos Verlag, Berlin, 2011.

[2] M. Adams, N. Bruce, W. Davies, R. Cain, P. Jennings, A. Carlyle, P. Cusack, K. Hume, and C. Plack. Soundwalking as a methodology for understanding soundscapes. In *Institute of Acoustics Spring Conference*, Reading, UK, 2008.

[3] S. M. Aglioti and M. Pazzaglia. Representing actions through their sound. *Experimental Brain Research*, 2010. Published online 04 July 2010. DOI 10.1007/s00221-010-2344-x.

[4] K. Alaerts, S. Swinnen, and N. Wenderoth. Interaction of sound and sight during action perception: Evidence for shared modality-dependent action representations. *Neuropsychologia*, 47(12):2593–2599, 2009.

[5] K. Aura, G. Lemaitre, and P. Susini. Verbal imitations of sound events enable recognition of the imitated

sounds. *Journal of the Acoustical Society of America*, 123:3414, 2008.

[6] M. Bain. Psychosonics, and the modulation of public space on subversive sonic techniques. In J. Seijdel and L. Melis, editors, *OPEN Sound: The Importance of the Auditory in Art and the Public Domain*. NAI, 2005.

[7] S. Barrass and G. Kramer. Using sonification. *Multimedia Systems*, 7(1):23–31, 1999.

[8] T. Bovermann, R. Tünnermann, and T. Hermann. Auditory augmentation. *International Journal on Ambient Computing and Intelligence (IJACI)*, 2(2):27–41, 2010.

[9] P. Cabe and J. Pittenger. Human sensitivity to acoustic information from vessel filling. *Journal of experimental psychology. Human perception and performance*, 26(1):313–324, 2000.

[10] Y. Chen and C. Spence. When hearing the bark helps to identify the dog: Semantically-congruent sounds modulate the identification of masked pictures. *Cognition*, 114(3):389–404, 2010.

[11] M. Chion. *Audio-vision: sound on screen*. Columbia University Press, 1994.

[12] M. Chion. *Le Son*. Editions Nathan, 1998.

[13] P. Cook. Principles for designing computer music controllers. In *Proceedings of the 2001 conference on New interfaces for musical expression*, pages 1–4. National University of Singapore, 2001.

[14] M. De Lucia, C. Camen, S. Clarke, and M. Murray. The role of actions in auditory object discrimination. *Neuroimage*, 48(2):475–485, 2009.

[15] S. Delle Monache, P. Polotti, and D. Rocchesso. A toolkit for explorations in sonic interaction design. In *Proceedings of the $5^t h$ Audio Mostly Conference: A Conference on Interaction with Sound*, pages 1:1–1:7, Piteå, Sweden, 2010. Association for Computer Machinery.

[16] K. Dobson. Blendie. In *Proceedings of the 5th conference on Designing interactive systems: processes, practices, methods, and techniques*, page 309. ACM, 2004.

[17] H. L. Dreyfus. Intelligence without representation – Merleau-Ponty's critique of mental representation: The relevance of phenomenology to scientific explanation. *Phenomenology and the Cognitive Sciences*, 1(4):367–383, 2002.

[18] I. Ekman and M. Rinott. Using vocal sketching for designing sonic interactions. In *Proceedings of the 8th ACM Conference on Designing Interactive Systems (DIS 2010)*, pages 123–131, Aarhus, Denmark, 2010. Association for Computing Machinery.

[19] M. Eriksson and R. Bresin. Improving running mechanisms by use of interactive sonification. In *Proceedings of the 3rd Interactive Sonification Workshop (ISon 2010)*, Stockholm, Sweden, April 2010.

[20] M. Fernström, E. Brazil, and L. Bannon. HCI design and interactive sonification for fingers and ears. *IEEE Multimedia*, 12(2):36–44, april-june 2005.

[21] K. Franinović and Y. Visell. New musical interfaces in context: sonic interaction design in the urban setting. In *Proceedings of the 7th International Conference on New Interfaces for Musical Expression (NIME 2007)*, pages 191–196. Association for Computing Machinery, 2007.

[22] K. Franinović and Y. Visell. Flops: Sonic and luminescent drinking glasses. In *Biennale Internationale du Design*. Cité du Design, Saint-Étienne, France, 2008.

[23] G. Galati, G. Committeri, G. Spitoni, T. Aprile, F. Di Russo, S. Pitzalis, and L. Pizzamiglio. A selective representation of the meaning of actions in the auditory mirror system. *Neuroimage*, 40(3):1274–1286, 2008.

[24] W. Gaver. What in the world do we hear?: An ecological approach to auditory event perception. *Ecological psychology*, 5(1):1–29, 1993.

[25] L. Gaye, L. E. Holmquist, F. Behrendt, and A. Tanaka. Mobile music technology: report on an emerging community. In *Proceedings of the 6th International Conference on New Interfaces for Musical Expression (NIME 2006)*, pages 22–25, Paris, France, 2006. IRCAM Centre Pompidou.

[26] M. Grassi and C. Casco. Audiovisual bounce-inducing effect: When sound congruence affects grouping in vision. *Attention, Perception, & Psychophysics*, 72(2):378, 2010.

[27] P. Green and L. Wei-Haas. The rapid development of user interfaces: Experience with the Wizard of Oz method. In *Human Factors and Ergonomics Society Annual Meeting Proceedings*, volume 29, pages 470–474. Human Factors and Ergonomics Society, 1985.

[28] T. Grosshauser and T. Hermann. Multimodal closed-loop human machine interaction. In R. Bresin, T. Hermann, and A. Hunt, editors, *Proceedings of the* 3rd *Interactive Sonification Workshop (ISon 2010)*, Stockholm, Sweden, 2010.

[29] R. Guski, U. Felscher-Suhr, and R. Schuemer. The concept of noise annoyance: how international experts see it. *Journal of Sound and Vibration*, 223(4):513–527, 1999.

[30] R. Guski and N. Troje. Audiovisual phenomenal causality. *Perception & Psychophysics*, 65(5):789–800, 2003.

[31] M. Hauenstein and T. Jenkin. Audio shaker. `http://www.tom-jenkins.net/projects/audioshaker.htm`.

[32] E. Hellier and J. Edworthy. The design and validation of attensons for a high workload environment. In N. A. Stanton and J. Edworthy, editors, *Human Factors in Auditory Warnings*. Ashgate Publishing Ltd., 1999.

[33] T. Hermann. Taxonomy and definitions for sonification and auditory display. In P. Susini and O. Warusfel, editors, *Proceedings* 14th *International Conference on Auditory Display (ICAD 2008)*, Paris, France, 2008. Institut de Recherche et de Coordination Acoustique Musique.

[34] D. Hong, T. Höllerer, M. Haller, H. Takemura, A. Cheok, G. Kim, M. Billinghurst, W. Woo, E. Hornecker, R.J.K. Jacob, C. Hummels, B. Ullmer, A. Schmidt, E. van den Hoven, and A. Mazalek Advances in Tangible Interaction and Ubiquitous Virtual Reality. *IEEE Pervasive Computing*, 7(2), pages 90–96, 2008.

[35] D. Hug. Investigating narrative and performative sound design strategies for interactive commodities. In S. Ystad, M. Aramaki, R. Kronland-Martinet, and K. Jensen, editors, *Auditory Display*, volume 5954 of *Lecture Notes in Computer Science*, pages 12–40. Springer Berlin / Heidelberg, 2010.

[36] C. Keysers, E. Kohler, M. Umiltà, L. Nanetti, L. Fogassi, and V. Gallese. Audiovisual mirror neurons and action recognition. *Experimental brain research*, 153(4):628–636, 2003.

[37] A. Lahav, E. Saltzman, and G. Schlaug. Action representation of sound: audiomotor recognition network while listening to newly acquired actions. *Journal of Neuroscience*, 27(2):308–314, 2007.

[38] B. Le Nindre and G. Guyader. Electrical vehicle sound quality: customer expectations and fears - crucial NVH stakes. In *Proceedings on the International Conference on Noise, Vibration and Harshness (NVH) of hybrid and electric vehicles*, Paris, France, 2008.

[39] G. Lemaitre, A. Dessein, P. Susini, and K. Aura. Vocal imitations and the identification of sound events. *Ecological Psychology*. in press.

[40] G. Lemaitre, O. Houix, K. Franinović, Y. Visell, and P. Susini. The Flops glass: a device to study the emotional reactions arising from sonic interactions. In F. Gouyon, Álvaro Barbosa, and X. Serra, editors, *Proceedings of the* 6th *Sound and Music Computing Conference (SMC 2009)*, Porto, Portugal, 2009.

[41] G. Lemaitre, O. Houix, Y. Visell, K. Franinović, N. Misdariis, and P. Susini. Toward the design and evaluation of continuous sound in tangible interfaces: The spinotron. *International Journal of Human-Computer Studies*, 67(11):976–993, 2009.

[42] G. Lemaitre, P. Susini, S. Winsberg, S. McAdams, and B. Letinturier. The sound quality of car horns: designing new representative sounds. *Acta Acustica united with Acustica*, 95(2):356–372, 2009.

[43] M. Leman. *Embodied music cognition and mediation technology*. The MIT Press, 2008.

[44] X. Li, R. Logan, and R. Pastore. Perception of acoustic source characteristics: Walking sounds. *The Journal of the Acoustical Society of America*, 90(6):3036–3049, 1991.

[45] D. Merrill and H. Raffle. The sound of touch. In *CHI '07 extended abstracts on Human Factors in Computing Systems*, pages 2807–2812, San Jose, CA, 2007. Association for Computing Machinery.

[46] D. Norman. *Emotional design: Why we love (or hate) everyday things*. Basic Books, 2004.

[47] S. O'Modhrain and G. Essl. An enactive approach to the design of new tangible musical instruments. *Organised Sound*, 11(3):285–296, 2006.

[48] L. Pizzamiglio, T. Aprile, G. Spitoni, S. Pitzalis, E. Bates, S. D'Amico, and F. Di Russo. Separate neural systems for processing action-or non-action-related sounds. *Neuroimage*, 24(3):852–861, 2005.

[49] P. Polotti, S. Delle Monache, S. Papetti, and D. Rocchesso. Gamelunch: forging a dining experience through

sound. In M. Czerwinski, A. M. Lund, and D. S. Tan, editors, *CHI '08 extended abstracts on Human Factors in Computing Systems*, pages 2281–2286, Florence, Italy, 2008. Association for Computing Machinery.

[50] M. Rath and D. Rocchesso. Continuous sonic feedback from a rolling ball. *IEEE MultiMedia*, 12(2):60–69, 2005.

[51] N. Ravaja, M. Turpeinen, T. Saari, S. Puttonen, and L. Keltikangas-Järvinen. The psychophysiology of James Bond: Phasic emotional responses to violent video game events. *Emotion*, 8(1):114–120, 2008.

[52] D. Rocchesso, P. Polotti, and S. Delle Monache. Designing continuous sonic interaction. *International Journal of Design*, 3(3):13–25, 2009.

[53] L. Rosenblum. Are hybrid cars too quiet? *Journal of the Acoustical Society of America*, 125:2744–2744, 2009.

[54] L. D. Rosenblum. *See what I'm saying: the extraordinary powers of our five senses.* W W. Norton & Company, Inc., New York, 2010.

[55] J. A. Russell and L. Feldman Barrett. Core affect, prototypical emotional episodes, and other things called emotion: dissecting the elephant. *Journal of personality and social psychology*, 76(5):805–819, 1999.

[56] R. M. Schafer. *The Soundscape: Our Sonic Environment and the Tuning of the World.* Destiny Books, 1994(1977).

[57] N. Schaffert, K. Mattes, and A. O. Effenberg. Listen to the boat motion: acoustic information for elite rowers. In R. Bresin, T. Hermann, and A. Hunt, editors, *Proceedings of the 3rd Interactive Sonification Workshop (ISon 2010)*, pages 31–37, Stockholm, Sweden, 2010.

[58] K. Scherer, E. Dan, and A. Flykt. What determines a feeling's position in affective space? A case for appraisal. *Cognition & Emotion*, 20(1):92–113, 2006.

[59] K. R. Scherer. What are emotions? And how can they be measured? *Social Science Information*, 44(4):695–729, 2005.

[60] P. Sengers and B. Gaver. Staying open to interpretation: engaging multiple meanings in design and evaluation. In *Proceedings of the 6th conference on Designing Interactive systems*, pages 99–108. ACM, 2006.

[61] X. Serra, M. Leman, and G. Widmer. A roadmap for sound and music computing. *The S2S Consortium*, 2007.

[62] E. Singer. Sonic banana: a novel bend-sensor-based midi controller. In F. Thibault, editor, *Proceedings of the 3rd International Conference on New Interfaces for Musical Expression (NIME 2003)*, pages 220–221, Montréal, Québec, Canada, 2003. McGill University, Faculty of Music.

[63] C. Spence and M. Zampini. Auditory contributions to multisensory product perception. *Acta Acustica united with Acustica*, 92(6):1009–1025, 2006.

[64] P. Susini, N. Misdariis, O. Houix, and G. Lemaitre. Does a "natural" sonic feedback affect perceived usability and emotion in the context of use of an ATM? In F. Gouyon, Álvaro Barbosa, and X. Serra, editors, *Proceedings of the 6th Sound and Music Computing Conference (SMC 2009)*, pages 207–212, Porto, Portugal, 2009.

[65] D. Västfjäll and M. Kleiner. Emotion in product sound design. In *Proceedings of the Journées du Design Sonore*, Paris, France, 2002.

[66] Y. Wand. A Proposal for a Formal Definition of the Design Concept. *In Design Requirements Engineering: A Ten-Year Perspective*, 14, 2009.

[67] H. Westerkamp. Soundwalking. *Sound Heritage*, 3(4):18–27, 1974.

Chapter 6

Evaluation of Auditory Display

Terri L. Bonebright and John H. Flowers

6.1 Chapter Overview

Evaluation of auditory displays is a crucial part of their design and implementation. Thus, the overriding purpose of this chapter is to provide novice researchers with basic information about research techniques appropriate for evaluating sound applications in addition to providing experienced perceptual researchers with examples of advanced techniques that they may wish to add to their toolkits. In this chapter, information is presented about general experimental procedures, data collection methods for evaluating perceptual qualities and relations among auditory stimuli, analysis techniques for quantitative data and distance data, and techniques for usability and active user testing. In perusing the information in this chapter, the reader is strongly urged to keep the following issues in mind.

First and foremost, all application development should have ongoing investigation of the perceptual aspects of the design from the beginning of the project (Salvendy, 1997; Sanders & McCormick, 1993; Schneiderman, 1998). It is an extreme waste of time and other resources to finish an auditory display and *then* have the target audience attempt to use it. Such mistakes in the design process used to be common in computer design, but the work of such individuals as Schneiderman (1998) in human-computer interaction work has made ongoing evaluation a regular part of computer software and hardware development for most companies. The same approach should be used for auditory display design as well.

Second, the reader should note that the choice of research method has to be intimately tied to the final goal of the project. If the project is designed to develop a full-scale sonification package for a specific group (for examples of such projects, see Childs, 2005; Valenzuela, Sansalone, Krumhansl, & Street, 1997), the researcher would need to use a variety of methods including *both* laboratory components and ecologically valid testing. For example, it might be shown in the laboratory experiments that a specific sound for rising indexes of financial data works better than another; however the researcher might find in the real-world application with stockbrokers that the winning sound from the lab may have spectral overlap with noises in the environment and therefore won't work for the target application. In this case, it would be ideal if sounds in the target environment were specified during an analysis and specification phase for the project prior to testing in the laboratory.

Third, it is important to note that even though this chapter extols the virtues of research techniques for developing good audio applications, experts should also use their own introspection and intuition, especially when beginning a project. Such expertise can be tremendously useful in narrowing down what might otherwise be a Herculean task for something as simple as determining which sounds might be most appropriate. But researchers should not rely on their expertise alone and must do actual testing to determine how well the application will work for the target audience. This can be illustrated most clearly when an expert in visualization techniques assumes that auditory displays can be developed using the same principles. Unfortunately, there are different perceptual properties that come to bear on building effective auditory display applications, such as limitations of sensory and short-term memory, that are less relevant to the design of visual displays.

As a final introductory point, the reader should be aware that it is not the intent of this chapter to *replace* any of the excellent references that are available on research design issues or data analysis techniques. Rather, the purpose is to provide an overview of the research process as it pertains specifically to auditory display design. Embedded within this overview are referrals to more detailed and in-depth work on each of the relevant topics. It is also hoped that this chapter will foster interdisciplinary collaboration among individuals who have expertise in each of the disciplines that contribute to auditory display design, such as cognitive and perceptual psychologists, psychoacousticians, musicians, computer scientists, and engineers, since this leads to the most rapid development of good applications.

6.2 General Experimental Procedures

In this section general information is presented about design issues pertinent to the investigation of perceptual characteristics of auditory stimuli. The first issue in designing an empirical study for sound applications is to have a clear idea of the goals for the specific auditory display of interest, which are then used to develop the questions to be considered in the study. It is important to emphasize that all experimental procedures must be developed within the context of the particular application and setting. Thus, each issue discussed in this section assumes that the context and the goal for the application are embedded within each decision step for setting up the study.

The second issue a researcher needs to consider is what types of data and statistical analyses are required to answer the questions of interest. One of the major problems experienced by novice researchers is that they fail to recognize that it is critical that data analysis techniques must be specified during the *design* stage, since they impact directly the type of data that should be collected, as well as other design considerations discussed in this chapter.

The following material on general experimental procedures moves from overarching concerns (e.g. experimenter and participant bias), to basic design topics (e.g. number and order of stimuli) and finishes with participant issues (e.g. participant selection). Unfortunately, the actual process is not linear in nature but resembles a recursive loop, since the researcher needs to adjust design parameters in relation to each other in order to develop a successful procedure. (See Keppel & Wickens, 2004 for a good general reference for research design for behavioral studies.)

6.2.1 Experimenter and Participant Bias

Experimenter effects occur when the investigators collecting the data either treat participants in experimental conditions differently or record data in a biased manner. Typically such bias happens when the experimenter has expectations about the probable or "desired" outcomes of the study and inadvertently impacts the participants in such a way that it modifies their responses. This is an especially crucial issue during usability and active use testing or when an investigator is conducting any type of interview procedure. It is noteworthy that investigators who are in a power hierarchy, such as graduate or undergraduate research assistants working with a professor, may be more prone to the effects of experimenter bias in general. Supervisors should talk openly about such problems with their data collection team as part of the training process. This should help minimize the effects of any previous knowledge about the expected results investigators carry with them into the experimental sessions, as well as to alleviate any perceived pressure to "please" the authority figure.

Experimenter bias interacts with the tendency for participants in experiments to want to be "good subjects", and as a consequence, they seek clues about what the "right" answer is, even if the investigator assures them that there is no such thing. Participants can be sensitive to these *demand characteristics* and provide feedback that reflects what they think the experimenter wants to have as the outcome. Obviously such bias on the part of both experimenters and participants is undesirable, and researchers can use a number of methods to reduce or eliminate these problems. For example, one common and quite effective practice for reducing demand characteristics is to have data collection performed by individuals who are "blind" to the hypotheses (and sometimes even the specific purposes) of the study. Another effective method is to automate the procedure as much as possible by using written or video recorded instructions and computerized testing.

6.2.2 Perceptual Limitations Relevant to Sound Perception

There are a number of cognitive and perceptual issues that are especially important for researchers interested in evaluating auditory displays. It is common for researchers new to the field to assume people's processing capabilities for sounds are very similar to their abilities for visual stimuli. Unfortunately, some fundamental differences between auditory and visual perception make this a dangerous and misleading assumption. Discussions of many of these critical differences between hearing and vision can be found in Bregman (1990), Handel (1989), Hass and Edworthy (2002), and McAdams and Bigand (1993) - sources which researchers and developers should be encouraged to read. Three aspects of auditory perception that place constraints on tasks and methods used to evaluate auditory displays are the transient nature of sounds, properties of memory for auditory events, and differences in the way attention is allocated in auditory as opposed to visual tasks.

Since sounds exist in time and are transient, unlike static visual displays that can be repeatedly inspected and "re-sampled over time" at the will of the observer, re-inspection of sound requires that it be replayed. Comparisons between sounds require that features of one sound be retained in memory while another is being heard, and/or that information about more than one sound be retained in memory at the same time . There are thus major limitations related to sensory memory, working memory, and long-term memory for sounds that are crucial to consider during the testing and design phases of a project. These limitations affect both the design of auditory display elements themselves, as well as how to go about effectively evaluating them. Specifically, these limitations constrain the optimum duration for a discrete auditory display presentation, the optimal duration between presentation of elements to be compared (the interstimulus interval [1]) , and the degree of control of the display that is given to a participant in an evaluation or a user study. For auditory display applications that present discrete "packages" of information by sound (e.g., earcons (see Chapter 14), auditory representations of discrete data samples, etc.) the designer usually has the ability to control display duration, and thus the determination of a duration that optimizes task performance should be one of the objectives of display evaluation. In designing and evaluating such applications participants or users will need to make comparisons between auditory displays (e.g., sorting tasks, similarity ratings). The effective duration of auditory sensory memory is an issue for making such comparisons; if displays or stimuli exceed 12 seconds or so, it is likely that memory for events at the beginning of the display will be degraded and the ability of participants to make reliable comparisons will be impaired. However, shortening the duration of a display of complex information runs the risk that perception of auditory patterns will be impaired because they are presented too rapidly. Thus there may be a three-way tradeoff between sensory memory, perception, and display complexity that designers need to consider and specifically investigate in designing such applications.

In most research designs, any task involving comparisons between auditory displays should be set up so that participants can repeat stimuli for as many times as they feel is necessary to make a good evaluation. The exception to this general rule is when the researcher desires to have an intuitive response, such as the almost reflexive response desired for an alarm; in such cases, the sounds should be limited to a single presentation. Additionally, if feasible, participants should be given control over the interstimulus interval, in order to ensure that there will be little interference between the perceptions of the stimuli. If it is necessary to have a fixed delay between display presentations, the interval should be long enough to allow perceptual separation between the displays, but not allow degradation of the sensory memory of the first display. A pilot study can be helpful to determine what seems to be a "comfortable" interstimulus interval for a given type of display - generally in the range of 0.5 to 4.0 seconds.

Evaluation of displays intended for on-line monitoring of continuous status information (e.g., industrial systems, patient vital signs in the operating room, etc.) present a somewhat different set of problems. The issue here is not the memory of the entire display but the detection of changes and patterns within the display which require action on the part of the observer. For these types of displays, most development research is concerned with determining optimal perceptual mappings between sound and data channels and how many streams of data to present (see Chapter 15). In such tasks, attention limitations are of particular importance, and these are generally assessed by measuring actual task performance by such measures as detection accuracy for "significant" events. However attentional capacity is also taxed

[1] An interstimulus interval is the amount of time between the offset of one stimulus and the onset of the following stimulus.

significantly in most auditory testing situations, even those involving comparisons of, or decisions about, "short" discrete auditory displays; therefore, the researcher should take extra care to make sure that participant fatigue does not impact the quality of the resulting data (see Chapter 4 for information on both perceptual and cognition issues).

Ideally, researchers testing auditory displays would benefit greatly from having basic data about perceptual abilities, including limitations, for auditory display elements in a fashion similar to the data that have been compiled in anthropometry research (Dreyfus, 1967; Roebuck, Kroemer, & Thomson, 1975). These data provide measures of hundreds of physical features of people that are used by industry to provide means and percentile groupings for manufacturing most of the products people use that are related to body size. General information about auditory perceptual abilities is available in a variety of journal papers and other manuscripts (i.e. Bregman (1990), Hass & Edworthy (2002), Handel (1989), McAdams & Bigand (1993), and Salvendy, 1997), but not in one complete comprehensive compilation with the necessary norms for the populations of interest. Such a guide of auditory perceptual parameters for auditory display researchers would allow the development of sound applications for specific groups in addition to the construction of sound applications that could provide a range of sounds that would work for the majority of individuals within a heterogeneous population.

6.2.3 Number and Order of Stimuli

While designing a study, researchers need to determine the appropriate number of stimuli and how these stimuli will be presented to the participants. Researchers and developers should carefully consider the issues of working memory and cognitive load when deciding how many stimulus attributes will be manipulated (e.g., pitch, intensity, etc.) and how many levels or values will be varied per attribute. In cases for which the investigator wishes to study basic perceptual abilities (as might be the case in exploratory stages of auditory display development), it may be preferable to err on the side of fewer rather than more stimuli in order to obtain useful data. On the other hand, in later stages of display development, in which the goal is to evaluate a display design in a real-world environment, it may be necessary to manipulate all applicable variables to determine how the display will perform.

Repeated stimuli may be added to the total number of stimuli to test subject reliability. Typically, a small number of randomly selected stimuli are repeated and randomly placed in the stimulus order, so that participants are unaware of the repeated trials. Data from these repeat trials are then used for cross correlation coefficients[2] to compute subject reliability. These correlation coefficients can then provide the researcher with information about which participants might be outliers since a low coefficient may indicate that the individual had a perceptual disability, did not take the task seriously, or did not understand the directions. Data from such participants are likely to provide an inaccurate picture of the perceptual response for the majority of the participants and lead to decisions about an auditory display that are misleading or incorrect.

Once the number of stimuli has been determined, the order of stimulus presentation should be considered. This is a particularly crucial issue for auditory stimuli since any stimulus

[2]More information on correlation can be found in section 6.4.1 and additional information about using such techniques for determining outliers in section 6.5.1

presented before another one has the possibility of changing the perception of the second stimulus. An example of this would be when a high amplitude, high frequency sound is presented directly before a sound that has low amplitude and low frequency. If the researcher is asking about basic information (such as perceived pitch or volume), the response for the second stimulus may be skewed from the exposure to the first sound. In studies where there are a very small number of stimuli (< 5), the best solution is to provide all possible orderings of stimuli. For most typical studies where the number of stimuli is too large to make all possible orders practical, the most effective method is to randomize the stimulus order, which distributes any order effects across participants and these are consequently averaged out in the composite data. Computer presentation allows for full randomization of stimuli across participants, but if stimuli must be presented in a fixed order (e.g., using pre-recorded audio media), then three or four randomly generated orders should be used.

6.2.4 Testing Conditions, Pilot Testing and Practice Trials

Decisions about the testing conditions under which data are collected should take into account the specific purpose of the study. For example, when conducting basic auditory perception research, it is essential to eliminate as many extraneous variables as possible (e.g., noise or visual stimuli) that could be distracting and to keep the environmental conditions constant across task conditions. On the other hand, for research projects designed to test the usability of a product for an industrial setting, the study should be conducted in the target environment. Regardless of the general testing conditions, instructions for the procedures should be carefully constructed and standardized in content and presentation for all participants.

The time it takes participants to complete an auditory display study is extremely important, since perceptual tasks tend to be demanding in terms of attention and vigilance, which can lead to participants becoming fatigued or losing motivation over the course of the session. As a general rule, most studies should have a limited task time of no more than 30 minutes, even though the complete session, including instructions, debriefing, practice trials, etc., might run for an hour or more. Even within a 30-minute session, pauses or breaks to help reduce fatigue can be included if deemed necessary from feedback during pilot sessions; however, if the task must be longer than 30 minutes, breaks should be built into the structure of the session. If a study consists of more than one hour of testing, it is advisable to consider breaking it up into multiple sessions, if possible. Researchers should keep in mind, however, that stretching a study across multiple sessions may produce greater risk that participants will change or adopt different strategies across sessions than they would within a single session. In some cases, the decision to have multiple sessions may be dictated by the participants in the targeted population. For example, if the researcher is working with students on a college campus or with individuals within a specific company, it may work quite well to ask them to commit to several sessions. Conversely, if the individuals must come to a location that is removed from their work or home, it may be easier to have them stay for an extended period of time rather than asking them to return for future testing.

Prior to formal data collection, pilot testing is strongly recommended to validate experimental procedures, to help ensure that the participants understand the instructions, and to test any equipment and software that will be used. This should include double-checking any data storage and back-up systems. A small number of participants (e.g., three to five) from the target population is usually sufficient for pilot testing; however, if problems are discovered in the procedures, additional pilot testing should be seriously considered. Novice researchers may feel that time spent piloting and debugging a procedure are not well spent; however, such testing may not only lead to higher quality data but may actually result in changes that make the data more readily interpretable.

A common practice that should be avoided is using colleagues or graduate student researchers for the pilot study. While such individuals should certainly be asked to provide feedback about research designs or questions on surveys, they should not be used in lieu of a sample from the participant pool. Normally, colleagues or collaborators will have additional experience and information that will allow them to read into questions information that may not actually appear, or they may know how the application is "supposed" to work. Thus, final feedback about the clarity of the procedure or survey questions can only be obtained from a sample of people from the target population, who in most instances will be inexperienced in terms of the sound application in question.

At the beginning of each experimental session, practice trials should be used to ensure that participants are familiar with the test procedures and that they have the opportunity to ask questions so that they understand the task. It is best if practice stimuli are similar, but not identical, to the actual stimuli used in the study. The optimal number of practice trials for a given study can be determined by considering previous research in the area, feedback from pilot testing, and the researcher's expertise. For some types of study, it may also be important for participants to first listen to the full set of stimuli if they will be asked to perform any type of comparative task (i.e., paired comparisons and sorting tasks). Exposure to the stimulus set assures that participants know the complete reference set of stimuli prior to judging the relations among members of the set. In some cases, such as those involving stimulus sets with relatively unfamiliar or complex information (e.g., auditory data displays), it may even be helpful to present sample auditory displays simultaneously with more familiar equivalent visual analogies (e.g. charts or graphs) to help familiarize the participants with the structure of the auditory displays they will be evaluating.

As a final general recommendation about experimental design, investigators should keep in mind that they are often seeking participants' *subjective* perceptions of the stimuli. In most cases, it follows that participants should be instructed to respond as they deem appropriate and that there are no absolutely right or wrong responses. Moreover, every attempt should be made to motivate participants to actively participate in the task, including appropriate remuneration. This may seem counterintuitive to the notion of the "detached" experimenter working within a laboratory setting, but it can have a large impact on the quality of the data procured from perceptual studies.

6.2.5 Ethical Treatment and Recruitment of Participants

Investigators who have limited experience with data collection from human participants should make sure that they are knowledgeable about issues relating to ethical treatment of subjects that are mandated by governmental and granting agencies within their countries, as well as human research policies specific to their research settings. In academic and research institutions in the United States there will usually be an institutional review board (IRB) that will have procedures clearly outlined for submitting applications to receive approval to conduct such studies[3]. Researchers at other types of institutions or settings that do not normally conduct research with human subjects should check with their institution and seriously consider collaborating with a colleague who has expertise in this area.

One of the most important considerations in designing an auditory display study is for the researcher to select or recruit participants that will be representative of the population that is targeted for use of the type of display being developed. Most of the time, researchers will be interested in the normal adult population with normal hearing. It is interesting to note, however, that very few studies actually include a hearing examination to verify whether the participants have hearing that falls within the normal range. With the increase in hearing deficits that have been documented due to environmental noise (Bauer, Korper, Neuberger, & Raber (1991) and the use of portable music devices (Biassoni et al. 2005; Meyer-Bisch, 1996), researchers should determine whether they need to include hearing testing or whether they may need to restrict the range and types of sounds they use for specific groups. Researchers may also be interested in designing auditory displays for specialized groups, such as children, the elderly, or people with visual impairments. In such cases, it is imperative that the participants reflect the relevant characteristics of the target population (for example, see Oren, Harding & Bonebright, 2008). It can be tempting for researchers to think that they can anticipate the needs of such groups, but this assumption should be quickly questioned. It is best if the research group includes at least one member of the desired target group as a consultant or full collaborator from the beginning of the project, if at all possible, in addition to actively recruiting individuals with the desired characteristics for the most valid test results.

It is also important to consider other general subject characteristics, such as gender, age, and type and level of relevant expertise that might impact the use of the auditory display. For example, there may be differences in the aesthetic value of certain sounds across age groups (see Chapter 7), or an expert user of a specific piece of equipment may be better able to accommodate the addition of sound. It is also important to keep in mind cultural differences that might impact the interpretation of a specific sound (Schueller, Bond, Fucci, Gunderson, & Vaz, 2004) or the perceived pleasantness of sounds (Breger, 1971). Researchers also need to consider that there are other individual differences within populations that may not be so readily apparent on the surface, but which may have dramatic impacts on participants' abilities to interact with an auditory display. For example, some individuals suffer from amusia, which is a disorder of pitch discrimination and melodic perceptual organization. Such individuals may appear normal in terms of performance on a standard hearing test that is based on simple detection of tones, yet be highly impaired in their ability to recognize melodies or detect changes and harmonic distortions in tone sequences that individuals with normal auditory ability can discriminate with ease (Marin & Perry, 1999). Recent studies (Hyde & Peretz, 2004; Peretz et al., 2002; Peretz & Hyde, 2003) suggest that approximately 4% of the population may have an inherited variety of amusia, while an additional (possibly larger proportion) may suffer from an acquired variety of amusia due to cortical injury related to stroke, trauma, or other pathological conditions (Sarkamo et al., 2009). Designers of auditory displays should thus recognize that just as color deficiency may prevent some individuals from effectively using certain advanced visualization designs, a similar circumstance may exist for the usability of auditory displays by a small proportion of

[3]The American Psychological Association (*www.apa.org*) or the National Institutes of Health (*www.nih.gov*) are good sources for information on ethical treatment of human subjects.

the population who have amusia or related deficits.

Another individual difference that researchers in this area have considered is musical ability, since it seems logical that musical expertise should have an impact on auditory perception, which would consequently affect how individuals interact with auditory displays. Studies that have included this variable have revealed inconsistent results for differences between musicians and non-musicians on basic sound perception tasks (i.e., Beauvois & Meddis, 1997; Neuhoff, Knight, & Wayand, 2002; van Zuijen, Sussman, Winkler, Naatanen, & Tervaniemi, 2005). One reason for these inconsistencies could be that the method for determining musical ability is not standardized. It is also possible that researchers interested in auditory display should be measuring some other construct, such as the basic sensitivity to sound qualities. This issue can only be settled through systematic research examining both basic musical ability and what types of auditory perception (such as auditory streaming, ability to follow simple tonal patterns, sensitivity to rhythm, etc.) are relevant for designing effective auditory displays. There have been some efforts to provide a better measure of auditory perception for elements specific to auditory displays (Edwards, Challis, Hankinson, & Pirie, 2000), but currently such tests have not been widely circulated or accepted within the field.

6.2.6 Sample Size and Power Analysis

Another important research design topic is the number of participants needed, called the sample size. Researchers need to consider the overall task context, which includes the number and type of stimuli, design type, and required statistical analyses to determine the appropriate sample size. These issues can be addressed by reviewing past research in the area to determine the number of participants used or pilot studies can be performed to help make this decision.

When researchers are interested in comparing stimuli, they must choose whether to use a *between groups* (different participants in each condition) or *within groups* (the same participants in all conditions) design. Within group designs have the advantage of needing fewer participants and of better statistical power since the variation due to differences in individual participants is statistically removed from the rest of the variance. Therefore, researchers typically make this decision by considering whether there would be any type of carry-over effect from one condition to the other. For example, in a study designed to investigate the effects of mappings for auditory graphs (such as signifying axis crossings with timbre changes versus momentary loudness changes), practice with one mapping would probably affect participants' ability to learn the other mapping scheme. In such circumstances, a within groups design will not work well, and a between groups design should be used.

Researchers should also consider performing a statistical procedure, called power analysis, which is designed to specify the number of participants needed to get a statistical result that allows for any real effect present to be detected. With insufficient power, results may not be significant simply due to the lack of sufficient sample size rather than that there is no effect to be found. There is also the possibility of having too many participants, which results in trivial effects revealed during statistical analysis, although the most likely error for researchers to make is to have a sample size that is too small rather than too large. The sample size needed for a given study depends on the type of statistical test that will be used, the number of stimuli to be tested, and the alpha level [4] that will be used to determine significance. There are a number of excellent resources available that present both conceptual background information and practical considerations for performing power analyses (Cohen, 1988; Kraemer & Thiemann, 1987; Lipsey, 1990, Lenth, 2001) and there are also a number of software packages, both commercial (i.e. SPSS and SAS) and freeware, that are available for use (Thomas, 1997). More complete information about design type and other related analysis and statistical topics is presented in sections 6.4 and 6.5.

6.3 Data Collection Methods for Evaluating Perceptual Qualities and Relationships among Auditory Stimuli

Five commonly used methods for exploring the perceptual qualities of auditory stimuli and their relationships to one another are identification tasks, attribute ratings, discrimination trials, dissimilarity ratings and sorting tasks (see Table 6.1 for a summary). As discussed in previous sections, the researcher should consider each technique in relation to the goals of the project as well as the desired analysis technique to determine which ones are appropriate. This discussion is presented to provide basic information about these techniques; further discussion about how these techniques fit within particular types of analyses will be presented in sections 6.4 and 6.5.

6.3.1 Identification Tasks

Identification tasks for auditory stimuli provide a measure of accuracy for determining whether participants can recognize and label sound stimuli. Normally such tasks provide data that show the percentage of participants who correctly identified the stimulus. Some researchers also collect reaction time data, which is assumed to be a measure of the amount of processing or cognitive effort it takes to complete the task. A short reaction time may indicate that the stimulus represents a well-known and/or quickly identifiable sound, or that a participant made "false starts". In contrast, a long reaction time may indicate that a sound is unfamiliar or that the participant has lost focus. If a researcher is testing the veracity of synthesized sounds, a long reaction time may indicate that the sound is not a convincing replication of the actual sound. Thus, it is suggested that researchers examine their data to determine whether the pattern of reaction times suggests that outliers are present or whether there is information in these data relevant to the stimulus quality.

Identification tasks for auditory displays include trials that require participants to listen to an auditory stimulus and respond either in a free-form or open-ended format with a written description or by selecting a response from a provided list. In some studies, it is best if the participants are allowed to play the sounds as many times as they desire with no time limit. In such cases, data can also be collected on the number of times the participant played each sound in addition to other measures. If the researcher wishes to obtain intuitive responses, participants are not allowed to change their responses and a time limit may also be imposed.

[4] The alpha level is the value set to determine if an obtained result is statistically significant or if it happened by chance. Typically alpha levels are set at .05 or lower, which minimizes the probability of rejecting the null hypothesis when it is true (called Type I error).

Task	Typical Measures	Typical Usage
Identification	▪ Accuracy (% correct) ▪ Reaction time for correct ID	▪ Design and selection of sounds that are perceptually distinct from each other or that are inherently "meaningful"
Attribute Ratings	▪ Rating scale (e.g. 1 to 7, where 1 = 'very unpleasant' to 7 = 'very pleasant')	▪ Discovery of relationships between perceptual and acoustic properties of sounds. ▪ "Labeling" dimensions that determine similarities and differences between sounds. ▪ Input data for factor analysis and other techniques for determining "structure" among a set of sounds.
Discrimination	▪ Accuracy (% correctly compared) ▪ Errors (number and type of incorrect responses)	▪ Design and selection of sounds that are perceptually distinct from each other.
Dissimilarity Ratings	▪ Numeric estimate of similarity between pairs of sounds. ▪ Dissimilarity or proximity matrix.	▪ To determine which sounds are highly similar (possibly confusable) or distinct. ▪ Input data for cluster analysis and MDS for determining perceptual "structure" of set of sounds.
Sorting	▪ Dissimilarity or proximity matrix	▪ To determine which sounds are highly similar (possibly confusable) or distinct. ▪ Input data for cluster analysis and MDS for determining perceptual "structure" of set of sounds.

Table 6.1: Summary table for data collection methods.

When the data collected for identification tasks are in an open-ended format, content analysis of the responses is required initially to determine if the meaning of the responses shows a pattern across participants. For example, when Miner (1998) asked subjects to identify a synthesized sound, the responses 'running river water', 'water running in a river', and 'river noise' were aggregated into a single term with a response count of three. The terms 'rain falling against a window' and 'rain' were not aggregated because the first term provided additional information that would be lost if it were combined with the simpler term 'rain'. It is important to note that even though this type of linguistic/semantic analysis can be conducted automatically with commercial and non-commercial packages, the researcher will still need to make fine distinctions manually in some cases as noted in the previous example. For both open-ended and fixed format responses, the resulting frequency data can be used to determine whether the participants correctly identified the sounds as well as determining which sounds were confused with each other. Such information can be especially useful for sound designers, since systematically confused sounds can be used as a basis to simplify and speed up the production of synthesized sounds for use in computer software and virtual reality environments (Cook, 2002 and see Chapter 9).

6.3.2 Attribute Ratings

Attribute ratings, also called semantic differential ratings, provide information about the perceptually salient qualities of auditory stimuli and are routinely used by investigators working with auditory displays. Researchers using attribute ratings are interested either in understanding the basic perceptual aspects of sound or in combining these data with other analysis techniques, such as factor analysis and multidimensional scaling (MDS), to provide a richer and more complete interpretation of the perceptual structure of the stimuli.

The researcher needs to determine what the appropriate attributes are depending on the type of stimuli used and the purposes of the sound application. Many times these will include a basic set of attributes, such as perceived loudness and pitch, although many other attributes can be used as well, such as roughness, annoyance, or pleasantness. The attributes also need to be clearly and consistently understood by the target population. Finally, the choice of the rating scale varies among researchers but typically semantic differential scales of 5, 7, or 9 points are preferred.

Analysis procedures for rating scale data consist of standard descriptive statistics as well as correlational analysis, analysis of variance, factor analysis, and as an additional measure for interpreting MDS solution spaces. A more complete discussion of these techniques will be presented in sections 6.4 and 6.5.

6.3.3 Discrimination Trials

For designing applications that use multiple auditory signals, it is important to determine if people can discriminate between the selected sounds and to measure the *extent* to which the sounds can be distinguished using a discrimination task. The procedure for a discrimination task requires participants to listen to two sequential stimuli (A and B), which are then followed by a third stimulus (X). Participants are then asked to determine if X is the same as A, B or neither of them (Ballas, 1993; Turnage, Bonebright, Buhman, & Flowers, 1996).

In the instructions, participants are informed that there will be a number of 'catch' trials on which the correct response would be neither. These trials are necessary to make sure that participants are attending to both stimuli A and B before making their judgments rather than adopting the simpler strategy of ignoring A, attending to B, and making a same-different judgment for the B-X pair (Garbin, 1988).

Basic analyses of data from this procedure consist of comparisons of correct responses and errors using descriptive statistics, such as means, standard deviations and ranges. Investigation of the types of errors for individual participants and for composite data from all participants can be examined for patterns that indicate perceptual similarity among the stimuli.

6.3.4 Dissimilarity Ratings

Dissimilarity rating [5], also referred to as proximity rating, paired comparison, or similarity judgment, is when participants provide a numerical assessment of dissimilarity for each possible pair of stimuli in the set. A typical dissimilarity rating task might instruct participants to "listen to each pair of sounds presented and to rate their degree of dissimilarity by using a 7-point rating scale where 1 = extremely similar and 7 = extremely dissimilar." It is also possible to have participants make a mark along a continuous line with labeled endpoints to indicate degree of similarity. A number of commercial data collection or survey software packages can be used for participants to enter such judgments. However, paper and pencil forms may also be used for participants to enter the ratings.

Regardless of how the rating data are recorded, considerable attention should be given to the manner in which the sound samples are presented. Comparing pairs of sounds presents some cognitive and perceptual issues that differ from those encountered with comparing *visual* displays. As discussed previously in this chapter, sounds must be listened to sequentially, which means that there is a memory component to the comparison task that would not be the case for simultaneously displayed visual stimuli. There may also be order effects, such that a similarity rating may be slightly different for the same pair, depending on which sound is played first. There are several choices for how one might present sound pairs to address these potential complications. One method of dealing with order effects is to present each pair twice - once in each order. The mean of the two ratings can be used as a participant's estimate of similarity between the pair of sound samples. If this procedure is followed, however, there will be a minimum of $N \cdot (N-1)$ ratings performed by each participant, where N is the number of sounds in the set. This procedure may produce a time consuming (and perhaps arduous) task if the number of stimuli is large. For example obtaining dissimilarity ratings for 50 different automobile horn samples would require presentation of 2450 pairs of horn toots to each participant. An alternative approach to dealing with potential order effects would involve randomizing the order of each pair and the order in which each pair is presented during the session for each participant, thereby cutting the number of pair presentations in half: $N \cdot (N-1)/2$ comparisons. A third alternative approach to addressing pair order effects involves allowing the participants to listen to each member of a pair in any order as many times as they wish before entering a dissimilarity rating. This can be achieved by presenting, on each trial, a pair of clickable icons that elicit each of the sound samples,

[5] For both dissimilarity ratings and sorting tasks, researchers should be mindful of differences among the stimuli, such as duration or amplitude that could perceptually overwhelm other attributes of the stimuli the researcher may wish to examine. In such cases, the stimuli could be equalized on the relevant dimensions.

along with an icon or button that presents a dissimilarity rating box that can be activated when a participant is ready to respond. Since participants may choose to listen to each sound more than once under this procedure, it may take slightly longer to complete than the fixed schedule with randomized order of pairs, and it does not give the researcher full control over the number of times participants are exposed to the stimuli. Investigators who plan to collect dissimilarity ratings among sound samples must thus weigh the costs and benefits of these alternative approaches for their particular research or development project (and perhaps also consider the option of assessing perceptual dissimilarity by using a sorting task, which will be described in the next section).

The data obtained from a dissimilarity rating task are typically configured into a "dissimilarity" matrix for each participant in which cell entries (assuming a 7-point rating scale) will vary between 1 and 7. Most computer programs for clustering or scaling such data require (or at least accept) a "lower left triangular" matrix (examples and more information about these techniques will be shown in section 6.5) for the input of such data, often called "proximity" data.

6.3.5 Sorting Tasks

An alternative method for obtaining perceptual distance or dissimilarity ratings among stimuli is a task in which participants sort a set of stimuli (typically 20-80 examples) into "piles" or "groups." Traditionally, such methods have been used for visual and tactile stimuli (Schiffman, Reynolds, & Young, 1981); however studies indicate their utility in investigating auditory stimuli as well (Bonebright, 1996, 1997; Flowers et al., 2001; Flowers & Grafel, 2002).

While sorting is not an activity normally associated with sounds, current technology makes it quite easy to collect sorting data on sound samples by presenting participants with a computer screen folder containing numbered or labeled icons that activate the presentation of a sound file. Participants are allowed to click on each icon as often as they wish to listen to it and to move the icons into different locations on the screen based upon their judgments of similarity until they are satisfied that they have formed meaningful groupings. The experimenter then records the group each stimulus was placed in (a process which could be automated by software that senses the screen position of the final icon locations). A *dissimilarity matrix* is generated for each participant by assigning the value "0" to each pair of stimuli that are sorted into the same pile, and the value of "1" to each stimulus pair that is sorted into a different pile. Logically, this is equivalent to obtaining dissimilarity ratings using a "two-point" rating scale for each participant, as opposed to the typical seven point scales used in dissimilarity rating tasks. As with actual dissimilarity ratings, one may sum these matrices across participants to obtain a group or "composite" dissimilarity matrix. Each cell entry of this composite matrix thus consists of an integer that is the count of how many participants assigned a particular stimulus pair to different piles. The composite matrix may be submitted for clustering or MDS procedures, and individual participant dissimilarity matrices may be reconfigured into linear vectors and submitted to reliability analysis programs. The authors of this chapter have developed simple software routines to perform this transformation. One version writes the lower triangular dissimilarity matrix (and sums these matrices across participants to provide a composite or group dissimilarity matrix), while the other version "stretches out" the individual participant dissimilarity ratings for each pair into a linear vector so that similarity of sorting patterns among subjects can be assessed by correlation and reliability analyses.

The following example (shown in Figure 6.1) illustrates the process of transforming sorting data into lower triangular dissimilarity matrices. Suppose there are three participants who each sort ten different sound samples (note that in an actual design scenario, there would likely be far more participants and stimuli). The three rows of ten digits on the left of Figure 6.1 represent the sorting data from each participant. The cell entries are the "pile number" in which that particular subject sorted each stimulus. For example, the first subject assigned stimuli #1 & #4 to pile 3, #2 and #3 to pile 1, #5, #6, and #10 to pile 4, and #7, #8 and #9 to pile 2. The "pile numbers" are arbitrary and need not correspond across participants, since the matrix and vector data only reflect whether each pair of stimuli were grouped together or not. Note that while the first two participants used four piles, the third participant only used three piles. For the first participant, pairs 1-4, 2-3, 5-6, 5-10, 6-10, 7-8, 7-9, and 8-9 should all receive a zero in the lower left triangular dissimilarity matrix (where the first number in the pair is the column and the second the row), and the remaining cell entries should be "ones". The right side of Figure 6.1 displays the three lower triangular dissimilarity matrices computed for these three participants, followed by the group (summed) matrix at the bottom. Note that in these matrices there is also the matrix diagonal (each entry a zero) since this is required by several popular data analysis packages, such as SPSS, when submitting dissimilarity data to perform multidimensional scaling or cluster analyses.

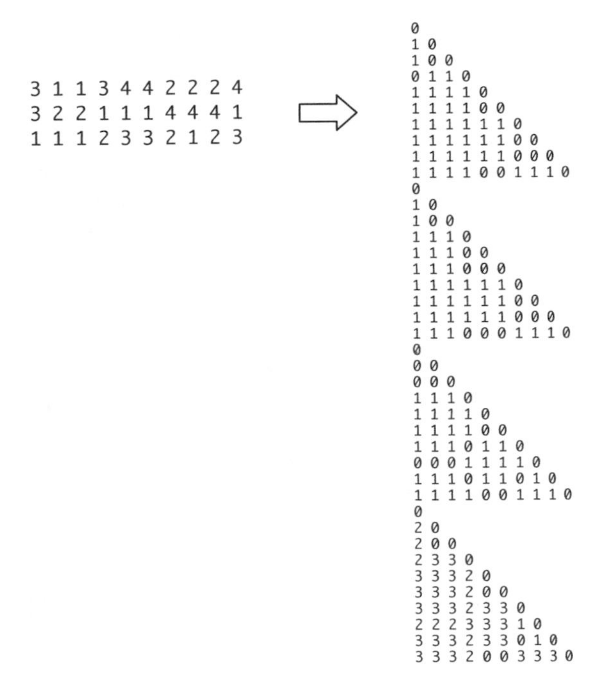

Figure 6.1: Transforming sorting data into individual and composite dissimilarity matrices.

Unless the participants hear all the stimuli first, it is probably best to allow at least two sorting trials for a set of stimuli, and to use the matrices generated by the final sorting for further analysis. The completion of the first sorting trial is necessary to familiarize each participant with the presence and range of variation among auditory attributes and features within the set of stimuli.

It should be noted, that when collecting sorting data, researchers sometimes place weak constraints on the number of groups (e.g., "no fewer than three, no more than eight"), and/or a restriction on the minimum number of stimuli (e.g., "2") that can be included in any group. In the opinion of the authors (and particularly for sonification design purposes), specific instructions of this type are of little value. If one stimulus is truly unique, participants should be allowed to indicate that. Additionally, failure to separate a relatively diverse of set of ten or more stimuli into at least three categories rarely happens.

One clear advantage of sorting tasks over dissimilarity rating tasks is the speed with which the data can be obtained. It is much quicker and much less tedious for participants to sort stimuli into piles that "go together" than to be presented with at least $N \cdot (N-1)/2$ numeric rating trials. Another advantage sorting tasks have in relation to dissimilarity rating tasks is that once the participants have finished sorting the stimuli into groups, they can be asked to label each of the categories. Such information can help the researcher understand what the participants *explicitly thought* they were doing, and it may also help in interpreting the results of the data analysis. However, it should be noted that participants may be using strategies of which they are *not* consciously aware, which the data analysis may be able to expose. When used in conjunction with techniques that correlate physical stimulus properties with positions in a multidimensional scaling plot or a cluster plot (examples of which will be discussed in section 6.5.2, below), labeling data may be quite instructive.

6.4 Analysis of Data Obtained from Identification, Attribute Rating, Discrimination, and Dissimilarity Rating Tasks

There are two categories of data analysis approaches, *correlation* based, and group or condition *comparison* based, which are often helpful in making decisions about the effectiveness of display properties. Correlation based analyses are performed when one wishes to determine the strength of relationship between two (usually continuous) quantitative variables (for example between pitch of a data stream in an auditory graph designed to display temperature, and observers' estimates of temperature). Group (or condition) comparison analyses are used when one wishes to see whether two or more different conditions produced different values (usually based on the mean) of some quantitative measure. For example, if one has two alternative designs for an alarm earcon in an industrial display, does one produce faster response times than the other? It is often the case that a designer will find it useful to employ *both* correlation based *and* comparison based analysis procedures during the process of designing an application involving auditory displays.

6.4.1 Correlation Analyses

While there are several different statistical measures of correlation, one of the most commonly used is the Pearson's correlation coefficient, r. In this analysis, two quantitative variables, labelled X and Y, are tested to determine whether there is a *linear*[6] relationship between them. The correlation coefficient r, provides information about the direction (whether X and Y vary in the same direction or vary in opposite directions, indicated by whether the computed value of r has a positive or negative value) and the strength of the relationship (a value of $+1.0$ or -1.0 indicate "perfect" linear relationships, and 0.0 indicates no relationship). For most designers of auditory displays, correlation analyses, by themselves, will not provide sufficient information for product evaluation without reliance on additional approaches such as group or condition comparisons. However, Pearson's correlation coefficient lies at the heart of more sophisticated analyses (such as regression analysis, factor analysis, cluster analysis, and MDS) that can be used effectively to determine the perceptual and acoustic qualities among sets of auditory stimuli. (For detailed statistical discussion of the Pearson's correlation coefficient and multiple regression readers should consult Cohen (2003) or Pedhazur (1997); for factor analysis, Gorsuch, 1983 or Brown, 2006 and for questions about multivariate analyses, Tabachnick & Fidell, 2006[7]).

6.4.2 Comparing Conditions or Groups: t-tests, ANOVA and Related Procedures

Most researchers designing auditory displays will wish to compare users' responses to individual sounds or to different display formats to determine which one would work best for a particular application. Such studies may use measures of *performance*, such as users' speed or accuracy in responding to an auditory display (see Bonebright & Nees, 2008 for an example study that uses both types of measures), or they may use subjective attribute ratings, or similarity ratings (for example to determine whether one set of auditory icons "matches" a set of visual icons better than another set of icons). The data from these studies is then submitted to an analysis technique that compares the means of performance measures or ratings among the conditions.

Two basic statistical procedures for evaluating differences between means are Analysis of Variance (ANOVA) and the t-test. Each of these has versions (with different computational techniques) suitable for comparing means among conditions in between-group and within-group designs (see section 6.2.6). The present discussion will focus on ANOVA since this technique is much more flexible and can test differences among multiple subject groups across multiple conditions while the t-test can only be used for testing differences between means from two conditions. Software routines for performing analyses using ANOVA and related techniques are available in a wide range of data analysis software packages (e.g., SAS, SPSS, R) and some limited capabilities for using these techniques are embedded among the "add on tools" in "professional" versions of popular spreadsheet programs such as Microsoft Excel. While a full discussion of ANOVA (and more advanced related techniques such as

[6]Researchers should keep in mind that there are other types of non-linear relationships between quantitative variables (such as quadratic), which will *not* result in a significant correlation coefficient. It is always valuable to plot the data points for a visual representation of the data to understand more fully the relationship. Please consult the references specified in the text for information on this and other issues related to the pitfalls of correlation analysis.

[7]There are a number of good references for multivariate statistics. The current recommendation for Tabachnick and Fidell is based on the applied nature and readability of their book. However, there are other references, such as Johnson, 2002; Johnson & Wichern, 1998; Hair, Tatham, Anderson, & Black, 1998, Abdi, Edelman, Valentin, & Dowling, 2009, that also provide excellent coverage of the topic.

MANOVA and ANCOVA) is beyond the scope of this chapter (interested readers should consult Tabachnick and Fidell, 2006 or one of the other references presented in footnote 7), a general discussion of the basic logic of statistically evaluating differences among means, as well as some of the common pitfalls encountered in applying and interpreting such analyses are in order.

The logic behind ANOVA and related statistical procedures is to compare the variance among the values (e.g., performance scores, ratings) within each group (condition), considered the *error variance*, with the variance *between* the groups or conditions (which is presumed to reflect both group differences and error variance). This comparison is done by computing a ratio, called an F-ratio. If the resulting F value differs more than would be "expected by chance", the "null hypothesis" that the means of the conditions or groups are the same is rejected, and the researcher then has some evidence that there are actual differences between the groups on the variable of interest. For a study that has more than two groups or conditions, follow-up analyses are needed to determine exactly where the differences lie since the "omnibus" F test only determines that there is a difference among the groups, but not *which* group differences are significant. Clearly, the researcher must also examine the values of the group or condition means to determine the *direction* of the differences, should the F value be significant. It is important to note that there are conflicting opinions about post hoc comparisons and their appropriate use. Researchers should check the previously mentioned references on multivariate statistics as well as consult with colleagues in their respective disciplines to determine which method is the best to use for publication purposes. It is also important to point out that display design and optimization decisions are not the same thing as pure scientific research being prepared for a research journal; thus, ultraconservative statistical tests may not be necessary.

6.4.3 Caveats When Using Techniques that Compare Groups

Even though comparing means by ANOVA and related techniques is relatively simple to perform with statistical software, there are pitfalls that should be recognized and avoided, particularly by researchers who have little or no prior experience with applying these techniques. One typical problem results from *failure to screen data* prior to performing mean comparisons. Not only are there common problems with data sets, such as missing data points or data that contain errors due to data entry or to equipment or software problems, but there are also more subtle issues that should be examined. Averaged data and particularly computations of variance are extremely sensitive to outliers, and if a data set contains them, the means can be either artificially inflated or deflated leading to finding a difference that doesn't really exist for the population of interest or not finding one that is there. There is also the possibility that the sample of participants may be made up of multiple populations, such as people who process sounds differently than others, or it could be that the outliers are participants who misunderstood the instructions. In the first case, it would be good for the researcher to be able to identify this sub-group so that appropriate accommodations can be made for them when they use the sound application. In the second case, typically indicated by a number of outliers who share no common pattern of responses, it is extremely difficult to determine if these occur due to general *perceptual* difficulties encountered with the displays, or to basic inattention to the task. Interviews or post testing surveys of individual participants may provide some guidance in this regard. In general, the presence of substantial lack of reliability among participants in responding to auditory displays should trigger a note of concern about whether a design has been adequately optimized. In the case where there is additional, clear empirical evidence that some participants did have difficulty understanding the task or were inattentive, the data can be removed from the data set before further analyses are performed and a full explanation for this action included in any manuscript written for publication purposes.

It is most important to note that neither data entry errors, nor presence of outlying data observations are likely to be easily discovered without an initial data screening. Screening can consist of simple visual inspection of data values in a table or spreadsheet if the number of data records is relatively small, but for larger data sets some type of software assisted screening should be considered. In some cases reliability analysis routines may be useful (an example will presented later in section 6.5.1), and it is possible that some types of visualization schemes, such as plotting condition profiles for each participant on a common plot to see if any visually "jump out", may also be helpful (Wegman, 2003). It should also be noted that even *sonification* of raw data values by mapping them to pitch (perhaps organized as profiles of observations from each participant) could be useful in pointing out anomalies in the data prior to formal statistical analyses.

Another pitfall researchers should be wary of is the difference between a *statistically significant* difference and a *practical* difference. If the analysis finds that the difference between the group means was significant, the researcher can assume that the difference most probably didn't happen by chance. But the probability value (typically set at less than .05) doesn't state what the actual *effect size* is. In order to determine this, additional statistical tests, such as η^2 or ω^2, which provide an estimate of the proportion of variance due to the differences in the conditions, need to be performed (Tabachnick & Fidell, 2006). However, even if there is a significant difference and the effect size is large, the difference between the means may not be *practically* large enough to matter when the sound application is used in a real world setting.

The final pitfall that researchers should keep in mind occurs when there are multiple comparisons being performed within a given study. Alpha inflation or *Familywise type I error* (FWER) occurs when each comparison performed has the probability of .05 that the null hypothesis was rejected when it should have been retained. For each additional analysis, the probability of committing this type of error increases by the amount of the probability value used. The issue of adjusting for alpha inflation is controversial, and there are a number of methods (such as Scheffe, Tukey, Dunnett, Bonferoni or Fisher tests) that can be used ranging in how conservative they are, that will correct for the type I error rate (Keppel & Wickens, 2004). Obviously, these corrections decrease the likelihood of finding a significant difference; however this is justified since the convention is to be conservative in terms of stating that differences exist.[8]

[8]There is a movement in a number of disciplines to use statistical techniques, such as Bayesian statistics, that do not have the disadvantages of null hypothesis significance testing (for a discussion of this issue, see Kruschke, 2010 or Wagenmakers,, Lodewyckx, Kuriyal and Grasman, 2010). However, statistical testing as described in this chapter is still the predominantly accepted method.

6.5 Using "Distance" Data Obtained by Dissimilarity Ratings, Sorting, and Other Tasks

Evaluation of the overall usability of an auditory display requires consideration of both the effectiveness of the *perceptual mappings* between sound and information that the designer intends to present, and the *reliability* of perception of the display among potential users. Perceptual mappings play a critical role in making sure that the listeners extract the desired information from the display. For example, if the designer wishes to present data values that are increasing, pitches that increase would be appropriate. However, if the designer also adds changes in loudness to this auditory stream, the interaction between changes in pitch and loudness may lead to "distorted" estimates of the magnitudes since changes in pitch can affect judgment of loudness and vice versa (see Neuhoff, Kramer, & Wayand, 2002). Such a display could be described as *reliably perceived*, since all the participants may perceive the graph in exactly the same way, but its ability to display the underlying information would be compromised. Alternatively, an auditory graph of data that appears to faithfully represent the structure of data to about 40% of users, but conveys little or no information to the remaining 60% (or, worse yet, conveys a totally *different* structure among a subset of users), would have serious *reliability* shortcomings, and thus its overall usability would also be low.

The use of data collection techniques that generate "perceived distance estimates" among auditory display elements can be used to address the issue of consistency of perception among users via reliability analysis, and produce descriptions of the actual perceptual relationships among the display elements via techniques such as cluster analysis and MDS. Solutions from clustering or MDS routines may then be examined to determine whether they meet the objectives for which the display is being designed. For example, if display elements are auditory graphs representing multivariate data, one can make statistical comparisons between values of variables in graphs included in different clusters, and/or one can use regression analysis to determine the relationship between numeric values of variables and the position of the graphs in an MDS structure (e.g., Flowers & Hauer, 1995). If the display elements are real or synthesized "product sounds", one can use such procedures to determine relationships between acoustical properties of sounds and user perceptions to guide design or predict consumer preferences.

There are several methods commonly used to assess the perceived "distance" or dissimilarity between stimuli for purposes of clustering or scaling. "Direct" methods include the use of dissimilarity ratings and sorting tasks, which were discussed in sections 6.3.4 and 6.3.5. Perceptual dissimilarity between stimuli can also be measured "indirectly" by computing it from attribute rating tasks, which were discussed in section 6.3.2. However perceptual dissimilarity measures can also be computed from measures of performance (speed or accuracy) from tasks requiring participants to make perceptual discriminations between different stimuli, such as same/different judgments of stimulus pairs, or speeded classification (e.g., "press the right key if you hear sound A; press the left key if you hear sound B"). The "direct" methods (dissimilarity rating and sorting) offer a considerable advantage in the speed of data collection and are probably preferable for most applications involving evaluation of auditory displays.

6.5.1 Using Reliability Analysis to Assess Dissimilarity Rating or Sorting Consistency among Participants

Reliability analysis is a technique typically used for the assessment of test items used in educational or psychometric tests. Presented here is the use of reliability analysis using sorting or dissimilarity rating data for obtaining a measure of "agreement" among the participants about the perceptual structure of a set of sounds that might be used in an auditory display. For this purpose, the "test items" are each participant's "stretched out dissimilarity matrix". There will be $N \cdot (N-1)/2$ of entries in each vector, where N is the number of stimuli that were sorted. Examples of such vectors could be generated by taking each of the individual matrices on the right side of Figure 6.1, eliminating the zeros that make up the diagonal, and then lining up the data in a separate column for each participant.

Reliability analysis routines, such as SPSS Reliabilities, compute several measures of reliability and scaling statistics, but for the present purposes, an overall measure of consistency among participants, the *Cronbach's alpha* (Cronbach, 1951) would be used. It would also be necessary to have a measure of consistency (correlation) between each participant's vector and the composite of the entire group of participants for detecting participants whose sorting patterns substantially depart from the overall group sorting pattern (outlier detection). For example, SPSS[9] provides a printout column titled "Item-Total Correlation" that presents this information for each participant (the participants are the "items" in the application), as well as a column titled "Alpha if Item Deleted" (a listing of what the overall reliability would be if a participant or "item" were to be excluded from the analysis). Alphas above 0.70 indicate quite reasonable agreement or consistency, with values above 0.80 providing a quite high level of confidence that there is a solid shared basis of judgments of similarity among the stimuli.

It is reasonable to expect on the basis of distribution of pitch discrimination impairment and other auditory deficiencies in the general population that a small percentage of participants will have difficulty with discriminating pitch changes normally encountered in music (Marin & Perry, 1999) or may simply fail to understand the nature of the sorting task. In practice, it is difficult and not overly useful to distinguish between these two types of participants. One can adopt a policy of excluding participants whose grouping patterns exhibit a negative correlation or a correlation of less than some small positive value (e.g., +0.05) with the remainder of the group (as indicated by the Item-total correlation) from inclusion in subsequent MDS or clustering analyses, on the basis that they are outliers and are not likely to be representative of the population for which the auditory displays will be used, particularly if the alpha after exclusion is substantial in size.

Reliability analysis of sorting patterns can also be useful as a general performance measure for making comparisons among different display designs or durations for purposes of optimizing display design. For example, Flowers & Grafel (2002) had participants sort two sets of auditory graphs representing monthly samples of climate data. One set, the "slow" displays, presented 31 days of weather observations in 14.4 seconds, while other ("fast") displays presented the same data in 7.2 seconds. Sorting reliability for the "slow" displays was substantially lower than for the "fast" displays (0.43 vs. 0.65), even though participants indicated a preference for the slow displays, and stated that the fast ones were "too fast to

[9]SPSS is only one of the commercial packages that can be used for the analysis specified here. Such an analysis can also be performed using freeware, such as R or Scilab.

perceive detail." Subsequent display evaluation, using displays of similar design but with an intermediate duration of 10.0 seconds produced higher sorting reliabilities ranging from 0.71 to 0.84. For these types of auditory time series graphs (which will be discussed in more detail in the next section), display duration clearly affected the consistency of sorting among users.

6.5.2 Inferring "Perceptual Structure" using "Distance" Data: Clustering and Multidimensional Scaling (MDS)

Once it has been ascertained through reliability analysis that participants agree, to a reasonable extent, about which sound samples are similar or dissimilar to each other, the application of clustering and/or MDS procedures can be used to generate displays of perceptual structure among the set of sounds being investigated. A full treatment of either cluster analysis or MDS procedures is beyond the scope of this chapter; readers interested in applying these procedures should consult one or more of the authoritative sources in these areas such as Borg & Groenen, 2005; Davison, 1992; Kruskal, 1977; Kruskal & Wish, 1978; Schiffman, et al., 1981; or Young & Hamer, 1987. However the following discussion should provide some basic guidelines about how these procedures can be used by investigators and designers of auditory displays. Both hierarchical clustering and MDS are data structure display techniques that analyze "distance" data, and provide a display that illustrates perceptual "distance" relationships among stimuli. Both techniques can be used in conjunction with either rating data or acoustical properties of the stimuli to show how these perceptual distance relationships relate to psychological (perceived) or physical stimulus attributes (see Davison, 1992).

To illustrate use of these techniques, examples of data analyses from a previously unpublished study of auditory "weather graph perception" conducted as a follow-up to the study of Flowers and Grafel (2002) are presented here. These data were generated by 30 participants who each sorted a set of 23 auditory graphs into perceptually similar groups. The auditory graphs displayed monthly samples of historical weather observations from Lincoln, Nebraska, obtained from the High Plains Regional Climate Center (www.hprcc.org). These 23 monthly records were selected to cover a representative range of variation in temperature and precipitation patterns typical of the Great Plains of the United States during warm season months across the historical period of 1934-2000 - a period during which substantial climate variation occurred. The auditory displays presented each day's high and low temperature as an alternating four note synthetic string MIDI stream for which pitch was mapped to temperature. On days in which precipitation occurred, a one to three note MIDI grand piano was imposed over the last half of the four note string sequences to indicate rainfall amount. (For additional details about the display format see Flowers, Whitwer, Grafel, & Kotan, 2001 and Flowers & Grafel, 2002).

The basic display output of a hierarchical clustering procedure is a "tree" structure (sometimes shown as an "icicle plot" or a "dendrogram" depending on one's display preferences). These displays depict clusters of stimuli that "belong together" under a hierarchical "agglomeration schedule" that adds stimuli to clusters and clusters to each other based on analysis of distance data. There are several choices among clustering algorithms used for determining the criteria for combining groups, and at what "level" the clusters or stimuli are combined. However the objectives of these algorithms are quite similar; in many cases the results they produce are also highly similar. Figure 6.2 displays a dendrogram created by SPSS using the weather sample sorting data and the *average linkage method*, which is a typical default clustering algorithm.[10]

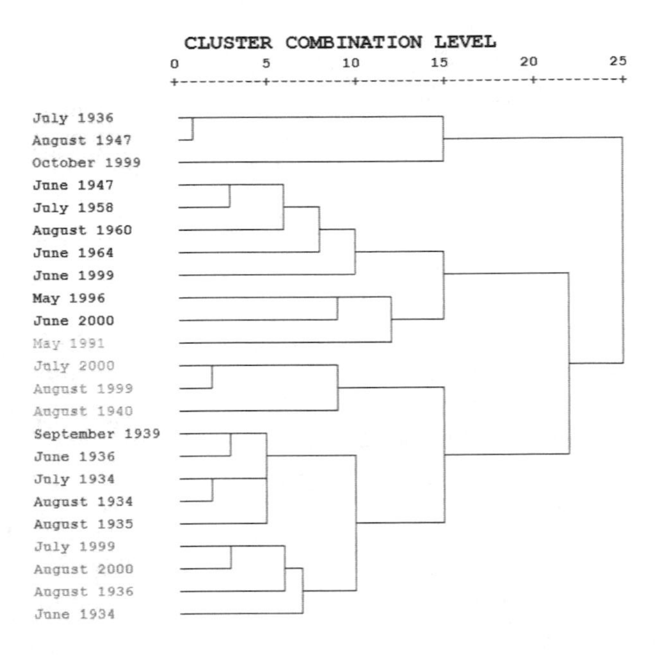

Figure 6.2: Dendogram of the cluster structure obtained from sorting patterns of auditory graphs of historical monthly weather patterns.

Determining the agglomeration level at which joined items should be considered "meaningful" to be treated as a group is a relatively subjective judgment. If the clusters that result after that judgment differ meaningfully in properties of the stimuli themselves, or correspond to additional rating or other performance data obtained with the stimuli, one gains confidence that the groupings reflect meaningful perceptual decisions on the part of the participants, and thus they can guide a variety of subsequent design decisions. Figure 6.2 was selected as an example since it happens to illustrate some "extremes" of what might happen (and *did* happen in this case) and to point to some interesting and informative data features. The visual overview of the agglomeration structure suggests three major groupings at about level 15, but with breaks among these groupings at levels 10 through 13 suggesting that a 5 "cluster" structure would be a reasonable description. However, there is one feature that stands out. One stimulus, the auditory display of the weather from October 1999 (sound example **S6.1**) does not combine with *any* group until level 15. This pattern of extreme late combination is suggestive of an *outlier* - a stimulus that does not belong to any group. Inspection of the weather properties for this month suggest that it was indeed meteorologically unique within the set of 23 monthly climate samples (additional sound examples are referenced directly prior to Figure 6.3). It was exceptionally dry (a trace of rain on each of three days), but quite cool. Coolness and dryness happen to be features that do not conjoin in the other stimuli in this set. Musically, the sonification of October 1999 consisted of an atypical low pitched

[10]For more details about different types of clustering analyses, see Johnson, 1967.

temperature stream with only five single high piano plinks representing rain. The two months with which it was combined, at the last resort, were August 1947 (sound example **S6.2**) and the dust bowl month of July 1936 (sound example **S6.3**). These were also months of exceptional drought and only three days of rain. But these two months also had *searing heat* (up to the all-time record 115 degrees Fahrenheit for the region) and would have produced a temperature stream averaging more than an octave higher in pitch throughout the 10-second display. Within the remaining 20 monthly weather samples there were both hot and cool months with either moderate or high amounts of precipitation, but no other cool and very dry months. So it "makes sense" that drought was probably the common attribute that determined October 1999's final admission to a cluster.

When clusters have been defined by a clustering routine, one may then inspect whether the clusters differ in terms of specific measurable properties of the stimuli or in terms of additional ratings of the stimuli obtained. Provided there are enough members of individual clusters to provide sufficient statistical power, traditional techniques such as ANOVA can be used for that purpose. In the present case, clusters differed significantly in terms of both total precipitation, and number of days on which precipitation occurred. When October 1999 was included in the analysis by clusters, there were no significant differences between clusters in temperature. However, with the removal of the October 1999 an overall significant effect of temperature was found that distinguished among the clusters as well as a pattern suggesting that participants were able to perceive the key meteorological properties of these different historical weather records by listening to them.

The objective of MDS procedures is to provide a *spatial* depiction of stimulus similarity relationships - typically in Euclidean space. MDS procedures use iterative algorithms to discover a spatial configuration of the stimuli that is compatible with at least the ordinal relationships among the dissimilarity measures among the stimuli - and to do so in a minimum number of Euclidean dimensions. How "compatible" a fit in a given number of dimensions (typically 2, 3, or sometimes 4 for perceptual stimuli) happens to be is usually assessed by at least one, and typically two measures of the "degree of fit" that has been achieved once the iterative routine has determined that it has "done its job". MDS computation routines such as ALSCAL[11] (Young & Lewyckyj, 1979) provide STRESS and R^2 as indices of discrepancy between distances among "optimally scaled" points and the positions produced by the final configuration (for a discussion of computational details see Kruskal & Wish, 1978). STRESS ranges between zero and one and is sometimes referred to as a measure of "badness of fit" since poor fits are associated with larger numbers. R^2 is a form of a multiple correlation coefficient – in this case between optimally scaled dissimilarities and the MDS model distances. It gets larger as the "fit" of the model improves, and it can be viewed, like other types of multiple correlations, as a "proportion of the variance" of the optimally scaled data that can be accounted for by the MDS solution. Good fit does not imply a *meaningful* solution however. The user should attempt to achieve a solution in the minimum number of dimensions that produces an acceptable level of fit, since using a large number of dimensions, may lead to small STRESS and large R^2 values, but a meaningless "degenerate" solution.

To illustrate an example of MDS applied to assessing perceptual structure of auditory display stimuli, the same example of weather data sonification used to illustrate clustering methods

[11]There are other MDS algorithms [for example, CLASCAL, see Winsberg & De Soete (1993) INDSCAL or MULTISCALE, see Young, 1984)] that can be used, although the discussion of their relevant advantages and disadvantages is beyond the scope of this chapter. See the MDS references for more information.

will be used. The SPSS ALSCAL routine was applied to the dissimilarity data from the sorting task obtained for monthly weather records that generated the cluster display previously shown in Figure 6.2. A "satisfactory" fit was obtained in two dimensions, stress $= 0.135$ and $R^2 = 0.923$. Figure 6.3 displays the spatial configuration in two dimensions, upon which rectangles encompassing the *cluster* groupings described in the earlier discussion of clustering are superimposed. Inspection of this display shows that October 1999 (sound example **S6.1**) again stands out as an outlier - farther apart from its neighbors in the cluster than any other month in the sample. However, the geometric relationships show that the other two members of that ill-defined cluster are spatially close to other brutally hot and almost as dry months such as August 1947 (sound example **S6.2**) and July 1936 (sound example **S6.3**). Notably cooler and very wet months, such as June 1947 (sound example **S6.4**), and May 1996 (sound example **S6.5**) are on the opposite (left) side of the display (Please refer to four additional sound files for more examples from the clusters in the MDS solution space - August 1960 (sound example **S6.6**), August 1940 (sound example **S6.7**), August 2000 (sound example **S6.8**), and July 1934 (sound example **S6.9**) as well as nine examples from fall and winter months that were not included in this study – December 2000 (sound example **S6.10**), December 2001 (sound example **S6.11**), December 1999 (sound example **S6.12**), February 1974 (sound example **S6.13**), January 2001 (sound example **S6.14**), January 1940 (sound example **S6.15**), January 1974 (sound example **S6.16**), November 1940 (sound example **S6.17**), and November 1985 (sound example **S6.18**).

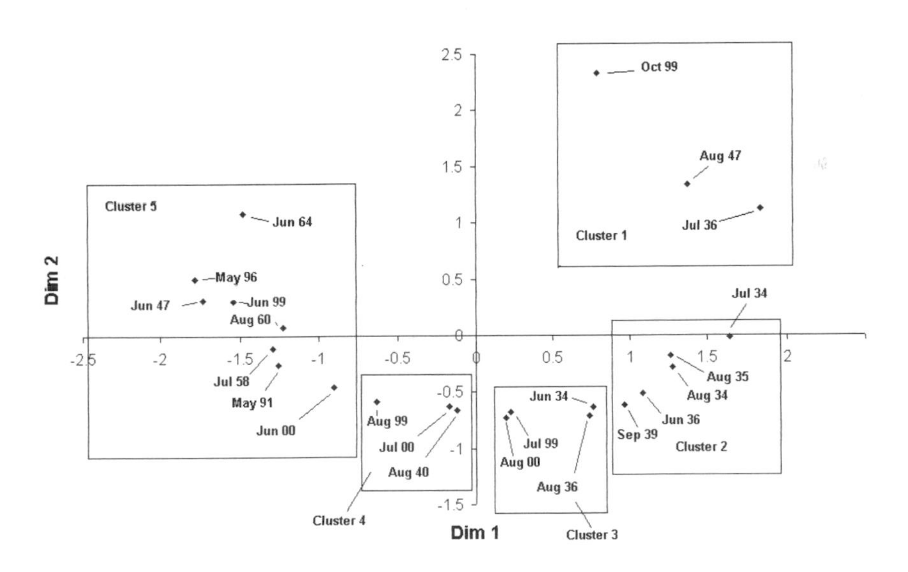

Figure 6.3: MDS configuration obtained from auditory weather graph sorting data.

It was previously mentioned that the mean temperature values, total precipitation, and number of days on which precipitation occurred differed significantly among the clusters (at least when the outlier October 1999 was excluded). Figure 6.3 clearly shows that the stimuli defined by these clusters appear in different spatial regions. With an MDS configuration, one can use multiple regression techniques to indicate the relationship between the positions of stimuli in the space defined by the MDS dimensions and some measured quantitative property of the stimuli, by using the MDS dimensions as predictors and the property value as the dependent measure. In the present case, regression can help identify regions of "wetness versus dryness", "many rainy days" versus "few rainy days", and "warm" versus "cool" through predicting total monthly precipitation, number of days on which rain fell, and mean temperature of the month, using the two MDS dimension scale values of each stimulus. The ratio of the beta weights of the two predictors defines the slope of the "best fit vector" for each of these predictors; thus one can draw a line, passing through the origin and use this computed slope to illustrate these relationships. Figure 6.4 displays such vectors. One is only justified in displaying property vectors in this manner if the result of the multiple regression analysis shows that MDS axes significantly predict the stimulus property being represented; in this case all three regressions were significantly predicted. The RSQ values listed on Figure 6.4 are the squared multiple correlation, or the proportion of variance accounted by the regression models. In this particular situation, one can infer that the experimental participants who listened to these auditory depictions of month-long samples of weather observations were indeed sensitive to the sonic representation of temperature and precipitation patterns.

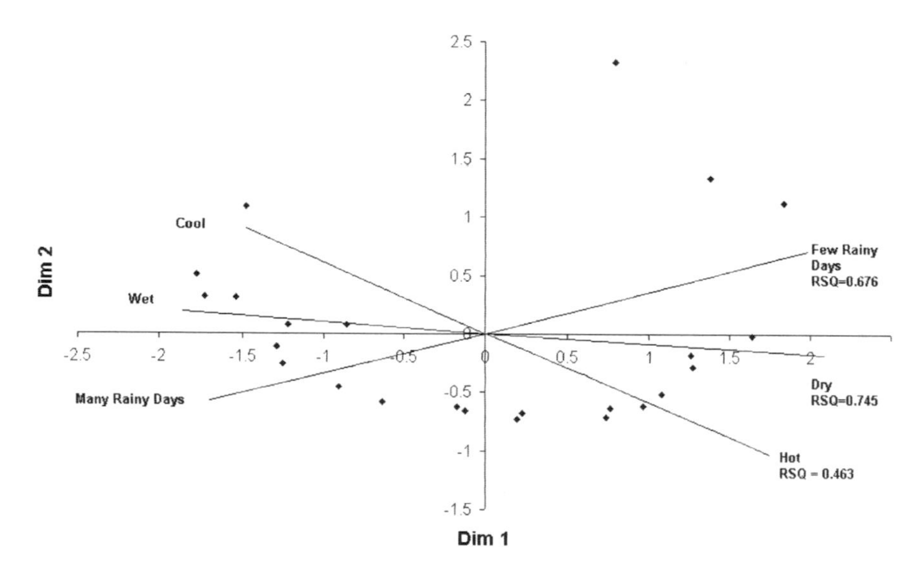

Figure 6.4: MDS configuration of auditory weather graph data with stimulus attribute vectors included.

In summary, the combination of clustering procedures with MDS and regression analyses based on stimulus attributes can provide a very useful set of exploratory and visualization tools for discovering perceptual relationships among auditory stimuli, and thereby guide choice or design of auditory display components for a wide range of applications, such as sonified data displays, auditory icons, earcons, status indicators, alarms, etc. These tools can guide discovery of which sounds are perceptually distinct from each other, and which have sufficient similarity that confusability might become an issue. The addition of regression procedures to MDS can help determine the relationships between subjective quality ratings of sounds, their acoustical properties, and their perceptual similarity structure (see Gygi, Kidd, & Watson, 2007 for another MDS analysis example used in the investigation of environmental sound perception).

6.6 Usability Testing Issues and Active Use Experimental Procedures

When considering auditory displays from a usability perspective, there are a number of issues that the designer and researcher must take into account. These issues are basic to all usability testing and include the time it takes the user to learn the application; the speed at which the user can accomplish tasks with the application; the number of errors that occur while using the application; the ease of retention over time for how to use the application; and the overall subjective satisfaction of the user with the application (Schneiderman, 1998).[12] These issues all point to the value of providing testing that promotes actual use of the auditory display by the target user population in the target environment early and often during the process. Conversely, it is also important that experts play a role so that they can use their knowledge to help winnow down options. For example computer scientists might evaluate the cost effectiveness of specific auditory displays in terms of computer processing power, while perceptual psychologists might consider the cognitive and perceptual abilities of human users in relation to the proposed display.

6.6.1 Testing in the Laboratory versus Testing in Real World Settings

It is also absolutely essential to consider the use of data collected in a laboratory setting in comparison with use in the target environment. Results of a strictly controlled experiment may suggest to a developer that a particular aspect of an auditory display "works well" because it produces statistically significant effects on performance that are in the desired direction. However, this does not indicate that the application will work well in a less controlled environment (e.g., a workplace that has noise that overpowers the sound display or a working environment that results in such sound displays aversely affecting co-workers); thus practical significance needs to be thoroughly examined. In addition, while participants in an experiment may be willing to endure multiple sessions for training purposes due to any compensation they might receive, users in a real environment must immediately see that the potential benefits outweigh any costs in learning or using the display, otherwise they may choose to simply disable the application.

Assessment of sound applications using active-use procedures emphasizes the actual use of the product or application in the "real-world" environment. Such techniques, including surveys, verbal protocols, focus groups and expert appraisals can be used both in the target environment or in a usability laboratory that is set up to provide a comparable environment to the one where the application will actually be used (Jordan, 2002; Nielsen, 1993). In these laboratories, participants can work with the display and provide feedback to the researchers. In this type of testing, it is imperative that the subjects realize that *they* are not being tested, but rather that it is the *application or product* that is under investigation.

[12] There are many useful references for usability testing. Schniederman's 1998 book is a good general reference, but the reader may also wish to consult Dumas and Redish (1993) or Nielsen (1993).

6.6.2 Surveys

Surveys can be designed to collect data before, during, and/or after the participant has worked with the application. For example, a researcher may ask individuals about their own expectations of how sound would work in a specific case, or what type of experience the participant has had with sound applications. In this case, the researcher wants to make sure that the participant is not biased by exposure to the target application. During the interaction with the application, participants may be required to provide specific responses that they might forget by the end of the session. However, most of the time, participants complete surveys after they have completed their interaction with the sound application. In this case, the survey serves the purpose of measuring the overall reactions to the application.

The questions in a survey are dictated by the particular application and concerns of the researchers. Demographic questions concerning age, gender, and other relevant personal characteristics should be selected carefully. For example, a researcher may find that women have a preference for a particular type of sound while men may prefer another. Obviously, this would be good to know and could result in the auditory display offering a variety of sounds in a "sound palette" (see Bonebright & Nees, 2008 for an example) to provide the best possible match of the application with the widest possible user audience. General questions about annoyance and distraction levels, overall satisfaction with the user interface, and whether the participant would use such a product would be particularly pertinent for sound applications. Finally, questions that are specific to the target application should be carefully prepared and selected to make sure that researchers have the information they desire. It should be strongly emphasized that construction of surveys can appear deceptively simple to someone who is uninitiated into this type of research. However, effectively wording questions for surveys takes experience and careful consideration of the target population. In addition to the obvious need of writing the questions clearly, using vocabulary that is familiar to the participants, and keeping questions as short as possible, survey items also need to be constructed to avoid leading or biasing questions. One common pitfall researchers make is to construct a questionnaire that is excessive in length. This should be avoided by carefully choosing items that will provide the necessary information to promote the design of the auditory display.

Responses to surveys can take a number of fixed response format items, such as rating scales, true or false questions, and check boxes for relevant properties, as well as free response options. However, particularly for the purposes of evaluation to guide design or refinement of a product, a good general guideline is to make more use of rating scale questions (e.g., 5, 7, or 9-point scales) rather than yes/no questions. Data from fixed response format items are easier to analyze, but free responses may provide a richer source of data. In many cases, a combination of fixed and open response items may provide the best balance for both the researcher's purpose and the ability of the participants to respond in a way that reflects their true opinions.

Surveys provide a relatively easy way to determine users' opinions about auditory displays, but they are not without shortcomings. For example users may react to the perceived demand characteristics of the research context, or they may also respond in ways that they believe are socially desirable. In both cases, the data provided by the participants does not reflect their true opinions or experiences and will lead to erroneous decisions about the effectiveness of the display. (For a good general reference for survey design and construction, see Bradburn,

Sudman, & Wansink, 2004 or Oppenheim, 1992).

6.6.3 Verbal Protocols

Verbal protocols require subjects to talk aloud while they work with an application. The participants' statements can be recorded and/or an experimenter can take notes and cue the participants to elaborate on their comments during the session. The advantages of this type of procedure are that participants do not need to rely on memory in order to report their responses at a later time and that participants can provide spontaneous comments about improvements or problems while they are working with the application. Some researchers have pairs of participants work together since this leads to more information for the researcher while the users explain aspects of the program to one another (Schneiderman, 1998). This approach may in fact lead to a more realistic evaluation of a sound application in two ways. First, when people learn a new application, many times they will have someone help them. Second, it could be especially informative for sound designers to determine whether the sound helps or hinders in what can be a social process.

In spite of the possible advantages of using verbal protocols to evaluate use of auditory displays, there are also potential disadvantages that should be considered before adopting this method. Some of these issues are general problems encountered with use of verbal protocols for evaluation of any product or process, while others are unique to (or perhaps even exacerbated by) situations involving evaluation of auditory displays. One general problem is often encountered when recording sessions by electronic means, such as videotaping or use of digital recording media. Use of passive recording methods can be falsely reassuring since a novice researcher will assume that this means that there is a permanent record of *all* aspects of the session. Unfortunately, the reality of using recording media is quite different from that expectation. For example, the verbal record can become obscured when the participant doesn't talk loudly enough, or the camera may be placed in such a way that there are important details that are not captured on the recording. It should also be noted that the recorded media will need to be coded at some point for analysis purposes, and while the researcher can choose to replay a section that was missed, the coding stage will still need to be completed at a later time than if it were done while the participant was interacting with the application. However, if the researcher chooses to have the session recorded by an investigator, it is extremely important to make sure that there is sufficient training so that investigators are consistent across sessions themselves and show a high degree of consistency with any other investigators working on the project. Finally, when examining the effectiveness of an auditory display with the participant talking about the experience, the researcher needs to be aware of any system sounds that might be missed due to the monologue of the participant.

It is important to note that verbal protocols were developed primarily for evaluation of computer software during usability studies (Virzi, 1992). To date, there has been limited use of this technique for auditory displays; therefore, researchers interested in trying this technique should keep in mind that the verbal protocol in addition to listening to an auditory task may have much larger effects on cognitive load and resource allocation than are seen when this technique is used for visual or text based scenarios.

6.6.4 Focus Groups

In focus groups, participants assemble with a discussion leader to provide reactions and opinions about a product (in this case, a sound application) that is being developed. The discussion leader typically has a list of items that he or she wishes to use as beginning discussion points. Such a list is normally generated by the researchers prior to the meeting to illuminate any of the facets they are considering trying or testing. However, it is also important to leave the conversation open so that the participants can bring up issues that are important to them that the researchers may not have anticipated. When working on sound applications, it would be most likely that focus groups would be conducted when the design was for a specific population, such as firefighters or physicians. In these cases, the researcher can gain valuable insight into the needs of the specific group that can then be considered during the subsequent design process. When conversation goes dry, prompts must not be leading so that the conversation is not biased toward a particular topic. Thus it is very important that discussion leaders be carefully trained.

Focus groups tend to consist of five to six participants so that there are not so many people that individuals become bored waiting for their turn nor that there are so few that there aren't enough voices to keep up the synergy. Individuals chosen for the group should also be carefully selected to make sure all constituents of the target group are involved. Finally, group dynamics must be managed well by the leader in order to get good information from all members.

Analysis of data from focus groups typically involves content analysis of the topics. In most cases, the discussion leader records the major points, as well as the emphasis placed on each, on a checklist type of format that leaves room to specify the topic and provide a rating of significance for the group. Once the data are collected, the content is analyzed for overlapping themes that can help further development of the display. Electronic methods of recording focus groups can also be used to assist with these analyses. However, some of the same caveats presented in the discussion of verbal protocols about the use of recorded media apply here as well. (See Jordan, 1998 or O'Donnell, Scobie, & Baxter, 1991 for further discussion of focus groups.)

6.6.5 Expert Appraisals

Enlisting the help of experts in relevant areas in which designers or researchers are not trained can and should be used when developing auditory displays. For example, a professional who works in sound synthesis will not necessarily have the expertise to make appropriate judgments about the human physical and cognitive limitations that are important to take into account when building a sound application. Or a researcher designing an application for visually impaired people will not necessarily be an expert on the types of needs of this particular population. In such cases an expert appraisal performed by a professional in the appropriate field can be used effectively to avoid pitfalls and streamline the entire process. One way an expert may perform an appraisal is to use a checklist when evaluating the proposed design for an auditory display. An example of this would be to have a perceptual psychologist check on a number of the known perimeters that can affect people's ability to use a display. It has also been shown in a number of usability studies that multiple experts contributing to the evaluation in their area of expertise can increase the benefits of this technique (Jeffries, Miller, Wharton, & Uyeda, 1991; Karat, Campbell, & Fiegel, 1992). This approach is particularly useful in the beginning stages of the project even before a prototype is built, but it should be used with other methods of obtaining information from the target population as mentioned previously in this section.

6.7 Conclusion

In concluding this chapter, it may be helpful to summarize some of the most important global "take home" messages:

- Researchers in auditory display need to use appropriate auditory design principles and good research methodology.
- Good design projects will likely use multiple research methods to provide sufficient information to produce a good display.
- The context for the auditory display and the target population must be included in the design of the display from the beginning of the process.
- Researchers should seriously consider using teams that include individuals with complementary training to assure that all aspects of auditory design are addressed.
- Human auditory perceptual abilities must be a central consideration for the development of auditory displays.
- Statistical techniques should be used when appropriate but should not replace real-world testing nor mitigate the practical significance of sound applications.
- Decisions about the appropriate statistics to use must take into account the ultimate goals of the project.

Finally, the authors wish to note that researchers working in the development of auditory displays have made great strides in applying appropriate techniques for evaluating the usefulness of such applications in a variety of contexts. Hopefully this chapter will further facilitate extension of these techniques into the discipline and will act as a catalyst and reference for both experienced and new researchers in this area.

Bibliography

[1] Abdi, H., Edelman, B., Valentin, D., & Dowling, W. J. (2009). Experimental design and analysis for psychology, Oxford, UK: Oxford University Press.

[2] Ballas, J. A. (1993). Common factors in the identification of an assortment of everyday sounds. *Journal of Experimental Psychology: Human perception and Performance 19*, 250-267.

[3] Bauer, P., Korpert, K., Neuberger, M., & Raber, A. (1991). Risk factors for hearing loss at different frequencies in a population of 47,388 noise-exposed workers. *Journal of the Acoustical Society of America, 90*, 3086-3098.

[4] Beauvois, M. W., & Meddis, R. (1997). Time delay of auditory stream biasing. *Perception & Psychophysics, 59*, 81-86.

[5] Biassoni, E. C., Serra, M. R., Richter, U., Joekes, S., Yacci, M. R., Carignani, J. A. et al. (2005). Recreational noise exposure and its effects on the hearing of adolescents. Part II: Development of hearing disorders. *International Journal of Audiology, 44*, 74-85.

[6] Bonebright, T. L. (1996). An investigation of data collection methods for auditory stimuli: Paired comparisons versus a computer sorting task. *Behavior Research Methods, Instruments, & Computers, 28*, 275-278.

[7] Bonebright, T. L. (1997). Vocal affect expression: A comparison of multidimensional scaling solutions for paired comparisons and computer sorting tasks using perceptual and acoustic measures. *Dissertation Abstracts International: Section B: The Sciences and Engineering, 57* (12-B), 7762.

[8] Bonebright, T. L., & Nees, M. A. (2008). Most earcons do not interfere with spoken passage comprehension. *Applied Cognitive Psychology, 23*, 431-445.

[9] Borg, I., & Groenen, P. (2005). *Modern multidimensional scaling: Theory and applications* (2nd ed.). New York, NY: Springer.

[10] Bradburn, N., Sudman, S., & Wansink, B. (2004). *Asking questions*. San Francisco: Jossey Bass.

[11] Breger, I. (1971). A cross-cultural study of auditory perception. *Journal of General Psychology, 85*, 315-316.

[12] Bregman, A. S. (1990). *Auditory scene analysis*. Cambridge, MA: MIT Press.

[13] Brown, T. A. (2006). *Confirmatory factor analysis for applied research*. New York, NY: Guilford Press.

[14] Childs, E. (2005). Auditory graphs of real-time data. *In Proceedings of the International Conference on Auditory Display (ICAD 2005)*. Limerick, Ireland.

[15] Cohen, J. (1988). *Statistical power analysis for the behavioral sciences* (2nd ed.). Hillsdale, NJ: Lawrence Erlbaum Associates.

[16] Cohen, J. (2003). *Applied multiple regression/correlation analysis for the behavioural sciences*. New York, NY: Psychology Press.

[17] Cook, P. R. (2002). *Real sound synthesis for interactive applications*. Natick, MA: AK Peters, Ltd.

[18] Cronbach, L. J. (1951). Coefficient alpha and the internal structure of tests. *Psychometrika. 16*, 297-334.

[19] Davison, M. L. (1992). *Multidimensional scaling*. Malabar, FL: Krieger Publishing.

[20] Dreyfus, W. (1976). *The measure of man: Human factors in design* (2nd ed.). New York, NY: Whitney Library of Design.

[21] Dumas, J., & Redish J. (1993). *A practical guide to usability testing*. Norwood, NJ: Ablex.

[22] Edwards A. D. N., Challis, B. P., Hankinson, J. C. K. & Pirie, F. L. (2000). Development of a standard test of musical ability for participants in auditory interface testing. *In Proceedings of the International Conference on Auditory Display (ICAD 2000)*, Atlanta GA.

[23] Flowers, J. H., & Grafel, D. C. (2002). Perception of daily sonified weather records. *Proceedings of the Human Factors and Ergonomics Society, 46th Annual Meeting*, 1579-1583.

[24] Flowers, J. H., & Hauer, T. A. (1995). Musical versus visual graphs: Cross-modal equivalence in perception of time series data. *Human Factors, 37*, 553-569.

[25] Flowers, J. H. Whitwer, L. E., Grafel, D. C. & Kotan, C. A. (2001). Sonification of daily weather records: Issues of perception, attention and memory in design choices. *In Proceedings of the International Conference on Auditory Display (ICAD 2001)*, Espoo, Finland.

[26] Garbin, C. P. (1988). Visual-haptic perceptual nonequivalence for shape information and its impact upon cross-modal performance. *Journal of Experimental Psychology: Human Perception and Performance 14*, 547-553.

[27] Gorsuch, R. L. (1983). *Factor analysis* (2nd ed.). Hillsdale, NJ: Lawrence Erlbaum Associates

[28] Gygi, B., Kidd, G. R., & Watson, C. S. (2007). Similarity and categorization of environmental sounds. *Perception & Psychophysics, 69*, 839-855.

[29] Hair, J. F., Tatham, R. L., Anderson, R. E., & Black, W. (1998). *Multivariate analysis* (5th ed.). Upper Saddle River, NJ: Prentice Hall.

[30] Handel, S. (1989). *Listening: An introduction to the perception of auditory events*. Cambridge MA: The MIT Press.

[31] Hass, E., & Edworthy, J. (Eds.). (2002). *Ergonomics of sound: Selections from human factors and ergonomics*

society annual meetings, 1985-2000. Human Factors and Ergonomics Society.

[32] Hyde, K. L., & Peretz, I. (2004). Brains that are out of tune but in time. *Psychological Science, 15*, 356-360.

[33] Jeffries, R., Miller, J. R. Wharton, C., & Uyeda, K. M. (1991). User interface evaluation in the real world: A comparison of four techniques, *Proceedings ACM CHI91 conference*, 119-124.

[34] Johnson, A. C. (2002). *Methods of multivariate analysis* (2nd ed.). New York, NY: Wiley-Interscience.

[35] Johnson, R. A., & Wichern, D. W. (2002). *Applied multivariate statistical analysis* (5th ed.). Upper Saddle River, NJ: Prentice Hall

[36] Johnson, S. C. (1967). Hierarchical clustering schemes. *Psychometrika, 32*, 241-254.

[37] Jordan, P. W. (1998). *An introduction to usability*. London, UK: Taylor & Francis.

[38] Jordan, P. W. (2002). *Designing pleasurable products*. London, UK: Taylor & Francis.

[39] Karat, C., Campbell, R., & Fiegel, T. (1992). Comparison of empirical testing and walkthrough methods in user interface evaluation. *Proceedings CHI92 Human Factors in Computing Systems, ACM*, New York, 397-404.

[40] Keppel, G., & Wickens, T. D. (2004). *Design and analysis: A researcher's handbook* (4th ed.). Upper Saddle River, NJ: Prentice Hall.

[41] Kraemer, H. C., & Thiemann S. (1987). *How many subjects?* London, UK: Sage Publications.

[42] Kruskal, J. B. (1977). The relationship between MDS and clustering. In J. V. Ryzin (Ed.) *Classification and clustering* (pp. 17-44). San Diego, CA: Academic Press.

[43] Kruskal, J. B., & Wish, M. (1978). *Multidimensional scaling*. Beverly Hills, CA: Sage.

[44] Kruschke, J. (2010). What to believe: Bayesian methods for data analysis. *Trends in Psychology, 14*, 293-300.

[45] Lenth, R. V. (2001). Some practical guidelines for effective sample size determination. *The American Statistician, 55*, 187-193.

[46] Lipsey, M. W. (1990). *Design sensitivity: Statistical power for experimental research*. Newbury Park, CA: Sage.

[47] Marin, O. S. M., & Perry, D. W. (1999). Neurological aspect of music perception and performance. In D. Deutsch (Ed.), *The psychology of music* (pp. 653-724). San Diego, CA: Academic Press.

[48] McAdams, S., & Bigand, E. (Eds.). (1993). *Thinking in sound: The cognitive psychology of human perception*. Oxford, UK: Clarendon Press.

[49] Meyer-Bisch, C. (1996). Epidemiological evaluation of hearing damage related to strongly amplified music (personal cassette players, discotheques, rock concerts)—High-definition audiometric survey on 1364 subjects. *Audiology, 35*, 121-142.

[50] Miner, N. E. (1998). Creating wavelet-based models for real-time synthesis of perceptually convincing environmental sounds. *Dissertation Abstracts International: Section B: The Sciences and Engineering 59*. (03-B), 1204.

[51] Neuhoff, J. G., Kramer, G., & Wayand, J. (2002). Pitch and loudness interact in auditory displays: Can the data get lost in the map? *Journal of Experimental Psychology: Applied, 8*, 17-25.

[52] Neuhoff, J. G., Knight, R., & Wayand, J. (2002) Pitch change, sonification, and musical expertise: Which way is up? *In Proceedings of the International Conference on Auditory Display (ICAD 2002)*, Kyoto, Japan.

[53] Nielsen, J. (1993). *Usability engineering*. San Diego, CA: Academic Press.

[54] O'Donnell, P. J., Scobie, G., & Baxter, I. (1991). The use of focus groups as an evaluation technique in HCI. In D. Diaper and N. Hammond (Eds.) *People and computers VI*, (pp. 211-224), Cambridge: Cambridge University Press.

[55] Oppenheim, A. N. (1992). *Questionnaire design, interviewing, and attitude measurement*. New York, NY: Pintner.

[56] Oren, M., Harding, C., & Bonebright, T. L. (2008). Design and usability testing of an audio platform game for players with visual impairments. *Journal of Visual Impairment and Blindness, 102,* 761-773.

[57] Pedhazur, J. (1997). *Multiple regression in behavioral research* (3rd ed.). Fort Worth, TX: Harcourt Brace.

[58] Peretz, I., Ayotte, J., Zatorre, R. J., Mehler, J., Ahad, P., Penhune, V. B., et al. (2002). Congenital amusia: A disorder of fine-grained pitch discrimination. *Neuron, 33*, 185-191.

[59] Peretz, I. & Hyde, K. L. (2003). What is specific to music processing? Insights from congenital amusia. *Trends in Cognitive Sciences, 7*, 362-367.

[60] Roebuck, J. A., Kroemer, K. H. E., and Thomson, W. G. (1975). *Engineering anthropometry methods*. New York, NY: Wiley.

[61] Salvendy, G. S. (1997). *Handbook of human factors and ergonomics* (2nd ed.). New York, NY: Wiley.

[62] Sanders, M. S., & McCormick, E. J. (1993). *Human factors in engineering and design* (7th ed.). New York, NY: McGraw-Hill.

[63] Sarkamo, T., Tervaniemi, M., Soinila, S., Autti, T. Silvennoinen, H. M., Laine, M., et al. (2009). Cognitive deficits associated with acquired amusia after stroke: A neuropsychological follow-up study, *Neuropsychologia, 47*, 2642-2651.

[64] Schiffman S. S., Reynolds M. L., & Young F. W. (1981). *Introduction to multidimensional scaling: Theory, methods and applications*. New York, NY: Academic Press.

[65] Schneiderman B. (1998). *Designing the user interface: Strategies for effective human computer interaction* (3rd ed.). Reading, MA: Addison-Wesley.

[66] Schueller, M., Bond, Z. S., Fucci, D., Gunderson, F. & Vaz, P. (2004). Possible influence of linguistic musical background on perceptual pitch-matching tasks: A pilot study. *Perceptual and Motor Skills, 99*, 421-428.

[67] Tabachnick, B. G., & Fidell, L. S. (2006). *Using multivariate statistics*. (5th ed.). Boston, MA: Allyn & Bacon.

[68] Thomas. L. (1997). A review of statistical power analysis software. *Bulletin of the Ecological Society of America, 78*, 126-139.

[69] Turnage, K. D., Bonebright, T. L., Buhman, D. C,, & Flowers, J. H. (1996). The effects of task demands on the equivalence of visual and auditory representations of periodic numerical data. *Behavior Research Methods, Instruments, & Computers, 28*, 270-274.

[70] van Zuijen T. L., Sussman, E., Winkler, I., Naatanen, R. & Tervaniemi, M. (2005). Auditory organization of sound sequences by a temporal or numerical regularity—A mismatch negativity study comparing musicians and non-musicians. *Cognitive Brain Research, 23*, 240-276.

[71] Valenzuela, M. L., Sansalone, M. J., Krumhansl, C. L., & Streett, W. B. (1997). Use of sound for the interpretation of impact echo signals. *In Proceedings of the International Conference on Auditory Displays (ICAD 1997),* Palo Alto, CA.

[72] Virzi, R. A. (1992). Refining the test phase of usability evaluation: How many subjects is enough? *Human Factors, 34*, 457-468.

[73] Wagenmakers, E., Lodewyckx, T., Kuriyal, H., & Grasman, R. (2010). Bayesian hypothesis testing for psychologists: A tutorial on the Savage-Dickey method. *Cognitive Psychology, 60*, 158-189.

[74] Wegman, E. J. (2003). Visual data mining. *Statistics in Medicine, 22*, 1383-1397.

[75] Winsberg, S., & De Soete, G. (1993). A latent class approach to fitting the weighted Euclidean model, CLASCAL. *Psychometrika, 58*, 315-330.

[76] Young, F. W. (1984). Scaling. *Annual review of psychology*, 35, 55-81.

[77] Young, F. W., & Hamer, R. M. (1987). *Multidimensional scaling: History, theory, and applications*. Mahwah, NJ: Lawrence Erlbaum Associates.

[78] Young, F. W., & Lewyckyj R. (1979). *ALSCAL-4: Users guide.* (2nd ed.). Chapel Hill, NC: Data Analysis and Theory Associates.

Chapter 7

Sonification Design and Aesthetics

Stephen Barrass and Paul Vickers

> We can forgive a man for making a useful thing as long as he does not admire it. The only excuse for making a useless thing is that one admires it intensely. All art is quite useless.
>
> — OSCAR WILDE, THE PICTURE OF DORIAN GRAY, 1890 [106]

> Form follows function. Form doesn't follow data. Data is incongruent by nature. Form follows a purpose, and in the case of Information Visualization, Form follows Revelation.
>
> — MANUEL LIMA, INFORMATION VISUALIZATION MANIFESTO, 2009 [58]

> The craft of composition is important to auditory display design. For example, a composer's skills can contribute to making auditory displays more pleasant and sonically integrated and so contribute significantly to the acceptance of such displays. There are clear parallels between the composer's role in AD and the graphic artist's role in data visualization. Improved aesthetics will likely reduce display fatigue. Similar conclusions can be reached about the benefits of a composer's skills to making displays more integrated, varied, defined, and less prone to rhythmic or melodic irritants.
>
> — GREGORY KRAMER, AUDITORY DISPLAY, 1994 [49, PP. 52–53]

Even in Beethoven's time the idea that music could be composed from extra-musical sources was not new; the Greeks composed with geometric ratios, and Mozart threw dice. In the 1930s Joseph Schillinger [85] proposed a "scientification" of music through a mathematical system that has been described as "a sort of computer music before the computer" [30]. With the invention of electroacoustic technologies Iannis Xenakis composed music from statistics and stochastic processes [109] (sound example **S7.1**). Computer music today is composed from fractal equations, cellular automata, neural networks, expert systems and other systematic rule-based systems, algorithms, and simulations [76]. Music is also composed from DNA sequences, financial indexes, internet traffic, Flickr images, Facebook connections, Twitter messages and just about anything in digital form. Generally, the composer is concerned with the musical experience, rather than the revelation of compositional materials. However, when the data or algorithm is made explicit it raises the question of whether some aspect of the phenomenon can be understood by listening to the piece. When the intention of the composer shifts to the revelation of the phenomenon, the work crosses into the realm of sonification. Until recently, sonification has mainly been the province of scientists, engineers, and technologists exploring the functional potential of synthetic sounds as a tool for observation and enquiry. These experiments have sometimes been criticized as unpleasant to listen to, and difficult to interpret. In many cases the most enlightening aspect of a sonification was the process of composing it.

This chapter proposes to address the issues of functionality and aesthetics in sonification by advocating a design-oriented approach that integrates scientific and artistic methods and techniques. Design is an iterative practice-based discipline involving cycles of hypothesis testing and critical evaluation that aims for solutions to specific problems in context. The chapter begins with a review of design methods and practice in sonification. The next section argues for a pragmatist information aesthetic that distinguishes sonification from computer music and psychoacoustics. The penultimate section addresses the issue of aesthetic design guidelines and metrics for sonification. The final section argues that the design approach can allow sonification to become a mass medium for the popular understanding and enjoyment of information in a non-verbal sonic form.

7.1 Background

A debate about whether music can have meaning beyond music itself has raged since the 18th century and continues to this day. Formalists argue that music is the most abstract of the arts and cannot represent anything beyond its own world of melody, harmony, dissonance, tension and resolution. Conversely, Beethoven's sixth symphony (the Pastoral) is often cited as an example of program music that has a narrative conveyed by the titles of the five movements and the music itself (sound example **S7.2**). The musical telling of the story about peasants dancing under a tree climaxes with a dynamic orchestral rendering of a thunderstorm that is a precursor to film sound today. Beethoven wrote a note in the margin of the manuscript that reads "*Mehr Ausdruck der Empfindung als Malerei*" ("More expressive of emotions than a painting") which "... marks the beginning of a conception of program music where the music does not merely convey a literary narrative through musical imitation of characteristic acoustic objects (think of Smetana's Moldau, for instance), but instead creates an imaginary drama or represents a poetic idea" [93, p. 287].

In the silent movie era it was common for a pianist to improvise an accompaniment to amplify the emotions and drama happening on the screen. The invention of the optical movie soundtrack allowed Foley recordings such as footsteps to be synchronized with events on the screen. Film sound designers soon began to explore and expand the functions of sound effects to convey off-screen events, cover cuts and scene transitions, signal flashbacks, and direct attention. A sophisticated theory of film sound developed and Chion [24], building upon Pierre Schaeffer's earlier work proposed three modes of listening: causal listening (attending to the source of the sound), semantic listening (attending to the meaning of a sound), and reduced listening (being concerned with the properties of the sound itself) [24] (see also section 18.2).[1]

The invention of magnetic audio tape ushered in a new era of musical innovation with *musique concrète*, cutup, reversal, looping, and many other possibilities. But musicians were not the only ones exploring the new affordances of tape. Seismologists sped up recordings of earth tremors to listen to sub-sonic events, and were able to distinguish earthquakes from underground nuclear tests [39]. The invention of analogue synthesizers allowed sounds to be controlled with knobs and sliders. Patterson synthesized sounds from the instruments in an aircraft cockpit, to study the effects of amplitude, frequency, tempo, and pulse patterns. His guidelines for cockpit warning and alarm sounds included onset and offset times of 30ms or more, limiting on-time to 100ms, patterns with at least five pulses, linking pulse rate with urgency, and limiting the vocabulary of symbolic sounds to seven [67]. Bly [16] studied the perception of multivariate data in sounds by mapping six-dimensional data to six characteristics of a synthesized tone (pitch, volume, duration, waveshape, attack and overtones). Her participants were able to classify data from the sonification as well as they could from a visual representation. The pioneering researchers in this area were brought together in 1992 by Gregory Kramer who founded the International Conference for Auditory Display (ICAD).[2] In the introduction to the proceedings of that meeting Albert Bregman outlined a near-future scenario in which an executive in a shoe company listens to sales data to hear trends over the past twelve months. Interestingly, this scenario remains futuristic, though not for technological reasons.[3] The participants at that first meeting introduced most of the sonification techniques that are current today, including audification [39], beacons [49], musical structure [62], gestalt stream-based heuristics [107], multivariate granular synthesis [89], and parameter mapping [81]. Scaletti [81] provided a "working definition of sonification" as "a mapping of numerically represented relations in some domain under study to relations in an acoustic domain for the purpose of interpreting, understanding, or communicating relations in the domain under study" [p. 224]. She classified sonification mappings by level of directness: level 0) audification, level 1) parameter mapping, and level 2) a mapping from one parameter to one or more other parameters. Kramer [49] arranged various techniques on a semiotic spectrum from analogic to symbolic, with audification at the analogic end, and parameter mapping in the middle. However, the suggestion that audification is more direct and intuitive is complicated by Hayward's observation that although seismic data could be readily understood, stock prices sounded like opaque noise because they are not constrained by the laws of physics [39]. Kramer [50] also observed that different auditory variables have different perceptual weightings, and suggested that psychoacoustic scaling could balance the perception of multivariate sonifications. Bly [17] presented an experiment in which three sonifications produced different understandings of the same data structure.

[1] It was Chion who enumerated the three modes of listening described above, but the name for reduced listening was first given by Schaeffer [82] in his treatment of the *quatre écoutes*.

[2] See `http://www.icad.org`.

[3] It is interesting to note that Bregman's scenario was already anticipated in fiction. In Douglas Adams' 1988 novel "Dirk Gently's Holistic Detective Agency" [1] the leading character made himself wealthy by devising a spreadsheet program that allowed company accounts to be represented musically [97].

7.2 Design

The field of sonification was consolidated by the attendance of more than a hundred researchers at the second ICAD in Santa Fe in 1994. The first session on Perceptual Issues followed on from the observations of the need for psychoacoustic underpinnings at the first ICAD. Watson and Kidd [104] discussed the cognitive aspects of auditory processing that included type of task, length of training, and complexity, and suggested that auditory science could provide principles for the design of effective displays. Smith et al. [88] psychometrically evaluated the perception of data structure from a granular synthesis technique. Barrass [6] described a psychometric scaling of sonification sequences modeled on the scaling of color sequences in scientific visualization. Other sessions were divided into topics of Spatialization, Systems Issues, Sound Generation, and Sonification and Speech Interfaces. The third ICAD at Xerox Parc in 1996 introduced a session on Design Issues in Auditory Displays in which Tkaczevski [95] presented an overview of aesthetic, technical, and musical issues in commercial sound design, and Back [3] introduced a sound design theory of micro-narratives. Walker and Kramer [102] presented an experiment in which participants interpreted an increase in the frequency of a tone as an increase in temperature, but as a decrease in size. In a further study, sighted participants interpreted an increase in frequency as representing more money, whilst visually impaired participants interpreted it as less [103]. They suggested this may have been because people in the visually impaired group were using a physical metaphor to interpret the sounds. Barrass [7] presented the TaDa task-oriented approach for designing sonifications described by the diagram in Figure 7.1. in which the upper facets of Task analysis and Data characterization specify the information requirements of the design brief, which is then realized by a perceptual representation and device rendering shown by the lower facets.

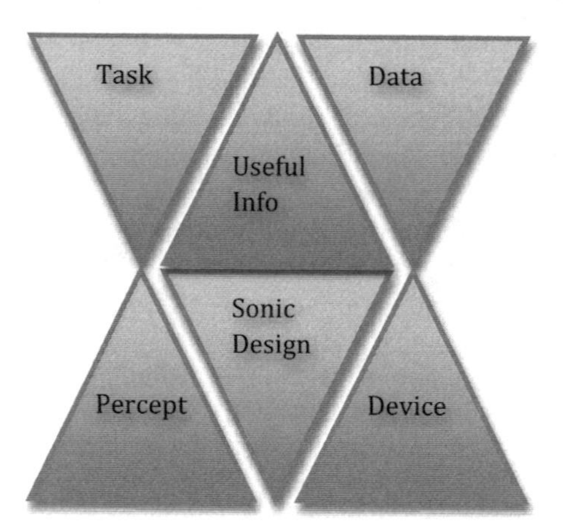

Figure 7.1: TaDa Sonification Design Process, reproduced from the original in Barrass [7]

The TaDa design approach reworks the definition of sonification to focus on functionality rather than representation: "Sonification is the design of sounds to support an information processing activity" [7]. This design-oriented definition augments the National Science Foundation White Paper on Sonification in which sonification is defined thus:

> ... the use of nonspeech audio to convey information. More specifically, sonification is the transformation of data relations into perceived relations in an acoustic signal for the purposes of facilitating communication or interpretation [51].

Another approach takes a view that exploration of a data set with sonification should be a more open-ended process. de Campo's Data Sonification Design Space Map (DSDSM) begins with the characterization of the data set in terms of size and dimensionality [28]. Techniques labeled audification, parameter mapping, filtering, textures, grain clouds, and model-based are shown as regions between the x-axis (number of data points) and y-axis (number of data dimensions). A third z-axis is used to describe the number of perceptual auditory streams produced by each technique. Techniques are grouped in regions labeled Discrete-Point, Continuous and Model-based. The Discrete-Point region contains note-based sonifications and parameter mappings that provide low levels of data transmission. The Continuous region contains Textures and Grain Clouds with higher densities and transitions to Audification. The Model-based region contains techniques that employ a mediating metaphor, such as a simulation of a gas-cloud or crystal growth [42]. The design process is shown by lines in the DSDSM that make implicit knowledge (often expressed as intuitive ad-hoc decisions) explicit and therefore available for reflection, discussion, and learning [28]. These lines depicting process can allow an understanding of the effect of decisions on the perceptual features of the sonification [29].

Methods for designing sounds from theatre, film, ethnographics, computer games, sound art, and architecture have also been introduced in sonification. Somers proposed that theatre sound provides a framework beyond theories of ecological sound and abstract music [91]. Saue [80] proposed a first person point of view for navigating spatial data sets. Macaulay and Crerar [59] employed ethnographic techniques to study auditory displays in an office environment. Cooley [25] took an art theoretic approach to argue that sonification had much to learn from the narrative qualities of computer game sound. In a study of an auditory display of the New York subway system, Rubin [79] concluded that future practice should include information design, sound design, and music as equal partners alongside the more traditional psychological methods. Design patterns, first developed in architecture [2] and applied more broadly in software engineering, were introduced to sonification by Barrass [9] and evaluated by Frauenberger in a study with a context space containing links to artifacts, examples, and problems [34].

7.2.1 Aesthetic awareness

In their call for art submissions for the ICAD conference in Japan in 2002 Rodney Berry and Noatoshi Osaka identified the need for more consideration of the aesthetic aspects of sonification highlighting the important role of aesthetic practice in the process of meaning-making that is sonification:

> In this year's ICAD we have included an art section in the hope that future ICADs might continue to explore some of the arguably less utilitarian aesthetic implications of auditory display. Due to budget and space restrictions, we could only manage to host one installation work this time. The work presented here is Acoustic Acclimation by Singapore-based artists and composers, Lulu

> Ong and Damien Lock who work together under the name Coscilia. The work itself is not a literal "aesthetically pleasing sonification of data-sets" kind of piece. Rather, Acoustic Acclimation explores the relationship between sound and meaning, together with how they combine to establish a sense of place. It is hoped that exposure to such works in future ICAD events might stimulate attendees' thinking about the crucial mapping stage of auditory display, and the interplay between data, information and meaning that concerns both scientists and artists. [15]

At the same conference Bob Sturm announced the release of a CD of sonifications of ocean buoy spectral data titled *Music from the Ocean* [94] (sound example **S7.3**). The proposal that sonification could be a musical experience was reiterated at ICAD 2003 in Boston where Marty Quinn released a CD of sonifications composed from data about the September 2001 attack on the World Trade Centre titled *For those who died* [72]. Barra et al. [5] explored ways to reduce listening fatigue by composing sonifications with "musical structure that's neutral with respect to the usual and conventional musical themes", inspired by futurist composer Luigi Russolo (1885–1947), Pierre Schaeffer's *musique concrète*, Edgard Varèse's *Poème Electronique* (1958) and John Cage's aleatoric compositions (e.g., *Music of Changes* (1951)) [4, p. 35]. The aesthetic potential of sonification as a medium has been developed by sound artists like Andrea Polli who made extensive use of sonification techniques in a public sound art installation on climate change [69] (sound example **S7.4**). Guillaume Potard's sonification of the Iraq Body Count site also demonstrates that sonification can be a political medium [70] (sound example **S7.5**). The growing attention to aesthetics in sonification was recognized by the introduction of a session on the Aesthetics of Auditory Displays at the ICAD conference in Sydney 2004 [12]. Vickers [98] reviewed long term process monitoring sonifications from an aesthetic perspective and called for sonification design to become more interdisciplinary and Leplâtre and McGregor [55] conducted an experiment in which it was found that the functional and aesthetic properties of auditory displays are not independent and the one impacts on the user's experience of the other.

The potential for sonification as a musical experience was tested by the introduction of a concert of sonifications that was ticketed to the general public and staged at the Sydney Opera House Studio [14]. The call for submissions for the Listening to the Mind Listening concert of EEG data asked for sonifications that were "musically satisfying" whilst also being "data driven" [13] (sound example **S7.6**). In their descriptions many composers wrote of meeting both criteria. Three used a notion of revelation or inherence, with a related idea that the data was musical in itself. One described the goal to be to "find naturally occurring rhythmic and musical structures in the data". Another also invoked Nature: "Nature itself creates the structure and balance upon which aesthetics are based. It stands to reason that data captured from such activity is naturally aesthetic when interpreted properly". At the same time, several identified the need to create or maintain musical "interest" and others noted that they selected or "highlighted" aspects that were more musically satisfying. Three recognized the duality of music and sonification as constraining, or even inherently conflicting. One wrote: "It is not to be expected that a sonification produced in a deterministic manner from the data will have any of the normal characteristics of a piece of music". Some contributors emphasized information and perception rather than music, and only a small subset used both musical and perceptual discourses. Several identified with non-music sound practices, using terms such as audio, soundscape, or composition rather than music to describe the results.

A second concert of sonifications was organized by Alberto de Campo for ICAD 2006 in London. "Global Music, The World by Ear" premiered eight sonifications of socio-economic data in an 8 speaker surround system at the Institute of Contemporary Arts [27].[4] The cross-fertilization between sonification and sound art was furthered by the organization of ICAD 2009 in parallel with the Re:New symposium on sound art which included three nights of performances in Copenhagen in 2009. A session on Sound Design Theory and Methods included a review of a workshop on design in sonification that highlighted the fact that knowledge in the field is currently focused on applications and techniques, and there is a need to consider users, environments and contextual issues more in the future [10]. Brazil and Fernström [19] reviewed a cross-section of subjective experience methods that are centered around early conceptual design. (See also Brazil's review of existing sonification design methods and frameworks [18].) Hug [44] presented a participatory design process narrative sound strategies from film and video game production. Fagerlönn and Liljedahl [32] described the AWESOME tool that enables users to become part of the sonification design process. Larsson [52] discussed the EARCONSAMPLER tool that was used in focus groups to help evaluate and improve the sound designs of auditory displays. Sessions on Design and Aesthetics, Philosophy, and Culture of Auditory Displays appeared on the agenda at ICAD 2010 in Washington. Straebel [93] provided a historic grounding that related sonification design to musical movements (especially Romanticism), concepts, and theories. Continuing the theme of participative design, Schertenleib and Barrass [84] introduced Web 2.0 concepts of community of practice, knowledge sharing, and cultural dynamics. Jeon [47] described an iterative sound design process used in industry whilst Vogt and Höldrich [101] discussed a metaphorical method that asked experts to imagine sounds to represent concepts from high energy physics as a basis for sonification design. Following the metaphor theme Fritz [35] proposed a design model based around the intersections of universally (culturally) perceived musical features. Goßman [36] worked from an ontological perspective to discuss the role of the human body as a mediator between external sounds and internal perceptions. Of particular interest here is the assertion that "the contribution of musicians, artists, composers etcetera is not so much in the area of creating aesthetic experiences related to the data, but in the expansion of cognitive models available to the actively exploring listener". The, by now, traditional ICAD concert was organized by Douglas Boyce on the theme "Sonic Discourse – Expression through Sound" with a program that included Spondike's "*Schnappschuss von der Erde*" which premiered at ICAD 2006, and Katharina Rosenberge's "Torsion" that establishes relationships between parabolic spirals found in sunflower heads and spectral analysis of the lowest octave of the piano. Other works emphasized the role of performers in musical performance as embodied techniques.[5]

The effects of aesthetic aspects of sonification have begun to be studied particularly in interactive sports and fitness applications. When a sine-wave sonification of the acceleration of a rowing skiff was played to elite athletes and coaches they commented that the sound was pleasing because it provided useful information that was difficult to see from a video [83]. However, Chris Henkelmann who was involved in a study of sonification on a rowing machine observed that a sine-wave sonification became annoying [40]. He hypothesized that computer music techniques, such as a timbre model and a formant synthesis, could improve the longer term experience. Some of these techniques were included in a study

[4]The full program, together with the audio tracks, may be heard at `http://www.dcs.qmul.ac.uk/research/imc/icad2006/proceedings/concert/index.html`.

[5]The full concert program is available at `http://web.me.com/douglasboyce123/icad/`.

of preferences between six different sonifications of kinetic data that included a sine-wave Sinification (sound example **S7.7**) [sic] pattern, a phase aligned formant Vocal pattern (sound example **S7.8**), a wind Metaphor pattern (sound example **S7.9**), a Musicification pattern using FM instruments (sound example **S7.10**), and a Gestalt stream-based sonification pattern (sound example **S7.11**) [11]. The participants could select between these sonifications on an iPod while involved in an outdoor recreational activity of their choice, such as walking, jogging, martial arts, or yoga. Selections between the sonifications were logged during the activity, and participants were interviewed about the experience afterwards. The interviews discovered a general preference for the Sinification and Musicification patterns and this corresponded with the data logs of time spent with each pattern. The interviews also revealed that the two most preferred patterns were also least preferred by some participants. It might be that recreational users prefer a more conventionally music-like experience whilst competitive athletes prefer more everyday informational sound. These observations show that aesthetics are as important as functionality, and the need to consider the expectations of the users and the context of use when designing a sonification.

The increasing interest in aesthetic dimensions in research studies and the development of sonification as an artistic medium have made it increasingly difficult to distinguish sonification from other practices. Hermann [41] sought to clarify the distinction by recasting the term to plant it firmly in the domain of scientific method by adding four conditions that a work should meet to be considered a sonification:

1. The sound reflects objective properties or relations in the input data.
2. The transformation is systematic. This means that there is a precise definition provided of how the data (and optional interactions) cause the sound to change.
3. The sonification is reproducible: given the same data and identical interactions (or triggers) the resulting sound has to be structurally identical.
4. The system can intentionally be used with different data, and also be used in repetition with the same data. [41]

However computer musicians use the same technologies, tools, and techniques to systematically synthesize sounds from data and algorithmic processes as sonification researchers, and vice-versa. The further statement that the "distinction between data and information is, as far as the above definition, irrelevant" [41], does not make sonification any more distinct. In this chapter we propose that it is the functional intention, rather than a systematic process, that sets sonification apart from other fields of sonic practice. Sonification is a rendering of data to sound with the purpose of allowing insight into the data and knowledge generation about the system from which the data is gathered. We propose that the defining feature of sonification is a pragmatic information aesthetic that combines the functionality of information design with the aesthetic sensibilities of the sonic arts. Casting sonification as purely scientific runs the risk of further polarizing C. P. Snow's [90] Two Cultures debate.[6]

[6]The Two Cultures is a reference to the existence of two separate cultures with little contact between them — one is based on the humanities and the other on the sciences [97, p. 2] a divide which James [46] described as a "psychotic bifurcation" [p. xiv]. James summarized the situation thus:

> In the modern age it is a basic assumption that music appeals directly to the soul and bypasses the brain altogether, while science operates in just the reverse fashion, confining itself to the realm of pure ratiocination and having no contact at all with the soul. Another way of stating this duality is to marshal on the side of music Oscar Wilde's dictum that 'All art is quite useless,' while postulating

7.2.2 What mapping?

The other motivation behind Hermann's recasting of the definition of sonification was the mapping question. Several definitions of sonification have been proposed over the past twenty years or so. For example, Scaletti [81], who gave one of the earliest, saw sonification as having two parts, one to do with information requirements and the other with information representations [8]. Scaletti provided the following definition:

> ... a mapping of numerically represented relations in some domain under study to relations in an acoustic domain for the purpose of interpreting, understanding, or communicating relations in the domain under study. [81, p. 224]

Barrass reconsidered Scaletti's definition of sonification from a design perspective by substituting the concept of 'usefulness' in place of 'interpretation' [8]. The resulting design-centric definition that sonification is the use of nonverbal sounds to convey useful information embraces both functionality and aesthetics, while sidestepping the thorny issues of veridical interpretation and objective communication. Usefulness allows for multiple sonifications of the same data for different purposes, and provides a basis for evaluation, iterative development, and theory building. This idea was taken up in the NSF Sonification Report of 1999, along with a fallback to a more succinct version of Scaletti's definition of sonification to give the current generally accepted definition:

> Sonification is the use of nonspeech audio to convey information. More specifically, sonification is the transformation of data relations into perceived relations in an acoustic signal for the purposes of facilitating communication or interpretation. [51]

Whilst this definition works very well for describing parameter mapping sonifications (where data drive (or 'play') the sound generating hardware; see Chapter 15) Hermann argued that it did not allow for model-based sonification (see Chapter 16) or other techniques not yet developed. In model-based sonification the data itself becomes the sound model and interaction tools are provided to allow the user to excite the model and thus generate sound which is thus itself a representation of the data. In this type of sonification it is the user that plays the data. Hermann states that model-based sonification allows us to "explore data by using sound in a way that is very different from a mapping" and that "structural information is holistically encoded into the sound signal, and is no longer a mere mapping of data to sound". However, even though the rendering of the data into sound takes a different form, there is still a mapping. Any time something is represented in a form external to itself, a mapping takes place; an object from a source domain is mapped to a corresponding object in the co-domain (or target domain). Sometimes the mappings are very obvious and transparent, as in parameter-mapped sonifications, but even model-based sonification involves mappings in this general sense as there are still transformation rules that determine how the data set and the interactions combine to produce sound which represents some state of the system. The mappings may not be simple, but mapping is still taking place.

Recognizing that all perceptualization (e.g., visualization and sonification) involves mapping in some form admits any possible number of mapping strategies whilst retaining the more

---

that science is the apotheosis of earthly usefulness, having no connection with anything that is not tangibly of this world. [p. xiii]

catholic definitions of Scaletti [81], Barrass [8], and the NSF report [51] and recognizing the potentially interdisciplinary nature of sonification design. Whilst sonification is undoubtedly used *within* scientific method the *design* of sonifications themselves must, we argue, remain an interdisciplinary endeavor as the Auditory Display community originally envisaged.

7.3 Aesthetics: sensuous perception

If sonification allows for (or even requires) interdisciplinary contributions we must consider the question of the role of artistic practice and wider aesthetic issues in sonification design. Sonification is a visualization activity in which sound is used to convey information about data sets.[7] Perhaps because of the novelty value in the early days of being able to make data go 'ping', many sonifications (including recent ones) have been created that are not particularly useful, useable, or meaningful. In the graphical visualization community a debate has been taking place in recent years over the role of function and its relationship with data art. Lima [58] set out the case against data-art-as-visualization thus:

> The criticism is slightly different from person to person, but it usually goes along these lines: "It's just visualization for the sake of visualization", "It's just eye-candy", "They all look the same".

It is instructive to consider the existing relationships between graphical visualization and art as the sonification field is experiencing similar tensions. The overall purpose of visualization is to shed light on the data being represented in order to allow meaning to be inferred. Information is data that has been given meaning and so without the meaning it remains only data. The process of meaning making can, of course, take place without the agency of a representation (we could begin examining the raw data looking for patterns) but sonification and visualization are concerned with the creation of representations of data that facilitate inference and meaning making. Often the forms of the representations are derived from the form of the underlying data [58] (indeed, de Campo's Data Sonification Design Space Map [28] was specifically devised to enable sonifications to be constructed in which hidden structures and patterns in the data emerge as perceptible sonic entities) but a foundational premise of design practice is that that form should follow function. Consider, for example, a beautiful piece like Radiohead's "House of Cards" video [73]. In Lima's view it ought not strictly to be considered information visualization as it provides no insight, it is pure spectacle. The value of the piece lies solely in its artistic properties as it does not fulfill the criterion of usefulness that visualizations must, it is argued, possess. We could marshall to Lima's side Redström who identifies a basic issue in interaction design aesthetics which is the question "of how through a certain design we aim to make a computational thing express what it can do through the way it presents itself to us through use over time" [75, p. 1]. Because the "purpose of visualization is insight, not pictures" [21, p. 6] so Redström puts the focus of aesthetics onto "expressions and expressiveness of things" [75, p. 2] and leads us to look at how material builds form through the logic underpinning those expressions. For example, on the subject of tangible interfaces Redström says:

[7] The classical definition of visualization is "the process of forming a mental image of some scene as described" [71, p. 320]. So, by visualization we mean the process by which mental images and cognitions (what we call *visualizations* are formed from the reading of external representations of data. Those representations may be visual, auditory, or even haptic. Data sets can be repositories of data, such as files, tables, etc. or real-time streams of data events such as would occur in a process monitoring application.

> ...it is not the fact that they are tangible that is the most crucial part of tangible user interfaces considered to comprise an interface design strategy, but how they aim to deal with the relation between appearance and functionality. [75, p. 15]

Wright et al. [108] suggest aesthetic experience should lie at the heart of how we think about human-computer interaction. This aesthetic-oriented view, they say, takes us beyond studying the way people interact with the technology we have designed and ends up influencing the way we design and build that technology.

7.3.1 Two *aesthetic turns*

Lima is in good and well-established company. William Morris [64] adjured us to have "nothing in your houses that you do not know to be useful, or believe to be beautiful". When Oscar Wilde proclaimed that all art is quite useless [106] this was not a dismissal of art as an irrelevance but an assertion that the utility of art lies not in terms of work to which it can be put but to its intrinsic aesthetic qualities and value; art *is*, tools *do* — this looks remarkably like another expression of the Two Cultures divide. And yet, product designers increasingly try to make tools that are also beautiful. This view would see the danger for visualization design being when the drive to instill beauty takes gets in the way of utility. Lima [58] argues strongly that "simply conveying data in a visual form, without shedding light on the portrayed subject, or even making it more complex, can only be considered a failure". If what we are building is neither very beautiful nor very useful, then we have, it would seem, failed altogether. What place, then, should aesthetics have in the work of sonification designers?

Aesthetics is commonly understood today to be the "philosophical study of art and the values, concepts, and kinds of experience associated with art such as taste, beauty, and the sublime" [45]. The word aesthetics stems from a broader Greek root having to do with perception and sense and, prior to the mid-eighteenth century aesthetics was a branch of philosophy concerned with perception by the senses.[8] Indeed, the word *anaesthetic* literally means the removal of feeling. Synaesthesia (same root) is the bringing together of the senses in perception (e.g. color-hearing). In the mid-eighteenth century a move began amongst German philosophers to consider these issues of taste, beauty, and the sublime. In 1750 Baumgarten defined aesthetics in terms of judging through or by sense. Through the work of Baumgarten's successors, Kant, Schiller, Schelling, and Hegel, by the end of the nineteenth century an *aesthetic turn* had taken place giving rise to our modern understanding of aesthetics which, according to Nake and Grabowski [65, p. 54], has beauty as a major focus.

The first turn

Rose-Coutre defines art as "purely and simply an aesthetic object that appeals to the senses in a certain way" [78, p. 5]. In Kantian philosophy, although the central questions are concerned with how we are able to make judgments of beauty, aesthetics occupies the realm of sensibility and aesthetic experience is "inexplicable without both an intuitive and a conceptual dimension" [20]. For Kant, perception and understanding are intertwined, even inseparable. Hegel, building upon Kant's work, defined art as a sensuous presentation of ideas, something that communicates concepts through our senses and our reason [26]. In Hegel's world, and somewhat in opposition to Wilde, art for art's sake is anathema; for him

[8] The etymological root of aesthetics is the Greek word αἰσθάνομαι meaning "I perceive, feel, sense" [38].

art was for beauty's sake as a sensuous (aesthetic) form of expressing truth; art's task "is the presentation of beauty and that beauty is a matter of content as well as form" [43].

The second turn

In recent years a second aesthetic turn has taken place in the fields of data visualization, data aesthetics, and Creative Commons. In the past five years or so there has been a popular uptake of computational tools, technologies, and processes that were previously only available to specialists, scientists, and engineers in centralized institutional labs such as those at NCSA, Nasa, CSIRO, etc. The development of open source or free access platforms such as Processing[9] and Many Eyes[10] has led to a much broader conceptualization and application of visualization in artistic media, advertising, DIY online culture, and communities that have a wide range of different goals, languages, and evaluative dimensions (e.g., affect, social significance, narrative, production quality, etc.) that are often grouped together under the umbrella term "aesthetics". The different sensibilities of the new designers and audiences in this "second wave" has led to a reassessment of visualization and debates about the differing principles used by first and second wave practitioners. For example, Lima's manifesto [58] is a clear example of the first wave in which functionality is of prime importance. Lima went as far as to describe himself as "a functionalist troubled by aesthetics."[11] For the first wave the inherent danger in visualization is summed up well by Carroll [22]:

> To some extent however this elegance, which makes data visualisation so immediately compelling, also represents a challenge. It's possible that the translation of data, networks and relationships into visual beauty becomes an end in itself and the field becomes a category of fine art. No harm in that perhaps. But as a strategist one wants not just to see data, but to hear its story. And it can seem that for some visualisations the aesthetic overpowers the story.[12]

"Second wavers", such as Vande Moere, on the other hand, have embraced aesthetics as a tool for visualization work and talk of "information aesthetics", "information aesthetic visualization", and "artistic data visualization" [96, 53]. For them, the second aesthetic turn provides the link between information visualization and data art and requires interdisciplinary practice. Very much in tune with Hegel and the first aesthetic turn, Lau and Vande Moere say that information aesthetics "adopts more interpretive mapping techniques to augment information visualization with extrinsic meaning, or considers functional aspects in visualization art to more effectively convey meanings underlying datasets" [53]. As an example of such interdisciplinary work in practice consider Keefe et al. [48] who described two interdisciplinary visualization projects in which computer scientists and artists worked together to build good representations. They propose a spectrum of representation (see Figure 7.2) at the left end of which lie those visualizations that we would normally label information art with more traditional information visualizations residing at the right hand end. The purpose of this spectrum is not to divide and categorize to help keep art and science and engineering apart but to show that both ends (and all points in between) are valid and meaningful expressions, and that the artist and the researcher should collaborate to develop new techniques and representations.

Figure 7.2 shows that systems with a tight connection to underlying data are highly indexical.

[9] `http://www.processing.org`
[10] `http://www.many-eyes.com`
[11] See Justin McMurrary's blog of 3 September, 2009 at madebymany.com: `http://tinyurl.com/5uqlwg6`.
[12] Jim Carroll made this statement in response to a talk by Manuel Lima at BBH Labs in 2009.

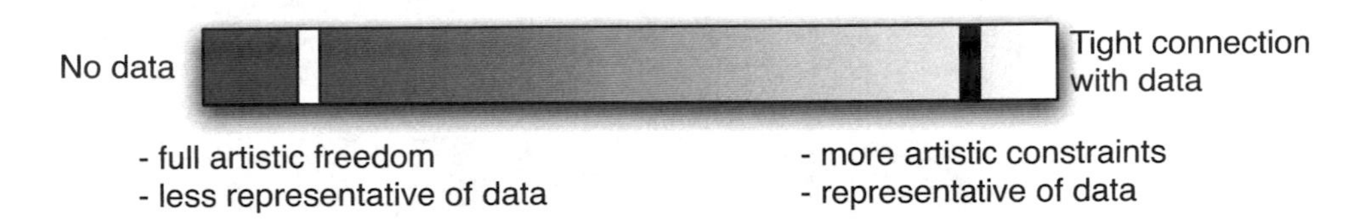

Figure 7.2: Indexicality in visualization (adapted from Keefe et al. [48]). The black and white bars indicate visualization tools operating at different ends of the representational continuum. The white bar is a system that is informed by underlying data but in which artistic freedom is the main driver. The black bar would be the position of a system in which artistic expression is much more tightly constrained with the focus being clear representation of a data set.

Vickers and Hogg [100] introduced to sonification discourse the concept of *indexicality*.[13] Something (a gesture, an utterance, a sign, etc.) that is indexical points to (indicates) some other thing that is external (an entity, an idea, etc.). In sonification practice indexicality becomes a measure of the arbitrariness of a mapping (in semiotic terms an indexical signifier is non-arbitrary and has a direct connection (physically or causally) to that which it is signifying [23]). In sonification it is the data that makes the sound (parameter-based sonification) or user interactions with the data that make the sound (model-based sonification). A sonification system exhibiting high indexicality is one in which the sound is derived directly from the data (for example, through the use of direct data-to-sound mappings). Low indexicality arises from more symbolic or interpretative mappings.

Keefe et al. [48] discovered that getting artists and visual designers to help with a visualization project at the design level from the outset is key and bears much more fruit than using them for "turning the knobs of existing visualization techniques" [p. 23]. Artists, they say, routinely "provide a unique source of visual insight and creativity for tackling difficult visual problems"; they do more than "merely making a picture pretty or clear for publication". For Keefe et al. the integration of function and aesthetics is a desirable challenge. It is the artist working within the tight constraints of programmatic data mappings and the computer scientist facing the issues of visual design that creates the opportunity for them to work together "to design novel visual techniques for exploring data and retesting hypotheses". For an example of this at work in sonification design, see Stallman et al. [92] who used a composer to help in the design of an auditory display for an automated telephone queue management application.

7.3.2 Aesthetics as a guide

Aesthetics or, specifically, aesthetic perception then, is a framework we use for making judgments about artistic works. Thanks to the aesthetic turns, when the word *aesthetic* is used in the same breath as sonification or auditory display it is often automatically assumed that one is talking about artistic qualities or properties. Just like the first-wavers would claim for visualization, sonification, the argument goes, belongs to science and engineering

[13]Indexicality is a concept from philosophy which is often used interchangeably with the linguistics term *deixis* and is also used in semiotic explanations of *sign*.

and we should not be discussing it as if it were art. The problem here though is that this is something of a false dichotomy predicated upon the assumption that art and science are somehow incompatible bedfellows. The issue here is that aesthetics is not synonymous with art. Aesthetics is about more than art, at its core it is about sensuous perception— we make aesthetic judgments every day about the products we buy (or don't buy), the clothes we wear, and the tools we use.

In recent times, as computer graphical user interfaces and interactive systems have become functionally richer and more impacted by graphic design, we are increasingly employing our aesthetic sense-making faculties in our engagement with them. Although aesthetics clearly plays a role in how we respond to the visual presentation of an interactive system, Graves Petersen et al. [37, p. 269] claim that it is a mistake to assume that aesthetics is restricted to visual impressions. Whitehead's claim that art "is the imposing of a pattern on experience, and our aesthetic enjoyment is recognition of the pattern" [105] suggests that whilst aesthetic judgment is required for enjoyment of art, the fact that patterns are involved means that there is potential for leveraging aesthetics in the design and use of visualization systems whose primary purpose is about gaining insight into data. Nake and Grabowski [65, p. 62] go as far as to say that because aesthetics is concerned with sensuous perception, questions of beauty are secondary. Graves Petersen et al. [37, p. 270] support this view by saying that those "who view the potential of aesthetics as the possibility to provide users with a pleasing visual appearance of products are leaving out much of the potential of aesthetics". They boldly claim that far from being an "added value" or even "an adhesive making things attractive" aesthetics is "an integral part of the understanding of an interactive system and its potential use" [p. 271].

In mathematics aesthetics has long been understood to play a vital role. Mathematicians strive to find simpler ways of describing an object or phenomenon. Sometimes this is for simplicity's sake, other times because the application of the simpler representation to a real-world problem makes the calculation easier or faster. Einstein's guiding principle was to seek mathematical beauty or simplicity. The physicist Paul Dirac took this idea even further in his "Principle of Mathematical Beauty". For Dirac, the more theories revealed about nature the more beautiful they would be; ugly theories could not be right. So, for mathematicians, truth and beauty are intertwined: beauty reveals truth and the truth is beautiful. But the point is not that mathematicians are seeking beauty for its own sake, but that the simple, that is, the beautiful, brings understanding more readily. To give a very practical example, metrics for aesthetics in graph drawing include the number of edge crossings (the fewer the better) and the amount of symmetry exhibited by the graph (the greater the better) [53]; both of these measures are associated with a graph's readability. So, aesthetics deals with judgment using the senses, and the easier the representation makes such judgments, the better the representation is. However, we must be careful not to assume that just because something is beautiful it is, therefore, interesting. In a discussion of his work on algorithms for generating low-complexity ('simple') art, Schmidhuber [86] says:

> Interest has to do with the unexpected. But not everything that is unexpected is interesting — just think of white noise. One reason for the interestingness (for some observers) of some of the pictures shown here may be that they exhibit unexpected structure. Certain aspects of these pictures are not only unexpected (for a typical observer), but unexpected in a regular, non-random way. [p. 102]

Just as Keefe et al. [48] recognized for visualization design, there is a tension in the design of auditory representations that requires aesthetic and artistic expression constrained by computational issues of data mapping. With regard to sonification design, Vickers [99] asserted:

> The larger questions of sonification design are concerned with issues of intrusiveness, distraction, listener fatigue, annoyance, display resolution and precision, comprehensibility of the sonification, and, perhaps *binding all these together, sonification aesthetics*. [emphasis added] [p. 57]

Indeed, Pedersen and Sokola [68] cited an impoverished aesthetic as being partly responsible for people growing quickly tired of the sonifications used in their Aroma system [99]. Kramer [49] was particularly frank:

> Gaver relates that SonicFinder was frequently disabled, Mynatt reports that poorly designed sounds degraded Mercator, and Kramer considers some of his sonification experiments downright ugly. [p. 52]

7.3.3 A pragmatist solution

If we can accept that aesthetics is not *only* about the art, when we consider sonification (and visualization more generally) we might go as far as saying that aesthetics isn't about the art at all.[14] By that we mean that thinking of aesthetics as being the framework for making decisions about artistic value and taste is unhelpful in this context because it limits what we can do and even diverts our thinking, thereby distracting us from considering what aesthetics can be used for: the design of effective sonifications that promote sense-making, understanding, and pattern recognition. Far from being the pinnacle of artistic expression, in sonification good aesthetic practice helps us to achieve ease of use which Manovich [61] describes as "anti-sublime". Being products of the first aesthetic turn the Romantics, Manovich points out, were concerned in their art with the sublime, with those phenomena and effects that go "beyond the limits of human senses and reason". Therefore, visualization systems are necessarily anti-sublime for their aim is to make representable the data sets underlying them.

The question, then, becomes how may aesthetics be applied or leveraged in the design of sonifications? For the mathematician aesthetics "involves concepts such as invariance, symmetry, parsimony, proportion, and harmony" [33, p. 9] and mathematics can be interrogated in the light of these factors. In physics aesthetics "is often linked to the use of symmetries to represent past generative states" [56, p. 307]. In sonification design we are presented with many of the same challenges that face designers of interactive computer systems who are trying to ensure a positive user experience. The problem is that one cannot design *a* user experience one can only design *for* user experience [87, p. 15]. In aesthetic terms this is the difference between analytic and pragmatist aesthetics. In Moore's [63] analytic view aesthetics exist as objects in their own right and are intuitively apprehended by a viewer [37]. In this paradigm the aesthetic properties arise when the artist or designer creates an artifact and they await being found by the viewer/user with the resultant implication that they

[14] Aesthetics is not about art any more than a painting is about the technology and chemistry of pigment design and manufacture, except that they are interdependent. Without the technology there is no art; without aesthetic input there is no meaningful or usable visualization.

have some objective reality. This parallels the view that a software designer can intend for a product to have a particular universally shared user experience. What the analytic view does not take into account are the socio-cultural factors that affect how an artifact is perceived [37], or *experienced* to use Dewey's [31] terminology (see also Macdonald [60]). Graves Petersen et al. [37] observe:

> Dewey insists that art and the aesthetic cannot be understood without full appreciation of their socio-historical dimensions ... that art is not an abstract, autonomously aesthetic notion, but something materially rooted in the real world and significantly structured by its socio economic and political factors. [p. 271]

Dewey's *pragmatist* stance recognizes that aesthetic experiences are the result of "the engagement of the whole embodied person in a situation" [108, p. 4]. This pragmatist aesthetics perspective reconciles us to the assertion that user experience may only be designed *for*, that we must do all we can to maximize the opportunities for meaningful dialogue with our sonifications, but recognizing that the experience will not be universal. Sonification engages the user in a sense-making process and as designers we need to remember that the user's interaction with the system "is based on not just the immediate sensational, but it builds upon earlier experiences as well as it draws upon the socio-cultural" [37, p. 272]. As Sharp et al. put it, "one cannot design a sensual experience, but only create the design features that can evoke it" [87, p. 15]. Wright et al. [108] suggest that because we cannot build the aesthetic experience (nor, in fact, significantly control the user's experience) our job as designers is to "provide resources through which users structure their experiences" [pp. 9–10].

In the pragmatist paradigm aesthetics is a kind of experience emerging from the interactions between the user and the context (including cultural and historical factors), and it is located in neither the artifact nor the viewer exclusively [108]. This pragmatist aesthetics takes into account that interaction is constructed as much by the user as by the designer and that the sense-making process involves not just cognitive skills but also "the sensual and emotional threads of experience situated in time and place" [108, p. 18]. In Kant's aesthetic worldview, the beauty of an object does not exist in the object itself but is a property that emerges as we respond to the object. For Kant, beauty was linked irrevocably to an object's form.[15]

In an application of pragmatist aesthetics to interaction design, Wright et al. [108] argued a need to place "felt life and human experience at the center of our theorizing and analysis". They observed:

> But putting aesthetic experience at the center of our theorizing about human-computer interaction is not just about how we analyze and evaluate people's interaction with technology; it affects the way we approach the design and making of digital artifacts. Our ... work, which has brought together software developers, electronics engineers, and contemporary jewelers, has provided a fertile ground for reflection on the process of interaction design and the way digital artifacts are framed within traditional HCI practice. [pp. 18–19]

They conclude that "if the key to good usability engineering is evaluation, then the key to

[15]For Kant, beauty was universal (or rather that which one would perceive as beautiful one would assume is a universal response even though it might not be in reality) but the perception of beauty is arrived at through a disinterested inspection. By that Kant means that the judgment is made independent of any interest we might have in the object, independent of its content, its moral or financial value, etc. The judgment of beauty is made only in terms of the object's form (its shape, its composition, its texture, etc.).

good aesthetic interaction design is understanding how the user makes sense of the artifact and his/her interactions with it at emotional, sensual, and intellectual levels". It becomes unhelpful to think about the aesthetics of artifacts in their own right as aesthetic potential is only realized through interaction or use which is dependent on context [37]. The pragmatist outlook also breaks the close bond between aesthetics and art thus providing "the basis for focusing on the aesthetics of interaction related to our everyday experiential qualities when engaging in and designing interactive systems" [37, p. 271]. The focus now shifts to how an artifact is appropriated into someone's life and how this is shaped by prior expectations, how the user's activities change to accommodate the technology, and they change the technology to assimilate it into their own world. The emphasis is on meaning in use: how the user's talk about technology changes, possibly even how the artifact ceases to become a topic of conversation, is a valuable source of data. One of the implications of this approach is that it takes place in situ and is oriented towards longer-term processes of change. Various forms of interpretive phenomenological analysis (IPA) are proving useful empirical techniques in this regard (for example, see Ní Chonchúir and McCarthy [66] and Light [57]). IPA is a psychologically-based qualitative research method that complements the more sociological grounded theory. Its aim is to gain insights into how a person experiences, or makes sense of, a phenomenon. Typically such phenomena would be of personal significance to the person being studied (e.g., changing job, moving house, starting a relationship, etc.) but IPA has also been used to study less personally-related phenomena such as using interactive computer systems or web sites. For instance, Ní Chonchúir and McCarthy [66] showed how IPA could get very personal insight into the user experience of Internet usage. Light [57] used IPA to study the experience of receiving phone calls to learn more about the issues that should be addressed in the design of mobile telephones. Traditional metric- and task-performance-based techniques have been used to measure sonification design factors such as accuracy, recall, precision, efficiency, etc. Whilst one could measure the improvement on performance of auditory displays that have been designed to maximize their aesthetics, aesthetic judgment itself remains primarily experiential and so we can envisage using qualitative tools like IPA not only to gain more understanding of how users experience sonifications, but to evaluate the aesthetic dimension more richly.

Raijmakers et al. [74] found that using a documentary film format to present personas of typical customers to product designers. They found that the films gave "access to incidental details that might or might not be important for design—the patients' activities, homes, aesthetic tastes, ways of expression, etc.—since these things made the personas "come alive" for them as characters who might use future products." If sonification is to move out of the lab and into the home, to become embedded in mainstream products, it is possible that radical techniques like this will enable us to get more understanding of the target user community.

7.4 Towards an aesthetic of sonification

If we admit the necessity of addressing aesthetic issues in sonification design and recognize that approaches such as pragmatist aesthetics offer useful frameworks for thinking about our aesthetic practice, the question still arises as to what *are* sonification aesthetics? What do they sound like? Are there some specific guidelines that, if codified, will guarantee (or at least offer the chance of) successful aesthetic expression? After all, areas such as graph theory and web design have established aesthetic metrics, sets of rules which, if followed, promise an easy-to-read graph or a usable web site. However, it has been observed that often many codified aesthetics are contradictory and so cannot all be achieved in one piece of work [71]. Furthermore, sonification is not a discrete singular discipline, it occupies space in perceptual psychology, computer science, engineering, sound design, and sonic art drawing to varying extents upon skills in all those fields (and others besides, no doubt). Sonification comes in many different styles using different sonic techniques each of which may have its own set of specific aesthetics. Take the case of musical renderings. If we draw on music

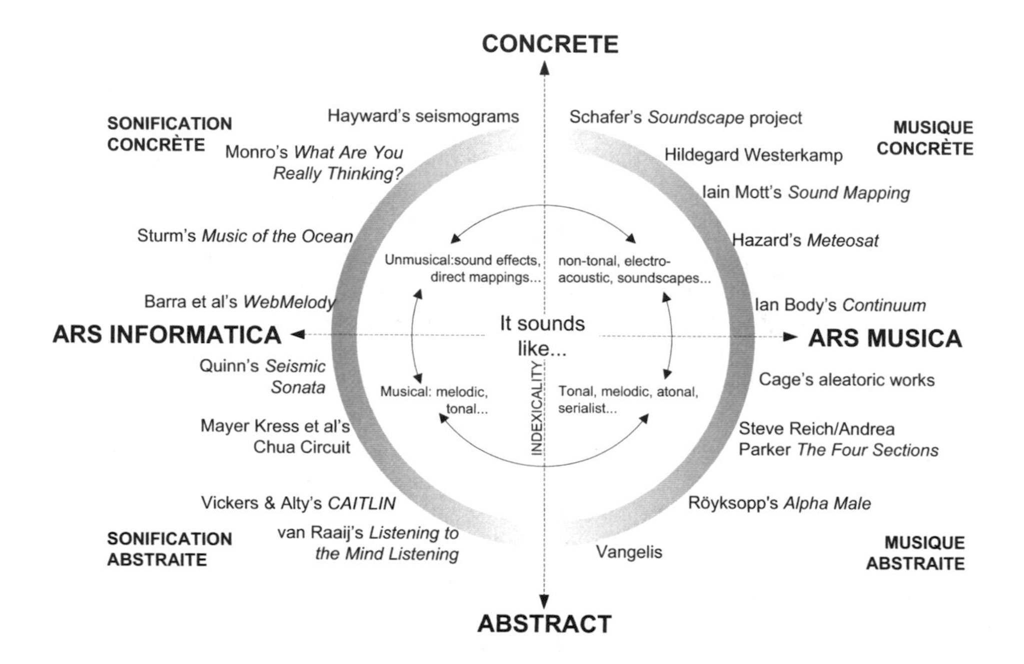

Figure 7.3: The Ars Musica — Ars Informatica Aesthetic Perspective Space (from Vickers and Hogg [100]).

practice for inspiration we see many different genres each with its own aesthetic rules. For example, Vickers and Hogg [100] suggested an aesthetic perspective space (see Figure 7.3) which associates sonifications with their closest analog in the musical world, the idea being that if a sonification is organized, say, along the lines of a piece of tonal music then it could draw upon the aesthetics of tonal musical composition (assuming an appropriate sub genre can be identified); likewise, a sonification that is organized like a piece of *musique concrète* could draw upon electroacoustic aesthetics. But each musical style has its own, quite distinct, aesthetic.

Sound design, arguably the field of sonic practice most closely related to sonification, is filled with practitioners who develop their own personal aesthetic rather than adhering to some definitive 'red book' of sound design. Sonification aesthetics is still in its infancy as far as detailed research goes and so we are not yet at a point where we can offer a definitive set of aesthetic guidelines. It is not even known whether any such set of guidelines is possible or, for that matter, even desirable. In reference to their work on aesthetic interaction design, Wright, Wallace, and McCarthy [108] said "nor does it seem sensible to talk of principles or guidelines for designing enchanting experiences".[16] Lavie and Tractinksy [54] observe that aesthetics is still marginalized within HCI, commenting that "readers of human-computer interaction textbooks can hardly find any reference to aesthetic considerations in design". They did, however, begin studies to discover what factors might make good measures of aesthetic quality in interactive systems (specifically, web sites). What we can offer at this stage, then, are some indications of where the interesting ground may lie and what aspects of auditory information rendering appear worthy of systematic and detailed attention in future research explorations. Addressing the aesthetics of large-scale information visualization systems, Quigley [71] identifies four different problems affecting the visualization of data using large graphs:

1. Graph drawing aesthetics
2. Computation
3. Screen space aesthetics
4. Cognitive load

He further breaks graph drawing aesthetics down into two subcategories, *drawing aesthetics* and *abstract representation aesthetics* both of which contain a number of organizing principles (such as the need to maximize symmetries, the avoidance of overlapping groups of nodes, etc.). These aesthetic principles are fairly tightly defined but relate only to a single visualization task, that of representing large data sets with graphs. Other visualization techniques will have their own aesthetic 'rules'. If we are to move towards such sets of rules for sonification we must first classify the different types of sonification practice. There are simple gross distinctions that can be made, for example between parameter-mapped sonifications and model-based sonifications, but even within these, as this volume attests, there are different representational techniques that can be used each of which is likely to have different aesthetics.

7.4.1 Aesthetic premises and oppositions

However, to see where the sonification aesthetics research focus might be placed, it is possible to offer some general areas which will affect aesthetic practice. In *Microsound* [77] Curtis Roads set out a collection of aesthetic premises and aesthetic oppositions that he found helpful to consider when composing in the granular synthesis domain. Some of the principles dealt with issues related to electronic music generally whilst others were concerned with the specific properties of the microsound domain. Whilst they do not especially inform sonification design practice, the very existence of an aesthetic philosophy for this relatively new area of music composition suggests that an undertaking to formulate an aesthetic philosophy for sonification might be fruitful. Perhaps more relevant to the subject at hand are the aesthetic oppositions which might serve as a basis for beginning the discussion about the aesthetic guidelines to which sonification designs might usefully adhere. Roads' ten oppositions are as follows:

1. Formalism versus intuitionism.
2. Coherence versus invention.

[16]Enchantment is a particular branch of experience which deals with the feeling of being caught up in something.

3. Spontaneity versus reflection.
4. Intervals versus morphologies.
5. Smoothness versus roughness.
6. Attraction versus repulsion in the time domain.
7. Parameter variation versus strategy variation.
8. Simplicity versus complexity.
9. Code versus grammar.
10. Sensation versus communication.

By stating these aesthetic dimensions in terms of opposites requires us to consider what is meant by sounds at either pole. For instance, when would smooth sounds be more suitable than rough sounds, and vice versa? To these oppositions we may add Leplâtre and McGregor's [55] basic aesthetic principles for sonification design: homogeneity of the design, temporal envelope, and sonic density. Leplâtre and McGregor found that "functional and aesthetic properties of auditory cannot be dealt with independently" and so to their and Roads' categories we might add low-level functional measures such as usefulness, usability, pleasantness, functionality, ergonomics, intuitiveness, learnability, and enjoyability (or, perhaps, annoyance). Some of these terms have analogs in the HCI/interaction design fields, though it should be noted that the trend in HCI is away from pure metrics and towards designing for user experience (hence the rise in phenomenological methods). As Roads noted, an aesthetic philosophy "is nothing more than a collection of ideas and preferences that inform the artist's decision-making" [77, p. 326] and so we must be careful not to treat as sacred any list of aesthetic guidelines. Even if aesthetics could be codified, they still require talent and skill to implement them; the talent must be innate and the skill must be taught or otherwise acquired. Any skilled practitioner also needs to know how and when it is appropriate to break the rules.

7.5 Where do we go from here?

To improve the aesthetics of our sonifications, then, we argue that first and foremost the designers of sonifications either need to be skilled in aesthetic thinking and practice or they need to work with someone who possesses such skills. We are beginning to see higher level university courses that embrace art and technology and which educate people to be literate and capable in both, and they show that technologists can learn aesthetic skills just as artists can learn to write code. But such courses are few and require a concerted will to think and work in an interdisciplinary way (which cuts against many university departmental structures and funding models). Until such a time as the majority of sonification designers possess aesthetic design skills we repeat Kramer's initial call for interdisciplinary work. Where Kramer called for the community to work with composers, the net needs to be cast wider to include sound designers and other sonic artists, all the while keeping our eyes on the goal which is to produce auditory representations that give insight into the data or realities they represent to enable inference and meaning making to take place.

In sonification and auditory display, where hard evidence of insights produced by the auditory representations is much less common than in graphical visualization, the integration of function and aesthetics is even more urgent and problematic, especially in the light of the strong affective and cultural aspects of sound that we have through musical education and experiences.

Sonification is becoming embedded in everyday objects and activities. This means that issues of desire, branding, emotion, and narrative will become increasingly important as they already have in graphical visualization. Where graphical visualization draws on graphic design these directions suggest that we can draw on sound design for commercial products (and toys) and film sound in the next era of ubiquitous everyday sonification where sonification becomes a commercial, domestic, consumer, mass medium. Whilst composers are not, of necessity, focused on functionality or accessibility to a broad audience, product designers and film sound designers are.[17] How does one design affective and persuasive sonifications? The question of beauty and its relationship to utility has been raised for both sonifications and graphical visualizations. This is where design thinking and aesthetic practice could help. Figure 7.4 shows that aesthetics (sensuous perception) is the common thread in sonic art and sonification and we contend that the wall between sonic art and sonification has been put up unnecessarily and that treating sonification as a truly interdisciplinary design process offers much scope for informing the work of the auditory display community as it matures and develops.

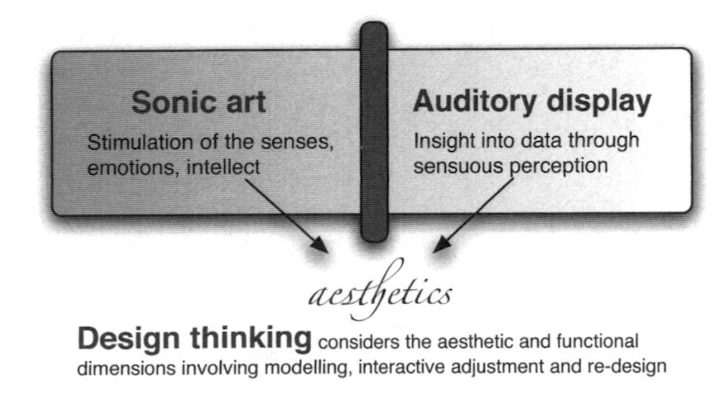

Figure 7.4: The wall between sonic art and sonification/auditory display is a false one. Aesthetics is a framework for working across the spectrum.

Despite the promise of sonification to provide new insights into data there is little to show in the way of scientific discoveries made through sonification in the past twenty years. A definition of sonification focusing on usefulness and enjoyment reconfigures sonification from an instrument solely for scientific enquiry into a mass medium for an audience with expectations of a functional and aesthetically satisfying experience. A design-centered approach also moves sonification on from engineering theories of information transmission to social theories of cultural communication. Developing this theme Schertenleib and Barrass [84] are developing the concept of sonification as a social medium through the Many Ears site for a community of practice in data sonification.[18] This site is modeled on the Many Eyes

[17] Of course, the popular music industry is predicated precisely upon appealing to a broad audience. However, there is nothing about musical composition *per se* that demands this.

[18] `http://www.many-ears.com`

site for shared visualization and discovery that combines facilities of a social networking site with online tools for graphing data.[19] Anyone can upload a data set, describe it, and make it available for others to visualize or download. The ease of use of the tools and the social features on Many Eyes have attracted a broad general audience who have produced unexpected political, recreational, cultural, and spiritual applications that differ markedly from conventional data analysis. The Many Ears project seeks to find out what will happen when data sonification is made more available as a mass medium. What new audiences will listen to sonifications? Who will create sonifications and for whom? What unexpected purposes will sonification be put to? [84]

Kramer's 1994 call (echoed a decade later by Vickers [98]) to include composers in the sonification design process [49] is as relevant today as it was then, and extends to sound artists, sound designers, film sound, and interactive product designers. At this stage it would appear that there is great potential for sonification to become a medium for communicating information about data sets to a broad music-listening audience who also have expectations of an aesthetically satisfying experience. A positive way forward is to adopt an approach that does not polarize art and science along some artificial simplistic dimension. Design thinking requires an approach that accepts that there are multiple constraints and multiple solutions in any problem domain. A good solution is one that addresses the requirements of the brief, which may be involve qualitative and quantitative aspects, and proper attention to the context and the audience. Auditory display is an exciting field at the intersection of future developments in music, design, and science and we look forward to the hearing the progress in these directions.

Bibliography

[1] Douglas Adams. *Dirk Gently's Holistic Detective Agency*. Pan, 1988.

[2] Christopher Alexander. *Timeless Way of Building*. Oxford University Press, 1979.

[3] Maribeth Back. Micro-narratives in sound design: Context, character, and caricature in waveform manipulation. In *ICAD 96 - Third International Conference on Auditory Display*, pages 75–80, Palo Alto, CA, 1996.

[4] Maria Barra, Tania Cillo, Antonio De Santis, Umberto F. Petrillo, Alberto Negro, and Vittorio Scarano. Multimodal monitoring of web servers. *IEEE Multimedia*, 9(3):32–41, 2002.

[5] Maria Barra, Tania Cillo, Antonio De Santis, Umberto Ferraro Petrillo, Alberto Negro, and Vittorio Scarano. Personal WebMelody: Customized sonification of web servers. In Jarmo Hiipakka, Nick Zacharov, and Tapio Takala, editors, *Proceedings of the 2001 International Conference on Auditory Display*, pages 1–9, Espoo, Finland, 29 July–1 August 2001. ICAD.

[6] Stephen Barrass. A perceptual framework for the auditory display of scientific data. In Gregory Kramer and Stuart Smith, editors, *ICAD '94 Second International Conference on Auditory Display*, pages 131–145, Santa Fe, NM, 1994. Santa Fe Institute.

[7] Stephen Barrass. TaDa! demonstrations of auditory information design. In Steven P. Frysinger and Gregory Kramer, editors, *ICAD '96 Third International Conference on Auditory Display*, pages 17–24, Palo Alto, 1996. Xerox PARC, Palo Alto, CA 94304.

[8] Stephen Barrass. *Auditory Information Design*. PhD thesis, Dept. Computer Science, Australian National University, Canberra, Australia, 1998.

[9] Stephen Barrass. Sonification from a design perspective: Keynote speech. In Eoin Brazil and Barbara

[19] `http://www-958.ibm.com/software/data/cognos/manyeyes/`

Shinn-Cunningham, editors, *ICAD '03 9th International Conference on Auditory Display*, Boston, MA, 2003. ICAD.

[10] Stephen Barrass and Christopher Frauenberger. A communal map of design in auditory display. In *Proc. 15th International Conference on Auditory Display*, Copenhagen, Denmark, 18–21 May 2009.

[11] Stephen Barrass, Nina Schaffert, and Tim Barrass. Probing preferences between six designs of interactive sonifications for recreational sports, health and fitness. In Roberto Bresin, Thomas Hermann, and Andy Hunt, editors, *ISon 2010: 3rd Interactive Sonification Workshop*, pages 23–29, Stockholm, Sweden, 7 April 2010. KTH.

[12] Stephen Barrass and Paul Vickers, editors. *ICAD 2004: Proceedings of the 10th Meeting of the International Conference on Auditory Display, 6-9 July*. Sydney, Australia, 6–9 July 2004.

[13] Stephen Barrass and Paul Vickers. Listening to the mind listening: Call for participation. `http://www.icad.org/websiteV2.0/Conferences/ICAD2004/call.htm#concert`, 2004.

[14] Stephen Barrass, Mitchell Whitelaw, and Freya Bailes. Listening to the mind listening: An analysis of sonification reviews, designs and correpondences. *Leonardo Music Journal*, 16:13–19, 2006.

[15] Rodney Berry and Naotoshi Osaka. Art gallery. In *ICAD 2002 – International Conference on Auditory Display*, 2002.

[16] Sara A. Bly. Presenting information in sound. In *CHI '82 Conference on Human Factors in Computing Systems*, The Proceedings of ACM-SIGCHI, pages 371–375. New York: ACM Press/Addison-Wesley, 1982.

[17] Sara A. Bly. Multivariate data mappings. In Gregory Kramer, editor, *Auditory Display*, volume XVIII of *Santa Fe Institute, Studies in the Sciences of Complexity Proceedings*, pages 405–416. Addison-Wesley, Reading, MA, 1994.

[18] Eoin Brazil. A review of methods and frameworks for sonic interaction design: Exploring existing approaches. In Sølvi Ystad, Mitsuko Aramaki, Richard Kronland-Martinet, and Kristoffer Jensen, editors, *Auditory Display*, volume 5954 of *Lecture Notes in Computer Science*, pages 41–67. Springer Berlin / Heidelberg, 2010.

[19] Eoin Brazil and Mikael Fernström. Subjective experience methods for early conceptual design of auditory display. In *Proc. 15th International Conference on Auditory Display*, Copenhagen, Denmark, 18–21 May 2009.

[20] Douglas Burnham. Kant's aesthetics. In James Fieser and Bradley Dowden, editors, *Internet Encyclopedia of Philosophy*. University of Tennessee, Last Accessed: December, 2010.

[21] Stuart K. Card, Jock Mackinlay, and Ben Shneiderman. Information visualization. In K. Card, Stuart, Jock Mackinlay, and Ben Shneiderman, editors, *Readings in Information Visualization: Using Vision to Think*, pages 1–34. Morgan Kaufman, 1999.

[22] Jim Carroll. From art to apps: Data visualisation finds a purpose. `http://bbh-labs.com/from-art-to-apps-data-visualisation-finds-a-purpose`, 2009.

[23] Daniel Chandler. *Semiotics: The Basics*. Routledge, 2 edition, 2007.

[24] Michel Chion. *Audio-Vision: Sound on Screen*. Columbia University Press, NY, 1994.

[25] Millicent Cooley. Sound + image in design. In Stephen A. Brewster and Alistair D. N. Edwards, editors, *ICAD '98 Fifth International Conference on Auditory Display*, Electronic Workshops in Computing, Glasgow, 1998. British Computer Society.

[26] Alain de Botton. "Are museums our new churches?", A Point of View. Prod. Adele Armstrong. BBC Radio 4, 28 January 2011.

[27] Alberto de Campo, editor. *Global Music – The World by Ear, the ICAD 2006 Concert*, London, UK, 20–23 June 2006.

[28] Alberto de Campo. Toward a data sonification design space map. In *13th International Conference on Auditory Display*, pages 342–347, Montréal, Canada, 26–29 June 2007.

[29] Alberto de Campo, Christopher Frauenberger, Katharina Vogt, Annette Wallisch, and Christian Dayé. Sonification as an interdisciplinary working process. In Tony Stockman, Louise Valgerður Nickerson, Christopher Frauenberger, Alistair D. N. Edwards, and Derek Brock, editors, *ICAD 2006 - The 12th Meeting*

of the International Conference on Auditory Display, pages 28–35, London, UK, 20–23 June 2006.

[30] Bruno Degazio. Nikola Tesla and Joseph Schillinger: The music of NT: The man who invented the twentieth century.

[31] John F. Dewey. *Art as Experience*. Perigee Books (orignally published in 1934), 2009.

[32] Johan Fagerlönn and Mats Liljedahl. Awesome sound design tool: A web based utility that invites end users into the audio design process. In *Proc. 15th International Conference on Auditory Display*, Copenhagen, Denmark, 18–21 May 2009.

[33] Paul A. Fishwick. An introduction to aesthetic computing. In Paul A. Fishwick, editor, *Aesthetic Computing*, LEONARDO, pages 3–28. MIT Press, Cambridge, MA, 2006.

[34] Christopher Frauenberger, Tony Stockman, and Marie-Luce Bourguet. Pattern design in the context space: PACO — a methodological framework for designing auditory display with patterns. In *Proceedings of the 14th Conference on Pattern Languages of Programs*, PLOP '07, pages 17:1–17:7, New York, NY, USA, 2007. ACM.

[35] Thomas Fritz. The anchor model of musical culture. In Eoin Brazil, editor, *16th International Conference on Auditory Display*, pages 141–144, Washington, DC, 9–15 June 2010. ICAD.

[36] Joachim Goßman. From metaphor to medium: Sonification as extension of our body. In Eoin Brazil, editor, *16th International Conference on Auditory Display*, pages 145–152, Washington, DC, 9–15 June 2010. ICAD.

[37] Marianne Graves Petersen, Ole Sejer Iversen, Peter Gall Krogh, and Martin Ludvigsen. Aesthetic interaction: A pragmatist's aesthetics of interactive systems. In *DIS '04: Proceedings of the 5th conference on Designing interactive systems*, pages 269–276, New York, NY, USA, 2004. ACM.

[38] Douglas Harper. Online etymology dictionary. `http://www.etymonline.com`.

[39] Chris Hayward. Listening to the Earth sing. In Gregory Kramer, editor, *Auditory Display*, volume XVIII of *Santa Fe Institute, Studies in the Sciences of Complexity Proceedings*, pages 369–404. Addison-Wesley, Reading, MA, 1994.

[40] Christoph Henkelmann. Improving the aesthetic quality of realtime motion data sonification. Technical Report CG-2007-4, Universität Bonn, October 2007.

[41] Thomas Hermann. Taxonomy and definitions for sonification and auditory display. In Brian Katz, editor, *Proc. 14th Int. Conf. Auditory Display (ICAD 2008)*, Paris, France, 24–27 June 2008. ICAD.

[42] Thomas Hermann and Helge Ritter. Crystallization sonification of high-dimensional datasets. *ACM Transactions on Applied Perception*, 2(4):550–558, October 2005.

[43] Stephen Houlgate. Hegel's aesthetics. In Edward N. Zalta, editor, *Stanford Encyclopedia of Philosophy*. The Metaphysics Research Lab Center for the Study of Language and Information Stanford University Stanford, CA 94305-4115, Last Accessed: February, 2011.

[44] Daniel Hug. Using a systematic design process to investigate narrative sound design strategies for interactive commodities. In *Proc. 15th International Conference on Auditory Display*, Copenhagen, Denmark, 18–21 May 2009.

[45] Alan Isaacs, Elizabeth Martin, Jonathan Law, Peter Blair, John Clark, and Amanda Isaacs, editors. *Oxford Encyclopedia*. Oxford University Press, 1998.

[46] Jamie James. *The Music of the Spheres*. Springer-Verlag, New York, 1 edition, 1993.

[47] Myounghoon Jeon. Two or three things you need to know about AUI design or designers. In Eoin Brazil, editor, *16th International Conference on Auditory Display*, pages 263–270, Washington, DC, 9–15 June 2010. ICAD.

[48] Daniel F. Keefe, David B. Karelitz, Eileen L. Vote, and David H. Laidlaw. Artistic collaboration in designing VR visualizations. *IEEE Computer Graphics and Applications*, 25(2):18–23, March–April 2005.

[49] Gregory Kramer. An introduction to auditory display. In Gregory Kramer, editor, *Auditory Display*, volume XVIII of *Santa Fe Institute, Studies in the Sciences of Complexity Proceedings*, pages 1–78. Addison-Wesley, Reading, MA, 1994.

[50] Gregory Kramer. Some organizing principles for representing data with sound. In Gregory Kramer, editor, *Auditory Display*, volume XVIII of *Santa Fe Institute, Studies in the Sciences of Complexity Proceedings*, pages 185–222. Addison-Wesley, Reading, MA, 1994.

[51] Gregory Kramer, Bruce Walker, Terri Bonebright, Perry Cook, John H. Flowers, Nadine Miner, and John Neuhoff. Sonification report: Status of the field and research agenda. Technical report, ICAD/NSF, 1999.

[52] Pontus Larsson. Earconsampler: A tool for designing emotional auditory driver-vehicle interfaces. In *Proc. 15th International Conference on Auditory Display*, Copenhagen, Denmark, 18–21 May 2009.

[53] Andrea Lau and Andrew Vande Moere. Towards a model of information aesthetics in information visualization. In *Proceedings of the 11th International Conference Information Visualization*, pages 87–92, Washington, DC, USA, 2007. IEEE Computer Society.

[54] Talia Lavie and Noam Tractinsky. Assessing dimensions of perceived visual aesthetics of web sites. *International Journal of Human-Computer Studies*, 60(3):269–298, 2004.

[55] Grégory Leplâtre and Iain McGregor. How to tackle auditory interface aesthetics? discussion and case study. In Stephen Barrass and Paul Vickers, editors, *ICAD 2004 – The Tenth Meeting of the International Conference on Auditory Display*, Sydney, 6–9 July 2004. ICAD.

[56] Michael Leyton. The foundations of aesthetics. In Paul A. Fishwick, editor, *Aesthetic Computing*, LEONARDO, pages 289–314. MIT Press, Cambridge, MA, 2006.

[57] Ann Light. Transports of delight? what the experience of receiving (mobile) phone calls can tell us about design. *Personal and Ubiquitous Computing*, 12:391–400, 2008.

[58] Manuel Lima. Visual complexity blog: Information visualization manifesto. http://www.visualcomplexity.com/vc/blog/?p=644, August 2009.

[59] Catriona Macaulay and Alison Crerar. 'observing' the workplace soundscape: Ethnography and auditory interface design. In Stephen A. Brewster and Alistair D. N. Edwards, editors, *ICAD '98 Fifth International Conference on Auditory Display*, Electronic Workshops in Computing, Glasgow, 1998. British Computer Society.

[60] Alastair S. Macdonald. The scenario of sensory encounter: Cultural factors in sensory-aesthetic experience. In William S. Green and Patrick W. Jordan, editors, *Pleasure With Products: Beyond Usability*, pages 113–124. Taylor & Francis Ltd, London, 2002.

[61] Lev Manovich. The anti-sublime ideal in data art. http://www.manovich.net/DOCS/data_art.doc.

[62] Gottfried Mayer-Kress, Robin Bargar, and Insook Choi. Musical structures in data from chaotic attractors. In Gregory Kramer, editor, *Auditory Display*, volume XVIII of *Santa Fe Institute, Studies in the Sciences of Complexity Proceedings*, pages 341–368. Addison-Wesley, Reading, MA, 1994.

[63] George Edward Moore. *Principia Ethica*. Cambridge University Press (reprint from 1903), 2 edition, 1993.

[64] "Morris, William" (1834–96). *The Oxford Dictionary of Quotations*. Oxford University Press, Revised 4th edition, 1996.

[65] Frieder Nake and Susanne Grabowski. The interface as sign and as aesthetic event. In Paul A. Fishwick, editor, *Aesthetic Computing*, LEONARDO, pages 53–70. MIT Press, Cambridge, MA, 2006.

[66] Mórna Ní Chonchúir and John McCarthy. The enchanting potential of technology: A dialogical case study of enchantment and the internet. *Personal and Ubiquitous Computing*, 12:401–409, 2008.

[67] Roy Patterson. *Guidelines for Auditory Warning Systems on Civil Aircraft*. Civil Aviation Authority, 1982.

[68] Elin Rønby Pedersen and Tomas Sokoler. Aroma: Abstract representation of presence supporting mutual awareness. In *CHI '97: Proceedings of the SIGCHI conference on Human factors in computing systems*, pages 51–58, New York, NY, USA, 1997. ACM Press.

[69] Andrea Polli, Glenn Van Knowe, and Chuck Vara. Atmospherics/weatherworks, the sonification of metereological data. http://www.andreapolli.com/studio/atmospherics/, 2002.

[70] Guillaume Potard. Guernica 2006. In Alberto de Campo, editor, *Global Music – The World by Ear, the ICAD 2006 Concert*, pages 210–216, London, UK, 20–23 June 2006.

[71] Aaron Quigley. Aesthetics of large-scale relational information visualization in practice. In Paul A. Fishwick, editor, *Aesthetic Computing*, LEONARDO, pages 315–333. MIT Press, Cambridge, MA, 2006.

[72] Marty Quinn, Wendy Quinn, and Bob Hatcher. For those who died: A 9/11 tribute. In Eoin Brazil and Barbara Shinn-Cunningham, editors, *ICAD '03 9th International Conference on Auditory Display*, pages 166–169, Boston, MA, 2003. ICAD.

[73] Radiohead. House of cards. `http://code.google.com/creative/radiohead/`, 2007.

[74] Bas Raijmakers, William W. Gaver, and Jon Bishay. Design documentaries: Inspiring design research through documentary film. In *DIS '06: Proceedings of the 6th conference on Designing Interactive systems*, pages 229–238, New York, NY, USA, 2006. ACM.

[75] Johan Redström. Tangled interaction: On the expressiveness of tangible user interfaces. *ACM Trans. Comput.-Hum. Interact.*, 15(4):1–17, 2008.

[76] Curtis Roads, editor. *The Computer Music Tutorial*. MIT Press, Cambridge, Massachusetts, 1998.

[77] Curtis Roads. *Microsound*. The MIT Press, 2004.

[78] Robert Rose-Coutré. Art as mimesis, aesthetic experience, and Orlan. *Q.ryptamine*, 1(2):4–6, February 2007.

[79] Benjamin U. Rubin. Audible information design in the new york city subway system: A case study. In Stephen A. Brewster and Alistair D. N. Edwards, editors, *ICAD '98 Fifth International Conference on Auditory Display*, Electronic Workshops in Computing, Glasgow, 1998. British Computer Society.

[80] Sigurd Saue. A model for interaction in exploratory sonification displays. In Perry R Cook, editor, *ICAD 2000 Sixth International Conference on Auditory Display*, pages 105–110, Atlanta, GA, 2000. International Community for Auditory Display.

[81] Carla Scaletti. Sound synthesis algorithms for auditory data representation. In Gregory Kramer, editor, *Auditory Display*, volume XVIII of *Santa Fe Institute, Studies in the Sciences of Complexity Proceedings*, pages 223–252. Addison-Wesley, Reading, MA, 1994.

[82] Pierre Schaeffer. *Traité Des Objets Musicaux*. Seuil, Paris, rev. edition, 1967.

[83] Nina Schaffert, Klaus Mattes, and Alfred O. Effenberg. A sound design for the purposes of movement optimisation in elite sport (using the example of rowing). In *Proc. 15th International Conference on Auditory Display*, Copenhagen, Denmark, 18–21 May 2009.

[84] Anton Schertenleib and Stephen Barrass. A social platform for information sonification: many-ears.com. In Eoin Brazil, editor, *16th International Conference on Auditory Display*, pages 295–299, Washington, DC, 9–15 June 2010. ICAD.

[85] Joseph Schillinger. *The Schillinger System of Musical Composition*. Carl Fischer, Inc., New York, NY, 1941.

[86] Jürgen Schmidhuber. Low-complexity art. *Leonardo*, 30(2):97–103, 1997.

[87] Helen Sharp, Yvonne Rogers, and Jenny Preece. *Interaction Design: Beyond Human-Computer Interaction*, volume 2. John Wiley & Sons Inc, 2007.

[88] Stuart Smith, Haim Levkowitz, Ronald M. Pickett, and Mark Torpey. A system for psychometric testing of auditory representations of scientific data. In Gregory Kramer and Stuart Smith, editors, *ICAD '94 Second International Conference on Auditory Display*, pages 217–230, Santa Fe, NM, 1994. Santa Fe Institute.

[89] Stuart Smith, Ronald M. Pickett, and Marian G. Williams. Environments for exploring auditory representations of multidimensional data. In Gregory Kramer, editor, *Auditory Display*, volume XVIII of *Santa Fe Institute, Studies in the Sciences of Complexity Proceedings*, pages 167–184. Addison-Wesley, Reading, MA, 1994.

[90] C. P. Snow. *The Two Cultures and the Scientific Revolution*. Cambridge University Press, 1959.

[91] Eric Somers. Abstract sound objects to expand the vocabulary of sound design for visual and theatrical media. In Perry R Cook, editor, *ICAD 2000 Sixth International Conference on Auditory Display*, pages 49–56, Atlanta, GA, 2000. International Community for Auditory Display.

[92] K. Stallmann, S. C. Peres, and P. Kortum. Auditory stimulus design: Musically informed. In Brian Katz, editor, *Proc. 14th Int. Conf. Auditory Display (ICAD 2008)*, Paris, France, 24–27 June 2008. ICAD.

[93] Volker Straebel. The sonification metaphor in instrumental music and sonification's romantic implications. In Eoin Brazil, editor, *16th International Conference on Auditory Display*, Washington, DC, 9–15 June 2010. ICAD.

[94] Bob L. Sturm. Music from the ocean, 2002.

[95] Alejandro Tkaczevski. Auditory interface problems and solutions for commercial multimedia products. In Steven P. Frysinger and Gregory Kramer, editors, *ICAD '96 Third International Conference on Auditory Display*, pages 81–84, Palo Alto, 1996. Xerox PARC, Palo Alto, CA 94304.

[96] Andrew Vande Moere. Information aesthetics. `http://infosthetics.com/`.

[97] Paul Vickers. *CAITLIN: Implementation of a Musical Program Auralisation System to Study the Effects on Debugging Tasks as Performed by Novice Pascal Programmers*. Ph.D. thesis, Loughborough University, Loughborough, Leicestershire, September 1999.

[98] Paul Vickers. External auditory representations of programs: Past, present, and future – an aesthetic perspective. In Stephen Barrass and Paul Vickers, editors, *ICAD 2004 – The Tenth Meeting of the International Conference on Auditory Display*, Sydney, 6–9 July 2004. ICAD.

[99] Paul Vickers. Lemma 4: Haptic input + auditory display = musical instrument? In David McGookin and Stephen Brewster, editors, *Haptic and Audio Interaction Design:First International Workshop, HAID 2006, Glasgow, UK, August 31 - September 1, 2006. Proceedings*, volume 4129/2006 of *Lecture Notes in Computer Science*, pages 56–67. Springer-Verlag, 2006.

[100] Paul Vickers and Bennett Hogg. Sonification abstraite/sonification concrète: An 'æsthetic perspective space' for classifying auditory displays in the ars musica domain. In Tony Stockman, Louise Valgerður Nickerson, Christopher Frauenberger, Alistair D. N. Edwards, and Derek Brock, editors, *ICAD 2006 - The 12th Meeting of the International Conference on Auditory Display*, pages 210–216, London, UK, 20–23 June 2006.

[101] Katharina Vogt and Robert Höldrich. Metaphoric sonification method — towards the acoustic standard model of particle physics. In Eoin Brazil, editor, *16th International Conference on Auditory Display*, pages 271–278, Washington, DC, 9–15 June 2010. ICAD.

[102] Bruce N. Walker and Gregory Kramer. Mappings and metaphors in auditory displays: An experimental assessment. In Steven P. Frysinger and Gregory Kramer, editors, *ICAD '96 Third International Conference on Auditory Display*, Palo Alto, 1996. Xerox PARC, Palo Alto, CA 94304.

[103] Bruce N. Walker and David M. Lane. Psychophysical scaling of sonification mappings: A comparision of visually impaired and sighted listeners. In Jarmo Hiipakka, Nick Zacharov, and Tapio Takala, editors, *ICAD 2001 7th International Conference on Auditory Display*, pages 90–94, Espoo, Finland, 29 July–1 August 2001. ICAD.

[104] Charles S. Watson and Gary R. Kidd. Factors in the design of effective auditory displays. In Gregory Kramer and Stuart Smith, editors, *ICAD '94 Second International Conference on Auditory Display*, pages 293–303, Santa Fe, NM, 1994. Santa Fe Institute.

[105] "Whitehead, Alfred North" (1861–1947). *The Oxford Dictionary of Quotations*. Oxford University Press, Revised 4th edition, 1996.

[106] Oscar Wilde. Preface to *The Picture of Dorian Gray*. Wordsworth Classics, 1992 (first pub. 1890).

[107] Sheila M. Williams. Perceptual principles in sound grouping. In Gregory Kramer, editor, *Auditory Display*, volume XVIII of *Santa Fe Institute, Studies in the Sciences of Complexity Proceedings*, pages 95–125. Addison-Wesley, Reading, MA, 1994.

[108] Peter Wright, Jayne Wallace, and John McCarthy. Aesthetics and experience-centered design. *ACM Trans. Comput.-Hum. Interact.*, 15(4):1–21, 2008.

[109] Iannis Xenakis. *Formalized Music: Thought and Mathematics in Composition*. Pendragon Press, Hillsdale, NJ.

Part II

Sonification Technology

Chapter 8

Statistical Sonification for Exploratory Data Analysis

Sam Ferguson, William Martens and Densil Cabrera

8.1 Introduction

At the time of writing, it is clear that more data is available than can be practically digested in a straightforward manner without some form of processing for the human observer. This problem is not a new one, but has been the subject of a great deal of practical investigation in many fields of inquiry. Where there is ready access to existing data, there have been a great many contributions from data analysts who have refined methods that span a wide range of applications, including the analysis of physical, biomedical, social, and economic data. A central concern has been the discovery of more or less hidden information in available data, and so statistical methods of data mining for 'the gold in there' have been a particular focus in these developments. A collection of tools that have been amassed in response to the need for such methods form a set that has been termed Exploratory Data Analysis [48], or EDA, which has become widely recognized as constituting a useful approach. The statistical methods employed in EDA are typically associated with graphical displays that seek to 'tease out' a structure in a dataset, and promote the understanding or falsification of hypothesized relationships between parameters in a dataset. Ultimately, these statistical methods culminate in the rendering of the resulting information for the human observer, to allow the substantial information processing capacity of human perceptual systems to be brought to bear on the problem, potentially adding the critical component in the successful exploration of datasets. While the most common output has been visual renderings of statistical data, a complementary (and sometimes clearly advantageous) approach has been to render the results of statistical analysis using sound. This chapter discusses the use of such sonification of statistical results, and for sake of comparison, the chapter includes analogous visual representations common in exploratory data analysis.

This chapter focuses on simple multi-dimensional datasets such as those that result from scientific experiments or measurements. Unfortunately, the scope of this chapter does not allow discussion of other types of multi-dimensional datasets, such as geographical information systems, or time or time-space organized data, each of which presents its own common problems and solutions.

8.1.1 From Visualization to Perceptualization

Treating visualization as a first choice for rendering the results of data analysis was common when the transmission of those results was primarily limited to paper and books. However, with the rise of many other communication methods and ubiquitous computing devices, it would seem better to consider the inherent suitability of each sensory modality and perceptual system for each problem, and then 'perceptualize' as appropriate. Indeed, devices with multiple input interface methods are becoming commonplace, and coordinated multimodal display shows promise when considering problem domains in which object recognition and scene analysis may be helpful.

Friedhoff's 'Visualization' monograph was the first comprehensive overview of computer-aided visualization of scientific data, and it redefined the term: 'Case studies suggest that visualization can be defined as the substitution of preconscious visual competencies for conscious thinking.' [28]. Just as is implied here for visualization applications, auditory information display can take advantage of preattentive, hard-wired processing resident in the physiology of the auditory system. Since this processing occurs without the application of conscious attention (it is 'preattentive'), the capacity of conscious thought is freed up for considering the meaning of the data, rather than cognizing its structure.

Multivariate data provides particular challenges for graphing. Chernoff notably used pictures of faces to represent a data point that varied in multiple dimensions – groups of observations with similar parameters would be seen as one type of face, while different data points would be seen as 'outsiders' [13]. Cleveland is commonly cited as providing the classic text on multi-dimensional data representation, as well as being involved with an important visualization software advance (Trellis graphics for S-Plus) [15].

Grinstein et al. [31] discussed the 'perceptualization' of scientific data, a term which may be used interchangeably with the more modern definition of 'visualization', although it is free of the sensory bias of the latter term. Ware [55] surveys the field of information visualization, a field distinct from scientific visualization, due to the non-physically organized nature of the information being visualized. While scientific visualization may seek to visualize, for instance, the physical shape of the tissue in and around a human organ, information visualization may wish to visualize the relationship between various causes of heart attacks.

8.1.2 Auditory Representations of Data

The auditory and visual modalities have different ecological purposes, and respond in different ways to stimuli in each domain [42]. The fundamental difference is physiological though – human eyes are designed to face forward, and although there is a broad angular range of visibility, the most sensitive part of the eye, the fovea, only focuses on the central part of the visual scene [55], while the ear is often used to monitor parts of the environment that the eye is not looking at currently. Eye movements and head movements are necessary to view any visual scene, and the ears often direct the eyes to the most important stimulus, rather than acting as a parallel information gathering system.

Auditory display methods have been applied in various fields. One area of widespread usage is auditory alert design, where auditory design flaws can have strong effects in various critical situations, such as air traffic control [10], warning sounds [20], and medical monitoring equipment [45] (see Chapter 19). Much research has focused on sonification for time-series or real-time monitoring of multiple data dimensions, such as for monitoring multiple sources of data in an anaesthesia context [23], stock market analysis [39], or EEG signals [36]. These types of signals are bound to time, and therefore sonification naturally is appropriate as sound is also bound to time, and expansions and contractions in time can be easily understood.

The early development of auditory data representations was surveyed by Frysinger [30], who highlights Pollack and Fick's early experiments [44] which were inspired by the advances made in information theory. They encoded information in a number of different manners and measured the bits transmitted by each method. They found that by encoding information in multiple dimensions simultaneously they were able to transmit more information than if the information was encoded unidimensionally. Frysinger also mentions Bly's 1982 work [6, 7], in which a number of auditory data representations were developed to allow the investigation of the Iris dataset [1]. Bly tested whether a participant could classify a single multivariate data point as one of three iris species accurately, based on learning from many representations of the measurements of each of the three irises (which are described in Section 8.2). Flowers and Hauer investigated auditory representations of statistical distributions, in terms of their shape, central tendency and variability, concluding that the information was transmitted easily using the auditory modality [27]. Later, Flowers et al. discussed an experiment on the visual and auditory modalities [25], again finding them to be equivalent in their efficacy for the evaluation of bivariate data correlations. However, Peres and Lane discussed experiments using auditory boxplots of datasets [43], and found that their respondents did not find auditory graphs easy to use, and the error rate regarding the statistical information presented did not improve with training as much as may be expected. They cautioned that this finding did not necessarily generalize to the entire auditory modality and may have been influenced by issues to do with the designs of the particular auditory graphs under investigation. Flowers described how, after 13 years of study and development, auditory data representation methods are still not common in modern desktop data analysis tools [24, 26].

Sonification has been defined in various ways, initially by Kramer: the process of transforming data to an attribute of sound. Recently, Hermann has expanded this definition, and has defined sonification in a more systematic manner, as a sound that: reflects a) objective properties or relations in the input data, has a b) systematic and c) reproducible transformation to sound, and can be d) used with different input data [35].

A common technique for sonification is *parameter mapping* which requires some kind of *mapping* of the data to the element of sound that is to represent it (see Chapter 15). Choosing that mapping is not a simple task [41], but Flowers [24] describes some simple strategies that produce useful outcomes, and Walker et al. have carried out fundamental research into strategies for mapping [52, 50], showing that certain types of data and polarities are more naturally mapped to particular sound attributes.

The representation of probability distributions has also been discussed by various authors, including Williamson and Murray-Smith [58, 59] who used granular synthesis as a method of displaying probability distributions that vary in time. Childs [14] discussed the use of probability distributions in Xenakis' composition *Achorripsis*, and Hermann [33] has investigated the sonification of Monte Carlo Chain Simulations.

Walker and Nees' research on auditory presentation of graphs has provided a description of data analysis tasks – they delineate trend analysis, pattern detection, pattern recognition, point estimation, and point comparison [53]. Harrar and Stockman described the effect of the presentation of data in discrete or continuous formats, finding that a continuous format was more effective at conveying a line graph overview as the complexity increased, but a discrete format was more effective for point estimation or comparison tasks [32]. De Campo has developed a sonification design space map to provide guidance on the appropriate sonification method (either discrete, continuous or model-based) for representing particular quantities and dimensionalities of data [11].

Hermann [34, 38] has introduced model-based sonification as a method distinct from parameter-mapping, whereby the data set is turned into a dynamic model to be explored interactively by the user, rather than sonifying the data directly. This method provides very task-specific and problem-specific tools to investigate high-dimensional data and is covered in Chapter 16. For a method that deals with large amounts of sequential univariate or time-series data, audification is a common choice, as discussed in Chapter 12. Ferguson and Cabrera [22, 21] have also extended exploratory data analysis sonification techniques to develop methods for sonifying the analysis of sound and music.

Perceptualization practice will gradually reveal when it is best to use auditory representation tools. Auditory representations can potentially extract patterns not previously discernible, and might make such patterns so obvious to the ear, that no-one will ever look for them with their eyes again. By capitalizing upon the inherently different capabilities of the human auditory system, invisible regularities can become audible, and complex temporal patterns can be "heard out" in what might appear to be noise.

8.2 Datasets and Data Analysis Methods

Tukey was possibly one of the first to prioritize visual representations for data analysis in his seminal work *Exploratory Data Analysis* [48]. He focused on the process of looking for patterns in data and finding hypotheses to test, rather than in testing the significance of presupposed hypotheses, thereby distinguishing *exploratory* data analysis from *confirmatory* data analysis. Through the use mainly of graphical methods, he showed how datasets could be summarized with either a small set of numbers or graphics that represented those numbers. In some situations (e.g. medical research), a confirmatory approach is common, where a hypothesis is asserted and statistical methods are used to test the hypothesis in a dataset drawn from an experimental procedure. In exploratory situations a hypothesis is not necessarily known in advance, and exploratory techniques may be used to find a 'clue' to the correct hypothesis to test based on a set of data. For a set of univariate observations there are several pieces of evidence that exploratory data analysis may find:

the *midpoint* of the data, described perhaps by the mean, mode or median or through some other measure of the central tendency of the data;

the *shape and spread* of the data, describing whether the data centres heavily on one particular value, or perhaps two or more values, or whether it is spread across its range;

the *range and ordering* of the data, describing the span between the highest and lowest values in the data, and the sorting of the data point values between these two extremes;

the *outliers* in the data, the data points that do not follow the data's general pattern and may indicate aberrations or perhaps significant points of interest;

the *relationships* between variables in the data, focusing on factors that explain or determine the variability in the data.

Overarching each of these pieces of evidence is the desire to understand any behaviors of, or structure in, the data, to form or discard hypotheses about the data, and to generally gain some kind of insight into the phenomena being observed. As a demonstration dataset the iris measurements of Anderson [1] will be analyzed. For each iris in the set there are four measurements, the sepal's length and width, and the petal's length and width. Fifty irises are measured in each of three species of iris, resulting in a total of 150 measurements.

8.2.1 Theoretical Frameworks for Data Analysis

Bertin [4, 5], one of the pioneers of interactive data analysis techniques, described a five-stage pattern of decision making in data analysis:

1. defining the problem;
2. defining the data table;
3. adopting a processing language;
4. processing the data, and;
5. interpreting, deciding or communicating.

For data analysis, Bertin developed the permutation matrix, an interactive graphical display that used rearrangeable cards. He argued that all representations of data are reducible to a single matrix (for examples and a review see [17]). Tufte was also heavily influential through his definition of the purpose of graphics as methods for 'reasoning about data', and highlighting of the importance of design features in the efficiency of information graphics [46, 47]. He draws a distinction between graphics whose data is distorted, imprisoned or obfuscated – and graphics which rapidly and usefully communicate the story the data tells. 'Above all else, show the data', is a maxim that shows his emphasis on both the quantity and priority of data, over 'non-data ink' – complex scales, grids or decorative elements of graphs, although Bateman et al. have shown that some types of decoration (such as elaborate borders, cartoon elements and 3-dimensional projections) can enhance a graphic's efficiency [3].

Another theoretical framework for statistical graphics that has already had far-reaching influence is Leland Wilkinson's *Grammar of Graphics* [57], which has been implemented in a software framework by Wickham [56]. This is one of the few conceptual frameworks to take a completely systematic object-oriented approach to designing graphics.

Ben Fry's *Computational Information Design* [29], presents a framework that attempts to link fields such as computer science, data mining, statistics, graphic design, and information visualization into a single integrated practice. He argues a 7-step process for collecting, managing and understanding data: 1. acquire, 2. parse, 3. filter, 4. mine, 5. represent, 6. refine, 7. interact. Crucial to this framework is software that can simplify the implementation of each operation, so that a single practitioner can practically undertake all of these steps, allowing the possibility of design iteration to incorporate many of the stages, and facilitating user interaction through dynamic alterations of the representation.

Barrass discusses the 'TaDa' design template in 'Auditory Information Design', a Ph.D. thesis [2]. The template attempts to delineate the *Ta*sk requirements, and the *Da*ta Characteristics. An important element of this design framework is that Barrass categorizes different types of data into a systematic data type and auditory relation taxonomy based on information design principles.

8.2.2 Probability

Some fundamental concepts are necessary in a discussion of statistical sonification. *Statistical probability* is a method of prediction based on inference from observation, rather than from induction. Data are in general understood as samples drawn from an unknown high-dimensional probability distribution. In one definition, Bulmer [9] describes a *statistical probability* as '...the limiting value of the relative frequency with which some event occurs.' This *limiting value* may be approximated by repeating an experiment n times and comparing that number with the number of times event A occurs, giving the probability $p(A)$ clearly as $\frac{n(A)}{n}$. As n increases, in most situations $p(A)$ moves closer to a particular value, reasonably assumed to be the limiting value described above. However, as no experiment may be repeated infinitely, the reasonableness of this assumption is strongly associated with the number of times the experiment is performed (n), and we can never know with absolute certainty what this limiting value is [9]. Using statistical probability, the way we infer that we have approximately a 50% chance of getting either a head or a tail when we toss a coin is by tossing that coin repeatedly and counting the two alternatives, rather than by inducing a probability through reasoning about the attributes of the coin and the throw.

8.2.3 Measures of Central Tendency

Once a set of observations has been obtained they can be quickly summarized using measures of their central tendency. The midpoint of a dataset is crucial and can be described in many ways, although a parsimonious approach is to use the *median*. The median is the middle point of the ranked data, or the mean of the two middle points if the dataset has an even count. The median implicitly describes the value at which the probability of a randomly drawn sample will fall either above or below it is equal. The *arithmetic mean* ($\overline{x}$) is another measure of central tendency that is useful for describing the midpoint of a set of data. It can be described mathematically as:

$$\overline{x} = \frac{x_1 + x_2 + ...x_n}{n} = \sum_{i=1}^{n} x_i/n \tag{1}$$

where n observations are termed x_1, x_2, ..., x_n [9]. Other methods for calculating the mean exist, usually used when the numeric scale is not linear: including the geometric mean and the harmonic mean. The *mode* is the measurement that is observed the most times in a

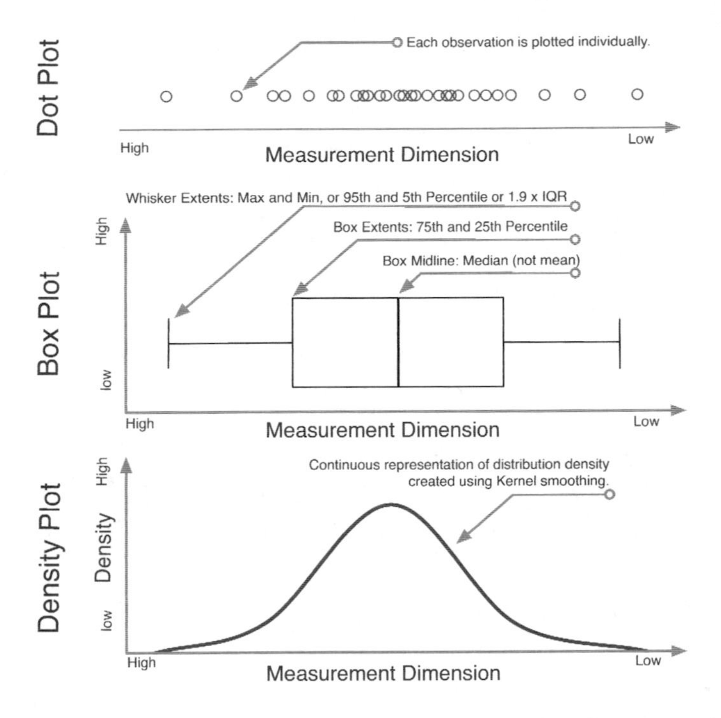

Figure 8.1: A distribution may be represented in several ways, but the purpose is to show the clustering and spread of a set of data points.

discrete distribution, or it is the point where the probability density function has the highest value in a continuous distribution.

8.2.4 Measures of Dispersion

Measures of dispersion allow us to build a more detailed summary of how the distribution of a dataset is shaped. Sorting a set of data points is an initial method for approaching a batch of observations. The top and bottom of this ranking are the *maximum* and *minimum*, or the extremes. The difference between these numbers is the *range* of the data. To provide a number that represents the dispersion of the distribution, one may take the mean and average all the absolute deviations from it, thus obtaining the mean absolute deviation. The standard deviation σ is similar, but uses the square root of the mean of the squared deviations from the mean, which makes the resulting number more comparable to the original measurements:

$$\sigma = \sqrt{\frac{1}{N}\sum_{i=1}^{N}(x_i - \overline{x})^2} \tag{2}$$

To provide more detail on the shape of the distribution we can use two more values. The central portion of the distribution is important to work out the spread of the data, and to summarize it we divide the data into 2 parts, the minimum to median, and the median to maximum. To find the 25th and 75th percentiles we then take the median of these 2 parts

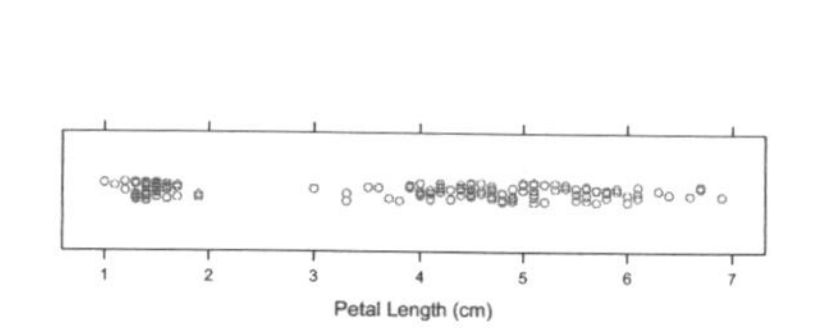

(a) A dotplot of the 150 measurements of Petal Length shows clear groupings.

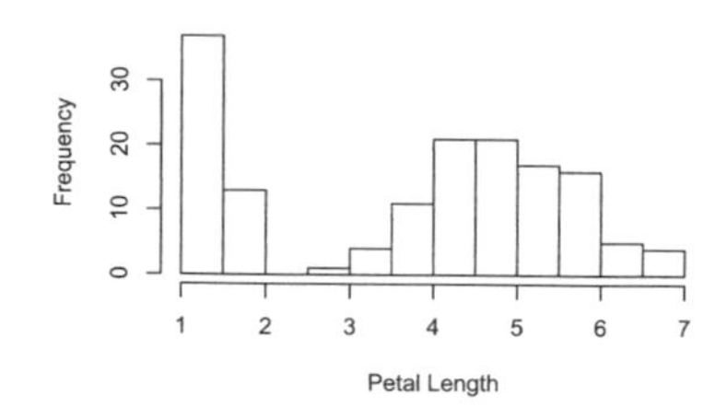

(b) A histogram can show the general shape within the Petal Length data.

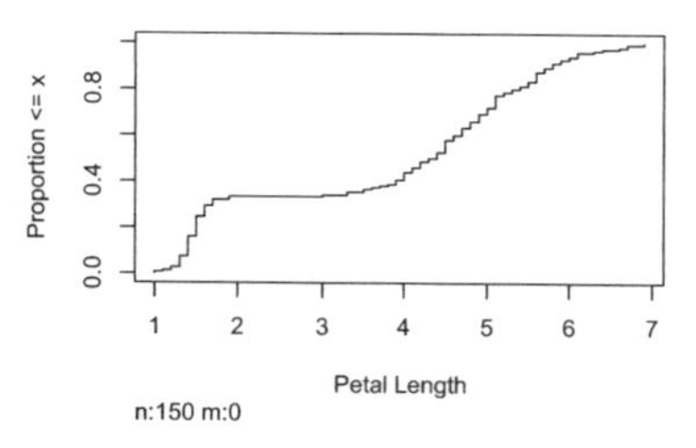

(c) A cumulative distribution function of the Petal Length measurements.

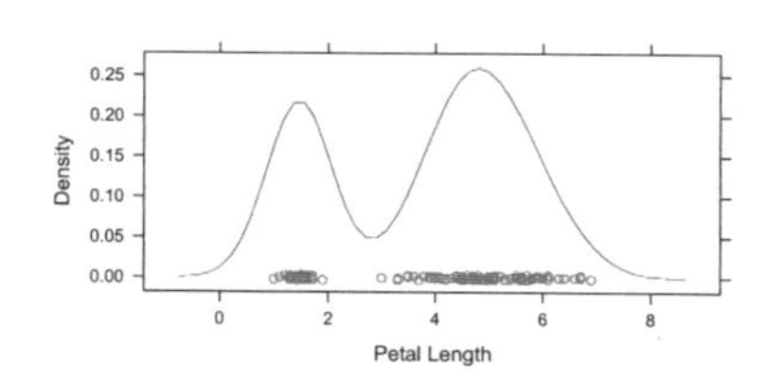

(d) A kernel smoothed estimation of the probability density function.

Figure 8.2: Various representations of a dimension of a dataset.

again. These numbers are also known as the 1st and 3rd quartiles, and the range between them is known as the interquartile range. A small interquartile range compared with the range denotes a distribution with high kurtosis (peakedness).

Tukey's *five number summary* presents the interquartile range, the extremes and the median to summarize a distribution, allowing distributions to be easily and quickly compared [48]. Tukey invented the boxplot by taking the five-number summary and representing it using a visual method. It uses a line in the middle for the median, a box around the line for the 25th to 75th percentile range (the inter-quartile range), and whiskers extending to maximum and minimum values (or sometimes these values may be the 95th and 5th percentile). Considering the batch of data as a shape, rather than a set of (five) numbers, can show the characteristics of the data distribution more clearly, especially if the distribution is not a typical unimodal bell shape (see Figure 8.1). A graph of the distribution demonstrates the ideas of the *range* of the data, the *midpoint* of the data, but also clearly shows well-defined aspects such as skew and kurtosis, as well as aspects that are not easily described and may be peculiar to the particular dataset being investigated.

Figure 8.2 shows four representations of the Petal Length variable in the iris dataset: a dotplot, a histogram, a cumulative distribution function, and a kernel density function. A histogram is a simple way of visualizing a distribution of data by using a set of bins across the data range and representing the number of observations that fall into the bin. A cumulative

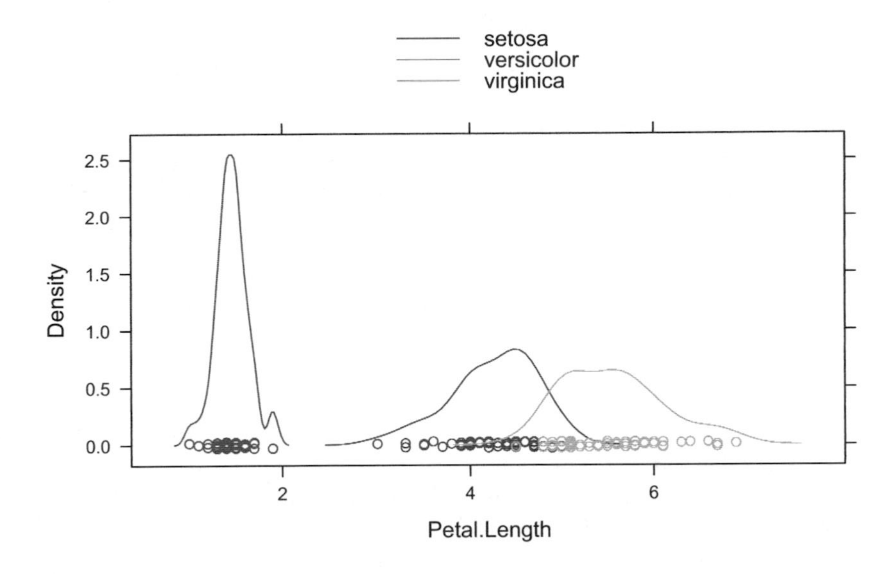

Figure 8.3: A kernel-smoothed estimation of the probability density function, grouped by the species of iris measured.

distribution function is another description of the same information. It is a graph of the probability (on the y-axis) of a random choice from the dataset being less than the value specified on the x-axis. A smoother representation of the probability density function may be obtained through the technique of kernel smoothing [54]. In this technique the distinct observations at each point are replaced by kernels, miniature symmetric unimodal probability density functions with a specified bandwidth. These kernels are then summed across the data range to produce a curve.

The distribution revealed by the kernel-smoothed probability density function (Figure 8.2(d)) does not exhibit the sort of unimodal shape that might be expected if the 150 data points were all sampled from a homogeneous population of iris flowers. Rather, there seems to be some basis for separating the data points into at least two groups, which is no surprise as we already know that multiple species of iris were measured. Although pattern classification could be based upon the *ground truth* that is already known about these data, it is also of particular interest whether membership of a given item in one of the three iris groups might be determined through exploration of the data. Before examining that possibility, the *ground truth* will be revealed here by dividing the single category into three categories based on the known iris species (Figure 8.3). The three color-coded curves in the figure show that there is one clearly separated group (graphed using a red curve), and two groups that overlap each other (the blue and green curves). What might be considered to be more typical of an exploratory data analysis process, and that which will be examined in greater depth from this point, is the attempt to assign group membership to each of the 150 items in a manner that is blind to the *ground truth* that is known about the iris data. That is the topic taken up in the next subsection.

8.2.5 Measures of Group Membership (*aka* Blind Pattern Classification)

Blind pattern classification is the process by which the items in a set of multivariate data may be sorted into different groups when there is no auxiliary information about the data that would aid in such a separation. What follows is an example based upon the four measurements that make up the iris dataset, even though the *ground truth* is known in this case.

The simplest approach to pattern classification would be to apply a *hard* clustering algorithm that merely assigns each of the 150 items into a number of groups. Of course, the number of groups may be unknown, and so the results of clustering into two, three, or more different groups may be compared to aid in deciding how many groups may be present. The most common approach to *hard* clustering, the so-called K-means clustering algorithm, takes the hypothesized number of groups, K, as an input parameter. If we hypothesize that three species of iris were measured, then the algorithm will iteratively seek a partitioning of the dataset into these three hypothetical groups. The process is to minimize the sum, over all groups, of the within-group sums of the distances between individual item values and the group centroids (which capture the group mean values on all four measurements as three unique points in a four-dimensional space). Of course, the K-means clustering result measures group membership only in the nominal sense, with a *hard* assignment of items to groups.

A more useful approach might be to determine how well each item fits into each of the groups, and such a determination is provided by a fuzzy partitioning algorithm. If again we hypothesize that three species of iris were measured, fuzzy partitioning will iteratively seek a partitioning of the dataset while calculating a group membership coefficient for each item in each of the three groups. Hence no *hard* clustering is enforced, but rather a partitioning in which membership is graded continuously, and quantified by three group membership coefficients taking values between 0 and 1. The result of a fuzzy partitioning of the iris measurements is shown in Figure 8.4. In the graph, group membership coefficients for all 150 items are plotted for only two of the three groups, termed here the red and the green (to be consistent with the color code used in the previous figure). Since the group membership coefficients sum to 1 across all three groups for each item, the value of the blue membership coefficient for each item is strictly determined by the values taken for the other two coefficients. To help visualize the 150 items' continuously-graded membership in the three groups, the plotting symbols in the figure were color-coded by treating the red, green, and blue group membership coefficients as an RGB color specification. Naturally, as the group membership values approach a value of 1, the color of the plotting symbol will become more saturated. The items fitting well in the red group are thus quite saturated, while those items that have highest blue-group membership values are not so far removed from the neutral grey that results when all three coefficient values equal 0.33.

Of course, the red group items can be classified as separate from the remaining items strictly in terms of the measurement on just one column of the four-column matrix that comprises the iris dataset, that column corresponding to the Petal-Length variable. However, the distribution of Petal-Length measurement values, which was examined from several perspectives above, does not enable the separation of the green and blue group items. Other methods that are based on multivariate analysis of the whole dataset may provide better results, as is discussed in the next subsection.

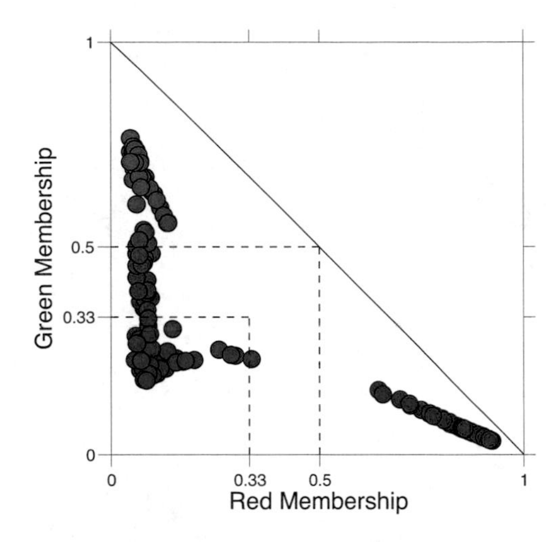

Figure 8.4: Results from fuzzy partitioning analysis on the 150 items in the iris dataset, using RGB color codes for plotting group membership coefficients.

8.2.6 Multivariate Data Exploration

It would be reasonable to assume that a multivariate analysis of the whole iris dataset might prove more effective in separating the 150 items into the above-hypothesized three groups, especially in comparison to an examination of the distribution of measurement values observed for a single variable. However, measurements for additional variables will only make substantial contributions in this regard if the values on those additional variables provide independent sources of information. As a quick check on this, a visual examination of two measurements at once will reveal how much independent information they might provide. Figure 8.5(a) is a scatterplot of measurement values available for each of the 150 items in the iris dataset on two of the variables, Petal-Width and Petal-Length. (Note that the plotting symbols in the figure are color coded to indicate the three known iris species that were measured.) It should be clear that there is a strong linear relationship between values on these two variables, and so there may be little independent information provided by measurements on the second variable. The fact that the two variables are highly correlated means that a good deal of the variance in the data is shared, and that shared variance might be represented by a projection of the items onto a single axis through the four-dimensional space defined by the four variables. The multivariate analytic technique that seeks out such a projection is Principal Component Analysis (*aka* PCA, see [19]).

PCA effectively rotates the axes in a multivariate space to find the principal axis along which the variance in the dataset is maximized, taking advantage of the covariance between all the variables. The analysis also finds a second axis, orthogonal to the first, that accounts for the greatest proportion of the remaining variance. In the case of the iris data, the scores calculated for each of the 150 items as projections onto each of these two axes, called principal component scores, are plotted in Figure 8.5(b). Scores on Principal Component 1 (PC1) separate the three groups well along the x-axis; however, scores on Principal Component 2 (PC2) do very little to further separate the three groups along the y-axis of

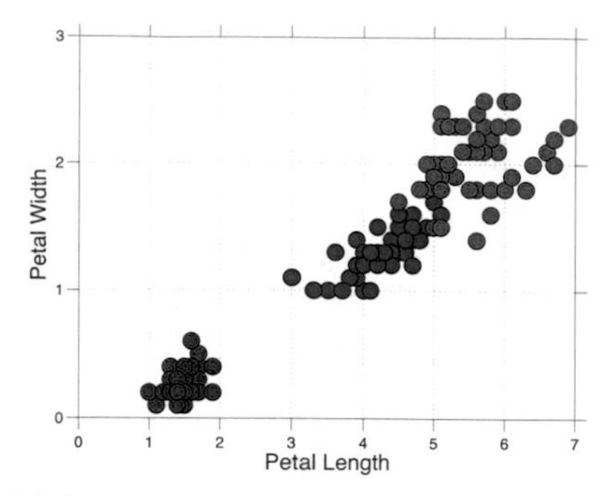

(a) Scatterplot of Petal-Width on Petal-Length measurement values for the 150 items in the iris dataset, using color codes indicating the hard partitions based on the *ground truth* about the three species of iris measured.

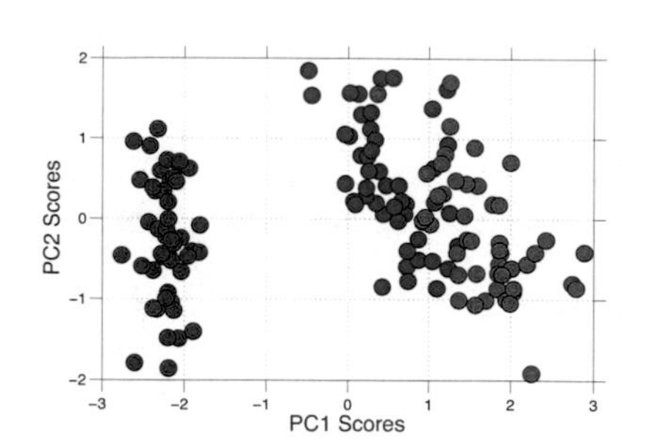

(b) Scatterplot of Principal Component 2 (PC2) scores on Principal Component 1 (PC1) scores for the 150 items in the iris dataset, again using color codes based on the *ground truth* about the three species of iris measured.

Figure 8.5: Principal Component Analysis rotates the axes of raw data in an attempt to find the most variance in a dataset.

the plot. This means that the blue group and green group items that inhabit the region of overlapping values would be difficult to classify, especially because in this exploration we assume that the species differences are not known. If it were a case of machine learning, in which the *ground truth* about the three iris species measured were known, rather than the exploratory data analysis that is under discussion here, then a linear discriminant analysis could be performed that would find a more optimal means of separating items from the three species [18]. There is some real value in evaluating the success of exploratory methods by comparing known categorization with categorization discovered blindly; however, for the introduction to statistical concepts that was deemed most relevant to this chapter, no further examination of such multivariate analytic techniques will be presented. Suffice it to say that the relative merits of visual and auditory information displays made possible employing such multivariate approaches are worth investigating in both exploratory and confirmatory analyses, though it is the former topic of exploratory sound analysis to which this chapter now turns.

8.3 Sonifications of Iris Dataset

This section presents several auditory representations of the iris dataset examined above from the visual perspective, and attempts to show the relative merit of the various univariate and multivariate representations that can be applied for exploratory data analysis. In this first subsection, the dataset is examined in a univariate manner, and this shows that groupings of items can be distinguished on the basis of their petal size measurements when sonically rendered just as when they are visually rendered (Figure 8.6).

8.3.1 Auditory Dotplot

Exploratory Data Analysis has a strong emphasis on the investigation of 'raw' data initially, rather than immediately calculating summaries or statistics. One visual method that can be used is the 'strip plot' or 'dotplot'. This presents one dimension (in this case, the measurements of Petal-Length) of the 150 data points along a single axis that locates the measured dimension numerically. This display is not primarily used for analytic tasks, but it possesses some characteristics that are very important for data analysis – rather than showing a statistic, it shows every data point directly, meaning the numerical Petal-Length values could be recreated from the visualization. This directness helps to create a clear impression of the data before it is transformed and summarized using statistical methods, and skipping this stage can sometimes be problematic.

A sonification method that mimics the useful attributes of a dotplot may be constructed in a variety of manners. It may map the data dimension to time, and simply represent the data through a series of short sounds. This sonification (sound example **S8.1**) scans across a data dimension using the time axis, rendering clusters, outliers and gaps in data audible. The use of short percussive sounds allows for a large number of them to be heard and assessed simultaneously.

8.3.2 Auditory Kernel Density Plot

In Figure 8.6 the kernel density plots additionally include a dotplot at the base of each set of axes. As described in section 8.2.4 this dotplot is summed (as kernels of a specified bandwidth) to produce the curve overlaid above it. A kernel density plot is very similar to a histogram in that it attempts to show the distribution of data points, but it has a couple of differences. It employs a more complex algorithm in its creation, rather than simply counting observations within a set of thresholds, and the units it uses are therefore not easily parsed – but it does create a curve rather than a bar-graph [54]. A histogram, by comparison, simply counts the number of data points within a number of bins and presents the bin counts as a bar graph. This means the algorithm is straightforward, and the units easy to understand, but the resolution can be fairly poor depending on the choice of bin width. In the auditory modality, however, a sonification algorithm can be simpler still.

A kernel density plot in the auditory modality can be created by mapping the sorted data points to a time axis based on the value of the data (with the data range used to normalize the data point's time-value along the time axis). This sonification achieves a similar type of summing to the kernel density plot, but in the auditory modality, through the rapid addition of multiple overlapping tones. The outcome is that groupings, and spaces between groupings, can be heard in the data sonification. This graphing technique describes the distribution of data, rather than a single summary of the data, and is easily extended to the auditory modality.

With so many overlapping data points we could give some consideration to the phase cancellation effects that might occur. Indeed, if many notes of identical frequencies were presented at the same time, some notes might increase the level more than the 3 dB expected with random phase relationships. However, it is very unlikely that two notes would be presented that have identical frequencies and spectra but are 180 degrees out of phase and

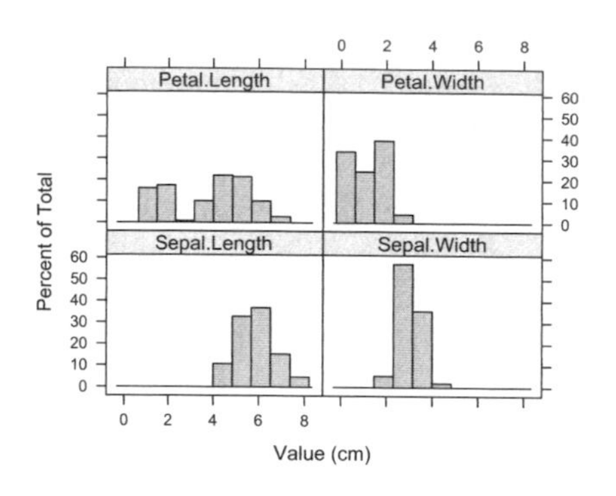

(a) Histograms for each of the four dimensions of the iris dataset.

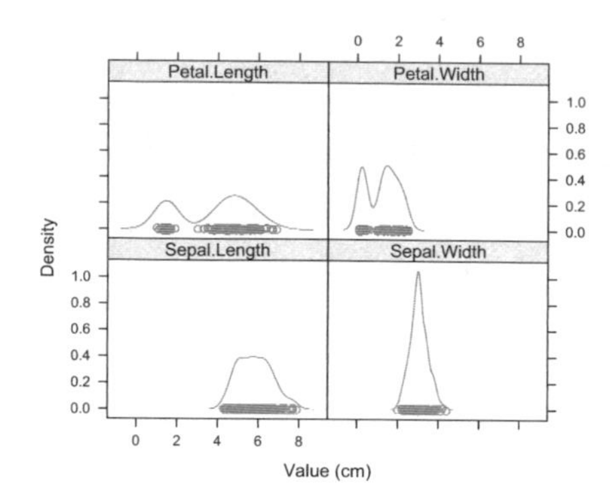

(b) Density plots for each of the four dimensions of the iris dataset.

Figure 8.6: Univariate representations of the dimensions of each dataset.

thereby cancel. A more likely scenario, with complex tonalities with temporal envelopes (created for instance using FM), is that the notes would be perceived as overlapping, but as two distinct elements – unless of course the temporal onset was simultaneous. The graphical technique that is used to avoid a similar effect is to use a specified amount of random jitter to ensure that graphical markers do not sit in exactly the same place, which is probably appropriate in this situation as well.

Sound example **S8.2** is an example of the Petal-Length data presented as an auditory kernel density plot. The first grouping is heard clearly separated from the other data points, and we can also make a rough guess of the number of data points we can hear. This first group happens to be the *setosa* species of iris, while the larger, longer group is made up of the two other iris species.

8.3.3 Auditory Boxplot

A boxplot is another common method for comparing different distributions of data (instead of assessing a single distribution as do the kernel density and histogram). Flowers et al. have discussed the representation of distributions through the use of boxplots (or arpeggio plots for their auditory counterpart) [26]. One way to represent these is to play each of the five numbers that form the boxplot in succession forming a type of arpeggio. Another method of summarizing the data is to randomly select and sonify data points from the dataset rapidly, creating a general impression of the range, density, and center of the dataset almost simultaneously. There is no need to stop sonifying when the number of data points has been reached, and selections could continue to be made indefinitely. This means that a very high rate of sonification can be achieved, summarizing the group very quickly, resulting in a stationary sonification of arbitrary length. This is very important when multiple groups of data are compared as the speed of comparison (and therefore the memory involved) is determined by the time it takes for each group to be sonified– using multiple short sonifications of each group can be done quickly using this method.

By choosing the range of data points selected through sorting the data and specifying a range, it is possible to listen to the interquartile range of the data, or only the median, maximum or minimum. These are useful techniques for comparing different groupings. This dataset can be sorted by the species of iris, into three groups of 50 flowers each. By using this sonification we can either summarize a group with a single measure of location, such as the median, or describe the distribution. By concatenating the sonifications of each of the groups we can compare the measures against each other rapidly, allowing an estimate of whether the groups can be readily separated. Figure 8.7 shows a traditional boxplot of the Petal-Length dimension of the dataset, which can be compared to sound example **S8.3** for the sonification method.

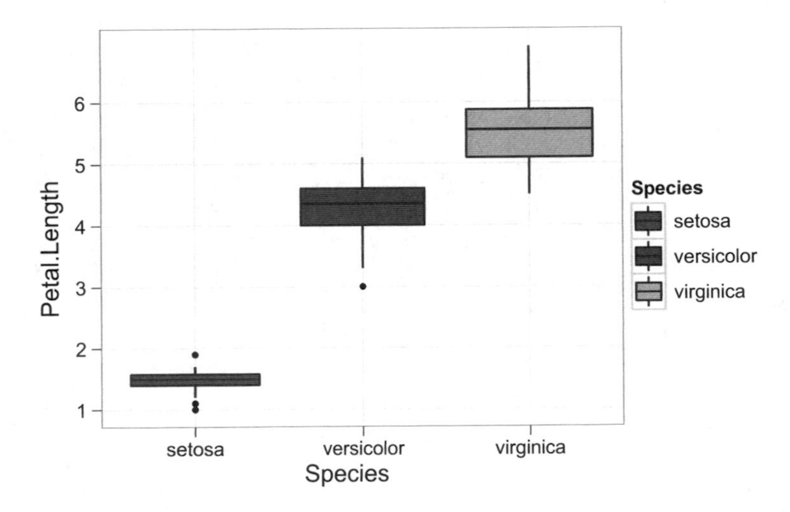

Figure 8.7: Boxplots represent distributions of data, and when different data distribution boxplots are presented adjacently they can be easily compared.

This sonification separates the three groups into different 'blocks' of sound, with successive narrowing of the range presented, from the full 95th to 5th percentile to 75-25 percentile and then the median only. Hence, this represents the range, interquartile range, and then the median of the data in succession. It is clear from the sonification as well as the graphic that the first group is relatively narrow in distribution, but the second and third are quite wide, and not as well separated.

8.3.4 Auditory Bivariate Scatterplot

Two-dimensional approaches can provide more differentiation than those using one dimension. In Figure 8.5(a) we see a bivariate scatterplot that compares two parameters. Using parameter mapping, we sonify the data again (sound example **S8.4**), this time using FM-synthesis (as introduced in chapter 9) to encode two data dimensions in one note. The petal length parameter is mapped to the pitch of the tone, while the petal width parameter is mapped to the modulation index for the tone. This method of representation allows the user to listen to the data and build an auditory representation quickly, internalizing it like an environmental sound for later recall. Single sounds can then be compared against the internalized sound for comparison and classification.

8.3.5 Multivariate Data Sonification

As suggested in the above discussion on visualization of the iris dataset, it has been assumed that a multivariate exploration of the whole iris dataset might prove more effective in separating the 150 items into three groups than would a univariate examination of measurement values. The same question should be asked here in comparing univariate, bivariate, and multivariate sonifications of the iris data.

The initial investigation of this idea is a demonstration of the difference between a Petal-Length sonification and some bivariate sonifications. The first bivariate sonification is strictly analogous (sound example **S8.4**) to the visual graph shown in Figure 8.5(a), the scatterplot of measurement values available for each of the 150 items in the iris dataset on Petal-Width and Petal-Length. Although there is some linear dependence between values on these two variables, there does seem to be a substantial amount of independent information provided by measurements on the second variable. The second bivariate sonification (sound example **S8.5**) is analogous to the visual graph shown in Figure 8.5(b), which plotted scores for each item on the first two principal components that were found when the iris dataset was submitted to PCA. As the PCA rotated the axes to maximize variance along the principal axis, and then to maximize the remaining variance along a second orthogonal axis, two auditory attributes of graded perceptual salience were applied in a PC-based sonification (as suggested by Hermann [34]). Comparing this PC-based sonification to the direct two-parameter sonification also presented here does not provide a convincing case for an advantage given the PC-based approach. To most listeners, the PC-based mapping does not produce a sonification that makes the distinction between groups any more audible than did the more straightforward two-parameter sonification. Might it be that correlation between item values on the original variates, when mapped to distinct sonification parameters, could be relatively more effective in the case of presenting the iris data, despite the inherent redundancy this would display?

Therefore, a four-dimensional sonification was created using measurement values on all four of the original variables (sound example **S8.6**). This sonification is related to the visualization using Chernoff's [13] faces that is illustrated in Figure 8.8. The sonification is not strictly analogous to the visualization, however, since the sonification allows individual items to be displayed in rapid succession, and also allows many repeat presentations to aid the observer in forming a concept of how sounds vary within each group of 50 items. A comparable succession of faces could be presented, but the more typical application of Chernoff's [13] faces is to present many faces simultaneously at plotting positions that are spread out in space. It may be that the opportunity to visually scan back and forth presents some distinct advantage over the strict temporal ordering of item sonifications followed in the sound example, but is is difficult to argue for a more general conclusion outside the context of the investigation of a particular dataset. Indeed, the authors are not aware of any empirical evaluation of the relative effectiveness of spatial vs. temporal distribution of faces for human discrimination of patterns in visualized data, let alone sonified data.

The four-dimensional sonification created for this discussion is not the first such sonification

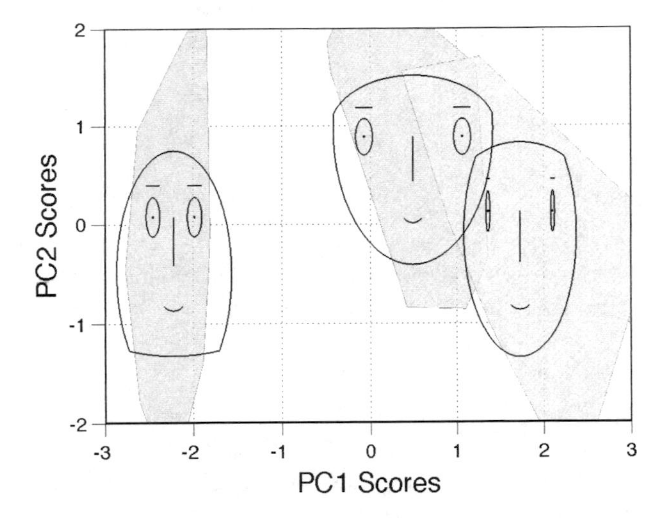

Figure 8.8: A Chernoff-face visualisation of the mean values for each of three groups of iris measured on four variables. Underlying each of the three Chernoff faces can be seen a color-coded patch illustrating the convex hull of principal component scores for the 50 items in each of the three groups contributing to the mean values visualized. Although the faces are constructed to show the four mean parameters as a four-dimensional visualization, each is positioned over the centroid of these patches in the bivariate plotting space to show the spread of the data in each of the three groups.

created to represent the iris dataset. In fact, about 30 years ago Sara Bly [7] presented a closely related sonification that was developed during her doctoral research. Her sonification used the following mapping from iris measurement variables to sound synthesis parameters: Variation in petal length was mapped to duration, petal width was mapped to waveshape, sepal length to pitch, and sepal width to volume. After a training session in which listeners heard example sonifications representing each of the three iris species, they were presented with 10 test items that could be more or less well classified. She reported that most casual observers could place her sonifications into the appropriate groups with few errors. No formal empirical evaluation of the success of the current sonification was undertaken, but informal listening tests suggest that similar good performance would be expected using the mapping chosen here, which is summarized as follows: The first two parameters, Petal-Length and Petal-Width, were mapped to the most elementary auditory attributes, pitch and duration, while the third and fourth parameters were mapped to timbral attributes, perhaps providing more subtle indication of parameter variation than the mappings for the first two parameters. These two timbral attributes could be heard as variations in tone coloration similar to that of vowel coloration changes characteristic of human speech. More specifically, the dataset values were mapped so as to move the synthesized tones through the vowel space defined by the first two formants of the human vocal tract, as follows: The measured Sepal-Length values modulated the resonant frequency of a lower-frequency formant filter, while Sepal-Width values were mapped to control the resonant frequency of a higher-frequency formant filter. Applying this co-ordinated pair of filters to the input signals that varied in pitch and duration resulted in tones that could be heard as perceptually rich and yet not overly complex, perhaps due to their speech-like character.

8.4 Discussion

Different priorities exist in the auditory modality, and an important difficulty for making sonification design decisions is the purpose of the representation. Sonification can be used for enabling blind and visually impaired people to use data representations, for 'ears-only' representations in high-workload monitoring environments, for multi-modal displays, and for representations where the auditory modality is more efficient than the visual.

Sonification is an important part of the process of developing next-generation data representations. The data representations described in this chapter could be used in isolation, but would also be appropriate for inclusion in a multi-modal interactive interface, to redundantly encode the information for better description. As data analysis slowly moves off the page and onto the computer screen, touch-screen or mobile phone, interfaces of this nature will become more important.

Interactivity is key to a high-quality representation, and sonification of many types benefits immensely from a strong interactive interface. Many of the representations described are designed as a single-pass presentation of an overview of the data, although many information visualizations employ an interactive zoom-and-filter technique. This technique could be appropriate for the interactive control of sonifications. The speed at which the data is presented is one of its strengths, allowing the sonification to be started and repeated rapidly, and therefore capable of responding to interaction in a time-appropriate manner while remaining audible. Natural user interfaces employing multi-touch technology have now appeared in many human-computer interaction situations, and sonification research has already started to address priorities associated with this form of data interface [37, 8, 49]. Interfaces such as these may be predecessors to widespread use of sonification exploratory data analysis.

8.4.1 Research Challenges or Continuing Difficulties?

Data sonification is possibly not as common as it could be for a few reasons. One that has been mentioned by many authors is logistical in nature – many practitioners have little access to, or cannot easily use, sonification software. While much of the available software (such as the *sonification sandbox* [51], or *SonEnvir* [12]) is free and easily available, it does not necessarily fit within a typical data analysis workflow, and is non-existent within most major statistical packages or spreadsheet programs [26].

Another logistical problem is that sonification is often time-bound, while data representations of all types usually need to be scanned in a non-linear fashion, as the user seeks to build a conception of the relationships between different points by making comparisons between data points, axes and labels. The eye is able to quickly move between multiple elements in the graph obtaining various pieces of information. Where the time-axis is employed as a mapping axis in a sonification, the sonification must be replayed and the particular elements to be compared must be listened for. This is analogous to taking a visual graph, wrapping it onto a cylinder and reading it by rotating the cylinder. The eye has a great advantage over the ear in this regard, as it is capable of scanning non-linearly, in a way that the ear cannot. Of course, it is not necessarily problematic that sonifications are time-bound, since the sensitivity of the human auditory system to variations in temporal patterns can be quite acute under many circumstances. For example, in highly interactive sonification interfaces, a variety of alternatives to the linear presentation format are available, and these user-driven interactive explorations can reveal the benefits of the auditory modality, with significant improvements resulting from the possibility of listening and re-listening, comparing and re-assessing the presented data (see Chapter 11 for more discussion of interactive sonification).

Open questions remain - there are not many methods that exist for sonification of datasets that simultaneously present more than five or six dimensions. This difficulty exists in the visual domain, and is often solved through multiple views of the same data presented and interacted with simultaneously (see `ggobi` for instance [16])– an auditory analogue faces obvious difficulties as the ear cannot ‘shift its gaze’ as easily as the eye. With careful interaction design however, similar or complementary possibilities may prove possible for sonification. For very large datasets the mapping of distinct auditory elements to each data dimension is not practical, and future research may investigate possible methods for developing methods that scale well for large numbers of data dimensions. Common methods used in the study of genetic data, for instance, use large permutation matrices in the form of a heat map, with sophisticated methods of interaction to highlight differences or outliers in the data (see Fry [29], Chapter 4 for a discussion). No auditory analogue yet exists that does not use some form of data reduction, but the auditory sense’s capability for processing large amounts of information seems well suited for this type of data.

Also, visualization can provide numerical indicators of values (e.g., numbers on axes), while it is difficult for sonification to be so specific. Auditory tick-marks (including defining the start of the sonification, time 0), and exploiting either physical or psychoacoustical scales can help, but in general, statistical sonification will often provide a general sense of the data distribution without providing the user with access to specific numeric values.

8.5 Conclusion and Caveat

This chapter has described previous and current methods for representing multivariate data through statistical sonification for the purposes of exploratory data analysis. It must be said that the current state of the art must be considered to be quite immature as yet, with many challenges for sonification research to tackle in the future. In fact, it might be proposed that the best approach to take in designing and developing statistical sonifications in particular would be one that includes critical evaluation of the results at each attempt. Indeed, in the early development of scientific visualization methods, such a summary of practical case studies did appear in the published collection entitled ‘Visual cues: Practical data visualization’ [40]. Perhaps the sonification case studies that are presented in the handbook in which this chapter appears provide a useful beginning for such an endeavor.

In the absence of a well-established paradigm representing the consensus of practitioners in this field, a strategy might be taken in which the effectiveness of sonifications, in contrast to visualizations, would be put directly under test, so that ineffective sonifications could most easily be rejected. Through such rigor practitioners may become confident that their attempts have real value; yet without such rigorous evaluation, less useful sonification approaches may be accepted as worthy examples to be followed, before they have been adequately examined.

Bibliography

[1] E. Anderson. The Irises of the Gaspe Peninsula. *Bulletin of the American Iris Society*, 59:2–5, 1935.

[2] S. Barrass. *Auditory Information Design*. Ph.D. Thesis, 1997.

[3] S. Bateman, R. L. Mandryk, C. Gutwin, A. Genest, D. McDine, and C. Brooks. Useful Junk?: The Effects of Visual Embellishment on Comprehension and Memorability of Charts. In *CHI '10: Proceedings of the 28th International Conference on Human Factors in Computing Systems*, pp. 2573–2582, Atlanta, Georgia, USA, 2010. ACM.

[4] J. Bertin. *Graphics and Graphic Information-processing*. de Gruyter, Berlin; New York, 1981.

[5] J. Bertin. *Semiology of Graphics : Diagrams, Networks, Maps*. University of Wisconsin Press, Madison, Wis., 1983.

[6] S. A. Bly. Presenting Information in Sound. In *CHI '82: Proceedings of the 1982 Conference on Human Factors in Computing Systems*, pp. 371–375, Gaithersburg, Maryland, United States, 1982. ACM.

[7] S. A. Bly. *Sound and Computer Information Presentation*. Ph.D. Thesis, 1982.

[8] T. Bovermann, T. Hermann, and H. Ritter. Tangible Data Scanning Sonification Model. In *Proceedings of the 12th International Conference on Auditory Display*, London, UK, 2006.

[9] M. G. Bulmer. *Principles of Statistics*. Dover Publications, New York, 1979.

[10] D. Cabrera, S. Ferguson, and G. Laing. Considerations Arising from the Development of Auditory Alerts for air traffic control consoles. In *Proceedings of the 12th International Conference on Auditory Display*, London, UK, 2006.

[11] A. de Campo. Towards a data sonification design space map. In *Proceedings of the 13th International Conference on Auditory Display*, Montreal, Canada, 2007.

[12] A. de Campo, C. Frauenberger, and R. Höldrich. Designing a generalized sonification environment. In *Proceedings of the 10th International Conference on Auditory Display*, Sydney, Australia, 2004.

[13] H. Chernoff. The use of faces to represent points in k-dimensional space graphically. *Journal of the American Statistical Association*, 68(342):361–368, 1973.

[14] E. Childs. Achorripsis: A sonification of probability distributions. In *Proceedings of the 8th International Conference on Auditory Display*, Kyoto, Japan, 2002.

[15] W. S. Cleveland. *The Elements of Graphing Data*. Wadsworth Publ. Co., Belmont, CA, USA, 1985.

[16] D. Cook and D. F. Swayne. *Interactive and Dynamic Graphics for Data Analysis: With R and GGobi*. Springer, New York, 2007.

[17] A. de Falguerolles, F. Friedrich, and G. Sawitzki. A tribute to J. Bertin's graphical data analysis. In *Softstat '97: the 9th Conference on the Scientific Use of Statistical Software*, Heidelberg, 1997.

[18] R. O. Duda, P. E. Hart, and D. G. Stork. *Pattern Classification*. Wiley-Interscience, New York, USA, 2001.

[19] G. H. Dunteman. *Principal components analysis*. SAGE Publications, Inc, Thousand Oaks, CA, USA, 1989.

[20] J. Edworthy, E. Hellier, K. Aldrich, and S. Loxley. Designing trend-monitoring sounds for helicopters: methodological issues and an application. *Journal of Experimental Psychology: Applied*, 10(4):203–218, 2004.

[21] S. Ferguson. *Exploratory Sound Analysis: Statistical Sonifications for the Investigation of Sound*. Ph.D. Thesis, The University of Sydney, 2009.

[22] S. Ferguson and D. Cabrera. Exploratory sound analysis: Sonifying data about sound. In *Proceedings of the*

14th International Conference on Auditory Display, Paris, France, 2008.

[23] W. T. Fitch and G. Kramer. Sonifying the body electric: Superiority of an auditory over a visual display in a complex, multivariate system. In G. Kramer, ed., *Auditory Display*, 18:307–325. Addison-Wesley, 1994.

[24] J. H. Flowers. Thirteen years of reflection on auditory graphing: Promises, pitfalls, and potential new directions. In *Proceedings of the 11th International Conference on Auditory Display*, Limerick, Ireland, 2005.

[25] J. H. Flowers, D. C. Buhman, and K. D. Turnage. Cross-modal equivalence of visual and auditory scatterplots for exploring bivariate data samples. *Human Factors*, 39(3):341–351, 1997.

[26] J. H. Flowers, D. C. Burman, and K. D. Turnage. Data sonification from the desktop: Should sound be part of standard data analysis software? *ACM Transactions on Applied Perception (TAP)*, 2(4):467–472, 2005.

[27] J. H. Flowers and T. A. Hauer. The ear's versus the eye's potential to assess characteristics of numeric data: Are we too visuocentric? *Behaviour Research Methods, Instruments & Computers*, 24(2):258–264, 1992.

[28] R. M. Friedhoff and W. Benzon. *Visualization. The second computer revolution*. Freeman, New York, NY, USA, 1989.

[29] B. Fry. *Computational Information Design*. Ph.D. Thesis, 2004.

[30] S. P. Frysinger. A brief history of auditory data representation to the 1980s. In *Proceedings of the 11th International Conference on Auditory Display*, Limerick, Ireland, 2005.

[31] G. Grinstein and S. Smith. The perceptualization of scientific data. *Proceedings of the SPIE/SPSE Conference on Electronic Imaging*, pp. 190–199, 1990.

[32] L. Harrar and T. Stockman. Designing auditory graph overviews: An examination of discrete vs. continuous sound and the influence of presentation speed. In *Proceedings of the 13th International Conference on Auditory Display*, Montréal, Canada, 2007.

[33] T. Hermann. Sonification of Markov Chain Monte Carlo Simulations. In *Proceedings of the 7th International Conference on Auditory Display*, Helsinki, Finland, 2001.

[34] T. Hermann. *Sonification for Exploratory Data Analysis*. Ph.D. Thesis, 2002.

[35] T. Hermann. Taxonomy and definitions for sonification and auditory display. In *Proceedings of the 2008 International Conference on Auditory Display*, Paris, France, 2008.

[36] T.Hermann, G. Baier, U. Stephani, and H. Ritter. Vocal Sonification of Pathologic EEG Features. In *Proceedings of the 12th International Conference on Auditory Display*, London, UK, 2006.

[37] T. Hermann, T. Bovermann, E. Riedenklau, and H. Ritter. Tangible Computing for Interactive Sonification of Multivariate Data. In *International Workshop on Interactive Sonification*, York, UK, 2007.

[38] T. Hermann and H. Ritter. Listen to your data: Model-based sonification for data analysis. In G. E. Lasker, ed., *Advances in intelligent computing and multimedia systems*, pp. 189–194. Int. Inst. for Advanced Studies in System research and cybernetics, Baden-Baden, Germany, 1999.

[39] P. Janata and E. Childs. Marketbuzz: Sonification of real-time financial data. In *Proceedings of the 10th International Conference on Auditory Display*, Sydney, Australia, 2004.

[40] P. Keller and M. Keller. *Visual cues : practical data visualization*. IEEE Computer Society Press ; IEEE Press, Los Alamitos, CA Piscataway, NJ, 1993.

[41] G. Kramer, B. N. Walker, T. Bonebright, P. R. Cook, J. H. Flowers, N. Miner, and J. G. Neuhoff. Sonification report: Status of the field and research agenda. Technical report, National Science Foundation, 1997.

[42] P. P. Lennox, T. Myatt, and J. M. Vaughan. From surround to true 3-d. In Audio Engineering Society, editor, *16th International Audio Engineering Society Conference on Spatial Sound Reproduction*, Rovaniemi, Finland, 1999. Audio Engineering Society.

[43] S. Camille Peres and D. M. Lane. Sonification of statistical graphs. In *Proceedings of the 9th International Conference on Auditory Display*, Boston, MA, USA, 2003.

[44] I. Pollack and L. Ficks. Information of elementary multidimensional auditory displays. *Journal of the Acoustical Society of America*, 26(1):155–8, 1954.

[45] P. Sanderson, A. Wee, E. Seah, and P. Lacherez. Auditory alarms, medical standards, and urgency. In

Proceedings of the 12th International Conference on Auditory Display,, London, UK, 2006.

[46] E. R. Tufte. *The Visual Display of Quantitative Information.* Graphics Press, Cheshire, Conn, 1983.

[47] E. R. Tufte. *Envisioning Information.* Graphics Press, Cheshire, Conn, 1992.

[48] J. W. Tukey. *Exploratory Data Analysis.* Addison-Wesley, Reading, Mass., 1977.

[49] R. Tünnermann and T. Hermann. Multi-touch interactions for model-based sonification. In *Proceedings of the 15th International Conference on Auditory Display*, Copenhagen, Denmark, 2009.

[50] B. N. Walker. Magnitude estimation of conceptual data dimensions for use in sonification. *Journal of Experimental Psychology: Applied*, 8(4):211–221, 2002.

[51] B. N. Walker and J. T. Cothran. Sonification sandbox: A graphical toolkit for auditory graphs. In *Proceedings of the 9th International Conference on Auditory Display*, Boston, MA, USA, 2003.

[52] B. N. Walker, G. Kramer, and D. M. Lane. Psychophysical scaling of sonification mappings. In *Proceedings of the International Conference on Auditory Display*, Atlanta, 2000.

[53] B. N. Walker and M. A. Nees. An agenda for research and development of multimodal graphs. In *Proceedings of the 11th International Conference on Auditory Display*, Limerick, Ireland, 2005.

[54] M. P. Wand and M. C. Jones. *Kernel Smoothing.* Monographs on Statistics and Applied Probability. Chapman & Hall, 1995.

[55] C. Ware. *Information visualization: Perception for design.* Morgan Kaufman, San Francisco, 2000.

[56] H. Wickham. *ggplot2: Elegant graphics for data analysis.* Springer, New York, 2009.

[57] L. Wilkinson. *The Grammar of Graphics (2nd Edition).* Springer, Berlin, 2005.

[58] J. Williamson and R. Murray-Smith. Granular synthesis for display of time-varying probability densities. In *International Workshop on Interactive Sonification*, Bielefeld, Germany, 2004.

[59] J. Williamson and R. Murray-Smith. Sonification of probabilistic feedback through granular synthesis. *IEEE Multimedia*, 12(2), pp. 5–52, 2005.

Chapter 9

Sound Synthesis for Auditory Display

Perry R. Cook

9.1 Introduction and Chapter Overview

Applications and research in auditory display require sound synthesis and manipulation algorithms that afford careful control over the sonic results. The long legacy of research in speech, computer music, acoustics, and human audio perception has yielded a wide variety of sound analysis/processing/synthesis algorithms that the auditory display designer may use. This chapter surveys algorithms and techniques for digital sound synthesis as related to auditory display.

Designers of auditory displays and systems employing sound at the user interface need flexible and powerful means to generate and process sounds. So the approach here is to present the basic technology behind each algorithm, then view it by examining the parameters it provides to the sound/interaction designer to allow manipulation of the final sonic result.

This chapter first walks through most of the major means for synthesizing audio, in descending order from non-parametric to highly parametric, eventually summarizing those in a concluding section. Techniques that begin with pre-recorded audio, then analyze, process, and resynthesize sound, will be followed by techniques that synthesize audio directly "from scratch" or models. But before we proceed, we will take a moment to clearly define "parametric."

9.2 Parametric vs. Non-Parametric Models

Here the word "parametric" is used as it is used in engineering "system identification" (fitting a model to the observable inputs and outputs of some system of interest). A "parametric" model is defined here as one that has a (relatively) few variable parameters that can be manipulated to change the interaction, sound, and perception. A highly parametric model of sound is the technique known as Linear Predictive Coding (LPC, discussed in a later section), which uses just a few numbers representing the spectral shape, and the (usually voice) source, to represent thousands of PCM samples. A close representation of an original block of samples can be resynthesized by running a source/filter model, informed by the extracted parameters. Further, we can synthesize longer or shorter (by just running the model slower or faster) while retaining all other parameter values such as pitch and spectral shape. We can also synthesize with higher or lower pitch, or turn sung vowels into whispers (less pitch, more noise), because the few parameters we have are meaningful and influential. As the LPC example points out, extracting a few powerful parameters from raw audio is closely related to audio compression and coding.

As a counter example, a non-parametric model of a particular sound would be the raw samples (called PCM as discussed in the next section), because the "model" has no small set of parameters that allows us to modify the sound in meaningful ways. The Fourier Transform (discussed at length two sections from now), while powerful for many reasons, is also a non-parametric model, in that it turns N time waveform samples into N frequency values, but those (equal number of) parameters don't allow us to manipulate the interaction, sound, and perception in a low-dimensional and flexible manner.

Of course, "parametricity" (not likely a word, but used here to represent how parametric a model/technique is), is relative, and a little tricky. If we find a way to represent 10,000 samples of 8-bit wave data with 9,000 bytes, maybe by just running those samples through a standard text symbol compressor such as WinZip, we will have reduced the size of the representation, but the 9,000 bytes aren't really parameters, since they do little to let us manipulate the "resynthesis." On the other hand, we could "code" every song released commercially in digital form by a fairly small and unique "tag" representing the serial number of the CD release, and the track number. This one small tag number is arguably not a "parameter", since even the slightest change in this number will yield a totally different recording of a totally different song.

Thus the definition of a "parameter" in this chapter is: **a (likely continuous) variable that, when changed slightly, yields slight changes in the synthesized sound, and when changed greatly, makes great changes.** The parametricity of the algorithm is determined based on the space of possible output sounds, relative to the number of such parameters. Herein lies flexible power for auditory display, because we can map data to those powerful parameters.

9.3 Digital Audio: The Basics of PCM

Digital audio signals are recorded by *sampling* analog (continuous in time and amplitude) signals at regular intervals in time, and then *quantizing* the amplitudes to discrete values. The process of sampling a waveform, holding the value, and quantizing the value to the nearest number that can be digitally represented (as a specific integer on a finite range of integers) is called Analog to Digital (A to D, or A/D) conversion [1]. A device that does A/D conversion is called an Analog to Digital Converter (ADC). Coding and representing waveforms in sampled digital form is called Pulse Code Modulation (PCM), and digital audio signals are often called PCM audio. The process of converting a sampled signal back into an analog signal is called Digital to Analog Conversion (D to A, or D/A), and the device which does this is called a Digital to Analog Converter (DAC). Low-pass filtering (smoothing the samples to remove unwanted high-frequencies) is necessary to reconstruct the sampled signal back into a smooth continuous time analog signal. This filtering is usually contained in the DAC hardware.

The time between successive samples is usually denoted as T. Sampling an analog signal first requires filtering it to remove unwanted high frequencies, to avoid "aliasing." Aliasing is caused by under-sampling frequencies higher than half the sample rate, causing them to not be accurately represented, as shown in Figure 9.1. The pre-filtering must eliminate all frequencies higher than half the sampling rate.

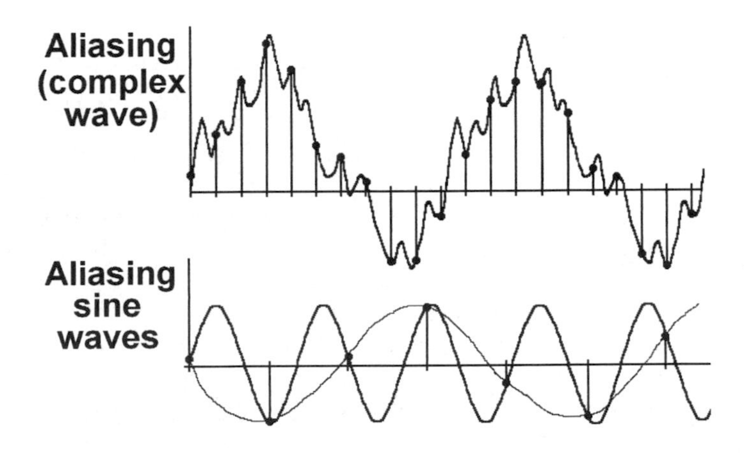

Figure 9.1: Because of inadequate sampling rate, aliasing causes important features to be lost.

The next step in Analog to Digital Conversion is to hold each waveform value steady for a period (using a Pulse) while a stable measurement can be made, then associating the analog value with a digital number (Coding). So PCM means to Modulate the analog signal with a Pulse, measure the value for that instant, then Code it into a digital number. Analog signals can have any of the infinity of real-numbered amplitude values. Since computers work with fixed word sizes (8-bit bytes, 16 bit words, etc.), digital signals can only have a finite number of amplitude values. In converting from analog to digital, rounding takes place and a given analog value must be quantized to the nearest digital value. The difference between quantization steps is called the quantum (not as in quantum physics or leaps, but that's just the Latin word for a fixed sized jump in value or magnitude). Sampling and quantization is shown in Figure 9.2. Note the errors introduced in some sample values due to the quantization process.

Humans can perceive frequencies from roughly 20 Hz to 20 kHz, thus requiring a minimum sampling rate of at least 40 kHz. Speech signals are often sampled at 8kHz ("telephone quality") or 11.025 kHz, while music is usually sampled at 44.1 kHz (the sampling rate used on audio Compact Disks), or 48 kHz. Some new formats allow for sampling rates of 96 kHz, and even 192 kHz.

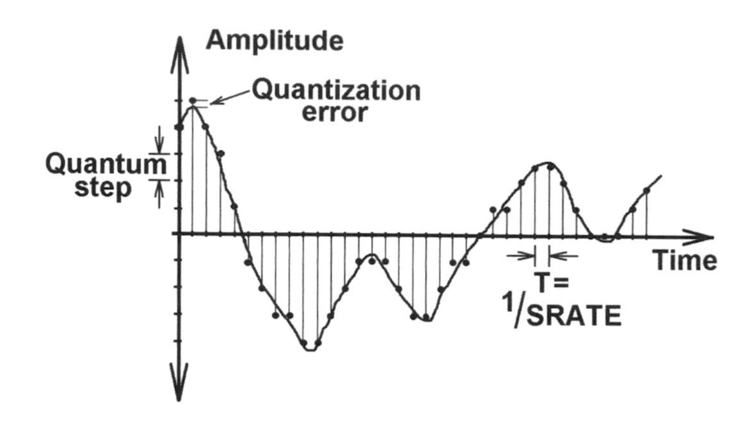

Figure 9.2: Linear sampling and quantization.

In a digital system, a fixed number of binary digits (bits) are used to sample the analog waveform, by quantizing it to the closest number that can be represented. This quantization is accomplished either by rounding to the quantum value nearest the actual analog value, or by truncation to the nearest quantum value less than or equal to the actual analog value. With uniform sampling in time, a properly band-limited signal can be exactly recovered provided that the sampling rate is twice the bandwidth or greater, but only if there is no quantization. When the signal values are rounded or truncated, the amplitude difference between the original signal and the quantized signal is lost forever. This can be viewed as an additive noise component upon reconstruction. Using the additive noise assumption gives an approximate best-case signal to quantization noise ratio (SNR) of approximately 6N dB, where N is the number of bits. Using this approximation implies that a 16 bit linear quantization system will exhibit an SNR of approximately 96 dB. 8 bit quantization exhibits a signal to quantization noise of approximately 48 dB. Each extra bit improves the signal to noise ratio by about 6 dB. Exact formulas for this are given in [1].

Most computer audio systems use two or three types of audio data words. As the data format used in Compact Disk systems, 16 bit (per channel) data is quite common. High definition formats allow for 24 bit samples. 8-bit data is common for speech data in PC and telephone systems, usually using methods of quantization that are non-linear. In non-linear quantization systems (mu-law or a-law) the quantum is smaller for small amplitudes, and larger for large amplitudes.

9.3.1 PCM (Wavetable, Sampling, Concatenative (Speech)) Synthesis

The majority of speech, music, and sound "synthesis" today is accomplished via the playback of stored PCM (Pulse Code Modulation) waveforms. Single-shot playback of entire segments of stored sounds is common for sound effects, narrations, prompts, segments of music, etc. Most high quality modern electronic music synthesizers, speech synthesis systems, and PC software systems for sound synthesis use pre-stored PCM as the basic data. This data is sometimes manipulated by filtering, pitch shifting, looping, and other means to yield the final output sound(s).

For speech, the most common synthesis technique is "concatenative" synthesis [2]. Concatenative phoneme synthesis relies on end-to-end splicing of roughly 40 (for English) pre-stored phonemes. Examples of vowel phonemes are /i/ as in beet, /I/ as in bit, /a/ as in father, /u/ as in boot, etc. Examples of nasals are /m/ as in mom, /n/ as in none, /ng/ as in sing, etc. Examples of fricative consonant phonemes are /s/ as in sit, /sh/ as in ship, /f/ as in fifty. Examples of voiced fricative consonants are /v/, /z/ (visualize). Examples of plosive consonants are /t/ as in tat, /p/ as in pop, /k/ as in kick, etc. Examples of voiced plosives include /d/, /b/, /g/ (dude, bob, & gag). Vowels and nasals are pitched periodic sounds, so the minimal required stored waveform is only one single period of each. Consonants require more storage because of their noisy (non-pitched, aperiodic) nature. Sound and movie examples **S9.1** and **S9.2** demonstrate concatenative voice/speech synthesis.

The quality of concatenative phoneme synthesis is generally considered quite low, due to the simplistic assumption that all of the pitched sounds (vowels, etc.) are purely periodic. Also, simply "gluing" /s/ /I/ and /ng/ together does not make for a high quality realistic synthesis of the word "sing." In actual speech, phonemes gradually blend to each other as the jaw, tongue, and other "articulators" move with time.

Accurately capturing the transitions between phonemes with PCM requires recording transitions from phoneme to phoneme, called "diphones". A concatenative diphone synthesizer blends together stored diphones. Examples of diphones include see, she, thee, and a subset of the roughly 40x40 possible combinations of phonemes. Much more storage is necessary for a diphone synthesizer, but the resulting increase in quality is significant.

Changing the playback sample rate on sampled sound results in a shift in pitch, time, and spectral shape. Many systems for recording, playback, processing, and synthesis of music, speech, or other sounds allow or require flexible control of pitch (sample rate). The most accurate pitch control is necessary for music synthesis. In sampling synthesis, this is accomplished by dynamic sample rate conversion (interpolation), which has three steps; band-limiting, interpolation, and re-sampling. The band-limiting is the same as is required for sampling, so if the new sample rate is lower than the original, frequencies higher than half the new rate must be removed.

Interpolation is the process of filling in the smooth waveform between existing samples, and can be accomplished by fitting line segments to the samples (not the best method, due to artifacts from the jagged edges), higher order curves (splines), or other means. The provably correct way (from engineering mathematics) to interpolate is to fit a *sinc* function to the samples, defined as:

$$\mathrm{sinc}(t/T) = \frac{\sin(\pi t/T)}{\pi t/T}, \quad \text{where } T = 1/\mathrm{SRATE} \tag{1}$$

The sinc function is the ideal reconstruction filter, but comes at a significant computational cost, so the designer of a high quality sample rate converter will choose an appropriately truncated sinc function [3] to meet the quality constraints of the system. Once the smooth waveform is reconstructed, it can then be re-sampled at the new target sample rate.

Note that stored voiced speech phonemes (pitched vowels are periodic, so can be stored as a single period and synthesized by looping that single period), or the phoneme components of diphones, can be shifted in pitch by sample rate conversion, allowing "prosody" to be imparted on synthesized speech. However, the pitch cannot be shifted too far in either direction, because the spectral properties shift accordingly, and the synthesized speaker can begin to sound like a chipmunk (shifted upward, therefore sounding like a smaller head size), or a giant (shifted downward, large head). PCM speech synthesis can be improved further by storing multi-samples, for different pitch ranges, genders, voice qualities, individual speaker voices, accents, etc.

9.3.2 Making PCM Parametric

In fact, storing the individual phonemes or diphones for speech synthesis is a form of hand-crafted parameterization. Noting that there are only 40 phonemes, and actually only a few hundred important diphones (any given language uses only a small subset of commonly occurring phoneme combinations), the index number of each phoneme/diphone can be considered a low-dimensional parameter. Combined with pitch shifting by interpolation, and the ability to loop phonemes for arbitrary lengths, a speech synthesizer becomes a form of parametric synthesizer. But from our definition, we desire *continuous* parameters that influence the sound as a reliable function of how much we perturb them.

For musical sounds, it is common to store only a loop, or wavetable, of the periodic component of a recorded sound waveform and play that loop back repeatedly. This is sometimes called "Wavetable Synthesis" [4], primarily in musical synthesis. In speech and other sound synthesis the more common term is "concatenative." For more realism, the attack or beginning portion of the recorded sound can be stored in addition to the periodic steady state part. Figure 9.3 shows the synthesis of a trumpet tone starting with an attack segment, followed by repetition of a periodic loop, ending with an enveloped decay (or release). "Envelope" is a synthesizer/computer music term for a time-varying change applied to a waveform amplitude, or other parameter. Envelopes are often described by four components; the *Attack Time*, the *Decay Time* ("decay" here means the initial decay down to the steady state segment), the *Sustain Level*, and the *Release Time* (final decay). Hence, envelopes are sometimes called *ADSR's*.

Originally called "Sampling Synthesis" in the music industry, any synthesis using stored PCM waveforms has now become commonly known as "Wavetable Synthesis". Filters are usually added to high-quality wavetable synthesis, allowing control of spectral brightness as a function of intensity, and to get more variety of sounds from a given set of samples. Thus making the model more parametric.

As discussed in the previous section, a given sample can be pitch-shifted only so far in either direction before it begins to sound unnatural. This can be dealt with by storing multiple recordings of the sound at different pitches, and switching or interpolating between these upon resynthesis. In music sampling and speech synthesizers, this is called "multi-sampling". Multi-sampling also might include the storage of separate samples for "loud" and "soft" sounds. Linear or other interpolation is used to blend the loudness of multi-samples as a function of the desired synthesized volume. This adds realism, for loudness is not simply a matter of amplitude or power; most sound sources exhibit spectral variations as a function of loudness due to driving energy and non-linearity. There is usually more high frequency energy ("brightness") in loud sounds than in soft sounds. Filters can also be used to add

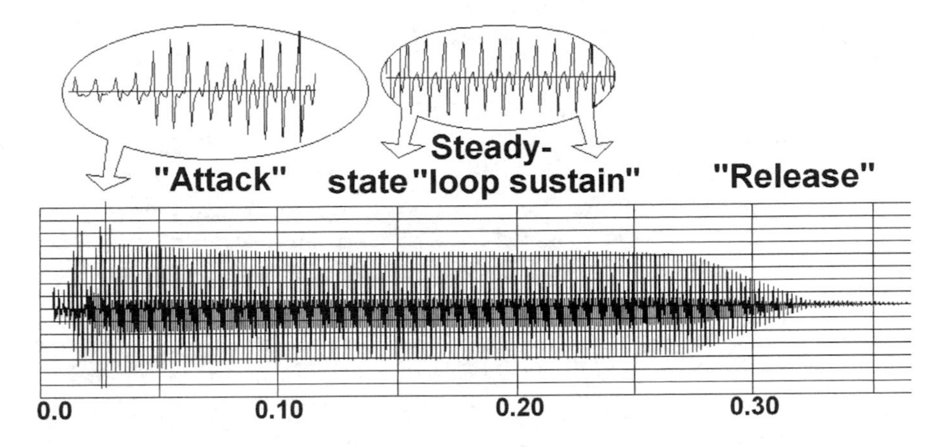

Figure 9.3: Wavetable synthesis of trumpet tone.

spectral variation.

A common tool used to describe the various components and steps of signal processing in performing digital music synthesis is the "synthesizer patch" (historically named from hooking various electrical components together using patch cords). In a patch, a set of fairly commonly agreed building blocks, called "unit generators" (also called modules, plug-ins, operators, op-codes, and other terms) are hooked together in a signal flow diagram. This historical [5] graphical method of describing signal generation and processing affords a visual representation that is easily printed in papers, textbooks, patents, etc. Further, graphical patching systems and languages have been important to the development and popularization of certain algorithms, and computer music in general. Figure 9.4 shows a PCM synthesizer patch with attack and loop wavetables whose amplitudes are controlled by an envelope generator, and a time-varying filter (also controlled by another envelope generator). As with all synthesis, panning (placement in stereo or more channels) can be controlled as an additional parameter.

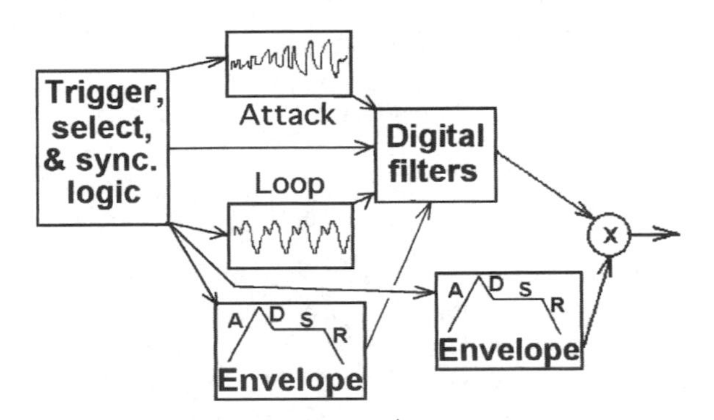

Figure 9.4: Block diagram of a wavetable synthesizer "patch", showing connections of unit generators such as wavetables, ADSR envelopes, digital filters, etc.

9.4 Fourier (Sinusoidal) "Synthesis"

Lots of sound-producing objects and systems exhibit sinusoidal modes, which are the natural oscillatory frequencies of any acoustical system. A plucked string might exhibit many modes, with the strength of each mode determined by the conditions of the terminations, and the nature of the excitation pluck (plucking at the end vs. the center). Striking a metal plate with a hammer excites many of the vibrational modes of the plate, determined by the shape of the plate, and by where it is struck. A singing voice, struck drum head, bowed violin string, struck bell, or blown trumpet exhibit oscillations characterized by a sum of sinusoids. The recognition of the fundamental nature of the sinusoid gives rise to a powerful model of sound synthesis based on summing up sinusoidal modes.

These modes have a very special relationship in the case of the plucked string, a singing voice, and some other limited systems, in that their frequencies are all integer multiples (at least approximately) of one basic sinusoid, called the "fundamental." This special series of sinusoids is called a "harmonic series", and lies at the basis of the "Fourier Series" representation of oscillations, waveforms, shapes, etc. The Fourier Series [6] solves many types of problems, including physical problems with boundary constraints, but is also applicable to any shape or function. Any periodic waveform (repeating over and over again) F_{per} can be transformed into a Fourier series, written as:

$$F_{\text{per}}(t) = a_0 + \sum_m [b_m \cos(2\pi f_0 m t) + c_m \sin(2\pi f_0 m t)] \tag{2}$$

Which states mathematically that any periodic function can be expressed as a sum of harmonically (integer multiples of some fundamental frequency) related sine and cosine functions, plus an arbitrary constant.

The limits of the summation are technically infinite, but we know that we can (and should) cut off frequencies at ½ the sampling frequency for digital signals. The a_0 term is a constant offset, or the average of the waveform. The b_m and c_m coefficients are the weights of the "m^{th} harmonic" cosine and sine terms. If the function $F_{\text{per}}(t)$ is purely "even" about $t = 0 \quad (F(-t) = F(t))$, only cosines are required to represent it, and all of the c_m terms would be zero. Similarly, if the function $F_{\text{per}}(t)$ is "odd" $(F(-t) = -F(t))$, only the c_m terms would be required. An arbitrary function $F_{\text{per}}(t)$ will require sinusoidal harmonics of arbitrary (but specific) amplitudes and phases. The magnitude A and phase θ of the mth harmonic in the Fourier Series can be found by:

$$A_m = \sqrt{b_m^2 + c_m^2} \tag{3}$$

$$\theta_m = \text{ArcTan}(c_m / b_m) \tag{4}$$

Phase is defined relative to the cosine, so if c_m is zero, θ_m is zero. As a brief example, Figure 9.5 shows the first few sinusoidal harmonics required to build up an approximation of a square wave. Note that due to symmetries only odd sine harmonics $(1, 3, 5, 7)$ are required. The amplitudes of the sine waves are expressed as $1/M$, where M is the harmonic number. Using more sines improves the approximation of the resulting synthesis, moving toward a pure square wave.

The process of determining the sine and cosine components of a signal or waveform is called "Fourier Analysis", or the "Fourier Transform". If the frequency variable is sampled (as is

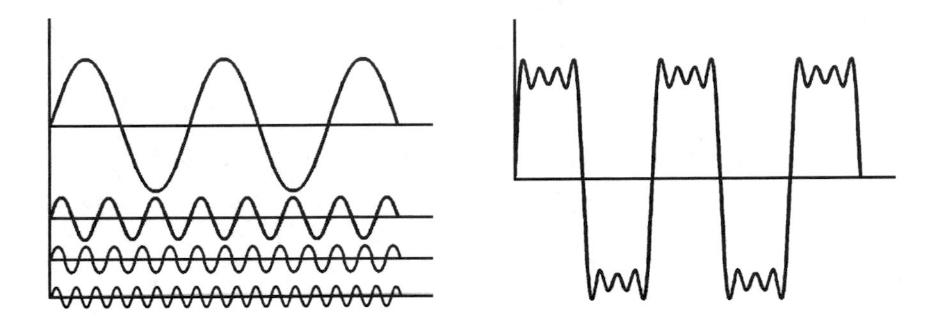

Figure 9.5: A sum of odd harmonics approximates a square wave.

the case in the Fourier Series, represented by m), and the time variable t is sampled as well (as it is in PCM waveform data, represented by n), then the Fourier Transform is called the "Discrete Fourier Transform", or DFT. The DFT is given by:

$$F(m) = \sum_{n=0}^{N-1} f(n) \left[\cos(2\pi mn/N) - j\sin(2\pi mn/N)\right] \tag{5}$$

Where N is the length (in samples) of the signal being analyzed. The inverse DFT (IDFT) is similar to the Fourier Series:

$$f(n) = \frac{1}{N} \sum_{m=0}^{N-1} F(m) \left[\cos(2\pi mn/N) + j\sin(2\pi mn/N)\right] \tag{6}$$

The convention is to use lower case for the time domain and upper case for the frequency domain. So $f(n)$ is the time-waveform (a sound for example), and $F(m)$ represents the spectral description.

The imaginary number $j = \sqrt{-1}$ is used to place the cosine and sine components in a unique mathematical arrangement, where odd ($x(-n) = -x(n)$) sine terms of the waveform are represented as imaginary components, and even ($x(-n) = x(n)$) cosine terms are represented as real components. This gives us a way of talking about the magnitude and phase in terms of the magnitude and phase of $F(m)$ (a complex number). There is a near-mystical expression of equality in mathematics known as Euler's Identity, which links trigonometry, exponential functions, and complex numbers in a single equation:

$$e^{j\theta} = \cos(\theta) + j\sin(\theta)\ . \tag{7}$$

We can use Euler's identity to write the DFT and IDFT in shorthand:

$$F(m) = \sum_{n=0}^{N-1} f(n) e^{-j2\pi mn/N} \tag{8}$$

$$f(n) = \frac{1}{N} \sum_{m=0}^{N-1} F(m) e^{j2\pi mn/N} \tag{9}$$

Converting the cosine/sine form to the complex exponential form allows lots of manipulations that would be difficult otherwise. But we can also write the DFT in real number terms as a form of the Fourier Series:

$$f(n) = \frac{1}{N} \sum_{m=0}^{N-1} F_b(n) \cos(2\pi mn/N) + F_c(n) \sin(2\pi mn/N) \tag{10}$$

where

$$F_b(m) = \sum_{n=0}^{N-1} f(n) \cos(2\pi mn/N) \tag{11}$$

$$F_c(m) = \sum_{n=0}^{N-1} -f(n) \sin(2\pi mn/N) \tag{12}$$

The Fast Fourier Transform (FFT) is a computationally efficient way of calculating the DFT. There are thousands of references on the FFT [6], and scores of implementations of it, so for our purposes we'll just say that it's lots more efficient than trying to compute the DFT directly from the definition. A well crafted FFT algorithm for real input data takes on the order of $N \log_2(N)$ multiply-adds to compute. Comparing this to the N^2 multiplies of the DFT, N doesn't have to be very big before the FFT is a winner. There are some downsides, such as the fact that FFTs can only be computed for signals whose lengths are exactly powers of 2, but the advantages of using it often outweigh the pesky power-of-two problems. Practically speaking, users of the FFT usually carve up signals into small chunks (powers of two), or "zero pad" a signal out to the next biggest power of two.

The Short-Time Fourier Transform (STFT) breaks up the signal into (usually overlapping) segments and applies the Fourier Transform to each segment individually [7]. By selecting the window size (length of the segments), and hop size (how far the window is advanced along the signal each step) to be perceptually relevant, the STFT can be thought of as a simple approximation of human audio perception. Figure 9.6 shows the waveform of the utterance of the word "synthesize", and some STFT spectra corresponding to windows at particular points in time.

9.4.1 Direct Fourier "Synthesis"

Fourier synthesis is essentially just the process of reconstructing the time domain waveform from the sines and cosines indicated by the Fourier Transform. In other words, it is the Inverse Fourier Transform. As such it is essentially the same as PCM synthesis, providing no meaningful parameters for transformations. There are ways to parameterize, however.

Using the Short Time Fourier Transform, the "Phase Vocoder" (VoiceCoder) [8, 9] processes sound by calculating and maintaining both magnitude and phase. The frequency bins (basis sinusoids) of the DFT can be viewed as narrowband filters, so the Fourier Transform of an input signal can be viewed as passing it through a bank of narrow band-pass filters. This means that on the order of hundreds to thousands of sub-bands are used.

The Phase Vocoder has found extensive use in computer music composition and sound design. Many interesting practical and artistic transformations can be accomplished using

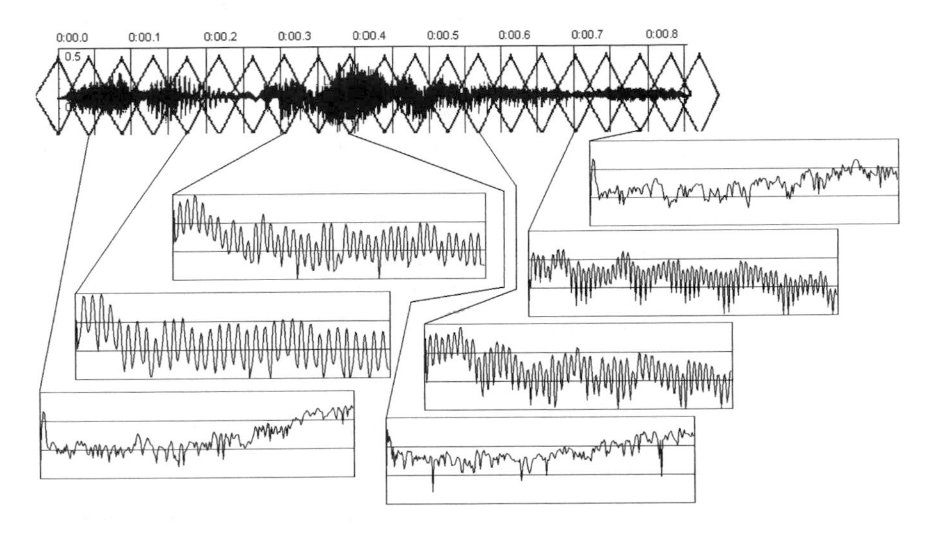

Figure 9.6: Some STFT frames of the word "synthesize".

the Phase Vocoder, including nearly artifact-free and independent time and pitch shifting, as demonstrated in sound example **S9.3**. A technique called "cross synthesis" assumes that one signal is the "analysis" signal. The time-varying magnitude spectrum of the analysis signal (usually smoothed in the frequency domain) is multiplied by the spectral frames of another "input" (or filtered) signal (often "brightened" by high frequency pre-emphasis), yielding a composite signal that has the attributes of both. Cross-synthesis has produced the sounds of talking cows, "morphs" between people and cats, trumpet/flute hybrids, etc.

These techniques are useful for analyzing and modifying sounds in some ways, but for auditory display, we can do more to make Fourier-related methods more parametric.

9.4.2 Making Fourier More Parametric

While the Fourier Transform is not parametric as directly implemented, we can use the Fourier Transform to extract useful parametric information about sounds. A brief list of audio "features" (also called descriptors) that can be extracted is:

- Gross Power in each window. If the audio stream suddenly gets louder or softer, then there is a high likelihood that something different is occurring. In speech recognition and some other tasks, however, we would like the classification to be loudness invariant (over some threshold used to determine if anyone is speaking).
- Spectral Centroid, which relates closely to the brightness of the sound, or the relative amounts of high and low frequency energy.
- Rolloff: Spectra almost always decrease in energy with increasing frequency. Rolloff is a measure of how rapidly, and is another important feature that captures more information about the brightness of an audio signal.
- Spectral Flux is the amount of frame-to-frame variance in the spectral shape. A steady

sound or texture will exhibit little spectral flux, while a modulating sound will exhibit more flux.

- Mel-Frequency Cepstral Coefficients, which are a compact (between 4 and 10 numbers) representation of spectral shape. LPC coefficients are sometimes used in this way as well.
- Low Energy is a feature, defined as the percentage of small analysis windows that contain less power than the average over the larger window that includes the smaller windows. This is a coarser time-scale version of flux, but computed only for energy.
- Zero-Crossing Rate is a simple measure of high frequency energy.
- Harmonicity is a measure of the "pitchyness" (and pitch) of a sound.
- Harmonics to Noise Ratio is a measure of the "breathiness" of a sound.
- Parametric Pitch Histogram is a multi-pitch estimate.
- Beat/Periodicity Histogram is a measure of beat (rhythm) strength and timing.

All of these can be extracted and used to understand, classify, and describe sounds, as is done in audio analysis, music information retrieval, content-based query, etc. [10]. However they are not sufficient for direct synthesis.

If we inspect the various spectra in Figure 9.6, we can note that the vowels exhibit harmonic spectra (clear, evenly spaced peaks corresponding to the harmonics of the pitched voice), while the consonants exhibit noisy spectra (no clear sinusoidal peaks). Recognizing that some sounds are well approximated/modeled by additive sine waves [11], while other sounds are essentially noisy, "spectral modeling" [12] breaks the sound into deterministic (sines) and stochastic (noise) components. Figure 9.7 shows a general Sines+Noise Additive Synthesis model, allowing us to control the amplitudes and frequencies of a number of sinusoidal oscillators, and model the noisy component with a noise source and a spectral shaping filter.

The beauty of this type of model is that it recognizes the dominant sinusoidal nature of many sounds, while still recognizing the noisy components that might be also present. More efficient and parametric representations, and many interesting modifications, can be made to the signal on resynthesis. For example, removing the harmonics from voiced speech, followed by resynthesizing with a scaled version of the noise residual, can result in the synthesis of whispered speech.

One further improvement to spectral modeling is the recognition [13] that there are often brief (impulsive) moments in sounds that are really too short in time to be adequately analyzed by spectrum analysis. Further, such moments in the signal usually corrupt the sinusoidal/noise analysis process. Such events, called transients, can be modeled other ways (often by simply keeping the stored PCM for that segment). As with Fourier synthesis, Spectral Modeling is most useful for transformation and modification of existing sounds. Indeed, some meaningful parameters on noise, spectral shape, and transient extraction could be exploited during resynthesis for auditory display. An excellent reference to Fourier and frequency domain techniques, and signal processing in general, can be found in [14].

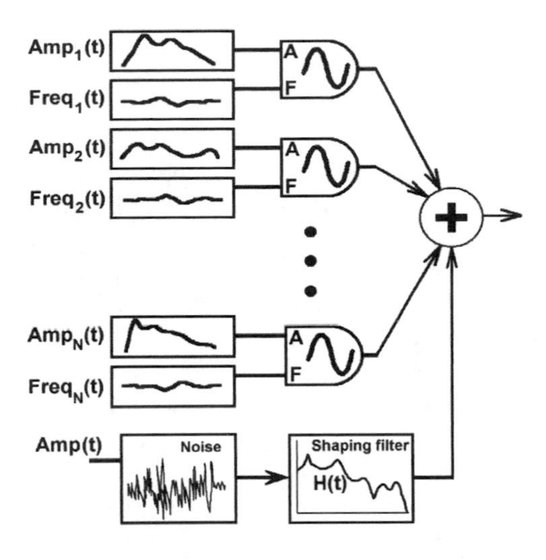

Figure 9.7: Sinusoidal model with filtered noise added for spectral modeling synthesis.

9.5 Modal (Damped Sinusoidal) Synthesis

The simplest physical system that does something acoustically (and musically) interesting is the mass-spring-damper [15]. The differential equation describing that system has a solution that is a single exponentially-decaying cosine wave. Another common system that behaves the same way is a pendulum under small displacements. The swinging back and forth of the pendulum follows the same exponentially-decaying cosine function. Yet one more system, the Helmholtz resonator (a large cavity, containing air, with a small long-necked opening, like a pop bottle), behaves like a mass-spring-damper system, with the same exponentially damped cosine behavior.

The equations describing the behavior of all of these systems, where m = mass, r = damping, and k = spring constant (restoring force) is:

$$\frac{d^2y}{dt^2} + \frac{r}{m}\frac{dy}{dt} + \frac{k}{m}y = 0 \tag{13}$$

$$y(t) = y_0 e^{(-rt/2m)} \cos\left(t\sqrt{\frac{k}{m} - \left(\frac{r}{2m}\right)^2}\right) \tag{14}$$

Where y is the displacement of the mass, dy/dt is the velocity of the mass, and d^2y/dt^2 is the acceleration of the mass. Of course, most systems that produce sound are more complex than the ideal mass-spring-damper system, or a pop bottle. And of course most sounds are more complex than a simple damped exponential sinusoid. Mathematical expressions of the physical forces (and thus the accelerations) can be written for nearly any system, but solving such equations is often difficult or impossible. Some systems have simple enough properties and geometries to allow an exact solution to be written out for their vibrational behavior. An ideal string under tension is one such system.

This section will present some graphical arguments and refers to the previous discussion of the Fourier Transform to further motivate the notion of sinusoids in real physical systems. The top of Figure 9.8 shows a string, lifted from a point in the center (half-way along its length). Below that is shown a set of sinusoidal "modes" that the center-plucked string vibration would have. These are spatial functions (sine as function of position along the string), but they also correspond to the natural frequencies of vibration of the string. At the bottom of Figure 9.8 is another set of modes that would not be possible with the center-plucked condition, because all of these "even" modes are restricted to have no vibration in the center of the string, and thus they could not contribute to the triangular shape of the 'initial central pluck condition' of the string. These conditions of no displacement, corresponding to the zero crossings of the sine functions, are called "nodes." Note that the end points are forced nodes of the plucked string system for all possible conditions of excitation.

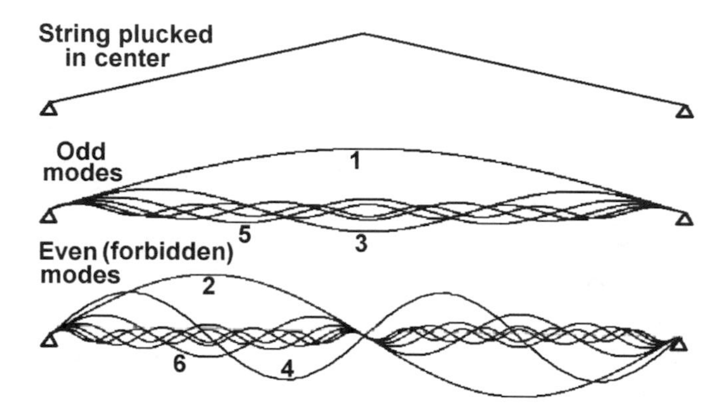

Figure 9.8: Plucked string (top). Center shows sinusoidal "modes" of vibration of a center-plucked string. Bottom shows the even modes, which would not be excited by the center-plucked condition. End points are called "nodes" of no-vibration.

Physical constraints on a system, such as the pinned ends of a string, and the center plucked initial shape, are known as "boundary conditions." Spatial sinusoidal solutions like those shown in Figure 9.8 are called "boundary solutions" (the legal sinusoidal modes of displacement and vibration) [16].

Just as Fourier Boundary methods (Fourier solutions taking into account physical limits and symmetries) can be used to solve the one-dimensional string, we can also extend boundary methods to two dimensions. Figure 9.9 shows the first few vibrational modes of a uniform square membrane. The little boxes at the lower left corners of each square modal shape depict the modes in a purely 2-dimensional way, showing lines corresponding to the spatial sinusoidal nodes (regions of no displacement vibration). Circular drum heads are more complex, but still exhibit a series of circular and radial modes of vibration. The square membrane modes are not integer-related inharmonic frequencies. In fact they obey the relationship:

$$f_{mn} = f_{11}\sqrt{(m^2 + n^2)/2} \tag{15}$$

where m and n range from 1 to (potentially) infinity, and f_{11} is $c/2L$ (speed of sound on the membrane divided by the square edge lengths).

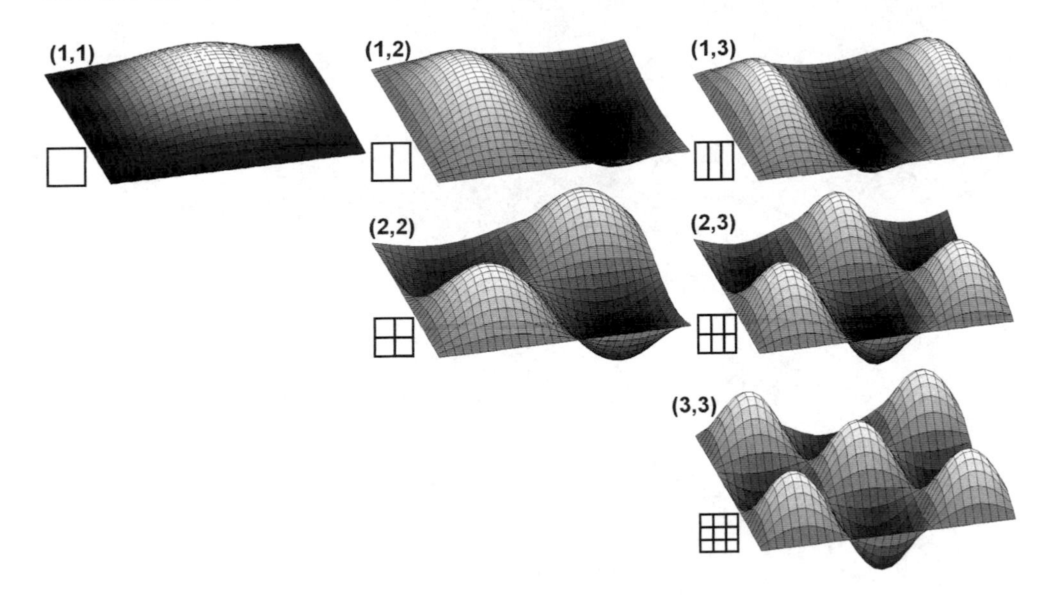

Figure 9.9: Square membrane vibration modes.

Unfortunately, circles, rectangles, and other simple geometries turn out to be the only ones for which the boundary conditions yield a closed-form solution in terms of spatial and temporal sinusoidal components. However, we can measure and model the modes of any system by computing the Fourier Transform of the sound it produces, and by looking for exponentially decaying sinusoidal components.

We can approximate the differential equation describing the mass-spring-damper system of Equation 13 by replacing the derivatives (velocity as the derivative of position, and acceleration as the 2nd derivative of position) with sampled time differences (normalized by the sampling interval T seconds). In doing so we arrive at an equation that is a recursion in past values of $y(n)$, the position variable:

$$\frac{y(n) - 2y(n-1) + y(n-2)}{T^2} + \frac{r}{m}\frac{y(n) - y(n-1)}{T} + \frac{k}{m}y(n) = 0 \qquad (16)$$

where $y(n)$ is the current value, $y(n-1)$ is the value one sample ago, and $y(n-2)$ is the twice-delayed sample. Note that if the values of mass, damping, spring constant, and sampling rate are constant, then the coefficients ($(2m+Tr)/(m+Tr+T^2k)$ for the single delay, and $m/(m+Tr+T^2k)$ for the twice delayed signal) applied to past y values are constant. DSP engineers would note that a standard Infinite Impulse Response (IIR) recursive filter as shown in Figure 9.10 can be used to implement Equation 16 (the Z^{-1} represents a single sample of delay). In fact, equation 16 (called the 2nd order 2-pole feedback filter by Digital Signal Processing engineers) can be used to generate an exponentially decaying sinusoid, called a "phasor" in DSP literature [17]. Here the term "filter" is used to mean anything that takes a signal as input, yields a signal as output, and does something interesting between (not strictly a requirement that it do something interesting, but why bother if not?). The connection between the 2nd order digital filter and the physical notion of a mode of vibration forms the basis for Modal Sound Synthesis [18], where a spectrally rich source such as an impulse or noise is used to excited modal filters to generate a variety of natural sounds.

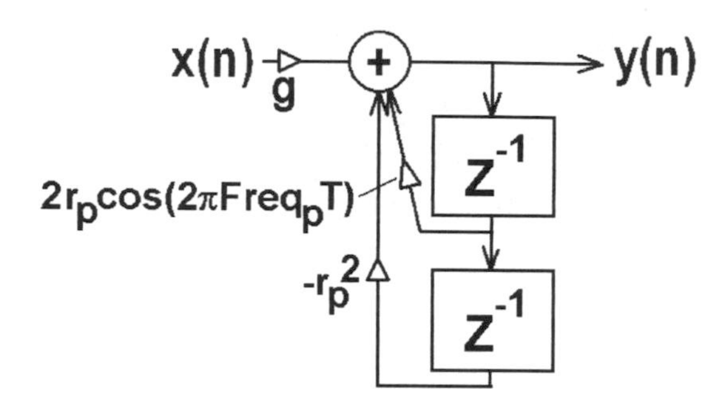

Figure 9.10: Two-Pole resonant filter

9.5.1 Making Modes Parametric

The extraction of modes, and synthesis using a resonant filter, can do a lot toward parameterizing many types of sounds. Stiff metal and glass objects, and some other systems tend to exhibit relatively few sinusoidal modes. In some cases, the location of excitation (striking or plucking) can be related to the excitation level of each mode (as was the case above with our center-plucked string). "Damping" in the system relates to the speed of decay of the exponentials describing each mode. High damping means rapid decay (as when we mute a guitar string). Thus, strike amplitude, strike location, and modal damping can become powerful parameters for controlling a modal synthesis model. The frequencies of the modes can be changed together, in groups, or separately, to yield different sonic results.

Figure 9.11 shows a general model for modal synthesis of struck/plucked objects, in which an impulsive excitation function is used to excite a number of filters that model the modes. Rules for controlling the modes as a function of strike position, striking object, changes in damping, and other physical constraints are included in the model. The flexibility of this simple model is demonstrated in sound example **S9.4**.

Modal synthesis is a powerful technique for auditory display, because we can control timbre, pitch, and time with a few "knobs". The nature of modal synthesis, where each sound begins with an impulsive excitation, and decays exponentially thereafter, lends it to alerts and alarms, and systems where rhythmic organization is an important part of the design of the auditory display.

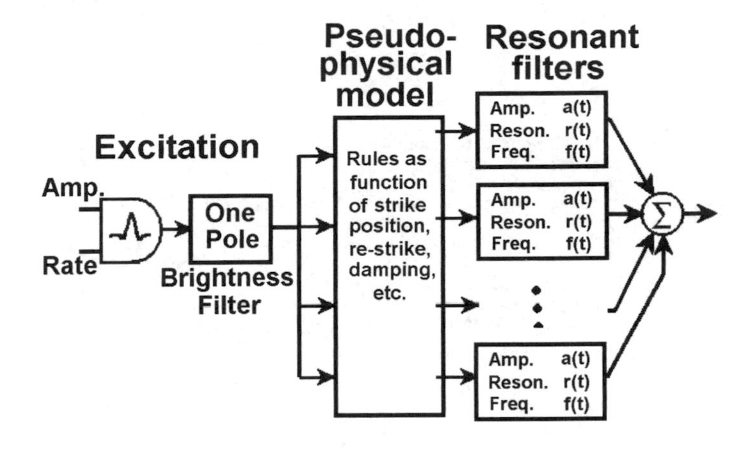

Figure 9.11: Flexible parametric modal synthesis algorithm.

9.6 Subtractive (Source-Filter) synthesis

Subtractive synthesis uses a complex source wave, such as an impulse, a periodic train of impulses, or white noise, to excite spectral-shaping filters. One of the earliest uses of electronic subtractive synthesis dates back to the 1920/30s, with the invention of the "Channel Vocoder" (for VOiceCODER) [19]. In this device, the spectrum is broken into sections called sub-bands, and the information in each sub-band is converted to a signal representing (generally slowly varying) power. The analyzed parameters are then stored or transmitted (potentially compressed) for reconstruction at another time or physical site. The parametric data representing the information in each sub-band can be manipulated in various ways, yielding transformations such as pitch or time shifting, spectral shaping, cross synthesis, and other effects. Figure 9.12 shows a block diagram of a channel vocoder. The detected envelopes serve as "control signals" for a bank of band-pass "synthesis filters" (identical to the "analysis filters" used to extract the sub-band envelopes). The synthesis filters have gain inputs that are fed by the analysis control signals.

When used to encode and process speech, the channel vocoder explicitly makes an assumption that the signal being modeled is a single human voice. The "source analysis" (upper left of Figure 10.12) block extracts parameters related to finer spectral details, such as whether the sound is pitched (vowel) or noisy (consonant or whispered). If the sound is pitched, the pitch is estimated. The overall energy in the signal is also estimated. These parameters become additional low-bandwidth control signals for the synthesizer. Intelligible speech can be synthesized using only a few hundred numbers per second. An example coding scheme might use 8 channel gains + pitch + power, per frame, at 40 frames per second, yielding a total of only 400 numbers per second. The channel vocoder, as designed for speech coding, does not generalize to arbitrary sounds, and fails horribly when the source parameters deviate from expected harmonicity, reasonable pitch range, etc. This can result in artifacts ranging from distortion, to rapid shifts in pitch and spectral peaks (often called "bubbling bells"). But the ideas of sub-band decomposition, envelope detection, and driving a synthesis filter bank with control signals give rise to many other interesting applications and implementations of

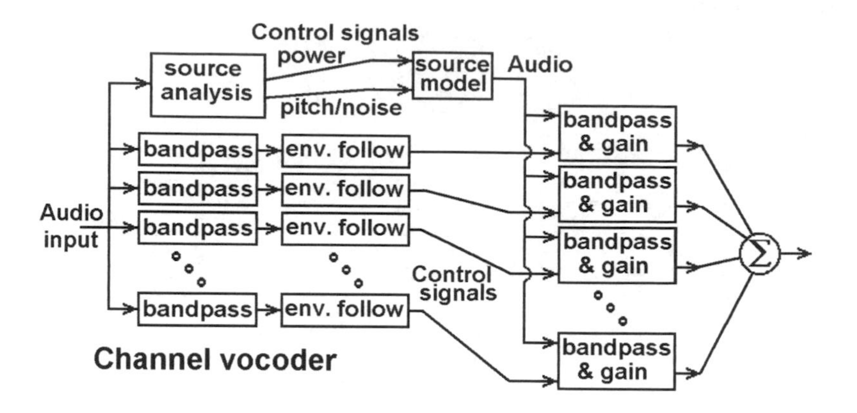

Figure 9.12: Channel vocoder block diagram.

the vocoder concepts. MPEG coding/compression, audio/speech analysis, and audio effects all use these ideas.

A family of filter-based frequency transforms known as "Wavelet Transforms" has been used for analysis and synthesis of sound. Instead of being based on steady sinusoids such as the Fourier Transform, Wavelet Transforms are based on the decomposition of signals into fairly arbitrary functions (called "wavelets") [20] with useful properties such as compactness (constrained in time or frequency), efficient computation, or other.

Some benefits of wavelet transforms over Fourier transforms are that they can be implemented using fairly arbitrary filter criteria, on a logarithmic frequency scale rather than a linear scale as in the DFT, and that time resolution can be a function of the frequency range of interest. This latter point means that we can say accurate things about high frequencies as well as low. This contrasts with the Fourier transform, which requires the analysis window width be the same for all frequencies, meaning that we must either average out lots of the interesting high-frequency time information in favor of being able to resolve low frequency information (large window), or opt for good time resolution (small window) at the expense of low-frequency resolution, or perform multiple transforms with different sized windows to catch both time and frequency details. There are a number of fast wavelet transform techniques that allow the sub-band decomposition to be accomplished in essentially $N \log_2(N)$ time, like the FFT.

While the channel vocoder, and other sub-band models, are interesting and useful for processing and compressing speech and other sounds, by itself the vocoder isn't strictly a synthesizer. Factoring out the voice source parameters and filter energies does reduce the sound to a few descriptive numbers, and these numbers can be modified to change the sound. But very few systems actually use a channel vocoder-like structure to perform synthesis. Spectral shaping of noise or arbitrary signals can be used for auditory display, and thus the channel vocoder ideas could be useful.

9.6.1 Linear Predictive Synthesis (Coding)

Modal Synthesis, as discussed before, is a form of Subtractive Synthesis, but the spectral characteristics of modes are sinusoidal, exhibiting very narrow spectral peaks. For modeling the gross peaks in a spectrum, which could correspond to weaker resonances, we can exploit the same two-pole resonance filters. This type of source-filter synthesis has been very popular for voice synthesis.

Having origins and applications in many different disciplines, Time Series Prediction is the task of estimating future sample values from prior samples. Linear Prediction is the task of estimating a future sample (usually the next in the time series) by forming a linear combination of some number of prior samples. Linear Predictive Coding (LPC) does this, and automatically extracts the gross spectral features by designing filters to match those, yielding a "source" that we can use to drive the filters [21, 22]. Figure 9.13 shows linear prediction in block diagram form (where each Z^{-1} box represents a sample of delay/memory). The difference equation for a linear predictor is:

$$y(n) = \hat{x}(n+1) = \sum_{i=0}^{m} a_i x(n-i) \tag{17}$$

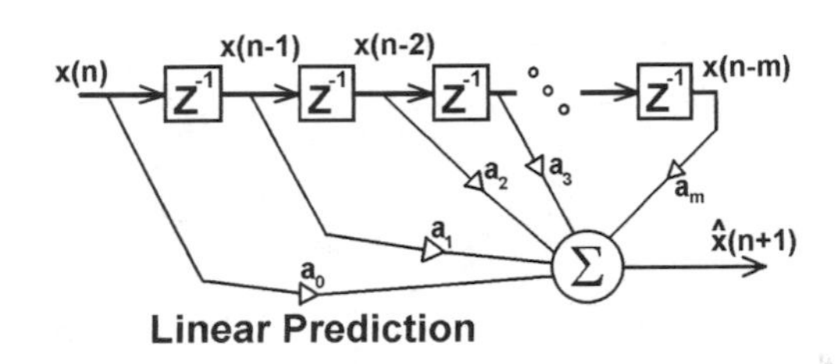

Figure 9.13: A linear prediction filter.

The task of linear prediction is to select the vector of predictor coefficients

$$A = [a_0, a_1, a_2, a_3, \ldots, a_m] \tag{18}$$

such that $\hat{x}(n+1)$ (the estimate) is as close as possible to $x(n+1)$ (the real sample) over a set of samples (often called a frame) $x(0)$ to $x(N-1)$. Usually "close as possible" is defined by minimizing the Mean Square Error (MSE):

$$\text{MSE} = \frac{1}{N} \sum_{n=1}^{N} [\hat{x}(n) - x(n)]^2 \tag{19}$$

Many methods exist for arriving at the predictor coefficients a_i which yield a minimum MSE. The most common method uses correlation or covariance data from each frame of samples to be predicted. The difference between the predicted and actual samples is called the "error" signal or "residual". The optimal coefficients form a digital filter. For low order LPC (delay order of 6–20 or so), the filter fits the coarse spectral features, and the residue contains the remaining part of the sound that cannot be linearly predicted. A common and popular use of

LPC is for speech analysis, synthesis, and compression. The reason for this is that the voice can be viewed as a "source-filter" model, where a spectrally rich input (pulses from the vocal folds or noise from turbulence) excites a filter (the resonances of the vocal tract). LPC is another form of spectral vocoder as discussed previously, but since LPC filters are not fixed in frequency or shape, fewer bands (than some vocoders) are needed to dynamically model the changing speech spectral shape.

LPC speech analysis/coding involves processing the signal in blocks and computing a set of filter coefficients for each block. Based on the slowly varying nature of speech sounds (the speech articulators can only move so fast), the coefficients are relatively stable for milliseconds at a time (typically 5-20ms is used in speech coders). If we store the coefficients and information about the residual signal for each block, we will have captured many of the essential aspects of the signal. Figure 9.14 shows an LPC fit to a speech spectrum. Note that the fit is better at the peak locations than in the valleys. This is due to the nature of the coefficient-computation mathematics, which performs a "least-squares error minimization criterion." Missing the mark on low-amplitude parts of the spectrum is not as important as missing it on high-amplitude parts. This is fortunate for audio signal modeling, in that the human auditory system is more sensitive to spectral peaks (poles, resonances), called "formants" in speech, than valleys (zeroes, anti-resonances).

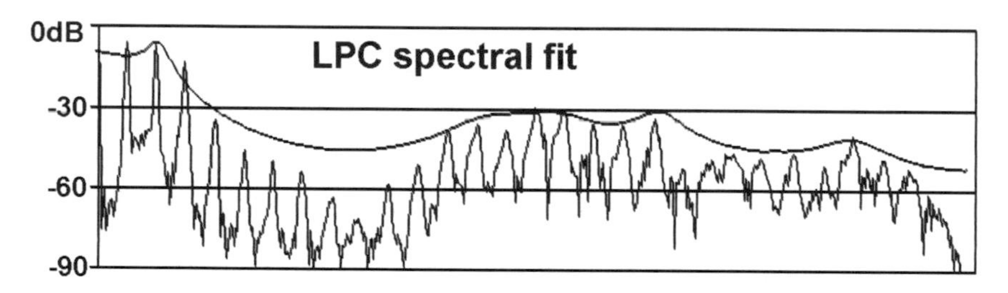

Figure 9.14: 10th order LPC filter fit to a voiced /u/ ("ooo") spectrum.

Once LPC has been performed on speech, inspecting the residual shows that it is often a stream of pulses for voiced speech, or white noise for unvoiced speech. Thus, if we store parameters about the residual, such as whether it is periodic pulses or noise, the frequency of the pulses, and the energy in the residual, then we can recreate a signal that is very close to the original. This is the basis of much modern speech compression. If a signal is entirely predictable using a linear combination of prior samples, and if the predictor filter is doing its job perfectly, we should be able to hook the output back to the input and let the filter predict the rest of the signal automatically. This form of filter, with feedback from output to input, is called "recursive." The recursive LPC reconstruction is sometimes called "all pole", referring to the high-gain "poles" corresponding to the primary resonances of the vocal tract. The poles do not capture all of the acoustic effects going on in speech, however, such as "zeroes" that are introduced in nasalization, aerodynamic effects, etc. However, as mentioned before, since our auditory systems are most sensitive to peaks (poles), LPC does a good job of capturing the most important aspects of speech spectra.

Any deviation of the predicted signal from the actual original signal will show up in the error signal, so if we excite the recursive LPC reconstruction filter with the residual signal itself, we can get back the original signal exactly. This is a form of what engineers call "identity analysis/resynthesis", performing "deconvolution" or "source-filter separation" to separate the source from the filter, and using the residue to excite the filter to arrive at the original signal.

9.6.2 The Parametric Nature of LPC

Using the parametric source model also allows for flexible time and pitch shifting, without modifying the basic timbre. The voiced pulse period can be modified, or the frame rate update of the filter coefficients can be modified, independently. So it is easy to speed up a speech sound while making the pitch lower, still retaining the basic spectral shapes of all vowels and consonants. Cross-synthesis can also be accomplished by replacing the excitation wave with an arbitrary sound, as shown in sound example **S9.5**.

In decomposing signals into a source and a filter, LPC can be a marvelous aid in analyzing and understanding some sound-producing systems. The recursive LPC reconstruction filter can be implemented in a variety of ways. Three different filter forms are commonly used to perform subtractive voice synthesis [23]. The filter can be implemented in series (cascade) as shown in Figure 9.15, factoring each resonance into a separate filter block with control over center frequency, width, and amplitude. The flexibility of the parallel formant model is demonstrated in sound and movie examples **S9.6** and **S9.7**. The filter can also be implemented in parallel

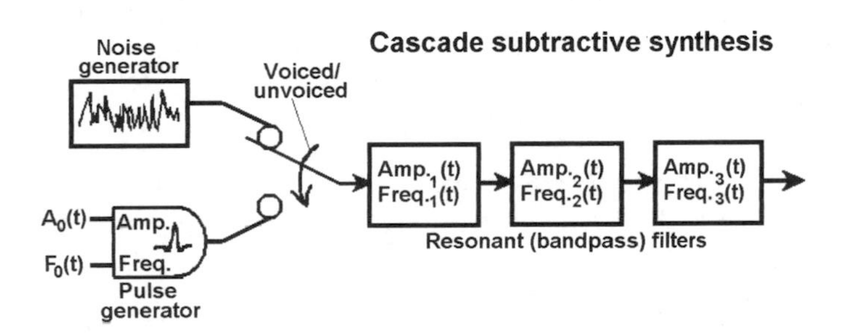

Figure 9.15: Cascade factored formant subtractive synthesizer.

(separate sub-band sections of the spectrum added together), as shown in Figure 9.16.

One additional implementation of the resonant filter is the ladder filter structure, which carries with it a notion of one-dimensional spatial propagation as well [24]. Figure 9.17 shows a ladder filter realization of an 8th order (output plus eight delayed versions of the output) IIR filter (Infinite Impulse Response, or feedback filter).

9.6.3 A Note on Parametric Analysis/Synthesis vs. Direct Synthesis

Note that most of our synthesis methods so far have relied (at least initially or in motivation) on analyzing or processing recorded sounds:

- PCM takes in a time-domain waveform and manipulates it directly;

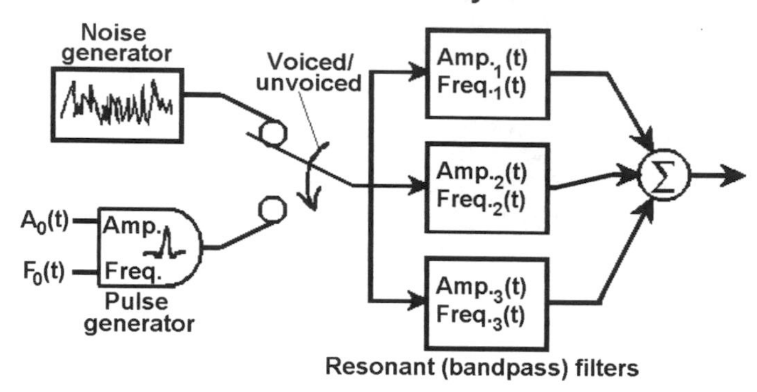

Figure 9.16: Parallel factored formant subtractive synthesizer.

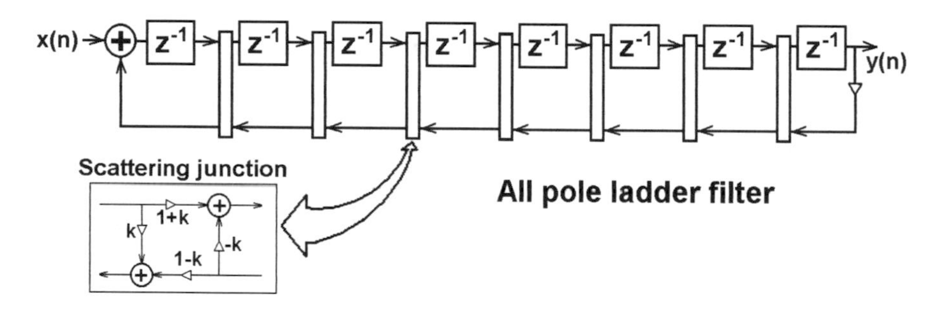

Figure 9.17: Ladder filter implementation of all-pole LPC filter.

- Fourier determines the sinusoidal components of a time-domain waveform;
- LPC determines the gross shape of a spectral filter and the source that, when driven through the filter, will yield an approximation of the original waveform.

As each technique was examined, ways were determined to extract or derive low(er)-order parameters for resynthesis that could be useful for auditory display. Based on this background knowledge and these techniques, the next sections look at methods for synthesizing directly from parameters, not necessarily relying on an original recording to be analyzed and manipulated.

9.7 Time Domain Formant Synthesis

FOFs (fonctions d'onde formantique, Formant Wave Functions) were created for voice synthesis using exponentially decaying sine waves, overlapped and added at the repetition period of the voice source [25]. Figure 9.18 depicts FOF synthesis of a vowel. FOFs are composed of a sinusoid at the formant center frequency, with an amplitude that rises rapidly upon excitation, then decays exponentially. The control parameters define the center frequency and bandwidth of the formant being modeled, and the rate at which the FOFs are generated and added determines the fundamental frequency of the voice.

Note that each individual FOF is a simple "wavelet" (local and compact wave both in frequency and time). FOFs provide essentially the same parameters as formant filters, but are implemented in the time domain.

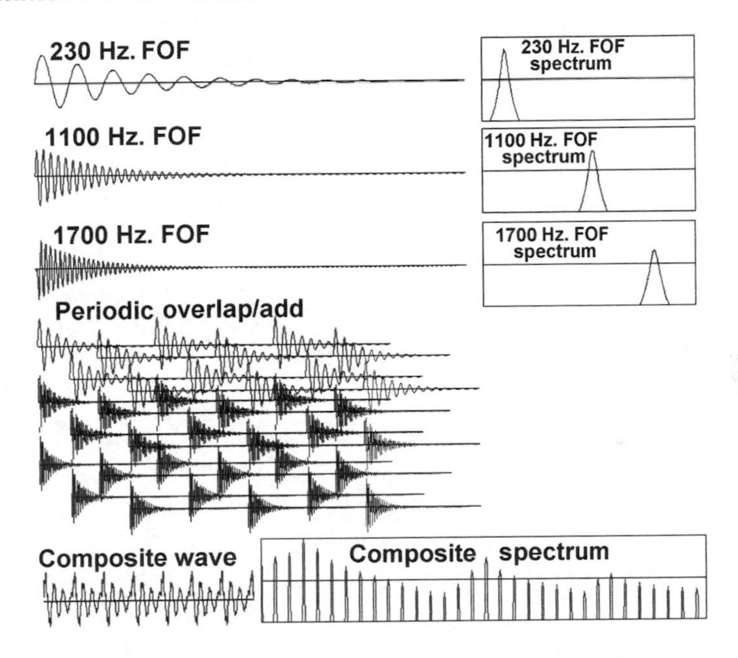

Figure 9.18: FOF synthesis of a vowel.

9.8 Waveshaping and FM Synthesis

Waveshaping synthesis involves warping a simple (usually a saw-tooth or sine wave) waveform with a non-linear function or lookup table [26, 27]. One popular form of waveshaping synthesis, called Frequency Modulation (FM), uses sine waves for both input and warping waveforms [28]. Frequency modulation relies on modulating the frequency of a simple periodic waveform with another simple periodic waveform. When the frequency of a sine wave of average frequency f_c (called the carrier wave), is modulated by another sine wave of frequency f_m (called the modulator wave), sinusoidal sidebands are created at frequencies equal to the carrier frequency plus and minus integer multiples of the modulator frequency. Figure 9.19 shows a block diagram for simple FM synthesis (one sinusoidal carrier and one sinusoidal modulator). Mathematically, FM is expressed as:

$$y(t) = \sin(2\pi t f_c + \Delta f_c \sin(2\pi t f_m)) \tag{20}$$

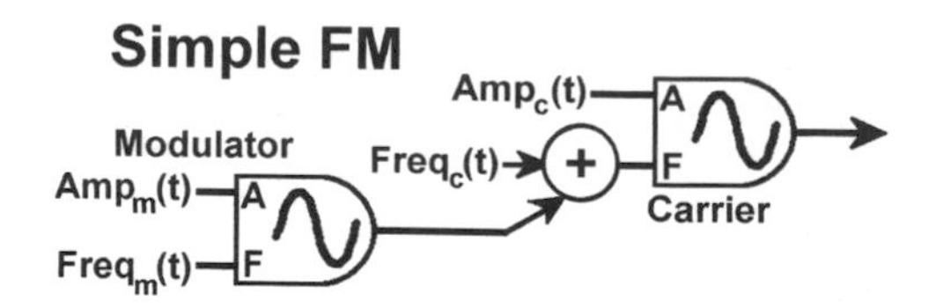

Figure 9.19: Simple FM (one carrier and one modulator sine wave) synthesis.

The index of modulation, I, is defined as $\Delta f_c / f_c$. Carson's rule (a rule of thumb) states that the number of significant bands on each side of the carrier frequency (sidebands) is roughly equal to $I + 2$. For example, a carrier sinusoid of frequency 600 Hz., a modulator sinusoid of frequency 100 Hz., and a modulation index of 3 would produce sinusoidal components of frequencies 600, {700, 500}, {800, 400}, {900, 300}, {1000, 200}, and {1100, 100} Hz. Inspecting these components reveals that a harmonic spectrum with 11 significant harmonics, based on a fundamental frequency of 100 Hz, can be produced by using only two sinusoidal generating functions. Figure 9.20 shows the spectrum of this synthesis. Sound example **S9.8** presents a series of FM-tones with increasing modulation index.

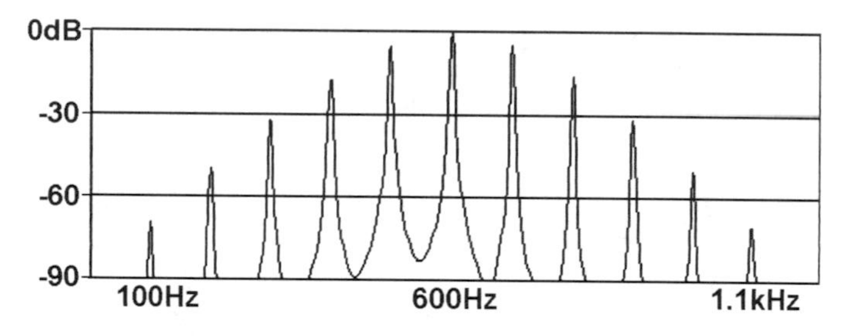

Figure 9.20: Simple FM with 600 Hz carrier, 100 Hz modulator, and index of modulation of 3.

Selecting carrier and modulator frequencies that are not related by simple integer ratios yields an inharmonic spectrum. For example, a carrier of 500 Hz, modulator of 273 Hz, and an index of 5 yields frequencies of 500 (carrier), 227, 46, 319, 592, 865, 1138, 1411 (negative sidebands), and 773, 1046, 1319, 1592, 1865, 2138, 2411 (positive sidebands). Figure 9.21 shows a spectrogram of this FM tone, as the index of modulation I is ramped from zero to 5. The synthesized waveforms at $I = 0$ and $I = 5$ are shown as well.

By setting the modulation index high enough, huge numbers of sidebands are generated, and the aliasing and addition of these results in noise. By careful selection of the component frequencies and index of modulation, and combining multiple carrier/modulator pairs, many spectra can be approximated using FM. The amplitudes and phases (described by Bessel functions) of the individual components cannot be independently controlled, however, so FM is not a truly generic sinusoidal, waveform, or spectral synthesis method.

Because of the extreme efficiency of FM (its ability to produce complex waveforms with the relatively small amounts of computer power to run a few oscillators) it became popular

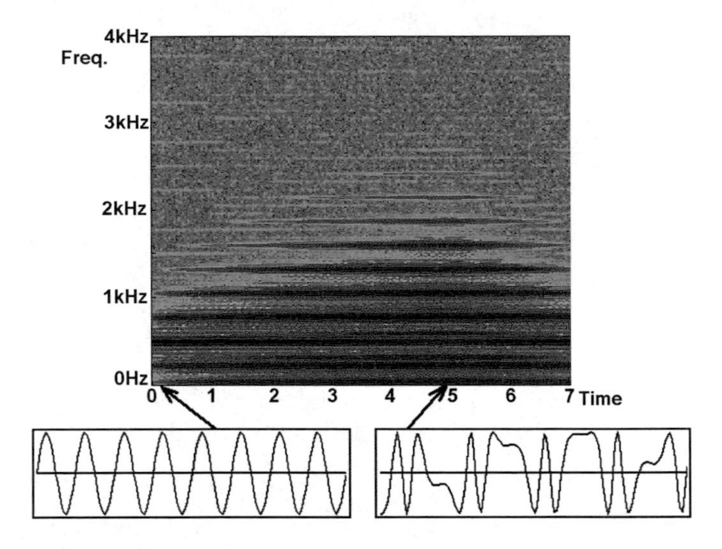

Figure 9.21: Inharmonic simple FM with 500 Hz carrier and 273 Hz modulator. The index of modulation is ramped from 0.0 to 5.0 then back to 0.0.

in the 1980s as a music synthesis algorithm. FM is sometimes used for auditory displays, partly due to popular commercial hardware, and partly due to the rich variety obtainable through manipulation of the few parameters. Carrier and Modulator frequencies determine harmonicity, inharmonicity, and pitch; the index of modulation determines spectral spread; and envelopes control time and spectral evolution. The sound/sonification designer must be careful with carrier/modulator ratio (inharmonicity), however, as often a small-seeming change can result in large categorical perceptual shifts in the resulting sound. Multiple carrier/modulator pairs lend more flexibility and more accurate spectral control. Using multiple carriers and modulators, connection topologies (algorithms) have been designed for the synthesis of complex sounds such as human voices [29], violins, brass instruments, percussion, etc.

9.9 Granular and PhISEM Synthesis

Much of classical physics can be modeled as objects interacting with each other. Lots of little objects are often called "particles." Granular synthesis involves cutting sound into "grains" (sonic particles) and reassembling them by adding, or mixing them back together [30]. The "grains" or "MicroSounds" [31] usually range in length from 10 to 100 ms. The reassembly can be systematic, but often granular synthesis involves randomized grain sizes, locations, and amplitudes. The transformed result usually bears some characteristics of the original sound, just as a mildly blended mixture of fruits still bears some attributes of the original fruits, as well as taking on new attributes due to the mixture. A FOF-Wavelet-related granular method is "Pulsar" synthesis [31]. Granular synthesis is mostly used as a music/composition type of signal processing, but some also take a more physically motivated viewpoint on sound "grains" [32].

The PhISEM (Physically Informed Stochastic Event Modeling) algorithm is based on pseudo-random overlapping and adding of parametrically synthesized sound grains [33]. At the heart of PhISEM algorithms are particle models, characterized by basic Newtonian equations governing the motion and collisions of point masses as can be found in any introductory physics textbook. By modeling the physical interactions of many particles by their statistical behavior, exhaustive calculation of the position, and velocity of each individual particle can be avoided. By factoring out the resonances of the system, the "wavelets" can be shortened to impulses or short bursts of exponentially decaying noise. The main PhISEM assumption is that the sound-producing particle collisions follow a common statistical process known as "Poisson", (exponential probability of waiting times between individual sounds), Another assumption is that the system energy decays exponentially (for example, the decay of the sound of a maraca after being shaken once). Figure 9.22 shows the PhISEM algorithm block diagram.

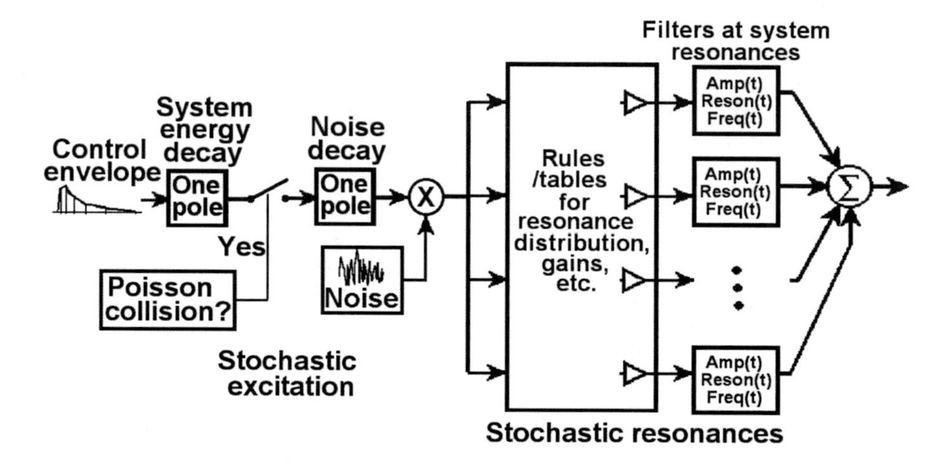

Figure 9.22: Complete PhISEM model showing stochastic resonances.

The PhISEM maraca synthesis algorithm requires only two random number calculations, two exponential decays, and one resonant filter calculation per sample. Other musical instruments that are quite similar to the maraca include the sekere and cabasa (afuche). Outside the realm of multi-cultural musical instruments, there are many real-world particle systems that exhibit one or two fixed resonances like the maraca. A bag/box of hard candy or gum, a salt shaker, a box of wooden matches, and gravel or leaves under walking feet all fit pretty well within this modeling technique.

In contrast to the maraca and guiro-like gourd resonator instruments, which exhibit one or two weak resonances, instruments such as the tambourine (timbrel) and sleigh bells use metal cymbals, coins, or bells suspended on a frame or stick. The interactions of the metal objects produce much more pronounced resonances than the maraca-type instruments, but the Poisson event and exponential system energy statistics are similar enough to justify the use of the PhISEM algorithm for synthesis. To implement these in PhISEM, more filters are used to model the individual partials, and at each collision, the resonant frequencies of the filters are randomly set to frequencies around the main resonances. Other sounds that can be modeled using stochastic filter resonances include bamboo wind chimes (related to a musical instrument as well in the Javanese anklung) [34].

Granular and particle models lend themselves well to continuous interactive auditory displays, where the parameters can be adjusted to modify the perceived "roughness", damping, size, number of objects, etc. Inspired by the work of Gaver's Sonic Finder [35], the earcons of Blattner [36] and auditory interfaces of Brewster [37], and others, Figure 9.23 shows a simple auditory display for desktop dragging and scrolling that uses PhISEM models to indicate whether the mouse is on the desktop (sonic "texture" of sand) or on the scrollbar of a web browser (tambourine model, with pitch mapped to location of the scrollbar in the window). This is demonstrated in movie example **S9.9**.

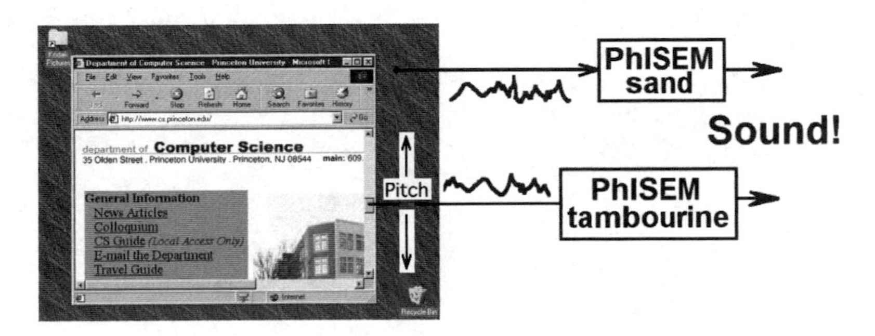

Figure 9.23: Sonically enhanced user interface.

9.10 Physical Modeling Synthesis

There is a simple differential equation that completely describes the motions of an ideal string under tension. Here it is, without derivation:

$$\frac{d^2y}{dx^2} = \frac{1}{c^2}\frac{d^2y}{dt^2} \tag{21}$$

The derivation and solution proof can be found in [16]. This equation (called "the wave equation") means that the acceleration (up and down) of any point on the string is equal to a constant times the curvature of the string at that point. The constant c is the speed of wave motion on the string, and is proportional to the square root of the string tension, and inversely proportional to the square root of the mass per unit length. This equation could be solved numerically, by sampling it in both time and space, and using the difference approximations for acceleration and curvature (much like was done with the mass-spring-damper system earlier). With boundary conditions (such as rigid terminations at each end), the solution of this equation could be expressed as a Fourier series, as was done earlier in graphical form (Figure 9.12). However, there is one more wonderfully simple solution to Equation 21, given by:

$$y(x,t) = y_l\left(t + \frac{x}{c}\right) + y_r\left(t - \frac{x}{c}\right) \tag{22}$$

This equation says that any vibration of the string can be expressed as a combination of two separate traveling waves, one traveling left (y_l) and one traveling right (y_r). They move at rate c, which is the speed of sound propagation on the string. For an ideal (no damping or stiffness) string, and ideally rigid boundaries at the ends, the wave reflects with an inversion at each end, and will travel back and forth indefinitely. This view of two traveling waves summing to make a displacement wave gives rise to the "Waveguide Filter" technique of modeling the vibrating string [38, 39]. Figure 9.24 shows a waveguide filter model of the ideal string. The two delay lines model the propagation of left and right going traveling waves. The conditions at the ends model the reflection of the traveling waves at the ends. The -1 on the left models the reflection with inversion of a displacement wave when it hits an ideally rigid termination (like a fret on a guitar neck). The -0.99 on the right reflection models the slight amount of loss that happens when the wave hits a termination that yields slightly (like the bridge of the guitar which couples the string motion to the body), and models all other losses the wave might experience (internal damping in the string, viscous losses as the string cuts the air, etc.) in making its round-trip path around the string.

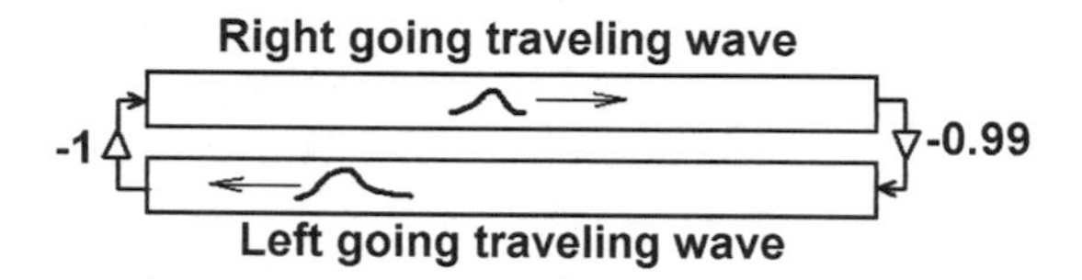

Figure 9.24: Waveguide string modeled as two delay lines.

Figure 9.25 shows the waveguide string as a digital filter block diagram. The $Z^{-P/2}$ blocks represent a delay equal to the time required for a wave to travel down the string. Thus a wave completes a round trip each P samples (down and back), which is the fundamental period of oscillation of the string, expressed in samples. Initial conditions can be injected into the string via the input $x(n)$. The output $y(n)$ would yield the right-going traveling wave component. Of course, neither of these conditions is actually physical in terms of the way a real string is plucked and listened to, but feeding the correct signal into x is identical to loading the delay lines with a pre-determined shape.

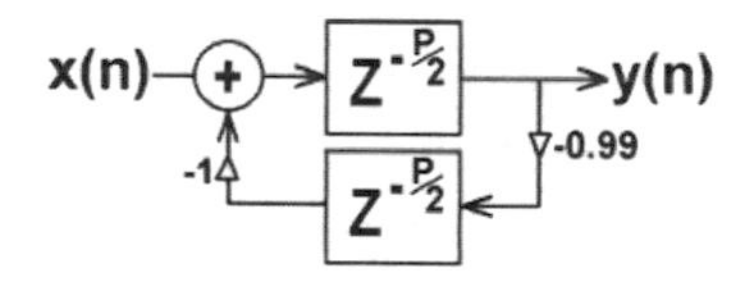

Figure 9.25: Digital filter view of waveguide string.

The impulse response and spectrum of the filter shown in Figure 9.25 is shown in Figure 9.26. As would be expected, the impulse response is an exponentially decaying train of pulses spaced $T = {}^{P}/_{\mathrm{SRate}}$ seconds apart, and the spectrum is a set of harmonics spaced $F_0 = {}^{1}/_{T}$ Hz apart. This type of filter response and spectrum is called a "comb filter", so named because of the comb-like appearance of the time domain impulse response, and of the frequency domain harmonics.

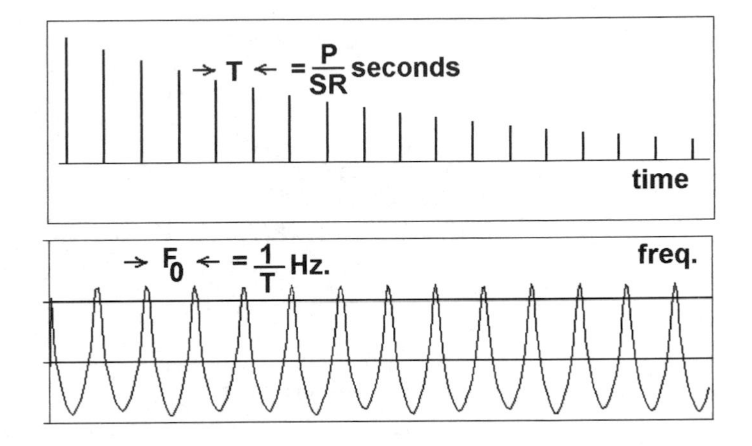

Figure 9.26: Impulse response and spectrum of comb (string) filter.

The two delay lines taken together are called a "waveguide filter." The sum of the contents of the two delay lines is the displacement of the string, and the difference of the contents of the two delay lines is the velocity of the string. If we wish to pluck the string, we simply need to load ½ of the initial string shape into each of the upper and lower delay lines. If we wish to strike the string, we would load in an initial velocity by entering a positive pulse into one delay line and a negative pulse into the other (difference = initial velocity, sum = initial position = 0). These conditions are shown in Figure 9.27.

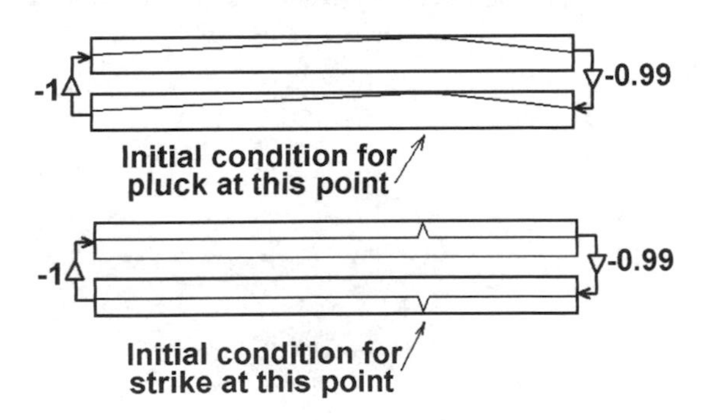

Figure 9.27: Waveguide pluck and strike initial conditions.

9.10.1 Making the String More Real (Parametric)

Figure 9.28 shows a relatively complete model of a plucked string using digital filters. The inverse comb filters model the nodal (rejected frequencies) effects of picking, and the output of an electrical pickup, emphasizing certain harmonics and forbidding others based on the pick (pickup) position [40]. Output channels for pickup position and body radiation are provided separately. A solid-body electric guitar would have no direct radiation and only pickup output(s), while a purely acoustic guitar would have no pickup output, but possibly a family of directional filters to model body radiation in different directions [41].

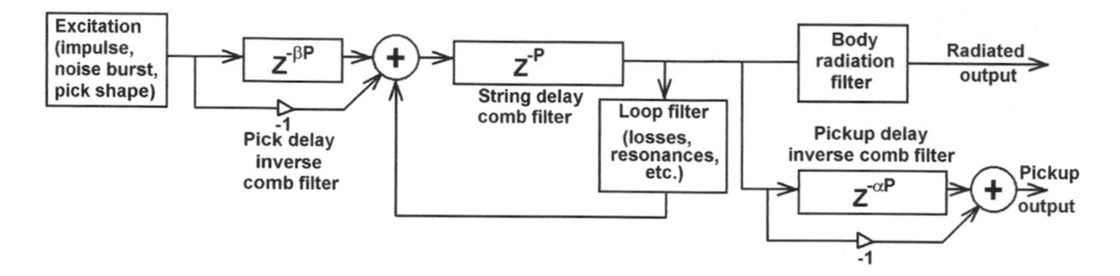

Figure 9.28: Fairly complete digital filter simulation of plucked string system.

9.10.2 Adding Stiffness

In an ideal string or membrane, the only restoring force is assumed to be the tension under which it is stretched. We can further refine solid systems such as strings and membranes to model more rigid objects, such as bars and plates, by noting that the more rigid objects exhibit internal restoring forces due to their stiffness. We know that if we bend a stiff string, it wants to return back to straightness even when there is no tension on the string. Cloth string or thread has almost no stiffness. Nylon and gut strings have some stiffness, but not as much as steel strings. Larger diameter strings have more stiffness than thin ones. In the musical world, piano strings exhibit the most stiffness. Stiffness results in the restoring force being higher (thus the speed of sound propagation as well) for high frequencies than for low. So the traveling wave solution is still true in stiff systems, but a frequency-dependent propagation speed is needed:

$$y(x,t) = y_l(t + {}^{x}/_{c(f)}) + y_r(t - {}^{x}/_{c(f)}) \qquad (23)$$

and the waveguide filter must be modified to simulate frequency-dependent delay, as shown in Figure 9.29.

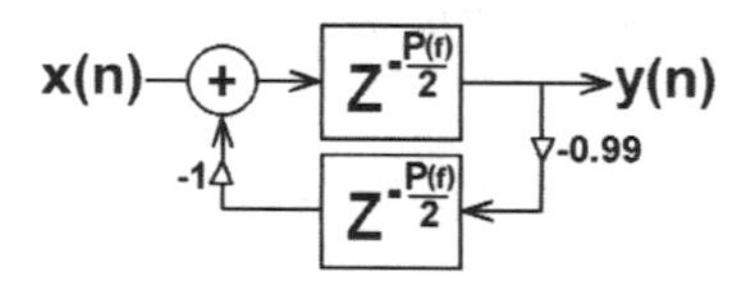

Figure 9.29: Stiffness-modified waveguide string filter.

For basic stiff strings, a function that predicts the frequencies of the partials has the form:

$$f_n = n f_0 (1 + Bn^2) \qquad (24)$$

where B is a number slightly greater than 0, equal to zero for perfect harmonicity (no stiffness), and increasing for increasing stiffness. This means that $P(f)$ should follow the inverse of the $\sqrt{1 + Bn^2}$ factor (round-trip time or period gets shorter with increasing frequency). Typical values of B are 0.00001 for guitar strings, and 0.004 or so for piano strings.

Unfortunately, implementing the $Z^{-P(f)/2}$ frequency-dependent delay function is not simple, especially for arbitrary functions of frequency. One way to implement the $P(f)$ function is by replacing each of the Z^{-1} with a first order all-pass (phase) filter, as shown in Figure 9.30 [40]. The first order all-pass filter has one pole and one zero, controlled by the same coefficient. The all-pass filter implements a frequency-dependent phase delay, but exhibits a gain of 1.0 for all frequencies. The coefficient α can take on values between -1.0 and 1.0. For $\alpha = 0$, the filter behaves as a standard unit delay. For $\alpha > 0.0$, the filter exhibits delays longer than one sample, increasingly long for higher frequencies. For $\alpha < 0.0$ the filter exhibits delays shorter than one sample, decreasingly so for high frequencies.

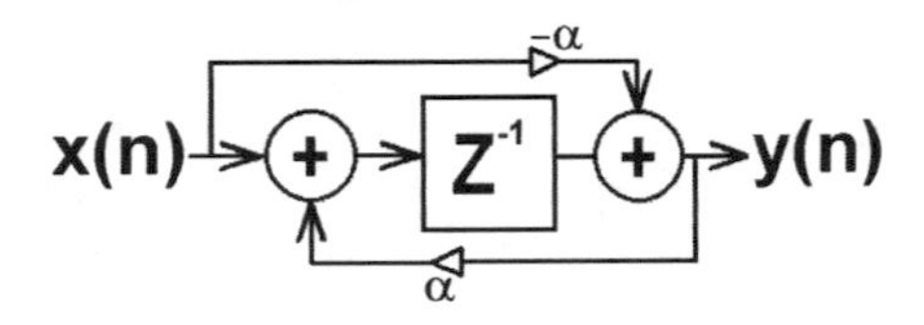

Figure 9.30: First-order all-pass filter.

It is much less efficient to implement a chain of all-pass filters than a simple delay line. But for weak stiffness it is possible that only a few all-pass sections will provide a good frequency-dependent delay. Another option is to implement a higher-order all-pass filter, designed to give the correct stretching of the upper frequencies, added to simple delay lines to give the correct longest bulk delay required.

For very stiff systems such as rigid bars, a single waveguide with all-pass filters is not adequate to give enough delay, or far too inefficient to calculate. A technique called "Banded Waveguides" employs sampling in time, space, and frequency to model stiff one-dimensional systems [42]. This can be viewed as a hybrid of modal and waveguide synthesis, in that each waveguide models the speed of sound in the region around each significant mode of the system. As an example, Figure 9.31 shows the spectrum of a struck marimba bar, with additional band-pass filters superimposed on the spectrum, centered at the three main modes. In the banded waveguide technique, each mode is modeled by a band-pass filter, plus a delay line to impose the correct round-trip delay, as shown in Figure 9.32.

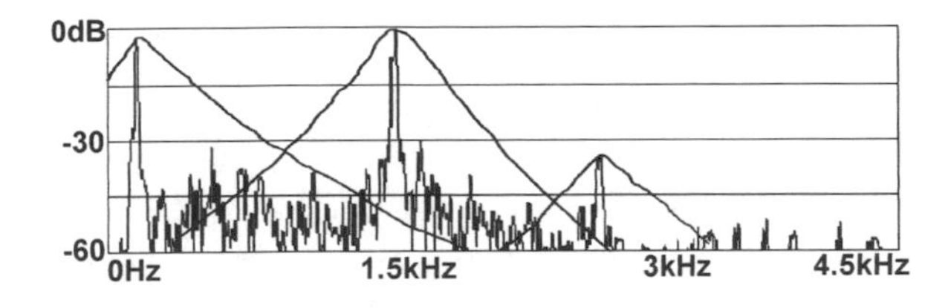

Figure 9.31: Banded decomposition of struck bar spectrum.

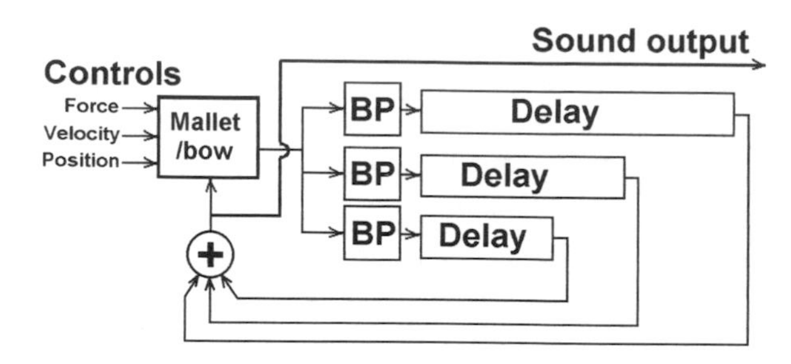

Figure 9.32: Banded waveguide model

9.10.3 Auditory Display with Strings and Bars

Plucked strings and banded waveguide models have many of the same advantages as modal synthesis, but usually with less computational cost. Pluck/strike location, damping, harmonicity/inharmonicity, and other parameters are easily manipulated to yield a wide variety of resulting sound. As an application example, Figure 9.33 shows the normalized stock prices of Red Hat Linux and Microsoft, for one year, February 2001–2002. It's pretty easy to see the trends in the stocks, but what if we wanted to track other information in addition to these curves? We might be interested in the daily volume of trading, and seemingly unrelated data like our own diastolic blood pressure during this period (to decide if it's really healthy to own these stocks). Figure 9.34 shows the five normalized curves consisting of two stock prices, two stock volumes, and one daily blood pressure measurement. It clearly becomes more difficult to tell what is going on.

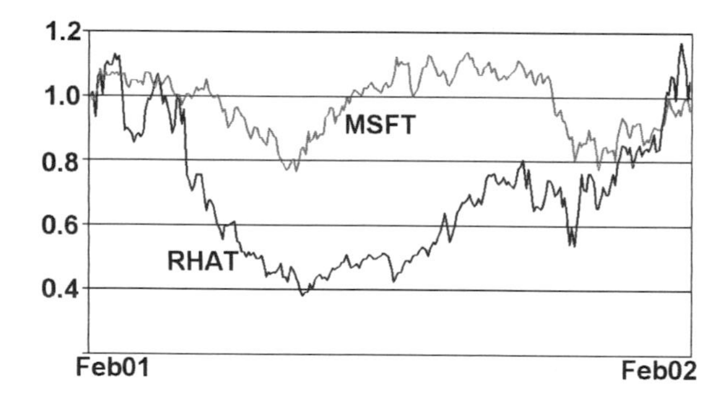

Figure 9.33: Stock prices, normalized to a $1 purchase on day 1.

Of course there are more sophisticated graphical means and techniques we could use to display this data. But some trends or patterns might emerge more quickly if we were to listen to the data. With a suitable auditory mapping of the data, we might be able to hear a lot more than we could see in a single glance. For example, the value of Red Hat could be mapped to the pitch of a plucked mandolin sound in the left speaker, with sound loudness controlled by trading volume (normalized so that even the minimum volume still makes a faint sound). Microsoft could be mapped to the pitch of a struck marimba sound in the right

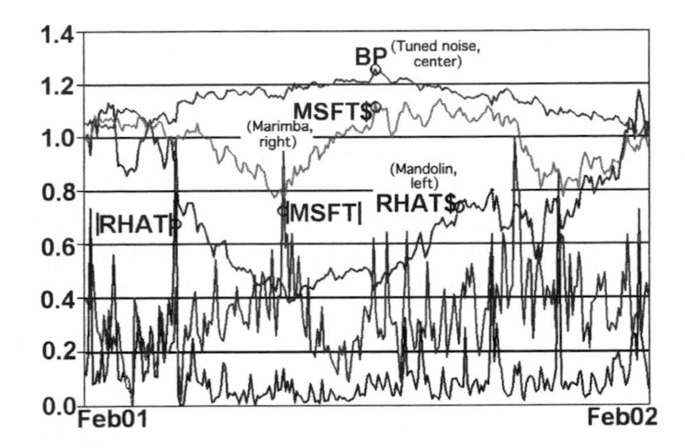

Figure 9.34: Normalized stock prices, plus volumes, plus the blood pressure of someone owning both stocks.

speaker, again with loudness controlled by trading volume. The pitch ranges are normalized so that the beginning prices on the first day of the graphs sound the same pitch. This way, on any day that the pitches are the same, our original (day 1) dollar investment in either stock would be worth the same. Finally, our normalized daily blood pressure could be mapped to the volume and pitch of a tuned noise sound, located in the center between the two speakers. Figure 9.35 shows the waveforms of these three signals. Of course, the point here is not to map visual data to visual data (waveforms), but rather to map to audio and listen to it as in sound example **S9.10**.

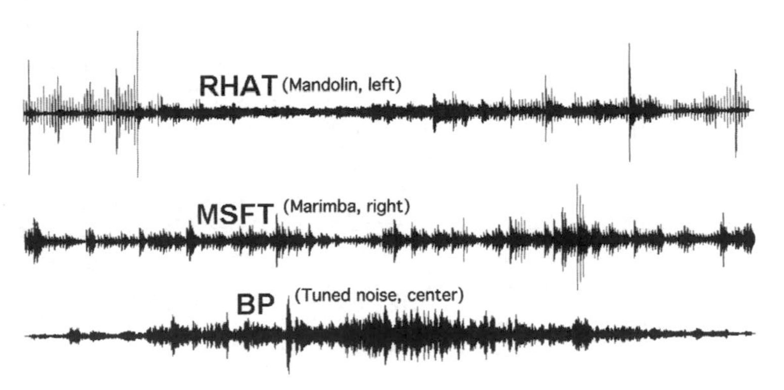

Figure 9.35: Figure 10.35 Audio waveforms of sonified stock prices, volumes, and blood pressure.

9.11 Non-Linear Physical Models

The physical models discussed so far are all linear, meaning that doubling the input excitation causes the output results to double. FM and waveshaping synthesis techniques are also spectral in nature, and non-linear, although not necessarily physical.

Many interesting interactions in the physical world, musical and non-musical, are non-linear. For example, adding a model of bowing friction allows the string model to be used for the violin and other bowed strings. This focused non-linearity is what is responsible for turning the steady linear motion of a bow into an oscillation of the string [43, 44]. The bow sticks to the string for a while, pulling it along, then the forces become too great and the string breaks away, flying back toward rest position. This process repeats, yielding a periodic oscillation. Figure 9.36 shows a simple bowed string model, in which string velocity is compared to bow velocity, then put through a nonlinear friction function controlled by bow force. The output of the nonlinear function is the velocity input back into the string.

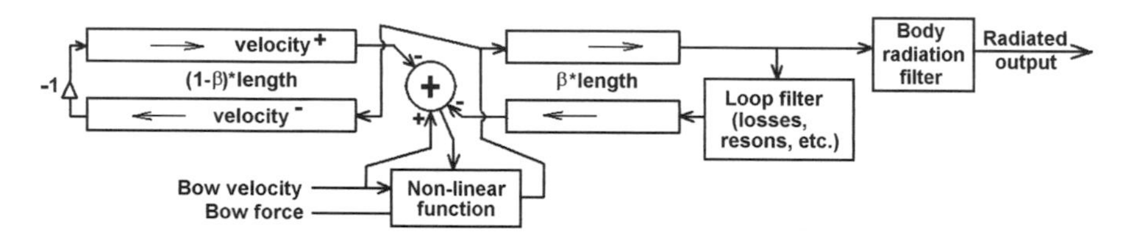

Figure 9.36: Bowed string model.

In mathematically describing the air within a cylindrical acoustic tube (like a trombone slide, clarinet bore, or human vocal tract), the defining equation is:

$$\frac{d^2P}{dx^2} = \frac{1}{c^2} \cdot \frac{d^2P}{dt^2} \tag{25}$$

which we would note has exactly the same form as Equation 21, except displacement y is replaced by pressure P. A very important paper in the history of physical modeling by [43] noted that many acoustical systems, especially musical instruments, can be characterized as a linear resonator, modeled by filters such as all-pole resonators or waveguides, and a single non-linear oscillator like the reed of the clarinet, the lips of the brass player, the jet of the flute, or the bow-string friction of the violin. Since the wave equation says that we can model a simple tube as a pair of bi-directional delay lines (waveguides), then we can build models using this simple structure. If we'd like to do something interesting with a tube, we could use it to build a flute or clarinet. Our simple clarinet model might look like the block diagram shown in Figure 9.37.

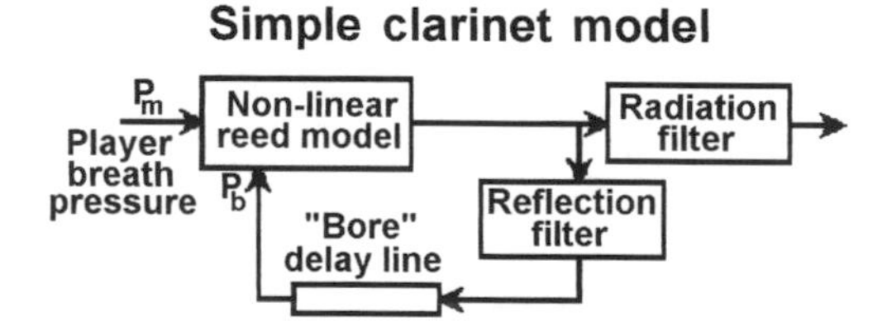

Figure 9.37: Simple clarinet model.

To model the reed, we assume that the mass of the reed is so small that the only thing that must be considered is the instantaneous force on the reed (spring). The pressure inside the bore P_b is the calculated pressure in our waveguide model, the mouth pressure P_m is an external control parameter representing the breath pressure inside the mouth of the player (see Figure 38(a)). The net force acting on the reed/spring can be calculated as the difference between the internal and external pressures, multiplied by the area of the reed (pressure is force per unit area). This can be used to calculate a reed opening position from the spring constant of the reed. From the reed opening, we can compute the amount of pressure that is allowed to leak into the bore from the player's mouth. If bore pressure is much greater than mouth pressure, the reed opens far. If mouth pressure is much greater than bore pressure, the reed slams shut. These two extreme conditions represent an asymmetric non-linearity in the reed response. Even a grossly simplified model of this non-linear spring action results in a pretty good model of a clarinet [44]. Figure 38(b) shows a plot of a simple reed reflection function (as seen from within the bore) as a function of differential pressure. Once this non-linear signal-dependent reflection coefficient is calculated (or looked up in a table), the right-going pressure injected into the bore can be calculated as $P_b^+ = \alpha P_b^- + (1 - \alpha) P_m$.

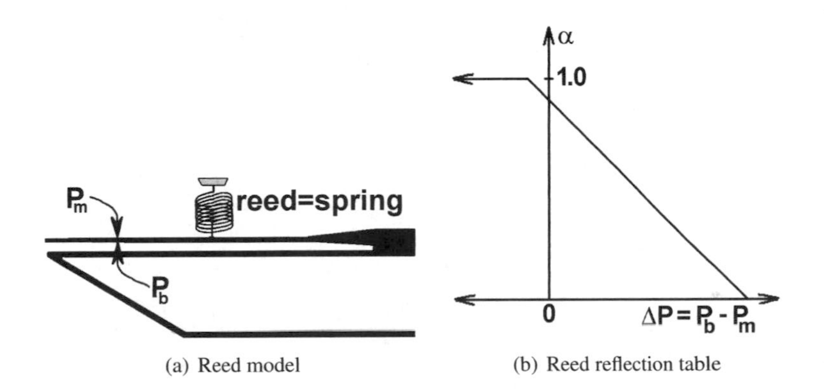

(a) Reed model (b) Reed reflection table

Figure 9.38: Reed model and reflection table

The clarinet is open at the bell end, and essentially closed at the reed end. This results in a reflection with inversion at the bell and a reflection without inversion (plus any added pressure from the mouth through the reed opening) at the reed end. These boundary conditions cause odd-harmonics to dominate in the clarinet spectrum, yielding a square-like wave as we constructed before using odd Fourier harmonics.

We noted that the ideal string equation and the ideal acoustic tube equation are essentially identical. Just as there are many refinements possible to the plucked string model to make it more realistic, there are many possible improvements for the clarinet model. Replacing the simple reed model with a variable mass-spring-damper allows the modeling of a lip reed as is found in brass instruments. Replacing the reed model with an air jet model allows the modeling of flute and recorder-like instruments. With all wind or friction (bowed) excited resonant systems, adding a little noise in the reed/jet/bow region adds greatly to the quality (and behavior) of the synthesized sound.

9.11.1 Auditory Display with Nonlinear Models

Nonlinear synthesis provides some interesting possibilities for auditory display. Since the parameters influence the resultant sound in physically meaningful ways, often the "intuition" for mapping data to parameters is much more natural than in abstract models. However, the nature of many non-linear systems is that a small change in a parameter might make a huge and unpredictable change in the behavior. Such is the case in blowing a clarinet just below the "speaking threshold", resulting in a noisy sound, but by increasing the blowing pressure slightly, the tone changes to a "warm" odd-harmonic oscillation. As any parent of a child studying violin knows too well, slight changes in bowing parameters make gross (literally) changes in the output sound (noisy, scratchy, pitched but irritating oscillation, beautiful sonorous singing quality).

So the interesting and physically meaningful behavior of many non-linear synthesis models is a double-edged sword; rich variety of sounds and responsiveness to small parameter changes, vs. unpredictability and non-linear mapping of parameters to output sound. For this reason, care should be taken in using such systems for reliable and repeatable auditory displays.

9.12 Synthesis for Auditory Display, Conclusion

There are other types of sound synthesis, such as random waveform and/or spectrum generation using genetic algorithms, fractals, neural networks, and other popular techniques that have been applied to a host of other problems in other domains [45]. These techniques can also be applied to the derivation and manipulation of parameters for parametric synthesis models. Scanned synthesis [46] is a hybrid of physical and wavetable synthesis, where a trajectory of a physical model running at one update rate is constantly scanned as a form of self-modifying wavetable. In a way this is a sonification in itself, where a physical process (not necessarily running at audio rate or generating audio itself) generates data that is scanned as a waveform.

Other projects involving the mapping of physical, pseudo-physical, or physically inspired simulations to synthesis range from sonifying a rolling ball or pouring water into a glass [47], to using sound textures to simulate the sound of swords/sticks traveling rapidly through the air [48]. These are examples of the mapping of physical process parameters to the control parameters of various synthesis techniques (like the PhISEM algorithm described above). Others have taken the approach of looking at the target sound to be made for games, sound effects, or other applications, and then deriving custom synthesis "patches" tailored to each sound or class of sound [49]. Again, these methods rely heavily on the basic synthesis methods described in this chapter. These are all examples of using one or more processes or models to control the parameters of sound synthesis, thus related to auditory display and sonification. Further, since many of these include a parametric model for generating synthesis parameters themselves, and since these models themselves have input parameters, they can be used in auditory displays, or for sonifying abstract data. These are examples of "mapping", which is covered at length in other chapters of this book.

On the simple side, just plain old pulses are interesting in many cases, such as the familiar Geiger counter (which could be viewed as a simple case of granular synthesis). Historically, computer researchers would attach an amplifier and speaker to a particular bit or set of bits inside a computer, and use the resulting pulse wave output to monitor and diagnose information about the behavior and state of the computer (a loop is easy to hear, access to memory, network traffic or collisions, all are possible to learn as direct sonifications). In fact, just placing an AM radio near a computer and tuning to certain frequencies allows the inner workings to be heard, as the sea of pulses at different rates generates Radio Frequency (RF) emissions. These simple and direct mappings rely mostly on the human ability to learn the sound of a process or state, rather than an explicit mapping of data to the parameters of a parametric synthesis algorithm.

On a more neuro-ecological note, the use of speech or speech-like sounds is perhaps the most powerful form of auditory display. Indeed so much of the legacy of sound synthesis comes from research on speech and communications, as our LPC, Formant, FOF, Vocoder algorithms point up. The danger, however, of using speech-like sounds is that they might trigger our linguistic "circuitry" and evoke lots of semantic, emotional, cultural, and other results, which could vary greatly from person to person, and culture to culture. Speech-motivated models are a very powerful tool for conveying even non-speech information, due to our sensitivity to pitch, quality, articulation, breathiness, etc. but designers must be cautious in using them.

Auditory display designers have a rich variety of techniques and tools at their disposal, and with the power of modern computers (even PDAs and cell phones), parametric synthesis is easily possible. The author hopes that more researchers and interaction designers will exploit the potential of synthesis in the future, rather than just using recorded PCM or "off the shelf" sounds.

Bibliography

[1] K. Pohlmann, *Principles of Digital Audio*, McGraw Hill, 2000.

[2] T. Dutoit, *An Introduction to Text-To-Speech Synthesis*, Kluwer Academic Publishers, 1996.

[3] J. O. Smith and P. Gossett, "A flexible sampling-rate conversion method". In *Proceedings of the International Conference on Acoustics, Speech, and Signal Processing, San Diego*, vol. 2, (New York), pp. 19.4.1–19.4.2, IEEE Press, 1984.

[4] R. Bristow-Johnson, "Wavetable Synthesis 101: a Fundamental Perspective" *AES 101 Conference*, Paper 4400, 1996.

[5] M. V. Mathews, *The Technology of Computer Music*. Cambridge, Massachusetts: MIT Press, 1969.

[6] R. Bracewell, *The Fourier Transform and its Applications*, McGraw Hill, 1999.

[7] R. Portnoff, "Time-scale modifications of speech based on short time Fourier analysis", *IEEE Transactions on Acoustics, Speech, and Signal Processing*, 29:3, pp. 374–390, 1981.

[8] A. Moorer, "The Use of the Phase Vocoder in Computer Music Applications". *Journal of the Audio Engineering Society*, 26 (1/2), pp. 42–45, 1978.

[9] M. Dolson, "The Phase Vocoder: A Tutorial", *Computer Music Journal*, 10 (4), pp. 14–27, 1986.

[10] *Proceedings of the International Symposium on Music Information Retrieval*, 2003-Present, website at `http://www.ismir.net`

[11] R. J. McAulay and T. Quatieri, "Speech Analysis/Synthesis Based on a Sinusoidal Representation", *IEEE Trans. ASSP*-34, pp. 744–754, 1986.

[12] X. Serra and J. O. Smith. "Spectral Modeling Synthesis: Sound Analysis/Synthesis Based on a Deterministic plus Stochastic Decomposition". *Computer Music Journal*, 14(4), pp. 12–24, 1990.

[13] T. Verma and T. Meng, "An Analysis/Synthesis Tool for Transient Signals that Allows a Flexible Sines+Transients+Noise Model for Audio", *1998 IEEE ICASSP-98*, pp. 77–78, Seattle, WA, 1998.

[14] U. Zölzer, Ed., *Digital Audio Effects*, John Wiley & Sons, 2002.

[15] P. Cook, *Real Sound Synthesis for Interactive Applications*, AK Peters LTD, 2002.

[16] P. Morse, *Vibration and Sound*, American Institute of Physics, for the Acoustical Society of America, 1986.

[17] K. Steiglitz, *A Digital Signal Processing Primer*, Addison Wesley, 1996.

[18] J. M. Adrien, *The missing link: Modal synthesis*. In: De Poli, Piccialli, and Roads, editors, *Representations of Musical Signals*, chapter 8, pp. 269–297. MIT Press, Cambridge, 1991.

[19] H. Dudley, 1939, "The Vocoder", Bell Laboratories Record, December, 1939. Reprinted in *IEEE Transactions on Acoustics, Speech and Signal Processing* ASSP-29(3):347–351, 1981.

[20] I. Daubechies "Orthonormal Bases of Compactly Supported Wavelets", Communications on Pure and Applied Math. Vol. 41, pp. 909–996, 1988.

[21] B. Atal, "Speech Analysis and Synthesis by Linear Prediction of the Speech Wave". Journal of the Acoustical Society of America V47, S65(A), 1970.

[22] A. Moorer, "The Use of Linear Prediction of Speech in Computer Music Applications", Journal of the Audio Engineering Society 27(3): pp. 134–140, 1979.

[23] D. Klatt, "Software for a Cascade/Parallel Formant Synthesizer", Journal of the Acoustical Society of America 67(3), pp. 971–995, 1980.

[24] J. L. Kelly Jr. and C. C. Lochbaum, "Speech Synthesis", in *Proceedings of the Fourth ICA*, Copenhagen, Denmark, pp. 1–4, 1962.

[25] X. Rodet, "Time-Domain Formant-Wave-Function Synthesis", *Computer Music Journal* 8 (3), pp. 9–14, 1984.

[26] M. Le Brun, "Digital Waveshaping Synthesis" in *Journal of the Audio Engineering Society*, 27(4), pp. 250–266, 1979.

[27] D. Arfib, "Digital synthesis of complex spectra by means of multiplication of non-linear distorted sine waves". *Journal of the Audio Engineering Society* 27: 10, pp. 757–768, 1979.

[28] J. Chowning, "The Synthesis of Complex Audio Spectra by Means of Frequency Modulation", Journal of the Audio Engineering Society 21(7), pp. 526–534, 1973; reprinted in Computer Music Journal 1(2), 1977; reprinted in *Foundations of Computer Music*, C. Roads and J. Strawn (eds.). MIT Press, 1985.

[29] J. Chowning, "Frequency Modulation Synthesis of the Singing Voice" in Mathews, M. V., and J. R. Pierce (eds), *Current Directions in Computer Music Research*, pp. 57–64,Cambridge, MIT Press, 1980.

[30] C. Roads, "Asynchronous Granular Synthesis" in G. De Poli, A. Piccialli, and C. Roads, eds. 1991. *Representations of Musical Signals*. Cambridge, Mass: The MIT Press, pp. 143–185, 1991.

[31] C. Roads, *Microsound*, Cambridge, MIT Press, 2001.

[32] C. Cadoz, A. Luciani and J. L. Florens. "CORDIS-ANIMA: A Modeling and Simulation System for Sound Image Synthesis-The General Formalization" *Computer Music Journal*, 17(1), pp. 23–29, 1993.

[33] P. Cook, "Physically Informed Sonic Modeling (PhISM): Percussion Instruments." In *Proceedings of the ICMC,* Hong Kong, pp. 228–231, 1996.

[34] P. Cook, "Physically Informed Sonic Modeling (PhISM): Synthesis of Percussive Sounds", *Computer Music Journal,* 21:3, pp. 38–49, 1997.

[35] W. Gaver, "The SonicFinder: An interface that uses auditory icons", *Human Computer Interaction*, 4(1), pp 67–94, 1989.

[36] M. Blattner, D. Sumikawa, & R. Greenberg, "Earcons and icons: Their structure and common design principles", *Human Computer Interaction*, 4(1), pp 11–44, 1989.

[37] S. Brewster, "Sonically-Enhanced Drag and Drop", *Proc. 5th Int. Conf. on Auditory Display (ICAD98)*, 1998.

[38] K. Karplus and A.Strong. "Digital synthesis of plucked-string and drum timbres". *Computer Music Journal*, 7(2), pp. 43–55, 1983.

[39] J. O. Smith, "Acoustic Modeling Using Digital Waveguides", in Roads et. al. (eds.), *Musical Signal Processing*, pp. 221–263, Netherlands, Swets and Zeitlinger, 1997.

[40] Jaffe, D. A. and Smith, J. O. Extensions of the Karplus-Strong plucked string algorithm. *Computer Music Journal*, 7(2), pp. 56–69, 1983.

[41] P. R. Cook, and D. Trueman, "Spherical Radiation from Stringed Instruments: Measured, Modeled, and Reproduced", *Journal of the Catgut Acoustical Society,* pp. 8–14, November 1999.

[42] George Essl and Perry Cook. Banded waveguides: Towards physical modeling of bar percussion instruments. In *Proc. 1999 Int. Computer Music Conf.*, pp. 321–324, Beijing, China,Computer Music Association, 1999.

[43] M. E. McIntyre, R. T. Schumacher, and J. Woodhouse, "On the oscillations of musical instruments." *J. Acoust. Soc. Am.*, 74(5), pp. 1325–1345, 1983.

[44] J. O. Smith, "Efficient simulation of the reed-bore and bow-string mechanisms", *Proceedings of the International Computer Music Conference*, pp. 275–280, 1986.

[45] J. Yuen and A. Horner, "Hybrid Sampling-Wavetable Synthesis with Genetic Algorithms", Journal of the Audio Engineering Society, 45:5, 1997.

[46] B. Verplank, M. Mathews, and R. Shaw, "Scanned synthesis", *Proceedings of the ICMC*, pp. 368–371, Berlin, 2000.

[47] D. Rocchesso and F. Fontana, (Eds.) *The Sounding Object*, `http://www.soundobject.org/SObBook`, 2003.

[48] Y. Dobashi, T. Yamamoto, and T. Nishita, "Real-Time Rendering of Aerodynamic Sound Using Sound Textures Based on Computational Fluid Dynamics", *Proc. ACM SIGGRAPH*, pp. 732–740, 2003.

[49] A. Farnell, *Designing Sound*, Cambridge, MIT Press, 2010.

Chapter 10

Laboratory Methods for Experimental Sonification

Till Bovermann, Julian Rohrhuber and Alberto de Campo

This chapter elaborates on sonification as an experimental method. It is based on the premise that there is no such thing as unconditional insight, no isolated discovery or invention; all research depends on methods. The understanding of their correct functioning depends on the context. Sonification as a relatively new ensemble of methods therefore requires the re-thinking and re-learning of commonly embraced understandings; a process that requires much experimentation.

Whoever has tried to understand something through sound knows that it opens up a maze full of both happy and unhappy surprises. For navigating this labyrinth, it is not sufficient to ask for the most effective tools to process data and output appropriate sounds through loudspeakers. Rather, sonification methods need to incrementally merge into the specific cultures of research, including learning, drafting, handling of complexity, and last but not least the communication within and between multiple communities. Sonification can be a great complement for creating multimodal approaches to interactive representation of data, models and processes, especially in contexts where phenomena are at stake that unfold in time, and where observation of parallel streams of events is desirable. The place where such a convergence may be found may be called a *sonification laboratory*, and this chapter discusses some aspects of its workings.

To begin with, what are the general requirements of such a working environment? A sonification laboratory must be flexible enough to allow for the development of new experimental methods for understanding phenomena through sound. It also must be a point of convergence between different methods, mindsets, and problem domains. Fortunately, today the core of such a laboratory is a computer, and in most cases its 'experimental equipment' is not hardware to be delivered by heavy duty vehicles, but is software which can be downloaded from online resources. This is convenient and flexible, but also a burden. It means that the division of labor between the development of tools, experiments, and theory cannot be taken for granted, and a given sonification toolset cannot be 'applied' without further knowledge;

within research, there is no such thing as 'applied sonification', as opposed to 'theoretical sonification'. Participants in sonification projects need to acquire some familiarity with both the relevant discipline and the methods of auditory display. Only once a suitable solution is found and has settled into regular usage, these complications disappear into the background, like the medical display of a patient's healthy pulse. Before this moment, both method and knowledge depend on each other like the proverbial chicken and egg. Because programming is an essential, but also sometimes intractable, part of developing sonifications, this chapter is dedicated to the software development aspect of sonification laboratory work. It begins with an indication of some common pitfalls and misconceptions. A number of sonification toolkits are discussed, together with music programming environments which can be useful for sonification research. The basics of programming are introduced with one such programming language, *SuperCollider*. Some basic sonification design issues are discussed in more detail, namely the relationship between time, order and sequence, and that between mapping and perception. Finally, four more complex cases of sonification designs are shown – vector spaces, trees, graphs, and algorithms – which may be helpful in the development process.

In order to allow both demonstration and discussion of complex and interesting cases, rather than comparing trivial examples between platforms, the examples are provided in a single computer language. In text-based languages, the program code also serves as precise readable documentation of the algorithms and the intentions behind them [17]. The examples given can therefore be implemented in other languages.

10.1 Programming as an interface between theory and laboratory practice

There is general agreement in the sonification community that the development of sonification methods requires the crossing of disciplinary boundaries. Just as the appropriate interpretation of visualized data requires training and theoretical background about the research questions under consideration, so does the interpretation of an auditory display. There are very few cases where sonification can just be applied as a standard tool without adaptation and understanding of its inner workings.

More knowledge, however, is required for productive work. This knowledge forms an intermediate stage, combining *know-how* and *know-why*. As laboratory studies have shown, the calibration and development of new means of display take up by far the most work in scientific research [24]. Both for arts and sciences, the conceptual re-thinking of methods and procedures is a constant activity. A *computer language* geared towards sound synthesis is a perfect medium for this kind of experimentation, as it can span the full scope from the development from first experiments to deeper investigations. It allows us to understand the non-trivial translations between data, theory, and perception, and permits a wider epistemic context (such as psychoacoustics, signal processing, and aesthetics) to be taken into account. Moreover, programming languages hold such knowledge in an operative form.

As algorithms are designed to specify processes, they dwell at the intersection between laboratory equipment and theory, as boundary objects that allow experimentation with different representation strategies. Some of what needs to be known in order to actively engage in the development and application of sonification methods is discussed in the subsequent sections in the form of generalized case studies.

10.1.1 Pitfalls and misconceptions

For clarification, this section discusses some common *pitfalls and misconceptions miscon-ceptions* that tend to surface in a sonification laboratory environment. Each section title describes a misunderstanding, which is then disentangled in the section which follows:

»Data is an immediate given« Today, measured and digitized data appears as one of the rocks upon which science is built, both for its abundance and its apparent solidity. A working scientist will however tend to emphasize the challenge of finding appropriate data material, and will, wherever required, doubt its relevance. In sonification, one of the clearest indications of the tentative character of data is the amount of working hours that goes into reading the file formats in which the data is encoded, and finding appropriate representations for them, i.e., data structures that make the data accessible in meaningful ways. In order to do this, a working understanding of the domain is indispensable.

»Sonification can only be applied to data.« Often sonification is treated as if it were a method applied to data only. However, sonification is just as much relevant for the understanding of processes and their changing inner state, models of such processes, and algorithms in general. Sonification may help to perceptualize changes of states as well as unknowns and background assumptions. Using the terminology by the German historian of science Rheinberger [24], we can say that it is the distinction between technical things (those effects and facts which we know about and which form the methodological background of the investigation) and epistemic things (those things which are the partly unknown objects of investigation) that makes up the essence of any research. In the course of experimentation, as we clarify the initially fuzzy understanding of what the object of interest is exactly the notion of what does or does not belong to the object to be sonified can change dramatically. To merely "apply sonification to data" without taking into account what it represents would mean to assume this process to be completed already. Thus, many other sources than the common static numerical data can be interesting objects for sonification research.

»Sonification provides intuitive and direct access.« To understand something not yet known requires bringing the right aspects to attention: theoretical or formal reasoning, experimental work, informal conversation, and methods of display, such as diagrams, photographic traces, or sonification. It is very common to assume that acoustic or visual displays provide us somehow with more immediate or intuitive access to the object of research. This is a common pitfall: every sonification (just like an image) may be *read* in very different ways, requires acquaintance with both the represented domain and its representation conventions, and implies theoretical assumptions in all fields involved (i.e., the research domain, acoustics, sonification, interaction design, and computer science). This pitfall can be avoided by not taking acoustic insight for granted. The sonification laboratory needs to allow us to gradually learn to listen for specific aspects of the sound and to judge them in relation to their origin together with the sonification method. In such a process, intuition changes, and understanding of the data under exploration is gained indirectly.

»Data "time" and sonification time are the same.« Deciding which sound events of a sonification happen close together in time is the most fundamental design decision:

temporal proximity is the strongest cue for perceptual grouping (see section 10.4.1). By sticking to a seemingly compelling order (data time must be mapped to sonification time), one loses the heuristic flexibility of really experimenting with orderings which may seem more far-fetched, but may actually reveal unexpected phenomena. It can be helpful to make the difference between *sonification time* and *domain time* explicit; one way to do this formally is to use a sonification variable $\mathring{t}$ as opposed to t. For a discussion of sonification variables, see section 10.4.5.

»Sound design is secondary, mappings are arbitrary.« For details to emerge in sonifications, perceptual salience of the acoustic phenomena of interest is essential and depends critically on psychoacoustically well-informed design. Furthermore, perception is sensitive to domain specific meanings, so finding convincing metaphors can substantially increase accessibility. Stephen Barrass' *ear benders* [2] provide many interesting examples. Finally, "aesthetic intentions" can be a source of problems. If one assumes that listeners will prefer hearing traditional musical instruments over more abstract sounds, then pitch differences will likely sound "wrong" rather than interesting. If one then designs the sonifications to be more "music-like" (e.g., by quantizing pitches to the tempered scale and rhythms to a regular grid), one loses essential details, introduces potentially misleading artefacts, and will likely still not end up with something that is worthwhile music. It seems more advisable here to create opportunities for practicing more open-minded listening, which may be both epistemically and aesthetically rewarding once one begins to read the sonification's details fluently.

10.2 Overview of languages and systems

The history of sonification is also a history of laboratory practice. In fact, within the research community, a number of sonification systems have been implemented and described since the 1980s. They all differ in scope of features and limitations, as they were designed as laboratory equipment, intended for different specialized contexts. These software systems should be taken as integral part of the amalgam of experimental and thought processes, as "reified theories" (a term coined by Bachelard [1]), or rather as a complex mix between observables, documents, practices, and conventions [14, p. 18]. Some systems are now historic, meaning they run on operating systems that are now obsolete, while others are in current use, and thus alive and well; most of them are toolkits meant for integration into other (usually visualization) applications. Few are really open and easily extensible; some are specialized for very particular types of datasets.

The following sections look at dedicated toolkits for sonification, then focus on mature sound and music programming environments, as they have turned out to be very useful platforms for fluid experimentation with sonification design alternatives.

10.2.1 Dedicated toolkits for sonification

xSonify has been developed at NASA [7]; it is based on Java, and runs as a web service[1]. It aims at making space physics data more easily accessible to visually impaired people. Considering that it requires data to be in a special format, and that it only features rather simplistic sonification approaches (here called 'modi'), it will likely only be used to play back NASA-prepared data and sonification designs.

The *Sonification Sandbox* [31] has intentionally limited range, but it covers that range well: Being written in Java, it is cross-platform; it generates MIDI output e.g., to be fed into any General MIDI synth (such as the internal synth on many sound cards). One can import data from CSV text files, and view these with visual graphs; a mapping editor lets users choose which data dimension to map to which sound parameter: Timbre (musical instruments), pitch (chromatic by default), amplitude, and (stereo) panning. One can select to hear an auditory reference grid (clicks) as context. It is very useful for learning basic concepts of parameter mapping sonification with simple data, and it may be sufficient for some auditory graph applications. Development is still continuing, as the release of version 6 (and later small updates) in 2010 shows.

Sandra Pauletto's toolkit for Sonification [21] is based on PureData and has been used for several application domains: Electromyography data for Physiotherapy [22], helicopter flight data, and others. While it supports some data types well, adapting it for new data is slow, mainly because PureData is not a general-purpose programming language where reader classes for data files are easier to write.

SonifYer [27] is a standalone application for OSX, as well as a forum run by the sonification research group at Berne University of the Arts[2]. In development for several years now, it supports sonification of EEG, fMRI, and seismological data, all with elaborate user interfaces. As sound algorithms, it provides audification and FM-based parameter mapping; users can tweak the settings of these, apply EQ, and create recordings of the sonifications created for their data of interest.

SoniPy is a recent and quite ambitious project, written in the Python language [33]. Its initial development push in 2007 looked very promising, and it takes a very comprehensive approach at all the elements the authors consider necessary for a sonification programming environment. It is an open source project and is hosted at sourceforge[3], and may well evolve into a powerful and interesting sonification system.

All these toolkits and applications are limited in different ways, based on resources for development available to their creators, and the applications envisioned for them. They tend to do well what they were intended for, and allow users quick access to experimenting with existing sonification designs with little learning effort.

While learning music and sound programming environments will require more effort, especially from users with little experience in doing creative work with sound and programming, they already provide rich and efficient possibilities for sound synthesis, spatialization, real-time control, and user interaction. Such systems can become extremely versatile tools for the sonification laboratory context by adding what is necessary for access to the data and its

[1] http://spdf.gsfc.nasa.gov/research/sonification
[2] http://sonifyer.org/
[3] http://sourceforge.net/projects/sonipy

domain. To provide some more background, an overview of the three main families of music programming environments follows.

10.2.2 Music and sound programming environments

Computer Music researchers have been developing a rich variety of tools and languages for creating sound and music structures and processes since the 1950s. Current music and sound programming environments offer many features that are directly useful for sonification purposes as well. Mainly, three big families of programs have evolved, and most other music programming systems are conceptually similar to one of them.

Offline synthesis: MusicN to CSound

MusicN languages originated in 1957/58 from the Music I program developed at Bell Labs by Max Mathews and others. Music IV [18] already featured many central concepts in computer music languages such as the idea of a Unit Generator (UGen) as the building block for audio processes (unit generators can be, for example, oscillators, noises, filters, delay lines, or envelopes). As the first widely used incarnation, Music V was written in FORTRAN and was thus relatively easy to port to new computer architectures, from where it spawned a large number of descendants.

The main strand of successors in this family is *CSound*, developed at MIT Media Lab beginning in 1985 [29], which has been very popular in academic as well as dance computer music. Its main approach is to use very reduced language dialects for orchestra files (consisting of descriptions of DSP processes called instruments), and score files (descriptions of sequences of events that each call one specific instrument with specific parameters at specific times). A large number of programs were developed as compositional front-ends in order to write score files based on algorithmic procedures, such as Cecilia [23], Cmix, Common Lisp Music, and others. CSound created a complete ecosystem of surrounding software.

CSound has a very wide range of unit generators and thus synthesis possibilities, and a strong community; the CSound Book demonstrates its scope impressively [4]. However, for sonification, it has a few substantial disadvantages. Even though it is text-based, it uses specialized dialects for music, and thus is not a full-featured programming language. Any control logic and domain-specific logic would have to be built into other languages or applications, while CSound could provide a sound synthesis back-end. Being originally designed for offline rendering, and not built for high-performance real-time demands, it is not an ideal choice for real-time synthesis either. One should emphasize however that CSound is being maintained well and is available on very many platforms.

Graphical patching: Max/FTS to Max/MSP(/Jitter) to PD/GEM

The second big family of music software began with Miller Puckette's work at IRCAM on Max/FTS in the mid-1980s, which later evolved into Opcode Max, which eventually became Cycling'74's Max/MSP/Jitter environment[4]. In the mid-1990s, Puckette began developing

[4]http://cycling74.com/products/maxmspjitter/

an open source program called PureData (Pd), later extended with a graphics system called GEM.[5] All these programs share a metaphor of "patching cables", with essentially static object allocation of both DSP and control graphs. This approach was never intended to be a full programming language, but a simple facility to allow connecting multiple DSP processes written in lower-level (and thus more efficient) languages. With Max/FTS, for example, the programs actually ran on proprietary DSP cards. Thus, the usual procedure for making patches for more complex ideas often entails writing new Max or Pd objects in C. While these can run very efficiently if well written, special expertise is required, and the development process is rather slow, and takes the developer out of the Pd environment, thus reducing the simplicity and transparency of development.

In terms of sound synthesis, Max/MSP has a much more limited palette than CSound, though a range of user-written MSP objects exist. Support for graphics with Jitter has become very powerful, and there is a recent development of the integration of Max/MSP into the digital audio environment Ableton Live. Both Max and Pd have a strong (and partially overlapping) user base; the Pd base is somewhat smaller, having started later than Max. While Max is commercial software with professional support by a company, Pd is open-source software maintained by a large user community. Max runs on Mac OS X and Windows, but not on Linux, while Pd runs on Linux, Windows, and OS X.

Real-time text-based environments: SuperCollider, ChucK

The SuperCollider language today is a full-fledged interpreted computer language which was designed for precise real-time control of sound synthesis, spatialization, and interaction on many different levels. As much of this chapter uses this language, it is discussed in detail in section 10.3.

The ChucK language has been written by Ge Wang and Perry Cook, starting in 2002. It is still under development, exploring specific notions such as being strongly-timed. Like SuperCollider, it is intended mainly as a music-specific environment. While being cross-platform, and having interfacing options similar to SC3 and Max, it currently features a considerably smaller palette of unit generator choices. One advantage of ChucK is that it allows very fine-grained control over time; both synthesis and control can have single-sample precision.

10.3 SuperCollider: Building blocks for a sonification laboratory

10.3.1 Overview of SuperCollider

The SuperCollider language and real-time rendering system results from the idea of merging both real-time synthesis and musical structure generation into a single environment, using the same language. Like Max/PD, it can be said to be an indirect descendant of MusicN and CSound. From SuperCollider 1 (SC1) written by James McCartney in 1996 [19], it has gone through three complete rewriting cycles, thus the current version SC3 is a very

[5] http://puredata.info/

mature system. In version 2 (SC2) it inherited much of its language characteristics from the Smalltalk language; in SC3 [20] the language and the synthesis engine were split into a client/server architecture, and many features from other languages such as APL and Ruby were adopted as options.

As a modern and fully-fledged text-based programming language, SuperCollider is a flexible environment for many uses, including sonification. Sound synthesis is very efficient, and the range of unit generators available is quite wide. SC3 provides a GUI system with a variety of interface widgets. Its main emphasis, however, is on stable real-time synthesis. Having become open-source with version 3, it has since flourished. Today, it has quite active developer and user communities. SC3 currently runs on OS X and Linux. There is also a less complete port to Windows.

10.3.2 Program architecture

SuperCollider is divided into two processes: the language (*sclang*, also referred to as *client*) and the sound rendering engine (*scsynth*, also referred to as *server*). These two systems connect to each other via the networking protocol OpenSoundControl (*OSC*).[6]

SuperCollider is an interpreted fully-featured programming language. While its architecture is modeled on Smalltalk, its syntax is more like C++. Key features of the language include its ability to express and realize timing very accurately, its rapid prototyping capabilities, and the algorithmic building blocks for musical and other time-based compositions.

In contrast to sclang, the server, scsynth, is a program with a fixed architecture that was designed for highly efficient real-time sound-rendering purposes. Sound processes are created by means of synthesis graphs, which are built from a dynamically loaded library of unit generators (UGens); signals can be routed on audio and control buses, and soundfiles and other data can be kept in buffers.

This two-fold implementation has major benefits. First, other applications can use the sound server for rendering audio; Second, it scales well to multiple machines/processor cores, i.e., scsynth can run on one or more autonomous machines; and Third, decoupling sclang and scserver makes both very stable.

However, there are also some drawbacks to take into account. Firstly, there is always network latency involved, i.e., real-time control of synthesis parameters is delayed by the (sometimes solely virtual) network interface. Secondly, the network interface introduces an artificial bottleneck for information transfer, which in turn makes it hard to operate directly on a per sample basis. Thirdly, there is no direct access to server memory from sclang. (On OS X, this is possible by using the internal server, so one can choose one's compromises.)

SuperCollider can be extended easily by writing new classes in the SC language. There is a large collection of such extension libraries called Quarks, which can be updated and installed from within SC3.[7] One can also write new Unit Generators, although a large collection of these is already available as sc3-plugins.[8]

[6] http://opensoundcontrol.org/
[7] See the Quarks help file for details
[8] http://sourceforge.net/projects/SC3 plugins/

10.3.3 Coding styles

Thanks to the scope of its class library and its flexible syntax, SuperCollider offers many techniques to render and control sounds, and a variety of styles of expressing ideas in code. This short overview describes the basics of two styles (object style and pattern style), and shows differences in the way to introduce sound dynamics depending on external processes (i.e., data sonification). For a more detailed introduction to SuperCollider as a sound rendering and control language, please refer to the *SuperCollider Book* [32]. This also features a dedicated chapter on sonification with SuperCollider.

Object style Object-style sound control hides the network-based communication between client and server with an object-oriented approach. All rendering of sound takes place within the synthesis server (scsynth). The atom of sound synthesis is the unit generator (Ugen) which produces samples depending on its input parameters. UGens form the constituents of a fixed structure derived from a high-level description, the `SynthDef`, in sclang:

```
1 SynthDef(\pulse, { // create a synth definition named "pulse"
2   |freq = 440, amp = 0.1| // controls that can be set at runtime
3     Out.ar( // create an outlet for the sound
4         0,      // on channel 0 (left)
5         Pulse.ar( // play a pulsing signal
6             freq // with the given frequency
7         ) * amp // multiply it by the amp factor to determine its volume
8     );
9 }).add; // add it to the pool of SynthDefs
```

In order to create a sound, we instantiate a `Synth` object parameterised by the SynthDef's name:

```
1 x = Synth(\pulse);
```

This does two things: firstly, it creates a synth object on the server which renders the sound described in the `pulse` synthesis definition, and secondly, it instantiates an object of type `Synth` on the client, a representation of the synth process on the server with which the language is able to control its parameters:

```
1 x.set(\freq, 936.236);  // set the frequency of the Synth
```

To stop the synthesis you can either evaluate

```
1 x.free;
```

or press the panic-button (hear sound example **S10.1**).[9] The latter will stop all synthesis processes, re-initialise the server, and stop all running tasks, whereas `x.free` properly releases only the synth process concerned and leaves everything else unaffected.

In this strategy, we can implement the simplest parameter mapping sonification possible in SuperCollider (see also section 10.4.2). Let's assume we have a dataset consisting of a one-dimensional array of numbers between 100 and 1000:

```
1 a = [ 191.73, 378.39, 649.01, 424.49, 883.94, 237.32, 677.15, 812.15 ];
```

[9] `<Cmd>-.` on OS X, `<Esc>` in gedit, `<Ctrl>-c <Ctrl>-s` in emacs, and `<alt>-.` on Windows.

With a construction called *Task*, a pauseable process that can run in parallel to the interactive shell, we are now able to step through this list and create a sound stream that changes its frequency according to the values in the list (hear sound example **S10.2**):

```
Task {
    // instantiate synth
    x = Synth(\pulse, [\freq, 20, \amp, 0]);
    0.1.wait;

    x.set(\amp, 0.1);          // turn up volume
    // step through the array
    a.do{|item| // go through each item in array a
        // set freq to current value
        x.set(\freq, item);

        // wait 0.1 seconds
        0.1.wait;
    };

    // remove synth
    x.free;
}.play;
```

The above SynthDef is continuous, i.e., it describes a sound that could continue forever. For many sound and sonification techniques, however, a sound with a pre-defined end is needed. This is done most simply with an envelope. It allows the generation of many very short sound events (sound grains). Such a grain can be defined as:

```
SynthDef(\sinegrain, {
  |out = 0, attack = 0.01, decay = 0.01, freq, pan = 0, amp = 0.5|

    var sound, env;

    // an amplitude envelope with fixed duration
    env = EnvGen.ar(Env.perc(attack, decay), doneAction: 2);

    // the underlying sound
    sound = FSinOsc.ar(freq);

    // use the envelope to control sound amplitude:
    sound = sound * (env * amp);

    // add stereo panning
    sound = Pan2.ar(sound, pan);

    // write to output bus
    Out.ar(out, sound)
}).add;
```

To render one such grain, we evaluate

```
Synth.grain(\sinegrain, [\freq, 4040, \pan, 1.0.rand2]);
```

Note that, in difference to the above example, the *grain* method creates an anonymous synth on the server, which cannot be modified while running. Thus, all its parameters are fixed when it is created. The grain is released automatically after the envelope is completed, i.e., the sound process stops and is removed from the server.

Using the dataset from above, a discrete parameter mapping sonification can be written like this (hear sound example **S10.3**):

```
Task {
    // step through the array
```

```
3       a.do{|item|
4           // create synth with freq parameter set to current value
5           // and set decay parameter to slightly overlap with next grain
6           Synth.grain(\sinegrain, [\freq, item, \attack, 0.001, \decay, 0.2]);
7
8           0.1.wait; // wait 0.1 seconds between grain onsets
9       };
10  }.play;
```

A third way to sonify a dataset is to first send it to a `Buffer` – a server-side storage for sequential data – and then use it as the source for dynamics control (hear sound example **S10.4**):

```
1   b = Buffer.loadCollection(
2       server: s,
3       collection: a,
4       numChannels: 1,
5       action: {"load completed".inform}
6   );
7
8   SynthDef(\bufferSon, {|out = 0, buf = 0, rate = 1, t_trig = 1, amp = 0.5|
9       var value, synthesis;
10
11      value = PlayBuf.ar(
12          numChannels: 1,
13          bufnum: buf,
14          rate: rate/SampleRate.ir,
15          trigger: t_trig,
16          loop: 0
17      );
18
19      synthesis = Saw.ar(value);
20
21      // write to outbus
22      Out.ar(out, synthesis * amp);
23  }).add;
24
25  x = Synth(\bufferSon, [\buf, b])
26
27  x.set(\rate, 5000); // set rate in samples per second
28  x.set(\t_trig, 1);  // start from beginning
29  x.free;             // free the synthesis process
```

This style is relatively easy to adapt for audification by removing the synthesis process and writing the data directly to the audio output:

```
1
2   SynthDef(\bufferAud, {|out = 0, buf = 0, rate = 1, t_trig = 1, amp = 0.5|
3
4       var synthesis = PlayBuf.ar(
5           numChannels: 1,
6           bufnum: buf,
7           rate: rate/SampleRate.ir,
8           trigger: t_trig,
9           loop: 0
10      );
11
12      // write to output bus
13      Out.ar(out, synthesis * amp)
14  }).add;
```

As the server's sample representation requires samples to be between -1.0 and 1.0, we have to make sure that the data is scaled accordingly. Also, a larger dataset is needed (see the chapter on audification, 12, for details). An artificially generated dataset might look like this:

```
a = {|i|cos(i**(sin(0.0175*i*i)))}!10000;
a.plot2; // show a graphical representation;
```

We can now load the dataset to the server and instantiate and control the synthesis process, just as we did in the example above (hear sound example **S10.5**):

```
b = Buffer.loadCollection(
    server: s,
    collection: a,
    numChannels: 1,
    action: {"load completed".inform}
);

// create synth
x = Synth(\bufferAud, [\buf, b, \rate, 44100]);

x.set(\t_trig, 1);             // restart
x.set(\rate, 200);             // adjust rate
x.set(\t_trig, 1, \rate, 400); // restart with adjusted rate
x.set(\t_trig, 1, \rate, 1500);

x.free;
```

Pattern style Patterns are a powerful option to generate and control sound synthesis processes in SuperCollider. A pattern is a high-level description of sequences of values that control a stream of sound events, which allows us to write, for example, a parameter mapping sonification in a way that also non-programmers can understand what is going on. Pattern-controlled synthesis is based on `Events`, defining a (predominately sonic) event with names and values for each parameter. Playing a single grain as defined in the object style paragraph then looks like this:

```
(instrument: \sinegrain, freq: 4040, pan: 1.0.rand2).play
```

When playing a pattern, it generates a sequence of events. The definition of the above discrete parameter mapping sonification in pattern style is (hear sound example **S10.6**):

```
a = [ 191.73, 378.39, 649.01, 424.49, 883.94, 237.32, 677.15, 812.15 ];
Pbind(
    \instrument, \sinegrain,
    \freq, Pseq( a ),  // a sequence  of the dataset a
    \attack, 0.001,    // and fixed values as desired
    \decay, 0.2,       // for the other parameters
    \dur, 0.1
).play
```

One benefit of the pattern style is that a wide range of these high-level controls already exist in the language. Let us assume the dataset under exploration is two-dimensional:

```
a = [
    [ 161.58, 395.14 ], [ 975.38, 918.96 ], [ 381.84, 293.27 ],
    [ 179.11, 146.75 ], [ 697.64, 439.80 ], [ 202.50, 571.75 ],
    [ 361.50, 985.79 ], [ 550.85, 767.34 ], [ 706.91, 901.56 ],
]
```

We can play the dataset by simply defining `a` with this dataset and evaluating the `Pbind` above. It results in two simultaneous streams of sound events, one for each pair (hear sound example **S10.7**). With a slight adjustment, we can even let the second data channel be played panned to the right (hear sound example **S10.8**):

```
Pbind(
    \instrument, \sinegrain,
    \freq, Pseq( a ),  // a sequence  of the data (a)
    \attack, 0.001,
    \decay, 0.2,
    \pan, [-1, 1],  // pan first channel to left output, second to right
    \dur, 0.1
).play
```

Comparison of styles

For modifying continuous sounds, and handling decisions unfolding in time very generally, 'tasks' are a very general and flexible tool. For creating streams from individual sounds, 'patterns' provide many options to express the implemented ideas in very concise terms. Depending on the context and personal preferences in thinking styles, one or other style may be better suited for the task at hand. The *Just In Time Programming Library* (JITLib) provides named proxies for tasks (Tdef), patterns (Pdef), and synths (Ndef), which allow to change running programs, simplify much technical administration, and thus can speed up development significantly.[10]

10.3.4 Interfacing

In this section, essential tools for loading data, recording the sonifications, and controlling the code from external processes are described. Due to the scope of this book, only the very essentials are covered. For a more in-depth overview on these themes, please consult the corresponding help pages, or the SuperCollider book [32].

Loading data Supposed, we have a dataset stored as comma-separated values (csv) in a text file called data.csv:

```
-0.49, 314.70,  964, 3.29
-0.27, 333.03,  979, 1.96
 0.11, 351.70, 1184, 5.18
-0.06, 117.13, 1261, 2.07
-0.02, 365.15,  897, 2.01
-0.03, 107.82, 1129, 2.24
-0.39, 342.26, 1232, 4.92
-0.29, 382.03,  993, 2.35
```

We can read these into SuperCollider with help of the CSVFileReader class:

```
a = CSVFileReader.readInterpret("data.csv");
a.postcs;  // post data
```

Each row of the dataset is now represented in SuperCollider as one array. These arrays are again collected in an enclosing array. A very simple sonification using the pattern method described in Section 10.3.3 looks like this:

```
// transpose the data representation
// now the inner arrays represent one row of the dataset
```

[10]For more information, see the JITLib help file, or the JITLib chapter in the SuperCollider book [32].

```
b = a.flop;

(
Pbind(
	\instrument, \sinegrain,
	\freq, Pseq([b[1], b[2]].flop, 2),
	\attack, 0.002,
	\decay, Pseq(b[3] * 0.1, inf),
	\pan, Pseq(b[0], inf),
	\dur, 0.1
).play
)
```

For very large data sets which are common in sonification it may be advisable to keep the data in a more efficiently readable format between sessions. For time series, such as EEG data, converting them to soundfiles will reduce load times considerably. For other cases, SuperCollider provides an archiving method for every object:

```
		// store data
a.writeArchive(path);
		// read data
a = Object.readArchive(path);
```

This can reduce load time by an order of two.

Recording sonifications SuperCollider provides easy and flexible ways to record real-time sonifications to soundfiles. Only the simplest case is covered here; please see the Server help file for more details.

```
		// start recording
s.record("/path/to/put/recording/test.wav");
		// run your sonification now ...
		// stop when done
s.stopRecording;
```

Control from external processes SuperCollider can be controlled from external applications by means of OpenSoundControl (OSC) [34]. Let us assume that an external program sends OSC messages in the following format to SC3[11]:

```
/data, iff 42 23.0 3.1415
```

You can set up a listener for this message with:

```
OSCresponder(nil, "/data", {|time, responder, message|
		"message % arrived at %\n".postf(message, time);
}).add;
```

We leave it as an exercise to the reader to integrate this into a sonification process. In-depth discussions of many sonification designs and their implementations in SC3 can be found in Bovermann [5] and de Campo [9].

[11]Note that SuperCollider's default port for incoming OSC messages is 57120.

10.4 Example laboratory workflows and guidelines for working on sonification designs

This section discusses many of the common concerns in creating, exploring and experimenting with sonification designs and how to integrate them in a laboratory workflow. Here, theoretical considerations alternate with examples that are generic enough to make it easy to adapt them to different contexts.

What is usually interesting about specific data sets is discovering the possible relationships between their constituents; some of these relations may be already established, whereas others may not yet be evident. Perceptualization is the systematic attempt to represent such relationships in data (or generally, objects under study) such that relationships between the constituents of the sensory rendering emerge in perception. This means that an observer notices *gestalts*, which may confirm or disprove hypotheses about relationships in the data. This process relies on human perceptual and cognitive abilities; most importantly that of organizing sensory events into larger groups. In auditory perception, this grouping of individual events depends on their perceptual parameters and their relationships, i.e., mainly inter-similarities and proximities.

In a successful sonification design, the relationships within the local dynamic sound structure (the proximal cues) allow a listener to infer insights into the data being sonified, effectively creating what can be considered distal cues. As there are very many possible variants of sonification design, finding those that can best be tuned to be very sensitive to the relationships of interest, however, is a nontrivial methodological problem.

The *Sonification Design Space Map (SDSM)* [8, 9] aims to help in the process of developing sonification designs. Put very briefly, while the working hypotheses evolve, as the sonification designs become more and more sophisticated, one repeatedly answers three questions:

1. How many data points are likely necessary for patterns to emerge perceptually?
2. How many and which data properties should be represented in the design?
3. How many parallel sound-generating streams should the design consist of?

Based on the answers, the SDSM recommends making sure the desired number of data points is rendered within a time window of 3–10 seconds (in order to fit within non-categorical echoic memory) [28], and it recommends suitable strategies (from Continuous, Discrete-Point, and Model-based approaches). As Figure 10.1 shows, changes in the answers correspond to movements of the current working location on the map: Zooming in to fewer data points for more detail moves it to the left, zooming out moves it to the right; displaying more data dimensions moves it up, while using more or fewer parallel sound streams moves it in the z-axis.

In practice, time spent exploring design alternatives is well spent, and helps by clarifying which (seemingly natural) implicit decisions are being taken as a design evolves. The process of exchange and discussion in a hypothetical research team, letting clearer questions evolve as the sonification designs become more and more sensitive to latent structures in the data, process or model under study, is of fundamental importance. It can be considered the equivalent of the common experience in laboratory work that much of the total work time is absorbed by setting up and calibrating equipment, until the experimental setup is fully

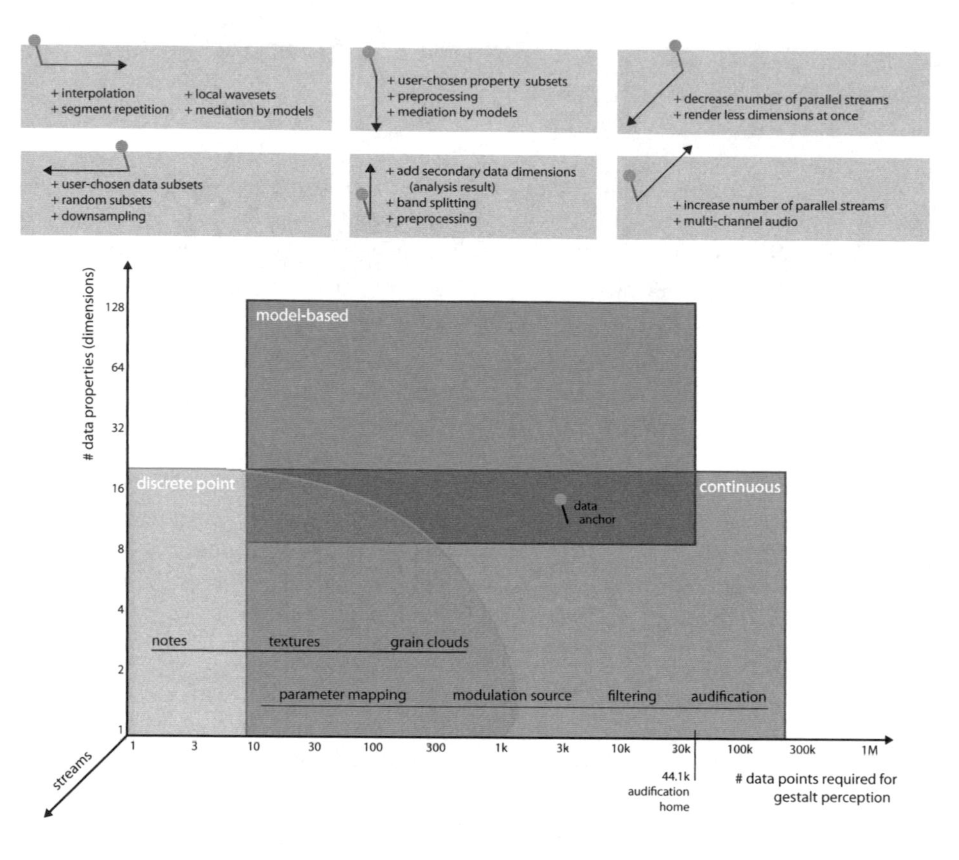

Figure 10.1: The Sonification Design Space Map.

"tuned", while by comparison, much less time is usually spent with actual measurement runs themselves. Such calibration processes may involve generating appropriate test data, as well as doing listening tests and training.

There now follow three sections explaining typical scenarios, in which data sonification workers may find themselves. As proximity in time is the property that creates the strongest perceptual grouping, especially in sound, the first section covers data ordering concepts and the handling of time in sonification. The second section discusses fundamental issues of mapping of data dimensions to sound properties via synthesis parameters, which requires taking perceptual principles into account. The later three sections address more complex cases, which raise more complex sets of questions.

10.4.1 Basics 1: Order, sequence, and time

In any data under study, we always need to decide which relations can be ordered and according to what criteria. Data from different geographic locations, for instance, may be ordered by longitude, latitude and/or altitude. Mapping a data order to a rendering order (such as altitude to a time sequence) means treating one dimension differently from the others.

As temporal order is the strongest cue for grouping events perceptually in the sonic domain, experimenting with mappings of different possible data orders to the temporal order of sounds can be very fruitful.

Example solutions

Here is a very simple example to demonstrate this constellation. Assume for simplicity that the domain data points are all single values, and they come in a two dimensional order, represented by an array:

```
a = [
    [ 0.97, 0.05, -0.22, 0.19, 0.53, -0.21, 0.54, 0.1, -0.35, 0.04 ],
    [ -0.07, 0.19,  0.67, 0.05, -0.91, 0.1,  -0.8, -0.21, 1, -0.17 ],
    [ 0.67, -0.05, -0.07, -0.05, 0.97, -0.65, -0.21, -0.8, 0.79, 0.75 ]
];
```

Two ordered dimensions are obvious, horizontal index, and vertical index, and a third one is implied: the magnitude of the individual numbers at each index pair. Depending on where this data came from, the dimensions may correlate with each other, and others may be implied, some of which may be unknown.

For experimentation, we define a very simple synthesis structure that creates percussive decaying sound events. This is just sufficient for mapping data values to the most sensitive perceptual property of sound - pitch - and experimenting with different ordering strategies.

```
SynthDef(\x, { |freq = 440, amp = 0.1, sustain = 1.0, out = 0|
    var sound = SinOsc.ar(freq);
    var env = EnvGen.kr(Env.perc(0.01, sustain, amp), doneAction: 2);
    Out.ar(out, sound * env);
}).add;
```

This sound event has four parameters: amplitude, sustain (duration of sound), frequency, and output channel number (assuming one uses a multichannel audio system).

As there is no inherent preferred ordering in the data, a beginning strategy would be to experiment with a number of possible orderings to develop a sense of familiarity with the data and its possibilities, and noting any interesting details that may emerge.

Relating time sequence with horizontal index, and frequency to the data value at that point, we can begin by playing only the first line of the data set (hear sound example **S10.9**):

```
// define a mapping from number value to frequency:
f = { |x| x.linexp(-1, 1, 250, 1000) };
Task {
    var line = a[0]; // first line of data
    line.do { |val|
        (instrument: \x, freq: f.value(val)).play;
        0.3.wait;
    }
}.play;
```

Next, we play all three lines, with a short pause between events and a longer pause between lines, to maintain the second order (hear sound example **S10.10**):

```
Task {
    a.do { |line|
        line.do { |val|
            (instrument: \x, freq: f.value(val)).play;
            0.1.wait;
    },
    0.3.wait;
}.play;
```

When we sort each line before playing it, the order in each line is replaced with order by magnitude (hear sound example **S10.11**):

```
Task {
    a.do { |line|
        line.copy.sort.do { |val|
            (instrument: \x, freq: f.value(val)).play;
            0.1.wait;
    };
    0.3.wait;
    }
}.play;
```

We play each line as one chord, so the order within each line becomes irrelevant (hear sound example **S10.12**):

```
Task {
    a.do { |line|
        line.do { |val|
            (instrument: \x, freq: f.value(val)).play; // no wait time here
    };
    0.3.wait;
}.play;
```

We can also use vertical order, and play a sequence of all columns (hear sound example **S10.13**):

```
Task {
    var cols = a.flop; // swap rows <-> columns
    cols.do { |col|
        col.do { |val|
            (instrument: \x, freq: f.value(val)).play;
            0.1.wait;     // comment out for 3-note chords
        };
        0.3.wait;
    };
}.play;
```

Finally, we play all values in ascending order (hear sound example **S10.14**):

```
Task {
    var all = a.flat.sort;
    all.do { |val|
        (instrument: \x, freq: f.value(val)).play;
        0.1.wait;
    };
}.play;
```

All these variants bring different aspects to the foreground: Hearing each line as a melody allows the listener to compare the overall shapes of the three lines. Hearing each column as a three note arpeggio permits comparing columns for similarities. Hearing each column as a chord brings similarity of the (unordered) sets of elements in each column into focus. Hearing each line sorted enables observation of what value ranges in each line values are denser or sparser. Sorting the entire set of values applies that observation to the entire dataset.

Discussion

It is productive to be aware of explicit orderable and un-orderable dimensions. Simple experiments help the designer to learn, to adjust, and to develop how these dimensions interrelate. Writing systematic variants of one experiment brings to the surface nuances that may become central evidence once discovered. For instance, with every new ordering, different structures might emerge. With unknown data, cultivating awareness of alternative orderings and data structures will help for fruitful experimentation and for learning to distinguish the impact of differences on a given sonification. Note that there are many psychoacoustic peculiarities in timing – for instance, parallel streams of sound may emerge or not dependent on tempo, and a series of events may fuse into a continuum.

10.4.2 Basics 2: Mapping and perception

Every sonification design involves decisions regarding how the subject of study determines audible aspects of the perceptible representation. It is thereby necessary to take into account the psychoacoustic and perceptual concepts underlying sound design decisions. Here, the discussion of these facts is very brief; for a more in-depth view see chapter 3, for a longer discussion of auditory dimensions see chapter 4, finally, for an introduction to mapping and scaling, see chapter 2.

Audible aspects of rendered sound may serve a number of different purposes:

1. Analogic display - a data dimension is mapped to a synthesis parameter which is easy to recognise and follow perceptually. Pitch is the most common choice here; timbral variation by modulation techniques is also well suited for creating rich, non-categorical variety.
2. Labelling a stream – this is needed for distinguishing categories, especially when several parallel streams are used. Many designers use instrumental timbres here; we find that spatial position is well suited as well, especially when using multiple loudspeakers as distinct physical sound sources.
3. Context information/orientation – this is the mapping non-data into the rendering, such as using clicks to represent a time grid, or creating pitch grids for reference.

Tuning the ranges of auditory display parameters plays a central role in parameter mapping sonification (see chapter 15), but indirectly it plays into all other approaches as well. Physical parameters, such as frequency and amplitude of a vibration, are often spoken of in identical terms to synthesis processes, as in the frequency and amplitude of an oscillator. They typically correspond to perceived sound properties, like pitch and loudness, but the correspondence is not always a simple one. First, we tend to perceive amounts of change relative to the absolute value; a change of 6% of frequency will sound like a tempered half-step in most of the audible frequency range. Second, small differences can be inaudible; the limit where half the test subjects say a pair of tones is the same and the other half says they are different is called the *just noticeable difference* (JND). The literature generally gives pitch JND as approximately

0.2 half-steps or about 1% frequency difference (degrading towards very high and very low pitches, and for very soft tones) and loudness differences of around 1 dB (again, worse for very soft sounds, and potentially finer for loud sounds). However, this will vary depending on the context: when designing sonifications, one can always create test data to learn which data differences will be audible with the current design. In the example in section 10.4.1, the numerical value of each data point was mapped to an exponential frequency range of 250 – 1000 Hz. In the first data row, the smallest difference between two values is 0.01. We may ask now whether this is audible with the given design (hear sound examples **S10.15** and **S10.16**):

```
// alternate the two close values
Task { loop {
    [0.53, 0.54].do { |val|
        (instrument: \x, freq: f.value(val)).play;
        0.1.wait;
    }
} }.play;

        // then switch between different mappings:
f = { |x| x.linexp(-1, 1, 250, 1000) }; // mapping as it was
f = { |x| x.linexp(-1, 1, 500, 1000) };  // narrower
f = { |x| x.linexp(-1, 1, 50, 10000) }; // much wider

        // run entire dataset with new mapping:
Task {
    a.do { |line|
        line.do { |val|
            (instrument: \x, freq: f.value(val)).play;
            0.1.wait;
        }
    };
    0.3.wait;
}.play;
```

When playing the entire dataset with the new wider mapping, a new problem emerges: the higher sounds appear louder than the lower ones. The human ear's perception of loudness of sine tones depends on their frequencies. This nonlinear sensitivity is measured experimentally in the equal loudness contours (see also chapter 3). In SC3, the UGen `AmpComp` models this: based on the frequency value, it generates boost or attenuation factors to balance the sound's loudness. The following SynthDef exemplifies its usage:

```
SynthDef(\x, { |freq = 440, amp = 0.1, sustain = 1.0, out = 0|
    var sound = SinOsc.ar(freq);
    var ampcomp = AmpComp.kr(freq.max(50));  // compensation factor
    var env = EnvGen.kr(Env.perc(0.01, sustain, amp), doneAction: 2);
    Out.ar(out, sound * ampcomp * env);
}).add;
```

So far also it is assumed that the sound events' duration (its sustain) is constant (at 1 second). This was just an arbitrary starting point; when one wants to render more sound events into the same time periods, shorter sounds have less overlap and thus produce a clearer sound shape. However, one loses resolution of pitch, because pitch perception becomes more vague with shorter sounds (hear sound example **S10.17**).

```
Task {
    a.do { |line|
        line.do { |val|
            (instrument: \x, freq: f.value(val), sustain: 0.3).play;
            0.1.wait;
        };
```

```
    };
    0.3.wait;
}.play;
```

We can also decide to assume that each of the three lines can have a different meaning; then we could map, for example, the values in the first line to frequency, the second to sustain, and the third to amplitude (hear sound example **S10.18**):

```
Task {
    var cols = a.flop; // swap rows <-> columns
    cols.do { |vals|
        var freq = vals[0].linexp(-1, 1, 300, 1000);
        var sustain = vals[1].linexp(-1, 1, 0.1, 1.0);
        var amp = vals[2].linexp(-1, 1, 0.03, 0.3);

        (instrument: \x,
            freq: freq,
            sustain: sustain,
            amp: amp
        ).play;
        0.2.wait;
    };
}.play;
```

Finally, we make a different set of assumptions, which leads to different meanings and mappings again: If we interpret the second line to be a comparable parameter to the first, and the third line to represent how important the contribution of the second line is, we can map the three lines to basic frequency, modulation frequency, and modulation depth (hear sound example **S10.19**):

```
SynthDef(\xmod, { |freq = 440, modfreq = 440, moddepth = 0,
    amp = 0.1, sustain = 1.0, out = 0|
    var mod = SinOsc.ar(modfreq) * moddepth;
    var sound = SinOsc.ar(freq, mod);
    var env = EnvGen.kr(Env.perc(0.01, sustain, amp), doneAction: 2);
    Out.ar(out, sound * env);
}).add;

Task {
    var cols = a.flop; // swap rows <-> columns
    cols.do { |vals|
        var freq = vals[0].linexp(-1, 1, 250, 1000);
        var modfreq = vals[1].linexp(-1, 1, 250, 1000);
        var moddepth = vals[2].linexp(-1, 1, 0.1, 4);
        (instrument: \xmod,
            modfreq: modfreq,
            moddepth: moddepth,
            freq: freq,
            sustain: 0.3,
            amp: 0.1
        ).postln.play;
        0.2.wait;
    };
}.play;
```

Discussion

Tuning display processes in such a way that they are easy to read perceptually is by no means trivial. Many synthesis and spatialization parameters behave in subtly or drastically different ways and lend themselves to different purposes in mappings. For example, while recurrent patterns (even if shifted and scaled) are relatively easy to discern when mapped to pitch, mapping them to loudness would make recognizing them more difficult, whereas mapping them to spatial positions would reduce the chances of false assignment.

In the sonification laboratory, it is good practice to test the audible representation, in much the same way as one would other methods. By creating or selecting well-understood test datasets, and verifying that the intended audience can confidently hear the expected level of perceptual detail, one can verify its basic viability for the context under study.

10.4.3 Vector spaces

While a given dataset is never entirely without semantics, it can be heuristically useful to abstract from what is known in order to identify new relations and thus build up a new semantic layer. One may quite often be confronted with the task of sonifying a numerical dataset that is embedded into a high-dimensional vector space, where axis descriptions were actively pruned. In other words, the orientation of the vector basis is arbitrary, thus carrying no particular meaning. We would like to be able to experiment with different approaches that take this arbitrariness into account.

Example solutions

Simple mapping approaches may consist of random mapping choices, picking one dimension for time ordering, and choosing others for mapping to control parameters of the sound display. For examples on these, see sections 10.4.1 and 10.4.2.

Principal Component Analysis (PCA) A related linearization method for datasets embedded into vector spaces is to find a linear transformation:

$$\hat{x}_j = \sum_i (x_j - \hat{o}) \cdot a_i$$

(with x_j data item, $\hat{x}_j$ transformed data item, $\hat{o}$ new origin, and a_i the i-th basis vector) that – based on either domain-specific knowledge or based on the actual dataset – makes sense. One option for deriving dataset-inherent knowledge is Principal Component Analysis (PCA) which returns basis vectors ordered by variances in their direction.[12] For more details on this approach, see chapter 8. The `pc1` method on `SequenceableCollection`, provided with the *MathLib* quark, calculates an estimation of the first principal component for a given, previously whitened dataset:

```
1 // a 2-d dataset with two quasi-gaussian distributions
2 d = {#[[-1, -0.5], [1, 0.5]].choose + ({0.95.gauss}!2)}!10000;
3 p = d.pc1; // first principal component
```

To estimate the next principal component, we have to subtract the fraction of the first one and do the estimation again:

```
1 f = d.collect{|x|
2     var proj;
```

[12] While nearly any dataset could be interpreted as a vector-space, PCA is more useful for high-dimensional data.

```
3      proj = ((v * x).sum * v); // projection of x to v
4      x-proj; // remove projection from data item
5  }
6  p = d.pc1; // second principal component
```

For computing all principal components of a multidimensional dataset, we can apply the process recursively:

```
1  // compute components for n-dimensional datasets
2  q = ();
3
4  q.data; // the data
5  q.dim = q.data.shape.last
6  q.subtractPC = {|q, data, pc|
7      var proj;
8      data.collect{|x|
9          proj = ((pc * x).sum * pc); // projection of x to v
10         x-proj; // remove projection from data item
11     };
12 }
13
14 // recursive function to calculate the steps needed
15 q.computePCs = {|q, data, pcs, dims|
16     var pc;
17
18     (dims > 1).if({
19         pc = data.pc1;
20         pcs[data.shape.last-dims] = pc;
21         pcs = q.computePCs(q.subtractPC(data, pc), pcs, dims-1);
22     }, {
23         pc = data.pc1;
24         pcs[data.shape.last-dims] = pc;
25     });
26     pcs;
27 }
28
29 // calculate and benchmark. This might take a while
30 {q.pcs = q.computePCs(q.data, 0!q.dim!q.dim, q.dim)}.bench;
```

This dimensional reduction, respectively dimension reorganization process alters the dataset's representation but not its (semantic) content. After this kind of pre-processing, strategies as introduced in Sections 10.4.1 and Section 10.4.2 become applicable again.

For more complex, statistical analysis methods, we suggest using tools like octave[13] or NumPy/SciPy[14] as they already implement well-tested methods for this, which otherwise have to be implemented and tested in SuperCollider.

Distance If the absolute data values are not of interest (e.g., because of the absence of a reference point), relative information might be worth sonifying. One such information is the distance between data items. It implicitly contains information like dataset density (both global and local), variance and outliers. The distance matrix for all data items can be computed by:

```
1  q = ();
2
3  // the dataset
4  q.data = {|dim = 4|
5      ({{1.0.rand}!dim + 5}!100) ++ ({{10.0.rand}!dim}!100)
6  }.value
```

[13] http://www.gnu.org/software/octave/
[14] www.scipy.org/

```
7
8
9  // function to compute the distance between two points
10 q.dist = {|q, a, b|
11     (a-b).squared.sum.sqrt;
12 }
13
14 // compute the distance matrix for the dataset
15 q.distanceMatrix = {|data|
16     var size = data.size;
17     var outMatrix = 0!size!size; // fill a matrix with zeroes
18     var dist;
19
20     data.do{|item, i|
21         i.do{|j|
22             dist = q.dist(item, data[j]);
23             outMatrix[i][j] = dist;
24             outMatrix[j][i] = dist;
25         }
26     };
27
28     outMatrix
29 }.value(q.data)
```

Since the resulting matrix can be recognised as an (undirected) graph, it can be sonified for example by methods as described in the upcoming section 10.4.4.

Model-based sonification A third set of methods to sonify vector spaces are model-based sonifications as introduced in chapter 16. Next is shown an example for a data sonogram [15], which uses the data and distance algorithm described in the previous paragraph.

```
1  SynthDef(\ping, {|freq = 2000, amp=1|
2      var src = Pulse.ar(freq);
3      var env = EnvGen.kr(Env.perc(0.0001, 0.01), 1, doneAction: 2) * amp;
4      Out.ar(0, (src * env) ! 2)
5  }).add;
6
7  (
8  q = q ? ();
9  // generate score of OSC messages, sort and play
10 q.createScore = {|q, dataset, rSpeed=1, impactPos|
11     var dist, onset, amp;
12
13     // set impactPos to a useful value - best by user interaction
14     impactPos = impactPos ? 0.0.dup(dataset.shape.last);
15
16
17         // for each data item, compute its distance from the impact center and
18         // create an event according to it in the score*/
19         // first ping represents impact
20     [[0, [\s_new, \ping, -1, 0, 0, \freq, 1500]]]
21         ++ dataset.collect{|row|
22                 // set onset time proportional to the distance
23             dist = q.dist(row, impactPos);
24             onset = dist * rSpeed;
25
26                 // compute amplitude according to distance from impactPos
27                 // less excitation > less amplitude
28             amp = dist.squared.reciprocal;
29
30                 // finally, create the event
31             [onset, [\s_new, \ping, -1, 0, 0, \amp, amp]];
32         };
33 };
34 )
```

```
35
36      // use the above defined function with a fixed position
37  q.scoreData = q.createScore(q.data, 2, 2!4);
38
39      // generate a score, sort, and play it
40  Score(q.scoreData).sort.play
41  )
```

In keeping with the scope of this chapter, this is quite a simplistic implementation. However, it can be easily extended to feature also additional values of a dataset, e.g., by mapping them to the frequency of the individual sonic grain. Also, it would be worth implementing a GUI representation, allowing the user to literally tap the dataset at various positions.

Discussion

While semantics are an inherent part of any dataset, it is sometimes beneficial to consciously neglect it. In the analysis of the insights gained, however, the semantics should be again considered in order to understand what the new structural findings could really mean. While keeping track of relevant details, methods like the ones discussed above permit a process of gradually shifting between structure and meaning and of moving between different domains.

10.4.4 Trees and graphs: towards sonifying higher order structures

Trees and graphs in general are characterised by the fact that they provide more than one way to access or traverse them. As we can typically reach a node from more than one other node, there is an inherent choice to be made. This also demonstrates that data is contextual and access not immediate and univocal. Traversing a graph can be a non-trivial task – as exemplified by Euler's famous problem from 1735, of how to cross all *Seven Bridges of Koenigsberg* only once, is a good example. Any grammatical structure, such as a computer program, or even this very sentence, implies graphs.

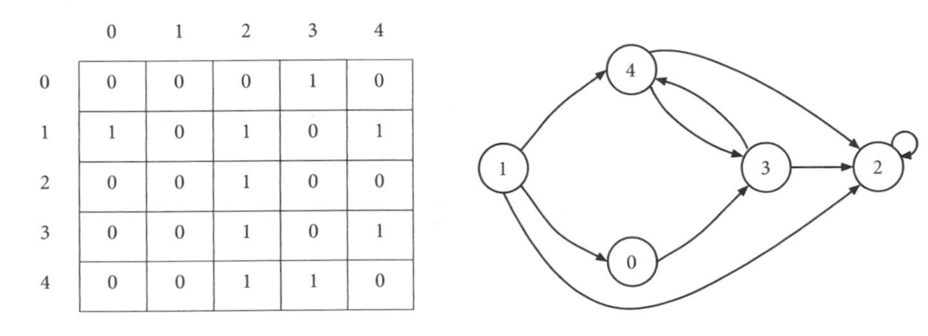

	0	1	2	3	4
0	0	0	0	1	0
1	1	0	1	0	1
2	0	0	1	0	0
3	0	0	1	0	1
4	0	0	1	1	0

Figure 10.2: Two representations of the same directed graph.

Example solutions

One way to specify a general graph is a simple two-dimensional table, where each entry represents a directed link between the start node (row number) and the end node (column number) (see Figure 10.2). One simple way to sonify such a graph would be to move from row to row and assign to each node X an n-dimensional sound event (e.g., a frequency spectrum) according to the set of nodes $S = \{X_0, X_1, \ldots X_n\}$ to which it is connected. This would sonify all links (vertices) S of each node. In order to also hear which node the connections belong to, the first example simply plays the starting node, then its connections, and the next node is separated by a longer pause (hear sound example **S10.20**).

```
q = ();

SynthDef(\x, { |freq = 440, amp = 0.1, sustain = 1.0, out = 0|
    var signal = SinOsc.ar(freq);
    var env = EnvGen.kr(Env.perc(0.01, sustain, amp), doneAction: 2);
    Out.ar(out, signal * env);
}).add;

q.graph = [
    [0, 0, 0, 1, 0],
    [1, 0, 1, 0, 1],
    [0, 0, 1, 0, 0],
    [0, 0, 1, 0, 1],
    [0, 0, 1, 1, 0]
];
    // arbitrary set of pitches to label nodes:
q.nodeNotes = (0..4) * 2.4; // equal tempered pentatonic

Task {
        // iterating over all nodes (order defaults to listing order)
    loop {
        q.graph.do { |arrows, i|
            var basenote = q.nodeNotes[i];
                // find the indices of the connected nodes:
            var indices = arrows.collect { |x, i| if(x > 0,  i, nil) };
                // keep only the connected indices (remove nils)
            var connectedIndices = indices.select { |x| x.notNil };
                // look up their pitches/note values
            var connectedNotes = q.nodeNotes[connectedIndices];
            (instrument: \x, note: basenote).play;
            0.15.wait;

            (instrument: \x, note: connectedNotes).play;
            0.45.wait;
    };
    0.5.wait;
    };
}.play;
```

Another way of displaying the structure is to follow the unidirectional connections: each node plays, then its connections, then one of the connections is chosen as the next starting node. While the first example looked at the connections "from above", this procedure remains faithful to the locality of connections as they appear "from inside" the graph (hear sound example **S10.21**).

```
Task {
    var playAndChoose = { |nodeIndex = 0|
        var indices = q.graph[nodeIndex].collect { |x, i| if(x > 0,  i, nil) };
        var connectedIndices = indices.select { |x| x.notNil };

        var basenote = q.nodeNotes[nodeIndex].postln;
```

```
7         var connectedNotes = q.nodeNotes[connectedIndices];
8
9         (instrument: \x, note: basenote).play;
10        0.15.wait;
11
12        (instrument: \x, note: connectedNotes).play;
13        0.3.wait;
14            // pick one connection and follow it
15        playAndChoose.value(connectedIndices.choose);
16    };
17    playAndChoose.value(q.size.rand); // start with a random node
18 }.play;
```

As node 2 only points to itself, the sonification quickly converges on a single sound that only connects to itself. If one changes node 2 to have more connections, other nodes become accessible again.

```
1     // connect node node 2 to node 4:
2 q.graph[2].put(4, 1); // nodes 3 and 4 become accessible
3 q.graph[2].put(4, 0); // disconnect 4 from 2 again
4
5 q.graph[2].put(1, 1); // connect 2 to 1: all nodes are accessible now
6 q.graph[2].put(1, 0);  // disconnect 1 from 2 again
```

Discussion

The examples given here were chosen for simplicity; certainly, more complex strategies can be imagined.

The last design discussed above could be extended by supporting *weighted connections* – they could be mapped to amplitude of the sounds representing the connected nodes, and to probability weights for choosing the next node among the available connections. One would likely want to include a halting condition for when a node has no further connections, maybe by introducing a longer pause, then beginning again with a randomly picked new node. These additions would allow monitoring a graph where connections gradually evolve over time. Adding information to the edge itself (here, a real number between 0 and 1) is one case of a *labeled graph*. Finally, this can be considered a sonification of an algorithm (as discussed in the next section 10.4.5): such a graph is a specification of a finite state machine or its statistical relative, a Markov chain. In a similar way, the syntactic structure of a program, which forms a graph, may be sonified.

A *tree* is a special case of a graph in which every vertex has a specific level. So instead of adding extra information to the edges, also the vertex may be augmented. In the case of a tree, we augment each vertex by adding a partial order that tells us whether this node is on a higher, a lower, or on the same level as any other node that it is connected to. This is important wherever there is some hierarchy or clustering which we try to investigate. For sonification, this additional information can be used to specify the way the graph is traversed (e.g., starting from the highest level and going down to the lowest first – *depth-first*, or covering every level first – *breadth-first*). Note that wherever one wants to guarantee that every edge is only sonified once, it is necessary to keep a list of edges already traversed. The order information need not solely inform the time order of sound events but also other parameters, such as pitch (see section 10.4.2). Also it is possible to sonify a graph without traversing it over time – it can also serve as a model for a synthesis tree directly [6].

So far we have discussed only directed and connected graphs. Undirected graphs may be represented by directed ones in which each pair of vertices is connected by two edges, so they don't pose new difficulties as such. For graphs that consist of several separate parts, we can first fully traverse all paths to find which parts exist. Then each subgraph can be sonified. Note that the first example, which uses the table of connections directly, works the same way with connected and unconnected graphs.

In general, graph traversal is a well covered problem in computer science, whose results offer a wide range of possibilities for graph sonification, most of which go far beyond anything covered here.

10.4.5 Algorithms: Sonifying causal and logical relations

There are cases in which we can represent the result of a process as a simple sequence of data points, but if we are interested in conditions and causality of processes, these often become part of what we want to sonify. So far, such causal and logical relations have only been implicit in the sound (in so far as they may become audible as a result of a successful sonification), but not explicit in the structure of the experiment. This is the next step to be taken, and next is shown how the sonification laboratory may provide methods to bring to the foreground the causal or logical relations between data.

By definition, within a computational system, causal relations are algorithmic relations and causal processes are computational processes. In a broad understanding, algorithms make up a reactive or interactive system, which can be described by a formal language. We may also take algorithms simply as systematic patterns of action. In a more narrow understanding, algorithms translate inputs via finite steps into definite outcomes.[15] Generally, we can say that – to the same degree that *cause and effect* are intertwined – algorithms connect one state and with the other in a specific manner. In such a way, they may serve as a way to represent natural laws, or definite relations between events. If we know how to sonify algorithms we have at our disposal the means to sonify data together with their theoretical context.

Up to this point we have already implicitly sonified algorithms: we have used algorithms to sonify data – they represented something like the transparent medium in which the relation between measured data points became apparent. It is this medium itself which becomes central now. This may happen on different levels.

The algorithm itself may be sonified:

1. in terms of its output (we call this *effective sonification* – treating it as a black box, equivalent algorithms are the same),
2. in terms of its internal steps (we call this *procedural sonification* – equivalent algorithms are different if they proceed differently),
3. in terms of the structure of its formal description (see section 10.4.4).

While often intertwined, we may think of different reasons why such a sonification may be significant:

1. we are interested in its mathematical properties,

[15]The term algorithm is ambiguous. For a useful discussion, see for instance [10].

2. it represents some of the assumed causal or logical chains between states of a system,
3. it reproduces some of the expected effects (simulation).

Examples for procedural and effective sonification of algorithms

For a demonstration of the difference between the first two approaches, effective and procedural sonification, there now follows a very simple example of the Euclidean Algorithm, which is still today an effective way to calculate the greatest common divisor of two whole numbers. For this we need only to repeatedly subtract the smaller number from the larger number or, slightly faster, obtain the rest of integer division (modulo).

In order to hear the relation between the input of the algorithm and its output, we sonify both the pair of its operands and the greatest common divisor (gcd) thus obtained as a chord of sine tones. We may call this *effective sonification* of an algorithm. As we only sonify its outcome, it should sound the same for different versions and even other algorithms that solve the same problem.

Two sets of numbers are provided below for whose pairs the gcd is calculated. We use a random set here. Each pair is presented together with its gcd as a chord of sine tones whose frequencies in Hertz are derived from a simple mapping function $g(x) = 100x$, whose offset guarantees that the lowest value $x = 1$ corresponds to an audible frequency (100 Hz) (hear sound example **S10.22**).

```
1  f = { |a, b|
2      var t;
3      while {
4          b != 0
5      } {
6          t = b;
7          b = a mod: b;
8          a = t;
9      };
10     a
11 };
12
13 g = { |i| i * 100 }; // define a mapping from  natural numbers to frequencies.
14
15 SynthDef(\x, { |freq = 440, amp = 0.1, sustain = 1.0, out = 0|
16     var signal = SinOsc.ar(freq) * AmpComp.ir(freq.max(50));
17     var env = EnvGen.kr(Env.perc(0.01, sustain, amp), doneAction: 2);
18     Out.ar(out, signal * env);
19 }).add;
20
21 Task {
22     var n = 64;
23     var a = { rrand (1, 100) } ! n; // a set n random numbers <= 100
24     var b = { rrand (1, 100) } ! n; // and a second dataset.
25     n.do { |i|
26         var x = a[i], y = b[i];
27         var gcd = f.value(x, y);
28         var nums = [x, y, gcd].postln; // two operands and the result
29         var freqs = g.value(nums);    // mapped to 3 freqs ...
30                     // in a chord of a sine grains
31             (instrument: \x, freq: freqs).play;
32             0.1.wait;
33     }
34
35 }.play
```

To realise a *procedural implementation* of the same algorithm, we have to access its internal variables. One way to do this is to evaluate a *callback function* at each iteration, passing the intermediate values back to where the gcd algorithm was called from. Within a co-routine like `Task`, the function may halt the algorithm for a moment in the middle (here 0.1 s), sonifying the intermediate steps. In our example, the intermediate gcd values are sonified as a pair of sine tones (hear sound example **S10.23**).

```
f = { |a, b, func|
    var t;
    while {
        b != 0
    } {         // return values before b can become 0
        func.value(a, b, t);
        t = b;
        b = a mod: b;
        a = t;
    };
};

// procedural sonification of the Euclidean algorithm
Task {
    var n = 64;
    var a = { rrand (1, 100) } ! n; // n random numbers <= 100.
    var b = { rrand (1, 100) } ! n; // and a second dataset.
    n.do { |i|
            f.value(a[i], b[i], { |a, b| // pass the
            var numbers = [a, b].postln;
                // a 2 note chord of sine grains
                 (instrument: \x, freq: g.value(numbers)).play;
                 0.1.wait; // halt briefly after each step
            });
            0.5.wait; // longer pause after each pair of operands
    }
}.play;
```

Operator based sonification

For the sonification laboratory it is essential to enable the researcher to easily move the border between measured data and theoretical background, so that tacit assumptions about either become evident. Integrating data and theory may help to develop both empirical data collection and the assumed laws that cause coherence. A sonification of a physical law, for instance, may be done by integrating the formal relations between entities into the sound generation algorithm (because this effectively maps a domain function to a sonification function, we call this *operator based sonification* [26, 30]).

An object falling from great height can, for simplicity, be sonified by assigning the height h to the frequency of a sine tone (assuming no air resistance and other effects): $y(t) = \sin(2\pi\theta t)$, where the phase $\theta = \int(h_0 - gt^2)dt$. For heights below 40 m, however, this sine wave is inaudible to the human ear ($f < 40$ Hz). Also, dependent on gravity, the fall may be too short. The sonification introduces scalings for appropriate parameter mapping (see section 10.4.2), changing the rate of change (duration) and the scaling of the sine frequency (k) (hear sound example **S10.24**).

```

(
SynthDef(\fall, { |h0 = 30, duration = 3, freqScale = 30|
    var y, law, integral, g, t, h, freq, phase, k;
    g = 9.81; // gravity constant
```

```
6      t = Line.ar(0, duration, duration); // advancing time (sec)
7      law = { |t| g * t.squared }; // Newtonian free fall
8      integral = { |x| Integrator.ar(x) * SampleDur.ir };
9      h = h0 - law.value(t); // changing height
10     freq = (max(h, 0) * freqScale); // stop at bottom, scale
11     phase = integral.(freq); // calculate sin phase
12     y = sin(2pi * phase);
13             // output sound - envelope frees synth when done
14     Out.ar(0, y * Linen.kr(h > 0, releaseTime:0.1, doneAction:2));
15   }).add;
16 );
17
18 Synth(\fall);
19
```

This however causes an ambiguity between values belonging to the sound algorithm $\sin(k2\pi t f)$ and those belonging to the sonification domain $h_0 - gt^2$. This simple example does not pose many problems, but for more complex attempts, it can be crucial to formally separate the domains more clearly.

This problem can be addressed by introducing *sonification variables* into the formalism that usually describes the domain. By superscribing variables that belong to the context of sonification by a *ring*,[16] sonification time is therefore distinguished as $\mathring{t}$ from the domain time variable t, and the audio signal y itself can be similarly marked as $\mathring{y}$. The above example's semantics become clearer: $\mathring{y}(\mathring{t}) = \sin(\mathring{k}2\pi\mathring{t}\int(h_0 - g\mathring{t})^2 d\mathring{t})$. All those variables which are introduced by the sonification are distinguishable, while all terms remain entirely explicit and do not lose their physical interpretation. For a discussion of sonification variables and operator based sonification, especially from quantum mechanics, see [30].

Integrating data and theory

For demonstrating how to combine this sonification of a physical law with experimental data, take a classical example in physics, namely Galileo Galilei's inclined plane experiment (from a fragment from 1604), which is classical also in the historiography of sonification. Before Stillman Drake's publications in the 1970s [13], it was assumed that Galileo measured time by means of a water clock. In a previously lost document, Drake surprisingly discovered evidence for a very different method. According to Drake, Galileo adjusted moveable gut frets on the inclined plane so that the ball touched them on its way down. These "detectors" could be moved until a regular rhythm could be heard, despite the accelerating motion of the ball.[17] This experiment has been reconstructed in various versions for didactic purposes, as well as for historical confirmation [25], [3], partly using adjustable strings or bells instead of gut frets.

The code below shows a simulation of the inclined plane experiment, in which the law of gravity ($s = gt^2$), only stated in this form later by Newton, is assumed. A list of distances (*pointsOfTouch*) is given at which the "detectors" are attached. Time is mapped as a linear parameter to the distance of the accelerating ball, and whenever this distance exceeds the

[16] In LaTeX, the little ring is written as \mathring{...}

[17] Drake's [12] more general discussion of Renaissance music provokes the idea of a possible continuity in the material culture of Mediterranean laboratory equipment: in a sense, this experiment is a remote relative of the Pythagorean monochord – just as tinkering with the latter allowed the discovery of an invariance in frequency ratios, the former helped to show the invariance in the law of gravity. At the time, many artists and theorists were Neopythagoreans, like Galileo's father [16].

distance of one of the detectors, it is triggered (hear sound example **S10.25**).

```
(
Ndef(\x, {
    var law, g = 9.81, angle;
    var pointsOfTouch;
    var ball, time, sound, grid;

//  pointsOfTouch = [1, 2, 3, 5, 8, 13, 21, 34]; // a wrong estimate
        // typical measured points by Riess et al (multiples of 3.1 cm):
    pointsOfTouch = [1, 4, 9, 16.1, 25.4, 35.5, 48.5, 63.7] * 0.031;

    angle = 1.9; // inclination of the plane in degrees
    law = { |t, gravity, angle|
        sin(angle / 360 * 2pi) * gravity * squared(t)
    };

    // linear "procession" of time:
    time = Line.ar(0, 60, 60);
        // distance of ball from origin is a function of time:
    ball = law.value(time, g, angle);

    sound = pointsOfTouch.collect { |distance, i|
        var passedPoint = ball > distance; // 0.0 if false, 1.0 if true
            // HPZ2: only a change from 0.0 to 1.0 triggers
        var trigger = HPZ2.ar(passedPoint);
            // simulate the ball hitting each gut fret
        Klank.ar(
        `[
            {exprand(100, 500) }    ! 5,
            { 1.0.rand }            ! 5,
            { exprand(0.02, 0.04) } ! 5
        ],
        Decay2.ar(trigger, 0.001, 0.01, PinkNoise.ar(1))
        )
    };

        // distribute points of touch in the stereo field from left to right
    Splay.ar(sound) * 10
        // optionally, add an acoustic reference grid
//  + Ringz.ar(HPZ2.ar(ball % (1/4)), 5000, 0.01);
}).play
)
```

A slightly richer, but of course historically less accurate variant of the above uses the method employed by Riess et al., in which it is not the frets that detect the rolling ball, but instrument strings (hear sound example **S10.26**).

```
Ndef(\x, {
    var law, g = 9.81, angle;
    var pointsOfTouch;
    var ball, time, sound, grid;

//  pointsOfTouch = [1, 2, 3, 5, 8, 13, 21, 34]; // a wrong estimate
        // typical measured points by Riess et al (multiples of 3.1 cm):
    pointsOfTouch = [1, 4, 9, 16.1, 25.4, 35.5, 48.5, 63.7] * 0.031;

    angle = 1.9; // inclination of the plane in degrees
    law = { |t, gravity, angle|
        sin(angle / 360 * 2pi) * gravity * squared(t)
    };

    // linear "procession" of time:
    time = Line.ar(0, 60, 60);
        // distance of ball from origin is a function of time:
    ball = law.value(time, g, angle);

```

```
20      sound = pointsOfTouch.collect { |distance, i|
21          var passedPoint = ball > distance; // 0.0 if false, 1.0 if true
22              // HPZ2: only a change from 0.0 to 1.0 triggers
23          var trigger = HPZ2.ar(passedPoint);
24              // using Galileo's father's music theory for tone intervals
25          var freq = 1040 * ((17/18) ** i);
26              // simple vibrating string model by comb filter.
27          CombL.ar(trigger, 0.1, 1/freq, 1.0 + 0.3.rand2)
28      };
29
30          // distribute points of touch in the stereo field from left to right
31      Splay.ar(sound) * 10
32          // optionally, add an acoustic reference grid
33  //  + Ringz.ar(HPZ2.ar(ball % (1/4)), 5000, 0.01);
34  }).play
```

Discussion

Evaluating the code above with varying settings for *pointsOfTouch*, and various angles, one can get an impression of the kind of precision possible in this setup. In a sense, it is much better suited than a visual reconstruction, because there are no visual clues that could distract the listening researcher. The whole example may serve as a starting point for quite different sonifications: it demonstrates how to acoustically relate a set of points with a continuous function. Replacing the *law* by another function would be a first step in the direction of an entirely different model.

As soon as one becomes aware of the fact that we do not only discover correlations in data, but also tacitly presume them – the border between auditory display and theory turns out to be porous. Correlations may either hint toward a causality in the domain or simply be a consequence of an artefact of sonification. Integrating data and theory more explicitly helps experimentation with this delimitation and the adjustment of it so that the sonification displays not only itself. The introduction of sonification variables may help both with a better understanding of the given tacit assumptions and with the task of finding a common language between disciplines, such as physics, mathematics, and sonification research.

As we have seen, one and the same algorithm gives rise to many different perspectives of sonification. Separating its "outside" (its effect) from its "inside" (its structure and procedural behavior) showed two extremes in this intricate spectrum.

Finally, this presentation gives a hint of an interesting way to approach the question *"What do we hear?"*. Necessarily, we hear a mix of the sonification method and its domain in every instance. Within the series of sound events in Galileo's experiment, for instance, we listen to the device (the frets and their arrangement) just as much as we listen to the law of gravity that determines the movement of the accelerating ball. Sonification research is interested in distal cues, outside of its own apparatus (which also produces proximal cues: see section 10.4). We try to hear the domain "through" the method, so to speak. There is an inside and an outside also to sonification.

Furthermore, data itself may be considered the external effect of an underlying hidden logic or causality, which is the actual subject of investigation. Unless surface data observation is sufficient for the task at hand, the "outside" of sonification is also the "inside" of the domain we are investigating.

Add to this the fact that today's sonification laboratory consists almost exclusively of algorithms, which may either be motivated by the sonic method or by the domain theory. Which "outside" are we finally listening to? How is the structure of the algorithm tied to the structure of the phenomenon displayed? What we face here is a veritable epistemological knot. This knot is implicit in the practice of sonification research, and each meaningful auditory display resolves it in one way or another.

10.5 Coda: back to the drawing board

This chapter has described a range of methods considered essential for a sonification laboratory, introduced SuperCollider, a programming language that is well suited for sonification research in lab conditions, and suggested guidelines for working on sonification designs and their implementations. This should provide useful orientation and context for developing appropriate methods for the acoustic perceptualization of knowledge.

However, depending on the domain under exploration, and the data concerned and its structures, each problem may need adaptations, or even the invention of new methods. This means a repeated return to the drawing board, revising not only the data, but also the equipment – calibrating sonification methods, programming interfaces, and discussing the implications of both approach and results. As sonification research involves multiple domains and communities, it is essential that all participants develop a vocabulary for cross-disciplinary communication and exchange; otherwise, the research effort will be less effective.

A sonification laboratory, being an ecosystem situated between and across various disciplines, should be capable of being both extremely precise and allowing leeway for rough sketching and productive errors. It is precisely this half-controlled continuum between purity and dirt [11] that makes a laboratory a place for discoveries. Doing sonification research means dealing with large numbers of notes, scribbles, software and implementation versions, incompatible data formats, and (potentially creative) misunderstandings. The clarity of a result is not usually present at the outset of this process – typically, the distinction between fact and artefact happens along the way.

Finally, a note on publishing, archiving and maintaining results: science thrives on generous open access to information, and the rate of change of computer technology constantly endangers useful working implementations. Sonification laboratory research does well to address both issues by adopting the traditions of open source software and literate programming. Publishing one's results along with the entire code, and documenting that code so clearly that re-implementations in new contexts become not just possible but actually practical makes one's research contributions much more valuable to the community, and will quite likely increase their lifetime considerably.

Bibliography

[1] G. Bachelard. *The Formation of the Scientific Mind. A Contribution to a Psychoanalysis of Objective Knowledge*. Clinamen Press, 2002 [1938].

[2] S. Barrass. *Auditory Information Design*. PhD thesis, Australian National University, 1997.

[3] F. Bevilacqua, G. Bonera, L. Borghi, A.D. Ambrosis, and CI Massara. Computer simulation and historical experiments. *European Journal of Physics*, 11:15, 1990.

[4] R. Boulanger. *The Csound Book: Perspectives in Software Synthesis, Sound Design, Signal Processing, and Programming*. MIT Press, Cambridge, MA, USA, 2000.

[5] T. Bovermann. *Tangible Auditory Interfaces. Combining Auditory Displays and Tangible Interfaces*. PhD thesis, Bielefeld University, 2009.

[6] T. Bovermann, J. Rohrhuber, and H. Ritter. Durcheinander. understanding clustering via interactive sonification. In *Proceedings of the 14th International Conference on Auditory Display*, Paris, France, June 2008.

[7] R. Candey, A. Schertenleib, and W. Diaz, Merced. xSonify: Sonification Tool for Space Physics. In *Proc. Int Conf. on Auditory Display (ICAD)*, London, UK, 2006.

[8] A. de Campo. Toward a Sonification Design Space Map. In *Proc. Int Conf. on Auditory Display (ICAD)*, Montreal, Canada, 2007.

[9] A. de Campo. *Science By Ear. An Interdisciplinary Approach to Sonifying Scientific Data*. PhD thesis, University for Music and Dramatic Arts Graz, Graz, Austria, 2009.

[10] E. Dietrich. Algorithm. In The MIT Encyclopedia of the Cognitive Sciences (Bradford Book), 1999.

[11] M. Douglas. *Purity and Danger*. Routledge and Kegan Paul, London, 1966.

[12] S. Drake. Renaissance music and experimental science. *Journal of the History of Ideas*, 31(4):483–500, 1970.

[13] S. Drake. The Role of Music in Galileo's Experiments. *Scientific American*, 232(8):98–104, June 1975.

[14] P. Galison. *Image & logic: A material culture of microphysics*. The University of Chicago Press, Chicago, 1997.

[15] T. Hermann and H. Ritter. Listen to your Data: Model-Based Sonification for Data Analysis. In *Advances in intelligent computing and multimedia systems*, pages 189–194, Baden-Baden, Germany, 1999. Int. Inst. for Advanced Studies in System research and cybernetics.

[16] C. Huffman. Pythagoreanism. In Edward N. Zalta, editor, *The Stanford Encyclopedia of Philosophy*. Summer 2010 edition, 2010.

[17] D. E. Knuth. *Literate Programming*. CSLI Lecture Notes, no. 27. Center for the Study of Language and Information, Stanford, California, 1992.

[18] M. Mathews and J. Miller. *Music IV programmer's manual*. Bell Telephone Laboratories, Murray Hill, NJ, USA, 1963.

[19] J. McCartney. Supercollider: A new real-time synthesis language. In *Proc. ICMC*, 1996.

[20] J. McCartney. Rethinking the computer music language: Supercollider. *Computer Music Journal*, 26(4):61–68, 2002.

[21] S. Pauletto and A. Hunt. A Toolkit for Interactive Sonification. In *Proceedings of ICAD 2004, Sydney*, 2004.

[22] S. Pauletto and A. Hunt. The Sonification of EMG data. In *Proceedings of the International Conference on Auditory Display (ICAD)*, London, UK, 2006.

[23] J. Piché and A. Burton. Cecilia: A Production Interface to Csound. *Computer Music Journal*, 22(2):52–55, 1998.

[24] H.-J. Rheinberger. *Experimentalsysteme und Epistemische Dinge (Experimental Systems and Epistemic Things)*. Suhrkamp, Germany, 2006.

[25] F. Riess, P. Heering, and D. Nawrath. Reconstructing Galileos Inclined Plane Experiments for Teaching Purposes. In *Proceedings of the International History, Philosophy, Sociology and Science Teaching Conference*, Leeds, 2005.

[26] J. Rohrhuber. $\mathring{S}$ – Introducing sonification variables . In *Proceedings of the Supercollider Symposium*, Berlin, 2010.

[27] A. Schoon and F. Dombois. Sonification in Music. In *International Conference on Auditory Display*, Copenhagen, 2009.

[28] B. Snyder. *Music and Memory*. MIT Press, 2000.

[29] B. Vercoe. *CSOUND: A Manual for the Audio Processing System and Supporting Programs*. M.I.T. Media

Laboratory, Cambridge, MA, USA, 1986.

[30] K. Vogt. *Sonification of Simulations in Computational Physics*. PhD thesis, Institute for Physics, Department of Theoretical Physics, University of Graz, Austria, Graz, 2010.

[31] B. Walker and J. T. Cothran. Sonification Sandbox: A Graphical Toolkit for Auditory Graphs. In *Proceedings of ICAD 2003, Boston*, 2003.

[32] S. Wilson, D. Cottle, and N. Collins, editors. *The SuperCollider Book*. MIT Press, Cambridge, MA, 2011.

[33] D. Worrall, M. Bylstra, S. Barrass, and R. Dean. SoniPy: The Design of an Extendable Software Framework for Sonification Research and Auditory Display. In *Proc. Int Conf. on Auditory Display (ICAD)*, Montreal, Canada, 2007.

[34] M. Wright and A. Freed. Open sound control: A new protocol for communicating with sound synthesizers. In *International Computer Music Conference*, pages 101–104, Thessaloniki, Hellas, 1997. International Computer Music Association.

Chapter 11

Interactive Sonification

Andy Hunt and Thomas Hermann

11.1 Chapter Overview

This chapter focuses on human interaction with sound. It looks at how human beings physically interact with the world, and how sonic feedback is part of this process. Musical instruments provide a rich heritage of interactive tools which allow humans to produce complex and expressive sounds. This chapter considers what can be learnt from the freedom of expression available with musical instruments when designing audio computing applications. It then describes how users can interact with computers in an interactive way in order to control the rendition of sound for the purposes of data analysis. Examples of such systems are provided to illustrate the possible applications of interactive sonification.

11.2 What is Interactive Sonification?

Not all sonification types demand interaction. For instance non-interactive sonification occurs in many alerting, monitoring and ambient information contexts, where sound may provide rich information independent of the user's actions. This chapter focuses on those situations where the user's attention is on the sound and the underlying data, and where it makes sense to consider the interaction in some detail, thinking about how the control of the system can be optimized.

Much of the theory, and many practical and relevant examples of such systems, are given and discussed in the proceedings of the Interactive Sonification workshops (in 2004, 2007, & 2010)[1] and published in special issues such as IEEE Multimedia [20].

At the first of those workshops Hermann and Hunt [13] defined this area of study as follows:

[1] ISon proceedings are available at www.interactive-sonification.org

> *"Interactive Sonification is the discipline of data exploration by interactively manipulating the data's transformation into sound."*

This chapter focuses on those situations where humans *interact* with a system that transforms data into sound. This forms an overlap between the topics of Sonification and Human Computer Interaction (HCI) (see Figure 11.1).

The general term *Auditory Display* is employed to describe the use of sound in computers to portray information. It covers not only the wide range of topics including alarm signals, earcons and sonification techniques, most of which are discussed by the International Community for Auditory Display (ICAD)[2], but also the actual display environment including the audio system, speakers, listening situation, etc.

Sonification is the more specific term used to describe the rendering of data sets as sound, or:

> "... *the transformation of data relations into perceived relations in an acoustic signal for the purposes of facilitating communication or interpretation"* [25]

or in a newer definition:

> *"Sonification is the data-dependent generation of sound, if the transformation is systematic, objective and reproducible, so that it can be used as scientific method"*[16]

Human Computer Interaction is defined as:

> "... *a discipline concerned with the design, evaluation and implementation of interactive computing systems for human use and with the study of major phenomena surrounding them."* [17]

In other words, this chapter considers the study of human beings interacting with computers to transform data into sound for the purposes of interpreting that data.

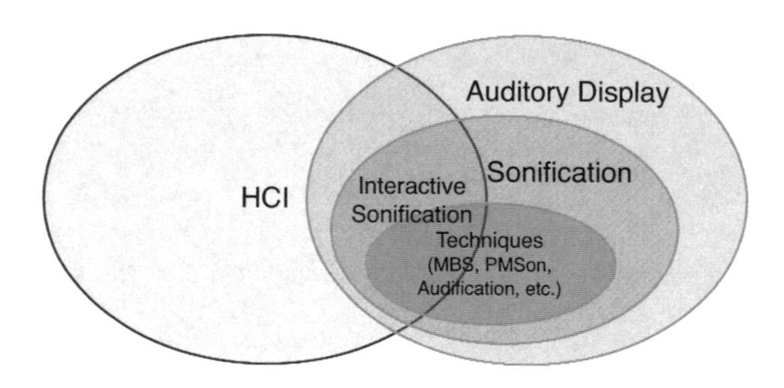

Figure 11.1: Topic Web for Interactive Sonification

The field of sonification has different aspects that can be studied, such as *(i)* the data transformation technique, the *(ii)* algorithmic implementation, or the *(iii)* interaction itself. The sub-topic of 'technique' is concerned with deciding the basic relation of data to its eventual

[2]ICAD: International Community for Auditory Display. Community website: `http://www.icad.org`

acoustic representation. Techniques such as Parameter Mapping Sonification (PMSon), Audification and Model-based Sonification (MBS)[3] are conceptually very different and are suited for different situations. The sub-topic of 'algorithmic implementation' concerns programming languages, computational issues, and performance. Both of the above sub-topics are complementary the third; that of interaction, which involves:

- the user,
- the user's needs,
- the user's actions in response to perceived sounds,
- the modes and means of controlling a sonification system,
- how the user and the sonification system form a closed loop
- how this loop impacts:
  - the ease of use,
  - ergonomics,
  - fun, and overall performance and experience.

By looking at sonification from the perspective of interaction we can gain ideas about how users can creatively access and manipulate the sonification process, to make the best use of it, and to adapt it to their own personal needs and interests. This raises questions about how responsive or real-time capable a system needs to be in order to match a person's natural or optimal interaction skills.

It furthermore stimulates questions about how humans more generally interact with sounding objects in everyday life and how such interaction skills have evolved to make use of real-world sound. The two most important yet very different sorts of interactions are:

1. physical manipulations, where sound is a highly informative *by-product* of human activity, and,
2. the *intentional* use of objects to actively create sound, as in musical instruments.

We would not typically regard musical instruments as interactive sonification devices because their primary function is to transform human *gestures* into sound for the purposes of *expression*. In contrast, Interactive Sonification systems transform *data* into sound (modulated and controlled by human gestures) for the purposes of data *analysis*. However, musical instruments do provide us with a range of tried and tested models of interaction with sound, which is why we study them in some detail in section 11.4.

The use of sound gives alternative insights into the data under examination. Until the mid-1990s the sheer computing power required to generate the sound output meant that, by necessity, the act of sonification was a non-interactive process. Data was loaded, parameters were selected, the algorithm set going, and some time later the sound emerged. Often in computing technology, when this time-lag is eliminated by improvements in processor speed, the *style* of interaction remains; and interaction is limited to setting parameters, then listening to a completed sound. This chapter therefore challenges designers to reconsider the complexity of the interaction, and to evaluate whether more continuous engagement by the

[3]see chapters 15, 12,16

user would be beneficial.

Because computers are good at processing data, there can be a tendency to expect the computer take on the bulk of the analysis work. Figure 11.2 portrays this situation graphically.

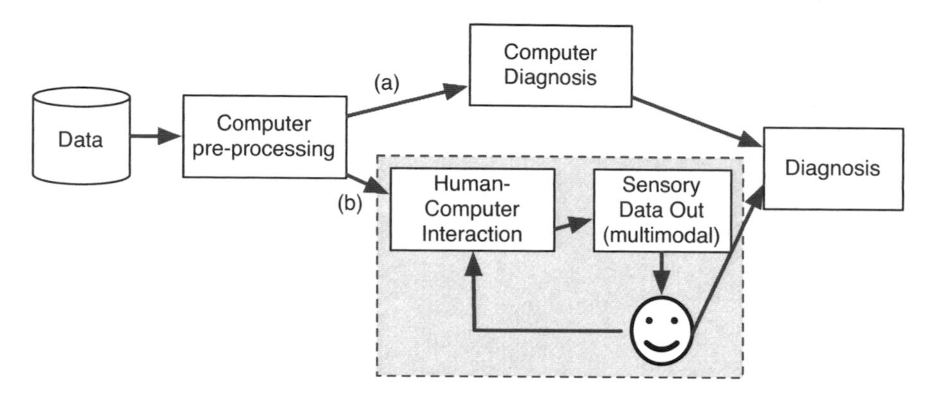

Figure 11.2: Different from (a) classical automated diagnosis and exploration schemes, Interactive Sonification (b) embeds the user in the loop to be in touch with the data.

Consider the field of medical diagnosis. Clinicians have been able to gather so much information from improved sensors that it appears that there can be *too much information* for them to process (long lists of numbers, or graphs which need too much scrolling to enable viewing at the correct resolution). The consequence of this is that diagnosis has been left for the computer to do, and often the expert human being is left out, as in route (a) on the diagram. However, another interpretation is possible; that the data is just not being converted in the correct way. Consider route (b) on the diagram. This represents the data being under the control of the expert human, who is searching it and sifting it, and coming to a human conclusion or diagnosis. Clearly both options are useful in different circumstances. Route (a) is particularly useful for those well-defined situations where the problem is well understood, whereas route (b) is perhaps the best way to explore unfamiliar data and to look for patterns that are not so well-defined. However, for route (b) to occur, the data needs to be rapidly portrayed to the user in large quantities, and that is where sonification excels.

The art of interactive sonification concerns the design of computer systems that enable human listeners to understand and interpret data by interacting with the data in a natural and effective manner. The next section looks at how to interact in such a manner by examining in some detail how humans interact with their everyday world.

11.3 Principles of Human Interaction

Before considering how humans interact with computers, this section takes some time to review how humans control and respond to their everyday environments.

11.3.1 Control loops

As human beings, from the moment we are born we begin to interact with the world. In fact a baby's first action in the world is to cry – to make a sound (hear sound example **S11.1**). As we grow we learn first how to control our bodies, and then how to interact with objects around us. The way that the world works – its physical laws, the constants and the variables – becomes coded into our developing brain.

We learn to take for granted that dropped objects fall to the ground, and that when we reach for an object we feel it and see it and hear it as we touch it. Watch a young child playing with a pile of bricks and you will notice how their movements develop by interacting with objects and obtaining from them instant and continuous feedback of their position, speed and texture (hear sound example **S11.2**). Hear the joy of two young children playing with pots and pans **S11.3**. Such control loops of human action and continuous feedback from the world become embedded deep within our mind-body system.

As we grow and learn to communicate with language, we begin to utilize another, very different, way of interacting with the world. Rather than directly manipulating an object and gaining instant feedback we instead use language to express our desires, so that others become involved in our loops. This is what is happening when a child asks his parent for a drink, and the parent supplies one. Listen to the young child make an early request for something from his mother **S11.4**. This method of interaction by command/request becomes increasingly predominant as we proceed through the education system. We will see in section 11.5 that much of our interaction with computers takes place via this command modality.

It matters too whether or not you are part of the control loop. Many passengers become travel sick whereas this condition rarely affects drivers. When you are controlling an object you know what to expect, as – by definition – you are initiating the reactions and can thus prepare your mental apparatus for the result. Maybe you have had the experience of being in a room where someone else is in charge of the TV remote control. You cannot believe how much they are 'playing around with it', driving to distraction everyone else in the room. However when *you* have it, everything is different, and you are 'simply seeing what's on the next channel' (hear sound example **S11.5**). It matters greatly whether you are in the control loop or not. Therefore, we should consider bringing more real-world interaction into our computing interfaces, by placing the human operator firmly in charge of a continuous control loop wherever possible and appropriate.

11.3.2 Control intimacy

A child playing with wooden blocks and a person operating a typical computer interface are both *interacting* with external objects. It is just that the *quality* of the interaction is different. The extent to which the interaction directly affects the object is one aspect of the *control intimacy* being exhibited; the other aspect being how well the human manages this control. Real-world objects seem to exhort us to spend time with them, and as we do, we subconsciously learn more about them, and master the skills of manipulating them until the control becomes almost automatic.

We are all aware of situations where we are controlling an object and almost forget that we are doing it. This was what was meant by the 'automatic' processes in the HCI literature [28]. Car drivers often report that they are shocked to find themselves at their destination, without knowing *how* they got there; even though the act of driving is an extremely complex interactive process. Many experienced performing musicians feel that their fingers are somehow playing the music by themselves. In musical performances, their minds appear to be concentrating on higher-level modes of expression, whilst their bodies are managing the physical act of manipulating the instrument. In fact, most musicians will recount the terrifying feeling of suddenly becoming *conscious* of what their fingers are doing, and as a result the performance grinds to a halt!

Csikszentmihalyi [4] called this type of disembodied interaction '*flow*'. He explains how it is found freely in children as they play, and less so in adult life. Certainly in many computer interfaces the flow is never allowed to happen, due to the constant choices, and the stop-start style of the interaction caused by the emphasis on reading words, processing and selecting options.

Let us review how we reached this situation – that of being highly adept at continuous interaction with our surroundings, yet inventing computer systems which interact with us in a comparatively stilted way.

11.3.3 Interacting with tools (from cavemen to computer users)

One of the defining attributes of human beings is that they appear to have always shaped their own environment by using tools - outside objects which are used to perform tasks which the unaided human body would find difficult or impossible. Since the dawn of history, humankind has fashioned objects out of found materials and has used them to change the world.

The earliest tools were purely physical - acting on other physical objects under manual control (listen to sound example **S11.6**). Through the ages the tools have become more sophisticated, using machine power and electricity to act on the world. In comparatively recent times (compared to the time-scale of human history) computers have allowed mental processes to be taken over by machines. This progression of tools is well summarized by Bongers [1] in Table 11.1.

This sort of development over time is usually regarded as a record of *progress*, but Bongers indicates that something has been *lost* along the way. Simple physical tools give us intimate and continuous tactile control as we use them, as well as implicit visual and audio feedback. The brain is free to concentrate on thinking about the task. Much of the task load itself is distributed from the brain to the human body. This body-brain system is very good at learning subtle and intricate control tasks given enough time. This is what is responsible for the centuries of fine craftsmanship, complex buildings, beautiful artwork and sublime music.

Having considered the astoundingly good results that are possible when humans use simple physical tools, we note that the opposite has often been the case with computer interfaces. Here the navigation of the interface itself can take up so much time, concentration and brain power that the task itself is often forgotten. But computer interfaces have been *designed* to do this. It should be noted that computers have been developing from a text-based command

Manual (objects)	Tools like a knife or hammer	Stone age
Mechanical (passive)	Levers, cogs, gears	
Mechanical (active)	Powered by steam, combustion engine	Industrial age
Electrical	Electricity, power and communication	
Electrical (analogue)	Modulation of electrical signals (vacuum tube, transistor)	Information age
Electronic (digital)	Integrated circuits	
Computer	Software	Digital age

Table 11.1: Stages of Human Tool development as described by Bongers [1]

interaction towards graphical interaction as we see it today, and the trend is clearly to include more flexible and direct methods of interaction, such as speech control, touch sensitive displays, malleable interfaces, and gestural controls. It seems that we are just about to rediscover the human interface richness we encounter in real-world interaction, and this trend needs to be continued and adapted to the peculiarities of interacting with virtual acoustic systems, such as those considered in sonification. We need to redesign interaction to allow truly intimate control over the data we are trying to explore.

11.3.4 Interacting with sound in control loops

Engineers use sound to deduce the internal state of engines (hear example **S11.7**) and complex machinery such as washing machines (**S11.8**). Sound warns us of dangers outside our relatively narrow field of view. It is also the medium by which much human communication takes place via speech and singing.

Whenever we interact with a physical object, sound is made. It confirms our initial contact with the object, but also tells us about its properties; whether it is solid or hollow, what material it is made of etc. The sound synchronizes with both our visual and tactile 'views' of the object. As we move the object, the sounds it makes give us continuous feedback about its state. Sound is a temporal indicator of the ongoing physical processes in the world around us.

The act of making sound may be satisfying to human beings precisely *because* they are in a very tightly responsive control loop. This does not by definition mean that *other* people find the sound satisfying. Think of times when a person mindlessly 'drums' his fingers on the table to help him think. He is part of the control loop, and so is expecting the moment-by-moment sonic response. The whole process often remains at the subconscious level, and he is unaware he is doing it. However, to other people in the vicinity (not in the loop) the sound can be intensely annoying, rather like the television remote control example mentioned earlier (hear, for instance, example **S11.9** and imagine working next to that person all day). Therefore, we see that there is something special about being the one to *initiate actions*, and receiving constant and immediate sonic results.

An observation about the individuality of interacting with sound became clear during the author's own experience of amateur radio operation . It is a common experience of radio hams that there is quite an art to tuning in the radio to pick out a particularly weak signal [30].

Somehow you need to be able to pick out the signal you are trying to listen to, in spite of the fact that there are much louder interfering signals nearby in the frequency spectrum, and background noise, and all manner of fluctuating signal levels and characteristics due to propagation conditions (hear example **S11.10**). To do this requires a fine balance with the tuning control, and the signal modulation controls, and sometimes even movement of the antenna. When two people are listening to the same radio signal, but only one is at the controls, it is quite common for the signal to be audible *only* to the person at the controls.

What can we learn from such an observation? Perhaps when a sound is made by a system, we ought to consider *who* the sound is intended for. Is it just for the person 'in the loop', since he is the one controlling the system parameters? Or, is the sound intended for everyone? Where data values are being portrayed as sound, for example in a hospital environment, it is important that everyone recognizes the sound. However, where the sound is being controlled interactively by a person, we might need to be aware that the operator could be inadvertently tuning the system for themselves. More complex sounds (which could appear as annoying or unpleasant) can be quite acceptable to people who are *in* the control loop.

The more general point to be inferred from the above example is that humans can use physical interaction to control the generation and modulation of sound in order to extract data from a noisy signal. *Interactive sonification appears to be a natural human diagnostic tool.*

Musical instruments are a special case of sound generating device where the main intention is that other people do indeed listen to the sound. Having said that, if you are sharing a house with someone practicing an instrument (particularly if the player is a beginner), the observation that it matters whether you are in control becomes obvious (hear example **S11.11**); the player can be engaged for hours, the listener is annoyed within minutes!

In the next section we look at the special case of human interaction with instruments in more detail.

11.4 Musical instruments – a 100,000 year case study

The sonic response of physical objects is so deeply ingrained in the human psyche that sound and music have been a fundamental part of every known human society. In this section, we take a closer look at human interaction with musical instruments; since much can be learned from this about what makes good quality real-time sonic interaction.

11.4.1 History

Musical instruments have been discovered by archaeologists which could be nearly 100,000 years old [5]. It seems that they are one of the earliest tools which humans developed.

Even by the second century B.C. quite complex mechanical devices were being invented [2], and through the ages many instruments were created with elements of automatic control. With the relatively recent blossoming of electronic recording and computer music, it seems that the development of musical instruments mirrors the development of tools that we saw in section 11.3.3.

Although new musical devices may be constantly invented, it is the repertoire of music written for the instruments that provides a sense of stability. For a piece of music to be performed, the instrument still needs to exist, and people need to be able to play it and to teach it to others.

However, such a long and rich history brings with it a perspective and longevity that we sometimes lack in the recent and rapidly changing world of computer interfaces. So let us look at the defining characteristics of this special form of sound interaction tool.

11.4.2 Characteristics

The common attributes of most acoustic musical instruments are as follows: [22]

- there is interaction with a physical object,
- co-ordinated hand and finger motions are crucial to the acoustic output,
- the acoustic reaction is instantaneous,
- the sound depends in complex ways on the detailed kinds of interaction (e.g., on simultaneous positions, velocities, accelerations, and pressures).

The physical interaction with the instrument causes an instantaneous acoustic reaction. This allows the player to utilize the everyday object manipulation skills developed throughout life. The player's energy is directly responsible for activating the sonic response of the system; when the player stops, the sound dies away. The mapping of system input to sonic output is complex (see section 11.4.3); many input parameters are cross-coupled, and connected in a non-linear manner to the sonic parameters. This can make an instrument difficult to play at first, but offers much scope for increased subtlety of control over time. As the player practices, he becomes better and better. This allows the control intimacy to increase to a level where the physical operation of the instrument becomes automatic. At this point the player often experiences the 'flow' of thinking at levels much higher than complex physical interface manipulations.

We should also not underestimate the importance of tactile feedback. Good performers will rarely look at their instrument, but will instead rely on the years of training, and the continuous feel of the instrument which is tightly coupled to the sound being produced. Human operators learn to wrap their mind-body system around the instrument to form a human-machine entity.

11.4.3 Mapping

One of the most notable facets of acoustic musical instruments is that they can take a long time to learn to play. This is partly because of the very complex ways in which controlling the instrument makes the sound. The physical nature of the instrument means that the playing interface and the sound generation apparatus are often subtly interwoven. The key on a flute is clearly the interface because the player controls it, but is also part of the sound generator because the key covers a hole which affects the vibrating air in the column of the instrument to make a different note.

Manufacturers of acoustic instruments shape materials to suit the human beings who will be playing them, *and* to generate the best possible sound. In other words, any 'mapping' of input device to sound generator is entirely implicit – it occurs because of the physics of the instrument.

With an electronic instrument the mapping needs to be *explicitly designed.* The input parameters coming from the player have to be 'mapped' onto the available control parameters for the sound generation device. A body of literature exists which examines this phenomenon in greater detail (see especially [21, 23, 31, 24]).

Many times it seems as if the mapping is made for engineering convenience. For example, maybe the positions of three available sliders are mapped onto the amplitude and frequency and wavetable of an oscillator modulating another oscillator without much thought as to whether this makes a suitable interface for the user. The studies above all show that considerable thought needs to be given to the mapping process, as it can affect not just the ease of performance, but also the very essence of whether the device can be considered as a useful musical instrument.

Empirical studies [22, 21] have shown that when there are several parameters for a human to control and monitor in real time, then a direct mapping interface performs very poorly, and a more complex one (which takes a while to learn) performs much better. It also seems that a simple, direct mapping such as the '3 sliders' example given above, does not *engage* the human player in the same way as a more complex mapping strategy. However, the more complex a mapping strategy, the longer it will take the human player to learn. And here we are faced with a fundamental problem: computers are often regarded as time-saving devices, yet if we are to interact with them in a subtle and meaningful way, then human users will need to spend time *practicing* the interface, in the same way that musicians spend hours practicing their instrument in order to gain complex control over their music.

A good compromise (or trade-off) should be found between (a) reducing the complexity of the interface so that it becomes very simple to use and easy to learn, and (b) enabling as much interaction bandwidth as possible, allowing the user to be in touch with the complexity of the data or sonification technique.

11.4.4 Instruments as exemplar interfaces

So, it appears that interfaces which are ultimately worthwhile and allow the user to transparently control complex end-products (sound or music) will require some practice, and may even be considered to be "too hard to control" at first.

However, we may also learn about user accessibility from different types of musical instrument. Some instruments are considered to be extremely expressive, such as the oboe (sound example **S11.12**), but they are almost impossible for a beginner to even make a sound. Other interfaces, such as the piano or guitar, make it quite easy for a beginner to play several notes (hear this intermediate player **S11.13**), but it still takes a long time to gain mastery over the instrument.

Therefore it seems from considering how people interact with musical instruments, that devices intended for sonic exploration need to have certain characteristics. These include:

- a real-time sonic response,
- a suitably complex control mapping, and

- tactile feedback tightly coupled to the sonic response.

11.5 A brief History of Human Computer Interaction

So far, this chapter has considered what musical instruments tell us about how interfaces have developed over thousands of years. On a human time-scale, by comparison, computer interfaces are mere new-born babies, yet they influence everything we do nowadays with data. So this section considers how computer interfaces began and how we reached those which exist today.

11.5.1 Early computer interfaces

In the earliest days of computing, there was no such thing as a 'user interface'. That concept did not exist. Computational machines were huge, power-consuming, cumbersome devices that were designed, built and operated by engineers. The operators knew how the machine worked because they had built it. If it went wrong they would crawl inside, locate and replace the affected component. The earliest all-electronic computer, built in 1943, was **ENIAC** – the Electronic Numerical Integrator And Computer (from which is coined the modern use of the word 'computer').

Much pioneering work in the 1960s and 1970s was carried out at universities and research institutes, and many of what we now consider to be modern interfaces were prototyped at this time. Only as computers became smaller, cheaper and available more widely was there any commercial need to study how the *user* felt about their machine, and how they interacted with it. Companies teamed up with university research labs to try and study what made a good interface, and so during the 1980s the market began to see a range of computing technologies which claimed to be 'user-friendly'. Full details of how interfaces and interaction styles developed can be found in Myers [27].

11.5.2 Command interfaces

A command-line interface consists of a typed series of instructions which the user types in to control the computer. The user is presented with a prompt such as:

```
C:>
```

Because this gives no information on what to type the user must learn the instructions before they can be used. Typically these commands allow users to move, copy and delete files, and to interact with the operating system in several other ways. The commands themselves have a complex syntax, for example:

```
cp *n.txt ..\outbox
```

which will copy all text files whose names end in the letter 'n' into a sibling directory called 'outbox' (one which exists as a different subdirectory of the parent directory).

Clearly users need to be rather knowledgeable about computer directory structures, as well as the existence of the command and its syntax. Errors were often met with unhelpful comments:

```
> cp *n.  txt \\outbox
syntax error
> help
c:\sys\op\sys\help.txt does not exist
> go away you stupid computer
syntax error
```

However, users with a good-to-expert degree of computing knowledge still find this kind of interface fast, flexible and indeed easier to get the job done than via the graphical interfaces that all but replaced them.

11.5.3 Graphical Interface devices

As computing display technology improved and became more affordable, computer design focused on portraying information and commands as graphical objects on the screen, with a mouse as the primary interaction device. This was the age of the 'user-friendly' computer, with the interface wars dominated by Apple's many successful computers, such as the Macintosh series. This paradigm of interaction was originally built upon Xerox PARC's pioneering experiments beginning in 1970 [18] and influenced by Ben Shneiderman's vision of *direct manipulation* [29], where graphical objects can be manipulated by the user with rapid feedback and reversibility.

Direct manipulation promised an era of easy-to-use interfaces that were visually intuitive. However, for many years, the reality was that computers were used for such a wide range of tasks that:

- not everything *could* be portrayed graphically
- this was very hard work for programmers.

Therefore what happened was that *Menus* were invented. Menus are simply lists of commands that are made visible to the users, so the users do not have to remember them. They have become ubiquitous in the world of computing, becoming known as WIMP (Windows, Icons, Menu, Pointer) interfaces. Some of the time they seem a reasonable way of proceeding (especially for beginners who do not know what commands are available in a piece of software). However, as soon as the user becomes competent (or maybe expert) menus often slow down the whole process.

Consider for a minute the fantastic flexibility of the human hands, and the intricate actions possible when coordinated with the eyes and ears. A mouse-menu system occupies the visual field of view, and requires the hand to coordinate a two-dimensional sweep across the desk, watching the visual feedback, waiting for the menu to appear, reading through the menu list (mostly a list of commands that are *not* wanted), requiring a further one-dimensional movement down (careful not to 'lose' the menu by going too far to the left) and finally a click. All this to select one command! It is a real waste of an amazing biological system.

We still have a long way to go in designing interfaces. However, at the time of writing, we find ourselves in period of revolution in the human-computer interfaces that are available to everyday users. This is exemplified by two trends: (a) alternative interaction paradigms for computer gaming, such as the wireless Nintendo Wii[4] system and the completely controller-free Xbox Kinect[5], and (b) the multi-touch-screen interface pioneered by Han [8], made commonplace via the iPhone[6], and now manifesting itself in a huge variety of tablet-style computer interfaces exemplified by the Apple iPad[7]. What all of these innovations are doing is making it acceptable (and *required* by commercial pressure) to move beyond the WIMP paradigm and explore a whole variety of more direct ways to control the data with which people are working. This is indeed an exciting time to be re-thinking how we interact with computers.

11.5.4 Contrast with Real-world interaction

Section 11.3.1 described how humans grow up interacting with the world in continuous control loops. Therefore it is hardly surprising that, later in life, we become rapidly frustrated with computer systems that engage with us in a very different and more limited manner. Here, too often, the interaction is dictated by the computer.

A prompt is given, or a list of options presented as icons or a menu. We have to choose from the selection offered by the computer at every stage of the process, and thus the interaction becomes a series of stilted prompt-choice cycles; a far cry from the way that we have learnt to interact with the everyday world.

It is as if we have designed our computer systems to always remain outside our control loop. We seem to expect them always to be under 'third-party' control; things to which we give instructions. Whilst this is completely acceptable for those situations where the computer needs to operate autonomously, the result is that we rarely gain the same intimacy of control with a computer as we do with objects in everyday life. A common observation is that much of our time working with computers is spent in navigating the interface, rather than completing the task.

11.5.5 Sound in Human Computer Interaction

Human computer interaction has never been a totally silent area. There are always sounds, at the very least the direct sound caused by the user interacting with the computer, such as the click sound accompanying each key-press on the keyboard, or sounds made while moving the computer mouse on its pad and clicking buttons (hear example **S11.14**). These sounds are mentioned here since they are typically forgotten, and regarded as irrelevant, since they are so ubiquitous during the interaction. Yet they are quite informative; they confirm the successful execution of elementary interactions. In fact we are often not aware of such utility unless we notice their absence. For instance, in modern cars it is technically possible to construct indicators (blinkers) without relays so that they are completely silent. The blinker sound was originally a technical artifact, but it is so useful that today artificial sonic replacements are actively produced in every car to fill the gap (hear example **S11.15**).

[4] http://www.nintendo.com/wii
[5] http://www.xbox.com/en-US/kinect
[6] http://www.apple.com/iphone
[7] http://www.apple.com/ipad

There are other artifact sounds which come from computers, such as the fan sound, and the sounds of disk drives, etc. Although they are by-products, they sometimes increase our awareness about what is going on. Technical developments (such as the replacement of moving drives with convenient solid-state devices – for instance USB memory sticks) have managed to remove the sound and as a side-effect also some useful information for the user.

However, sound has also been actively introduced as a communication channel between computers and users. At first tiny beep sounds indicated errors or signals (think of bar-code scanners in cash registers), and later characteristic sounds such as an operating system startup sound, or interaction sounds (to indicate execution of activities, e.g., clicking on an icon, or drawing attention to a newly opened window (listen to sound example **S11.16**). The *SonicFinder* [6] was the first successful interface of this type and gave rise to the explicit introduction of Auditory Icons. Earcons are also frequently used for this sort of communication (see chapters 14 and 13).

Most of these sonic elements differ from real-world interaction sounds in two regards: Firstly these sounds do not deliver *continuous feedback* to the user's actions: most feedback sounds are event-like, played on occurrence of a condition independent of the ongoing activity. This is a huge contrast to our everyday experience of sound where we hear lots of continuous sound feedback, for instance as we continuously move a glass on a table (hear example **S11.17**). Secondly, sounds in most typical computer work do not have an analogous component. In real-world interactions, by contrast, we frequently experience a direct coupling of sound attributes to the detailed sort of interaction or the objects involved: e.g., interaction sounds depend on the size or hardness of the objects. Such functions could easily also be used for human-computer interaction, and in fact this is what is meant by *Parameterized Auditory Icons* (as introduced by Gaver [6]). So there is a great deal of unexploited potential for improving the ergonomics of sonic human computer interaction even within the context of mouse-based interactions and the graphical computer desktop metaphor.

11.6 Interacting with Sonification

As stated in section 11.5.4, many prominent paradigms of computer interaction prevent control intimacy from developing. This section examines how to re-introduce interaction into the art of making sound.

Now that computers can run fast enough to generate sound in real-time, we should re-design our data-to-sound algorithms to take advantage of the rich possibilities of continuous human interaction. How can we facilitate a 'flow' experience of data sonification to take place?

The following four sub-sections focus in turn on how interaction can be used and developed in the fields of:

- Auditory Icons and Earcons
- Audification
- Parameter Mapping Sonification, and
- Model-based Sonification.

11.6.1 Interaction in Auditory Icons and Earcons

Auditory Icons and Earcons play an important role in many human-computer and human-machine interfaces and interactions. They signal discrete events, such as:

- the successful accomplishment of an activity,
- informing the user about an error state, or
- reporting a new event (such as an incoming mail sound signal).

Let us take a look at those Earcons and Auditory Icons which occur in direct response to a user's activity. Examples are the file deletion sound played after dragging a file icon to the trashcan on the computer desktop, or the warning beep sound played by a car's computer on starting the engine before fastening a safety belt (hear example **S11.18**).

These sounds appear to be interactive because they happen at the same time as the activity which triggers them. However, the interactivity is limited to this 'trigger coincidence', whereas other bindings which are usually encountered in real-world interaction are missing. If, for instance, a sound is caused in response to a physical interaction, it is an expected natural occurrence that the energy or the properties of the constituent objects influence the sounds.

Extrapolating this to the abovementioned examples could mean that the deletion of a small file might sound "smaller" than the deletion of a huge directory. Or when turning the car key quicker, the "buckle-up" warning sounds more intensive (hear example **S11.19**). Such bindings are nowadays easy to implement and it may be that such adaptations increase the acceptance of such sonic feedback and the device as a whole.

Yet there is another aspect where interaction is limited in Auditory Icons and Earcons. Typically, the sounds are played until they come to an end. In contrast, typical physical interactions, even punctual ones, can be manipulated, interrupted, and even stopped by the nature of the physical interaction. Such modes of interaction are missing, which is maybe not important if the auditory messages are quite short (e.g., less than 300ms), yet with longer earcons some additional annoyance can be connected with this lack of interaction.

Finally, and surprisingly disturbing to human listeners, we typically encounter a total lack of sound variability in most existing notification sounds. Every event sounds 100% identical, since it is just the playback of the same sound file. In real-world interactions, every interaction sounds different; even switching the light on and off produces subtly different sounds each time. Sound example **S11.20** contains eight clicks of a switch (real life) followed by 8 repeated samples of a single click. On careful listening the difference is quite noticeable. Although these differences may seem negligible, this can influence the overall acceptance of sound in interacting with a machine, as humans are finely tuned to expect unique and variable sonic events. As sonification is a scientific technique we would certainly expect that the same data, when repeated, should result in identical sounds. However the meaning of *identity* may or should be different from *sound-sample identity*, as discussed by Hermann [16]. Otherwise even subtractive synthesis using filtered noise would not fulfill reproducibility, as each sonification rendering of identical data would be different at the sample level. Subtle variability, as it occurs in the sounds of everyday object manipulation, indeed encodes subtleties in the interaction. There are interactive sonification methods, such as MBS (see chapter 16), which allow a similar variability and richness of the sound to emerge on repeated interaction.

Parameterized Auditory Icons (mentioned in section 11.5.5) are a good step towards the increase of information and interactivity in symbolic data sonification. One way to advance interactivity in auditory icons (as a direct signal following a button press) even further would be to use real-time physical modeling and a continuous influence of the sound based on the detailed user's actions, for instance measured by force-sensitive controllers instead of simple buttons.

11.6.2 Interaction in Audification

Audification (see chapter 12) is the most direct type of sonification technique. It plays ordered data values directly by converting them to instantaneous sound pressure levels. Typically audification is a non-interactive technique, and interaction occurs only at the point where the user starts the audification process. The audification of a data set is often rendered as a sound file, and this opens standard sound file interaction techniques available in music playing user interfaces, such as play, stop, pause, and sometimes also forward and rewind. Besides interactions associated with the actual rendering of the data as sound, we might also consider interactions which occur *prior* to the rendering, such as the selection of the start and end item in the dataset or additional processing such as compression.

A real-world analogue to audification is the gramophone, where data values can be imagined as represented in the form of groove. A more physical analogy might be scratching on a surface where the surface profile represents data values. Thinking about this physical analogy helps us to consider the sort of interactions that users would quite naturally and intuitively perform:

- moving their hand back and forth,
  - at different velocities and pressures,
  - using different interaction points (the fingernails, or the full hand),
- moving backwards and forwards, or
- using two-handed scratching to simultaneously compare different parts of the surface.

Interestingly we imagine complex spatial physical movement of our hands for control, while navigation in the computer is often just clicking on a forward or backward button. The abovementioned interactions are actually perfectly possible to implement in computer systems. For example by using a pressure sensitive touchpad (such as a Wacom Pen Tablet[8]) interaction can be easily provided that even allows pressure sensitive scratching of data for real-time interactive audification. However, as long as there is only one interaction point (such as with a single mouse pointer or a tablet with only a single pen), the user can only experience one position at a time, whereas in everyday life we can easily scratch on several spatially separated locations with our two hands. The developments in force-sensitive multi-touch surfaces (mentioned above) will probably overcome this limitation and open new opportunities for interactive sonification.

Hermann and Paschalidou [15] demonstrated extended interaction possibilities with a biman-

[8] `http://www.wacom.com`

ual gestural interface to control audification, using one hand to navigate the audification and the other to interactively control filter frequencies. Such developments show that audification bears unexploited potential for increased interaction and that further exploratory quests may benefit by making use of this potential. Many parameters can be adjusted, such as compression, filter frequencies, dynamic compression ratios, interpolation type, time-stretching coefficients and the like. Many audification systems rely on interrupted interactions, so that the parameter values are adjusted by sliders or text fields, and then the rendition of the audification is triggered and the sound is heard. The main effect is that the adjustment of parameters becomes a time-consuming and stilted process. With powerful real-time interactive sound rendition engines such as SuperCollider, it is no problem to adjust such parameters *while* the audification plays in a loop, opening up extended interaction possibilities.

11.6.3 Interaction in Parameter Mapping Sonification

Perhaps the most common sonification strategy is *parameter mapping sonification (PMSon)*[9], which involves the mapping of data features onto acoustic parameters of sonic events (such as pitch, level, duration, and onset time). There are two types of PMSon: *discrete PMSon* involves creating for each data vector a sound event, whereas *continuous PMSon* means that acoustic parameters of a continuous sound stream are changed by the data. In both cases, the sonification time of the events is a very salient parameter, and therefore often a key data feature is mapped to time. If the data are themselves time-stamped, it is straightforward to map the time value onto the sonification time. The resulting sound track is then normally listened to without interruption so that the evolution in time can be understood, and this removes direct interactivity. Certainly the listener may navigate the time axis in a similar way to audio 'tape' interactions (pause, play, forward/backward), yet such interactions are quite rudimentary. If, however, sonification time remains unmapped, i.e., data features are mapped only onto other sonic features of the sound stream such as pitch, brightness, level or spatial panning, it is much easier to provide interaction modes to the user.

We can discern two types of interactions: (i) *interactive data selection*, which means controlling what subset of the data set under exploration is to be sonified and (ii) *mapping interactions*, which means adjusting either the mappings or mapping-related parameters (i.e., ranges, scaling laws, etc.). Certainly, both interactions can go hand in hand. **Importantly, the sound of such an interactive sonification will only make sense to the one who is in the control-loop**, since others do not know whether sound changes are due to the data or due to parameter changes performed in the interaction loop.

Even if sonification time is occupied by mapping from a data feature, there is the possibility having some *pseudo-interactivity*, namely if the whole sonification is relatively short: if the presentation of the whole data set lasts only a few seconds, an interaction *loop* can be constituted by repeated triggering of the sonification. Since the short sonic pattern fits into short-term auditory memory, the listener can make comparisons in mind and judge whether the changes of mapping parameters or data selection has improved their understanding of the data. In our experience it is helpful to work with such short sonification units of a few seconds so that the interaction is heard as fast as possible after the control has been adjusted.

Let us discuss the above interaction types for PMSon in more detail:

[9]see chapter 15

(i) Interactive data selection can be achieved for instance by running the sonification program with different subsets of the data. However, a very intuitive and direct form of interacting with the data is to provide a visual user interface where the data set can be visually inspected. The basic technique is known in visualization as *brushing*: interactions in one scatter plot (or another visualization type) cause highlighting of display elements in a coupled visualization next to it. Here we propose the term *Sonic Brushing*, where selections in a single visual display cause selected data to be presented in real-time by sonification. A practical implementation can be done with a user-adjustable *Aura* (as introduced/used by Maidin & Fernström [26] for direct navigation of musical tunes), an audio selection circle. Only data points that fall within the Aura's scope are represented sonically. For instance, while moving the Aura with a mouse pointer as shown in Figure 11.3 all points that enter/leave the Aura will be sonified, and on a mouse click a sonification of all selected data could be rendered and played. This brushing technique has been demonstrated in [11] using gestural interaction on top of an interactive table surface and a self-organizing map as a visual display. Providing

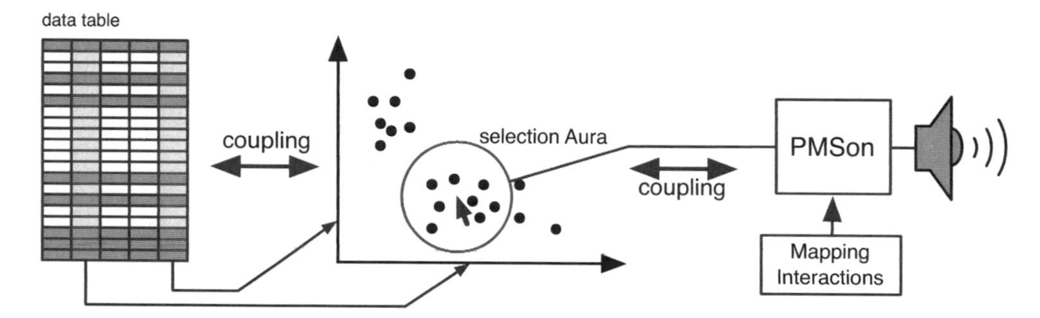

Figure 11.3: *Brushing* for coupled interactive data exploration. The figure depicts a coupled table viewer and scatter plot. The user moves the Aura to select a subset of data items (colored grey in the table) to be fed into the Parameter Mapping sonification (PMSon).

that time is not used for the mapping itself, this interaction would be very direct, since a stationary soundscape would be updated at very low latency after any selection or mapping change.

ii) Mapping interactions are the second type of interactive control, where the mapping itself and mapping parameters are changed by the user. However, it is quite difficult to program a good interface with which to influence the mapping. This often results in a demand for too much knowledge about the software system (e.g., Pure Data or SuperCollider)[10] for the user to change the mapping in an intuitive way. One solution would be to connect the mapping parameters to tangible controllers, such as the faders of a sound mixer board, novel controllers, or simple sliders in a Graphical User Interface (GUI), or even number boxes. This has been done in many sonification toolkits yet a new GUI is needed for every new combination of data, mapping, and synthesizer. As with Interactive Data Selection (above), if sonification time is used within the mapping, it is a good practice to render short, looped sonifications, so that the effect of mapping and parameter changes become clear within

[10]For details on Pd and SuperCollider, the most widespread programming systems suitable for interactive sonification on standard computers and also on mobile devices, see chapter 10.

the next few seconds at most. A parameter-mapping sonification program, written in the SuperCollider language, is provided on the accompanying book website to give an example of such looped interactive sonification. The example code might be a useful starting point for readers to interactively optimize mappings. The example video **S11.21** demonstrates how such closed-loop interaction helps in finding useful sonifications.

Evolutionary algorithms and genetic programming allow the user to be freed from the need to have any explicit knowledge of the synthesizer or the mapping, as demonstrated in [9]. In this example a recommender system generates a couple of new parameter mapping sonifications of the same data using different mappings. As usual in evolutionary algorithms these offspring sonifications are called children and mutation is the principle to determine the mapping. The user simply listens to these and provides a rating, such as relevance on a scale from 0 to 1. This is similar to evolution but here a good rating provides the conditions for the survival of those mappings which the user finds informative. Also, this keeps the user's mental load free for focusing on the sounds without being burdened by mapping details. With a few additional sliders the user can adjust whether the artificial evolution of new sonifications should be more focused on exploring new terrain of the mapping space or be more focused on maximizing the rating.

Sonification example **S11.22** demonstrates a series of parameter-mapping sonifications during such a user-directed exploration sequence for the Iris data set, a 4-dimensional data set containing geometrical features of Iris plants discussed in detail in chapter 8. The Iris data set contains three classes, two of which are slightly interconnected. During the evolutionary mapping optimization process, the user aimed to discover mappings where the clustering structure can be discerned. In the series of sonifications it can be heard how the clustering structure becomes more and more audible, showing that this procedure was helpful in discovering structure in the data.

The two approaches highlighted in this section with code and sound clips are examples of how parameter mapping sonification could be made more interactive. They show quite practically how parameter adjustment can be made more seamless in a continuously updated closed loop.

11.6.4 Interaction in Model-Based Sonification

Since interaction with objects is something we are already familiar with from real-world interactions and manipulations, it is advisable to customize interaction with sonification systems so that humans can rely on their already existing intuitive interaction skills. *Model-Based Sonification* (MBS) is a technique that starts with a linkage between interactions and sonifications that is similar to the linkage between interaction and sound in the real-world. MBS is described in detail in chapter 16. To give a brief summary: in MBS, the data set is used to configure a dynamic system equipped with given dynamical laws and initial conditions. Excitatory interactions to this dynamic model cause acoustic responses, which convey information about structural aspects of the data.

Concerning interaction there is one particular difference between MBS and PMSon: in MBS, the default mode of interaction is by excitation to elicit sonic responses. Interaction is thereby naturally built-in from design, whereas it needs to be added to PMSon as an extra step. More details on MBS are given chapter 16.

In the real world we most frequently interact with objects directly with our hands or by using tools. Our hands permit very flexible or high-dimensional control (i.e., involving many degrees of freedom (DOF)) and we have in most cases continuous control (i.e., we can control the applied pressure on an object continuously). In contrast to these characteristics, we often find that everyday computer interactions are rather *low-dimensional* (e.g., sliders, buttons with just one DOF) and *discretized* (e.g., drop-down menus, radio buttons, or on/off choices). Furthermore, an important variable to characterize interactions is the *directness*: the more direct an interaction is, the lower the latency until the effect becomes perceivable.

According to the descriptors *directness* and *dimensionality* we can categorize interactions in an imaginary 2D space, which we might term the *Interaction landscape*, as shown in Figure 11.4. Real-world interactions often show up in the upper right corner while computer interactions are mostly found at lower directness and dimensionality. From the imbalance between natural and computer interaction it becomes clear to what direction interaction needs to continue to develop in order to meet humans' expectations: interactions that exploit the unique interaction potential given by (bi-)manual interaction, including object interactions such as squeezing, deforming, etc.

MBS provides a 'conceptual glue' about how to bind such excitatory patterns to useful changes in dynamic systems so that meaningful sounds occur as result. Current sonification models already demonstrate interactions such as spatially resolved hitting / knocking, squeezing, shaking and twisting/deforming. The audio-haptic ball interface [12] provided an early interface to bridge the gap between our manual intelligence and sonification models. Equipped with acceleration sensors and force sensitive resistors for each finger, it allowed the real-time sensing and performance of the abovementioned interactions. Within a few years, sensors have become widely available in modern smart phones allowing MBS to be brought into everyday experience. The *shoogle* system [32] is a good example of this: the user shakes a mobile phone to query for incoming text messages, which sound – according to the sonification model – as grains in a box (see example video **S11.23**).

Even if no particular interfaces are available, the metaphors delivered by MBS are helpful for supporting interaction since they connect with the human expectation about what should happen after an interaction.

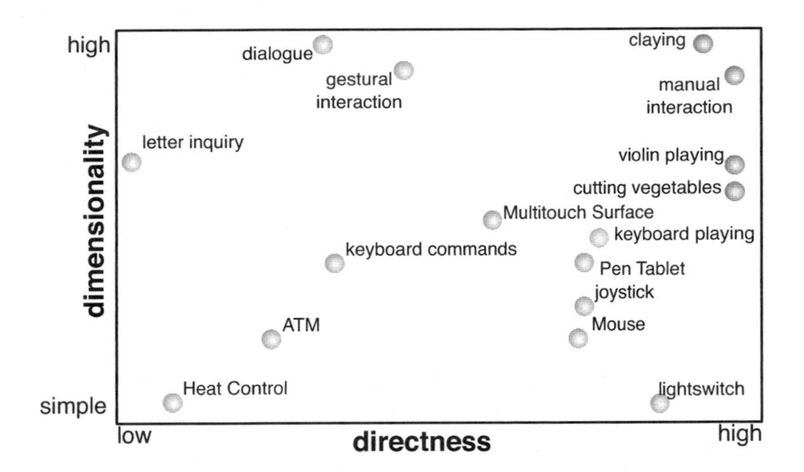

Figure 11.4: Interaction landscape, showing various interactions both in the real-world and with computer interfaces, organized roughly by directness (low latency) and dimensionality (degrees of freedom and continuity of control): Red dots are physical interactions in the real-world: unbeaten in directness and dimensionality.

11.7 Guidelines & Research Agenda for Interactive Sonification

Interactive Sonification offers a relevant perspective on how we interact with computer systems and how a tight control loop matters when we explore data by listening. This section focuses the abovementioned aspects into some guidelines for the design of interactive sonification systems that respect basic underlying mechanisms observable from real-world interaction.

11.7.1 Multiplicity of Sonic Views

A sonification delivers a single isolated 'sonic view' on the data set. Analogous to visual perception, where we need to see a scene from two angles (with our two eyes) in order to extract 3D information, we probably need several different sonic views in order to truly understand a system from its sound. In the visual interpretation of everyday objects such as cups or sculptures, we fail, or at least have much more difficulty, in grasping the 3D structure from a single view alone. By walking around or changing perspective we naturally (interactively) acquire different views which we utilize to inform our understanding. Likewise several sonic views may be helpful, if we assume the sonification to be an analogy – a projection from complex data spaces onto a linear audio signal. Changing perspective is then the equivalent of changing the parameters of the transformation, i.e., the mapping. Where the information is less complex, a single sonification may already suffice to communicate the complete information.

How can we acquire *sonic* views? We interactively query the world. Each footstep is a question to the world, each object manipulation, such as putting a cup on a table, generates a sonic view of the objects involved. This leads to the following guideline: think of the sonification as only one piece in the puzzle. Make it simple and seamless to collect other sonic views. The best way to do so is to consider sonifications that fit well into short-term memory, like short contact sounds of physical objects, rendered with low latency in response to manual/physical interactions.

11.7.2 Multi-modal Displays

In real-world situations we almost always receive feedback in various modalities. For instance, if we put a cup on the table we collect the combined tactile, auditory and visual perception of the physical event, possibly even accompanied with a temperature perception as we touch or release the cup (hear again example **S11.17**). Our perceptual systems are tuned to inter-relate the information between these channels. In order to support these mechanisms which are highly adapted and trained since birth, it is important to create displays that do not present information streams in isolation. Furthermore, it does not help to arbitrarily combine visual, auditory etc. information streams. Instead they need to be coupled by the same (or similar) mechanism which couples perceptual units in the real world: the underlying unity of physical processes.

11.7.3 Balanced Interaction

In everyday interaction sound is often not the single or primary information stream, and in fact it is sometimes only a *by-product* of interaction. Many sonification approaches feature the sound so prominently that they neglect its relation to other modalities. In fact, looking at how we use our sensual perceptions together in different tasks we notice that we distribute our attention to different components of the multimodal stimulus, depending on the task and other factors. Furthermore we reassign our attention with learning or expertise. For example, during the early process of learning to play a musical piece we may mainly struggle with the visual score and our tactile coordination on the instrument, using sound only as secondary feedback. However, when we are finally performing or improvising we attend mainly to the resulting sound (and to the much more abstract features of the music to do with emotion or expression). As a guideline, consider sonification as *additional component* in a mixture of sensory signals and ask what task or activity would be similar in character in real-world tasks? How would you use your senses in this situation? What can you learn from that for the use of sound in the sonification scenario to be designed?

11.7.4 Human Learning Capabilities

Where data sets are particularly complex, or the user does not know the structure of the data they are looking for, then maybe a more flexible interface is called for, analogous to that found on a musical instrument. Such interactions first need some practice and the process of becoming familiar with the interface. Consider how long it takes to learn to play a violin.

Designers of such interfaces perhaps should consider how to engage the user in practice and learning. This is possibly best achieved by creating sonifications which contain information on multiple levels: a coarse level gives useful information even when the interaction is not mastered well; whereas a more subtle information level may be accessed with growing interaction competence, which furthermore motivates the user to engage in the interaction and in learning.

11.7.5 Interaction Ergonomics

It is advisable to respect the bindings between physical actions and acoustic reactions that we have been familiar with since birth – or are possibly even coded into our sensory organs and brains. We expect louder sounds when exciting a system more strongly; we expect systems to sound higher when they are under more tension (e.g., guitar strings); we expect sound to fade out once we stop putting energy into the system. This listening skill of interpreting sound as caused by a underlying process is referred to as *causal listening* by Chion [3] and *everyday listening* by Hermann [10]. The guideline is to respect natural physical coherences and to be aware that interfaces that deviate from them may give decreased performance by not connecting the users so well with physically expected linkages. Model-Based Sonification here again shows particular advantages since it fulfills the bindings almost automatically if a suitable model has been designed and dynamical laws are chosen that are similar to real-life physics.

The above guidelines may be a bit unspecific since they are so generic, but it should be straightforward to apply them as questions to be considered when creating a new interactive sonification system. The guidelines are mostly the result of the authors' personal experiences over several years with designing, programming and using interactive sonifications in various application contexts, and in detailed discussions with other researchers.

However, they call for further investigation and research, which leads to relevant research questions to be addressed in the future. It will be important to develop a scheme for the evaluation of interactive sonification systems and to understand how humans allocate and adapt their perceptual and manipulation resources in order to accomplish a task. Furthermore we need to understand more about how users build up expertise and how the level of expertise influences the formation of automaticity and delegation of control. Only then can we start to investigate the positive effects in comparative studies between different designs that either stick or deviate from the guidelines in different ways.

The challenge is huge; there are infinitely many possibilities, techniques, multi-modal mixtures, tasks, etc. to be investigated. We are far away from a coherent theory of multi-modal sonification-based interactive exploration.

Despite this gap in theoretical underpinning, it will be a useful pathway to take to develop standards for interaction with sonification, both concerning methods and interfaces, and to format these standards in a modular form (e.g., interaction patterns) so that they can be easily be reused. Just as we have become familiar with the GUI and the mouse, we need to find sufficiently effective interactive-sonification methods that we are willing to develop a routine for their regular use.

11.8 Conclusions

This chapter has looked at the ways in which humans naturally interact with real-world objects, and musical instruments, and has compared this with the various methods of interacting with computer systems. With this focus on interaction, it has reviewed the main methods of sonification and considered how they can be configured to provide the user with best possible interface. It has proposed some general guidelines for how sonification systems of the future can be enhanced by providing a multimodal, ergonomic and balanced interaction experience for the user.

Extrapolating from the recent progress in the field of interactive sonification (and furthermore considering the evolution of new interfaces for continuous real-time interaction and the increased overall awareness and interest in multimodal displays) it is an interesting exercise to forecast how we will interact with computers in 2050, and guess what role interactive sonification might have by then.

If the standard desktop computer survives, and is not replaced by pervasive / wearable augmented-reality devices or pads, it will probably immerse the user much more into information spaces. With immersive 3D graphics, and fluent interaction using strong physical interaction metaphors, files and folders become virtual graspable units that the user can physically interact with in the information space which interweaves and overlaps with our real-world physical space. Interaction will probably be highly multimodal, supported by tactile sensing and haptic emitters in unobtrusive interaction gloves, and using latency-free audio-visual-haptic feedback. The auditory component will consist of rendered (as opposed to replayed) interaction sounds, using established sonification models to communicate gross and subtle information about the data under investigation.

Tangible interactions with physical objects and gestural interaction with visualized scenes will be possible and this will allow humans to make use of their flexible bimanual interaction modes for navigating and manipulating information spaces. Sound will be as ubiquitous, informative and complex as it is in the real world, and sonification will have evolved to a degree that interaction sounds are rather quiet, transient, and tightly correlated to the interaction. Sound will be quite possibly dynamically projected towards the user's ears via directional sonic beams, so that the auditory information is both private and does not disturb others. However, for increased privacy, earphones or bone conduction headphones will still be in use. Auditory Interaction will become a strongly bidirectional interface, allowing the user not only to communicate verbally with the computer, but also to use his/her vocal tract to query information non-verbally, or to filter and select patterns in sonifications. Sonification will furthermore help to reduce the barriers that today's information technology often puts up for people with visual disabilities. The multimodal interaction will be more physical, demanding more healthy physical activity from the user, and being less cognitively exhausting than current computer work. In summary, the computer of the future will respect much more the modes of multimodal perception and action that humans are biologically equipped with.

It is an exciting time to contribute to the dynamic field of HCI in light of the many opportunities of how sound, and particularly interactive sonification, can help to better bridge the gap between complex information spaces and our own perceptual systems.

Bibliography

[1] Bongers, B., (2004). "Interaction with our electronic environment an ecological approach to physical interface design", Cahier Book series (no 34) of the Faculty of Journalism and Communication, Hogeschool van Utrecht, April 2004 ISBN 90-77575-02-2

[2] Burns, K., (1997). History of electronic and computer music including automatic instruments and composition machines, available on-line at: http://www.djmaquiavelo.com/History.html

[3] Chion, M., (1994). Audio-Vision: Sound on Screen. New York, NY: Columbia University press

[4] Csikszentmihalyi, M., (1991). Flow: The Psychology of Optimal Experience, Harper Perennial, ISBN: 0060920432

[5] Fink, B., (1997). Neanderthal Flute, *Scientific-American* (Sept.,'97); available on-line at: http://www.greenwych.ca/fl-compl.htm

[6] Gaver, W. W., (1989). The SonicFinder, An Interface That Uses Auditory Icons, http://cms.gold.ac.uk/media/02gaver.sonicFinder.89.pdfandDOI:10.1207/s15327051hci0401_3, Human-Computer Interaction, vol. 4, nr. 1, pp. 67–94

[7] Gaver, W. W., (1994). Synthesizing Auditory Icons. In Proceedings of INTERCHI '93. New York: ACM.

[8] Han, J.Y., (2006). Multi-touch interaction research, Proceeding SIGGRAPH '06 ACM SIGGRAPH 2006

[9] Hermann, T., Bunte, K., and Ritter, H., (2007). "Relevance-based interactive optimization of sonification," in Proceedings of the 13th International Conference on Auditory Display (B. Martens, ed.), (Montreal, Canada), pp. 461–467, International Community for Auditory Display (ICAD)

[10] Hermann, T. (2002). Sonification for Exploratory Data Analysis. PhD thesis, Bielefeld University, Bielefeld, Germany.

[11] Hermann, T.; Henning, T. & Ritter, H. (2004). In: Camurri, A. & Volpe, G. (Eds.) Gesture Desk – An Integrated Multi-modal Gestural Workplace for Sonification Gesture-Based Communication in Human-Computer Interaction, 5th International Gesture Workshop, GW 2003 Genova, Italy, April 15–17, 2003, Selected Revised Papers, Springer, 2004, 2915/2004, pp. 369–379

[12] Hermann, T., Krause, J., & Ritter, H., (2002). "Real-time control of sonification models with an audio-haptic interface," Proceedings of the International Conference on Auditory Display, 2002, pp. 82–86.

[13] Hermann, T. & Hunt, A. (2004). The Discipline of Interactive Sonification. Proceedings of the Int. Workshop on Interactive Sonification, Bielefeld, Germany, http://www.interactive-sonification.org

[14] Hermann, T., & Hunt, A., (2005). An Introduction to Interactive Sonification, Guest Editors introduction, IEEE Multimedia Special Issue on Interactive Sonification, pp. 20–26

[15] Hermann, T., Paschalidou, S., Beckmann, D., & Ritter, H., (2005). "Gestural interactions for multi-parameter audio control and audification," in Gesture in Human-Computer Interaction and Simulation: 6th International Gesture Workshop, GW 2005, Berder Island, France, May 18–20, 2005, Revised Selected Papers (S. Gibet, N. Courty, and J.-F. Kamp, eds.), vol. 3881/2006 of Lecture Notes in Computer Science, (Berlin, Heidelberg), pp. 335–338, Springer

[16] Hermann, T., (2008). "Taxonomy and definitions for sonification and auditory display", Proceedings of the International Conference on Auditory Display, ICAD

[17] Hewett, T. T. (ed), (1992). ACM SIGCHI Curricula for Human-Computer Interaction, chapter 2. Available online at: http://old.sigchi.org/cdg/index.html

[18] Hiltzik, M., (1999). Dealers of Lightning: Xerox PARC and the Dawn of the Computer Age, HarperCollins, ISBN:0887308910

[19] Hunt, A., & Hermann, T., (2004). "The importance of interaction in sonification", Proceedings of the International Conference on Auditory Display, Sydney, Australia, July 6–9, 2004. http://www.icad.org/websiteV2.0/Conferences/ICAD2004/papers/hunt_hermann.pdf

[20] Hunt, A., & Hermann, T., (eds.) (2005). "Interactive Sonification", IEEE Multimedia special issue, Apr–June 2005.

[21] Hunt, A., Wanderley, M, & Kirk, P. R., (2000). "Towards a model for instrumental mapping in expert musical

interaction", Proc. International Computer Music Conference, pp. 209–211, San Francisco, CA, International Computer Music Association.

[22] Hunt, A., (1999). "Radical User Interfaces for real-time musical control", PhD Thesis, University of York, UK.

[23] Hunt, A., & Wanderley, M., (eds.) (2000). Mapping of control variables to musical variables. Interactive Systems and Instrument Design in Music, Working Group, http://www.igmusic.org

[24] Hunt, A., & Wanderley, M., (2002). "Mapping performer parameters to synthesis engines", Organised Sound, Vol 7, No.2, pp. 97–108, Cambridge University Press.

[25] Kramer, G. (chair). (1999). NSF Sonification Report: Status of the field and research agenda, 1999. http://www.icad.org/node/400

[26] Maidin, D. & Fernström, M., (2000). "The best of two worlds: Retrieving and Browsing", Proc. DAFX-00, Italy

[27] Myers, B., (1998). "A Brief History of Human Computer Interaction Technology". ACM interactions. Vol. 5, no. 2, March, 1998. pp. 44–54. http://www.cs.cmu.edu/~amulet/papers/uihistory.tr.html

[28] Schneider, W. & Shiffrin, R. M., (1977). "Controlled and automatic human information processing: II. Perceptual learning, automatic attending, and a general theory", Psychol. Rev. 84: 1–66

[29] Shneiderman, B. (1983). "Direct Manipulation: A Step Beyond Programming Languages", IEEE Computer, 16(8) pp. 57–67

[30] Silver, W., (2007). "DX-ing - Contacting those faraway places", in The ARRL Operating Manual for Radio Amateurs, ARRL, 2007. Available on Google Books at http://books.google.co.uk/books?id=YeDTpxHt37kC&lpg=SA6--PA9&ots=h1SOz_Ynko&dq=tuning%20in%20to%20DX%20signals&pg=PP1#v=onepage&q=tuning%20in%20to%20DX%20signals&f=false

[31] Wanderley, M.M. (ed.), (2002). Special issue on "Mapping in Computer Music", *Organised Sound*, Vol 7, No. 2, Cambridge University Press.

[32] Williamson, J., Murray-Smith, R., Hughes, S. (2007). "Shoogle: excitatory multimodal interaction on mobile devices", Proc. SIGCHI conf. on Human factors in computing systems.

Part III

Sonification Techniques

Chapter 12

Audification

Florian Dombois and Gerhard Eckel

12.1 Introduction

Music is ephemeral. It intrudes into reality for a moment, but escapes it in the next. Music has a form, which is expressed over time, but cannot be touched. Evidently, music is difficult to archive, and two main techniques have challenged its transience: (i) the *score*, as a code for instructing instrumentalists or other sound generators for later re-enactment, and (ii) the *recording*, as the acoustic complement to photography, which registers the sound wave at a specific point of listening. Both techniques have their advantages and disadvantages since they cannot exactly repeat the original, but they both open a particular perspective on the original sound and disclose aspects perhaps otherwise not to be heard. The first approach stresses, for example, more the tones and their symbolic value, whereas the second traces the exact physical wave in an analog manner. In sonification, one can find these two perspectives too: (i) the technique of *parameter mapping* (see chapter 15), and (ii) the technique of *audification*. This chapter concentrates on the second.

Gregory Kramer defines in his book *Auditory Display*: "The direct playback of data samples I refer to as 'audification"' [32, p. xxvii]. And as a later update of this definition: "Audification is the direct translation of a data waveform into sound." [60, p. 152]. The series of data might not even belong to the sound domain. A common way of displaying this visually would be a Cartesian graph. If the visualized data have a wave-like shape, e.g., an EEG signal, audification would mean to attribute their values to air pressure, and transferring the result to a loudspeaker, whereby the data then become audible. The aim behind this media shift, as in all sonification techniques, is that the other mode of representation discloses or makes emerge aspects of the data that might not have been discovered before. It is a direct alternative approach to visualization, since all abstract data series can be either visualized or sonified. So one might define: *Audification is a technique of making sense of data by interpreting any kind of one-dimensional signal (or of a two-dimensional signal-like data set) as amplitude over time and playing it back on a loudspeaker for the purpose of listening.* And since all data end up in a loudspeaker, audification is essentially a continuous, non-digital interpretation of data sets.

In audification one can distinguish between different types of data that result in different types of sounds. Often, sound recordings themselves have already been named "audification" if they have been shifted in pitch. Therefore, we want to include with our definition above all data sets that can be listened to, i.e. also all sound recordings themselves. We see four groups of data (see Fig. 12.1, withdrawing more and more from the audio-context: (i) sound recording data, (ii) general acoustical data, (iii) physical data, and (iv) abstract data.

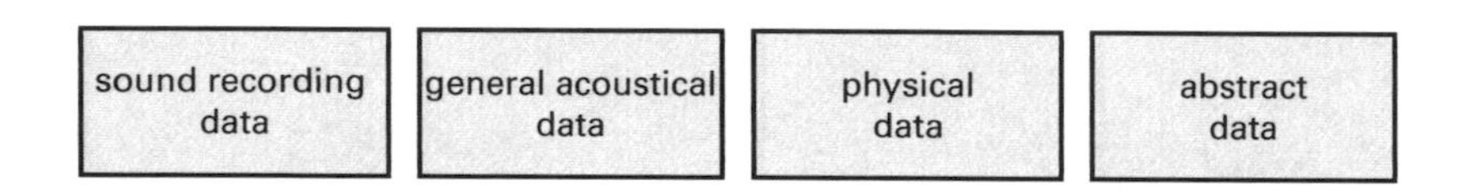

Figure 12.1: Classification of data for audification

(i) Sound Recording Data: The first group of data, to which audification can be applied, are sound recordings, which are today normally sampled digitally as series of numbers. Taking these time series one could say that every CD-Player has an audification module, which is the Digital-to-Analog (DA)-converter transforming the series of data points into a continuous sound signal. Now, usually there is little special from the viewpoint of sonification about listening to sound recordings themselves. This becomes different when sound recordings are amplified, thereby revealing unheard aspects in the recordings. And it becomes even more interesting when the recordings are time-compressed or -stretched. For example, ultrasonic signals such as bat calls are inaudible to the human ear unless they are transposed (sound examples **S12.1**, **S12.2**). The change of playback speed is then certainly more than a gimmick, and audification can function in this context as an acoustic micro- or telescope.

(ii) General Acoustical Data: All kinds of measurements in elastomechanics, which follow the same physical laws as an acoustic wave, constitute a major area of interest for audification. In particular, vibrational data of mechanical waves are easily accessible by listening to their audification. From applying our ears to a railroad rail, a mast or a human belly, from using sounding-boards, stethoscopes or sonar, we are familiar with interpreting mechanical waves acoustically. And, even though they are always a combination of compressional and transversal waves, the character of mechanical waves is usually preserved when being reduced to a one-dimensional audio signal. Changing the playback speed also usually proves to be of minor influence on the plausibility of the resulting sound. This is especially evident in Auditory Seismology, where seismograms are often audified with acceleration factors of 2,000 and more (sound example **S12.3**) [15].

(iii) Physical Data: There are measurements of other physical processes outside the mechanical domain that can be audified too. But these data, for example electromagnetic waves, usually lack acoustic familiarity with our daily hearing. The different velocities of wave-propagation or of the dimensions of refraction and reflection effects etc. result in a new soundscape unfamiliar to human experience. Therefore, one has to be careful with

interpretation; e.g., EEG data (sound example **S12.4**) of several electrodes around a head cannot simply be compared to a similar arrangement of microphones within a room.

(iv) Abstract Data: The lack of acoustic familiarity may worsen when using abstract data for audification which do not stem from a physical system. Examples of this non-physical data might be stock market data, or when listening to a fax-machine (sound example **S12.5**) or a computer-modem at the telephone (sound example **S12.6**). Not all wave-like shapes in abstract data conform to the wave equation, therefore interpreting those audified signals usually takes more time for the user to become habituated. Nevertheless, non-acoustic and abstract data can easily be audified when they are arranged as a time series.[1]

Audification is the simplest technique of sonification and is therefore often used as a first approach to a new area of investigation, but then mostly neglected in the further development of sonification projects. This chapter hopes to show that there are many reasons not to undervalue the potential of audification, especially when adding various acoustic conditioning techniques as described in Section 12.3. It also hopes to widen the scope for development towards more interactivity in the use of audification parameters.[2] The remainder of the chapter is organized as follows:

Section 12.2 gives a brief introduction to the history of audification from the 19th century until the first ICAD in 1992. Here, the focus is not only on the history of science, but also some examples from the history of music and art. Section 12.3 is dedicated to the technical side of audification and tries to unfold many possibilities for optimizing the acoustic result of audification. Audification can seem to be a simple reformatting procedure only at a first glance. Instead, it can be much more sophisticated when extended to the art of sound processing. Section 12.4 summarizes the areas in which audification is used today, especially referring to the ICAD and its papers. Section 12.5 gives some rules of thumb, how and when to use, and what to expect from the audification of a data set. Finally, Section 12.6 outlines the suggested next steps that need to be taken in order to advance the application of audification.

12.2 Brief Historical Overview (before ICAD, 1800-1991)

Three inventions from the 19th century are of great importance to the history of sonification: (i) the *telephone,* invented by Bell in 1876, (ii) the *phonograph,* invented by Edison in 1877, and (iii) *radiotelegraphy,* developed by Marconi in 1895. The transformation of sound waves into electric signals, and vice versa, started here, and the development of the loudspeaker began as a side product of the telephone. These tools for registering and displaying sound gave rise not only to a new era of listening,[3] but also to the research field of audification. If we take the "Time Axis Manipulation" of *sound recording data* as the simplest form of intentional audification, we find Edison demonstrating this technique already in 1878 in New York [24, p. 27f.] and in the 1890s the Columbia Phonograph Company suggested reversing the direction of playback as an inspiration for new melodies. [24, p. 52].

As well as reproduction, mediation is an important part of audification. The whole idea of

[1] This classification of data in four groups could be developed further and one could discern, for example, in each of the four between continuous and discrete or digital datasets (cf. Section 12.3), etc.

[2] Cf. [29, p. 21] and [28, p. 6]

[3] Cf. [58] for a profound investigation of how much these inventions changed our relation to hearing in general.

data-driven sound, which is the key concept of sonification, could only be made possible by the introduction of a medium, which, as electricity or an engraved curve, also makes it possible for all forms of data to be displayed in the sound domain. By extending the human auditory sense with technology, the process of listening to data can be thought of as involving data, conversion, display and perception.

12.2.1 Listening to the measurement in science

In science, visualization has played the dominant role for centuries, whereas sound and listening to natural phenomena has always been under suspicion. Nevertheless, there are a few exceptions. One of the very early uses of scientific audification of *general acoustical data* (cf. Section 12.1), even before the electrical age, was that made by the French doctor R. T. H. Laënnec who, in 1819, invented the stethoscope [35]. This auditory device, which is still in use, had a great career as an instrument for medical diagnosis, especially after being redesigned by Georg Philip Camman, and is one of the few important examples of an accepted  scientific device using audio[4] (sound example **S12.7**). Auenbrugger later added "interactivity" to it by introducing percussion and gave listening into the human body another boom of success [2]. This has not stopped, and we find similar instruments handling mechanical waves even in today's plumbers' equipment to track leaking conduits ("Hördosen").

In audifying *physical data*, as introduced above, the earliest examples date back as early as 1878, when a series of papers was published about connecting muscle cells with the newly invented telephone.[5] These groundbreaking publications have not been considered by the ICAD community, so far as we know, but, for example, J. Bernstein and C. Schönlein, in their paper of 1881[8], describe nothing less than how they studied the reaction frequencies of muscle cells and the transmitting qualities of the cells as what they call "muscle telephone" [8] (p. 27) by listening to its audification. The first audification of nerve currents was published a little later by Nikolai Evgenievic Wedenskii, in 1883, also using the loudspeaker of a telephone as an audio display of physiological processes [61]. Later, in 1934, a few years after Hans Berger's famous publication of EEG waves, E.D. Adrian and B.H.C. Matthews proved his experiments also using audification [1]. Another successful example of audifying non-acoustic data is the Geiger counter, which was invented by Hans Geiger in 1908 and developed further in 1928, resulting in the Geiger-Müller tube, which is still used today (sound example **S12.8**).

Almost at the same time is found what is probably the first use of Time Axis Manipulation in the scientific context: US scientists applied new methods of sound processing to the echo-locating sounds of bats and, in 1924, released a record with transposed bat recordings now audible for the human ear [31, p. 152]. Also, the technique of sonar (SOund Navigation And Ranging) dates from this time, even though the idea apparently can already be found in Leonardo da Vinci's manuscripts [33, p. 29]. It was developed during World War I in Great Britain to track submarines. The use of the Vocoder as a sound encoding device seems to be worth mentioning here. In the SIGSALY system for secure voice transmission

[4]There are several training websites for doctors to learn auscultation, e.g. `http://www.wilkes.med.ucla.edu` (accessed Jan 30 2011)

[5]The first paper, we found, is [26]. For more material cf. [17] and [18]. There is also an ongoing research project, "History of Sonification" by Andi Schoon and Axel Volmar, who have found about 30 more articles in that early era. First publications are [57], [59] and [56], but please watch out for their coming publications.

developed in 1942-1943, the Vocoder would encrypt the speech signal by another sound signal, for example with a noise recording. The transmitted signal would then sound like random noise, but at the receiving station the same noise sequence would be used to decode the speech signal, thus making the voice recognizable again [31, p. 78]. Depending on the encrypting signal the encrypted voice can become of interesting fashion, an effect that has been used in many science-fiction movies (sound example **S12.9**) and pop songs (sound example **S12.10**).

We have already mentioned that inventions of new technologies and media have a major influence on scientific development. One important improvement for audification research happened when audio tape was developed by Fritz Pfleumer in 1928, on paper, and later in 1935 by BASF, on plastic tape. This new material was soon used as a common data storage medium, and already Alan Turing seemed to have thought of magnetic audio tape as storing material for the upcoming computer. Using a recording and a playback head, allowing backward and forward play, magnetic tape was the start for "every thinkable data manipulation".[6] This new sound recording material transformed the temporal phenomenon of sound, even more explicitly than the wax roll, into a trace on a substrate – time is linearly represented in space. By manipulating the substrate (e.g., cutting and splicing a tape), sound can be transformed out-of-time, and Time Axis Manipulation becomes feasible through changing the reading speed of the substrate. This might explain the great success of audio tape after World War II and its intense use in all kinds of application areas. Among others, in the 1950s, seismologists started to write seismological recordings directly on magnetic tape in an audio format and several researchers listened to their data first for entertainment, we have been told, but then figured out that this method is especially valuable for detecting signals of seismic events in noisy records. The first scientific paper on audifying seismic recordings is "Seismometer Sounds" by S. D. Speeth, in 1961, where he describes testing audification for signal discrimination between natural quakes and atomic explosions (cf. [55]; see also [21]).

12.2.2 Making the inaudible audible: Audification in the arts

Interestingly enough, audification not only has its roots in science but also very much in the arts. The first text that is known on an audification of *abstract data* can be found with Rainer Maria Rilke, a German poet, dating back to 1919. "Ur-Geräusch" (Primal Sound) [52] is a short essay reflecting on the form of the coronal suture of a skull, imagining the shape translated into a sound. Another important text was written by László Moholy-Nagy, in discussion with Piet Mondrian in 1923, where he wants to compose New Music by etching sound curves directly on the record [43], which is nothing less than an audification of graphical lines, long before the computer was invented. Ideas similar to that of etching a disc were brought up by introducing soundtracks to film in the 1920s. Oskar Fischinger, in 1932, in reaction to the famous "Farbe-Ton-Kongreß" in Hamburg[7], started to investigate painting ornaments directly on the soundtrack of a film (see Figure 12.2).[8] This technique resulted in new synthetic sounds, very much like those from electric synthesizers, and produced a huge

[6] [31, p.165] ("jede erdenkliche Manipulation an Daten")

[7] In 1927 and 1930 the first two conferences on the relation between color and tone were held in Hamburg, Germany, that had a major impact on the development of synaesthetic art.

[8] First published in *Deutsche Allgemeine Zeitung* 8.7.1932; cf. [44, pp. 42-44]

press reaction in Europe, the US, and even in Japan. Nevertheless, Fischinger unfortunately could not get a grant for further research and released only a few recordings, which at least were well-received by John Cage and Edgard Varèse.

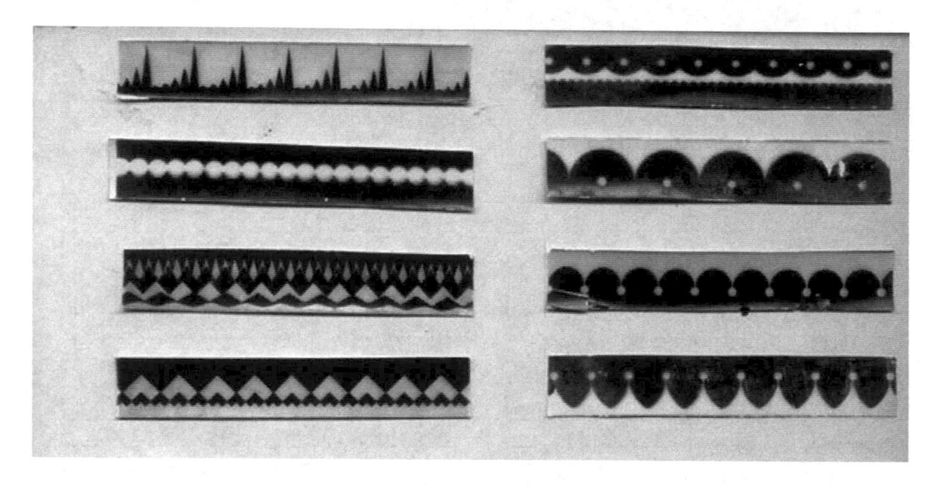

Figure 12.2: Detail from Oskar Fischinger's studies of sounding ornaments.

Another genealogic line follows composers of New Music using new technology for their compositions. For example in 1922 Darius Milhaud began to experiment with "vocal transformation by phonograph speed change" [54, p. 68] in Paris and continued it over the next 5 years. The Twenties also brought a wealth of early electronic instruments — including the Theremin (sound example **S12.11**), the Ondes Martenot (sound example **S12.12**) and

the Trautonium (sound example **S12.13**), which were designed to reproduce microtonal sounds. The historical beginnings of electronic music can also be interpreted as a history of audification, because all electronic instruments use electric processes audified on a loudspeaker. This chapter does not describe further this rich story of further inventions of data processing with acoustic results (i.e., the history of electronic music), since it moves too far away from the core of audification and must defer here to the existing literature.

But at least two more developments seem worth mentioning here:

(i) In 1948 Pierre Schaeffer developed his idea of "musique concrète" which was the furious starting point for composing with recorded material and sound samples in music. Naturally, all forms of manipulation and conditioning of the acoustic material were developed subsequently (Time Axis Manipulation, all sorts of transpositions, reverse playing, filtering etc.), first mainly with audio tape and record players, and later with the computer.

(ii) Another influential development was the artists' interest in the unheard sound cosmos. For example Jean Cocteau's reaction to the discoveries of ultrasounds was an enthusiastic conjuration of yelling fish filling up the sea with noise.[9] And many projects, like the LP record

"BAT" of Wolfgang Müller (1989) (sound example **S12.14**) with down-pitched ultrasounds,

[9] Cf. [10, p. 36f.]: "Die Welt des Tons ist durch die noch unbekannte Welt des Ultraschalls bereichert worden. Wir werden erfahren, daß die Fische schreien, daß die Meere von Lärm erfüllt sind, und wir werden wissen, daß die Leere bevölkert ist von realistischen Geistern, in deren Augen wir ebenfalls Geister sind."

received their attention in the art world because they displayed another world beside ours, so to say making the inaudible audible.

12.3 Methods of Audification

As with any sonification technique, the overall goal of data audification is to enable us to listen to the results of scientific measurements or simulations in order to make sense of them. As described in the introduction of this chapter, the particularity of audification lies in the fact that the data analysis is delegated almost completely to the human auditory sense. In the case of audification, we therefore try to minimize the transformations of data prior to listening in order to be able to perceive them as far as possible in their "raw" state (i.e., in the form they have been acquired). This chapter refers to these transformations as *signal conditioning* to underline the difference with parameter mapping sonification.

12.3.1 Analytic Listening

Audification is especially useful in cases where numerical data analysis methods fail or are significantly outperformed by the analytical capabilities of the human auditory system. In these cases, it cannot be decided which aspects of the data contain the information we may be interested in. By engaging with the data in a process of analytic listening, patterns may emerge which are otherwise undetectable. Listening to data audifications is a demanding task that needs training and experience. Any approach to audification has to support this task in the best possible way. This implies that all stages of the audification process have to be made explicit such that the listener can determine their influence on the perception of the data at any time. It is crucial to understand which aspects of a sound may stem from which stage of the audification process. This is the only way to distinguish features in the data set from artifacts inevitably introduced by any process of observation.

12.3.2 The Audification Process Model

The various technical aspects of audification are presented in Figure 12.3 according to a model of a typical audification process, which may be divided into three sequential stages: 1) data acquisition, 2) signal conditioning, and 3) sound projection. Each of these stages will be discussed in detail below once the overall constraints informing this process have been clarified. The goal of the audification process is to format the data in such a way that it is best exposed to the analytical capabilities of the human auditory system. The overall design of the audification process and the choices to be taken at each of its stages depend on a) the characteristics of the auditory system, b) the characteristics of the data set, and c) the questions that drive our analysis (cf. Fig. 12.3).

Characteristics of the Human Auditory System

The characteristics of the human auditory system most relevant to audification are its frequency range, frequency resolution, temporal resolution, dynamic range, masking effects, and

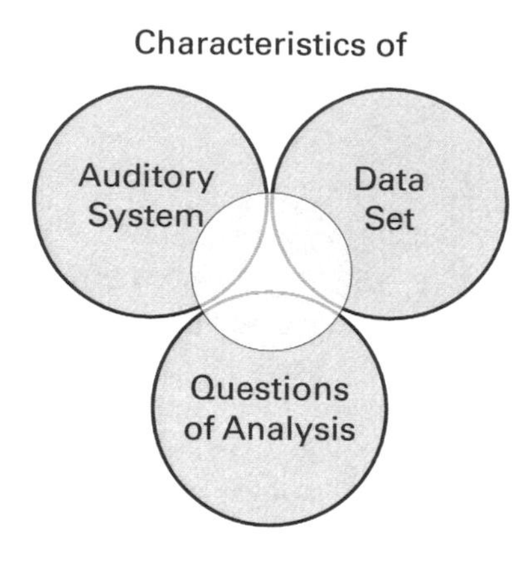

Figure 12.3: Aspects Informing the Audification Process

the different types of detection and discrimination tasks the auditory system has specialized in during its evolution (see chapters 3 and 4 for details).

The ear is sensitive over a *frequency range* of about 20 Hz to 20 kHz. Especially the upper limit of this range varies significantly with age and gender. Young individuals may be able to detect signals far above 20 kHz. Due to age-related hearing loss (presbycusis) the upper limit drops as much as an octave with progressing age, but there are significant differences between individuals. Noise-induced hearing impairment may further reduce the upper limit. It seems relatively safe to assume that even older individuals with typical presbycusis and no additional impairment hear up to 10 kHz. In practical applications the lower end of the frequency range is limited by the typical reproduction system by yet another octave. Thus, from the theoretical bandwidth of 10 octaves, only 8 can be used in practical applications (40 Hz – 10 kHz). The conditioning of a data set has to respect these bandwidth limitations, i.e., interesting signal components have to be transformed into this range. The *frequency resolution* of the ear is about 4 Hz in the middle range (1 – 2 kHz), i.e., smaller changes in frequency in a data set cannot be detected. The *temporal resolution* with which the ear can detect individual events in a signal is linked to the lower limit of the frequency range and lies somewhere between 20 and 50 ms. This limitation has to be taken into account if temporal structures in a data set are supposed to be detected through audification.

The *dynamic range* of the human auditory system (see chapter 3) is capable of covering an extent of about 120 dB at middle frequencies (around 1 kHz). This range reflects the level difference between the threshold of hearing (0 dB SPL[10] @ 1 kHz) and the threshold of pain (120 dB SPL @ 1 kHz). Evidently, this range is not available in practical applications. The lower end is limited by background noise (as much as 50 dBA[11] for a typical office space or as little as 20 dBA for a professional studio), and the upper end by the threshold of comfort (around 100 dBA). This limits the usable dynamic range in a practical application to about

[10]SPL stands for Sound Pressure Level. This is an absolute level measured in decibels (dB) refering to the threshold of hearing at 1 kHz, which is defined as the sound pressure of 20 μPa RMS = 0 dB SPL.

[11]The postfix A indicates an A-weighted sound pressure level generally used for noise measurements.

50 to 80 dB. If a data set exhibits a substantially larger dynamic range, the conditioning will have to include dynamic compression.

Another very important characteristic of the human ear is *masking*, occurring in different forms, which may render certain signal components inaudible in the presence of others. These effects are difficult to quantify, and this chapter can only give an overview of the typical situations where masking occurs. If two sounds occur simultaneously and one is masked by the other, this effect is referred to as *simultaneous masking*. A sound close in frequency to a more intense sound is more likely to be masked than if it is far apart in frequency. Masking not only occurs in the frequency domain but also in time. There are two types of *temporal masking*. A weak sound occurring soon after the end of a more intense sound is masked by the more intense sound (forward masking). Even a weak sound appearing just before a more intense sound can be masked by the higher intensity sound (backward masking). The amount of masking occurring in a particular signal depends on the structure of the signal and can only be predicted by a psychoacoustic model (cf. chapter 3). Since the masking effects are frequency- and time-dependent, different forms of conditioning (especially various forms of Time Axis Manipulation) may dramatically change the amount of masking that occurs in a data set.

Besides the characteristics discussed so far, the ear exhibits a number of very specialized capacities to detect and discriminate signals. These capabilities evolved out of any animal's need to determine the sources of sounds in their environment for reasons of survival. Thus, the ear is specialized in grouping elements in complex sounds and attributing these groups to sources. Typically, the grouping is based on discovering structural invariants in the time-frequency patterns of sound. Such invariants are the result of the physical constraints that vibrating objects obey, and which result in clear signatures in the sound. The ear can be thought of as constantly trying to build and verify these grouping hypotheses. This is why it can also deal with situations quite efficiently where the clear signatures of mechanically produced sounds are missing. The ability to scrutinize abstract sound through analytical listening and make sense of it can be considered the basis of data audification. The perceptual strategies listeners employ in this process are generally referred to as Auditory Scene Analysis [9], which is concerned with sequential and simultaneous integration (perceptual fusion) and the various interactions between these two basic types of perceptual grouping.

Characteristics of the Data set

The decisions to be taken at the different stages of the audification process also depend, to a large extent, on the type of data to be explored. As described in the introduction of this chapter, we can distinguish four types of data used in audification. In all cases, it is important to know how the data sets were acquired and which aspects of which processes they are meant to represent. In the first three cases, the data is derived from physical processes, implying that some kind of sensor[12] is used for data acquisition. Usually, these sensors produce an electric current, which has to be amplified and which is–nowadays–directly converted to a digital signal (digitization, A/D conversion) and stored in this form. The quality of a digital signal (i.e., how accurately it represents the physical property measured) mainly depends on the performance of the sensor, the amplifier, and the A/D converter.

[12] A sensor is a transducer which converts a physical condition into an analog electric signal (e.g., microphone, hydrophone, seismometer, thermometer, barometer, EEG electrode, VLF receiver, image sensor).

Every data acquisition process is afflicted with the introduction of artifacts. Typical artifacts are thermal noise[13], mains hum, RF interference, as well as linear and non-linear distortion. If an analog storage medium is used before digitization (e.g., a tape recorder), yet another source of artifacts is introduced. Apart from the technical artifacts introduced by the data acquisition process, other disturbing signal components are usually present in the acquired data (e.g., different types of environmental noise such as ocean waves and traffic noise in seismograms, or as muscle signals and DC offsets and drifts in EEG registrations).

In the audification process the decision needs to be made whether to remove what is considered an artifact before listening to the data, or if this task is delegated to the auditory system, which is often much better in doing so than a pre-applied conditioning algorithm. As discussed above, artifacts are only a problem for the analytical listening process if they are not identified as such, and if they cannot be attributed to a defined stage of the audification process. Therefore, the listener's ears need training to clearly identify the artifacts of each stage (e.g., environmental noise, sensor non-linearities, amplifier noise[14], mains hum, quantization noise, aliasing, data compression artifacts). This is comparatively easy in the case of acquiring data from a physical process. If the data to be audified stems from an "abstract" process, such as a numerical simulation or a collection of numerical data representing social or economic quantities (e.g., stock market data), it is much more difficult to decide what is an artifact. In this case, it is important to understand the underlying process as much as possible (e.g., estimate its bandwidth in order to choose an adequate sampling rate). If it is not possible to obtain the necessary insight, then the listener has to keep in mind that the acquired data may appear substantially obscured.

Questions that Drive the Analysis

Another important aspect informing the audification process lies in the questions to be answered through listening to the data. There are situations where this question cannot be answered precisely and thus audification is used exactly for that reason – because it is not yet known what is being searched for. In this case, users want to quickly play a potentially very large data set and then refine their choice of conditioning depending on what they find. Such an approach may be useful when employing audification to quality control of data. Another task may consist in trying to categorize data quickly. This may require listening to the same data set with different conditioning options before being able to take the decision to which category a data set belongs. A typical question to be answered by audification is whether there is some structure in a very noisy data set. Also, in this case, trying a wide range of conditioning options (e.g., different filters and Time Axis Manipulations) may help to answer this question. If users are looking for temporal segmentation of the data, they may benefit from using conditioning options that enhance the contrast in the data, or they may choose a Time Axis Manipulation that does not affect the pitch. If the goal is to find structures in the frequency domain, the signal can be conditioned such that tonal signal components are amplified and noise is rejected – a kind of spectral contrast enhancement, for example implemented as a multi-band noise gate. From this brief overview of the possible questions

[13]Typically induced directly by Brownian motion in resistors

[14]From recent efforts in quality control of biomedical signals we know that in certain frequency bands the SNR of EEG signals is less than 20 dB due to amplifier noise. Training the ear to distinguish the amplifier noise from the neuronal noise is essential in this situation.

that drive data audification it can be seen that special signal conditioning tools may need to be developed in particular cases. But there exists at least a set of general conditioning tools (described below), which has proven to be useful in most audification tasks.

12.3.3 Stages of the Audification Process

The quality of an audification is determined to a large extent by the quality of the signal to be displayed as sound. This is why the systemic constraints of digital signal representation are discussed in some detail in the section on *data acquisition*. Although this stage may not be under the control of the person performing the audification, it is essential to fully understand what happens at this stage for reasons of transparency. Evidently, the *signal conditioning* stage has another important influence on the signal quality. Most of the audification examples accessible on the Internet today suffer from severe distortions introduced at this stage, showing that there is little or no awareness of this problem. In the audio domain, the problems of data acquisition (sound recording) and conditioning (analog and digital audio signal processing) are much more easily detected because in this case listening to the signals is the main purpose. In other domains such problems are usually only discovered once audification is used. Another oft-neglected aspect of the audification process concerns the interface to our ears – the *sound projection* (just think of the typical speakers used with desktop or laptop computers, or the computer-induced background noise in an average office space). According to the proverb "a chain is only as strong as its weakest link", this stage is as important as the preceding ones.

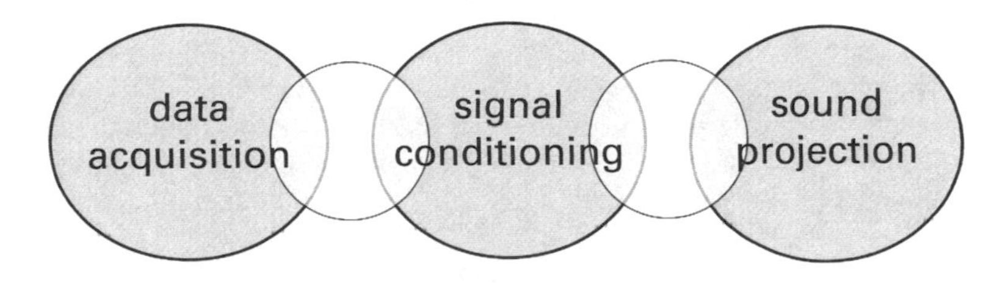

Figure 12.4: Stages of the Audification Process

The Data Acquisition Stage

Not every data set is suitable for audification. An important prerequisite for audification is that the data has the form of – or can be transformed into – a signal. Since input and output data of all audification processes are signals, this section now briefly recalls the foundations of information theory and the sampling process. Most signals of interest for audification are modeled as functions of time or position. A function is a relation, where each element of a set, called *domain* (or input), is associated with a unique element of another set, called *codomain* (or output). The set of all actual output values of a function is called its range. As an example, consider an acoustic signal. Its domain is time and its codomain is air pressure. Its range are all the air pressure values that can be recorded, for instance, with a given microphone.

There are analog and digital signals. Analog signals are continuous, whereas digital signals are discrete and quantized. Analog signals can be converted to digital signals through a process called sampling. Electric signals produced by sensors are discretized and quantized by an analog-to-digital converter (ADC) which implements the sampling process producing the digital signal. Digital signals can also be produced directly by numerical simulations. In both cases the sampling theorem has to be respected. It states: when sampling a signal in time, the sampling frequency must be greater than twice the bandwidth of the input signal in order to be able to reconstruct the original correctly from the sampled version (this reconstruction is essentially what we are doing when audifying data). If this condition is not met, signal components with a frequency above the Nyquist frequency (i.e., half the sampling frequency) will be mirrored at the Nyquist frequency. This artifact is called aliasing or foldover and introduces components into the digital signal which are not present in the analog signal. Similar effects arise in numerical simulations when the time resolution of the simulation is chosen inadequately. Aliasing can be avoided by setting the sampling frequency to cover the full bandwidth of a signal or by limiting the bandwidth with a low-pass filter (anti-aliasing filter) (see chapter 9). High-quality ADCs are equipped with such filters, thus aliasing is usually not a big problem with sampled analog signals. It is much more difficult to avoid (or even to detect) with digitally produced signals, as it is sometimes very hard to estimate the bandwidth of the numerical process. Sparsely sampled data sets, which often also exhibit jitter and missing values, such as historic barometric or temperature data, pose a particular problem to audification. The audible artifacts resulting from the different violations of the sampling theorem need special attention on a case-by-case basis, as a generalization of these effects is impossible.

Another source of errors in digital signals is the quantization of the sample values which adds quantization noise to the signal. The signal-to-noise ratio (SNR) is a measure of the signal quality that can be achieved with a certain level of quantization. It depends on the number of bits used to encode the sample value and can be computed with the following formula (assuming PCM[15] encoding):

$$\text{SNR} = N \cdot 20 \cdot \log_{10}(2), \text{ where } N \text{ is the number of bits.} \tag{1}$$

Consider the following example: according to this formula, a 16-bit digital audio signal has a theoretical SNR of 96.33 dB. But this SNR is only reached for a signal with the highest representable level, i.e., when all bits are used for the quantization. Signal parts with lower levels also have a lower SNR. This is one of the reasons why audio signal encoding moved from 16 to 24 bits in the last decade. This results not only in a better SNR, but also more headroom when recording. Headroom is the margin between the average recording level and the maximum recordable level. A larger headroom prevents signal clipping caused by unexpected signals peaks. Clipping occurs when the signal to be encoded exceeds the quantization range. Like aliasing, clipping also adds signal components to the digital signal which are not present in the analog signal (non-linear distortion). All these aspects of sampling discussed for audio signal here apply equally well to any other type of signal acquisition (e.g., seismic or electromagnetic signals).

[15]Pulse-Code Modulation (PCM) is the standard method used to digitally represent sampled analog signals.

The Signal Conditioning Stage

In the following paragraphs a minimum set of standard signal conditioning tools is introduced in a tutorial style. At the signal conditioning stage it can be assumed that the acquired data is represented as a digital signal with a defined sampling rate sr, a bit-depth q. The first example assumes an EEG registration with $sr = 250$ Hz, $q = 16$ bits.

The Best Conditioning is No Conditioning In the simplest case, no signal conditioning at all is needed to audify a data set. For our EEG signal, this would mean playing it back unchanged at a standard audio sampling rate. As the audio rate is significantly higher than the data rate, this operation would time-compress the signal by a factor equal to the quotient of audio and data rate. For an audio rate of 44.1 kHz, the compression factor would amount to $44100/250 = 176.4$, and so 1 minute of EEG data could be listened to in about 3.4 seconds. Such an audification would typically be useful in a screening or quality control task, for example looking for special events (e.g., electrodes loosening or an epileptic seizure) in a large data set (e.g., in a 24 h registration, which could be scanned in a little over 8 minutes). Evidently, all frequencies in the EEG signals would be transposed up by the same factor (i.e., about 7 1/2 octaves). This means that, for instance the so-called *alpha waves*, which occur in the EEG signal in a frequency range of 8–13 Hz, would become audible in a range of about 1.4–2.3 kHz, which happens to fall into the region where the human ear is most sensitive.

Next Best is Resampling To illustrate a simple case of conditioning by resampling, imagine that the source data are elephant calls, recorded with a microphone and a DAT recorder. As one would expect, typical elephant calls have a very low fundamental frequency (between about 15 and 25 Hz). In order to decide how many individuals are calling on the recording, it must be transposed to a frequency range in which our ears are able to separate the calls. At the original speed, it is only possible to hear an unclear mumbling which all sounds alike. If the sounds had been recorded on an analog tape recorder, the tape could have been played back at double speed and the problem would have been (almost) solved. Eventually, it turns out that a transposition by a factor of 3 moves the elephants' fundamental and the formant frequencies into a range convenient for our ears to easily attribute the calls to different individuals. In the digital domain such a transposition is accomplished by a process called resampling or bandlimited interpolation[16]. This process, which is based on the sampling theorem[17] and the cardinal series, may add samples to a signal by means of interpolation or remove samples from it by reducing its bandwidth and resampling it. In the second case, information is lost since fast fluctuations in the signal (high frequencies) have to be suppressed to avoid aliasing (as discussed above). On the above recording this would concern the birdcalls that were recorded together with the elephants. As they have very high fundamental frequencies, they would have folded over if the bandwidth of the signal were not reduced by low-pass filtering before decimating it. The information loss in the upper part of the spectrum is not critical in this case since the partials of the elephant calls do not reach that far up and we are not interested in the birds.

[16]cf. `http://ccrma.stanford.edu/~jos/resample/resample.html` (accessed Jan 30 2011)

[17]cf. `http://en.wikipedia.org/wiki/Nyquist-Shannon_sampling_theorem` (accessed Jan 30 2011)

Filtering is Useful and Mostly Harmless Filtering is a general-purpose conditioning tool useful in various situations. It introduces linear distortion in the signal, i.e., it changes the levels of existing signal components but does not add new components as non-linear distortion does. As an example, imagine that the abovementioned EEG signal was recorded in Europe, where the power lines are operated at a frequency of 50 Hz. With EEG registration it is almost impossible to avoid recording mains hum together with the brain waves. In the earlier example, the artifact would appear as a very annoying sinusoidal parasite at 8820 Hz (50 Hz $\times$ 176.4). Applying a notch filter[18] tuned to this frequency will remove the artifact without interfering too much with other frequency regions. We could solve this problem also by deciding to ignore all frequencies above 40 Hz, which is a relatively sensible assumption for most EEG signals. In this case, we would use a higher-order low-pass filter with a cut-off frequency of 40 Hz. This would attenuate frequencies above 40 Hz (the attenuation having reached -3 dB at the cut-off frequency). Often, filtering is combined with resampling. By resampling the EEG signal by a factor of 3, the original bandwidth of 125 Hz would be reduced to 41.6 Hz. Playing this signal at 44.1 kHz results in a time-compression factor of almost 530 (i.e., a 24 h registration could be auditioned in less than 3 minutes – assuming a young pair of ears that can still hear up to 20 kHz). High-pass filters are used to limit the lower end of the frequency range of a signal. This is typically needed to remove a DC offset or drift from the signal. Another very interesting use of filtering is to control the masking effect. When removing strong signal components with one or more band-pass filters, lower-level components in the surroundings of the attenuated regions will exceed the altered masking threshold and will thus become audible.

Compress Only if there is No Other Way Dynamic compression is needed when signals show large level variations and when the very loud and the very soft parts should both be made audible. As dynamic compression adds non-linear distortion to the signal, it should only be used if there is no other way (e.g., improving the listening conditions). A typical compressor reduces the dynamic range of a signal if it becomes louder than a set *threshold* value. The amount of level reduction is usually determined by a *ratio* control. For instance, with a ratio of 5:1, if the input level is 5 dB above threshold, the signal will be reduced so that the output level will be only 1 dB above threshold. Compressors usually have controls to set the speed with which they respond to changes in input level (known as *attack time*) and how fast they return to neutral operation once the input level falls below threshold (known as *release time*). Dynamic compression may change the character of a signal in subtle to quite drastic ways depending on the settings used. A typical case for dynamic compression in audification involves seismic signals, which exhibit a very large dynamic range and are therefore usually quantized with 24 or even 32 bits (representing 144 resp. 192 dB of theoretical dynamic range). In rare cases dynamic expansion – the inverse process of dynamic compression – may also be useful when a signal exhibits only very small differences in level and a contrast enhancement is needed to better detect changes.

Use Special Tools in Special Cases Only Besides the conditioning techniques described so far, any imaginable signal-processing algorithm may prove useful for a particular audification task. The simplest imaginable conditioning can be seen in time reversal. Due to

[18] A notch filter (also known as band-stop or band-rejection filter) is a special kind of filter that suppresses frequencies only in an extremely small frequency range.

the asymmetry of temporal masking, the listener may detect more (or other) details in the time-reversed version of a signal. Most of the common audio engineering tools such as gates, noise reduction, frequency shifters, etc. may be adequate in special cases. Reverberation, for instance, another common audio engineering tool, may be helpful in situations where very short transient signals are to be audified. Reverberation prolongs the transients in a way that is rather neutral (if a good algorithm with a defined frequency response is used) and which is familiar to our ears. This effect is perceived very dramatically when using headphones for sound projection, since then the influence of room acoustics of the reproduction space is suppressed and different transients may sound quite alike. Once reverberation is added, they can be distinguished much more easily. Non-linear techniques with drastic effects on the signal, such as ring modulation, which is sometimes used for frequency doubling [25], should best be avoided. The spectral side-effects of these techniques are very hard to control and cannot be predicted easily by the human auditory system. But sometimes even such techniques are useful as in heterodyning bat detectors, which use them due to real-time constraints[19].

Making Use of Advanced Spectral Processing Tools Analysis/resynthesis algorithms, such as the phase vocoder [20] and its recent improvements [50, 12], are an interesting class of signal conditioning tools. The phase vocoder allows for time stretching without affecting the pitch, and pitch scaling without affecting the signal duration. In the case of signal resampling discussed above, time and pitch manipulations are always linked (time compression scales pitches up, time stretching scales pitches down). Despite the improvements of the phase vocoder algorithms, they still produce audible artifacts, but they can be easily detected as such. A special case of improvement concerns the treatment of transient portions in a signal, which cannot be time-stretched by definition. The algorithm [53] detects the transients and treats them specially. Independent control of time and pitch is a powerful tool for audification as the signal can be adapted to the characteristics of the auditory system. Signals with rhythmical structures too fast to perceive in detail can be slowed down without making them unperceivable due to low pitch. The spectral structure of a signal can be transposed to regions where the ear is especially efficient in building gestalts without changing the temporal structure of the signal.

The Sound Projection Stage

In the sound projection stage the conditioned signals may be mapped spatially. This is of interest when audifying more than one channel at a time and when the data set exhibits a spatial structure which should be displayed. Another reason for spatial rendering is to exploit the listener's spatial hearing capabilities in order to assist in the auditory gestalt formation, e.g., through spatial unmasking and spatial auditory scene analysis.

The achievable quality of spatial rendering depends on the sound projection setup and the algorithms used to drive a particular setup. Evidently, the best rendering quality is reached if one speaker is used per signal, in which case rendering consists of a simple assignment of signals to loudspeaker channels. It is best to distinguish between rendering for various configurations of loudspeakers as well as for stereo headphones. Rendering for headphones

[19]Cf. `http://www.bats.org.uk/pages/bat_detectors.html` (accessed Jan 30 2011)

may use binaural synthesis employing head-related transfer functions (HRTF) and eventually room acoustic modeling to position sound sources in a virtual sound scene. The localization quality of a binaural display can be enhanced by employing a head tracker to compensate for the user's head movements. Tracked binaural synthesis allows for a stable rendering of sound source locations (i.e., the sound scene does not move with the user's head movement). Under special conditions, binaural rendering may also be used with stereo loudspeaker projection. In this case, the location of the user's head is constrained to a small region. A typical situation for this type of projection is being seated in front of a computer screen. In order to improve localization of sources from behind the user, a second pair of loudspeakers may be used, placed behind the user's head. In this case, an extension of binaural rendering using crosstalk cancellation has to be employed. If head tracking is available, the crosstalk cancellation can be made dynamic, thus enlarging the available sweet spot considerably [36].

Various rendering techniques for multichannel loudspeaker setups are available ranging from simple panning techniques (e.g., Vector Base Amplitude Panning / VBAP [51]) via Higher-Order Ambisonics (HOA [38]) to wave field synthesis (WFS [7]). WFS is a technique which requires a large number of loudspeakers (up to hundreds) but can achieve a very high quality of rendering – but at a very high cost. HOA and simpler amplitude panning techniques are used for loudspeaker setups in two dimensions (rendering of sources in a plane around the listener, e.g., with a ring of 5 or 8 speakers) and three dimensions (rendering of sources with elevation, e.g., a dome of 24 speakers). Standard formats such as stereo or 5.1 surround may also be used for reasons of compatibility and availability.

12.4 Audification now (1992-today)

The first ICAD in 1992, organized by Gregory Kramer, was not only a kick-off for the succeeding conferences, but also the formation of sonification research as a new discipline. It is therefore appropriate to assume a *caesura* here and to give the audification research after the first ICAD another section in this chapter.

In 1992, audification as a sonification technique received its name [32, p. xxvii] and its definition was quickly refined.[20] As a result professional investigation improved and empirically reliable audification research could develop, especially in the context of the yearly ICAD. Nevertheless, papers explicitly on audification are – as we will see – rare. Even today, audification is usually used only as a mock-up for sonification research, and is not described (or only without much detail) in publications of the ICAD proceedings. Much more vivid are the amateurs' applications of audification to all kinds of data, mostly for science popularization or amusement. This boom of auditory bricolage certainly relates to the development of the computer as an audiovisual display. Notably, the function of audification as a gimmick in mathematical visualization software, such as *Mathematica* ("Play[...]") or *MathTrax* should not be underestimated.

The following paragraphs summarize the few serious scientific works on audification and try to give a little overview of further ideas and applications in different areas of less scientific claim.

[20] Cf. also the *Sonification Report* of 1997, `http://www.icad.org/websiteV2.0/References/nsf.html`, accessed Jan 30 2011 and [27, pp. 35-40]

12.4.1 Scientific Examples

The most recent profound investigation of audification as a technique was carried out by Sandra Pauletto and Andy Hunt (2005), in which they tried to understand the advantages and disadvantages of audification in general [48]. There are many other papers which compare auditory and visual representations and also deal briefly with audification, although they usually do not go into detail. As representative of this type of work, the reader's attention is drawn to [37] and [6].

(i) Medicine: From the abovementioned examples of early audifications (cf. section 12.2), the stethoscope is still in use with great success, even though it functions no longer as a scientific proof, but more as a demonstration of argument. The idea of audifying EEG and nerve measurements developed in the 19th century was almost forgotten for more than a century but is now making a comeback: people such as Gerold Baier and Thomas Hermann are seriously investigating EEG sonifications, even though the technique of audification plays a minor role compared to parameter mapping (sound example **S12.15**).[21] One special publication in this field worth mentioning was delivered by Jesus Olivan, Bob Kemp and Marco Roessen in 2004 [47], in which they used audification to investigate EEG sleep recordings. Besides EEG data, heart rate variability has also attracted the interest of sonification researchers using audification [5].

(ii) Seismology: One area where audification finds a highly promising application is seismology. Chris Hayward brought up the topic at the first ICAD in 1992 (sound example **S12.16**) [25]. Co-operating closely with Gregory Kramer, he carried out an extensive and diligent investigation of the topic and was the first to examine the overall potential of audification in the area of seismics and seismology. Today, there are people still researching in the area, such as Frank Scherbaum from the University of Potsdam or Florian Dombois (sound example **S12.17**), who established what is known today as "Auditory Seismology".[22] And in 2008, Meier and Saranti presented some sonic explorations with seismic data [42].

(iii) Physics: Apart from Hayward's, there are two more examples of the use of audification in the proceedings of ICAD 1992: one on computational fluid dynamics data by McCabe and Rangwalla [41], and one on chaotic attractors by Mayer-Kress, Bargar and Insook [40]. The propositions of these authors have not been followed too far, as far as is known. An exception was the research group of Hans Diebner at ZKM Karlsruhe (D), who audified chaos of all kinds, displaying it at some of the art shows of ZKM. A more famous example of successful audification in Physics is the discovery of quantum oscillations in He-3 atoms, where Pereverzev et al., in 1997, found the searched-for frequency by listening directly to measured data [49]. Another interesting application in the nano sector was presented in 2004 at the Interactive Sonification workshop in Bielefeld, where Martini et al. used audification for "fishing" single atoms [39].

(iv) Stock market: In the area of *abstract data*, the stock market has a major attraction for obvious reasons. Unfortunately, these research results are usually not published. S. P. Frysinger's work of 1990 used audification of stock market data [23], which has been

[21] Cf. [4] and [3]; cf. also [27]; a lot of listening examples can be found at `http://www.sonifyer.org/sound/eeg/` (accessed Jan 30 2011)

[22] For an overview of the history of Auditory Seismology see [15]. For further research look at [13], [14] or [16]. An updated publication list and several sound examples can be found at `http://www.auditory-seismology.org` (accessed Jan 30 2011).

evaluated by Keith Nesbitt and Stephen Barrass [45, 46]. And David Worrall published a paper with a fine overview on the area of sonifying stock market data [62].

(v) Statistics: There was also some interesting work done in the area of high order statistics by Frauenberger, de Campo and Eckel, analyzing statistical properties of time-series data by auditory means [22]. Audification was mainly used to judge skewness and kurtosis.

All in all, one can say that audification has become, over the last ten years, quite common in the scientific community, due to computer programs such as *Mathematica* or *MathTrax*, that easily audify all kinds of data. Nevertheless, the number of researchers that assume audification as a reasonable research method is still small. It is seen more as a nice gimmick in the popularization of science, and here one can find innumerable websites in all scientific domains:

For example, famous are the transposed whale chantings that have been recorded and sold as CDs all over the world. A project, "The dark side of the cell", also received a lot of publicity because membrane oscillations of living cells had been audified.[23] In astronomy, NASA has used audification over the last several years to portray their results in a novel way in order to attract the interest of the general public. In 2004, when Cassini flew in and out of Saturn's ring plane, the press release also contained an audification of radio and plasma wave measurements.[24] Similar audifications can also be found of Titan or general solar wind registrations.[25] NASA even installed an online web radio of real-time audifications of VLF recordings at the Marshall Space Flight Center as part of their teaching program, INSPIRE.[26] Also, astrophysicists, such as Mark Whittle, used audio to successfully market their big bang calculations not only in the scientific community.[27] And there are sound-lovers, like Don Gurnett, who are publishing all kinds of sonifications of astrophysical data on the web.[28] According to Tim O'Brien, astronomers are increasingly listening into stars and other space sounds because "[i]t's interesting in itself [and i]t's also scientifically useful."[29] In Geophysics, we also find educational uses of audification. The USGS and John Louie set up two reasonable websites with audifications of earthquake registrations.[30]

12.4.2 Artistic Examples

The interest in creating new sounds, especially in computer music, led many musicians into audification of all kinds of data. But this section highlights only a very few examples of

[23] `http://www.darksideofcell.info` (accessed Jan 30 2011)

[24] `http://www1.nasa.gov/mission_pages/cassini/multimedia/pia06410.html` (accessed Jan 30 2011)

[25] `http://www.nasa.gov/vision/universe/solarsystem/voyager-sound.html` or `http://www.esa.int/SPECIALS/Cassini-Huygens/SEM85Q71Y3E_0.html` (accessed Jan 30 2011)

[26] `http://www.spaceweather.com/glossary/inspire.html` resp. `http://science.nasa.gov/headlines/y2001/ast19jan_1.htm`. Further VLF-recordings can be found at `http://www-pw.physics.uiowa.edu/plasma-wave/istp/polar/magnetosound.html` (accessed Jan 30 2011)

[27] `http://www.astro.virginia.edu/~dmw8f/index.php`, see also `http://staff.washington.edu/seymour/altvw104.html` (accessed Jan 30 2011)

[28] `http://www-pw.physics.uiowa.edu/space-audio/` (accessed Jan 30 2011)

[29] `http://news.bbc.co.uk/2/hi/science/nature/7687286.stm` (accessed Jan 30 2011)

[30] John Louie `http://crack.seismo.unr.edu/ftp/pub/louie/sounds/` and USGS `http://quake.wr.usgs.gov/info/listen/`. See also the amateur site of Mauro Mariotti `http://mariottim.interfree.it/doc12_e.htm` (accessed Jan 30 2011)

intentional use of audification in the last years, where the resulting sounds have been used without major aesthetic manipulations. Left out is the huge body of works which transform mechanical waves directly into sound waves without any transposition.

An interesting project is "According to Scripture", by Paul DeMarini, who revitalized, in 2002, a series of visual waveform diagrams from the 19th century, that were not registered on a phonograph but drawn directly on paper. Among others, one finds here E.W. Scripture's famous notations from 1853–1890 digitized and reaudified, giving the ear access to the oldest sound registrations ever made.

Christina Kubisch's famous project, "Electrical Walks" (sound example **S12.18**), was first shown in 2004. The visitor receives a headphone that audifies electromagnetic induction from the surroundings, and a city map with a suggested tour.[31] An enormous sound space is opened up by the prepared headphone that restructures the topology of the city.[32]

In 2004, the Australian group *radioqualia* displayed their piece "Radio Astronomy" at the Ars Electronica Festival in Linz (A), where real-time VLF recordings could be listened to. It was a network project, working together with the Windward Community College Radio Observatory in Hawaii, USA, NASA's Radio Jove network, the Ventspils International Radio Astronomy Centre in Latvia and the cultural center RIXC from Riga, Latvia.

Under the motto "art as research", several sound installations of seismological data by Florian Dombois have been shown at Cologne Gallery Haferkamp in 2003 and 2006[33] and [34], and also at Gallery gelbe MUSIK in Berlin 2009 etc. Here, one could listen, for example in "Circum Pacific" (sound example **S12.19**), to five seismic stations monitoring the seismic activity around the pacific plate.

Also at Ars Electronica, the following year 2005, "G-Player" (sound example **S12.20**) was shown, an artwork by Jens Brand from Germany. He uses a topographic model of the earth and imagines satellites behaving like the needle of a record player, so that a topographic cross section (following the flight route) can be directly audified and listened to [30, p. 342f.].

12.5 Conclusion: What audification should be used for

Before applying any sonification technique to a data set, the focus of interest should be considered. Sonification is not a "universal remedy", and if, for example, a visualization has been successful, why invest in another data transformation? One should be aware that listening is quite different from looking, and one should therefore approach a data set acoustically mainly in those cases where visualization usually fails.

Within the sonification techniques, audification is surely the most direct and simple to handle. No sound engines are needed, no instruments, no libraries of samples, no acoustic inputs. Audification, therefore, always bears a fundamental surprise, and the characteristics of the acoustic results are usually hard to foresee. But not every data set is suitable; it should fulfill some preconditions:

1. **number of samples:** Audification requires large quantities of data. The usual resolu-

[31] Cf. the interview in [11].
[32] `http://www.cabinetmagazine.org/issues/21/kubisch.php` (accessed Jan 30 2011)
[33] Cf. `http://www.rachelhaferkamp.eu` (accessed Jan 30 2011)

tion of an audio track is 44100 samples per second. To apply audification to a data set with listenable output, it should contain at least a few thousands of samples.

2. **wave-like signals:** Audification always needs at least one-dimensional data sets interpreted as a time series signal. The data should not have too many breaks or dropouts, and a round curve-like shape usually delivers the best output. The audification approach is most promising when dealing with data following physical laws, especially from elastomechanics. In these cases, audification can be assumed as an extension of the ear, conquering frequencies outside the usual range of perception, comparable to thermo- or x-ray-photography.
3. **complexity:** The ear can cope easily with complex sounds, and audification affords this opportunity much more easily than any other sonification technique. One should, therefore, always supply audification first with all the complexity that is in the data before doing any reduction or filtering.
4. **signal to noise ratio:** The cocktail-party-effect – i.e., the ability to focus one's listening attention on a single talker among a mixture of conversations and background noises, ignoring other conversations – is quite well known. Due to this effect, audification is a good approach for finding hidden signals in a noisy data record, assuming the audified data to be a mixture of unknown sources. The ear is trained to separate different sources, and therefore audification can lead to the discovery of unexpected signals or structural invariants in the time-frequency patterns of sound or all kinds of implausible artifacts that are interspersing.
5. **subtle changes:** Slow changes in data characteristics (e.g., frequencies or rhythms) can be obtained mostly easily due to the high flow rate of samples, whereas driftings are often difficult to recognize and often get lost by high-pass filtering.
6. **simultaneousness of rhythmical patterns:** Cross-correlations usually need a lot of calculation time and are visually difficult to obtain. Here, audification gives good access to follow several signals simultaneously and can demonstrate whether the rhythm is in or out of synchrony.
7. **data screening:** A suitable area for audification are all kinds of screening tasks. It is a good approach for getting a quick overview of the characteristics of different data sets.
8. **classification:** The ear structures signals very differently from the eye. One can, therefore, use audification also for all questions of classification.

12.6 Towards Better Audification Tools

Until recently, software tools that support the process of analytical listening to the degree required for a successful application of data audification have been lacking. This section presents a few general guidelines for the development of such tools. The recent past has seen several initiatives to create an integrated software environment for sonification.[34] As the field

[34]SonEnvir at IEM in Graz (`http://sonenvir.at/`), Denkgeräusche at HKB in Bern (`http://www.sonifyer.org` see also [17]), NASA's xSonify (`http://spdf.gsfc.nasa.gov/research/sonification/sonification_software.html`), Georgia Tech's Sonification Sandbox `http://sonify.psych.gatech.edu/research/sonification_sandbox/`. (All accessed Jan 30 2011)

of sonification is vast, and audification still tends to be undervalued, the particular needs of listening to data as directly as possible are only taken into account to a small degree in these development projects.

As the focus is on listening, all graphical data representations supporting this task have to be part of a general-purpose audification tool, under the condition that they can be switched off, i.e., made invisible, in order not to bias listening. Our tool has to allow us to grow with it, to improve our listening skills by using it and to enable us to extend it if we reach the limits of the possibilities foreseen by the developers. One of the most important features of the tool in question is the possibility of quickly checking ad hoc hypotheses about the problem under examination, as users develop in the process of analytical listening. Transparency of operation and the highest possible implementation quality of the employed signal processing algorithms (e.g., resampling) are the other essential features.

Programmability should be integrated on the level of an extension language, either in textual form, as for instance in SuperCollider,[35] or in graphical form like in Max[36] (or best in both). Extensibility is important to develop custom signal conditioning algorithms and for adding data import routines, as they are needed, when audifying data sets in formats not yet supported by the audification tool. As a central element, the tool has to include a multichannel time domain and frequency domain signal editor (with features as in AudioSculpt[37] and STx[38]). It goes without saying that another important feature of our ideal tool is interactivity. Changing conditioning parameters in real time and efficiently browsing large data sets are important prerequisites to work efficiently with audification.

Once a more powerful software environment is settled the real work can start: audifying all kinds of data. We think, that mechanical waves are most promising and especially natural frequencies. Listening to ultrasonic resonances of, for example, fruits, vegetables, cheese[39] up to infrasonic as in sculptures, buildings, bridges, planetary bodies etc.[40]. There is a whole cosmos of neglected sounds, that wait to be investigated.

Bibliography

[1] Adrian, E. D., & Matthews, B. H. C. (1934). The Berger Rhythm: Potential Changes from the Occipital Lobes in Man. *Brain* 57:355–385.

[2] Auenbrugger, L. (1936). On the Percussion of the Chest (transl. by John Forbes). *Bulletin of the History of Medicine* 4:373–404.

[3] Baier, G. (2001). *Rhythmus. Tanz in Körper und Gehirn*. Hamburg: Rowohlt.

[4] Baier, G., Hermann, Th., Lara, O. M. & Müller, M. (2005). Using Sonification to Detect Weak Cross-Correlations in Coupled Excitable Systems. Paper read at ICAD 2005, July 6–9, at Limerick, Ireland.

[5] Ballora, M., Pennycook, B. W., & Glass, L. (2000). Audification of Heart Rhythms in Csound. In *The Csound Book: Perspectives in Software Synthesis, Sound Design, Signal Processing, and Programming*, edited by R. Belanger. Cambridge (MA): MIT Press.

[35] `http://supercollider.sourceforge.net/` (accessed Jan 30 2011)

[36] `http://cycling74.com/products/maxmspjitter/` (accessed Jan 30 2011)

[37] cf. `http://forumnet.ircam.fr/349.html?\&L=1` (accessed Jan 30 2011)

[38] cf. `http://www.kfs.oeaw.ac.at/content/blogsection/11/443/lang,8859-1/` (accessed Jan 30 2011)

[39] cf. `http://www.sonifyer.org/sound/kartoffel/` or `http://www.sonifyer.org/sound/ausserdem/?id=27`

[40] cf. e.g. `http://www.klangkunstpreis.de/preise_2010.php` or several examples in [34]

[6] Barrass, S., & Kramer, G. (1999). Using Sonification. *Multimedia Systems* 7:23–31.

[7] Berkhout, A. J., De Vries, D. & Vogel, P. (1993). Acoustic Control by Wave Field Synthesis. In *Journal of the Acoustical Society of America*, 93:2764–2778

[8] Bernstein, J., & C. Schönlein. (1881). Telephonische Wahrnehmung der Schwankungen des Muskelstromes bei der Contraction. *Sitzungsberichte der Naturforschenden Gesellschaft zu Halle*:18–27.

[9] Bregman, A. S. (1990). *Auditory Scene Analysis: The Perceptual Organization of Sound.* Cambridge, Massachusetts: The MIT Press.

[10] Cocteau, J. (1983). *Kino und Poesie. Notizen.* Frankfurt a. M.: Ullstein.

[11] Cox, Ch. (2006). Invisible Cities: An Interview with Christina Kubisch. In *Cabinet Magazine* 21

[12] Dolson, M. & Laroche, J. (1999). Improved phase vocoder time-scale modification of audio. *IEEE Transactions on Speech and Audio Processing* 7.3:323–332.

[13] Dombois, F. (1999). *Earthquake Sounds. Volume 1: Kobe 16.1.1995, 20:46 ut.* St. Augustin. (Audio CD)

[14] Dombois, F. (2001). Using Audification in Planetary Seismology. Paper read at 2001 International Conference on Auditory Display, July 29–August 1, at Espoo.

[15] Dombois, F. (2002). Auditory Seismology: On Free Oscillations, Focal Mechanisms, Explosions and Synthetic Seismograms. Paper read at 2002 International Conference on Auditory Display, July 2–5, at Kyoto.

[16] Dombois, F. (2006). Reflektierte Phantasie. Vom Erfinden und Erkennen, insbesondere in der Seismologie. *Paragrana* Beiheft 1:109–119.

[17] Dombois, F. (2008a). Sonifikation: Ein Plädoyer, dem naturwissenschaftlichen Verfahren eine kulturhistorische Einschätzung zukommen zu lassen. In *Acoustic Turn* edited by Petra Maria Meyer. Paderborn: Fink, 91–100.

[18] Dombois, F. (2008b). The 'Muscle Telephone': The Undiscovered Start of Audification in the 1870s. In *Sounds of Science – Schall im Labor (1800–1930)* edited by Julia Kursell. Berlin: Max Planck Institute for the History of Science, Preprint 346:41–45.

[19] Dombois, F., Brodwolf, O., Friedli, O., Rennert, I. & König, Th. (2008). SONIFYER – A concept, a Software, a Platform. Paper read at 2008 International Conference on Auditory Display, June 24–27, at Paris.

[20] Flanagan, J. L. & Golden, R. M. (1966). Phase vocoder, *Bell Syst. Tech. J.* 45:1493–1509.

[21] Frantti, G. E., & Leverault, L. A. (1965). Auditory Discrimination of Seismic Signals from Earthquakes and Explosions. *Bulletin of the Seismological Society of America* 55:1–25.

[22] Frauenberger, Ch., de Campo, A. & Eckel, G. (2007). Analysing Time Series Data. Paper read at 2007 International Conference on Auditory Display, June 26–29, at Montreal, Canada.

[23] Frysinger, S. P. (1990). Applied Research in Auditory Data Representation. Paper read at SPIE/SPSE Symposium on Electronic Imaging, February, at Springfield (VA).

[24] Gelatt, R. (1977). *The Fabulous Phonograph 1877–1977.* 2 ed. London: Cassell.

[25] Hayward, Ch. (1994). Listening to the Earth Sing. In *Auditory Display. Sonification, Audification, and Auditory Interfaces*, edited by G. Kramer. Reading: Addison-Wesley, 369–404.

[26] Hermann, L. (1878). Ueber electrophysiologische Verwendung des Telephons. *Archiv für die gesamte Physiologie des Menschen und der Tiere* 16:504–509.

[27] Hermann, Th. (2002). *Sonification for exploratory data analysis*, PhD thesis, Bielefeld University, Bielefeld, available via `http://sonification.de/publications/Hermann2002-SFE`

[28] Hermann, Th., & Hunt, A. (2004). The Discipline of Interactive Sonification. Proc. Int. Workshop on Interactive Sonification (ISon 2004), Bielefeld, `http://interactive-sonification.org/ISon2004/proceedings/papers/HermannHunt2004-TDO.pdf`

[29] Hermann, Th., & Hunt, A. (2005). An Introduction to Interactive Sonification. *IEEE Multimedia* 12 (2):20–24.

[30] Kiefer, P. (ed.). (2010). *Klangräume der Kunst.* Heidelberg: Kehrer.

[31] Kittler, F. (1986). *Grammophon. Film. Typewriter.* Berlin: Brinkmann & Bose.

[32] Kramer, G. (1994). An Introduction to Auditory Display. In *Auditory Display: Sonification, Audification and*

Auditory Interfaces, edited by G. Kramer. Reading (MA): Addison-Wesley.

[33] Kramer, G. (ed.). (1994). *Auditory Display: Sonification, Audification and Auditory Interfaces*. Reading (MA): Addison-Wesley.

[34] Kunsthalle Bern (ed.). (2010). *Florian Dombois: What Are the Places of Danger. Works 1999-2009*. Berlin: argobooks.

[35] Laënnec, R. Th. H. (1830). *A Treatise on the Diseases of the Chest and on Mediate Auscultation*. Translated by J. Forbes. 3 ed. New York: Wood, Collins and Hannay.

[36] Lentz, T., Assenmacher, I., & Sokoll, J. (2005). Performance of Spatial Audio Using Dynamic Cross-Talk Cancellation. Proceedings of the 119th Audio Engineering Society Convention, NY.

[37] Madhyastha, T. M., & Reed, D. A. (1995). Data Sonification: Do You See What I Hear? *IEEE Software* 12 (2):45–56.

[38] Malham, D. G. (1999). Higher order Ambisonic systems for the spatialisation of sound. In *Proceedings of the International Computer Music conference, Beijing* 484–487.

[39] Martini, J., Hermann, Th., Anselmetti, D., & Ritter, H. (2004). Interactive Sonification for exploring Single Molecule Properties with AFM based Force Spectroscopy. Paper read at International Workshop on Interactive Sonification, January, 8, at Bielefeld.

[40] Mayer-Kress, G., Bargar, R., & Insook, Ch. (1994). Musical structures in Data from chaotic attractors. In *Auditory Display: Sonification, Audification and Auditory Interfaces*, edited by G. Kramer. Reading, Massachusetts, USA: Addison-Wesley.

[41] McCabe, K., & Rangwalla, A. (1994). Auditory Display of Computational Fluid Dynamics Data. In *Auditory Display: Sonification, Audification and Auditory Interfaces*, edited by G. Kramer. Reading, Massachusetts, USA: Addison-Wesley.

[42] Meier, M., & Saranti, A. (2008). Sonic Explorations with Earthquake Data. Paper read at ICAD 2008, 24.–27.6.2008, at Paris, France.

[43] Moholy-Nagy, L. (1923). Neue Gestaltung in der Musik. *Der Sturm* Juli:102–106.

[44] Moritz, W. (2004). *Optical Poetry. The Life and Work of Oskar Fischinger*. Bloomington: Indiana University Press.

[45] Nesbitt, K., & Barrass, S. (2002). Evaluation of a Multimodal Sonification and Visualisation of Depth of Market Stock Data. Paper read at 2002 International Conference on Auditory Display (ICAD), at Kyoto.

[46] Nesbitt, K., & Barrass, S. (2004). Finding Trading Patterns in Stock Market Data. *IEEE Computer Graphics* 24 (5):45–55.

[47] Olivan, J., Kemp, B. & Roessen, M. (2004). Easy listening to sleep recordings: tools and examples. *Sleep Medicine* 5:601–603.

[48] Pauletto, S., & Hunt, A. (2005). A Comparison of Audio & Visual Analysis of Complex Time-Series Data Sets. Paper read at ICAD 2005, 6.–9.7.2005, at Limerick, Ireland.

[49] Pereverzev, S. V., Loshak, A., Backhaus, S, Davis, J. C., & Packard, R. (1997). Quantum oscillations between two weakly coupled reservoirs of superfluid He-3. *Nature* 388:449–451.

[50] Puckette, M.(1995). Phase-locked vocoder. *Proc. IEEE ASSP Conference on Applications of Signal Processing to Audio and Acoustics* (Mohonk, N.Y.).

[51] Pulkki, V. (1997). Virtual sound source positioning using vector base amplitude panning. In *Journal of the Audio Engineering Society*, 45(6):456–466.

[52] Rilke, R. M. (1987). Ur-Geräusch (1919). In *Sämtliche Werke Bd. 6: Die Aufzeichungen des Malte Laurids Bridge. Prosa 1906–1926*. Frankfurt a. M.: Insel.

[53] Röbel, A. (2003). A new approach to transient processing in the phase vocoder, Proc. of the 6th Int. Conference on Digital Audio Effects (DAFx-03), London, UK.

[54] Russcol, H. (1972). *The Liberation of Sound. An introduction to electronic music*. Englewood Cliffs (NJ): Prentice Hall.

[55] Speeth, S. D. (1961). Seismometer Sounds. *Journal of the Acoustical Society of America* 33:909–916.

[56] Schoon, A. (2011). Datenkörper aushorchen. *kunsttexte.de / Auditive Perspektiven* 2.

[57] Schoon, A., & Dombois, F. (2009). Sonification in Music. Paper read at ICAD 2009, 18.–22.5.2009, Copenhagen, Denmark.

[58] Sterne, J. (2003). *The Audible Past: Cultural origins of sound reproduction*. Durham: Duke University Press.

[59] Volmar, A. (2010). Listening to the Body Electric. Electrophysiology and the Telephone in the Late 19th century. *The Virtual Laboratory,* edited by Max-Planck-Institute for the History of Science (ISSN 1866-4784), http://vlp.mpiwg-berlin.mpg.de/references?id=art76 (accessed Aug 9 2011).

[60] Walker, B., & Kramer, G. (2004). Ecological Psychoacoustics and Auditory Display. In *Ecological psychoacoustics*, edited by J. G. Neuhoff. San Diego: Elsevier.

[61] Wedenskii, N. (1883). Die telephonischen Wirkungen des erregten Nerven. *Centralblatt für die medicinischen Wissenschaften* 31 (26):465-468.

[62] Worrall, D. (2008). The Use of Sonic Articulation in Identifying Correlation in Capital Market Trading Data. Paper read at ICAD 2009, 18.–22.6.2009, at Copenhagen, Denmark.

Chapter 13

Auditory Icons

Eoin Brazil and Mikael Fernström

In the early 1980s the first explorations of the sound capabilities of personal computers appeared. While the desktop user interface metaphor emerged as the major visual user interface paradigm, for example Bill Gaver experimented with adding sounds [25]. He called the sounds, *auditory icons*, as they were the auditory equivalent of the visual icons used in the desktop metaphor. The rationale behind Gaver's work was his interpretation of Gibson's ecological theory of visual perception [28] adapted and applied for the design of auditory user interfaces.

Auditory icons mimic everyday non-speech sounds that we might be familiar with from our everyday experience of the real world, hence the meaning of the sounds seldom has to be learnt as they metaphorically draw upon our previous experiences. For example, deleting a document might be represented by the sound of crumpling a piece of paper; an application error may be represented by the sound of breaking glass or a similar destructive sound. Gaver's work on the *SonicFinder* extended the Apple operating system's file management application *Finder* (an integral part of Apple's desktop metaphor) using auditory icons that possessed limited parametric control. The strength of the SonicFinder was that it reinforced the desktop user interface metaphor, which enhanced the illusion that the components of the system were tangible objects that could be directly manipulated.

13.1 Auditory icons and the ecological approach

One way to understand how we can pick up information from our environment and perform actions in that environment is the concept of *affordances*. This concept was originally introduced by James Gibson to describe the relationship between an organism and its environment that potentially allows the organism to carry out actions [28]. Gibson's work was situated in theories of perception, in particular visual perception. Don Norman adapted the term for the context of human-machine interactions [38].

13.1.1 Auditory affordances

The work initiated by VanDerveer [51] on the formulation of an ecological approach to acoustics was continued by Gaver [21] who included theories of human-computer interaction, resulting in the development of the SonicFinder application. He refined the affordance concept in the context of human use of technology [23]. Sound in a user interface can afford users information about the success of an action. It can also support collaboration between users and may be used to supplement visual information.

For auditory icons, the following definition of auditory affordance applies:

> "The fact that sounds are available to the ear implies that information about the sound-producing events is also present. This information is available in the form of higher-level relations among the physical parameters of the sounds that correspond to attributes of their sources. Of particular importance is that these relations remain invariant over other transformations of the sounds if the corresponding source attribute is also unchanging, and change if the source attribute changes. The perceptual system "picks up" this information, actively seeking it and attuning itself to its presence. In particular, the perceptual system is sensitive to "affordances", information specifying the functional relations between the source and the listener. These affordances are partly responsible for the significance of the sounds." [21, p. 20]

Stanton and Edworthy refined the auditory affordances concept, in the context of auditory warnings [47], discussing the concept as something that is perceived but also learnt in a social and cultural context. Additional perspectives on the concept of affordances are discussed in Neuhoff's book on Ecological Psychoacoustics [37]. The ecological psychoacoustics approach uses experiments focused on probing the low level perceptual dimensions of sounds to assist in "*designing new sounds for representing complex information structures*" [30, p. 3160–3161].

13.2 Auditory icons and events

Auditory icons aim to provide an intuitive linkage between the metaphorical model worlds of computer applications by sonically representing objects and events in applications, using sounds that are likely to be familiar to users from their everyday life (sound examples **S13.1**, **S13.9**, **S13.21**)[1]. There are, of course, objects and events that do not have any corresponding sound in the real world and in such cases other forms of iconic representations may be considered (sound example **S13.22**). A summary of approaches on how to link a computer event to an everyday world event is provided by Brazil [10]. Approaches include the use of earcons (see chapter 14), or creative sound design practices for the design of a sound of something that does not exist in the real world. An example of this can be seen in the classic movie from 1977, *Star Wars* [40], where Ben Burtt had to create a new sound for the *light sabers* used by the Jedi knights in the movie. A second example by Ben Burtt from this movie was the R2D2 droid which he stated was "*50% of the droid's voice is generated electronically; the rest is a combination and blending of water pipes, whistles, and*

[1]For a full list of sound examples, see Table 13.1

vocalizations" (sound example **S13.24**). Sound effect design [17] is increasingly using open source software to design ordinary, everyday sounds.

Film sound design often uses the layering of sounds where concrete identifiable sounds are combined with each other or with more abstract sounds to add a richer meaning to the simpler sounds. In the case of Burtt, he created sound effects using familiar animal or machinery sound to ensure was elements of the sound that were recognizable; he then used acoustic manipulations (pitch shifting, filtering, time stretching, amplitude enveloping, etc.) to provide fantastic objects such as *light sabers* with the necessary amount of familiarity and credibility to the listener [54]. Film and media are increasingly influencing sound design, particularly for narrative sounds in areas such as interactive objects [32]. A fundamental difference between auditory icons and earcons is that earcons can be considered to be arbitrary symbolic representations while auditory icons can be regarded as analogical representations.

Description	**Example**	**Type**
Water, splashing on tiled floor	**S13.1**	Recorded
Water, flowing in the River Shannon	**S13.2**	Recorded
Water, filling a plastic bottle	**S13.3**	Recorded
Water, filling a plastic bottle, Sound Object Model, Cartoonification	**S13.4**	Synthesized
Water, filling a plastic bottle, Sound Object Model, Cartoonification	**S13.5**	Synthesized
Water, filling a hand wash basin	**S13.6**	Recorded
Water, dripping tap	**S13.7**	Recorded
Water, boiling	**S13.8**	Recorded
Walking on tarmac	**S13.9**	Recorded
Walking on gravel	**S13.10**	Recorded
Walking down stairs	**S13.11**	Recorded
Vodhran	**S13.12**	Synthesized
Hammering a nail into wood	**S13.13**	Recorded
Sawing a plank of wood	**S13.14**	Recorded
Shoogle	**S13.15**	Synthesized
Dropping one rubber ball on wooden floor	**S13.16**	Recorded
Dropping two rubber balls on wooden floor	**S13.17**	Recorded
Breaking three glasses	**S13.18**	Recorded
Car, starting and idling	**S13.19**	Recorded
Closing a drawer	**S13.20**	Recorded
Closing a door	**S13.21**	Recorded
Software defined buttons in a audio-only user interface	**S13.22**	Synthesized
Sonifying the Body Electric, from Fitch & Kramer ICAD' 92	**S13.23**	Synthesized
R2D2 Droid from StarWars inspired by Ben Burtt's sound design	**S13.24**	Synthesized

Table 13.1: Sound examples for auditory icons: some sounds are discussed in the chapter, additional examples are provided for inspiration.

13.3 Applications using auditory icons

The list of applications in this section is by no means exhaustive, instead a few key examples were chosen to demonstrate what auditory icons can be used for and how they work.

13.3.1 The Sonic Finder

The SonicFinder mapped qualities and quantities of events occurring within a computer to perceptible attributes of everyday sounds. It was the first user interface using auditory icons and was designed as an extension to the existing Finder (file manger) application of Apple's Macintosh operating system. The Finder application was the file manager on the Macintosh and used for organizing, manipulating, creating and deleting files. SonicFinder used digitized recordings of sounds (sound examples **S13.4**, **S13.20**) that were played when the system was used. Most of the user's actions were represented by auditory icons. The complete list of mappings for the SonicFinder is shown in Table 13.2 on page 328. Gaver [22] claimed that the intuitive mappings of auditory icons resulted in an increased feeling of engagement with the metaphorical world of the computer. The SonicFinder application was informally evaluated, and in general, received a positive response. A major challenge for the system was the size of sound files, since data storage and distribution media were very limited at the time [2].

Event to Sound Mappings for the SonicFinder	
Computer Finder Event	***Auditory Icon***
Objects	
Selection	*Hitting Sound*
Type (file, application, folder, disk, trash)	Sound Source (wood, metal, etc.)
Size	Pitch
Opening	*Whooshing Sound*
Size of opened object	Pitch
Dragging	*Scraping Sound*
Size	Pitch
Location (window or desk)	Sound type (bandwidth)
Possible Drop-In ?	Disk, folder, or trashcan selection sound
Drop-In	*Noise of object landing*
Amount in destination	Pitch
Copying	*Pouring sound*
Amount completed	Pitch
Windows	
Selection	*Clink*
Dragging	*Scraping*
Growing	*Clink*
Window size	Pitch
Scrolling	*Tick sound*
Underlying surface size	Rate
Trashcan	
Drop-in	*Crash*
Empty	*Crunch*

Table 13.2: Mappings used in the SonicFinder [22].

There were a number of issues with the mapping in SonicFinder. The first problem occurred when selecting or increasing the size a window in the visual user interface. Physical windows

[2] 1 to 4 MB RAM, 20 MB hard disk, 800 KB floppy disks

in the real world open slowly but windows in a graphical user interface typically zoom in or pop out quickly. This meant that using real sounds of a window's opening or closing would be inappropriate. The SonicFinder used '*whooshing*' sounds, which highlighted the potential for creative sound effects as alternatives when mappings based on real sounds were difficult to find. Another issue was the auditory icon for copying, where a pouring liquid sound was used; to represent the progress of the copy operation rather than the concept of copying.

13.3.2 SoundShark

Following on from the design of the SonicFinder, Gaver worked with a colleague to create the *SoundShark* [26] application. This expanded upon an earlier multiprocessing, collaborative environment, *SharedARK* [46]; by adding auditory icons to create a new system. SharedARK was a collaborative application designed as a virtual physics laboratory for distance education [46]. The auditory icons represented user interactions, ongoing processes and modes. They were designed to support navigation and to improve awareness of other users' activities in the system. Auditory icons were used to represent the activity of ongoing processes even when not within a visible window or view on the screen. This improved co-ordination between collaborators who could not see each other but who could still hear each other via the application. The distance between a user's cursor in the system and the source of a sound was represented by changing the relative loudness and low-pass filtering of the auditory icons. System modes were represented by low volume background sounds.

13.3.3 ARKola simulation

The ARKola simulation was an exploration of how auditory icons can be used to facilitate collaboration between people controlling a process [27]. It simulated the operation of a soft drink bottling plant with a single assembly line with nine *machines* for the different processes involved in the bottling processes from cooking, bottling, provision of supplies, to the financial tracking of the processes. The simulation was designed so that the graphical representation would require two full screens, i.e. a single screen represented half of the processes. The idea was to see how well an auditory display worked for monitoring a process and acting upon events arising in the simulation, including machine breakdown. Each machine had a unique auditory icon representing its function. The rate for each machine was represented by repetition rate of the sounds for the particular machine, while problems were signaled using various alarm sounds such as breaking glass and overflowing liquid.

This system used up to 14 simultaneous auditory icons, designed to maximize discrimination and to be semantically related to the events they represented. The plant's auditory display created a dynamic soundscape enabling users to understand the complex process of the simulated plant. The system was evaluated to explore the simulation as a collaborative process between the two participants, each focused on half of the processes. The results found that sound led to improved collaboration as the participants could directly hear the activity status of their partner's half of the plant.

13.3.4 Sonification of vital signs

Fitch and Kramer developed a simulator with eight continuously changing variables representing different vital signs of a patient. They found that subjects (students who had received a short training session as anesthesiologists) performed faster and with fewer errors when using auditory display compared to visual display, especially with multivariate changes [19]. The auditory display method used was a combination of iconic sounds and symbolic sounds. The mappings used in this system (sound example **S13.23**) are shown in Table 13.3.4.

Physiological variable	**Mapped to**
Heart rate	Rate of heart-like sound
Breathing	Rate of breath-like sound
CO2 level	Change in timbre of heart sound
Body temperature	Center-frequency of breath sound
Systolic blood pressure	Change of pitch of heart sound
AV dissociation	Random modulation of A-pulse in heart sound
Fibrillation	Random modulation of both A and V pulses in heart sound
Reflex	High FM tone on/off

Table 13.3: The mapping used by Fitch and Kramer [19] for the training system for anesthesiologists.

13.3.5 Mobile devices

Shoogle [55] was an experimental application developed for mobile devices such as smartphones. The inspirational metaphor for the design was from the action of shaking containers or objects to determine if they were full or empty and to get an approximation for the amount

in the container (sound example **S13.15**). A real-life example of this sort if behavior is where a box of a matches is shaken to determine if it is empty, if there are a few matches inside or if it is full. Received messages (such as SMS or email) were represented by bouncing sounds, and one of the mappings modeled ceramic marbles bouncing around in a metal box. When the user handled or moved the mobile device, data from the built-in sensors for acceleration was used to excite sound object models with the number of sound objects being mapped to the number of messages. A further aspect of this mapping used different timbres (i.e., the sound of different materials) to represent the sender's grouping or domain (e.g., colleague, friend, family, unknown) with larger objects (low-pitched) being mapped to longer messages and smaller objects (high-pitched) being mapped to shorter messages. By shaking the mobile device, users could estimate how many messages they had and the size of messages. Shoogle demonstrated an *eyes-free* interface responding to users' gestures through an interactive sonification using auditory icons.

13.4 Designing auditory icons

Auditory icons can be created, generated and controlled in different ways. In the simplest case, a recording of the particular everyday sound is made and stored as a sound file for use in the application when the signified event occurs. This approach can be extended to numerous sounds, either with sequential single sounds or with parallel multiple sounds being played to represent complex events and actions within a user interface. These recordings give a high-fidelity exact reproduction of the sound, however the lack of variety in the reproduction raises the potential of annoyance and limits the amount of information that can be conveyed. Beyond the simple case, applications can use either synthesized sounds or multiple versions of a sound recording to represent multiple levels of a parameter within the user interface.

13.4.1 Parametric auditory icons

As everyday sounds are quite expressive and can be used to communicate multiple dimensions simultaneously, there is the possibility of creating and using parametric auditory icons, e.g., changing sizes of objects, or the rate of pouring a liquid into a container. Simple forms of parameterization include the changing of loudness, varying the playback rate (changing the pitch) or lowpass filtering. This can be achieved in a limited fashion by processing the playback of the audio file to represent objects and events of different size and distance. However, this only works in a narrow range as the sounds processed in such a fashion start to sound unnatural or even lose identifiability when, for example, playback rate is changed too much. Another approach to developing parametric auditory icons is to utilize different recordings of the event, e.g., bouncing a small ball, a medium sized ball and a large ball (sound examples **S13.16**, **S13.17**). A *size* parameter can be used to determine which of the sound files to play. One downside of this approach is that it requires additional storage due to the necessity of multiple files and there may be interpolation issues between parameter values, particularly when combining multiple sound files with processing of the audio file.

13.4.2 Synthesizing auditory icons

Sound synthesis is another approach to creating auditory icons, which is sub-divided into two dominant styles. A signal-based approach, with homomorphic spectral modulations to mimic everyday sounds [43, 44], or a physical modeling approach simulating the propagation of acoustic energy through a model of an object [45, 50, 49]. This approach is based on the use of real-time mathematical simulations or models of real-world events. As these simulations are executed in real-time, it is possible to change the parameters of the simulated event. An example might be where the model represents the sound of a marble rolling across a table and where the size or speed of the marble can be varied through its parameters with the sound changing accordingly. For a further discussion of sound synthesis for auditory display, see chapter 9.

Auditory icons can be understood within two categories of sound object, as either *fully formed objects* or as *evolutionary objects* [12]. In the case of *fully formed objects*, the variables are all known at the instantiation of the object so that when the sound is produced it occurs from beginning to end, for example like playing back a record. This means that the sound is in essence immutable once it has been instantiated. This differs from evolutionary objects where the variables controlling the properties of the sound can be updated whilst the sound is playing. An analogy for these categories can be made by considering the difference between hitting a glass with a spoon and the filling of a container with a liquid. The hitting action creates a sound but you have no control of this sound after the hitting action whilst in the case of the filling action you can change the rate of pouring continuously. This separation can be in terms of a discrete sound versus a continuous sound. The view of an auditory icon as an evolutionary object raises the argument as to whether this type of sound would be better classified under the heading of an *interaction sonification*, see chapter 11.

An example of this kind of hybrid system was the *Ballancer* experiment [41, 39], shown in Figure 13.1. It used parametric control of an *evolutionary object* for the sound model in a simulated task of balancing of a virtual ball on a real stick. The sound of the *evolutionary object* mimicked the sound of a rolling steel ball, hence it was an iconic representation. The equilibrium task explored by the Ballancer system showed that a well designed sound object model could improve performance and the illusion of substance in continuous interaction tasks.

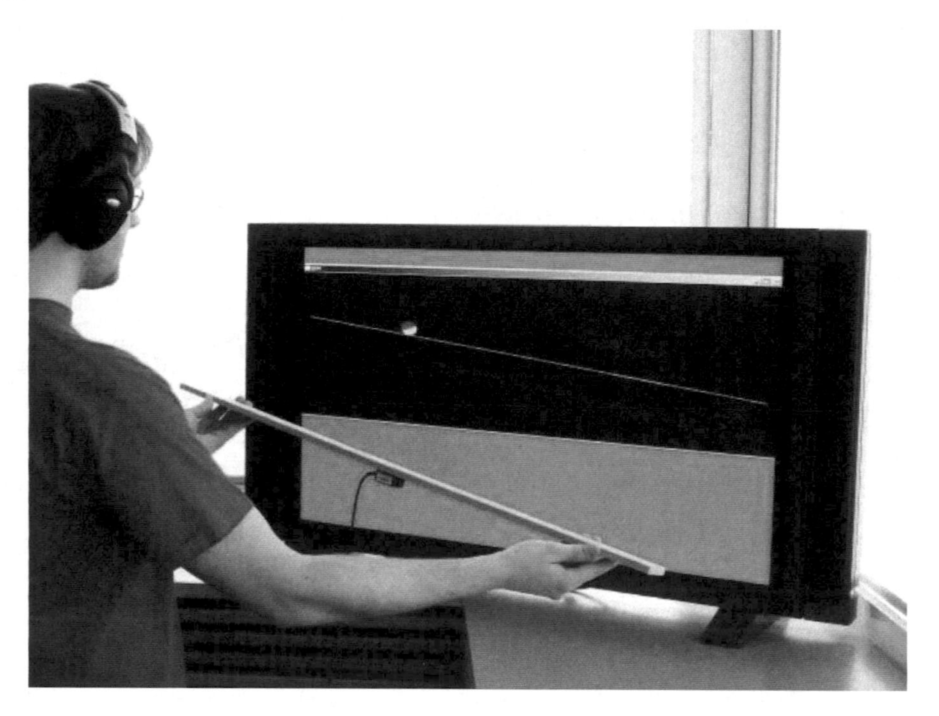

Figure 13.1: The Ballancer auditory equilibrium task. The screen displays a virtual ball and auditory feedback helps a user position the virtual ball in the specified spot on the stick. The physical stick is used to control the movement of the virtual ball.

It has been shown that it is possible to use synthesized parametric auditory icons to create a user interface without any direct visual component [18]. Using a touch sensitive tablet, Fernström et al designed an experiment where users were to find invisible software defined buttons (soft-buttons). The soft-buttons in this experiment reacted when a user's fingers swept over the area occupied by a button, with a friction-like sound being produced, including a faint click when crossing over the edge of a button area. The auditory icons with the friction sound object model responded continuously and directly to the user's actions, giving the participants a pesudo-haptic experience of touching buttons. When a user's fingers were between buttons, no sound was played. The participants in the experiment could easily find the buttons and draw the shape of the button layout.

Cartoonification of auditory icons

All objects resonate if excited by an external force, e.g. hitting, scraping or rolling. The resonances in objects are modal, i.e. the acoustic energy propagating through the object is reflected between edges and surfaces of the object. Complex and compound objects have more modes [1]. It has been shown that it is possible to reduce the number of modes when modeling a sound object while retaining an acceptable degree of indentifiability, as long as the macro-temporal patterns remain intact [29] [53]. This implies that cartoonification of sound object models [24] can make the models more computationally efficient and may potentially make these sounds more distinguishable than *real sounds* being used as auditory icons. This type of caricaturization can be seen as ignoring aspects of a sound object model whilst emphasizing other aspects of the same model. The aim is to improve recognition by exaggerating those features of the model to further distinguish these types of sound model from real-world sounds (sound examples **S13.12** and **S13.22**).

13.4.3 Choosing sounds for auditory icons

The main issue with auditory icons is that they have to be easily identifiable and understood as everyday sounds. At first, it may appear to be easy but many everyday sounds can be heard differently depending on their context. For example, the sound of frying and rain may sound similar. If we mix the sound of rain or frying with the sound of a clashing plate and cutlery, the subjective context is more likely to be a kitchen and frying becomes a more likely response in a listening test. Mynatt [35] discussed the recognition problem when choosing sounds for the interface. It is an art with many hidden dangers and dependent upon the skills of the designer. She developed a set of guidelines for designing auditory icons suggesting four factors: identifiability, conceptual mapping, physical parameters, and user preference, that influence the usability of auditory icons [36, p. 71].

1. Choose short sounds that have a wide bandwidth, and where length, intensity, and sampling quality are controlled. The set of sounds should represent the variety and meaning needed for the anticipated design space.
2. Evaluate the identifiability of the auditory cues using free-form answers.
3. Evaluate the learnability of the auditory cues that are not readily identified.
4. Test possible conceptual mappings for the auditory cues.
5. Evaluate possible sets of auditory icons for potential problems with masking, discriminability and conflicting mappings
6. Conduct usability experiments with interfaces using the derived auditory icons.

In addition to these guidelines, more generalized design advice was given, such as

> "... advocated evaluating auditory cues independently from the intended interface. These experiments were useful in that they highlighted general design guidelines. Like many interface design tasks, it is difficult to design components of the interface separate from the context of the interface. One reason this statement is true for designing auditory icons is that the icons must be designed as a cohesive set ... The design guidelines for controlling the length and "complexity" of sounds are useful when comparing relative differences between sounds. Another difficulty is choosing sounds for similar concepts in the user interface ... There is little chance that a successful set of icons would result from designing the auditory icons independently of each other." [36, p. 87]

13.4.4 Methods and frameworks designing auditory icons

Beyond Mynatt's guidelines, there are several methods and frameworks available to help auditory display designers select auditory icons for better identification, to elicit more meaningful mappings for the auditory icon; and to determine what type of expression (i.e., recorded, parameterized, synthesized, or hybrid) is best used within the auditory display. A review of relevant methods covering issues such as subjective experience, sound identification, confusion of sounds, cognition of sound and pragmatic mental models can assist in designing better auditory icons. James Ballas and his collaborators investigated a number of factors that have significant influence over how we identify brief everyday sounds in listening tests [7, 5, 3, 4]. Based on listening tests, they showed how a *Measure of Causal Uncertainty* could be calculated to rank sounds in terms of identifiability and they found that the subjective context for sounds affected identifiability [6]. Additional approaches for the subjective classification of sounds include *similarity ratings/scaling* [9, 42] and *sonic maps* combined with '*ear-witness accounts*' [14]. Stephen Barrass [8] developed a general framework for designing auditory display systems and applications, *TaDa* (Task and Data analysis of information requirements). A traditional task analysis was carried out followed by data characterization. He then suggested a case-based tool, *EarBenders* to investigate potentially semantic links to the application domain, i.e., construction of an interface metaphor. This was based on Erickson's work [16], using storytelling as a way to describe tasks. Barrass created a database with short stories with everyday observations of situations where sounds were significant to a number of tasks. When searching for a suitable sound to represent a user activity, a good starting point was to find a description in the database matching the intended description. Other approaches have been proposed, for example *paco* design patterns [20]; and methodologies for designing emotional interactive artifacts [15] or functional artifacts [52]. The *repertory grid* technique can be used to '*build up mental maps of the clients' world in their own words*' [48], with similarity ratings/scaling methods used for exploring attributes or perceptual space for a set of stimuli. The *similarity scaling technique* [11] (a derivative of the *similarity rating technique* [34]) presents sounds in listening tests and uses multidimensional scaling or sorting, rather than a single dimension at a time, to get similarity ratings. Several of these methods can be combined to complement each other, as proposed by Brazil [10], for a better understanding of the design space.

Another domain that can provide insights and techniques for auditory icon design is cinematographic sound design. This has inspired sonic methodologies including Back's micro narratives [2] and Hug's design oriented approach [32], using narratives to support the design of interactive sonic artifacts. Cinematic sound design concepts such as '*ergo audition*" regard sound making as an expressive act [13], the experience of hearing oneself acting and the acoustic manifestation of this influence on the world. These ideas can assist in moving beyond "*magical*" or anthropomorphized interfaces of procedural interaction styles [31].

13.5 Conclusion

This chapter has explained what auditory icons are, and how they can be designed, evaluated and applied. Auditory icons draw upon our familiarity with sounds from our everyday experience of the real world. In their simplest form, these can be a small number of recorded everyday sounds that represent user actions with metaphorical objects in a user interface. The next form or level of auditory icon allows for greater granularity of representation through the ability to display levels; this is achieved through the use of recorded or synthesized parametric auditory icons. The most complex and expressive form of auditory icons, moves the sound into the domain of interactive sonification where a continuous representation is possible, allowing complex user gestures and processes to be displayed. There are a number of methods to test identifiability and how well interface metaphors and mappings work. The growth of mobile computing and auditory displays in this context has been supported by a renewed interest in techniques and approaches [10, 33] to determining the meaning attributed by listeners to sounds. Many of these methods can be used in combination for a deeper understanding of the design space for auditory icons. They can deepen the understanding of the salient perceptual and cognitive aspects of the sounds for the specific context; in turn this can help create more meaningful mappings and auditory icons.

Auditory icons were historically used to complement graphical user interfaces, to reinforce a desktop metaphor. However, since then, they have been applied across a range of interfaces and domains, most recently in the areas of mobile and wearable computing. The growth in ubiquity and ubiquitous forms of computing will demand new interaction mechanisms and methods for the control of such interactions (e.g., multi-touch, free gestures). Auditory icons are likely to find even more use to support and augment these new interfaces as they can provide intuitive, yet complex mappings between the sound and the action or interface.

Bibliography

[1] F. Avanzini, M. Rath, D. Rocchesso, L. Ottaviani, and F. Fontana. *The sounding object*, chapter ??Low-level models: resonators, interactions, surface textures. Mondo Estremo, Firenze, Italy, 2003.

[2] M. Back and D. Des. Micro-narratives in sound design: Context, character, and caricature in waveform manipulation. In *International Conference on Auditory Display ICAD-96*, 1996.

[3] J. A. Ballas. Common factors in the identification of an assortment of brief everyday sounds. *Experimental Psychology: Human Perception and Performance*, 19(2):250–267, 1993.

[4] J. A. Ballas. The interpretation of natural sounds in the cockpit. In N.A. Stanton and J. Edworthy, editors, *Human Factors in Auditory Warnings*, pages 91–112. Ashgate, Aldershot, Hants, UK, 1999.

[5] J. A. Ballas and J. H. Howard. Interpreting the language of environmental sounds. *Environment and Behaviour*, 1(9):91–114, 1987.

[6] J. A. Ballas and R. T. Mullins. Effects of context on the identification of everyday sounds. *Human Performance*, 4:199–219, 1991.

[7] J. A. Ballas, M. J. Sliwinsky, and J. P. Harding. Uncertainty and response time in identifying non-speech sounds. *Acoustical Society of America*, 79, 1986.

[8] S. Barrass. *Auditory Information Design*. Phd thesis, Australian National University, 1997.

[9] T. Bonebright. Perceptual structure of everyday sounds: A multidimensional scaling approach. In J. Hiipakka, N. Zacharov, and T. Takala, editors, *ICAD 2002*, pages 73–78, Helsinki, Finland, 2001. Laboratory of Acoustics and Audio Signal Processing and the Telecommunications Software and Multimedia Laboratory, Univeristy of Helsinki, Espoo, Finland.

[10] E. Brazil. A review of methods and frameworks for sonic interaction design: Exploring existing approaches. *LNCS - Proceedings of the Auditory Display Conference 2009*, 5954:41–67, 2010.

[11] E. Brazil, M. Fernström, and L. Ottaviani. A new experimental technique for gathering similarity ratings for sounds. In *ICAD 2003*, pages 238–242, 2003.

[12] W. Buxton, W. Gaver, and S. Bly. *Auditory interfaces: the Use of non-speech audio at the interface*. Unpublished manuscript, 1994.

[13] M. Chion. *Audio-Vision: Sound on Screen*. Columbia University Press, 1st edition, 1994.

[14] G. W. Coleman. *The Sonic Mapping Tool*. Phd thesis, University of Dundee, August 2008.

[15] A. DeWitt and R. Bresin. Sound design for affective interactive. *Proc. ACH 2007 LNCS 4738*, pages 523–533, 2007.

[16] T. Erickson. Design as storytelling. *Interactions*, 3(4):30–35, 1996.

[17] Andy Farnell. *Designing Sound*. M.I.T. Press, 1st edition, 2010.

[18] J. M. Fernström, E. Brazil, and L. Bannon. HCI design and interactive sonification for fingers and ears. *IEEE Multimedia*, 12(2):36–44, 2005.

[19] W. T. Fitch and G. Kramer. Sonifying the body electric: Superiority of an auditory over visual display in a complex, multivariate system. In G. Kramer, editor, *Auditory Display: Sonification, Audification and Auditory interfaces*, pages 307–326. Addison-Wesley Publishing Company, Reading, MA, USA, 1994.

[20] C. Frauenberger and T. Stockman. Auditory display design—an investigation of a design pattern approach. *International Journal of Human-Computer Studies*, 67(11):907–922, November 2009.

[21] W. W. Gaver. *Everyday Listening and Auditory Icons*. Phd thesis, Cognitive Science and Psychology, San Diego, University of California, 1988.

[22] W. W. Gaver. The Sonic Finder: An interface that uses Auditory Icons the use of non-speech audio at the interface. In *ACM CHI '89*, Austin, Texas, USA, 1989. ACM Press.

[23] W. W. Gaver. Technology affordances. In *CHI'91*, pages 79–84, New Orleans, Louisiana, USA, 1991. ACM Press.

[24] W. W. Gaver. How do we hear in the world? explorations of ecological acoustics. *Ecological Psychology*, 5(4):285–313, 1993.

[25] W. W. Gaver. Using and Creating Auditory Icons. In G. Kramer, editor, *Auditory Display: Sonification, Audification and Auditory interfaces*, pages 417–446. Addison-Wesley Publishing Company, Reading, MA, USA, 1994.

[26] W. W. Gaver and R. Smith. Auditory Icons in large-scale collaborative environments. In D. Diaper, editor, *INTERACT'90*, pages 735–740. Elsevier Science Publishers B.V. (North-Holland), 1990.

[27] W. W. Gaver, R. Smith, and T. O'Shea. Effective sounds in complex systems: the arkola simulation. In *CHI'91*, pages 85–90, New Orleans, Louisiana, USA, 1991. ACM Press.

[28] J. J. Gibson. *The Ecological Approach to Visual Perception*. Lawrence Erlbaum Associates Inc. Publishers, Hillsdale, NJ, USA, 10th (1986) edition, 1979.

[29] B. Gygi. *Factors in the Identification of Environmental Sounds*. Phd thesis, Dept. of Psychology, Indiana University, 2002.

[30] B. Gygi and V. Shafiro. General functions and specific applications of environmental sound research. *Frontiers in Bioscience*, 12:3152–3166, May 2007.

[31] D. Hug. Genie in a bottle: Object-sound reconfigurations for interactive commodities. In *Proc. of Audio Mostly Conference*, pages 56–63, Pitea, Sweden 2008.

[32] D. Hug. Using a systematic design process to investigate narrative sound design strategies for interactive commodities. In *ICAD 2009*, pages 19–26, Copenhagen, Denmark, 2009.

[33] G. Lemaitre, O. Houix, N. Misdariis, and P. Susini. Listener expertise and sound identification influence the categorization of environmental sounds. *Journal of Experimental Psychology: Applied*, 16(1):16–32, March 2010.

[34] S. McAdams, S. Winsberg, S. Donnadieu, G. D. Soete, and J. Krimphoff. Perceptual scaling of synthesized musical timbres: common dimensions, specificities and latent subject classes. *Psychological Research*, 58:177–192, 1995.

[35] E. D. Mynatt. Designing with Auditory Icons. In G. Kramer and S. Smith, editors, *Second International Conference on Auditory Display (ICAD '94)*, pages 109–119, Santa Fe, New Mexico, 1994. Santa Fe Institute.

[36] E. D. Mynatt. *Transforming Graphical Interfaces into Auditory Interfaces*. Phd thesis, Georgia Institute of Technology, 1995.

[37] J. Neuhoff. *Ecologial Psychoacoustics*, chapter Ecologial Psychoacoustics: Introduction and History, pages 1–13. Elsevier, London, 1st edition, 2004.

[38] D. Norman. *The Psychology of Everyday Things*. Basic Books Inc., NY, USA, 1988.

[39] M. Rath and D. Rocchesso. Continuous sonic feedback from a rolling ball. *IEEE MultiMedia*, 12(2):60–69, 2005.

[40] J. W. Rinzler. *The Sounds of Star Wars*. Chronicle Books, San Francisco, CA, USA, 2010.

[41] D. Rocchesso. Physically-based Sounding Objects, as we develop them today. *Journal of New Music Research*, 33(3):305–313, 2004.

[42] G. P. Scavone, S. Lakatos, and P. R. Cook. Knowledge acquisition by listeners in a source learning task using physical models. *J. Acoustic Society of America*, 107(5):2817–2818, May 2000.

[43] X. Serra. *A system for sound analysis/transformation/synthesis based on a deterministic plus stochastic decomposition*. Phd thesis, Dept. of Music, CCRMA, Stanford University, 1989.

[44] X. Serra. *Musical Sound Modeling with Sinusoids plus Noise*, pages 91–122. Swets and Zeitlinger, 1997.

[45] J. O. III Smith. Physical modeling using digital waveguides. *Computer Music*, 4(16):74–91, 1992.

[46] R. Smith. A prototype futuristic technology for distance education. In *Proceedings of the NATO Advanced Workshop on New Directions in Educational Technology*, Cranfield, UK, 1988.

[47] N. Stanton and J. Edworthy. Auditory affordances in the intensive treatment unit. *Applied ergonomics*, 5(29):389–394, 1998.

[48] O. Tomico, M. Pifarré, and J. Lloveras. Needs, desires and fantasies: techniques for analyzing user interaction from a subjective experience point of view. In *Proceedings of NordiCHI 2006*, pages 4–9. COST294-MAUSE, ACM Press, 2006.

[49] K. van den Doel, P. Kry, and D. K. Pai. FoleyAutomatic: Physically-based sound effects for interactive simulation and animation. In *Pric. of SIGGRAPH 2001*, pages 537–544, Los Angeles, CA, USA, 2001. ACM Press/ACM SIGGRAPH.

[50] K. van den Doel and D. K. Pai. Synthesis of shape dependent sounds with physical modeling. In *Proceedings of the International Conference on Auditory Displays 1996*, 1996.

[51] N. J. Vanderveer. *Ecological Acoustics: Human Perception of Environmental Sounds*. Phd thesis, University Of Cornell, 1979.

[52] Y. Visell, K. Franinovic, and J. Scott. Closing the loop of sound evaluation and design (closed) deliverable 3.2 experimental sonic objects: Concepts, development, and prototypes. FP6-NEST-PATH project no: 29085 Project Deliverable 3.2, HGKZ (Zurich), 2008.

[53] W. Warren and R. Verbrugge. Auditory perception of breaking and bouncing events: A case study in ecological acoustics. *Auditory perception of breaking and bouncing events: A case study in ecological acoustics*, 10(5):704–712, 1984.

[54] W. Whittington. *Sound Design and Science Fiction.* University of Texas Press, Austin, Texas, USA, 2007.

[55] J. Williamson, R. Murray-Smith, and S. Hughes. Shoogle: Excitatory multimodal interaction on mobile devices. In *Proceedings of the SIGCHI conference on Human factors in computing systems*, pages 121–124, San Jose, California, USA, 2007. ACM Press.

Chapter 14

Earcons

David McGookin and Stephen Brewster

14.1 Introduction

In Chapter 13 Auditory Icons were introduced. These short, environmental sounds are useful to represent iconic information about operations and actions in a user interface. Auditory Icons require there to be an existing relationship between the sound and its meaning, something that may not always exist. In such cases, it may be better to employ Earcons. Blattner *et al.* [5] defined Earcons as: *"non-verbal audio messages used in the user-computer interface to provide information to the user about some computer object, operation, or interaction"*. Brewster [8] further refined this definition as: *"abstract, synthetic tones that can be used in structured combinations to create auditory messages"*. More concretely, Earcons can be thought of as short, structured musical messages, where different musical properties of sound are associated with different parameters of the data being communicated. The key difference between these and Auditory Icons is that there is no assumption of an existing relationship between the sound and the information that it represents. This relationship must, at least initially, be learned. Auditory Icons and Earcons are complementary in an auditory display; both may be useful in the same situations, but with different advantages and disadvantages (see Section 14.4 for more discussion on this).

Although Earcons have only been developed and employed in human computer interfaces over the last twenty years, their core features are much older. In the 19th century American soldiers used bugles to broadcast orders and information to troops in camps and on the battlefield. Before the advent of radios, bugles were a primary means of widely distributing orders. Different melodies represented different orders and information: the mail arriving, mealtime or that the camp was under attack [61]. The use of an auditory display allowed clear broadcast over a wide area and, due to the arbitrary nature of the sound and its meaning, a certain security in communication.

Today Earcons are used in a variety of places. Onboard aircraft for example, it is often necessary for cabin crew to communicate with each other or to be alerted to events requiring attention (e.g., a passenger requesting assistance). In such situations a sequence of tones is played whose meaning has been learned by the crew. These Earcons allow discrete communication without disturbing passengers. Mobile telephones represent another example use of Earcons. Mobile manufacturers typically provide the ability to associate different ringtones with different callers or groups of callers. These allow the recipient to be aware of who is calling without the need to visually attend to the device. Since the callers cannot be known by the manufacturer, each user must make their own mappings between the ringtones and callers. This means that the overall relationship between the ringtones and the callers is essentially arbitrary.

14.2 Initial Earcon Research

Earcons were initially proposed by Blattner, Sumikawa and Greenberg [5]. Their work built upon existing research in auditory warnings for safety critical applications such as intensive care units [50], as well as existing visual icon research by Marcus [38]. They proposed that Earcons be composed of motives: *"brief successions of pitches arranged to produce a rhythmic and tonal pattern sufficiently distinct to function as an individual recognizable entity"*. Motives have long been used in music. The composer, Sergei Prokofiev, employed leitmotifs in his composition "Peter and the Wolf" to indicate the different characters in the story. Blattner *et al.* [5] proposed that using motives allowed messages to be constructed systematically. Systematic combination or manipulation of motives would change their meaning. More radical changes increased the dissimilarity between the motives, allowing them to be grouped into families (e.g., a set of motives representing computer errors could be classed as a family). In their definition, rhythm and pitch were fixed parameters of the motive, and motives with different rhythms and pitches represented different families. Blattner *et al.* [5] proposed that the motives be manipulated using commonly understood musical principles, such as changes in timbre, dynamics and register to form variants and related members of the motive family. By learning the ways in which these Earcons were manipulated, the use of systematic motive manipulation could ease learnability. Blattner, Sumikawa and Greenberg [5] proposed four different ways in which motives could be manipulated to form families of Earcons: One-element Earcons, Compound Earcons, Transformational Earcons and Hierarchical Earcons. The following sections look at each of these in turn.

14.2.1 One-Element Earcons

One-element Earcons are the simplest type and can be used to communicate a single parameter of information. They may be only a single pitch or have rhythmic qualities. In either case, the one-element Earcon, unlike the other three types, cannot be further decomposed [5]. In many ways, one-element Earcons are like non-parameterized Auditory Icons, except they use abstract sounds whose meaning must be learned as opposed to the intuitive meaning of Auditory Icons. One-element Earcons are analogous to the SMS arrival sound on mobile telephones. Whilst there may be different sounds to indicate messages from different types of people, work, home etc., there is no structured relationship between the sounds. Each relationship is unique and its meaning must be individually learned. For large datasets, or in cases where more than one parameter of the data must be communicated, the number of sounds, and mappings of data to those sounds, could become extremely large. The following three types of Earcon attempt to provide solutions to such situations.

14.2.2 Compound Earcons

Compound Earcons are formed by concatenating one-element Earcons, or any other form, together to create more meaningful messages (see Figure 14.1). In many ways they are analogous to forming a phrase out of words, where one-element Earcons represent words and compound Earcons represent phrases. For example, three one-element Earcons representing "save", "open" and "file" can form compound Earcons by being played sequentially to form Earcons for the "open file" and "save file" operations [17] (hear sound examples **S14.1** - **S14.2**). When using compound Earcons it is important to consider Earcon length as the messages can easily become too long to be usable. Blattner *et al.* [5] proposed that each motive should be composed of no more than four individual notes so as to balance between excessive length and forming a melodic pattern.

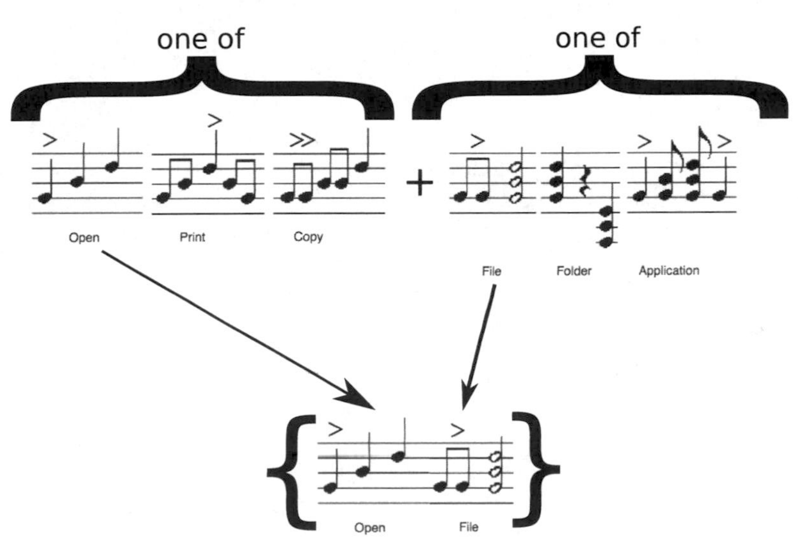

Figure 14.1: An example of how compound Earcons can be formed to create richer messages. The operation and object Earcons can be compounded to create multiple different messages. Adapted from Brewster [8].

14.2.3 Transformational Earcons

Transformational Earcon families are constructed around a "grammar" or set of rules, where there exists a consistent set of structured symbolic mappings from individual data parameters (such as file type) to individual sound attributes (such as timbre). Specific values of data parameters (e.g., a paint file) are then mapped to specific values of the corresponding auditory attribute (e.g., a piano timbre). Figure 14.2 shows how a set of transformational Earcons representing theme park rides is constructed. Each Earcon represents a theme park ride encoding three data parameters (type of ride, intensity and cost). Each timbre, melody and presentation register can be "mixed and matched" to create a set of 27 individual Earcons that represent all combinations of the data parameters (hear sound examples **S14.3**, **S14.4**, **S14.5**, **S14.6**, **S14.7**).

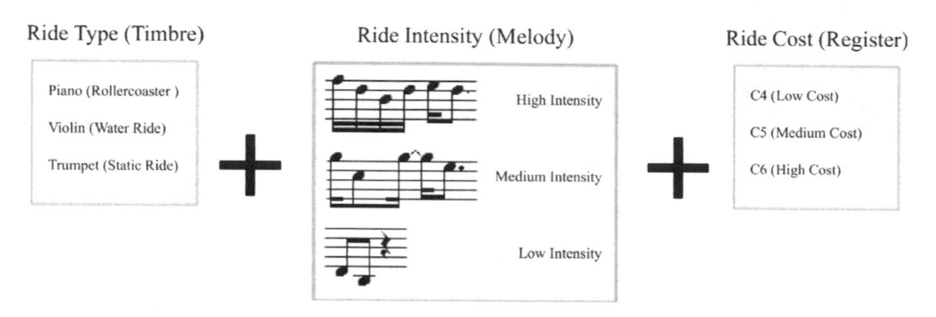

Figure 14.2: An example family of transformational Earcons representing attributes of theme park rides. Each data attribute is mapped to an auditory parameter. Adapted from McGookin [44]. Individual values for Type, Intensity and Cost can be "mixed and matched" to create a set of 27 unique Earcons.

Due to the consistency in the mappings used, a large set of complex auditory messages can be represented by a small set of rules. This makes learning of those messages easier for users to undertake. It is only necessary to learn the rules by which auditory parameters are mapped to data attributes in order to understand the Earcons. This is unlike one-element Earcons, where each data item has its own sound without any structured relationship between all of the sounds and the data they represent, and thus must be learnt individually.

14.2.4 Hierarchical Earcons

Hierarchical Earcons are similar to transformational Earcons as they are constructed around a set of rules. Each Earcon is a node in a tree and each node inherits all of the properties of the nodes above it in the tree. Hence, an un-pitched rhythm might represent an error, the next level may alter the pitch of that rhythm to represent the type of error, etc. This is summarised in Figure 14.3. Since the Earcons can be long at the terminal nodes of the tree, Blattner *et al.* [5] proposed that the principles of transformational Earcons (see Section 14.2.3) could be used to shorten hierarchical Earcons for "expert users", so that only the last part of the Earcon would be played.

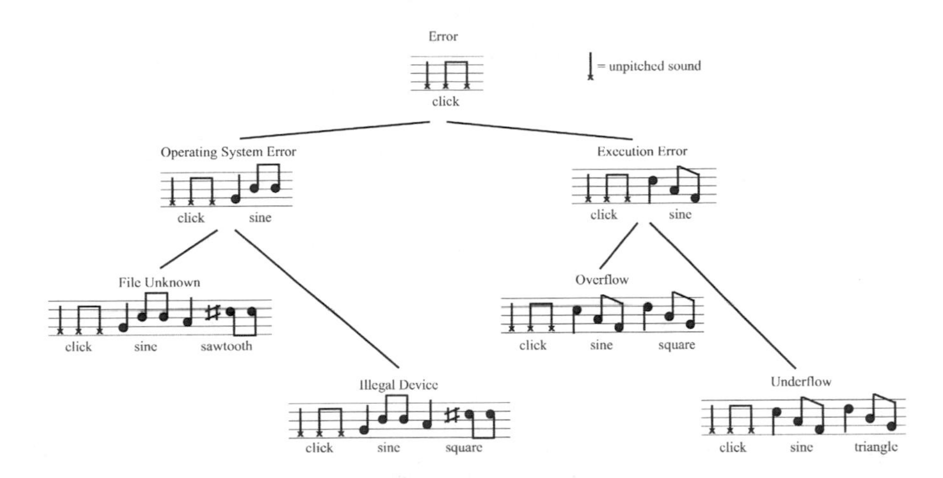

Figure 14.3: An overview of the rules used to construct a family of hierarchical Earcons representing computer error messages. Taken from Blattner *et al.* [5].

14.3 Creating Earcons

Design is of key importance in the effective application of Earcons. Earcons that have not been carefully designed are likely to pose problems in being distinguishable and identifiable by users. Fortunately, work on Earcons has led to design guidelines that can be employed to create effective Earcon families.

14.3.1 Earcon Design

Although Blattner, Sumikawa and Greenberg [5] proposed guidelines for how Earcons should be designed, they performed very little work on the validation of those guidelines. Work by Brewster, Wright and Edwards [17] sought to validate and improve the guidelines through empirical research studies. They designed two families of compound Earcons: one representing files in a graphical user interface and another representing operations that could be performed on those files. Brewster, Wright and Edwards compared three different designs of the two Earcon sets. One design was based on the guidelines of Blattner, Sumikawa and Greenberg [5] (Simple Set), using rhythm, pitch structure (see Figure 14.4), register and sinusoidal timbres to encode information about the files and operations. Another design was based on these guidelines (Musical Set), but with some modifications, notably the use of musical timbres rather than sinusoidal tones, as these were considered to improve performance. The third design (Control Set) avoided the use of rhythm to encode information. Instead, only musical timbre and pitch were used. This was considered to be similar to the simple beeps that computer systems made at the time. An outline of the encoding of information is shown in Table 14.1.

Participants were trained on the Earcon sets by having the relationship described, learning the names of what each Earcon represented and listening to the Earcons three times. Participants

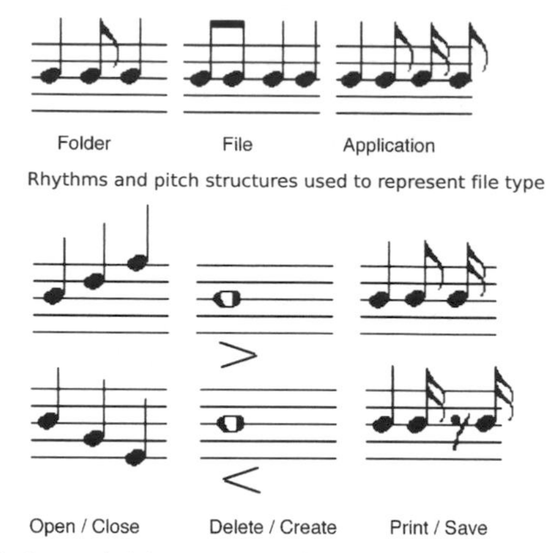

Figure 14.4: Rhythms and pitch structures used by Brewster, Wright and Edwards [17].

	Musical Set	Simple Set	Control Set
Application Family	Timbre (Musical)	Timbre (Sinusoidal)	Timbre (Musical)
File Type	Rhythm	Rhythm	Register
Instance	Register	Register	Register

Table 14.1: Overview of the Earcon sets used by Brewster, Wright and Edwards [17].

were then tested by having the Earcons played back in a random order. Participants had to write down all the information that they could remember about each Earcon. Brewster, Wright and Edwards found Earcons that represented operations over files were significantly better identified than those which represented the file, with recognition rates of 80% for the musical set. On analysis they found that the timbre of the Earcon had been well identified for both the file and operation Earcons from the musical set, but the rhythm for the operation Earcons had been significantly better identified due to their more complex intra-Earcon pitch structure (see Figure 14.4). Results also showed that the application family had been significantly better identified when represented using a musical timbre than when the sinusoidal waveforms suggested by Blattner, Sumikawa and Greenberg [5] were used. They also identified that the simple control sounds, which had application family identification rates as high as the musical sounds, had significantly worse identification for file type, which was represented by pitch rather than rhythm.

On the basis of these results Brewster, Wright and Edwards [17] redesigned the musical set of Earcons to incorporate greater differences between the rhythms used; varying the number of notes and the pitch structure of those notes. Additionally, based on data by Patterson [49], register changes of at least an octave were introduced to differentiate Earcons that had the same file type and application. A repeat of the evaluation showed a significant improvement in Earcon identification between the original and revised musical Earcon sets. From this work, Brewster, Wright and Edwards derived updated guidelines [18] for the effective design of Earcons. These augment and clarify those proposed by Blattner, Sumikawa and Greenberg [5]. Amongst the key guidelines were the use of musical timbre instead of the sinusoidal tones proposed by Blattner, Sumikawa and Greenberg [5], avoiding the use of pitch as the sole means of encoding information, and leaving a gap (of at least 0.1 seconds) when playing Earcons consecutively to ensure that they are identified as two, rather than one, Earcon (see Table 14.2). However, the more important theme emerging from the guidelines was that for any auditory attribute used to encode information as part of an Earcon, better recognition and identification could be gained by ensuring the maximum possible differences between those attributes. Timbres used should come from different musical families (e.g., brass and organ), and strongly dissimilar rhythms and intra-Earcon pitch structures should be used. Their guidelines constrain the size of an Earcon family to around three different values for each of the three main auditory attributes (timbre, melody and pitch). Additionally, as few musical instruments have a range that covers the several octaves necessary for using register, great care needs to be taken in the selection of the values for those auditory attributes. However, even with a constrained set, Brewster, Raty and Kortekangas [16] have shown that a hierarchical set of 27 Earcons can be effective at representing menu items in a telephone based menu system.

There is also scope to incorporate more subtle musical elements within Earcon design to increase the size of the hierarchy. Leplâtre [35] evaluated a simulated mobile telephone menu system which incorporated more sophisticated musical principles, such as attack, sustain and harmonic progression, to indicate the level and node the user was at in the menu hierarchy. He found that the application of hierarchical Earcons significantly reduced the number of errors users made when executing given tasks. However, the role of the Earcons in this case was to act as indicators of the position in the menu hierarchy, and as such augment the visual interface, rather than be explicitly learned or recalled themselves. More work is required to ascertain how densely encoded information in Earcons can be, and the full role that more subtle musical elements can play. We should also note that it is possible to combine auditory attributes and Earcon types. Watson [64], through a combination of pitch and duration, created a set of nine Earcons to represent blood pressure. These Earcons could be combined with each other to form compound earcons that allowed comparison with standard or prior readings of blood pressure. This created a set of 6,561 possible Earcon pairs. In evaluation, Watson found that blood pressure could be monitored to a high degree of accuracy. When errors did occur, they were more often found to be a misreading of an Earcon component as the next highest or lowest pressure level. These values were distinguished by pitch, which as already discussed, is a weak link in Earcon design.

Less work has been carried out on what data parameters should be mapped to what sound attributes. Walker and Kramer [62] distinguish between categorical sound dimensions (such as timbre) as opposed to sound dimensions that are on a linear continuum (such as pitch). They argue that it is more appropriate for data parameters to be mapped to sound parameters that share the same dimension type. For example, representing different car manufacturers in an Earcon would be best done by timbre, whereas representing car sales by those manufacturers would be done by mapping sales to register or pitch. Timbre, like car manufacturer, is a categorical dimension. Car sales, like musical pitch, exist on a clear continuum from low to high. They also note that the polarity of the mapping, such as increasing sales being represented by increasing pitch, should be appropriate to the context. It may not always be the case that a higher pitch should represent a larger data parameter. Although Earcons can have strictly arbitrary mappings between the sound and its meaning, it is important to ensure that any design does not violate pre-existing mappings that may exist. For example, Lemmens *et al.* [34] investigated whether playing a major or minor musical chord affected user reaction time in a picture selection task. Major chords are musically considered as positive, whereas minor chords are considered as negative. Users were presented with a picture of either an animal or a musical instrument and had to press a button marked 'yes' or 'no' in response to the question: "The picture is of an animal?". Lemmens *et al.* hypothesized that congruent images and chords (i.e., the image was that of an animal and a major chord was played) would result in faster reaction times from users. However, they found the opposite was true. They suggest a number of possible explanations, such as pictures of musical instruments being more strongly associated with the musical sounds than with the images of animals due to their musical character. The questions where the user was asked if the picture was of a musical instrument had a greater impact in the results. More work is required to fully understand this result. However, as noted by Walker and Kramer [62], pre-determined associations and cultural conventions in Earcon design is an area where more work is required.

Auditory Attribute	**Guidelines on Use**
Timbre	Musical timbre should be used rather than sinusoidal tones. Different data values should be encoded with timbres that are easily distinguishable. When Earcons with the same timbre encoded attribute are concurrently presented present each with different instruments from the same musical family.
Register	If listeners must make absolute judgments, then register should not be used. In other cases it should encode the least important data attribute. Large differences (two or three octaves) should be used between the registers used.
Pitch	Complex intra-Earcon pitch structures are effective in differentiating Earcons if used along with rhythm or another parameter. The maximum pitch used should be no higher than 5kHz (four octaves above C3) and no lower than 125Hz-150Hz (the octave of C4) so that the sounds are not easily masked and are within the hearing range of most listeners [49].
Rhythm	Rhythms used should be as different as possible. Combining rhythm with pitch is effective and provides greater differences between the Earcons.
Intensity	Intensity should not be used on its own to differentiate Earcons. Listeners are poor at making judgments in loudness.
Spatial Location	Spatial separation is useful in allowing Earcons from different families to be distinguished. A small number of spatial locations can be used to encode an additional parameter of data. In cases where Earcons from the same family are concurrently presented, spatialisation should be used to improve identification.
Timing	Short gaps should be included between Earcons. Where compound Earcons are used, gaps of at least 0.1 second should be introduced between the end and start of each component. When Earcons are concurrently presented, at least 300ms should separate Earcons start times.

Table 14.2: General guidelines on how auditory attributes should be employed in Earcon design. Adapted from Brewster, Wright and Edwards [18].

14.3.2 Concurrent Earcons

When using compound Earcons, the time to present them can be lengthy. This can make it difficult for the sound to keep pace with interaction in a human computer interface. Brewster, Wright and Edwards [18] carried out a study to determine if the time taken to present such Earcons could be reduced by presenting each part in parallel. They found that two sets of compound Earcons representing information about file types and operations, each encoding three parameters of information (the same as those discussed in Section 14.3.1), could be presented simultaneously, and identified with similar accuracy than when presented consecutively. In order to achieve these results the Earcons from each family had to differ in timbre, and each part of the Earcon was maximally stereo panned to ensure it was treated as a separate auditory stream [7]. Again, the key recommendation was to incorporate gross differences between the two Earcons to make them discriminable.

In many situations however, such gross separation may not be possible, such as when Earcons from the same family must be concurrently presented. For example, if timbre is mapped to some data attribute it cannot be grossly modified if two Earcons with the same timbre-encoded attribute are concurrently presented. Brewster, Wright and Edwards [18] concurrently presented two Earcons from different families that varied in acoustic parameters. An auditory display designer may wish to present multiple overlapping sound sources. The ARKola simulation by Gaver [26] exploited the temporal, overlapping qualities of Auditory Icons to indicate how processes were collaborating in a virtual bottling plant. Sawhney and Schmandt [58] used multiple overlapping instances of auditory feedback, including Earcons, speech and Auditory Icons, to present status and news information over a wearable auditory display. Where such displays include Earcons, it is possible that Earcons from the same family will be played together and overlap.

McGookin and Brewster [41] evaluated a transformational set of Earcons to determine how they performed when concurrently presented. In an identification task participants heard between one and four Earcons formed from the same Earcon family. Each Earcon encoded three parameters of a theme park ride: type (represented by timbre), ride intensity (represented by melody) and cost (represented by presentation register). The Earcons were designed in accordance with the guidelines of Brewster, Wright and Edwards [18]. McGookin and Brewster [41] found that the proportion of Earcons, and their attributes, that could be identified significantly decreased as the number of Earcons presented increased (see Figure 14.5).

McGookin and Brewster carried out further experiments to try to improve identification when Earcons were simultaneously presented [42]. They found that when Earcons with the same timbre encoded attribute were concurrently presented, the use of different timbres from the same musical family (e.g., acoustic grand piano and electric grand piano) improved identification of the melody encoded attribute and, as such, overall Earcon identification performance was significantly improved. A similar significant effect was found by staggering the onset-to-onset time of concurrently presented Earcons by at least 300ms. Additional studies presenting four Earcons in a spatialised auditory environment also improved identification [43]. Performance was significantly improved, but that improvement was not large. Earcons, at least those from the same family, are sensitive to being presented concurrently to a much greater extent than Auditory Icons [6]. Therefore, concurrently presenting Earcons from the same set should be avoided wherever possible.

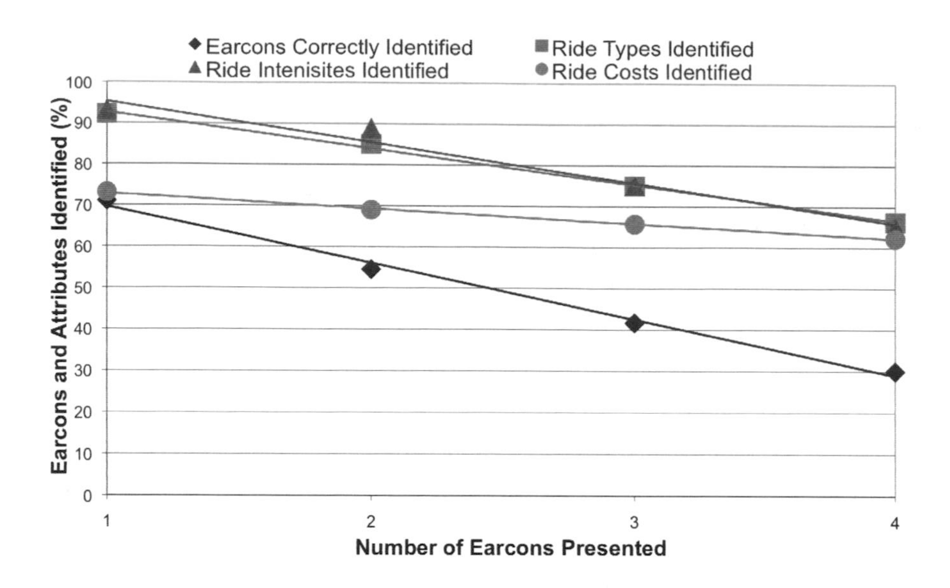

Figure 14.5: Graph showing how the proportion of Earcons and Earcon attributes correctly identified is affected by the number of concurrently presented Earcons. Taken from McGookin [43].

14.3.3 Learning and Understanding Earcons

It is important to design Earcons in such a way that they are effective means of communication, but a no less important feature is to consider how users will learn Earcons and the data mappings contained within them. As already stated, Earcons have abstract mappings between sound and what that sound represents, so some training is essential if users are to understand the intended meaning of an Earcon.

Garzonis *et al.* [25] compared "intuitiveness" in understanding Earcons and Auditory Icons. They found that although high recognition rates were quickly obtained with Auditory Icons, Earcons, without training, had recognition rates of around 10%. During subsequent phases of their study participants were given feedback on the accuracy of identification, and as such Earcon identification rates, as well as Auditory Icon identification rates, rose, with 90% of Earcons being correctly identified. McGee-Lennon, Wolters and McBryan [39] found that for simple Earcons designed to represent notifications in home reminder systems, requiring participants to correctly identify each Earcon once before starting the study led to similar recognition rates for the Earcons as for speech representing the same reminders.

The design of Earcons can also influence their learnability. The structure inherent in hierarchical and transformational Earcons, and the rules by which sound attributes and data parameters are mapped to each other, can be exploited to make Earcons easier to learn. Brewster [10] found that participants could correctly determine information about an Earcon even when that Earcon had an attribute changed that the participant had not learned (e.g., a new rhythm or timbre). In Leplâtre and Brewster's [35] study on hierarchical Earcons for telephone-based menus, participants could correctly determine the position of an Earcon in the hierarchy that had not been explicitly learned. Therefore participants could exploit the rules that the Earcons were formed from to infer information about them. Hankinson and Edwards [29] went further and applied a musical grammar to a designed set of compound Earcons. They introduced musical modifications such that combinations of Earcons that were valid within the grammar were aesthetically pleasing. If the user attempted to carry out an illegal operation, such as printing a disk drive, the auditory feedback from the combined Earcons would sound discordant and alert the user to the mistake. Hankinson and Edwards found that participants were able to quickly learn this grammar, making it a solid foundation to augment an existing set of Earcons with new examples.

Whilst work shows that the structure of an Earcon family can aid users in learning, it is also important to consider how, and in what way, users will initially learn the Earcon family. Brewster [10] compared several techniques for learning Earcons. These ranged from training that might be provided "out of the box", with only a verbal description of the Earcons, through to personalized training, where the experimenter played and described the Earcon structure as well as providing five minutes for self-guided training listening to the Earcons. Results showed that personalized training was the most effective, allowing significantly more Earcons to be correctly identified than when participants were given only a training sheet. However, he identified that the ability to listen to the Earcons was the significant factor in achieving good identification performance. After around 5 minutes of training, identification performance was around 80%. This is a common range reported for many Earcon identification studies, but it is usually based on an initial 5-10 minute training phase. Hoggan and Brewster [30], in the design of cross-modal icons – Earcons that can either be presented aurally or as Tactons [13] (tactile icons) – carried out a longitudinal study of user performance and retention rates. They found that a set of transformational Earcons encoding three data attributes, timbre (four values), rhythm (three values) and spatial location (three values), creating a set of 36 distinct Earcons, reached 100% identification accuracy after four 10 minute training sessions. In a comparison of Earcons, Auditory Icons and Spearcons (speeded up text-to-speech) to represent menu entries, Palladino and Walker [48] found that participants had to listen to a set of 30 hierarchical Earcons between five and seven times before perfect recall was obtained.

In conclusion, Earcons can be easily designed to represent a large number of multi-parameter data items, but care must be taken in the choice of auditory attributes to ensure that they can effectively communicate data. However, even when Earcons are well designed, consideration must be given to how users will be trained to interpret them. From the work discussed, if no training is provided, identification will be poor and users will become frustrated, perceiving the Earcons to be annoying. However, training does not need to be extensive, as even 5-10 minutes allows users to achieve high levels of identification.

14.4 Earcons and Auditory Icons

14.4.1 The Semiotics of Earcons

When designers consider the use of a set of Earcons to communicate information in an auditory display, they may also consider the use of Auditory Icons to fulfil the same role. As discussed in Chapter 13, Auditory Icons are: *"Everyday sounds mapped to computer events by analogy with everyday sound-producing events"* [28]. Both Earcons and Auditory Icons seek to fill the same role in an auditory display and considerable effort has been made to compare and contrast them over the years. It has largely been found that the advantages of Earcons are the disadvantages of Auditory Icons and *vice versa*. An understanding of the similarities and differences between these two forms of auditory cue is important in understanding when to employ them. In addition, a clear understanding allows us to consider other cues that are neither Earcon nor Auditory Icon.

Auditory Icons have been shown to be easier to learn than Earcons with good identification performance in the absence of training [25]. This is due to the mapping between the sound and the data represented being "intuitive". A carefully designed set of Auditory Icons requires little training. In many cases however, it can be hard to identify suitable sounds for all of the information that would be desired, and thus it becomes difficult to create a good set of Auditory Icons. It is possible to parameterize Auditory Icons in a similar way to Earcons [27], but in practice this can be difficult. Although the use of Auditory Icons may be seen to be the better choice, there is evidence that in some cases Earcons may be more appropriate. Sikora, Roberts and Murray [59] and Roberts and Sikora [53] compared environmental and musical sounds in terms of agreement of function, appropriateness and pleasantness. They found that the musical sounds were rated as more appropriate and pleasant than the real world sounds. Both sound types were rated as less pleasant and appropriate than speech. They also found that many users would turn off the environmental sounds if they were played multiple times a day in a business context, meaning that Earcons may be more appropriate.

The relative advantages and disadvantages of Earcons and Auditory Icons can be better understood when we consider that both are audible examples of semiotic signs. A full discussion of semiotics is beyond the scope of this chapter, but a basic consideration is important to understand why Earcons without a meaning are just sounds.

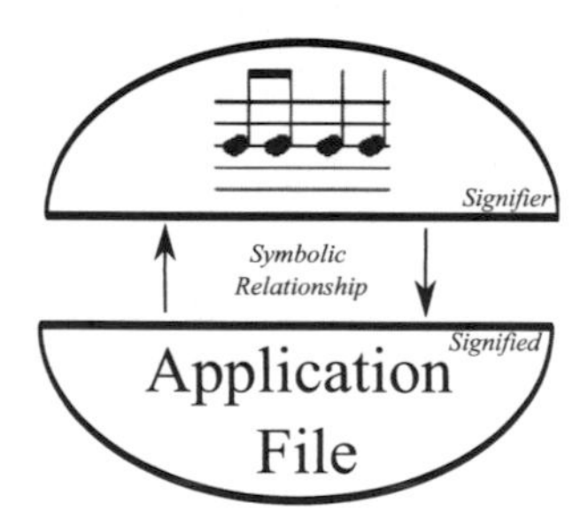

Figure 14.6: An Earcon illustrated as a sign, showing both the signifier, signified and relationship between the two.

Signs can be considered to be composed of two separate, but related components [19]: the signifier and the signified (see Figure 14.6). The signifier is the physical manifestation representing the sign. A visual example would be a road sign, whereas an aural example would be a musical motive or the sound of breaking glass. Related to the signifier is the signified, which as the name suggests, represents the meaning of the signifier. For a road sign, this may be that the driver should be aware of falling rocks or should reduce speed because of sharp bends in the road ahead. In a computer interface a musical motive or breaking glass signifier may signify that the current application has stopped working and has been closed by the operating system. In either case, the sign (Earcon or Auditory Icon) is composed of both the signifier and the signified; without either part it cannot be considered a sign. A musical motive is not an Earcon without a signified component and an Auditory Icon is only a sound effect. This is because it is the relationship between the signifier and signified that determines the sign as an Earcon or Auditory Icon. To be strictly classified as an Earcon this relationship should be arbitrary, hence the need to learn the Earcons before use. For an Auditory Icon there should be some cultural or other pre-existing relationship between the signifier and signified [1] [2]. Peirce [51] determined that there were three broad categories of relationship between the signifier and the signified: iconic, indexical and symbolic. Each of these categories represents a progressively less obvious or pre-existing relationship between the signifier and signified. Conventionally, Auditory Icons exist towards the iconic end and Earcons exist at the symbolic end, with some overlap in the middle (see Figure 14.7).

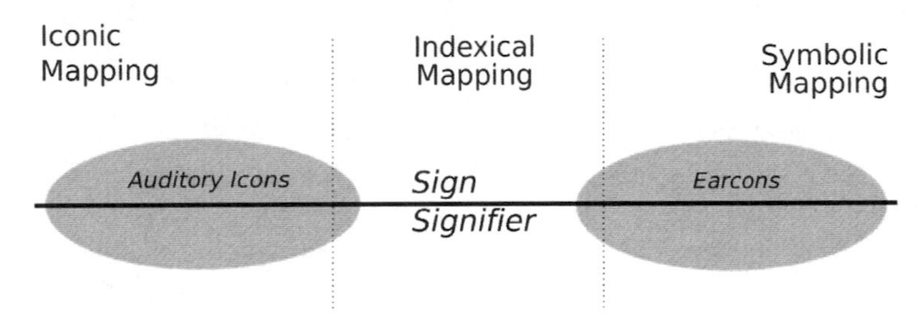

Figure 14.7: An illustration of the way in which Auditory Icons and Earcons relate to each other and the semiotic concept of signs.

14.4.2 Alternative Auditory Cues

From Figure 14.7, we might conclude that all forms of auditory notification are either Earcons or Auditory Icons. However, recent work has indicated that there are other auditory notifications that are neither. Ma, Fellbaum and Cook [37] for example, propose an auditory language composed of Auditory Icons that can be combined to form phrases for communication amongst people with language disabilities. Walker, Nance and Lindsay [63] proposed Spearcons, synthesised speech that is speeded up to the extent that it is no longer heard as speech. This unique auditory fingerprint can be used to represent items in a menu structure and in initial tests has shown that faster performance can be obtained over both Auditory Icons and Earcons, with significantly less training required. Palladino and Walker [48] discuss how Spearcons could incorporate more symbolic parameterisation, such as using voices of different genders to encode information.

A second example of alternative approaches comes from Isaacs, Walendowski and Ranganthan [32]. They considered the use of self- selected song extracts to uniquely identify a

[1] The original definition of Earcons by Blattner *et al.* [5], discusses representational Earcons which are analogous to Auditory Icons. However, these have never gained widespread acceptance and the term Auditory Icons has stuck.

[2] In the work of Sikora, Roberts and Murray [59] previously discussed, no data parameters were mapped to the sounds used. Therefore, according to our definition of Earcons, we do not consider their work to involve Earcons, and by the same definition Auditory Icons.

user in their Hubbub instant messaging system. Song extracts acted as parts of compound Earcons, with more conventional Earcons being compounded to form rich status messages. This approach has also been investigated by Dingler, Lindsay and Walker [24]. They used Earcons compounded with Auditory Icons to overcome the training issues with Earcons and the parameterisation problems with Auditory Icons. The use of existing music as a notification cue has also been investigated by McGee-Lennon *et al.* [40] in the study of Musicons. Musicons are short snippets (approximately 0.5 seconds long) of existing songs or musical pieces. These are then used to act as auditory reminders. In a study comparing Musicons to speech, Musicons were comparable in both speed and accuracy of identification. This was in spite of the participants receiving very limited opportunity to listen to the cues. McGee-Lennon *et al.* identified that the cultural knowledge inherent in knowing the music used was instrumental in the recognition rate. Musicons exploit an indexical-like relationship between the sound and its meaning in the reminder context. For example, the theme from the "Friends" TV show was associated with closing the door. Participants associated the closing of doors as characters entered or exited scenes to help recall the mapping.

In conclusion, Earcons and Auditory Icons are only two reasonably well defined and understood areas on a scale that is capable of containing a very much larger number of auditory cues. Little work investigating these other cues, which combine the advantages of both Earcons and Auditory Icons, has been carried out. This is a clear area for future investigation.

14.5 Using Earcons

Earcons have been applied in many contexts. This section presents three of the main areas, providing a clear understanding of where Earcons have been used and the situations and contexts where they have proven to be useful.

14.5.1 Sonically Enhanced Widgets

An initial source of Earcon application was in the augmentation of visual widgets in a graphical user interface. Many operating systems use on-screen widgets, such as buttons, scrollbars and menus, which are activated when the user releases the mouse button when the cursor is over the widget. If the user "slips off" the widget a selection error occurs. It can take time for the user to realize the error and correct it. The addition of Earcons can more quickly communicate such problems and provide feedback.

Brewster and Clarke [14] compared a drawing package application to one that had been augmented with Earcons. They added a relatively simple set of one-element Earcons to the tool palette. A marimba timbre was used to represent the default tool, and all other tools were represented by a trumpet timbre. These were played when the user either single or double clicked on the drawing area, or when the user selected a different tool in the palette. Brewster and Clarke found that the number of errors made in the drawing tasks that participants carried out was significantly reduced when the Earcons were played. Crucially, they found that workload was not significantly increased by the Earcons. Brewster [9] also investigated how drag and drop could be improved via the application of Earcons. Gaver [26] noted: *"A common problem in hitting such targets comes when the object, but not the cursor, is positioned over the target. In this situation, dropping the object does not place it inside the target, but instead positions it so that it obscures the target further"*. Brewster applied different Earcons to each stage of the drag and drop operation. Again, these were one-element Earcons. A reed organ timbre with a single note rhythm was used to indicate that the object had been moved onto the target destination. This Earcon was repeated until the user moved the object off the target or released the mouse button. If the user released the Earcon over the target, the same Earcon with a tinkle bell was played. If the user released the object when the cursor was not on the target, the same Earcon was played but with an orchestral hit timbre. This was designed to be more attention grabbing and alert the user to an error. In both complex (cluttered interfaces with multiple objects and targets) and simple interfaces, the application of the Earcons significantly reduced the time taken to complete drag and drop tasks over the same interface without the Earcons. Again, subjective workload was significantly reduced. With the addition of auditory feedback to visual menus already discussed in Section 14.3.3 and further work by Barfield, Rosenberg and Levasseur [3], Earcons can have a significantly positive impact on operations in a graphical user interface.

The examples presented however, deal only with cases where the Earcons are providing information about the primary task in the user interface and augment existing visual feedback. Earcons can also be useful in situations where they inform the user of background operations, or the status of objects out of the user's visual field. An obvious example is a scrollbar. Scrollbars tend to be operated as users scan a long visual document, either by using the mouse scroll wheel or by grabbing and dragging the scroll indicator with the mouse. In either event, the user is focusing on the document area rather than the scrollbar. This can make it difficult to keep track of the location in the document. If the user clicks off the scroll indicator his or her position in the document can quickly jump to an undesired location. Brewster [8] applied Earcons to assist in these two events. A low intensity continuous tone played on an electric organ timbre was presented just above the audible threshold. As the user moved between pages this tone was audibly increased and played with either a low or high pitch dependent on whether the user scrolled up or down. These augmented a scaling of low to high pitches that were mapped to the length of the document and played with the same timbre. Thus as the user scrolled up, the tones increased in frequency and *vice versa*. The user was therefore made aware of both distance travelled as well as page boundary changes. In a comparison with a visual scrollbar, Brewster [8] found that although there was no significant reduction in time or number of errors when using the auditory scrollbar, there was a significant reduction in subjective user workload. Similar studies [20] have been carried out on progress bars (hear sound example **S14.8**), where the user may wish to remain aware of an ongoing operation but not constantly visually monitor it. Beaudouin-Lafon and Conversey [4] proposed the use of Sheppard-Risset tones for auditory progress bars. These are auditory illusions which appear to be able to infinitely increase or decrease in pitch. Unfortunately, Beaudouin-Lafon and Conversey did not evaluate their technique. Isaacs, Wilendowski and Ranganthan [32], in a five month study on the use of Earcons to represent messages and status of users in their Hubbub instant messaging system, found that the Earcons helped users to develop an awareness of the activities of other users.

The work discussed here shows that simple Earcons can offer significant improvement to interactions in a graphical user interface and, as discussed in previous sections, Earcons are capable of encoding a great deal more information. However, Pacey and MacGregor [47], in a study comparing different auditory progress bar designs identified that there was a trade-off between performance and the complexity of the auditory cues (and thus the information they encoded). More complex auditory cues reduced the performance of participants on the primary task and increased annoyance. It is therefore important that only the information that needs to be communicated is encoded, rather than the maximum information possible.

A way to deal with these issues is to provide a comprehensive toolkit for developers to use. In existing graphical user interfaces developers do not have to program the look and feel of each button or how a scrollbar should work. Why should it be necessary to do so when using Earcons? Crease, Brewster and Gray [22] developed a toolkit of user interface widgets that were sensitive to the resources available for both the device and its context. The display of these widgets could be dynamically altered given the current resources. For example, an application running on a desktop computer system would present a progress bar visually, but on a mobile device would switch to an audio representation due to the reduction in screen space. Indeed, this may occur simply by reducing the visual size of the progress bar on the desktop. Below a certain size, the system would introduce an auditory progress bar. In such a way programmers would not have to worry about developing particular Earcons for individual applications.

14.5.2 Mobile and Ubiquitous Interaction

As computing has moved from desktops and laptops – tied to the wall with power and communication cables – towards small, mobile, always connected devices that a user has constant access to, consideration has been given to the role Earcons can play in improving interaction. Mobile devices have improved in both power and functionality over recent years, but they still suffer from a number of limitations such as a small visual display which requires on-screen elements to be reduced in size. Brewster and Cryer [11] considered how Earcons could be applied to number entry tasks on a mobile device. Simple Earcons were developed to signify the states of interaction with virtual buttons presented on a touchscreen, such as whether users had slipped off without selecting a button. Although only simple Earcons were used, these increased the number of four digit codes that could be accurately typed by participants. As the size of the buttons was decreased, from 16x16 pixels to 4x4 pixels, the use of Earcons allowed significantly more codes to be entered. Further work [12] found that Earcon enhanced buttons could allow a similar number of codes to be entered as non-sound enhanced buttons that were twice the size when the user was walking. Hoggan *et al.* [31] compared performance on text entry tasks when the user was sitting on a moving subway train, with either no button feedback or button feedback provided by Earcons. They found that the application of auditory feedback, even in the constantly changing environment of a moving subway car, allowed a significantly higher number of correctly typed phrases to be entered. Although during periods of very high background noise, audio was found to be less effective than tactile feedback.

As already stated (see Section 14.3.1), Leplâtre and Brewster [36] have evaluated the use of Earcons in providing context in mobile telephone hierarchies. They found that the use of Earcons to provide information about the current node in the hierarchy allowed users to complete tasks with fewer button clicks, and more quickly, than when no sound was provided. Further work showed that Earcons could enhance the monitoring of background tasks [21] such as download progress indicators and stock market trades [15]. Work by Ronkainen and Marila [55] investigated the use of simple Earcons to indicate key taps and timeouts for multitap input on mobile phone keypads. They found that these allowed for faster and more accurate text entry. In conclusion, there is strong evidence that the reduction in visual display space, prevalent in mobile devices, can be offset by the effective application of Earcons.

The work already discussed mostly considers the augmentation of mobile applications with Earcons to improve performance: allowing the size of on-screen elements to be reduced, or providing awareness of errors in interaction. More recent developments in mobile technology have considered how interaction can be performed without a visual display and how technology can be embedded in the physical environment. In both cases, communicating from the device to the user needs to be in a multimodal manner as a visual display is either not available, or cannot be constantly attended to.

Ronkainen [54] has considered how Earcons in a mobile telephone can alert a user to services or features in the environment that the phone can interact with. For example, a mobile phone could alert the user of a nearby vending machine that will allow the phone to purchase items electronically, rather than through the insertion of coins. Ronkainen identified several classes of notification, each of which he represented with an Earcon. With consideration of the discussion of Section 14.4, Ronkainen attempted to exploit previous perceptual understanding of sounds, such as using a faster rhythm to represent a more urgent notification. Work by Jung [33] has considered how Earcons can be embedded in an existing ambient soundscape to provide unobtrusive notifications. He proposed that Earcons would be less obtrusive than environmental sounds - partly due to the similarity of the Earcons to the musical soundscape in which they were embedded. Jung asked participants to complete a number of simple mathematical problems and identify when an auditory notification had been provided. Note that he did not associate information with the Earcons (although they did have to be learned) and thus the task was merely to detect that the cue had been played, rather than identify the information it contained. He found that both Auditory Icons and Earcons had similar identification rates of around 80%, but that it took longer for participants to become aware of the Earcons. This validated his assumption that, in this context, Earcons would be more suitable for less important notifications. This shows that Earcons do not have to be attention grabbing and can be used for more subtle notifications if required.

In the above cases it is not critical that users attend to alerts or notifications, and as discussed by Jung, the ability to ignore events is equally important. Earcons have also been applied in cases where the need to quickly understand the data encoded is important. Watson and Gill [65] have been working on providing better situational awareness by encoding the systolic and diastolic blood pressure of patients in an operating theatre. Such monitoring is important, but can be challenging due to the "eyes busy" nature of the environment. Watson and Gill found that users had a high level of performance at identifying mean blood pressure when engaged in a simple mathematical distracter task. The consecutive presentation of beacon Earcons, sonifying prior pressure readings with Earcons for current blood pressure, allowed participants to understand how pressure changed over time. However, Watson and Gill note that the true level of task performance with Earcons could only be determined when compared against a visual monitor with an experienced operator. In other cases, Earcons can be used when it is necessary that a reminder is presented, but that the reminder may be of a sensitive nature either due to its content or that it is necessary at all. Patients with dementia may need to be reminded to take complex medication, to switch-off dangerous appliances or even to remember to eat and drink. Sainz-Salces, England and Vickers [56] made an early attempt to encode information about home reminders for the elderly with Earcons. This has been extended by McGee-Lennon, Wolters and McBryan [39], who identified that

Earcons were found to be less obtrusive when compared to synthetic speech for the same messages. However, it remains an open question as to the usefulness of reminders in such an environment, as to date, no *in situ* studies have been performed.

14.5.3 Visual Impairment

A third area where Earcons have been applied is to increase access to user interfaces and computer-based data for people who are blind or visually impaired. Pleis and Lucas [52] report that there are 18.7 million such people in the United States. For these users, accessing a computer system can be challenging (see a more detailed description in Chapter 17). Interaction often occurs via the use of screen readers. Users move through the on-screen elements using a large number of keyboard shortcuts, with speech feedback used to read out names or data values. This works well for text-based information, but it is less useful for graphical or highly structured data.

Alty and Rigas [1] developed the AUDIOGRAPH system, which was designed to allow blind and visually impaired users to access graphical diagrams. Their system made extensive use of Earcons to indicate operations that had been performed by the users, such as expanding or contracting a shape, or undoing a previously carried out operation. Of note, Rigas and Alty incorporated more iconic (demonstrative) mappings [3] into their Earcon set. The 'expand' Earcon was constructed from a melody that aurally appeared to expand, with the 'contract' Earcon aurally contracting. The 'undo' Earcon was created by playing a motive with an "error" and then playing the motive without the error. Rigas and Alty mention that: *"At first hearing users were baffled by this, but on hearing the explanation they understood it immediately and had no further trouble recognizing it"*. To communicate information, pixels were sonified as the user moved over them. Two timbres were used to represent the x and y axes, with increasing pitch used to represent distance from the origin. By using this method, blind participants were able to succesfully draw the shapes that were presented [2]. A closely related system has been developed by Murphy *et al.* [45] to allow blind and visually impaired users to obtain a spatial overview of a Web page. Using a force-feedback mouse, auditory information, including Earcons, was presented as the user moved over on-screen elements.

Another important area is in the browsing and manipulation of algebraic equations. Mathematical equations are terse, with an unambiguous meaning. However, when presented through synthetic speech, this unambiguous meaning can be confused due to the time taken to present the equation. Stevens, Edwards and Harling [60], in addition to incorporating prosody changes in the way equations were spoken to allow terms to be better understood, considered how Earcons could be used to provide a quick overview of the expressions, and provide context during more detailed analysis. Their algebra Earcons, rather than trying to provide information about the contents of the algebraic expression, provided structural information. For example, the number of terms, sub-terms, and the type of each sub-expression were communicated through the Earcons. Rhythm was used to denote the number of terms in each expression, with one note per term. Superscripts (such as x^2) were played with an increase in pitch, and sub-expressions were played in a slightly lower register. Operands such as +, =, etc. were played with a different timbre. In this way the Earcons formed an overview of how the equation was structured and aided the user in understanding it. Stevens, Edwards

[3] Alty and Rigas [1] denote these as metaphorical mappings. We use the semiotic definition which regards the relationship as iconic [19].

and Harling [60] found that listeners were able to identify the structure of mathematical equations with high accuracy for both simple and more complex expressions.

A final point to note is that there is a great similarity between designing interfaces for visually impaired people and for more general mobile interaction. Sun shining on the display and the general reduction in visual display size can, temporarily at least, visually impair the user. As Newell [46] notes: *"a particular person's capabilities are specified to a substantial degree by the environment within which he or she has to operate"*. Therefore much of the work on applying Earcons to mobile and ubiquitous scenarios discussed in Section 14.5.2 could also be applied here. Yalla and Walker [66] for example, are considering how Earcons can be used to communicate information about scrollbars and operations for people with a visual impairment.

14.6 Future Directions

Although there has been a lot of work carried out to consider both the design and application of Earcons, there are still areas that remain largely unaddressed in the literature. The first area is in the design and implementation of more iconic Earcons, whose meanings are demonstrative or implicit. The work discussed in Section 14.3 takes a very "pure" approach to Earcon design, ensuring that the mapping between the sound and data is initially abstract. Learning must always be performed for such Earcons, with the emphasis being that listening to the Earcons before use is important. However, there are many cases, such as with consumer electronic devices, where users would not be trained on Earcons before use. This practically restricts such Earcons to specialised environments where training is mandatory, or that Earcons used must be few or simple so that they can easily be picked up. This may be one of the reasons why the work discussed in Section 14.5 largely focuses on one-element Earcons. However, researchers such as Leplâtre and Brewster [35] and Alty and Rigas [1] have shown that Earcons can be designed to be implicitly, rather than explicitly, learned; exploiting more musical properties of the sound to allow for more metaphorical mappings between the sound and data. As with much Earcon research, the influence of music and musicianship is lacking. A full investigation into how it can be best exploited is a ripe area for future research.

Another area for investigation is in the use of hybrid displays: those that combine both Auditory Icons and Earcons. Most research carried out has been to compare Earcons to Auditory Icons (see Section 14.4). Less work has been carried out to see how they can be best combined within an interface. Work has shown that this is possible, such as in the auditory presentation of social networks [23, 32, 57], but no detailed studies have been carried out to consider the issues of combining Earcons and Auditory Icons in such a way. For example, if Auditory Icons are used as parts of compound Earcons, can the easy-to-learn attributes of Auditory Icons be combined with the easy parameterization of Earcons? Related to this is the discussion of Section 14.4 and the other forms of auditory mapping that exist between data and sound. It is impossible to derive the number of permutations and different types of auditory cues that could be developed. Walker, Nance and Lindsay's [63] work on Spearcons and McGee-Lennon *et al's* work on Musicons [40] are the only two examples that seek to explore the world beyond Earcons and Auditory Icons in any real detail.

A final area is the use of Earcons beyond the concept of an auditory display. Tactons [13] are structured tactile messages that are formed via the same basic manipulations of motives as

Earcons. Presented through vibration motors, they can effectively communicate the same information as Earcons but through a different modality. More recently these have been combined to produce cross-modal icons [31]. These are again formed from the same basic manipulations, but can be presented through both audio and tactile modalities. Training users on tactile presentation means that they will be able to identify cues presented via audio and *vice versa* [30]. This means that presentation modality can be dynamically changed based on the current environment. For example, in a noisy environment the tactile modality may be best, whereas if the user has the device in a pocket, audio may be more suitable.

14.7 Conclusions

This chapter has discussed Earcons, short structured auditory messages that can be used to effectively communicate information in a human-computer interface. As well as describing Earcons, there has also been practical guidance for their design, implementation and use. As shown, Earcons can be of benefit in a wide range of applications, from simple augmentations of desktop widgets to sophisticated auditory interfaces for social media browsing. However, they must be well designed, and users must be trained in their use, at least to some extent. More subtle musical principles and metaphorical designs can be incorporated to allow for implicit learning of Earcons to occur. This is of benefit in situations where no explicit training is provided. In such situations simple Earcons that can be quickly "picked up" are more likely to be employed, such as the examples from Section 14.5.2.

In conclusion, Earcons offer a valuable means of communicating multi-parameter data in an auditory display. They are both easy to create, with effective guidelines for creation, and easy to learn, with a body of work illustrating the effective ways to train users. Their continued use, twenty years after being first proposed, is testament to their durability, and they continue to be one of the cornerstones onto which effective auditory displays are constructed.

Bibliography

[1] James L. Alty and Dimitrios I. Rigas. Communicating graphical information to blind users using music: The role of context. In *Proceedings of CHI 98*, volume 1, pages 574–581, Los Angeles, USA, 1998. ACM Press.

[2] James L. Alty and Dimitrios I. Rigas. Exploring the use of structured musical stimuli to communicate simple diagrams: The role of context. *International Journal of Human-Computer Studies*, 62:21–40, 2005.

[3] Woodrow Barfield, Craig Rosenberg, and Gerald Levasseur. The use of icons, earcons and commands in the design of an online hierarchical menu. *IEEE Transactions on Professional Communication*, 34(2):101–108, 1991.

[4] Michel Beaudoun-Lafon and Stephane Conversy. Auditory illusions for audio feedback. In *Proceedings of CHI 96*, volume 2, pages 299–300, Vancouver, Canada, 1996. ACM Press.

[5] Meera M. Blattner, Denise A. Sumikawa, and Robert M. Greenberg. Earcons and icons: Their structure and common design principles. *Human Computer Interaction*, 4(1):11–44, 1989.

[6] Eoin Brazil, M. Fernström, and John Bowers. Exploring concurrent auditory icon recognition. In *Proceedings of ICAD 2009*, volume 1, pages 1–4, Copenhagen, Denmark, 2009. ICAD.

[7] Albert S Bregman. *Auditory Scene Analysis*. MIT Press, Cambridge, Massachusetts, 1994.

[8] Stephen A. Brewster. *Providing a structured method for integrating non-speech audio into human-computer interfaces*. PhD thesis, 1994.

[9] Stephen A. Brewster. Sonically-enhanced drag and drop. In *Proceedings of ICAD 98*, Glasgow, UK, 1998. ICAD.

[10] Stephen A. Brewster. Using non-speech sounds to provide navigation cues. *ACM Transactions on CHI*, 5(2):224–259, 1998.

[11] Stephen A. Brewster. Sound in the interface to a mobile computer. In *Proceedings of HCI International'99*, volume 1, pages 43–47, Munich, Germany, 1999. Erlbaum Associates.

[12] Stephen A. Brewster. Overcoming the lack of screen space on mobile computers. *Personal and Ubiquitous Computing*, 6(3):188–205, 2002.

[13] Stephen A. Brewster and Lorna M. Brown. Tactons: structured tactile messages for non-visual information display. In *Proceedings of the fifth conference on Australasian user interface*, volume 28, pages 15–23, Dunedin, New Zealand, 2004. Australian Computer Society, Inc.

[14] Stephen A. Brewster and Catherine V. Clarke. The design and evaluation of a sonically-enhanced tool palette. In *Proceedings of ICAD 97*, Atlanta, Georgia, 1997. ICAD.

[15] Stephen A. Brewster and Robin Murray. Presenting dynamic information on mobile computers. *Personal Technologies*, 2(4):209–212, 2000.

[16] Stephen A. Brewster, Veli P. Raty, and Atte Kortekangas. Earcons as a method of providing navigational cues in a menu hierarchy. In *Proceedings of HCI 96*, volume 1, pages 167–183, London, UK, 1996. Springer.

[17] Stephen A. Brewster, Peter C. Wright, and Alistair D. N. Edwards. A detailed investigation into the effectiveness of earcons. In Gregory Kramer, editor, *Auditory Display: Sonification, Audification, and Auditory Interfaces*, pages 471–498. Addison-Wesley, Reading, Massachusetts, 1994.

[18] Stephen A. Brewster, Peter C. Wright, and Alistair D. N. Edwards. Experimentally derived guidelines for the creation of earcons. In *Proceedings of BCS-HCI 95*, volume 2, pages 155–159, Huddersfield, UK, 1995. Springer.

[19] Daniel Chandler. *Semiotics: the basics*. Routledge, New York, 2002.

[20] Murray Crease and Stephen A Brewster. Making progress with sounds - the design and evaluation of an audio progress bar. In *Proceedings of ICAD 98*, Glasgow, UK, 1998. ICAD.

[21] Murray Crease and Stephen A Brewster. Scope for progress - monitoring background tasks with sound. In *Proceedings of Interact 99*, volume 2, pages 21–22, Edinburgh, UK, 1999. BCS.

[22] Murray Crease, Stephen A. Brewster, and Philip Gray. Caring, sharing widgets: a toolkit of sensitive widgets. In *Proceedings of BCS-HCI 2000*, volume 1, pages 257–270, Sunderland, UK, 2000. Springer.

[23] Tilman Dingler and Stephen A. Brewster. Audiofeeds - a mobile auditory application for monitoring online activities. In *Proceedings of ACM Multimedia 2010*, volume 1, pages 1067–1070, Florence, Italy, 2010. ACM.

[24] Tilman Dingler, Jeffrey Lindsay, and Bruce N. Walker. Learnability of sound cues for environmental features: Auditory icons, earcons, spearcons and speech. In *ICAD 2008*, pages 1–6, Paris, France, 2008. ICAD.

[25] Stavros Garzonis, Simon Jones, Tim Jay, and Eamonn O'Neill. Auditory icon and earcon mobile service notifications: intuitiveness, learnability, memorability and preference. In *Proceedings of the CHI 2009*, volume 1, pages 1513–1522, Boston, MA, USA, 2009. ACM. 1518932 1513-1522.

[26] William W. Gaver. The sonicfinder: An interface that uses auditory icons. *Human Computer Interaction*, 4(1):67–94, 1989.

[27] William W. Gaver. Synthesizing auditory icons. In *Proceedings of INTERCHI'93*, volume 1, pages 228–235, Amsterdam, The Netherlands, 1993. ACM Press.

[28] William W. Gaver. Auditory interfaces. In Martin G Helander, Thomas K Landauer, and Prasad V Prabhu, editors, *Handbook of Human-Computer Interaction*, pages 1003–1041. Elsevier, Amsterdam, 2nd edition, 1997.

[29] John C K Hankinson and Alistair D N Edwards. Musical phrase-structured audio communication. In *Proceedings of ICAD 2000*, volume 1, Atlanta, Georgia, 2000. ICAD.

[30] Eve E. Hoggan and Stephen A. Brewster. Crosstrainer: testing the use of multimodal interfaces in situ. In *Proceedings of CHI 2010*, volume 1, pages 333–342, Atlanta, Georgia, USA, 2010. ACM.

[31] Eve E. Hoggan, Andrew Crossan, Stephen A. Brewster, and Topi Kaaresoja. Audio or tactile feedback: which modality when? In *Proceedings of CHI 2009*, volume 1, pages 2253–2256, Boston, MA, USA, 2009. ACM.

[32] Ellen Isaacs, Alan Walendowski, and Dipti Ranganthan. Hubbub: A sound-enhanced mobile instant messenger that supports awareness and opportunistic interactions. In *Proceedings of CHI 2002*, volume 1, pages 179–186, Minneapolis, USA, 2002. ACM Press.

[33] Ralf Jung. Ambience for auditory displays: Embedded musical instruments as peripheral audio cues. In *Proceedings of ICAD 2008*, volume 1, Paris, France, 2008. ICAD.

[34] Paul M. C. Lemmens, A. De Hann, G. P. Van Galen, and R. G. J. Meulenbroek. Emotionally charged earcons reveal affective congruency effects. *Ergonomics*, 50(12):2017–2025, 2007.

[35] Gregory Leplâtre. *The design and evaluation of non-speech sounds to support navigation in restricted display devices.* PhD thesis, 2002.

[36] Gregory Leplâtre and Stephen A Brewster. Designing non-speech sounds to support navigation in mobile phone menus. In *Proceedings of ICAD 2000*, pages 190–199, Atlanta, GA, 2000. ICAD.

[37] Xiaojuan Ma, Christiane Fellbaum, and Perry Cook. Soundnet: Investigating a language composed of environmental sounds. In *Proceedings of CHI 2010*, volume 1, pages 1945–1954, Atlanta, USA, 2010. ACM Press.

[38] Aron Marcus. Corporate identity for iconic interface design: The graphic design perspective. *Computer Graphics and Applications*, 4(12):24–32, 1984.

[39] Marylin R. McGee-Lennon, Maria Wolters, and Tony McBryan. Audio reminders in the home environment. In *Proceedings of ICAD 2007*, pages 437–444, Montreal, Canada, 2007. ICAD.

[40] Marylin R. McGee-Lennon, Maria K. Wolters, Ross McLachlan, Stephen Brewster, and Cordy Hall. Name that tune: Musicons as reminders in the home. In *Proceedings of CHI 2011*, volume 2, Vancouver, Canada, 2011. ACM Press.

[41] David K. McGookin and Stephen A. Brewster. An investigation into the identification of concurrently presented earcons. In *Proceedings of ICAD 2003*, pages 42–46, Boston, Massachusetts, 2003. ICAD.

[42] David K. McGookin and Stephen A. Brewster. Empirically derived guidelines for the presentation of concurrent earcons. In *Proceedings of BCS-HCI 2004*, volume 2, pages 65–68, Leeds, UK, 2004. BCS.

[43] David K. McGookin and Stephen A. Brewster. Space the final frontearcon: The identification of concurrently presented earcons in a synthetic spatialised auditory environment. In *Proceedings of ICAD 2004*, Sydney, Australia, 2004. ICAD.

[44] David K. McGookin and Stephen A. Brewster. Understanding concurrent earcons: Applying auditory scene analysis principles to concurrent earcon recognition. *ACM Transactions on Applied Perception*, 1(2):130–155, 2004.

[45] Emma Murphy, Ravi Kuber, Philip Strain, and Graham McAllister. Developing sounds for a multimodal interface: Conveying spatial information to visually impaired web users. In *Proceedings of ICAD 2007*, volume 1, pages 348–355, Montreal, Canada, 2007. ICAD.

[46] Alan F. Newell. Extra-ordinary human-computer interaction. In Alistair D N Edwards, editor, *Extra-Ordinary Human-Computer Interaction: Interfaces for Users with Disabilites*, pages 3–18. Cambridge University Press, Cambridge, 1995.

[47] Margaret Pacey and Carolyn MacGregor. Auditory cues for monitoring background process: A comparative evaluation. In *Proceedings of Interact 2001*, volume 1, pages 174–181. IEEE, 2001.

[48] Dianne K. Palladino and Bruce N. Walker. Learning rates for auditory menus enhanced with spearcons versus earcons. In *Proceedings of ICAD 2007*, volume 1, pages 274–279, Montreal, Canada, 2007. ICAD.

[49] Roy D. Patterson. Guidelines for the design of auditory warning sounds. *Institute of Acoustics*, 11(5):17–25, 1989.

[50] Roy D. Patterson, Judy Edworthy, and Michael J. Shailer. Alarm sounds for medical equipment in intensive care areas and operating theatres. Technical report, Institute of Sound and Vibration Research, June 1986.

[51] Charles S. Peirce. *The Essential Peirce: Selected Philosophical Writings.* Indiana University Press, Bloomington, IN, 1998.

[52] J.R. Pleis and J.W. Lucas. Provisonal report: Summary health statistics of u.s. adults: National health interview survey,2008. *Vital Health Statistics*, 10(242), 2008.

[53] Linda A. Roberts and Cynthia A. Sikora. Optimizing feedback signals for multimedia devices: Earcons vs auditory icons vs speech. In *Proceedings of IEA 97*, volume 2, pages 224–226, Tampere, Finland, 1997. IEA.

[54] Sami Ronkainen. Earcons in motion - defining language for an intelligent mobile device. In *Proceedings of ICAD 2001*, volume 1, pages 126–131, Espoo, Finland, 2001. ICAD.

[55] Sami Ronkainen and Juha Marila. Effects of auditory feedback on multitap text input using standard telephone keypad. In *Proceedings of ICAD 2002*, volume 1, pages 1–5, Kyoto, Japan, 2002. ICAD.

[56] Fausto J. Sainz-Salces, David England, and Paul Vickers. Household appliances control device for the elderly. In *Proceedings of ICAD 2003*, volume 1, pages 224–227, Boston, MA, 2003. ICAD.

[57] Nitin Sawhney and Chris Schmandt. Nomadic radio: Scaleable and contextual notification for wearable audio messaging. In *Proceedings of CHI 99*, volume 1, pages 96–103, Pittsburgh, Pennsylvania, 1999. ACM Press.

[58] Nitin Sawhney and Chris Schmandt. Nomadic radio: Speech and audio interaction for contextual messaging in nomadic environments. *ACM Transactions on CHI*, 7(3):353–383, 2000.

[59] Cynthia A. Sikora, Linda Roberts, and La Tondra Murray. Musical vs. real world feedback signals. In *Proceedings of CHI 95*, volume 2, pages 220–221, Denver, USA., 1995. ACM Press.

[60] Robert D. Stevens, Alistair D. N. Edwards, and Philip A. Harling. Access to mathematics for visually disabled students through multimodal interaction. *Human Computer Interaction*, 12:47–92, 1997.

[61] Emory Upton. *A New System of Infantry Tactics, Double and Single Rank*. Appleton, New York, 1867.

[62] Bruce N. Walker and Gregory Kramer. Ecological psychoacoustics and auditory displays: Hearing, grouping and meaning making. In John G. Neuhoff, editor, *Ecological Psychoacoustics*. Elsevier Academic Press, San Diego, 2004.

[63] Bruce N. Walker, Amanda Nance, and Jeffrey Lindsay. Spearcons: Speech-based earcons improve navigation performance in auditory menus. In *Proceedings of ICAD 2006*, volume 1, pages 63–68, London, UK, 2006. ICAD.

[64] Marcus Watson. Scalable earcons: Bridging the gap between intermittent and continuous displays. In *Proceedings of ICAD 2006*, pages 59–62, London, UK, 2006. ICAD.

[65] Marcus Watson and Toby Gill. Earcon for intermittent information in monitoring enviornments. In *Proceedings of OZCHI 2004*, volume 1, pages 1–4, Wollongong, Australia, 2004. Ergonomics Society of Australia.

[66] Pavani Yalla and Bruce N. Walker. Advanced auditory menus: Design and evaluation of auditory scroll bars. In *Proceedings of Assets 2008*, volume 1, pages 105–112, Halifax, Canada, 2008. ACM Press.

Chapter 15

Parameter Mapping Sonification

Florian Grond, Jonathan Berger

15.1 Introduction

Parameter Mapping Sonification (PMSon) involves the association of information with auditory parameters for the purpose of data display. Since sound is inherently multidimensional, PMSon is – at least in principle – particularly well suited for displaying multivariate data. PMSon has been used in a wide range of application areas.

The idiophonic acoustical output of material, electrical or chemical interaction (or the direct conversion of such interactions to electroacoustic signals), can often be used as a direct means of data interpretation. Consider, for example, the sound produced from the state change of water in a whistling tea kettle as it approaches boiling point. However, unlike the boiling tea kettle, the output signal that we wish to monitor may be beyond perceptual limen – whether below difference thresholds, or beyond of the limits of human hearing. Furthermore, the direct acoustic signal typically integrates a number of contributory factors, where attending to select attributes (while discounting others) may be desirable. Using PMSon a predetermined intermediary association can be established between one or more attributes of the information under scrutiny and its resulting auditory display. This association can, when necessary, be scaled to adapt to perceptual features and constraints of human hearing in order to optimize interpretive potential.

Our whistling tea kettle, arguably, generates an awful lot of sound for a simple binary signal whose purpose is to alert the user to remove the kettle from the stove and pour the water into the teacup. A simpler, albeit perhaps less charming, auditory signal might be achieved by monitoring the output of a thermometer measuring the water temperature in the tea kettle, and mapping the numeric output to a sound synthesis parameter. A simple mapping, for example, would link temperature to frequency, pitch or, perhaps a more obvious and explicit auditory signal (example **S15.1**). Rather than simply hearing when the target temperature is reached one might wish to listen to the continuous change of the rising water temperature (sound example **S15.2**), or, perhaps, to hear selective temperatures at various times during the heating process as in sound example **S15.3** which identifies the five traditional stages of Chinese tea preparation.

Therein lies both power and problem. Specifically, the enormous range of interpretive mapping decisions provides equally enormous opportunities to create an appropriate auditory display for a particular desired purpose. However, the wide variety of mapping possibilities poses a challenge in terms of consistency and comprehensibility, a challenge that has, for visual data mapping, been attenuated by evolution and the a-temporal nature of the display.

While mapping temperature to frequency or pitch is an intuitive and effective means of monitoring a critical stage in the process of tea preparation, it is useful to consider what information is lost in replacing the direct auditory output of the reed in the kettle with this simple mapping. For one thing, the kettle's whistle is, for many, a sound whose meaning is broadly, indeed universally, understood. For many, the sound carries a positive emotional valence. In addition, there is an anticipatory and musical component in the transformation from noise through variable and unsteady frequency to a relatively stable frequency[1]. Thus, while PMSon may offer precision and efficiency of display, it is important to consider the degree to which the sonification is intuitive along with aesthetic and emotive issues that may be desirable and even vital in the auditory display. In the following sections we consider these factors.

Sound example **S15.2** displays the continuously monitored temperature change of the water in the tea kettle by directly mapping temperature to frequency (i.e., 100 F = 100 Hz) at a relatively high sampling rate. The mapping provides both the numerical and temporal precision needed for the task at hand. However, unlike in sound example **S15.1**, in which the display is set only to sound at the boiling point (and thus, using virtually any mapped sound, an intuitive and effective result is produced), this display lacks an intuitive representation of context and goal. A simple solution might be to add a reference tone representing the boiling point that sounds together with the sonification as in sound example **S15.4**. This reference tone is analogous to a grid line on a graph. In the absence of this reference, the listener would be required to have absolute pitch and to know beforehand that 212 F is, in fact the boiling temperature of water [2]. Implicit in these examples is the notion that the sound parameters employed for mapping may be physical (for example, frequency) or psychophysical (pitch)[3].

Effective PMSon often involves some compromise between *intuitive*, *pleasant*, and *precise* display characteristics. For example, in sound example **S15.3**, temperature is polled at discrete time points and mapped to pitch. As opposed to the auditory 'line graph' of sound example **S15.2**, this example is analogous to a scatter plot. As is the case with visual graphs there are benefits and deficiencies in every approach. Discrete data polling using sounds and intervening silences that result in a display that is neither overly obtrusive nor exceedingly habituating is as much an art as a science. In sound example **S15.5** the tea water temperature is mapped to quantized frequencies corresponding to a pentatonic scale with pitch classes [C,

[1]The fact, that the idiophonic acoustic output is often already music with lots of variations and details, full of information in its own right, is addressed by the emerging field of auditory augmentation (see Bovermann et al. [11], and Grond et al. [33])

[2]In fact, since 212 F, or better 212 Hz, this falls between G#3 and A3, even absolute pitch might not be sufficient.

[3]As discussed further on, mappings may additionally be complexes in the time and/or frequency domains such as timbre, chords, melodic or rhythmic patterns

D, E, G and A] on the musical scale. The lowest pitch sounded is C2, 212 F is set to C5, and all other sounded pitches are scaled to their closest match within this range. Although the temperature is sampled regularly, the number of data points sonified is determined by the temperature, such that the higher the temperature, more (and higher) pitches are sounded until the boiling point is reached and the pitch is repeated every 200 ms. Sounded at a volume level that is neither masked by the environment, nor far above the masking threshold, and using a recognizable timbre with a relatively high degree of harmonicity, the resulting display is generally deemed pleasant, unobtrusive, and effective. Compound these intuitive but relatively arbitrary mapping decisions by the number of variable parameters comprising audio, let alone by adding additional dimensions to be displayed and the inherent strengths and challenges of implementing effective PMSon become apparent.

In addition to selecting particular increments for mapping data, there is a wide range of available sound parameters, and virtually limitless scaling strategies from which to choose. The optimal mapping is thus – even in this elementary case – the result of various considerations such as what are the task related aspects of the data to sonify, how much of this data is useful for auditory display (i.e., the granularity of the display), what sound dimension and parameter to use, and how to scale the data to the chosen parameter? A systematic analysis of these considerations is depicted in the diagram in Figure 15.1 and will be discussed in the following sections.

15.2 Data Features

In Figure 15.1, the field *data features*, encapsulating *data domain* and the *data preparation*, represents the stage in which the parameter-mapping sonification (PMSon) design involves objective thought regarding the data. Although we will present the process of PMSon design formally and in discrete segments, the bi-directionality of the design flow at the *data preparation* stage, suggests that the process is influenced at all stages by decisions regarding sound synthesis parameters.

15.2.1 Data Domain

The *data domain* describes all properties directly related to the data. Initial categorization of the data would identify *channels* or *dimensions* and whether the data are *continuous or discrete*. These aspects and their implications for the design process have been discussed in the *design space map* by deCampo in [21] and are treated in depth in the theory chapter chapter 2.

Data of a continuous variable that is densely sampled in time can potentially translate into smoothly interpolating auditory streams. Here sound has the potential to create the impression of analog continuity, in contrast to the inherently discrete nature of digital data. The ordering of the data is translated one to one (that is, one data point mapped to one signal parameter setting) into auditory features. In contrast, data that do not represent an continuos variable but rather a statistic process may fluctuate dramatically, and thus approaches such as granular synthesis may be appropriate. For this category in particular, but also the case for continuous data, attention must be paid to data format properties (such as data sampling rate) which, although unrelated to the data content, influence selection and design of mapping strategies.

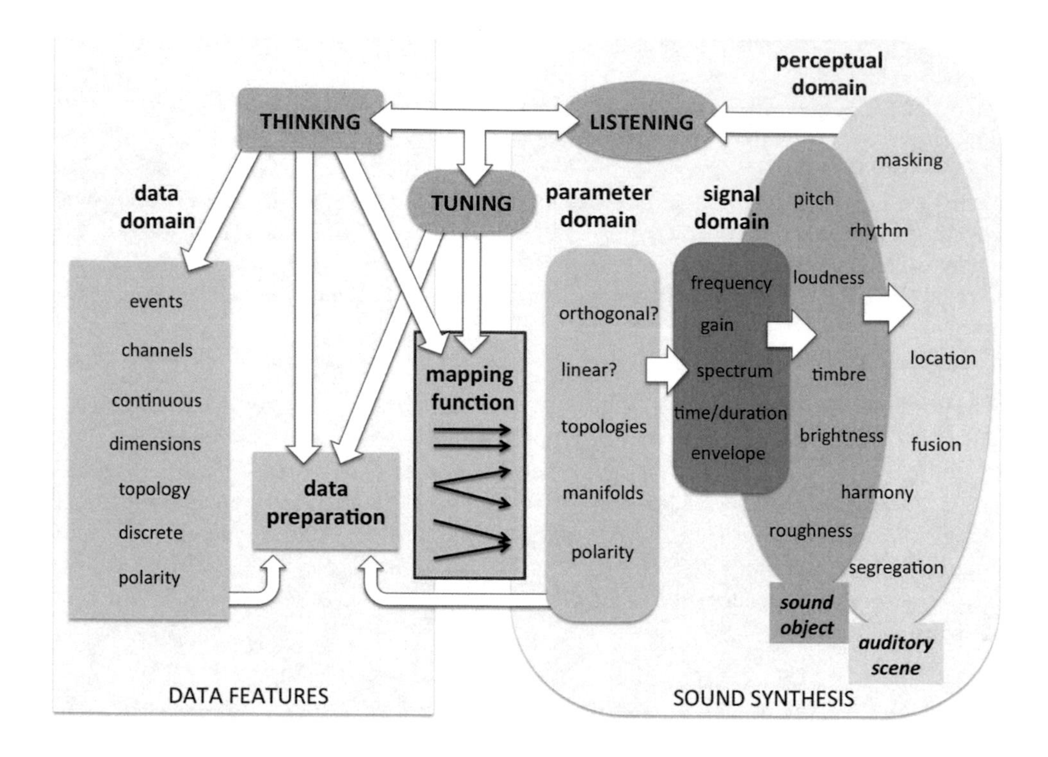

Figure 15.1: Map for a general design process of PMSon. Effective PMSon involves translating data features (left) into sound synthesis parameters (right). In this diagram, this process is divided between data and numerical control in grey, and sound and auditory factors in blue. Although data processing (rectangle) can be handled with rigorous objectivity (grey rectangle), human perception (green oval) imposes a subjective component to the process. As the figure suggests, the design of PMSon involves the interplay of, and the conscious intervention in both the data and the signal domains. Integrating both worlds is key in creating effective sonification.

If these format properties are reflected in the sound, the sonification runs the risk of producing audible display artifacts that compromise the veracity and efficacy of the auditory display. A third category involves data sets in which the data points are not equidistant. These are sometimes encoded in the original data, through non-equidistant timestamps. They can also be extracted from continuous or discrete data series as identified *events*, e.g. maxima, turning points or zero crossings, in the data preparation process.

Findings for this preliminary data assessment have important consequences for the subsequent mapping decisions. It is further important to identify *topological* structures in the data. Are they, for instance, organized on a ring or any other particular manifold shape structure which might suggest a particular spatialization of the auditory stream? Depending upon the outcome of this analysis, questions of *polarity* in the data can be addressed. A similar structuring approach for the design process has been developed by Barrass in [2].

15.2.2 Data Preparation

Following an objective data assessment, PMSon design involves a *data preparation* step which is not only influenced by the data structure but also by the inherent structure of the available sound synthesis parameters selected from the *parameter domain*. It is here that perception influences decisions. If, for instance, a selected sound synthesis parameter creates a perceptual categorical boundary [4] the sonification designer might apply this perceptual feature to a dichotomic data feature.

Appropriate data preparation is key for successful PMSon, particularly when the data set is multivariate and – as is typically the case – demands dimension reduction. Methods for dimension reduction can include principal component analysis (PCA) [45], self organizing maps (SOM) [35] among other methods. The purpose of dimension reduction for sonification is twofold: First, the available dimensions of synthesis parameters are often of limited orthogonality and of highly varying saliency and hence need to be used as efficiently as possible. Second, dimensionality reduction in the data domain must be considered in terms of resulting noise or perceptible distortion in the auditory display. Imagine several data channels that correlate in a complex manner but are each very noisy. Only after the information in the signal is properly separated from the noise can the correlation become perceivable in the sonification.

In contrast to data reduction at the *data preparation* stage, for scalar time series, *data preparation* may involve calculating derivatives as complementary information to map. Derivatives can, for example, be interesting data channels for expressing movement as changes in posture. In this case, the velocity as one indicator of the movement energy can be then expressed appropriately through the sound signal energy.

Data preparation can also consist of extracting events (e.g., extrema, zero crossings, intersections of data channels) from either discrete or continuous data, as a way to 'condense' the data substrate and adapt it to perceptual constraints.

The *data preparation* step can also help to objectify mapping decisions later on in the process. Principle Component Sonification [36] for example, determines the information content of data dimensions and maps them to sound parameters ordered by salience. This means that the data channel with most information (deviating from Gaussian distribution) is mapped to the most salient sound synthesis parameter.

Both the analysis of the *data domain* and analysis of the *data preparation* must be based on conscious and rational decisions. This is why we find on the side of the data features the human activity *thinking*. This might seem self-evident, but during the design process one has to constantly switch between careful *thinking* and concentrated *listening*. Awareness about which modality is necessary to move one step forward in the design process is crucial.

15.3 Connecting Data and Sound

The next field, the *mapping function* is the essence of what defines PMSon, which is to map data features to sound synthesis parameters in order to understand structures in the data. In PMSon the mapping function poses two challenges: The first is a proper formalization of

[4] for example, a sharp transition from a pitched to an unpitched source filtered by a spectral envelope

the transfer function which connects the *data domain* of hard facts with the somewhat more elusive perceptual domain. The second challenge is to find a good mapping topology, i.e., how the sets of *data features* and *synthesis parameters* should be linked in order to achieve a perceptually valid result.

15.3.1 The Transfer Function

A formalization of the mapping function can be found by Hermann in [36] and is here briefly recapitulated: Given a d-dimensional dataset $\{\vec{x}_1, ..., \vec{x}_n\}$, $\vec{x}_i \in \mathbb{R}^d$, an acoustic event is described by a signal generation function $f : \mathbb{R}^{m+1} \rightarrow \mathbb{R}^q$ which computes a q-channel sound signal $s(t) = f(\vec{p}; t)$ as a function of time. $\vec{p}$ is an m-dimensional vector of acoustic attributes which are parameters of the signal generator. q is the number of dimensions available through the spatialization setup. A PMSon is then computed by:

$$s(t) = \sum_{i=1}^{N} f(g(\vec{x}_i), t) \qquad (1)$$

where $g : \mathbb{R}^d \rightarrow \mathbb{R}^m$ is the parameter mapping function. For the (rare) ideal situation of a perceptually linearly scaling synthesis parameter, the linear mapping with a clipping to min/max values in the attribute domain can be used as shown in Figure 15.2. In less than ideal cases mapping functions can take various forms such as linear, exponential, sigmoid, or step functions. For the case of continuous data-stream sonifications, for instance, mapping functions can effectively shape amplitude and spectral envelopes and hence give the sonification a distinct articulation in time.

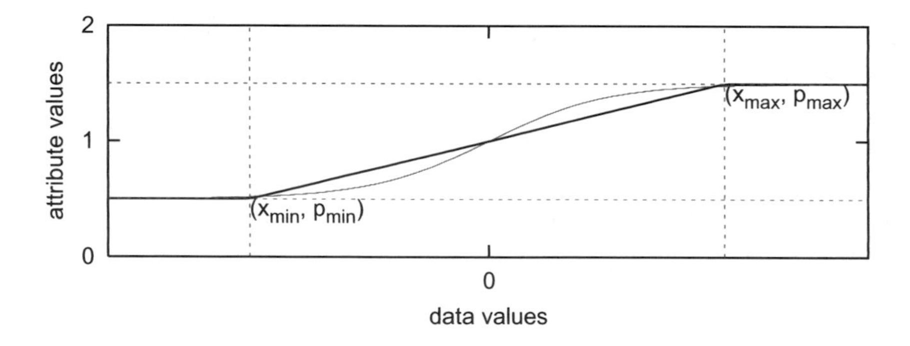

Figure 15.2: Typical piecewise linear transfer function (black line), the red line depicts an alternative sigmoidal mapping, Figure from [36].

Hermann further suggests a simpler representation as shown in Table 15.1. In this notation '_'means that mapping limits in the data domain are extracted from the smallest and greatest values from the data themselves. Instead of extreme data values, quantiles are often a good choice in order to make a sonification robust against outliers. The formalization of the mapping and its notation as proposed by Hermann makes this key aspect of PMSon explicit and transparent. Its equation does not, however, directly account for the rate of progress from one data point to the next. In a time-based medium like sound, a change in playback speed

data feature		sound synthesis parameter
$\text{datafeature}_1[\_,\_]$	$\rightarrow$	onset [10 ms, 20 ms]
$\text{datafeature}_2[\_,\_]$	$\rightarrow$	freq [50 midinote, 52 midinote]
$\text{datafeature}_3[\_,\_]$	$\rightarrow$	level [−18 dBV, 0 dBV]

Table 15.1: A readable textual representation for ease of interpreting the mapping of a sonification, where each parameter in the target domain is given a meaningful name.

affects many, if not all, perceptual characteristics of the display simultaneously and is hence a very important parameter.

15.3.2 Towards a General Formalization

Time, which is inherent to sonification, is well integrated in a formalism introduced by Rohrhuber [61] which introduces mixed expressions that intertwine sonification method and domain formalism. This formalism is not restricted to parameter mapping but also encompasses model-based sonifications and audification, see chapter 12 and 16. All sonification specific variables and operators are denoted with a ring, e.g., $\mathring{t}$ for sonification time. This ensures that a change in time in the data domain is not confused with the time that elapses while the sonification is being listened to. The sonification operator $\mathring{S}$ is further defined as $\mathring{S} = A\langle\xi\rangle \rightarrow \mathring{A}\langle\xi, \mathring{t}, \mathring{y}\rangle$, where A stands for a function, relation or other term, ξ is a set of domain variables (corresponding to the data domain) and $\mathring{y}$ is the sound signal of the sonification. The formalism is also discussed in chapter 2. The generality and rigor of this formalism comes at the cost of the typically steep learning curve associated with understanding abstract mathematical notation. Its intention and strength is to tighten expert domain mathematical formalism with synthesis processes. However it describes a transformation into the signal domain which, contrary to Table 15.1, does not contain perceptual qualities. It might be interesting for future extensions to consider the explicit integration of necessary perceptual corrections in the sonification operator $\mathring{S}$.

It is important to keep in mind that western music notation evolved over centuries in response to progressive refinements and changes in musical textures, timbre, pitch and rhythmic attributes. Thus music notation is, in a sense, a mapping from visual representation into pitch, loudness, duration and timbre instrumental space with time progressing according to an independent tempo map. The challenge which is genuine to sonification is that notations need to be functional and practical like the ones in music. However, they equally need to address the more abstract signal domain without losing sight of their ultimate purpose, which is to map and represent data rather than musical ideas.

15.4 Mapping Topology

The mapping function as shown in its readable form in Table 15.1 shows a mixture of mappings from data features into either a) strictly signal related parameters (onset in milliseconds), or b) hybrid mixtures i.e., mapping into level (expressed as dBV) or frequency (scaled as MIDI notes). This mixture of signal-related and perceptual- related categories and units expresses the two big challenges from opposite directions in the formalization of the mapping function: Are the signal synthesis parameters perceptually valid? Can the perceptual categories be smoothly varied and well expressed by synthesis parameters? One way to address this challenge is to choose a proper mapping topology.

'one-to-one' mapping

'One-to-one' mappings (as shown in Table 15.1) can, strictly, only be mappings to parameters of the signal domain since the parameters in the perceptual domain are generally not independent. The *Principal Component Mapping*, explained in the *data preparation* step on page 367, also falls into this category. Here, data channels are ordered with respect to their information content which are then assigned in a 'one-to-one' mapping scheme to the synthesis parameters with respect to their perceptual saliency (although in the case of *Principal Component Mapping*, the'one-to-one' approach is more related to the idea of parsimony in the use of perceptual dimensions).

'one-to-many' mapping

Mapping one data feature to several synthesis parameters ('one-to-many' mapping, also known as *divergent mapping* was first introduced by Kramer in [48]. The motivation behind this approach is to account for the fact that idiophonic objects usually change their sound characteristics in several aspects at the same time when varying, for instance, the amount of energy input. In [29] this method was expanded by scaling the ranges in the perceptual domain. This can be of particular interest if variations must be made noticeable in the display from small to large scales .

data feature		**sound synthesis parameter**
$\text{datafeature}_1[0, 30]$	$\rightarrow$	Δ gain [−90 dBV, 0 dBV]
$\text{datafeature}_1[20, 50]$	$\rightarrow$	between unvoiced and voiced
$\text{datafeature}_1[40, 70]$	$\rightarrow$	blends between the vowels [a:] and [i:]
$\text{datafeature}_1[60, 90]$	$\rightarrow$	fundamental freq 82 to 116 Hz
$\text{datafeature}_1[80, 100]$	$\rightarrow$	brightening of the vowel

Table 15.2: 'one-to-many' mapping table; the ranges in the data domain are percentiles of the variable for the given data set.

Table 15.2 gives an example where a single variable – here a variable from the Rössler attractor – is mapped to 5 different acoustic parameters in the example all ranges overlap so that the evolution of the different sound parameters builds a single sound stream. The spectrogram in Figure 15.3 depicts the resulting complex sound using this mapping strategy, which corresponds to the sound example **S15.6**. The transition between unvoiced and voiced and the transition between vowels are both audible and visually identifiable. One interesting aspect of this mapping strategy is that the mapping range of all parameters in the sound domain can be kept small but the perceived sum of all effects yields a highly differentiable result of a continuous auditory stream.

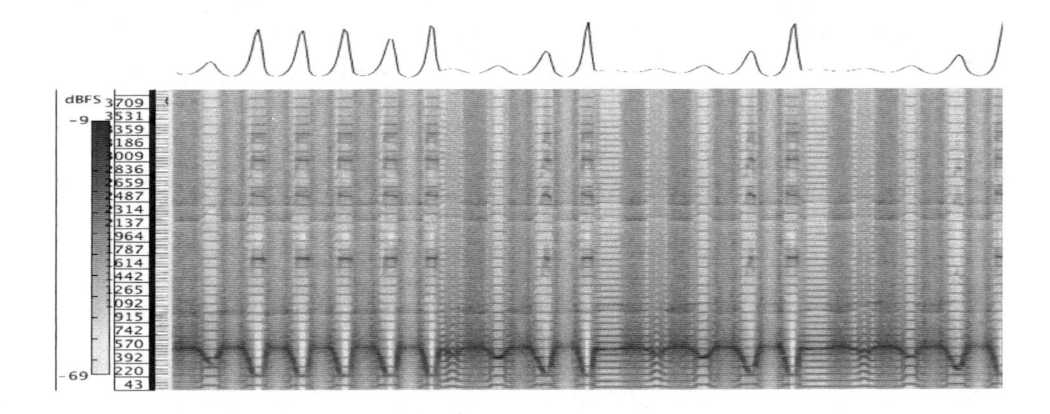

Figure 15.3: Spectrogram showing the described *one to many* mapping; on top is shown the logarithmic time series of the z variable. See also [29]

'many-to-one' mapping

'Many-to-one' mapping, also known as *convergent mapping*, can indirectly occur through the perceptual interdependence of sound synthesis parameters. Further, when mapping two or more data features to physical modeling processes, their variation can change perceptual properties such that a 'many-to-one' mapping is implicitly present. Convergent mappings have been explored mostly in gesture to sound synthesis mappings for signal models by Rovan et al. in [62]. The authors discuss and situate this approach in the field of new musical interfaces and identify this strategy as a determinant factor in expressivity control. Hunt et al. extended these definitions [43, 44]. The main goal in musical interface research is expressivity and fine control of musical material, see VanNort [71], and not necessarily the use of sound to highlight possibly hidden relationships in data sets. However, it is commonly agreed that it is important to have a perceivable relationship between gestures and sounds in performances. With regard to musical performance, expressivity is obviously an important issue for new musical interfaces, but its function is less clear for the purpose of representing data through sonification, since expressivity remains mostly a performer related category.

15.5 Signal and Sound

The *parameter domain* depicted in Figure 15.1 has already been touched upon by the discussion of the data preparation step and the mapping function. Sound synthesis parameters are part of the formalization of the mapping function and are in practical terms often controllable arguments of the signal generating units. Despite their role in the operational aspects in PMSon they also equally belong to the perceptual side and their proper scaling is part of the sound design expertise. In Table 15.1 we find, for instance, that some sound synthesis parameters can be expressed in perceptually approximatively linear units (e.g., MIDI notes, dBV), and hence a linear transfer function can be formulated. These perceptually linear units have then to be mapped into units of the signal domain (i.e., base frequency, amplitude). The corresponding mapping functions that take these perceptual relations into account can be sometimes made explicit. For instance, the relationship between loudness perception and frequency is captured in isophonic curves and hence psychoacoustic amplitude compensation can be applied. Within the context of PMSon this problem was demonstrated by Neuhoff et al. in [55]. For sounds of complex timbres however, this loudness compensation in the signal domain can become difficult and within PMSon appropriate compensations have still to be integrated. Nonetheless, if perceptual anchors can be defined in a sufficiently smooth manifold of non-orthogonal perceptual subspaces, data mining techniques can be applied to interpolate between those anchors. Related concepts have been explored for the formant space in [38] and for a smooth transition within sets of earcons and sets of auditory icons in [20].

15.5.1 Parameter and Signal Domain

In a strict sense the *parameter domain* belongs only to the *signal domain* in Figure 15.1 since parameters influence only signal generating processes as described in the formalism above. Furthermore, although we can rely on a good deal of empirical psychoacoustic formulae, the *perceptual domain* is ultimately reliant on the interactive process of listening, assessing, and refining parameter mappings. The arrangement in the diagram however reflects the challenge a PMSon designer is faced with which is the problem of representation in creative processes: An experienced sound designer or composer generally has a preconception of how the resultant signal should sound *before* it is generated and listened at. This imagined result belongs to the *perceptual domain* which demystifies the common notion of *inexplicable intuition* by presupposing that a conceptual auditory model is, through experience, created by the PMSon designer. Although this preconception is a strong guiding principle, the sound has to be heard for further evaluation and refinement.

15.5.2 Perceptual Domain

The psychophysical limits of just-noticeable differences (JND), masking, thresholds of hearing in the frequency, amplitude, and time domains, necessitates the constant interaction between 'thinking' and 'listening'. Thus, in Figure 15.1, the *signal domain* and the *perceptual domain* overlap partially, which is because not all aspects of a signal can be heard due to restrictions in the audible range and masking phenomena.

Here it is again interesting to compare sonification with classical music and composition practices, where mapping often refers to instrumentation or orchestration. Both are tightly coupled to the perceptual domain. For instance, a note in a score represents pitch and not frequency; a single pitch mapped simultaneously to a clarinet and oboe represents a single novel timbre combination rather than two discrete sounds. This musical approach can typically be found in MIDI based sonifications, where perceptual categories and scaling is contained in the signal product of the output sample which is triggered by a particular MIDI command. Because of the indexical nature of the signal in this case (i.e., being the trace of a calibrated instrument), creating a MIDI sample sonification effectively generates a quasi-musical score. Hence the signal category does not play the same role compared to sound synthesis through signal generators. MIDI-based sonification avoids (or skirts around) some perceptual challenges. However, its limitations deprive PMSon of the potentially high dimensional parameter space of sound synthesis methods. Nonetheless, the comparison of such mappings to musical scores suggests the efficacy of building sound synthesis units as perceptually calibrated 'mini instruments' whenever possible. The perceptual calibration can then be tested before preprocessed data are integrated in the PMSon.

The Sound Object

The sound taxonomy introduced in Pierre Schaeffer's classic work *Traité des objets musicaux* [63, 16] provides useful concepts for PMSon design. The first level of the *perceptual domain* can be considered as the abstract *object sonore* constituted by idealized *reduced listening*, a listening mode without causal and semantic connotations.

In practical terms this particularly applies when data features are mapped to mediating synthesis processes such as physical modeling. Here the target space into which data features are mapped often has parameters that describe material properties of simulated idiophonic objects, such as stiffness or mass. Through the dynamics of the physical model they control a complex combination of signal properties which integrate to perceptual categories such as timbre.

The Auditory Scene

Whereas the *object sonore* and its related theory describes the basic sounding units of a sonification design, principles of Auditory Scene Analysis (ASA), which describes how the human auditory system organizes sound into perceptually meaningful elements [12], allow us to account for their interplay. ASA describes how streams of sounds are grouped or segregated – vital considerations for effective sonification. Efficient PMSon design must take into account the cognitive load of users and their available capacities for a given task. One must, for example, be certain to avoid mappings in which three distinct data features are perceived as two auditory streams due to the pitch or timbral proximity of two of the three features. In addition to fusion, segregation and various Gestalt-based grouping principles, PMSon design must take into account auditory masking issues both with respect to the mutual influence of all sounds within the auditory display itself or the environment within which it is sounded.

ASA principles provide a framework where sound parameters, spatialization, harmonicity and their perceptual effects on masking, stream fusion, and segregation can be critically assessed. For literature and an in-depth discussion on stream-based sonification we refer to Worrall [83]. *Masking, stream fusion and segregation* are difficult challenges but parameter mapping can be considered as the sonification technique which allows for the most flexible intervention by the designer in order to address them individually. In audification for instance masking effects can only be explored by changing the playback-speed, which has at the same time implications on all other perceptual levels.

15.6 Listening, Thinking, Tuning

The projected result of a sound design idea is then evaluated through the three fields of human activities: *listening*, *tuning*, and *thinking*, as depicted in the diagram in Figure 15.1.

Listening in this context means a real listening mode as opposed to an ideal one. This is why the arrow from *listening* to *thinking* in Figure 15.1 is bi-directional.

A common pitfall in developing a PMSon is convincing oneself that variations in a sound are clearly audible and informative (since sonification designers know what they are seeking to hear) when, in fact, these variations are not at all salient to the naive listener. Interactively changing the mapping function, or exploratory *Tuning* the design can effectively avoid this trap. Yet, tuning that is undirected or uncontrolled runs the risk of leading the sonification designer to frustration and failure.

The iterative *tuning* process as an inherent element of PMSon has been investigated by methods of tangible computing through Bovermann et al. in [9, 10]. In their work a tangible desk was used to support an externalized representation of the parameter space while exploring various configurations within. These works fall into the category of interactive sonifications which are further described in chapter 11.

In [39] the *tuning* process and its evolutionary dynamics was partly operationalized by integrating it within a numerical optimization algorithm. This algorithm ensures a systematic and hence potentially efficient exploration of the parameter space. In numerical optimizations usually a cost-function is formulated, which defines a potential surface on which existing optima must be found. Since the functionality of sonification can ultimately only be assessed on the perceptual level the cost-function was substituted with user ratings and their comparison.

These examples show that *tuning* as an important aspect of PMSon has been identified as an interesting field of research which has also prompted the need for *just in time* (JIT) sonification tools that facilitate a quick evaluation cycle of idea, implementation, and listening. Related techiques are described in chapter 11. The function of tuning thereby is not only restricted to improve a PMSon in absolute terms but can also support exploratory data analysis, since different structures in the data may require different parameter sets to be discovered. *Tuning* however is an activity that can counteract learning efforts where parameter invariance is essential for building up internal representations.

15.7 Integrating Perception in PMSon

In the previous description of the PMSon design diagram, the constant tension between, phenomena of perception and formalization of the data-to-signal transformation has been encountered in several situations. However, auditory display has been successfully applied to various problems and in various circumstances. This unresolved tension is the reason why there is no generic way of designing a multipurpose PMSon for auditory display. This tension also exists, in principle, for visualizations but appears as less problematic in this modality. Why is this so? And what can we learn from this apparent advantage of visualization?

The visual modality is biased through our predominantly visual culture and the connection of knowledge with visual rather than auditory representation. It is further important that at least in 2D, structures such as correlations in data can be transparently visualized because of orthogonal (x, y) display dimensions and magnitude can be absolutely accessed through explicit scales. From a cultural/historic standpoint, the introduction of central perspective has further introduced a working mathematical formalization of some perceptional aspects of human vision. This operationalization contributes greatly to its credibility (and myth) as an objective knowledge representing medium.

Due to the complexity of the human auditory sense, which is in terms of the sonic qualities potentially multidimensional, but at the same time on the signal level only a "2-channel" time-based medium, formalizations of perceptual aspects and their operationalization for PMSon are considerably more difficult. In the next section some possible integration of perceptual corrections is discussed.

Proactive Corrections in the Mapping Function

In the ideal case, concerns regarding perception could be explicitly integrated in the formalized mapping functions as correcting terms. Possible sound aspects of intervention belong first and foremost to the perceptual field of the *sound object*. Imagine a PMSon that maps data solely to pure sine tones. In this case psychoacoustic amplitude compensation can be applied, which adapts the amplitude of a given frequency according to equal-loudness contours. In a similar way the Mel scale can be used to account for the nonlinear pitch frequency relation.

In the perceptional field of the *auditory scene* mapping aspects related to spatialization are one field where, in the case of HRTF rendering measurements of idealized head and torso shapes and their contribution to the sound perception, are well formalized. Here even individual perceptional features can be explicitly taken into account in the case of personalized HRTF.

Retroactive optimization through Tuning of the Mapping Function

Some aspects of the *auditory scene* are so difficult to operationalize that they cannot be addressed in the mapping function in the first step. In this case it is necessary to retroactively optimize the PMSon. Psychoacoustic models can provide an appropriate measure to optimize a sonification, and can be applied post fact.An interesting future question is whether quantifiable perceptual criteria lead to sonic optima that are pleasant enough to be functional for an auditory display. In the case of *masking*, however, this approach could ensure that PMSon can be ruled out where the data is obscured in the auditory display.

In the field of *qualitative perceptual effects* human intervention imposes the aesthetic judgment and criteria to a sonification. Therefore it can only partly be operationalized as described by Hermann et al. in [39], and criteria have to be formulated in order to quantify qualitative aspects of a PMSon. Attempts to formulate these high-level criteria for sonifications have been made by Vogt in [74].

Ultimately, the principal tenets of effective sonification – design of displays that are intuitive, learned with relatively little effort, able to draw attention to salient data features, and able to maintain interest and minimize satiation – involve both perceptual and, to some degree, aesthetic considerations. Such considerations are often arrived at through retroactive tuning.

Post-production Interventions

The equivalent of this step in music production is *mastering*, an activity with a particular level of expertise in auditory perception. Mastering processes that can be relevant for sonification include equalization and compression. Since these processes are prone to compromise the veracity of the sonification they should, ideally, be considered in the mapping function or sound synthesis processes rather than applied post factum. For equalization this should be done by proper psychoacoustic corrections. With respect to a signal compressor, the data can be scaled during the preparation step based on statistical properties (mean, median, variance, quantiles), which formalizes the display in a transparent way. However, compression might be appropriate when data magnitudes cannot be properly estimated in advance as is the case with online monitoring of data.

15.8 Auditory graphs

Auditory graphs are the auditory equivalent of mapping data to visual plots, graphs and charts. Going back to the tea ceremony from the introduction, mapping the continuous change in the kettle's water temperature to frequency is the attempt to produce the auditory equivalent of a line graph. The field of application for auditory graphs is usually as assistive technology for visually impaired or blind users, which is treated in depth in chapter 17. Auditory graph implementations are often clear cases of PMSon. For scatterplots, however, data sonograms have been introduced [41] which belong to MBS but usually involve a PMSon element, which is the categorial labeling of the data points as tone color. In the next paragraphs, we summarize a review over auditory graphs from [32] with a focus on PMSon design considerations.

Foundational work was laid in 1985 by Mansur et al. [51] who presented the first system for the creation of computer-generated sound patterns of two-dimensional line graphs, which they called *sound graphs*. Their goal was to provide the blind and visually impaired with a display for line graphs that allowed for a holistic understanding similarly to those of sighted users. The sonification approach employed by the authors was mapping the Ordinate values to pitch, varying it continuously while progressing in time along the x-axis. In their study the authors compared a prototype system with tactile-graph methods and found that mathematical concepts such as symmetry, monotonicity, and the slopes of lines could be determined quickly.

After presenting a longitudinal research approach in [8], Bonebright suggested an agenda for auditory graphs in [7]. The main agenda items were effectiveness, role of memory and attention, and longitudinal studies of learning.

A good overview of the field of auditory graphs can be found in Stockman et al. [68] The authors outline issues of a research agenda for design and evaluation which are comparisons, understanding and recall of sonified charts. Further, they introduce the 'multiple views' paradigm and how to design intuitive representations the same information using different levels of detail. Interactivity as a way to explore auditory graphs was also addressed as an important point of the proposed research agenda.

Harrar et al. [34] address the need to look at the mode of presentation and what effects changing the presentation parameters of auditory graphs have on the user's ability to gain an overview or to identify specific graph characteristics. Amongst other aspects they discuss the influence of playback speed in which the auditory graph is presented.

A first conceptual model of auditory graph comprehension can be found with Nees et al. in [54]. The authors motivate the necessity for their extensive review of the research field by pointing out that:

> *Auditory graph design and implementation often has been subject to criticisms of arbitrary or atheoretical decision-making processes in both research and application.*

Nees et al. attempt to make links to the relevant literature on basic auditory perception, and support a number of common design practices. The general design process diagram for PMSon from Figure 15.1 provides a supporting structure for design decisions. Nees et al. mention that information in an auditory graph has been occasionally mapped to spectral characteristics such as brightness, although they restrict their discussion of mapping strategies to frequency/pitch mappings which are still the most popular and dominant approaches. The conceptual model as presented in [54] also analyzes the role of the auditory context in an auditory graph. This includes concepts of tick-marks and the indication of quadrant transition. As a further auditory element the idea of an acoustic bounding box was introduced in [30].

Flowers gives (in [25]) an overview on auditory graphs and the state of the art in 2005. From a PMSon perspective, this is particularly interesting in that it lists not only successful mapping strategies but also those which failed. Some of the mappings that worked contained: pitch coding of numeric value, manipulating loudness changes in pitch-mapped stream as contextual cues and signal critical events, choosing distinct timbres to minimize stream confusions and unwanted grouping, sequential comparisons of sonified data. Approaches that failed were either due to the complex nature of loudness perception which cannot be used to represent an important continuous variable, or due to grouping if the simultaneous presentation of some continuous variables were of similar timbres, or if too many simultaneous continuous variables were presented at the same time using pitch mapping. Flowers also lists what we need to know more about, which are the effects of stream timbre and patterning on perceptual grouping, and the representation of multiple variables in a single auditory stream.

Despite the fact that research questions in the field of auditory graphs are far from being conclusively answered, applications that make use of auditory graphs are available. From the perspective of assistive technologies, the Java program MathTrax [64] [5] must be mentioned since its interface is tailored to typical requirements for the blind and partially sighted users, see chapter 17. It allows for an interactive exploration of the sonified function and presents visual, acoustic and descriptive information which can be all accessed through screen readers. Derivatives of a function can be sonified parallel to the function values. In comparison to the impressive features of the user interface, the sonification relies on a basic parameter mapping of function value to frequency/pitch. Mathtrax also allows the user to play the sonification of several functions – e.g., function values and their derivatives – at the same time.

From an *auditory scene* analysis standpoint the PMSon from MathTrax splits into two or several auditory streams exhibiting different varying pitch contours, whose integrated

[5] http://prime.jsc.nasa.gov/mathtrax/

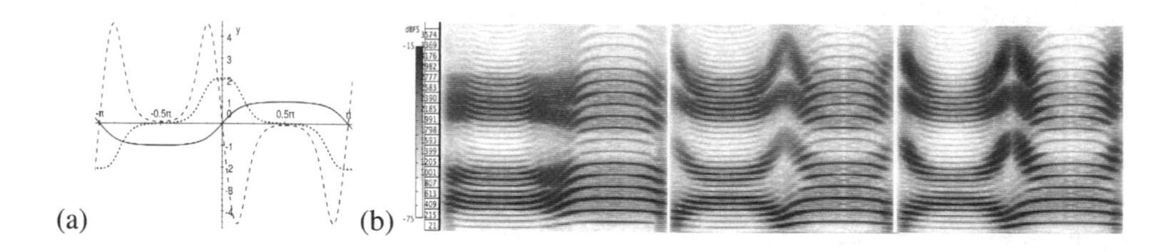

Figure 15.4: (a) function (continuous line) to be represented by sound, first derivative (narrow dashed) and second derivative (wide dashed). (b) 3 different PMSons in an auditory graph with increasing complexity, similar to the multiple views paradigm. Left: function value to pitch. Middle: additional vowel transition for slope. Right: additional brightness for curvature

interpretation requires developed listening skills; a problematic choice for auditory graphs as discussed in [25].

If we look at the particular case of mathematical function sonification we need to analyze what contributes to the understanding of visual plots, i.e., the recognition of shapes based on principles of visual gestalt perception. These shapes are a perceptual unit and do not fall apart into constituent elements like function value and slope. Auditory graphs with this perceptual goal in mind were designed by Grond et al. in [32]. The integration of the first two derivatives of the function of interest into the mapping should help to create individual sonic fingerprints for each position on the x-axis, so that the function properties could be assessed at any x position as a steady sound independently of a progression in time, i.e., along the x-axis. The mapping of function value to pitch was an orthodox choice. The slope was additionally mapped to the vowel transition [a:]–[i:] and the curvature to vowel brightness. This resulted in a single auditory stream of articulated timbre variations according to function shapes. A combined spectrogram of both stereo channels can be seen in Figure 15.4, which corresponds to sound example **S15.7**. An evaluation of this mapping strategy demonstrated the improved auditory contrast between similar functions over simple pitch mapping strategies. However this contrast was considerably weaker in an interactive exploration context, which suggests that this approach did not succeed in capturing the spatial and time-invariant aspects of visual shapes.

The results of over 25 years of research illustrate the potential of PMSon for auditory graphs as well as the need for further research and development of methods to include some key properties of their visual counterparts in order to maximize the potential to convey information through sound.

15.9 Vowel / Formant based PMSon

As stated above the principal tenets of effective sonification include the ability to communicate information with intuitive and easily learned auditory cues. An example of intuitive mapping would be the use of a physical model of an acoustic generator in which the parameter mapping creates a correspondence between change in data and change in sound. An interesting and effective example of this is the sonification of a rolling ball by Rath et al. [60]

and Lagrange et al. [49] (see also chapter 13).

However, not all sonification scenarios have a correspondingly intuitive physical model. Sonification of highly dimensional data by directly mapping data to synthesis parameters is thus often limited by the lack of an auditory model that will ensure coherent and intuitive sonic results. One promising direction is the use of vocal tract models with the rationale that recognition and categorization of vowels is a highly developed feature of human auditory perception. When identifying dissimilar sounds such as human vowels, the human auditory system is most sensitive to peaks in the signal spectrum. These resonant peaks in the spectrum are called formants (Figure 15.5).

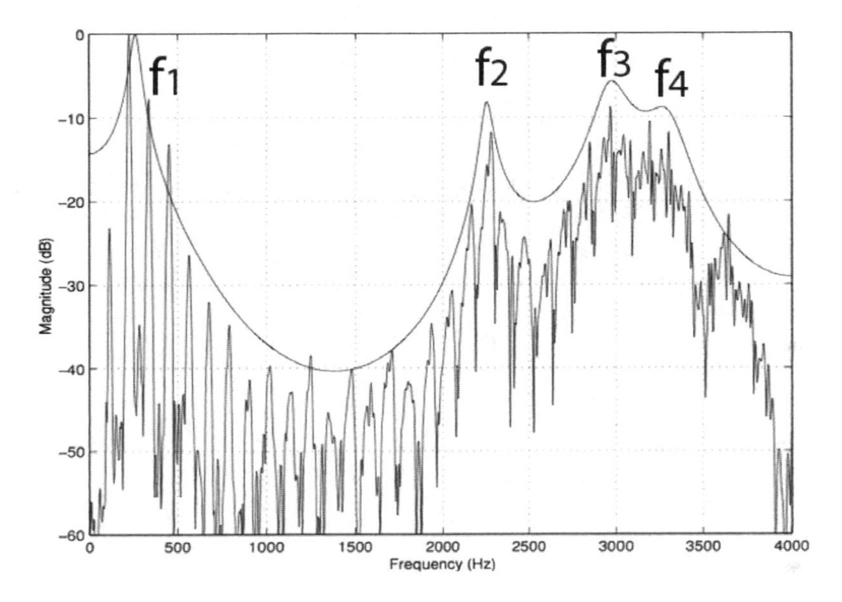

Figure 15.5: Spectrum of vocal utterance of the vowel /i/ as in 'team'. The smooth line enveloping the lower spectrum corresponds to the vocal tract transfer function. The resonant peaks (f_1, ..., f_4) of this curve are called formants.

Formant frequencies for human vowels vary according to age and gender, as well as the type of vowel uttered. In 1952, Peterson and Barney [57] measured formant frequencies of human vowels and suggested control methods that are well-suited for adaptation for PMSon.

Existing physical models include a formant filter model based on Klatt [46], and a physical model based on the pioneering work of Cook [19] and have been implemented for sonification including Cassidy et al. who used vowel synthesis to improve the diagnosis of colon tissue [15], and Hermann et al., who experimented with vowel-based sonification for the diagnostics of epilepsy via EEG signals [37].

Other approaches that produce formant structures easily categorizable into discrete vowel sounds include implementations of the tristimulus model [58] and FM-based formant synthesis [18]. Each of these methods has been used in sonification to good effect [26, 4]. Grond et al. [29] developed a potent and easily adaptable set of vowel synthesis tools using the SuperCollider programming language.

15.10 Features of PMSon

The potential benefits of PMSon include the possibility of highly effective multivariate displays and the potential to represent both physical and data spaces in a variety of ways that allow for exploration and observation *monitoring* (See also chapter 18). PMSon is potentially useful in a wide range of applications from data mining to assistive technology for the visually impaired. Computational speeds of real-time sound synthesis systems, efficient and flexible network protocols (for example, [84]), and a number of parameter mapping software tools like the *Sonification sandbox* [78] and *SONArt* [5] make real-time PMSon quite feasible for a host of applications. (See also chapter 10.)

Coming back to our introductory example of the boiling teapot, assuming appropriate means of transmission and reception one could as readily monitor the sensor in a teapot while in Qandahar as when sitting beside it in Palo Alto. Using sensors, whether proximate or remote, to transmit data, PMSon has been implemented to display diverse tasks including navigation, kinematic tracking, medical, environmental, geophysical, oceanographic and astrophysical sensing. In addition to numerical datasets, PMSon has been used to sonify static and moving images. Sonification of human movement, for example, is used in diagnostic and rehabilitative medicine [81], and athletic training (including golf, rowing, iceskating, and tai-chi) [47, 76].

15.10.1 Multidimensional mapping for multivariate display

Even the most compact representations of sound are multidimensional. Depending upon context, the contributory dimensions that make up a sound can alternately be heard as an integrated whole or segregated into independent components. Humans are adept at processing and interpreting some degree of multiple simultaneously sounding signals. Thus PMSon seems naturally amenable to auditory display of multivariate data.

Contributing factors in the perception of sound include frequency, duration, magnitude, spectrum, and spatial orientation and their musical correlates, pitch, rhythm and tempo, dynamics and timbre. Time-variant attributes include onset character and spectral flux, as well as characteristics that involve the integration of multiple parameters such as distance which integrates loudness, spectral character and reverberation. The ability to effectively simulate a sound moving through illusory space adds yet another perceptual cue, potentially useful in auditory display. Most importantly, the perceptual features of stream segregation and integration in auditory scene analysis provide a basis for sonification that can, within perceptual limits, display data polyphonically as discrete data streams mapped to independent sounds, or monophonically, as single sounds in which one or more data dimensions are mapped to sound dimensions.

The use of digitally synthesized sound to display multidimensional data was used by Max Matthews in the early 1970s with the mapping of three dimensions of a five-dimensional dataset to three sound parameters, while presenting the remaining two dimensions on a standard scatterplot. The mappings were pitch (150–700 Hz, quantized as chromatic pitches), spectral content using additive synthesis, and amplitude modulation varying the amplitude of a 15 Hz modulator. Experiments with a number of scientific datasets suggested the approach to be effective.

One of the earliest validating experiments of PMSon was Yeung's work on pattern recognition by audio representation of multivariate analytical data [86] in which data was mapped to register, loudness, decay time, left-right localization, duration and period of silence. Yeung allowed subjects two training sessions before they performed a classification task involving the presence of metals in a sample. Accuracy in the classification test was remarkably high.

Since that time the promise and allure of PMSon of data has been of persistent interest to a growing community of researchers. Principles of PMSon design have been proposed notably Brewster et al. [13], Walker and Kramer [79] and Barass [3]. However standard practices have yet to be broadly adopted.

15.10.2 Exploratory, observational and hybrid PMSon

Broadly speaking, PMSon can be used for two purposes, spatial exploration of data or *exploratory PMSon*, and monitoring sequential data (*observational PMSon*). *Exploratory PMSon* can be useful both for physical and virtual data spaces. *Observational PMSon* is useful with time series or sequential data streams. In some instances exploratory sonification is done on temporal data. We refer to this exploratory-observational approach as *hybrid PMSon*. From the user perspective the primary distinction between exploratory and observational PMSon is contingent upon the mode of inquiry which can either be sequential, which we refer to as *scanning*, or non-sequential, which we term *probing* [85]. Model-based sonification (MBS, see chapter 16) of which, some examples employ PMSon, provides a means of systematically sonifying data in the absence of an inherent time domain.

Exploratory PMSon

Effective mapping of spatial or relational information to time is key for useful exploratory sonification. Geometric display, that is, representing the properties and relations of magnitudes in a data vector, can be used to navigate and probe for diagnostic or analytical purposes. Exploratory PMSon has been used to display a wide range of complex data including EEG [40], turbulence [6], meteorological data [59], hyperspectral imaging [15], and geophysical data, as well as providing a means to display hierarchical data such as trees and graphs [72].

Auditory exploration of spatial states is routine in many everyday tasks. We tap on plaster-covered sheet rockhouse walls to determine whether there is a cross-beam or hollow space behind the surface. We slap a watermelon to determine whether it will be sweet and ripe. A physician listens to respiration, circulation and digestion using an archaic technology to evaluate respiratory state of her patient [6]. Analogous PMSon methods allow the user to probe vectors of data with the purpose of locating a target sound that closely matches a predetermined source sound. If the source sound is readily categorizable and the range of differentiation in the auditory mappings is sufficiently wide then it is quite possible to train a user to locate and identify particular data states. For example, a vocal tract model in which particular data states were anchored to specific phoneme sounds provided a remarkably intuitive sonification scheme [15].

[6] For a fascinating historical statement on the impact of Laennec and Auenbrugger see Edward Otis' 1898 Presidential speech to the Massachusetts' Medical Society [56]

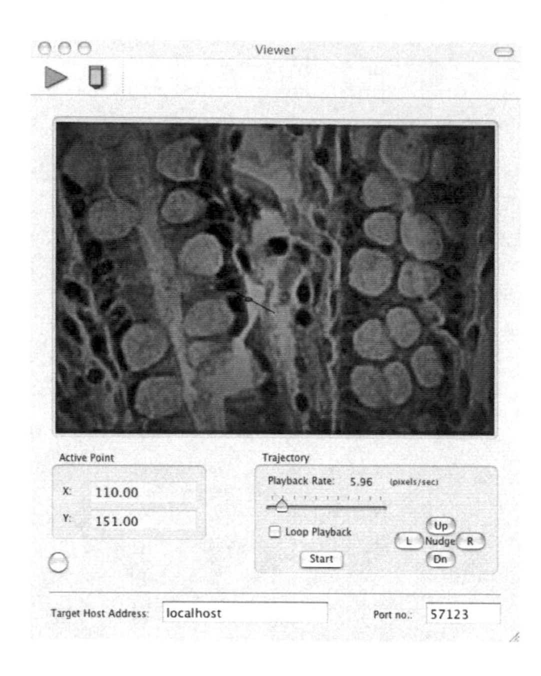

Figure 15.6: Viewer applet with Hyperspectral image of colon cell in window

In the area of auditory display, indeed in virtually all sonic situations, time is not just yet another parameter but rather the principal dimension to which all other auditory parameters address. While this may conform well to representing sequential or time series information streams, it demands particular consideration when representing spatial or geometric data. This challenge is particularly true when it is useful to have a synoptic overview of data.

A-temporal exploration may integrate scanning with *probing* data in which arbitrary points in a dataset are selectively sonified. Probing is typically necessary in diagnostic situations [15].

Figure 15.6 illustrates the user interface of a data viewer that provides a hybrid PMSon approach incorporating both data probing and data scanning. In this example a high dimensional hyperspectral image of a colon cell is represented visually as a two-dimensional surface with each data coordinate represented as a digital color pixel (Fig. 15.6). The vector at any data point can be probed by positioning the cursor over the associated pixel and clicking the mouse. By holding down the mouse button and dragging, the user can access all the data that is encountered during the scan. Alternatively, pre-set scan paths can be generated from a given data pixel. Accessed data is then packaged as an OSC message and sent to a user-designated port for sonification. The application allows the user to scan arbitrarily long sequences to gain a general auditory representation of local regions by dragging the cursor across the image, to hear a particular trajectory (for example a spiral generating from the originating pixel) or to probe a particular data point by clicking the mouse on a particular pixel [15]. Sound examples **S15.8**, **S15.9** and **S15.10** provide examples of, respectively, auditory probe (a), a linear scan (b) and a spiral originating from the cursor position (c)[7]. In examples **S15.9** and **S15.10** the spectral components of each vector associated with the selected pixel are scaled and mapped to frequencies in the audio range. A number of other approaches to probing data for exploratory sonification have been developed. These include the placement of *virtual microphones* [27, 23], the use of simulated sonar and the use of an adapted waveguide mesh [15].

While typically useful in the display of individual data points, probing can also be used to characterize data clusters. In [24] a digital waveguide mesh is used as a framework for PMSon where the variable parameter is not a specific sound parameter but rather the impedance values at junctures of the mesh. By creating an impulse at an arbitrary point along the mesh and converting the output to sound different configurations of data will create distinctly different auditory displays.

Spatial exploration

In addition to the potential benefits of the inherent temporality of hearing in terms of sonification, the human auditory system's acuity to determine the source and distance of a sound based on temporal, spectral, and amplitude cues suggests enormous potential for exploring data through spatial exploration. Spatial exploration with auditory display has been an area of considerable interest for decades [65]. The goals range from blind navigation through real or virtual space to exploration of geo-referenced data.

A number of areas of sound and music research have driven much of the work in auditory spatial exploration. These include: computational modeling of auditory localization with head-related transfer functions (HRTFs) [1], music composition using simulated spatial placement and movement of sound in illusory space [17], industry-driven efforts in developing compelling spatial simulations using binaural headsets and speakers, and research in understanding the neural mechanisms of localization (e.g., [22]).

Zhao et al. [87] describe four *Auditory Information Seeking Principles* (AISP) for PMSon modeled after methods of visual information searching. These include:

- gist – the capability of hearing an essential summary of data in a given region,
- navigate – the ability to scan sequential orderings of data,
- filter – the capability to selectively seek data according to specific criteria, and
- details on demand – the ability to obtain details of one or more data collections for comparison

AISP was shown to be a useful design approach for data sonification. Geo-referenced data sonification was demonstrated to be effective in distinguishing five-category value distribution patterns referring to a geographical map of the United States (Figure 15.7).

The mapping of two- or three-dimensional data onto a spatial auditory display is limited by a number of psychoacoustic factors (this is, in fact, true of any sonification using spatialization). Localization by humans is influenced by the type of signal and the familiarity of the sound. In addition, sensitivity to azimuth, confusion of placement in front or behind, and other

[7] For additional references on data scanning options for sonification of high-dimensional spaces see [28] and [75]

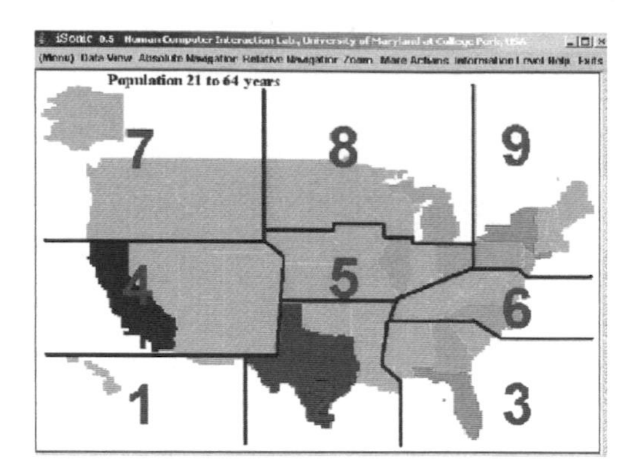

Figure 15.7: Blind navigation – PMSon for maps and other spatially oriented displays. Zhao et al. [88], for example, describe YMap, an auditory interactive exploratory display of choropleth maps. (courtesy Zhao et al.)

aspects of auditory spatial perception limit the specificity with which space can be used in PMSon.

In some instances extremely simple spatial mappings produce extremely effective means for orientation and exploration. The Lorenz system of radio beam auditory navigation, for example, provided a successful means for pilots to find their way in low visibility situations by using Morse code like dots and dashes – one sound for each directional beam broadcast from independent transmitters strategically located as a means of orientation. The antennas were fairly directional, projecting their signals slightly to each side of the runway centerline, one slightly left, the other slightly right. In the intersecting area of the transmissions the dots of the one signal and the dashes of the other combined to create a steady tone[8].

Subsequent ongoing research in the use of illusory spatial cues for training and navigational assistance have introduced more complex and sophisticated mappings with work incorporating HRTFs to create three-dimensional sound fields. However, the efficacy of using spatial information for sonifying statistical information remains inconclusive.

15.10.3 Observational PMSon

The purpose of observational PMSon is to facilitate exploration of sequential data. In some cases attention to the evolving character of the data requires conscious engagement, while in other instances, an auditory signal of exceptional cases or conditions might be sufficient for a useful display. In the former case, *engaged attentive observation* demands focus on the sonification in order to interpret the data. Conversely, *selectively-attentive observation* is a passive-attentive observational situation in which the behavior of the data draws attention to

[8] A development of the Lorenz system called the *Knickebein* was implemented as an auditory orienting method for German pilots on night time bombing raids. British intelligence ultimately discovered the method and countered it by broadcasting randomly placed transmissions of dots and dashes. The Germans and British engaged in what is called by some historians, the *battle of the beams* as new methods and different frequencies were tried and discovered.

a feature of the auditory scene that otherwise may or may not demand attention. For an in depth discussion of monitoring in sonification we refer to chapter 18. Sound example **S15.11** demonstrates a means of anchoring data by scaling to a particular phoneme [4]. In this example, a particular trend in stock market data is detectable by mapping the sound to a frequency modulation synthesis algorithm that can produce vocal formants. This mapping allowed the sonification task to be described as simply 'you are seeking for a condition displayed when you hear the sung vowel [o] in the sonification'.

15.11 Design Challenges of PMSon

We have thus far discussed PMSon of time-series data (sonified in or out of 'real-time'), as well as PMSon for non-temporal data in which information is converted into the time-domain for auditory display. Processing methods might involve numeric conversion, data filtering, or re-sampling, as well as mapping considerations of parameter association, polarity, scaling, among other decisions.

However, as Lumsden and Brewster noted, the fact remains that sounds are often used in ad hoc and ineffective ways [50]. It is important to note that the ineffective use of sound could well have as much to do with flaws in preparation of the data. The term *signal conditioning* has been used to describe signal processing performed on data for auditory display purposes. These methods include re-sampling, filtering and compression. It is useful to consider three aspects of parameter association: the polarity of mapping, dimensional scaling and the need for contextual and referential sounds to orient the listener to an effective auditory display [66].

15.11.1 Polarity

Polarity [77] is the directional mapping of data to a sound parameter. That is, whether, for instance, an increasing number sequence should be mapped to increasing or decreasing numbers in the mapped parameter. Whereas mapping the polarity of the rising temperature of the teapot thermometer to rising pitch seems intuitive enough, some studies suggest that inverting the polarity of pitch to pressure (that is, decreasing pressure mapped to increasing pitch) is preferable [79, 67]. One confounding factor in polarity is a lack of consistency in listeners' mental models. Even musical parameter mappings are not always as intuitive as generally thought. When one of the authors' daughter started studying the 'cello she confused pitch direction and the verbal descriptions of 'higher' and 'lower'. This polarity reversal reflected the physical nature of producing progressively 'higher' notes on a given string by moving the hand downwards.

15.11.2 Scale

It is often, indeed typically, necessary to scale from a data domain that transcends the perceptual limits of hearing such that data values must be scaled to adapt to a perceptually useful range of a particular auditory parameter.

Returning to our tea kettle example, while the unscaled mapping of numeric change between temperature and pitch as the water heat rises from 106 F to 212 F seems appropriate enough to be expressed as a musical octave, it may not serve the purpose of the sonification, which may, for example wish to delineate the range between the threshold of pain (130 F) and the boiling point by designating an octave relationship between these points. If the purpose of sonification is comparative (say, for example, one wishes to hear the temporal difference between boiling under varying atmospheric pressures, or perhaps, hear the differences between the boiling points of water, cesium (613 F), tellurium (989 F), and antimony (1750 F). These hypothetical analytical tasks exemplify the comparative analytical mode of PMSon. Another PMSon approach involves the integration of data into a single perceptual group (as opposed to discrete segregated data display).

The likelihood is that numeric changes in a data set may have insufficient or inappropriate number ranges to be directly mapped to frequency change. Scaling the values is typically needed to effectively represent data in sound. Furthermore, categorical boundaries such as scale degrees are often useful in providing effective sonification. Octave equivalence is a fundamental aspect of music perception in a wide swath of cultures. Therefore, it would seem that a representation of doubling of a numerical value as the doubling of frequency would be an appropriate auditory mapping. However, this introduces problems of scaling as well as of multiplicity.

The perceptual challenges of scaling lie in the absence of uniformity and wide divergence of perceptual acuity. For example, judgment of the relative loudness of two sounds is both difficult and subject to contextual and other biases such as pitch range [52] and duration. In the case of loudness perception, the ranges and limits of hearing are fairly well understood. Human hearing is remarkably sensitive. We can hear a sound whose energy is so weak as to move the tympanic membrane a small fraction of the diameter of a hydrogen molecule, while the dynamic range of hearing is in the order of a hundred trillion to one. However the inability to categorize loudness and the highly contextual nature of loudness perception make it virtually impossible to systematically map data to loudness for anything but relatively crude associations. These issues are exacerbated by masking phenomena, habituation and satiation of repeated auditory stimuli, and the startle response, all of which can diminish the effectiveness of PMSon. That said, frequency-dependent amplitude compensation according to psychoacoustically derived guidelines offer a means to appropriately map to loudness.

15.11.3 Context

Just as the use of axes and tick marks in visual graphs provide a referential context, it is often essential to provide an auditory reference in PMSon. The challenge is to provide additional contextual information without overtaxing or cluttering the auditory scene.

It seems that such context can be very beneficial, but only when it adds information, and not just clutter, to the display [67]. Since most users have little experience with sonification, training is critical. It remains to be determined how best to provide instruction in sonification comprehension. As stated earlier, intuitive and easily learned sounds allow for exceptional situations in which the instruction to the listener may be simple and unencumbered. For example, in vowel-based sonification one might instruct the user to listen for a particular vowel sound which represents a particular condition in the data.

The greatest challenges of PMSon involve the potential for ambiguity in data interpretation and the risk of incomprehensibility. Furthermore the lack of standards and ubiquity in mapping strategies often makes sonification research akin to working on the tower of Babel. The challenge remains to establish which mappings optimally represent particular data types.

Common agreement on what constitutes the best representative sounds for display, and the decisions of what data dimension will map to one or more attributes of the selected display sounds remain elusive. There are, broadly speaking, two approaches: *symbolic* mapping of data to sound (in which case the level of abstraction between the sound and the information displayed is highly variable) and *iconic* mapping. A third method, specifically, the use of vocal sounds, is considered here as a special case.

In Walker and Kramer's 1996 study of parameter mapping [79], subjects controlled operations of a simulated crystal factory with the task of monitoring temperature, pressure, size and rate using sonification to track the parameters simultaneously. Responses and reaction times were monitored by instructing subjects to take action in response to auditory cues. These included turning on a cooling fan when the heat was rising and turning on a pump when pressure fell. Loudness, pitch, tempo and onset rise-time were the auditory parameters used. The mappings were interchanged in different experiment trials permuting the mappings in what the researchers thought would be graduated from most intuitive to seemingly random mappings seen in Table 15.3.

representation	**temperature**	**pressure**	**rate**	**size**
intuitive	pitch	onset	tempo	loudness
okay	loudness	pitch	onset	tempo
bad	onset	tempo	loudness	pitch
random	tempo	loudness	pitch	onset

Table 15.3: Crystal factory mappings (from Walker & Kramer [79])

Surprisingly, both accuracy and reaction time measurements suggested that the *Bad* and *Random* mappings were more effective than the *Intuitive* or *Okay* groups. Furthermore, changing the polarity radically altered these assessments. Ultimately, the most effective mappings that were found are compiled in Table 15.4:

temperature	loudness
rate	pitch
onset	size
ineffective	tempo

Table 15.4: Effective mappings in Walker and Kramer's 1996 study

Another specific challenge for PMSon is the lack of a consistent set of perceptual guidelines to distinguish when context sounds are critical and when they are unnecessary or even obtrusive additions to the auditory display.

In some instances changes in one dimension affect perception in another (for example, consider how pitch affects loudness perception). Furthermore, many aspects of human hearing are nonlinear and thus the degree of sensitivity is highly variable across auditory dimensions. For example, humans can hear in the order of ten octaves. Discriminatory difference limen however differ dramatically across this range. Whereas frequency differences as small as one Hertz can be detected around one kHz the JND at 100 Hz is about three Hertz.

Principles of auditory scene analysis [12] provide the groundwork for perceptually valid sonification, particularly as it relates to streaming. Understanding the likelihood for formation of perceptual streams is key in building displays that will reliably convey the information to the listener [82]. Conversely, unanticipated perceptual grouping can undermine the effectiveness of auditory display, even when the stimulus was designed to be aurally interesting and appealing. A comparative study of psychophysical scaling of PMSon on sighted and visually impaired listeners suggests that establishing perceptual magnitude scales of sonification is further complicated by inconsistency across the two populations [80].

A particular challenge to effective PMSon is the absence of a perceptual metric for timbre. As demonstrated with synthetic phoneme-based sonification [4], timbre can be highly effective in revealing data. Within the sonification community relative timbre assessment has been used as a basis for data representation including the utilization of vowel space as an intuitive categorical space [15], crystallization sonification [42], and numerous applications in which traditional musical instrument sounds were used to represent data. Alas, since there is currently no perceptually valid metric that quantifies the distance between two particular sounds, the use of timbre is restricted to rather broad classes of sounds and necessarily lacks the nuance and fine gradations needed for auditory display of highly detailed data. Recent research on timbre space models [70] while promising, remain incomplete and inconclusive.

There is no consistent guiding set of principles regarding how to train listeners for a sonification task. Effective sonification must be intuitive and easily learned. Training should optimally be a minimal investment in time. The issue of training in how to comprehend and interpret a sonification is addressed in [66]. It is useful to consider how auditory display in direct audification tasks such as medical auscultation is taught.

While visualization has considerable groundwork in this regard, the perceptual reliability of parameter mapping sonification has piecemeal evidence.

There is empirical support for the ability to interpret a sonified trace of a two-dimensional line graph with one or two data dimensions [14]. Considerable evidence also supports the ability to trace the sonified borders of two-dimensional geometric shapes [73].

15.12 Synthesis and signal processing methods used in PMSon

Most existing PMSon applications use frequency, time, intensity, and, in a broad sense, timbre (often as MIDI-based instrument mapping) as principal mapping parameters. However, it is useful to consider synthesis and parameter mappings whose auditory results are not necessarily directly identifiable as a basic parameter, but that nonetheless may have identifiable characteristic auditory display properties. For more information on sound synthesis ins sonification see also chapter 9.

Additive Synthesis

The basic parameters of additive synthesis include the ratio between the fundamental and each harmonic or inharmonic spectral component the complex sound, the peak amplitude of each component, and the amplitude envelope of each partial. Additional parameters can include frequency skew, and periodic and aperiodic vibrato rates and frequencies for each partial. Sound examples **S15.12**, **S15.13**, **S15.14**, and **S15.15** demonstrate the potentially effective use of additive synthesis parameter mapping on high dimensional data[9]. In the first example **S15.12** of this set, a sequence of two 12-dimensional vectors are mapped to twelve partials of a harmonic complex such that the data values dictate the relative amplitudes of each partial. Although it is impossible to interpret the timbral result of the complex in terms of the specific mappings, patterns of relationships based, for instance, on the degree of spectral *brightness* are easily distinguished.

In sound example **S15.13** the vector elements are mapped to the rise time of each partial's amplitude envelope. In examples **S15.14** and **S15.15**, the two example vectors are mapped to frequency ratios. In sound example **S15.14** these relationships are scaled such that there is a linear relationship between the matrix values and the mapped frequency. The mapping produces a complex sound that is primarily inharmonic. In sound example **S15.15** the values are scaled such that the tendency gravitates toward harmonic complexes. Note that data points sharing harmonic relationships fuse while elements that dissonate with the fundamental are noticeable. As evident in these examples, parameter mapping strategies such as these are most useful when thought of as task-specific decisions rather than as general solutions for all sonification needs. Furthermore, these examples demonstrate the critical role of scaling data. How the data is scaled dictates the degree to which the sonification expresses the internal relationships between the vector elements.

Nonlinear synthesis

The advantage of additive synthesis is the direct correspondence between each instrument parameter and the effect upon the resultant sound. Nonlinear methods such as frequency modulation have less direct and intuitive correspondences. However, parameters such as modulation index and depth can be used effectively in creating identifiable timbre categories particularly when they are oriented toward highly characteristic sounds such as vowels and phonemes. The biggest drawback here is that, given the limited number of control parameters deems this method less amenable to representing highly dimensional data.

Resonators

Filter parameters provide the opportunity to sculpt an input sound whether noise or otherwise by controlling such parameters as the center frequency, bandwidth, peak amplitude and Q of frequency components. In sound examples **S15.16** and **S15.17** subtractive synthesis is used to sonify stock market data from two respective years (2000 and 2001) such that the

[9]The vectors in the examples are [.322 .959 .251 .301 .078 .405 .077 .209 .101 .095 .173 .221] and [.121 .073 .595 .008 .709 .177 .405 .278 .991 .651 .029 .922] respectively. These are vectors from two adjacent pixels of the hyperspectral dataset described elsewhere in this chapter, data scaled appropriately for amplitude and frequency mapping.

resonances express formant regions producing vowel like sounds that identify particular trends[4].

Physical models

As mentioned previously, physical models, in which the physical properties and mechanics of an acoustic sound generator is modeled using a variety of digital filtering methods provide a means to map a combination of data to create complex timbral characters (such as vowels and phonemes). Mapping data to one or more of the filter controls is a potentially effective method of PMSon. Physical models ranging from adaptations of the simple Karplus-Strong plucked string model to more complex models such as Cook's physical model of the vocal tract have been effectively used for parameter mapping sonification. An example of phyisical model based sonification is the use of a two dimensional waveguide mesh to sonify proximity to data clusters **S15.18** and **S15.19**. In these examples, coded by Greg Sell, the popular game *Battleship* is simulated such that the 'ship' coordinates are mapped to the scatter functions on junctions along the two dimensional mesh. The proximity of a probe to target data clusters, as well as characteristics of the cluster are audible.

Spectral mapping sonification

Resonators and other digital filter instruments provide a means of mapping multiple time-variant data streams to a sound complex such that the timbral distortion of the sound can provide useful information about pattern embedded in the data. Another approach to this goal is spectral mapping sonification [40] in which the frequency bands of the short-term Fourier transform of each time series (in this case EEG data channels) is mapped to a waveform oscillator with parametric control of frequency and amplitude such that the frequency band is mapped to the time-variant oscillator frequency and the energy is mapped to its amplitude. By 'tuning' the bands to a musically consonant interval (in this case a perfect fifth) the resulting sonification is perceptually pleasing and characteristics of the overall EEG are evident.

15.13 Artistic applications of PMSon

Finally we consider the use of PMSon for artistic creation (sometimes referred to as *musification* which includes sonifying a-temporal datasets (DNA, for instance, has inspired numerous musical works), time-based datasets such as solar activity [10], tides, and meteorological records [11].

The use of geometric relationships and mathematical processes have inspired composers for centuries. Famous examples include the use of the architectural proportions of Brunelleschi's Basilica di Santa Maria del Fiore in Dufay's *Nuper Rosarum Flores (1436)* , and Xenakis'

[10] examples can be found in *Sol (2005)* [31] `http://www.sol-sol.de` and *Brilliant Noise (2006)* by semiconductor `http://www.semiconductorfilms.com` last checked: 20/08/2011

[11] see for instance *Flood Tide (2008)* and *Hour Angle (2008)* by Eacott: `http://www.informal.org` last checked: 20/08/2011

mapping of statistical and stochastic processes to sound in *Metastasis (1965)* and other works [12].

Another approach to PMSon for musical purposes is the mapping of geographical coordinates to sound. Larry Austin's *Canadian Coastlines: Canonic Fractals for Musicians and Computer Band* (1981) in which the traced coastal contour dictated compositional data. A more recent creative example of PMSon is Jonathan Berger's *Jiyeh* (2008)[13], (of which there are two versions, one for eight channel computer-generated audio, and another for violin, percussion, cimbalom, and string orchestra). *Jiyeh* maps the contours of oil dispersion patterns from a catastrophic oil spill in the Mediterranean Sea. Using a sequence of satellite images, the spread of the oil over a period of time was, in the electroacoustic version, sonified and scaled both temporally and spatially to provide a sense of the enormity of the environmental event.

PMSon has been used in sonifying biofeedback. Alvin Lucier's work *Music for Solo Performer (1965)* sounds the composer's brain activity through electroencephalogram scalp electrodes and used the speaker output to resonate percussion instruments (an indirect mapping into instrumental space that, while musically effective renders only limited meaningful information).

Electroencephalography has provided source material for a number of composers. Knapp et al. at the Music, Sensors, and Emotion (MUuSE) lab at the Sonic Art Research Centre (SARC) at Queen's University Belfast [14], continually develop and explore tools to monitor various electrical activities of the body through electroencephalogram, electrooculogram, electrocardiogram, and electromyogram, and translate this data into MIDI control messages and other musical signal representations.

Image based sonifications used for artistic purposes have been realized with the software SONART [5]. Tanka has created various works that translate images into sounds amongst which *Bondage (2004)* [69] maps an image to spectral sound features. A similar approach can be found in Grond's *Along the Line (2008)* [15], which explores space filling curves as a means for spectral mapping.

A great deal of works involving PMSon in motion-tracked dance [53] have been created, and a number of robust software environments such as Isadora [16] have been developed to support these efforts. A number of sound installations use environmental data including Berger's *Echos of Light and Time (2000)* which continually polled the intensity of sunlight and temperature over 18 months in a collaboration with sculptor Dale Chihuly. Halbig's *Antarktika (2006)* [17] translates ice-core data reflecting the climatic development of our planet into the score for a string quartet. Chafe's *Tomato Quintet (2007,2011)* [18] sonifies the ripening process of tomatoes. The ripening process mapped carbon dioxide, temperature and light

[12] For a collection of 'historic' sonifications and mappings see: `http://locusonus.org/nmsat` last checked: 20/08/2011

[13] `http://ccrma.stanford.edu/~brg/jiyeh/` last checked: 20/08/2011

[14] `http://www.somasa.qub.ac.uk/~MuSE/?cat=1` last chacked: 20/08/2011

[15] `http://www.grond.at/html/projects/along_the_line/along_the_line.htm` last checked: 20/08/2011

[16] `http://www.troikatronix.com/isadora.html` last checked: 20/08/2011

[17] `http://www.antarktika.at` last checked: 20/08/2011

[18] `https://ccrma.stanford.edu/~cc/shtml/2007tomatoQuintet.shtml` last checked: 20/08/2011

readings from sensors in each vat to synthesis and processing parameters. Subsequently, the duration of the resulting sonification was accelerated to different time scales.

15.14 Conclusion

Virtually every domain of research has been marked by an explosive rise in data. Scientific and diagnostic data has become not only more voluminous but also far more complex. The need for effective exploratory analysis of that data bears new demands for novel methods of interpretation and representation.

Mapping of a data parameter to an auditory parameter is, in some cases, the most appropriate means of representing a data trend. This is particularly true when the purpose of the display is to represent general trends. The rise and fall of temperature or commodity prices, for example, can be readily heard. Furthermore, simultaneous display of multiple trends can often be effectively represented by segregating the mode of representation by distinguishing data sets according to basic principles of auditory scene analysis. For example, we can distinguish between temperature changes in two locations by spatially segregating the two sets. Similarly, we can retain coherent independent auditory scenes by representing two data sets with highly contrasting timbres.

The degree to which we can maintain perceived integrity of multiple simultaneously displayed data sets is dependent partly upon the degree of parameter segregation maintained and to individual abilities.

As new tools and mappings emerge PMSon will undoubtedly continue to seek new methods of auditory display. Among the challenges to be sought is to reach beyond auditory representation of the gestalts of curves (as in auditory graphs) to represent shapes in higher dimensional data sets.

The opportunities and challenges of PMSon are complicated by the inherent perceptual entanglement of sound synthesis and processing parameters, which constitute the two big challenges in the formalization of the mapping function. Establishing perceptually valid mappings between data and signal synthesis and/or processing parameters is key. This perceptual validity includes not only clearly audible categorical boundaries but also the ability to express data with variation in parameter changes that produce appropriately smooth transitions. Thus, the 'punchline' of this chapter is that it is crucial to fully understand the mutual influences between data preparation and the selected sound synthesis method, in order to design a successful auditory display with PMSon.

The PMSon design process diagram in Figure 15.1 is meant to provide the sonification designer with practical guidelines. We hope this diagram helps to identify and to address all challenges when conceptualizing, programming, tuning, reimplementing, and last but not least listening to the sounds of a PMSon.

Bibliography

[1] N. H. Adams and G. H. Wakefield. State-space synthesis of virtual auditory space state-space synthesis of virtual auditory space state-space synthesis of virtual auditory space. *IEEE Transactions on Audio, Speech and Language Processing*, 16(5):881–890, July 2008.

[2] S. Barrass. *Auditory Information Design*. PhD thesis, Australian National University, 1997.

[3] S. Barrass. Some golden rules for designing auditory displays. In R. Boulanger, editor, *Csound Textbook – Perspectives in Software Synthesis, Sound Design, Signal Processing and Programming*. MIT Press, 1998.

[4] O. Ben-Tal and J. Berger. Creative aspects of sonification. *Leonardo*, 37(3):229–232, 2004.

[5] O. Ben-Tal, J. Berger, P. R. Cook, M. Daniels, and G. Scavone. SonART: The sonification application research toolbox. In R. Nakatsu and H. Kawahara, editors, *Proceedings of the 8th International Conference on Auditory Display (ICAD2002)*, Kyoto, Japan, July 2-5 2002. Advanced Telecommunications Research Institute (ATR), Kyoto, Japan.

[6] M. M. Blattner, R. M. Greenberg, and M. Kamegai. Listening to turbulence: An example of scientific audiolization. In M. M. Blattner and R. Dannenberg, editors, *Multimedia Interface Design*, pages 87–107. Addison-Wesley, 1992.

[7] T. L. Bonebright. A suggested agenda for auditory graph research. In E. Brazil, editor, *Proceedings of the 11th International Conference on Auditory Display (ICAD 2005)*, pages 398–402, Limerick, Ireland, 2005. Department of Computer Science and Information Systems, University of Limerick.

[8] T. L. Bonebright, M. A. Nees, T. T. Connerley, and G. R. McCain. Testing the effectiveness of sonified graphs for education: A programmatic research project. In *Proc. Int. Conf. Auditory Display*, pages 62–66. Citeseer, 2001.

[9] T. Bovermann, T. Hermann, and H. Ritter. AudioDB. Get in Touch With Sound. In *Proceedings of the 14th International Conference on Auditory Display (ICAD 2008)*, Paris, France, 2008. inproceedings.

[10] T. Bovermann, J. Rohrhuber, and H. Ritter. Durcheinander. understanding clustering via interactive sonification. In *Proceedings of the 14th International Conference on Auditory Display (ICAD 2008)*, Paris, France, 2008. inproceedings.

[11] T. Bovermann, R. Tünnermann, and T. Hermann. Auditory augmentation. *International Journal of Ambient Computing and Intelligence (IJACI)*, 2(2):27–41, 2010.

[12] A. S. Bregman. Auditory scene analysis: The perceptual organization of sound. *MITPress*, September 1994.

[13] S. A. Brewster. *Providing a structured method for integrating non-speech audio into human-computer interfaces*. PhD thesis, University of York, York, UK, 1994.

[14] L. M. Brown and S. A. Brewster. Drawing by Ear: Interpreting Sonified Line Graphs. In *Proceedings of the 9th International Conference on Auditory Display (ICAD 2002)*, Boston, Mass., USA, 2003.

[15] R. J. Cassidy, J. Berger, K. Lee, M. Maggioni, and R. R. Coifman. Auditory display of hyperspectral colon tissue images using vocal synthesis models. In S. Barrass and P. Vickers, editors, *Proceedings of the 10th International Conference on Auditory Display (ICAD 2004)*, Sydney, Australia, 2004.

[16] M. Chion. *Guide To Sound Objects: Pierre Schaeffer and Musical Research, trans. John Dack and Christine North*. Éditions Buchet/Chastel, 1983.

[17] J. M. Chowning. The simulation of moving sound sources. *Computer Music Journal*, 1(3):48–52, June 1977.

[18] J. M. Chowning. Frequency modulation synthesis of the singing voice. In M. V. Mathews and J. R. Pierce, editors, *Current Directions in Computer Music Research Cambridge*, pages 57–63. MIT Press, 1980.

[19] P. R. Cook. *Identification of Control Parameters in an Articulatory Vocal Tract Model, with Applications to the Synthesis of Singing (CCRMA thesis)*. PhD thesis, Elec. Engineering Dept., Stanford University (CCRMA), Dec, 1990.

[20] A. Csapo and P. Baranyi. Perceptua interpolation and open-ended extrapolation of auditory icons and earcons. In *Proceedings of the 17th International Conference on Auditory Display (ICAD 2011)*, June 20 – 24 2011.

[21] A. deCampo. Toward a data sonification design space map. In Gary P Scavone, editor, *Proceedings of the 13th International Conference on Auditory Display*, pages 342–347, Montreal, Canada, 2007. Schulich School of Music, McGill University, Schulich School of Music, McGill University.

[22] S. Devore and B. Delgutte. Effects of reverberation on the directional sensitivity of auditory neurons across the tonotopic axis: Influences of interaural time and level differences. *The Journal of Neuroscience*, 30(23):7826–7837, June 2010.

[23] N. Diniz, M. Demey, and M. Leman. An interactive framework for multilevel sonification. In B. Bresin, T. Hermann, and A. Hunt, editors, *Proceedings of ISon 2010, 3rd Interactive Sonification Workshop, KTH, Stockholm, Sweden, April 7, 2010*, 2010.

[24] J. Feng, D. Yi, R. Krishna, S. Guo, and Buchanan-Wollaston V. Listen to genes: Dealing with microarray data in the frequency domain. *PLoS ON*, 4:e5098, April 2009.

[25] J. H. Flowers. Thirteen years of reflection on auditory graphing: Promises, pitfalls, and potential new directions. In *Proceedings of ICAD 05-Eleventh Meeting of the International Conference on Auditory Display,*, Ireland, Limerick, July 6–9 2005.

[26] J. Fox, J. Carlisle, and J. Berger. Sonimime: sonification of fine motor skills. In *Proceedings of the 11th Meeting of the International Conference on Auditory Display (ICAD 2005)*, Limerick, Ireland, July 6–9 2005.

[27] M. Gröhn and T. Takala. Magicmikes – multiple aerial probes for sonification of spatial datasets. In G. Kramer and S. Smith, editors, *Proceedings of the 2nd International Conference on Auditory Display*, pages 271–272, Santa Fe, New Mexico, 1994.

[28] F. Grond. Organized Data for Organized Sound Space Filling Curves in Sonification. In G. P. Scavone, editor, *Proceedings of the 13th International Conference on Auditory Display*, pages 476–482, Montreal, Canada, 2007. Schulich School of Music, McGill University, ICAD.

[29] F. Grond, T. Bovermann, and T. Hermann. A supercollider class for vowel synthesis and its use for sonification. In D. Worrall, editor, *Proceedings of the 17th International Conference on Auditory Display (ICAD 2011)*, Budapest, Hungary, June 20–24 2011. OPAKFI.

[30] F. Grond, T. Droßard, and T. Hermann. SonicFunction Experiments with a Functionbrowser for the Blind. In *Proceedings of the 16th International Conference on Auditory Display (ICAD 2010)*, pages 15–21, Washington D.C., 2010. ICAD.

[31] F. Grond, F. Halbig, J. M. Jensen, and T. Lausten. *SOL Expo Nr.16*. ISSN 1600-8499. Esbjerg Kunstmuseum, Esbjerg, Denmark, 2005.

[32] F. Grond and T. Hermann. Singing function, exploring auditory graphs with a vowel based sonification. *Journal on Multimodal User Interfaces*, 2011 in press.

[33] F. Grond, O. Kramer, and T. Hermann. Interactive sonification monitoring in evolutionary optimization. In D. Worrall, editor, *Proceedings of the 17th International Conference on Auditory Display (ICAD 2011)*, Budapest, Hungary, June 20 - 24 2011. OPAKFI.

[34] L. Harrar and T. Stockman. Designing auditory graph overviews: An examination of discrete vs. continuous sound and the influence of presentation speed. In G. P. Scavone, editor, *Proceedings of the 13th International Conference on Auditory Display (ICAD2007)*, pages 299–305, Montreal, Canada, 2007. Schulich School of Music, McGill University.

[35] S. Haykin. *Neural networks - A comprehensive foundation (2nd ed.)*, chapter Chapter 9 "Self-organizing maps". Prentice-Hall, 1999.

[36] T. Hermann. *Sonification for Exploratory Data Analysis*. PhD thesis, Bielefeld University, 2002.

[37] T. Hermann, G. Baier, U. Stephani, and H. Ritter. Vocal sonification of pathologic EEG features. In T. Stockman, editor, *Proceedings of the 12th International Conference on Auditory Display*, pages 158–163, London, UK, 06 2006. International Community for Auditory Display (ICAD), Department of Computer Science, Queen Mary, University of London UK.

[38] T. Hermann, G. Baier, U. Stephani, and H. Ritter. Kernel Regression Mapping for Vocal EEG Sonification. In *Proceedings of the 14th International Conference on Auditory Display*, Paris, France, 2008.

[39] T. Hermann, K. Bunte, and H. Ritter. Relevance-based interactive optimization of sonification. In W. Martens, editor, *Proceedings of the 13th International Conference on Auditory Display*, pages 461–467, Montreal, Canada, 06 2007. International Community for Auditory Display (ICAD), ICAD.

[40] T. Hermann, P. Meinicke, H. Bekel, H. Ritter, H. Müller, and S. Weiss. Sonification for EEG Data Analysis. In R. Nakatsu and H. Kawahara, editors, *Proceedings of the 2002 International Conference on Auditory Display*, pages 37–41, Kyoto, Japan, July 2002. International Community for Auditory Display (ICAD), ICAD.

[41] T. Hermann and H. Ritter. Listen to your data: Model-based sonification for data analysis. In G. E. Lasker, editor, *Advances in intelligent computing and multimedia systems*, pages 189–194, Baden-Baden, Germany,

08 1999. Int. Inst. for Advanced Studies in System research and cybernetics.

[42] T. Hermann and H. Ritter. Crystallization sonification of high-dimensional datasets. *ACM Trans. Applied Perception*, 2(4):550–558, 10 2005.

[43] A. Hunt and M. M. Wanderley. Mapping performer parameters to synthesis engines. *Invited article for Organised Sound, special issue on Mapping*, 7(2):97–108, 2002.

[44] A. D. Hunt, M. Paradis, and M. M. Wanderley. The importance of parameter mapping in electronic instrument design. *Invited paper for the Journal of New Music Research, SWETS, special issue on New Interfaces for Musical Performance and Interaction,*, 32(4):429–440, December 2003 2003.

[45] I. T. Jolliffe. *Principal Component Analysis*. Springer-Verlag, 1986.

[46] D. H. Klatt. Software for a cascade/parallel formant synthesizer. *Journal of the Acoustical Society of America*, 67(3):971–995, March 1980.

[47] M. Kleiman-Weiner and J. Berger. The sound of one arm swinging: A model for multidimensional auditory display of physical motion. In *Proceedings of the 12th International Conference on Auditory Display, London, UK*, 2006.

[48] G. Kramer. *Auditory Display: Sonification, Audification, and Auditory Interfaces*. Perseus Publishing, 1993.

[49] M. Lagrange, G. Scavone, and P. Depalle. Time-domain analysis / synthesis of the excitation signal in a source / filter model of contact sounds. In *Proceedings of the 14th International Conference on Auditory Display*, Paris, France, 2008. inproceedings.

[50] J. Lumsden, S. A. Brewster, M. Crease, and P. Gray. Guidelines for audio-enhancement of graphical user interface widgets. In *Proceedings of People and Computers XVI Memorable Yet Invisible: Human Computer Interaction 2002 (HCI-2002)*, London, 2002.

[51] D. L. Mansur, M. M. Blattner, and K. I. Joy. Sound graphs: A numerical data analysis method for the blind. *Journal of Medical Systems*, 9(3), 1985.

[52] B. C. J. Moore. Chapter thirteen. loudness, pitch and timbre. In E. B. Goldstein, editor, *Blackwell Handbook of Sensation and Perception*, page 388. Blackwell, 2005.

[53] R. Morales-Manzanares, E. F. Morales, Dannenberg R., and J. Berger. Sicib: An interactive music composition system using body movements. *Computer Music Journal*, 25(2):25–36, 2001.

[54] M. A. Nees and B. N. Walker. Listener, task, and auditory graph: Toward a conceptual model of auditory graph comprehension. In Gary P. Scavone, editor, *Proceedings of the 13th International Conference on Auditory Display (ICAD2007)*, pages 266–273, Montreal, Canada, 2007. Schulich School of Music, McGill University.

[55] J. Neuhoff, J. Wayand, and G. Kramer. Pitch and loudness interact in auditory displays: Can the data get lost in the map? *Journal of Experimental Psychology: Applied*, 8(1):17–25, 2002.

[56] E. O. Otis. President's Address. Auenbrugger and Laennec, the Discoverers of Percussion and Auscultation. *Trans Am Climatol Assoc.*, 14:1–23, 1898.

[57] G. E. Peterson and H. L. Barn. Control methods used in a study of the vowels. *Journal of the Acoustical Society of America*, 24:175–184, 1952.

[58] H. Pollard and E. Janson. A tristimulus method for the specification of musical timbre. *Acoustica*, 51:162–171, 1982.

[59] A. Polli. Atmospherics/weather works: A spatialized meteorological data sonification project. *Leonardo*, 38(1):31–36, 2005.

[60] M. Rath and D. Rocchesso. Continuous sonic feedback from a rolling ball. *IEEE Interactive Sonification*, 2005.

[61] J. Rohrhuber. $\mathring{S}$ – Introducing sonification variables. In *Proceedings of the Supercollider Symposium*, 2010.

[62] J. B. Rovan, M. M. Wanderley, S. Dubnov, and P. Depalle. Instrumental gestural mapping strategies as expressivity determinants in computer music performance. In *Proceedings of the Kansei – The Technology of Emotion Workshop*, pages 68–73, Genova – Italy, 1997.

[63] P. Schaeffer. *Traité des objets musicaux*. Le Seuil, Paris, 1997.

[64] R. Shelton, S. Smith, T. Hodgson, and D. Dexter. Mathtrax website: `http://prime.jsc.nasa.gov/MathTrax/index.html`.

[65] B. G. Shinn-Cunningham and T. Streeter. Spatial auditory display: Commentary on shinn-cunningham et al. icad 2001. *ACM Trans. Applied Perception*, 2(4):426–429, 2005.

[66] D. R. Smith and B. Walker. Effects of auditory context cues and training on performance of a point estimation sonification task. *Applied Cognitive Psychology*, 19:1065–1087, 2005.

[67] D. R. Smith and B. N. Walker. Tick-marks, axes, and labels: The effects of adding context to auditory graphs. In *Proceedings of the 8th International Conference on Auditory Display (ICAD2002)*, pages 362–367, Kyoto, Japan, 2002.

[68] T. Stockman, L. V. Nickerson, and G. Hind. Auditory graphs: A summary of current experience and towards a research agenda. In Eoin Brazil, editor, *Proceedings of the 11th International Conference on Auditory Display (ICAD 2005)*, pages 420–422, Limerick, Ireland, 2005. Department of Computer Science and Information Systems, University of Limerick.

[69] A. Tanaka. The sound of photographic image. *Artificial Intelligence and Society*, 27(2), 2012.

[70] H. Terasawa, M. Slaney, and J. Berger. Perceptual distance in timbre space. In Eoin Brazil, editor, *Proceedings of the 11th Meeting of the International Conference on Auditory Display (ICAD2005)*, pages 61–68, Limerick, Ireland, 2005.

[71] D. Van Nort. *Modular and Adaptive Control of Sound Processing*. PhD thesis, McGill University, 2010.

[72] P. Vickers. Whither And Wherefore The Auditory Graph? Abstractions & Aesthetics In Auditory And Sonified Graphs. In E. Brazil, editor, *Proceedings of ICAD 05-Eleventh Meeting of the International Conference on Auditory Display*, pages 423–442. ICAD, 2005.

[73] P. Vickers and J. L. Alty. Towards some organising principles for musical program auralisations. In S. A. Brewster and A. D. N. Edwards, editors, *Proceedings of the 5th International Conference on Auditory Display*, Glasgow, Scotland, 1998.

[74] K. Vogt. A quantitative evaluation approach to sonification. In D. Worrall, editor, *Proceedings of the 17th International Conference on Auditory Display (ICAD-2011)*, Budapest, Hungary, June 20 – 24 2011. OPAKFI.

[75] K. Vogt, T. Bovermann, P. Huber, and A. deCampo. Exploration of 4d-data spaces. sonification in lattice qcd. In *Proceedings of the 14th International Conference on Auditory Display*, Paris, France, 2008. inproceedings.

[76] K. Vogt, D. Pirro, I. Kobenz, R. Höldrich, and G. Eckel. Physiosonic - movement sonification as auditory feedback. In *Proceedings of the 15th International Conference on Auditory Display, Copenhagen, Denmark*, 2009.

[77] B. N. Walker. *Magnitude estimation of conceptual data dimensions for use in sonification*. PhD thesis, Rice University, Houston, TX., 2000.

[78] B. N. Walker and J. T. Cothran. Sonification sandbox: A graphical toolkit for auditory graphs. In E. Brazil and B. Shinn-Cunningham, editors, *Proceedings of the 9th International Conference on Auditory Display (ICAD2003)*, pages 161–163, Boston, USA, 2003. Boston University Publications Production Department.

[79] B. N. Walker and G. Kramer. Mappings and metaphors in auditory displays: An experimental assessment. In *Proceedings of the Third International Conference on Auditory Display ICAD96*, Palo Alto, CA, USA, 1996.

[80] B. N. Walker and D. M. Lane. Psychophysical scaling of sonification mappings: A comparison of visually impaired and sighted listeners. In *Proceedings of the Seventh International Conference on Auditory Display ICAD2001*, pages 90–94., 2001.

[81] I. Wallis, T. Ingalls, T. Rikakis, L. Olsen, Y. Chen, W. Xu, and H. Sundaram. Real-time sonification of movement for an immersive stroke rehabilitation environment. In *Proceedings of the 13th International Conference on Auditory Display, Montréal, Canada*, 2007.

[82] S. Williams. Perceptual principles in sound grouping. In G. Kramer, editor, *Auditory Display: Sonification, Audification and Auditory Interfaces - SFI Studies in the Sciences of Complexity*, volume XVIII, Reading, MA., 1994. Addison-Wesley.

[83] D. Worrall. *Sonification and Information - Concepts, Instruments and Techniques*. PhD thesis, University of Canberra, 2009.

[84] M. Wright, A. Freed, and A. Momeni. Opensound control: State of the art 2003. In *Proceedings of the 2003 International Conference on New Interfaces for Musical Expression (NIME*, Montreal, Quebec, Canada, 2003.

[85] W. Seung Yeo and J. Berger. A framework for designing image sonification methods. In *Proceedings of ICAD 05-Eleventh Meeting of the International Conference on Auditory Display*, Limerick, Ireland, 2005.

[86] E. Yeung. Pattern recognition by audio representation of multivariate analytical data. *Analytical Chemistry*, 52(7):1120–1123, 1980.

[87] H. Zhao, C. Plaisant, B. Schneiderman, and R. Duraiswami. Sonification of geo-referended data for auditory information seeking: Design principle and pilot study. In *Proceedings of the 10th International Conference on Auditory Display (ICAD 2004)*, 2004.

[88] H. Zhao, B. Shneiderman, C. Plaisant, D. N. Zotkin, and R. Duraiswami. Improving accessibility and usability of geo-referenced statistical data. In *Proc. of the Digital Government Research Conference*, page 147, 2003.

Chapter 16

Model-Based Sonification

Thomas Hermann

16.1 Introduction

Almost every human activity in the world is accompanied with an acoustic response. Interaction in the world typically provides us with rich feedback about the nature of the involved materials, as well as the strength and type of contact. It is stunning that, despite the ubiquity of action-driven informative sounds, we have tended to limit traditional computer interfaces to visual-only displays. *Model-Based Sonification* is a sonification technique that takes a particular look at how acoustic responses are generated in response to the user's actions, and offers a framework to govern how these insights can be carried over to data sonification. As a result, Model-Based Sonification demands the creation of *processes* that involve the data in a systematic way, and that are capable of evolving in time to generate an acoustic signal. A *sonification model* is the set of instructions for the creation of such a "virtual sound-capable system" and for how to interact with it. Sonification models remain typically silent in the absence of excitation, and start to change according to their dynamics only when a user interacts with them. The acoustic response, or sonification, is directly linked to the temporal evolution of the model.

Model-Based Sonification has been introduced by the author [14] and was elaborated in more detail [10]. Several sonification models have been developed since then [21, 27, 4, 23, 18, 3, 16, 12, 15, 10, 11, 13, 14], which give examples for model design, exploration tasks in the context of exploratory data analysis and interaction modes.

This chapter gives a full introduction to Model-Based Sonification (MBS), including its definition, some design guidelines, description of selected sonification models and a discussion of the benefits and problems of MBS in general. Since MBS is a *conceptually* different approach than Audification and Parameter Mapping Sonification, its relation to these will be addressed in detail. Finally, a research agenda for MBS is formulated.

16.1.1 Listening modes

A very helpful experiment to understand how the human auditory system works, is to play a short example sound and ask listeners to describe in as much detail as they can what they  have heard. As experiment the reader might try this now with sound example **S16.1**. Please stop reading here until you have listened to the sound, and be as accurate as possible and write down keywords of your description. Done?

Most listeners will now have characterized the sound by a guess of what *source* or *action* might have caused the sound. For instance, you may have described the sound as 'somebody is coughing', 'surely a male, and not a child', 'it sounds like bronchitis', and so on. Such descriptions are very typical and we are not aware of how dominating this source-identification default is. Let us call this listening mode *everyday listening*.

There is, however an alternative way to characterize the example, as 'a sequence of 7 noise bursts', 'their roughness and loudness decreases', 'they form a certain rhythmical pattern', and so on, characterizing the sound by its acoustic shape, its rhythm, harmony, melody, pattern, structure, etc. Such a description is just as valid as the one given from everyday listening, only the focus is different: rather than focussing on the signified it describes the sign itself. [1]. Let us call this *musical listening*. We can indeed experience our world with 'other ears' just by purposefully changing our listening mode.

Obviously, our brain and auditory system is capable of operating in different modes, and 'everyday listening' is the dominant or default mode. This is possibly because an accurate and quick sound source identification was evolutionarily advantageous since it enabled quick and correct *reaction*, e.g., to choose to flight or fight [17]. This argumentation would at least explain why our brain is specifically good at interpreting the sound source and source characteristics with a focus on the appropriate reaction rather than on conscious reflection.

There is yet another mode of listening, which we may call *analytical everyday listening, see listening, modes of*: this is the conscious use of all listening skills to distinguish and analyze an object under investigation. To give some examples, think of the task of determining the contents of an opaque box by shaking it, or the task of diagnosing a malfunctioning car engine from the sounds it makes. Such analytical listening is a 'diagnostic' use of listening, and thus most inspiring to be used for sonification.

The above list of listening modes is certainly incomplete. For instance the particular modes of listening to language and speech sounds have not been mentioned, or the enjoyment mode when listening to music. A discussion on listening modes can be found in [7] and [17]. Model-Based Sonification addresses our everyday listening and analytical listening skills. In the following section we categorize functions and contexts of sounds in order to better understand how information is encoded into sounds in our physical world.

16.1.2 Sound and Information

The sounds that we have heard in our lives can be categorized in the following classes:

Passive sounds: sounds that come from an external source, not directly caused by our own activity. These sounds give us information about the environment (e.g., a sense of

[1] more on semiotics can be found in chapter 18

where we are), and may direct our attention or even alert us.

Active sounds: sounds that are created in the course of physical activity, which directly accompany the owner's actions. Examples are the rustle of clothes while moving, the clip-clop of footsteps, the soft hiss of breathing, or contact sounds in response to direct or indirect manipulation of physical objects.

There is no strict separation between these classes as, for instance, actions may cause passive sounds. Also, other people's active sounds are indeed passive sounds for us as listeners. Most active sounds are a by-product of the activity and not its goal. As a special case we can identify *intentional active sounds* as active sounds where the subject has performed the (inter-)action intentionally in order to create the sound. Playing musical instruments, shaking an opaque box in order to learn about its content by listening, and clapping the hands to understand the surrounding reverberation characteristics are some examples for such intentional interactions.

Language sounds and musical sounds are highly specific to a cultural tradition, and the relation between the sounds and their meaning are largely learned or memorized bindings. The semantics of sound on the more basic level of environmental sounds and interaction sounds, however, is more universal. Sonification techniques that rely on sounds which the typical human is likely to have encountered are likely to be more culturally independent. For this reason, we now take a closer look at how information is encoded in real-world acoustics or physical sounds.

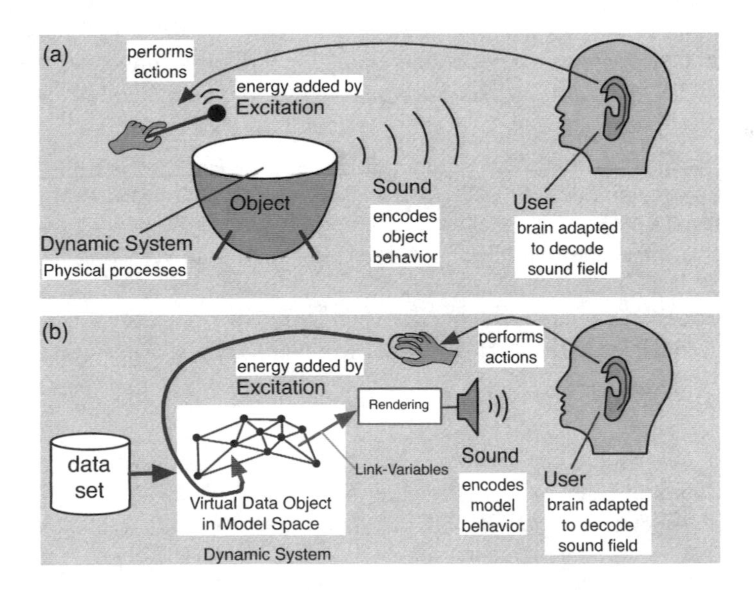

Figure 16.1: (a) Sonic loop in physical interaction: the user is tightly integrated into a closed-loop. The brain is adapted to interpret sonic patterns for source properties and to explore sound dependencies for the given excitation. (b) shows the modifications from real-world sonic loops for a typical Model-Based Sonification, as explained in the text.

Fig. 16.1 (a) illustrates a sonic loop from physical interaction with an object to the interpretation of the sound in the listener's mind. As starting point let us investigate the relation between the sound-capable object and the listener: audible sound is simply a pattern of vibration in a suitable frequency range (20 Hz to 16000 Hz), and we typically perceive sound since it is transported to our ears via sound waves that propagate through air. The detailed pattern of pressure variation, however, is a direct but complex image (or copy) of the vibrating object. The encoding of information into the wave field is neither unique nor invertible: different objects may lead to identical sound signals. Equally, an identical object may under repeated interaction also create slightly different sound responses. How do source properties then relate to the sound signals' There is unfortunately no simple answer to that question. A structural change of the physical object will typically lead to completely different sound signals, so that we may assume that the source properties are holistically encoded into the sound wave field. In addition, the sound will change sensitively with any change of the interaction.

It seems hopeless and overly complex to quickly obtain an inverse estimation of source properties from such distributed information. Bregman compares it to the task of estimating the number of ships, their position and velocity by simply looking at the fluctuations in the water waves in a pair of one meter long channels dug into the beach at the edge of a pond [6]. It would seem impossible to answer these questions from visually observing the water levels going up and down. However, the example is an analogy for human listening with the channels representing our ear canals. With our auditory systems we find that such inverse mappings (which infer source properties via incoming sound signals) are perfectly feasible, and our listening system has even been optimized to infer source-relevant information. Experiments have, for instance, demonstrated that material, size and rolling speed of balls on a surface can be perceived quite accurately [19]. For sonification, we can thus hope to exploit these *inverse mapping skills* to understand systems and in turn the underlying data. Physical systems as shown in Fig. 16.1 typically possess dissipative elements. Internal friction and the radiation of sound waves cause energy loss which makes physical systems converge towards a state of equilibrium. Since in this state there is no more vibration, the sound fades to silence. Often systems are excited and thus perturbed actively from their state of equilibrium by our own interaction. We can think of interaction as actively querying the object, which answers with sounds. Since we can reproduce sounds by repeated interaction, and thereby understand the systematic changes in sound that correlate with our change of excitation, we can gradually build up a mental representation which enables the miraculous inverse mapping from sound to interaction.

In summary, everyday sounds often stem from a closed-loop system where interactions are followed by physical/acoustic reactions which then lead to auditory perceptions and their cognitive interpretation. The human is tightly embedded in the loop and assesses source properties via the indirect *holistic* encoding in action-driven sounds.

16.1.3 Conclusions for Sonification

If we take the abovementioned observations from real-world sonic interactions seriously, there are several consequences for inherently interactive data sonifications:

ubiquity: almost every interaction with data should be accompanied by sound (as almost any interaction with the world causes some sound).

invariance of binding mechanism: the sound-producing laws should be invariant and structurally independent of the actual data – in the same way that the laws of physics and their invariance means that we can understand different objects in the world by attending to their sounds when we interact with them.

immediate response: sonifications should deliver an immediate (real-time) response to the interaction since this is the action-perception pattern we are familiar with from real-world interaction. The brain is tuned to interpret sound in this way; it is even optimized to associate synchronization between different modalities, e.g. our proprioception, visual changes and correlating acoustic patterns.

sonic variability: sonifications should depend on a subtle level on the interaction and data, in the same way that real-world sounds are never strictly identical at the sample level on repeated interaction, but depend very much on the actual dynamic state and the details of excitation.

information richness: sonifications should be 'non-trivial'. In other words they should be complex and rich on different layers of information. This is similar to the way that everyday sounds are complex, due to nonlinearities in the physical systems which produce them. It seems that the human brain expects this 'non-trivialness' and values it highly. If it is missing, the sounds may be perceived as boring, or just may not connect as well as possible with our auditory listening skills.

Model-Based Sonification offers a framework for the creation of sonification models which *automatically* behave according to these requirements, which underly sound generation in the real-world, as depicted in Fig. 16.1. How this is achieved is described in detail in the following section.

16.2 Definition of Model-Based Sonification

Model-Based Sonification (MBS) is defined as the general term for all concrete sonification techniques that make use of dynamic models which mathematically describe the evolution of a system in time, parameterize and configure them during initialization with the available data and offer interaction/excitation modes to the user as the interface to actively query sonic responses which depend systematically upon the temporal evolution model. In this section we will review the different 'ingredients' or elements of this complex and lengthy definition step-by-step. Hopefully this will clarify what is meant and how MBS is generally different from mapping sonification.

Model-Based Sonification (MBS) is the general framework or paradigm for how to define, design and implement specific, task-oriented sonification techniques. A specific design or instance obtained with MBS is called a *sonification model*. Model-Based Sonifications draw inspiration from physics, yet the designer is free to specify otherwise and may even invent non-physical dynamic models. A good procedure for the design of sonification models according to the MBS framework is given by the step-by step definition of the following six components: *setup*, *dynamics*, *excitation*, *initial state*, *link-variables*, and *listener characteristics*, which will be described in turn.

These steps are illustrated by using a simple MBS sonification model called *data sonograms*. In a nutshell, the data sonogram sonification model allows the user to excite a shock wave in data space that slowly propagates spherically. The wave-front in turn excites mass-spring systems attached at locations specified by each data point's coordinates . Fig. 16.3 on page 409 illustrates this setup. Using this sonification users can experience the spatial organization of data and how data density changes relative to the shock wave excitation center. While this sonification model is helpful for a MBS tutorial it should be emphasized that it is only one particular example model – other models can be structurally very different, as will hopefully become clear in section 16.3.

Model-Based Sonification mediates between data and sound by means of a dynamic model. The data neither determine the sound signal (as in audification) nor features of the sound (as in parameter mapping sonification), but instead they determine *the architecture of a 'dynamic' model* which in turn generates sound. Thereby MBS introduces the model space between the data space and the sound space, as depicted in Fig. 16.2.

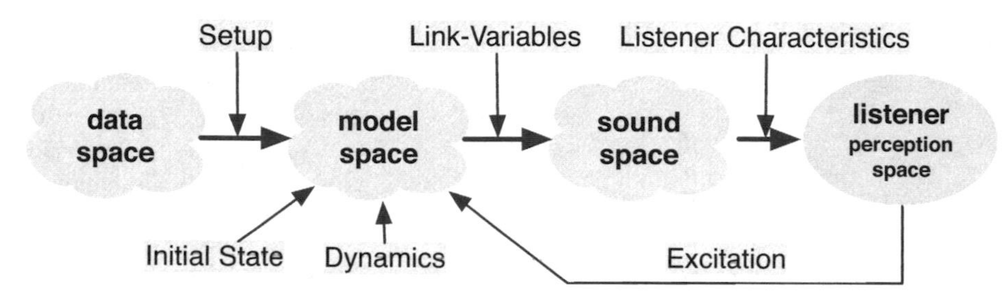

Figure 16.2: Transformations from data space via model space to sound space and to the listener's perception space. The elementary model specification steps are depicted at the location where they provide their specification.

16.2.1 MBS step 1: Model Setup

The model setup determines how data define the configuration of a dynamic system with internal degrees of freedom. The setup bridges the gap between the immaterial, abstract and static world of high-dimensional data and the more tangible world of a dynamic model where elements move in time and thereby cause the sound. It is helpful to distinguish between the data space and the model space. This may become clearer with a concrete example.

For example, assume that a d-dimensional data set with N records is given. The data set can then be represented as a table of N rows with d columns where each column is a feature and each row an instance or sample. In a census data set the columns could for instance be 'income', 'size', or 'sex', and the rows would be different persons. The cell values would then hold information such as 'this particular person's income in euros' etc. A frequently used representation in mathematics is that the data set defines a cloud of N points in the d-dimensional feature vector space, using the feature values as coordinates. So we can imagine the data space as a mathematical vector space.

With this representation in mind, a spatial setup of the model space is tempting. In the example of data sonograms the model setup is defined so that point masses are attached to springs so that they can oscillate or collide with each other. In this case the model space is also spatial, and its dimension may be chosen by the designer. For a d-dimensional data set, however, it is straightforward to create a model space of same dimensionality. Still, it remains the question of how the data vectors should determine these mass-spring elements in model space. Data sonograms use the data vector coordinates as the location vectors of the point to which the spring is attached. Of course, there are manifold different possibilities of how to connect the data space and the model space, examples of which are given in section 16.3.

16.2.2 MBS step 2: Model Dynamics

The ultimate goal is to get a sound signal which is a sonic representation of the data under analysis. Since sound evolves in time it makes sense to introduce a temporal evolution to the model, called *model dynamics*. More precisely, dynamics refers to the equations of motion that describe how the system's state vector changes in time, how the next state $\vec{s}(t + \Delta t)$ is computed from the actual state $\vec{s}(t)$. Since we are dealing with a mathematical description of the model, the equations of motion are usually specified as differential equations, similar to the equations of motion that describe how a mechanical system changes with time. Certainly other laws from electrodynamics, chemistry, or even machine learning are sometimes useful.

For the data sonogram model example where point masses are attached to springs, we need to specify how to update the position and velocities of each mass when the springs exert an actual force to the mass. The dynamics are given by equations $\vec{s}(t + \Delta t) = f(\vec{s}(t))$ which are inspired from physics and the mechanics of spring-mass systems.

Models may need several mechanisms of dynamic behavior. For the data sonogram model, for instance, we need dynamical laws that describe how excitation causes shock waves and how these waves propagate in model space, or how they interact with mass-spring systems. Other mechanisms such as energy flow are presented in section 16.3.

Physical principles such as kinetic and potential energy, and furthermore dissipation mechanisms such as friction, and specifically principles from acoustics provide rich inspiration on how to introduce dynamics that create a specific qualitative behavior. Not only do model developers need to specify the equations of motion, but most dynamical laws demand the inclusion of certain parameters that need to be adjusted. The parameter choice seems to be a source of arbitrariness, yet this is not really a problem if the parameters remain unchanged whatever data set is explored. Then the listener can adapt to the specific sonifications that are implied with the given dynamics and parameter settings.

In addition, the number of parameters is normally much lower than those needed for the specification of a parameter mapping sonification, and furthermore they also have a clear 'physical' meaning with respect to the model, which makes it easy to understand how their change affects the sound. This will be elaborated later in section 16.7.

16.2.3 MBS step 3: Model Excitation

Excitation is a key element in MBS, since it defines how the users interact with the model.

In acoustic systems physical objects (e.g., a bell) eventually come to rest in a state of equilibrium without external excitation. In a similar way the dynamics of sonification models often contain a term which leads to a system state of equilibrium. Excitatory interactions allow the users to feed new energy into the system and in turn users experience the acoustic reaction as a direct response. Firstly this prevents never-ending sound which would be annoying after some time. Secondly it enables the users to bring in rich manual interaction skills to examine a system. Think for instance of how many ways there are to shake, squeeze, tilt, incline, hit, etc. an opaque box to probe its content; such interaction can then be defined for use in interacting with sonification models.

Formally, excitation can be modeled as an external force in the equations of motion, which depends on the state of controllers or input devices. In the data sonogram example, a mouse click triggers a shock wave in model space, but other interactions are possible. For instance, the shaking of the mouse could inject energy into all spring-mass systems within a certain radius simultaneously.

Excitation type can range from elementary triggering (e.g., a mouse click or keystroke) through more detailed punctual interactions such as hitting a surface at specific locations with a certain velocity, to continuous interactions such as squeezing, shaking, rubbing or deforming controllers or tangible objects. Certainly, a mixture of these interactions may occur, depending on the interfaces used.

The better the metaphor binds interaction to the sonification model, the more the users will be capable of developing intuition about model properties, and understanding how these manifest in the resulting sonic response. Therefore, the specification of excitation cannot be done without keeping in mind the bigger picture and the idea of the sonification model.

Besides the mandatory excitation modes, there may be additional interface-to-model couplings that allow users to influence the dynamics. In real life a bottle filled partially with water sounds different when hit at various locations while changing the bottle's orientation. In a similar vein it may make sense, for instance, to allow the user to excite the sonification model at one location while controlling other parameters by rotating or squeezing a controller etc. Such excitation via parameter-rich interfaces brings the users more tightly in touch with the model and allows them to make use of their already available interaction competence from real-world interactions.

16.2.4 MBS step 4: Initial State

The initial state describes the configuration of the sonification model directly after setup. One's first thought might be that this has already happened during the Model Setup phase, yet that merely defines the system and how data are used to determine the architecture of the sonification model. For instance, in the data sonogram sonification model, the data vectors determine the location that the springs are attached to, whereas the initial state would determine the initial location and velocities of the point masses. In other words the Setup phase actually creates the model and then the initialization stage puts it into position ready for the first user interaction.

Normally, the designer knows – from insight into the equations of motion – the equilibrium state and initializes the system accordingly. If this is not possible, that is not a problem since the model will anyway relax from a random initial state to an equilibrium state, assuming that there is built-in dissipation. To prevent disturbingly loud noises, however, it is strongly advisable in this case to mute the audio output until the system has relaxed a bit.

16.2.5 MBS step 5: Model Link-Variables

Link-variables are the 'glue' which connects the model's dynamic processes to sound as shown in Fig. 16.2. In the most straightforward manner, the model's state variables can be used directly as a sound signal, which would be a good and direct analogy to how sound is generated in real-world acoustic systems. Think, for instance, of a drum head whose movement describes more or less one-to-one the sound signal that propagates to the ear. Expressed in terms of sonification techniques such a direct connection of a dynamic state variable with a sound signal could be called *audification* of the model dynamics. Sometimes it is more useful to condense several state variables $x_1(t), \ldots, x_n(t)$ into a single sound signal $s(t)$ by means of a feature function $s(t) = f(x_1(t), \ldots, x_n(t))$. For instance, in the particle trajectory sonification model explained in section 16.3 the kinetic energy of each particle is used as a link variable for the sound signal.

For some sonification models, the designer may consider the linking of state variables in a more complex or indirect way to the sound signal. For instance, the designer might want to map the overall model energy to the sound level. Such explicit parameter mappings can occur in MBS model design, and even help to make sound computation more efficient, yet they introduce a level of arbitrariness and the need for explanation which MBS design principles suggest keeping at a minimum.

One main problem of Model-Based Sonification is that the computation of tens of thousands of update steps necessary to generate even one second of a sonification is complex and time-intensive and even with current computing power in 2011 this is beyond real-time rendition even for moderately large systems. The reason is that the equations of motion may be coupled and demand the computation of the distances to all elements (e.g., masses in the model space) for each single update step of each mass, which leads to an explosion of the number of operations with increasing number of elements. However, real-time computation is crucial for MBS to tightly close the interaction loop. For that reason, implementation shortcuts are often used, which decouple the model update from the sound signal generation to some extent. For the data sonogram example, instead of computing the detailed motion of the mass-spring-system at 44100 steps per second[2], it may suffice to compute the average energy of a mass spring system at 50 Hz and to apply sample-based interpolation between successive amplitude values of an appropriately tuned sine generator. The result may be an acceptable approximation of the *real* model output with a reduced number of operations per second. Similar implementation shortcuts are necessary for many sonification models to reach real-time computability, yet it is most likely that with increasing computing power in a few years they can be minimized or avoided. Actually, while such shortcut procedures may be fine on first sight, they may just cut out subtleties in the sound signals which our ears demand and are tuned to pick up. More examples for implementation shortcuts will be given in section 16.3.

[2]to render CD quality signals at 44100 Hz sampling frequency

16.2.6 MBS step 6: Listener Characteristics

In everyday interaction with sounding objects and environments we either experience an object as a single sound source (e.g., knocking on a melon), or we experience ourselves embedded into a distributed soundscape (e.g., birds in the forest). In the same sense there are sonification models where the suitable metaphor is that the model forms a single sounding object or that the users are located and embedded in a space with the model elements around them. Let us distinguish these types as *microscopic* vs. *macroscopic* sonification models.

Listener Characteristics addresses all issues related to location, orientation or distance between sound sources (link-variables) and the user/listener. Spatial (macroscopic) models usually demand a more complex rendition and sound output, either using multi-channel audio systems or HTRF-convolution [3]. Furthermore they may need head-tracking to achieve a truly convincing spatial model experience. In contrast, the microscopic sonification models are much simpler yet may nonetheless deliver the majority of the information. The metaphor is that the whole model becomes a single sounding object.

For the data sonogram sonification model, the listener is assumed to be located at the shock wave center, so this is a macroscopic sonification model. In a stereo sound setup, it makes sense to play spring-mass sound contributions with stereo panning using the orientation of the spring-mass system relative to the user.

16.3 Sonification Models

The MBS framework is very open, i.e. it enables very different model specifications using very different sources of inspiration. Before providing general guidelines for MBS design in section 16.4, it is helpful to briefly review some existing sonification models. This section gives such an overview, where the model definition steps (setup, dynamics, excitation, etc.) are explained as compact and figuratively as possible. Mathematical details can be found in the referenced articles. However, sound examples are provided and are briefly discussed to bring this section to life.

16.3.1 The Data Sonogram Sonification Model

This model (see Fig. 16.3) has already been used as tutorial example in section 16.2. In summary, the *model setup* is to use one mass-spring system per data vector in a model space of the same dimensionality as the data space, each spring being attached at positions given by the data vector. The user interacts with a scatter plot of the data set and excites shock waves that spherically propagate through the model space. The shock wave speed can be adjusted - typical values for full traversal through the model space are 2 - 5 seconds. The shock wave front, as it passes, displaces mass-spring elements from their equilibrium state and these oscillate with some damping around their position according to the given equations of motion. The resulting sum of all mass-spring displacements constitutes the sonification which is roughly spatialized in stereo around the listener who is imagined to rest at the shock

[3]HRTF = Head-related transfer functions

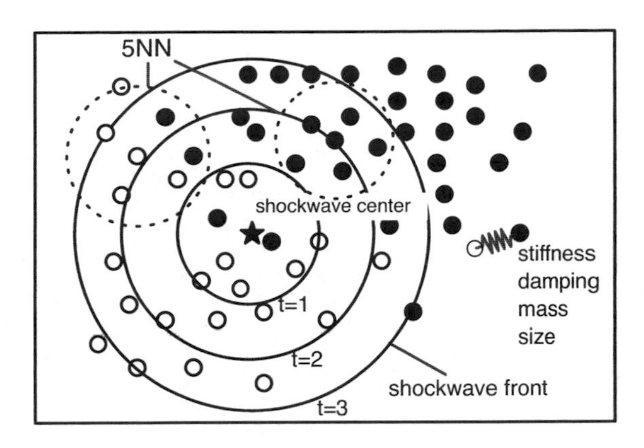

Figure 16.3: Data Sonogram Model Space

wave center. Both mouse clicks and multitouch displays have been used as interfaces to excite the system [14, 27].

Standard Data Sonograms provide information about the data density along a spherical sweep. But task-specific refinements of the model allow specific features such as the class label in data from classification problems to be used to control physical properties of the system, e.g. the stiffness or damping of the individual springs. In general, MBS allows for the definition of individual physical properties at hand of 'local' features. For instance, if the local class mixing entropy[4] among the nearest neighbors of each data point determines the spring stiffness, regions in the data space where different classes overlap will sound higher pitched since the higher local entropy leads to stiffer springs. This may be coined a 'class-border sensitive data sonogram' and it may be useful to quickly assess whether data from classification problems are separable or not. Data sonograms generally support an understanding of the clustering structure of data.

Sound examples **S16.2** are typical data sonograms for clustered data sets. More details on these examples can be found in [10].

16.3.2 Tangible Data Scanning

In Tangible Data Scanning (TDS), data points are represented by localized mass-spring systems just as in the Data Sonogram model as shown in Fig. 16.4). However, now the data are embedded into the 3D-space around the user. Thereby the model is mainly useful for 3D data, or for 3D projections of data. In contrast to data sonograms, interaction is very direct: the user moves a planar object such as a cardboard sheet as an interaction tool which is tracked by a motion capture system. Whenever the surface intersects a mass-spring system in the model space, the latter is excited and oscillates around its position. Even if the sound is played as monophonic audio, the directness allows the user to build up a mental model about the spatial data distribution. It suddenly makes sense to refer to the cluster 'down left around my left knee', or 'in *that* corner of the room'. Similar to Data Sonograms, modified /

[4] which is high when neighboring data points belong to different classes

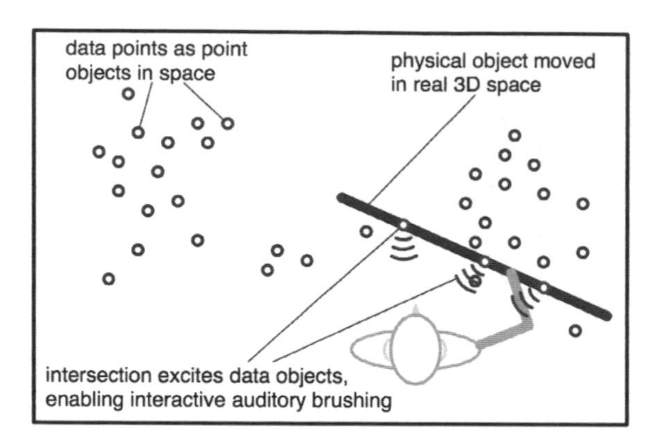

Figure 16.4: Tangible Data Scanning

derived models can use more elaborated definitions of how physical properties depend on local features. Interaction video **S16.3** illustrates a scanning of the space using a clustered data set (Iris data set containing three clusters). More details are reported in [4].

16.3.3 Principal Curve Sonification

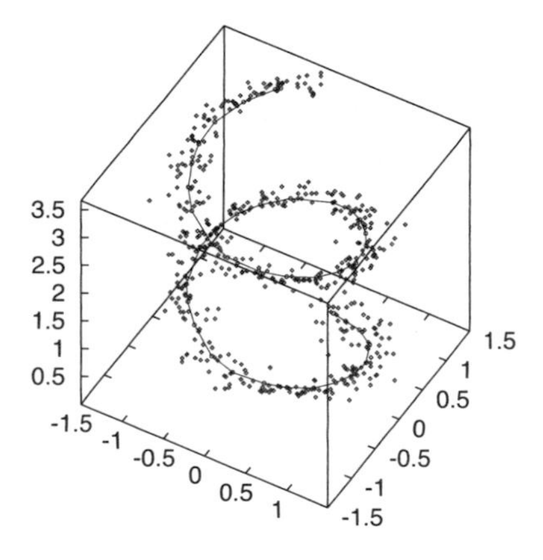

Figure 16.5: PCS for a spiral data set: noise structure along the spiral is difficult to see but easy to hear using PCS

The principal curve (PC) is a machine learning technique to compute a smooth path through a data set which passes nearby all data points [20, 9]. In this sonification model (see Fig. 16.5), each data point in the data space corresponds to a sound source in the model space which may contribute to a continuous overall soundscape, or just be silent. The interaction mode is that the users move along the curve through data space and hear only those data points that project onto their location on the curve. Alternatively, passing along the data points excites the sound sources. As a result, principal curve sonification (PCS) serializes high-dimensional

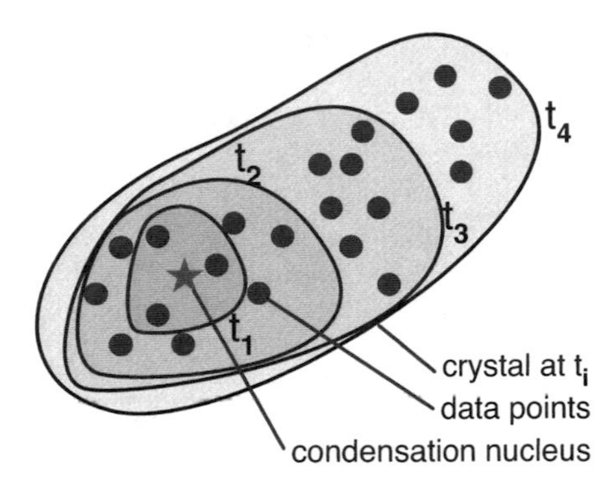

Figure 16.6: DCS crystal growth: crystal hull at various times.

data into a time-organized sequence where movement in space along the curve becomes the main mode of experiencing the data.

This model is very suitable for understanding the clustering structure of data since typically the PC passes once through all clusters. The sound example **S16.4** presents a PCS of a data set where the data are distributed along a noisy spiral: density modulations along the spiral become more easily heard than they can be perceived visually, see [13] for details.

16.3.4 Data Crystallization Sonification

The Data Crystallization Sonification (DCS) is inspired by the chemical process of crystal growth, here applied to the agglomerative inclusion of data points into a growing 'data crystal'. The model is a spatial one: data points specify the locations of 'molecules' in the model space as depicted in Fig. 16.6. These molecules are fixed and never move during the whole procedure. Excitation is done by setting a condensation nucleus, e.g., by clicking the mouse somewhere in the scatter plot. Molecules are then included with increasing distance from this center into a growing 'data crystal'. The metaphor is that the inclusion of a molecule sets free some energy which contributes to the overall vibration energy of the growing data crystal. The crystal's modes of oscillation are not defined in analogy to physics, but instead use the covariance matrix of the data set at each growth step as follows: the eigenvalues determine the harmonic series while the overall variance determines the size and thereby the fundamental frequency of the sound. During growth thereby the pitch drops whereas the brightness signature modulates. Understanding the mathematics helps to better understand the implications of sound changes and to interpret the sound as a fingerprint of the data crystal. Nonetheless, patterns can be discerned, characterized and compared even without this specific knowledge. The technique is suitable for discovering the clustering structure of data and particularly the local dimensionality structure of clusters in data sets. Sound examples **S16.5** illustrate typical sonifications, and more detailed explanations are given in [18].

16.3.5 Particle Trajectory Sonification Model

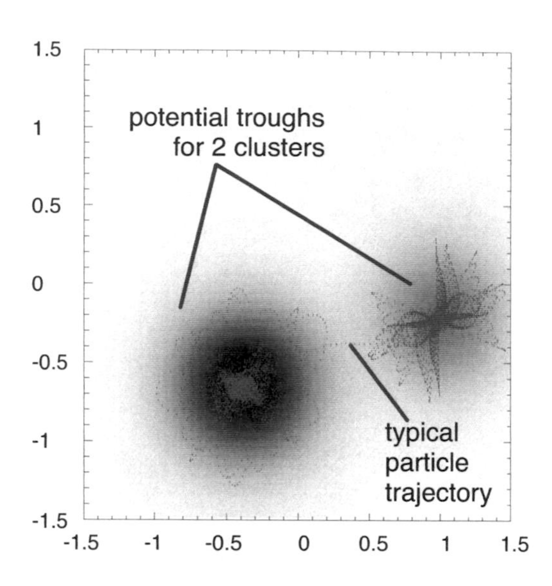

Figure 16.7: PTSM: 2D-potential and a particle for large σ, smooth $V(\vec{x})$

The Particle Trajectory Sonification Model (PTSM) demonstrates how MBS can *holistically* encode information into sound in a way which goes beyond what would be attainable with Parameter Mapping Sonification. For that reason we will discuss the model in more detail. In a nutshell, the model space is a d-dimensional vector space, the same as the data space. Data vectors determine coordinates of point masses in model space, which contribute to an overall 'gravitational' potential function $V(\vec{x})$. There are no dynamic elements connected to these fixed masses so the model remains silent so far.

The model of the universe is a useful metaphor for this, and we can imagine data points as stars that are fixed in space. Additional particles are now introduced to probe the model. They move fast in the 'data universe' according to the laws of mechanics. Staying with the metaphor of the universe, these are like comets as shown in Fig. 16.7.

As potential function, instead of a Coulomb potential, here an inverse Gaussian function $\phi_\alpha(\vec{x}) = -\mathcal{N}\exp(-(\vec{x} - \vec{x}^\alpha)^2/(2\sigma^2))$ is used, where σ controls the width of the potential trough, $\vec{x}^\alpha$ is the position of mass α, and $\mathcal{N}$ is a normalization constant. In the overall potential $V(\vec{x}) = \sum_\alpha \phi_\alpha(\vec{x})$ each particle moves according to Newton's law $m_p\vec{a}(t) = -\nabla_x V(\vec{x}(t)) - R\vec{v}(t)$, where $\vec{a}$ is the acceleration, $\vec{v}$ the velocity, R a friction constant and m_p the particle mass. As a result each particle moves on a deterministic trajectory through the data space. Collisions with other particles or masses are excluded. Finally, due to the friction term, each particle comes to rest at a local minimum of $V(\vec{x})$, which cannot be determined in advance. The link-variables are the instantaneous particle energies $W_i(t) = m_p\vec{v}_i(t)^2$, and their sum represents the overall sonification.

So what can be heard? In the beginning, a particle has enough energy to move freely in the data space, attracted by the data masses, moving on rather chaotic trajectories. This translates to rather noisy sounds. With energy loss, the particle is captured within a cluster (in the metaphor: a galaxy) and finally comes to rest at a minimum of V. The oscillations depend on V and change over time with decaying energy, providing an implicit and partial idea of the data distribution. While a single particle gives limited information, a number of particles create a qualitative sonic image of the data universe structure. Sound examples **S16.6** are single particle sounds.

Excitation in this model means either the injection of a bunch of particles into the model space, or the excitation of existing particles by giving them an impact. Depending on the excitation type, different interfaces can be used, ranging from a mouse click in a plot window for triggering particle injection, to shaking the mouse or other controllers such as an audio-haptic ball interface [12] to inject energy.

An important parameter for understanding the data distribution is the potential width σ: at large values the particles move in a very smooth Gaussian potential; decreasing σ lets more detail appear, first clustering structure, and finally potential troughs around each data point. Thereby the overall sound of the particles depends strongly on σ and this parameter can be offered as control to the user for interactive adjustment. For instance, with an audio-haptic interface [12] it is intuitive to use the squeeze force to control $1/\sigma$. Sound examples **S16.7** are sweeps while decreasing σ. The first example is for a data set consisting of three clusters. Stable pitches occur during decay at middle values of σ corresponding to well-shaped potential troughs at clusters. The second data set is only a single Gaussian distribution without further substructure, and in turn this pitch structure is absent in the sonification.

The primary analysis task of the model is to make perceptible the homogeneity and clustering shape of high-dimensional data. The structure can be understood from stable sonic pitch plateaus and noisy patterns during the transitions between these modes. Timbre complexity is obviously very high and there is no explicit definition of a synthesizer or sound generator. Data points are not explicitly responsible for sound structure. In contrast, data points contribute to the overall potential function and thereby contribute in a complex way to a holistic encoding of information into the sound wave field. Obviously the human auditory system can pick up structural properties, and we are likely to adapt further during sustained use of the model since the sound signal possesses the expected complexity and richness we are familiar with from contact sounds and noises in natural environments.

16.3.6 Growing Neural Gas Sonification Model

Growing Neural Gas (GNG) is a method for computing a topology-preserving graph representation of reduced complexity for a given high-dimensional data set [8]. For the GNG sonification (GNGS) model, the setup consists of the GNG graph trained with the data (see Fig. 16.8). The nodes of the graph are called neurons and can be imagined as points in the data space. For the model setup, an energy level variable is associated with each neuron.

The dynamics of the model operate on two levels: first, via an equation which determines how energy flows along graph edges to neighboring neurons; second, via different equations of motion for neurons to generate sound depending on their local properties (i.e. energy, graph connectivity structure). The model is excited by injecting energy into a neuron, e.g., by touching the location in a visual representation. The equations of motions spread the initially concentrated energy throughout the connected sub-patch of the GNG. Each

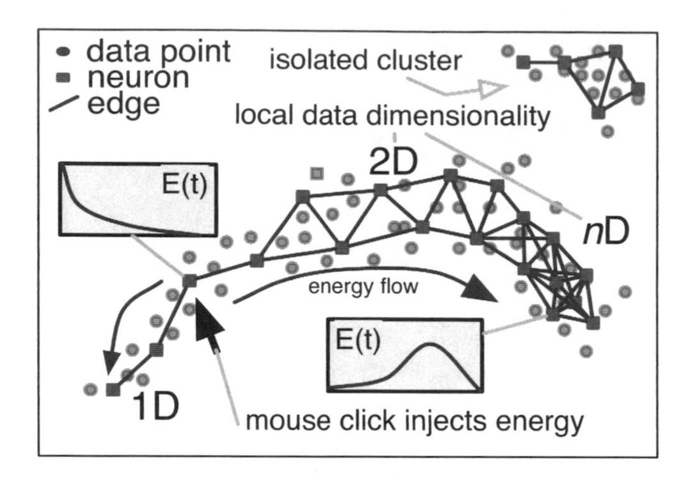

Figure 16.8: GNGS: energy flows through the network.

neuron contributes its sound to the sonification, allowing the perception of graph structure by listening.

What is the sound of a neuron? Assume that edges fix a neuron at its location, so the more edges there are, the higher the net restoring force, and qualitatively, the higher the frequency this neuron oscillates around its position. Following this logic, each neuron generates a sine wave, its energy determining the amplitude, and the number of edges influencing the stiffness and in turn the frequency. As a result, the overall connectivity of the structure becomes audible while energy spreads in the graphs. An important characteristics of GNG graphs is that the edge number at each neuron roughly scales with the local (intrinsic) dimensionality of the data. Thereby the sonification is an *implicit representation* of intrinsic dimensionality, an important feature for modelling and data analysis.

Exciting the GNG at different locations allows the user to perceive, at first, the local properties, then later the average properties of connected GNG patches. As a promising alternative use to excitation, the sonification can be rendered while the GNG grows. This allows the user to perceive the progress of adaptation and even to hear at what point overfitting sets in. Overfitting means that the graph merely describes the randomness of the data instead of the underlying relationship. This shows that MBS is not only useful for active exploration, but is also a suitable technique for process monitoring applications., see chapter 18.

Sonification examples **S16.8** show that clusters of different intrinsic dimension[5] sound differently when energy is injected into one of their neurons: note that higher-dimensional distributions automatically sound more brilliant without this feature having been mapped or computed explicitly during any part of the model construction. Sonification example video **S16.9** shows a sonified GNG growth process. You can hear how the structural hypothesis changes during learning. More details are provided at [16].

---

[5]degrees of freedom to span the volume, e.g. 1d is a curved line, 2d a twisted plane, 3d a volume, etc.

The reader may also check sonification models omitted here such as Shoogle for shaking text messages [30], the Local Heat exploration model [3], Data bubbles [23], Markov-chain Monte Carlo sonification [11], data solids [12], Multitouch GNGS [28, 21], and for scatter plot exploration for visually impaired people using active tangible objects [24].

16.4 MBS Use and Design Guidelines

How can designers quickly create useful sonification models for a certain task? This section provides guidelines, (a) to decide whether to use MBS at all and if so, (b) how and why to design new models and (c) how best to use MBS.

When to use MBS The motivation is to gain a rapid understanding of what is of interest in the data. If the data are organized in time (e.g., multivariate time series data), it is in most cases straightforward to maintain this dimension and to consider Audification or Parameter-Mapping Sonification. In the latter case it is important to consider if the available features can be meaningfully mapped to acoustic features, allowing the user to experience the temporal evolution in an informative way. MBS is rarely used for time-indexed data.

If, however, there is no time index, nor any other unique continuous feature for temporal organization of the sonification, it may appear unclear how to proceed. To give an example, the *Glass Identification Data Set*[6] contains 10 different physical properties such as refractive index, and chemical analyses such as Na, Mg, Al,... (in weight percents) of different types of glass samples (from buildings, vehicles and containers). The challenge is to identify the glass (or a new unseen glass sample) correctly from its features. In this example there is no time axis! Furthermore, the dimensionality is too high to understand the structure from looking at scatter plots. In this case, mapping all the features to acoustic features would be difficult. 10 meaningful acoustic features would be needed, which is quite challenging. The next problem is that there are infinite possibilities for the mapping, so the question arises of how to map what feature to what acoustic parameter? Any mapping will give an arbitrary sonic image, and it is highly likely that only the features mapped to event onset and pitch will mainly attract the listener's attention.

In such situations, MBS can be very useful. Think of MBS as a kind of tool box, each tool designed for a specific analysis goal. In the same way as you would not use pliers or a screwdriver to hammer a nail into the wall, each sonification model has been (or should be) developed to support a specific analysis task. The task is often so general that it abstracts largely from the concrete data. If, for instance, the task is to detect linear dependencies, a model would be applicable (and ignorant) to whether the data are chemical compound ratios or stock prices or census data features. The sonification model gives structure-specific information, which is good since it may inspire analysts to find new ideas for modelling the data or visualizing them in a way not thought about before. Similar to a motor mechanic who naturally listens to the engine sound before checking part by part for malfunctions, MBS may help data analysts to understand more quickly what's going on and in what direction to proceed with analysis. For example, if you discover linear dependencies you would certainly apply principal component analysis. If you discover clustering, you would proceed with

[6]see `http://archive.ics.uci.edu/ml/datasets/Glass+Identification`

clustering algorithms etc.

The currently existing sonification models are not strongly optimized for such specific tasks. Moreover, they typically allow the user to perceive additional aspects beyond the main objective. In our glass data example, the data sonogram sonification model using the class-entropy-based stiffness control explained on page 409 provides sounds that allow us to understand how strong the different classes of the glass probes overlap, or whether these classes can be nicely separated. After some interactions, particularly when paired with an interactive data selection to filter out some glass types, you may get a good idea what glass types are easily separable .

The fact that MBS is independent of the concrete data semantics is also a big advantage for another reason: the sound patterns remain stable over many uses with many different data sets. Therefore users can build up knowledge and experience in how specific structures sound.

How to Design Sonification Models If you want to create a new sonification model, the first question should be what is the main analysis task, or what type of pattern or structure should become apparent from listening. Taking a task-centered view helps the designer to focus on the relevant features. For example, assume that the goal was to hear whether the data set contains outliers.

Outliers are data points which are far away from the rest of the distribution, often due to erroneous data. They are sometimes difficult to recognize in multivariate data. Think of a census data set where females provide the information "x =age" and "y = number of children". $x = 12$ is not an outlier, nor is $y = 3$. Yet the tuple $(x = 3, y = 12)$ is certainly impossible and must be an outlier. So in order to detect outliers it is not enough to look at single features.

Here is one way to invent a sonification model for outlier detection. We could start by the following observation: outliers typically have few nearest neighbors in data space. So, if we create a dynamic system whose properties depend on this neighborhood emptiness we would obtain sounds where outliers stand out. For instance, we could represent each data point by a mass-spring system and define that the distance to cover the 5 nearest neighbors in data space determines the spring stiffness. After excitation of the masses, the 'outlier candidates' would sound at very high pitch and perceptually stand out. However, data points in a sparse region might also cause similarly high-pitched sounds. Thus, pitch is not necessarily an indication for outliers. Also, if the data space is rescaled, all stiffnesses increase and all oscillations sound higher pitched. One solution is to take the *relative* size of the 5-nearest neighbor sphere, so as to divide the radius by the standard deviation of the complete data set, etc. This should give an idea how the model could evolve further at the next design steps.

However, we could alternatively start from a completely different angle. Assume we connect d guitar strings from each data point in the d-dimensional data set to the points $\vec{x}_i$ that are the center of the k nearest neighbors if we would leave the ith vector component out. We could then send wind through the model space, or hit the whole model and as a result those data points which have long strings will contribute very low-frequency percussive sounds.

It is difficult to imagine what this model would actually sound like, yet certainly it would be possible to iteratively optimize a model to be both satisfying to use and informative. Perhaps, after a series of model inventions and refinements we would arrive at a quite suitable model to perceive outliers. The important point is that the models provide *analog* information and leave the inference and interpretation to the user. This is in contrast to procedures where a detector algorithm simply finds outliers and signals the result since then the user is detached from the analytical and decision-making process. The simpler and easier to understand the dynamic system, the better it will be for users to learn interactively how sound relates to patterns. The hope is that useful sonification models will - by being used - at some point in time become a *standard* tool for a given task, and are then effortlessly understood and routinely used to accelerate data analysis.

However, the designer may lack a concrete idea of what structure the sonification model will work best with, and may start from a random design seed. This probably bears a higher risk of creating useless models, but may eventually offer a higher chance of discovering something really unexpected and new. In the end, it is the utility of the sound to better understand the data which decides if sonification models 'survive' and will be used.

16.4.1 Metaphors for Sonification Model Design

Metaphors are very helpful both for the design process and the user. Some examples are the "shaking objects in a box" metaphor as used with the audio-haptic ball sonification model in [12] or in shoogle [30], or the "moving particles in a data universe" metaphor or the growing data crystal metaphor presented earlier.

To give an example let us start with a metaphor of 'dropping water' for the model design. Going back to the outlier detection sonification model considered in the previous section, we could imagine data points to be little pinholes through which water drips every second. Certainly we need to invent a law to describe in what direction the drops fall (e.g., they could fall towards the plane spanned by the first two principal components of the data distribution) and what sound they make when they touch this plane (i.e. what is the sound rendering process for this virtual water? - will it sound like real water drops?). The metaphor of dripping opens up ideas for new models. It might even inspire new interaction ideas, e.g., squeezing a tube interface to press more drops through the pinholes. If the metaphor works well, we may even consider ideas about how we can shape the dynamics so that the sonification is more similar in perceptual qualities to what we, as the designer, would have expected.

In summary, metaphors are useful both for the design and interpretation of MBS. However, the underlying coherence in a sonification model is usually stronger than just a metaphor (which works in some aspects but fails in others). The model is not a metaphor but has its own logic and consistency – the metaphor, however can be helpful for speeding up design and learning.

16.4.2 Task-oriented templates

Model-Based Sonifications abstract from the application-specific details of the data and are ignorant to the semantics. In other words it does not matter whether data come from chemistry, biology, economy, etc., when used in MBS the focus is on the data's structural properties. This makes MBS a bit more complex to understand and use, but it increases reusability.

The model developer's goal is to have a powerful toolbox of sonification models for whatever structure could potentially be of interest and to quickly explore a new data set with these 'interaction tools' to rapidly understand what's going on. This does *not* replace further investigation, but informs analysts so that their choices regarding their next steps are better rooted in experience. In the same way as there are several types of screwdrivers for similar screws, there may be several sonification models for similar tasks. It will also be a matter of personal preference, taste, or familiarity as to which sonification model works well for whom.

16.4.3 Model optimization

Sonification models are dynamic systems, and these typically contain a number of control variables that determine the detailed dynamics. These parameters need to be adjusted and tuned. However, this tuning is normally only done once by the model designer, so that the model can be applied *without any changes* by the user to arbitrary data sets. Sometimes a few parameters are provided to the users as interactive controls. The data sonogram sonification model for instance allows the user to control the propagation speed of the shock wave. This is useful for moving between very quick scans for rapid comparison of regions and slow spatial scans to attend to spatial density patterns.

Typically the number of parameters is low, compared to the many parameters to be adjusted when working with parameter mapping sonifications of d-dimensional data onto a p-dimensional synthesizer. This reduction of complexity on the side of the parameters goes hand in hand with the additional benefit that the model parameters are meaningful since the users can relate these to their internal imagination of what is going on.

16.5 Interaction in Model-Based Sonification

Interaction is an important part in MBS because MBS is interactive 'by-design' through the necessary excitation of the model. The general motivation for the importance of interaction is given in chapter 11 where some interaction modes are also explained.

The main purpose of interaction is to put energy into the dynamic system. As a result the system develops in time which causes the sound. A strong advantage of this approach is that interaction binds different modalities together. For example, if we excite a sonification model by knocking on some visualized data points using a multitouch display, we obtain a coupled audio-visual-haptic response and media synchronization helps us to relate the different media to each other and to bind them into multimodal units. Importantly, media synchronization does not need to be programmed explicitly, it emerges naturally from the coherence of the model.

Interaction furthermore enables the users to bring in their highly developed manual interaction skills which they have built up since birth: interaction in the real-world is far more complex than our typical interaction with computer interfaces such as mouse and keyboard. Think for example of the richness of interaction while shaking a box to find out what is inside, or while sculpting with clay. Model-Based Sonification aims to connect to such complex interaction abilities.

All sorts of interactions which we perform with real-world objects are candidates for MBS. Examples are scratching, rubbing, hitting, plucking, squeezing, deforming, stretching, bending, touching, etc. Interactions can be organized into the continuum between contact interactions and continuous interactions.

Contact Interactions are interactions where there is a very short energy transfer to the system. If we tap on a melon to hear whether it is matured, or if we knock on a wall to hear whether it is hollow or solid, we use contact interactions. For sonification models the implementation of these interactions can be as simple as using a mouse click in a scatter plot or as complex as using a multitouch surface equipped with contact microphones to sense details of the contact interaction. In objects such as mobile phones, acceleration sensors allow the measurement of contact interactions.

Continuous Interactions are those where the interaction progresses and changes while sound is being generated. Stroking, rubbing or scratching a surface are examples. Practically, they can be sensed by spatially resolved sensors such as touch-sensitive screens or tactile mats of sufficiently high resolution [1]. However, continuous interactions may also be non-excitatory, which means that they only manipulate the system (e.g., rotating an object or squeezing it) without putting energy into it. For example, imagine how a drum head interaction sound changes while the user's other hand moves or changes the pressure at a different position. In this way continuous interactions may control MBS parameters.

16.6 Applications

Model-Based Sonification was introduced as a framework to turn immaterial, non-sounding data sets into something that is sound-capable, so the primary applications are in the area of exploratory data analysis. However, MBS may also be useful in other fields as will be outlined briefly in the following sections.

Exploratory Data Analysis The best data mining 'machine' for the task of discovering and identifying hidden patterns and structures in complex data is the human brain. Our sensory organs and neural networks in the brain are excellent at making sense of the signals we encounter in the world, and allow us to recognize trees, cars, buildings, objects from the signals that come in via our eyes, ears and other sensory channels. However, as highly adapted as the brain is to make sense of structures as they appear in the world, it is bad at finding patterns in huge tables of numbers, which is the most direct representation of data. For this reason, there is the need to bridge the gap between the data spaces (mathematical vector spaces filled with data points) and our brain's preferred perceptual spaces. Model-Based Sonifications offer interaction-based mediators that turn data spaces into model spaces that are capable of creating sound.

The main capability that our brain offers here is *automatic concept formation*: the brain processes the sensory stimuli, automatically discovers patterns and instantiates categories to organize the perceived signals. In machine learning this is called 'symbol grounding', the transition from sub-symbolic signals to symbols. Here is a good opportunity to connect this to Kramer's continuum from analogic to symbolic displays as a means of categorizing auditory displays (see p. 23 in this volume): In exploratory data analysis we do not want to extract symbols (recognized patterns) from the data and represent them by auditory symbols, we rather want to turn the data into complex *analogic* representations which are suitable for the brain to discover patterns and symbols.

The key requirement to enable this learning process is the *invariance* of the binding between the data and the sound. Examples for that have been given in the sonification models discussed in the previous sections. In principle, all sorts of structures can be subject to sonification model design, such as outlier detection, local intrinsic dimensionality, clustering structure, separability of classes, multi-scale structure, and rhythmical patterns (e.g. where data points are aligned on a grid). Furthermore sonification models can also support meta-tasks such as determining how robust a mathematical model is in explaining the data (generalization), or when and how during the training of a machine learning model overfitting sets in.

For cluster analysis, the GNG sonification model, the particle trajectory sonification model, the data sonogram model and the tangible data scanning offer basic tools. For understanding the topology and intrinsic data dimensionality, the GNG sonification model and the data crystallization sonification model can be used. For understanding multi-scale structures, the growth process sonification of the GNG sonification models, and the particle trajectory sonification model (while controlling the bandwidth parameter σ) can be used. For understanding the separability of classes in classification problems, the data sonograms with class-entropy-based spring stiffness may be used. These models are just starting points and hopefully in the future more powerful and optimized sonification models will be developed for specific data exploration tasks.

Augmenting Human Computer Interaction Model-Based Sonification could in future make positive contributions to HCI, for instance, to create more informative, acoustically complex and situation-specific interaction sounds in Computer Desktop interaction. MBS could be used as a principal mechanism to couple any user interaction to acoustic responses, e.g., on the desktop computer or in virtual reality (VR) systems. For instance, a mouse click action could excite the GUI element clicked (buttons, widgets, background, icons, or link) and the resulting sound could help us to be more aware of where we clicked, and what the state of that element is. For instance, a frequently activated link could sound less fresh. There would be a rich, action-dependent informative soundscape while interacting with the computer, similar to the complex and analogous dependencies of real-world interaction sounds,. Furthermore MBS could enhance continuous interaction such as dragging the mouse while holding an object, using a slider, shaking icons with the mouse, or probing objects by knocking on them with a mouse click. Particularly in Virtual and Augmented Reality (VR/AR) where there is often no haptic or tactile feedback when interacting with objects, Model-Based Sonification can create some of the tactile information by sound while adding relevant data-driven information.

Process Monitoring In Model-Based Sonification, the excitation is normally done by the user. If we modify this basic idea so that changes in the data do not only change the model setup, but also provide some excitation, we obtain a sonification model which generates sound without user interaction, and which may be quite useful for process monitoring. Basic ideas for using sonification models for process monitoring have already been given with the

GNGS (section 16.3.6, p. 413) where the adaptation process of a growing neural gas has been used both to excite the sonification model and to configure it.

Auditory Augmentation and Ambient Information Model-Based Sonification also bears the potential for mixed-reality applications that support human activity and provide an ambient information display. Imagine for instance that each time you press a key while typing, in addition to the physical key sound, you also hear an additional sound resulting from the excitation of a sonification model. For example you could hear by a subtle overlapped cue how much space is left in a twitter message or SMS. Sonification models are just the right approach for such action-coupled information displays and would naturally extend the information value of interaction sounds. In [5] we have outlined techniques for augmented acoustics using contact microphones as detectors. Taking such signals as the excitation of a sonification model is the next step.

16.7 Discussion

Model-Based Sonification has been introduced as a mediator between data and sound. Dynamic models bridge the gap between non-sounding numbers and acoustic responses in a different manner to other sonification techniques such as parameter mapping sonification or audification. This section points out the most relevant differences, benefits and drawbacks of this technique compared to other approaches. Much more research in the form of comparative studies is needed to substantiate the claims, which here emerge mainly from long experience and qualitative observations.

Generality of Sonification Models From the brief overview of sonification models in section 16.3 it should have become clear that models are abstract: they are ignorant to the semantics or meaning of the data features, but only demand a certain generic structure. For instance most sonification models can be used independent of the data source, the data dimensionality or the number of data points in the data set and only demand that the data can be represented as a point cloud in a vector space.

Suitability for data that have no time argument Most sonification models have been defined for data sets where there is no time argument in the data, simply because in this case it is most difficult to specify in a canonic way what should be mapped to sonification time. The models also allow us to treat different dimensions equally, without any particular emphasis of one dimension as would happen in parameter mapping sonification due to the different saliency of acoustic parameters.

Dimensionality and Cardinality Independence Model-Based Sonifications can be defined and designed so that they operate on data of any size and dimension. Dimensionality independence is a particularly nice feature since it allows for reusing a model without modification in other contexts. This is in contrast to Parameter Mapping Sonification which requires that for each data set there must be selected a new set of mapping variables onto acoustic features, and a fresh decision about what to do with the remaining unmapped variables.

Learning MBS offers three benefits compared to mapping sonifications concerning learnability and interpretation. Firstly, Model-Based Sonifications address our everyday listening skills which we naturally use to understand everyday interaction sounds when we identify objects and their characteristics. In contrast, the interpretation of mapping sonifications requires more explicit knowledge of the mapping and musical structures to infer meaning from sound. Secondly, MBS sounds are rather stable in structure when using the sonification model with different data sets. This simply gives the user more opportunities to 'tune in' and to learn the 'language of the sound'. In contrast, for mapping sonification, usually you have a new independent mapping and sound structure for different data domains. Thirdly, MBS is interactive by design, naturally allowing the user to connect changes in interaction with changes in sound. Also, users can adapt their exploratory actions immediately as their understanding of the data changes.

Auditory Gestalt Formation Model-Based Sonification aims to provide an analogous auditory data representation according to the continuum definition of Kramer [22]. This analogous representation is particularly useful for auditory gestalt formation since it uses the same mechanisms which encode information into a sound wave as in real-world sound generation. Our listening system is evolutionarily prepared for detecting and conceptualizing gestalts from these kinds of signals.

Ergonomics From the author's experience, the following reasons seem to show that MBS may positively support human well-being and overall system performance. Firstly, since sonification models create sound only after excitation, the sound will be less annoying than sonifications which fill the soundscape decoupled from the user's initiative: they are integrated into a closed-loop (see chapter 11). In addition, interaction sounds accompany the user's actions, so MBS matches their expectations. Secondly, MBS enriches otherwise artificially soundless environments so that the information load is distributed on several perceptual channels. This may reduce fatigue and furthermore engage users into the work process. Thirdly, MBS may increase awareness of the data and actions, thereby helping to avoid misinterpretations or errors. Finally, MBS offers rich and more complex interaction modes such as shaking, scratching, squeezing, hitting a sonification model, for instance by using special interfaces and controllers beyond the mouse and keyboard. This turns data exploration into a much more comprehensive human activity and may also positively impact the healthiness of the work place.

Complexity of sound Sonification models which evolve according to dynamic laws are likely to render sounds which are otherwise intentionally difficult to synthesize. Depending on the model, they may possess a complexity and richness which exceeds the capacity of parameter-mapping sonification sounds. Since the concrete sound depends on the details of the interaction, every sonification will sound slightly different – similar to the way it is impossible to reproduce the signal-identical sound by plucking a real guitar string. However, our ears appreciate this variability and it does not hinder the auditory system to discover the relevant structures behind the 'signal surface' of the sound.

Reusability MBS sonification models are tools, designed to deliver interaction-driven task-specific information. They can be (and often are) defined to operate on a larger class of problems such as 'all data sets which can be represented as point cloud in an Euclidean vector space', or 'all data sets that represent variable distributions on a 2D surface' etc. This makes the sonification highly reusable without the need to adjust any parameters. MBS is a 'design once – use many times' paradigm. Only the developer needs to work hard; it should be simple for the users.

Intuitive Parameters MBS sonification models usually introduce some parameters within the model implementation. Examples are the shock wave velocity of propagation in data sonograms, the energy decay rate in GNG sonification model, etc. These parameters are either specified by the designer, or provided as interactive controls to the users. In the latter case, these parameters will be intuitive controls for users who understand the model. Generally, MBS provides fewer parameters than parameter mapping sonification where both the mapping of data to sound and the parameter ranges are variable. In addition, MBS parameters are often more meaningful since they refer to a physical process that can be imagined.

The Problems of Computational Complexity Sonification models can be extremely demanding in terms of computation. This is especially true for models where the degrees of freedom (e.g. number of moving particles) influence each other so that the number of operations scales quadratically or worse with the number of data points. Since MBS constructs virtual sounding objects from the data, their sound synthesis is as complex as the numeric physical modelling of acoustic instruments, and full-quality rendering of this may exceed the available computer power for many years. There are two alternative ways to address this problem: (i) model simplification, i.e. to invent implementation shortcuts that yield coarsely the expected signals without requiring full numeric simulation, and (ii) model analysis, i.e. using modal analysis from physics or other tricks that enable the efficient computation of the full resolution sound.

16.7.1 Model-Based Sonification vs. Parameter Mapping Sonification

The discussion has pointed out that MBS is quite different from parameter-mapping sonification (PMS). MBS creates dynamic models that are capable of rendering sound themselves whereas PMS maps data values to sound attributes and actively synthesizes the sounds. MBS is interaction-driven whereas interactivity needs to be added artificially in PMS. MBS needs only a few parameters whereas PMS typically needs a more complex mapping specification. MBS addresses everyday listening whereas PMS addresses musical listening. MBS is a 'design-once-use-many' paradigm whereas parameter mapping sonifications need to be set up for each individual data set.

Can we interpret MBS as parameter mapping sonification? On first sight it may appear so in some models. For instance, is the data sonogram model not just a mapping sonification where distance from the shock wave center is mapped to onset? In fact this could be one of the implementation shortcuts to practically implement the model for real-time operation. However, even if mapping is used in MBS for practical reasons such as a more efficient implementation, the model dictates exactly how to map. This may be called model-induced parameter mapping. MBS is also different in character: it can lead to 'holistic' representations, as for instance shown in the particle trajectory sonification model, which parameter mapping cannot create.

Can we understand MBS as audification? On first sight this may appear so as well: For instance, the particle trajectory sonification model is – concerning the rendering – an audification of state variables, specifically the particles' kinetic energies. Yet MBS is not an audification of the data under examination.

Finally, there are two other sources of confusion. Firstly, physical models have become popular for rendering sound signals. If such a physical model is used within a parameter-mapping sonification, this is not a MBS. On the other hand, MBS does not necessarily imply the use of physical modeling synthesis. Secondly, Kramer's *virtual engine* approach, where data are mapped to controllers of a dynamic system [22] is different from MBS despite the fact that a dynamic model is used: again, still the concept of mapping connects data and (in this case a more complex) synthesizer. In MBS, however, the data is not 'playing' the instrument, but the data set itself 'becomes' the instrument and the playing is left to the users. The sonification techniques may appear to lack clear borders, depending on how they are looked at, yet the approaches have their own place. In conclusion MBS is a new category qualitatively different from parameter mapping sonification and audification.

16.7.2 Model-Based Sonification and Physical Modeling

Physical modeling has become a major trend in modern sound synthesis for achieving complex, natural and interesting sounds. The structural vicinity to MBS motivates the question as to how methods from this field can be used for MBS. Few selected examples provide pointers to the relation.

There is a body of research on Sounding Objects [25], which provides assistance for the creation of physics-based models and for controlling their parameters in order to achieve continuous controlled events or interactive systems using these models. These methods are powerful for the generation of parameterized auditory icons (see Ch. 13), yet they can also be used for MBS. A sonification model would be the result if the data set under analysis would determine aspects of the model configuration.

There are also systems developed for music control and synthesis that offer inspiration and useful methods for MBS: for instance Cordis-Anima [26] is a sound synthesis engine, mainly used for music creation, but also capable of visual animation or multimodal simulations. It numerically integrates dynamic processes, e.g., using mechanical interactions, and furthermore it provides the means to excite the physical system via force-feedback gestural controllers. If the mechanical system was determined and set up from the data under analysis (Model Setup) Cordis-Anima would render Model-Based Sonifications.

Scanned Synthesis [2, 29] is a sound synthesis technique which also uses a dynamic system and its temporal evolution to shape sound. Different from simulated acoustics, here the model (e.g., a simulated spring) is scanned cyclically at audio rate to create the audio signal, allowing excitation and interaction to shape dynamic timbre evolutions at a lower control rate. Scanned Synthesis offers an interesting approach to mediate between the model's configuration and the resulting sound, giving inspiration for future sonification models to come.

16.8 Conclusion

This chapter has introduced Model-Based Sonification as a sonification technique that mediates between data spaces and the sound space by means of dynamic interactive models. Starting from an analysis of listening modes, we discovered the potential of human listening to make sense of sound wave fields that represent dynamic processes. This led to the definition of MBS as a paradigm, and *sonification models* as concrete task-centered designs, which need a specification of setup, dynamics, excitation, initial conditions, link-variables and listener characteristics. Various sonification models have been explained and demonstrated. From this background, guidelines for the use and design of MBS sonification models have been formulated. After highlighting interaction and the main application fields, the benefits and problems have been analyzed.

MBS research is still in its infancy. The next step will be to create a toolbox of optimized sonification models for many different tasks, and a good tutorial on how to apply, use, and learn them. For this it will be helpful to have an atlas of reference sonifications for certain structures so that the users can faster assess the structure in the data. Currently existing sonification models are just the first examples and possibly far from optimal. We hope for an evolution where many models will be invented, used, refined or rejected; working towards a set of good standard sonification models tuned to certain tasks. These models will perhaps become as stable and widely understood as pie charts or scatter plots are in visualization. This process will go hand in hand with the evolution of interfaces that allow us to use our skilled manual interactions to manipulate information spaces.

A research agenda for MBS includes, besides the development of the abovementioned MBS toolbox: research into ways of implementing the models so that they can be used for larger data sets with limited computation power; research into how best to interweave MBS with standard visual interfaces and the workflow of data analysts; and finally how to evaluate MBS and how to assess its effects on performance, flow, fatigue, depth of understanding, acceptance, etc. In summary, Model-Based Sonification opens up new opportunities for interactive HCI and multimodal data exploration, and will over time find its way into standard user interfaces.

Bibliography

[1] J. Anlauff, T. Hermann, T. Großhauser, and J. Cooperstock. Modular tactiles for sonic interactions with smart environments. In Altinoy, Jekosch, Brewster, editors, *Proceedings of the 4. International Workshop on Haptic and Auditory Design (HAID 09)*, volume 2 of *Lecture Notes on Computer Science*, pp. 26–27, Dresden, Germany, 09 2009. Springer.

[2] R. Boulanger, P. Smaragdis, and J. ffitch. Scanned synthesis: An introduction and demonstration of a new synthesis and signal processing technique. In *Proc. ICMC 2000*, Berlin, 2000. ICMC.

[3] T. Bovermann, T. Hermann, and H. Ritter. The local heat exploration model for interactive sonification. In E. Brazil (ed.), *Proceedings of the International Conference on Auditory Display (ICAD 2005)*, pp. 85–91, Limerick, Ireland, 07 2005. ICAD, International Community for Auditory Display.

[4] T. Bovermann, T. Hermann, and H. Ritter. Tangible data scanning sonification model. In T. Stockman (ed.), *Proceedings of the International Conference on Auditory Display (ICAD 2006)*, pp. 77–82, London, UK, 06 2006. International Community for Auditory Display (ICAD), Department of Computer Science, Queen Mary, University of London.

[5] T. Bovermann, R. Tünnermann, and T. Hermann. Auditory augmentation. *International Journal on Ambient Computing and Intelligence (IJACI)*, 2(2):27–41, 2010.

[6] A. Bregman. *Auditory Scene Analysis: The Perceptual Organization of Sound.* MIT Press, Cambrigde Massachusetts, 1990.

[7] M. Chion. *Audio-Vision: Sound on Screen.* Columbia University press, New York, NY, 1994.

[8] B. Fritzke. A growing neural gas network learns topologies. In G. Tesauro, D. Touretzky, and T. Leen, editors, *Advances in Neural Information Processing Systems*, volume 7, pp. 625–632. The MIT Press, 1995.

[9] T. Hastie and W. Stuetzle. Principal curves. *Journal of the American Statistical Association*, 84:502–516, 1989.

[10] T. Hermann. *Sonification for Exploratory Data Analysis.* PhD thesis, Bielefeld University, Bielefeld, Germany, 02/2002, `http://sonification.de/publications/Hermann2002-SFE`

[11] T. Hermann, M. H. Hansen, and H. Ritter. Sonification of Markov-chain Monte Carlo simulations. In J. Hiipakka, N. Zacharov, and T. Takala, (eds.), *Proceedings of 7th International Conference on Auditory Display*, pp. 208–216, Helsinki University of Technology, 2001.

[12] T. Hermann, J. Krause, and H. Ritter. Real-time control of sonification models with an audio-haptic interface. In R. Nakatsu and H. Kawahara, editors, *Proceedings of the International Conference on Auditory Display*, pp. 82–86, Kyoto, Japan, 2002. International Community for Auditory Display (ICAD), ICAD.

[13] T. Hermann, P. Meinicke, and H. Ritter. Principal curve sonification. In P. R. Cook, editor, *Proceedings of the International Conference on Auditory Display*, pp. 81–86. ICAD, International Community for Auditory Display (ICAD), 2000.

[14] T. Hermann and H. Ritter. Listen to your data: Model-based sonification for data analysis. In G. E. Lasker, editor, *Advances in intelligent computing and multimedia systems*, pp. 189–194, Baden-Baden, Germany, 1999. Int. Inst. for Advanced Studies in System research and cybernetics.

[15] T. Hermann and H. Ritter. Crystallization sonification of high-dimensional datasets. In R. Nakatsu and H. Kawahara, editors, *Proceedings of the International Conference on Auditory Display*, pp. 76–81, Kyoto, Japan, 2002. International Community for Auditory Display (ICAD), ICAD.

[16] T. Hermann and H. Ritter. Neural gas sonification – growing adaptive interfaces for interacting with data. In E. Banissi and K. Börner, (eds.), *IV '04: Proceedings of the Information Visualisation, Eighth International Conference on (IV'04)*, pp. 871–878, Washington, DC, USA, 2004. IEEE CNF, IEEE Computer Society.

[17] T. Hermann and H. Ritter. Sound and meaning in auditory data display. *Proceedings of the IEEE (Special Issue on Engineering and Music – Supervisory Control and Auditory Communication)*, 92(4):730–741, 2004.

[18] T. Hermann and H. Ritter. Crystallization sonification of high-dimensional datasets. *ACM Trans. Applied Perception*, 2(4):550–558, 2005.

[19] M.J. Houben, A. Kohlrausch, and D. J. Hermes. The contribution of spectral and temporal information to the auditory perception of the size and speed of rolling balls. *Acta Acustica united with Acustica*, 91(6):1007–1015, 2005.

[20] B. Kégl, A. Krzyzak, T. Linder, and K. Zeger. Learning and design of principal curves. *IEEE Transaction on Pattern Analysis and Machine Intelligence*, 22(3):281–297, 2000.

[21] L. Kolbe, R. Tünnermann, and T. Hermann. Growing neural gas sonification model for interactive surfaces. In Bresin, (ed.), *Proceedings of the 3rd Interactive Sonification Workshop (ISon 2010)*, Stockholm, 2010. ISon, KTH.

[22] G. Kramer. An introduction to auditory display. In G. Kramer, editor, *Auditory Display*. Addison-Wesley,

1994.

[23] M. Milczynski, T. Hermann, T. Bovermann, and H. Ritter. A malleable device with applications to sonification-based data exploration. In T. Stockman (ed.), *Proceedings of the International Conference on Auditory Display (ICAD 2006)*, pp. 69–76, London, UK, 2006. ICAD, Department of Computer Science, Queen Mary, University of London.

[24] E. Riedenklau, T. Hermann, and H. Ritter. Tangible active objects and interactive sonification as a scatter plot alternative for the visually impaired. In *Proc. of the 16th Int. Conference on Auditory Display (ICAD-2010), June 9-15, 2010, Washington, D.C, USA*, pp. 1–7. 2010.

[25] D. Rocchesso, R. Bresin, and M. Fernström. Sounding objects. *IEEE Multimedia*, 10(2):42–52, 2003.

[26] O. Tache and C. Cadoz. Organizing mass-interaction physical models: The cordis-anima musical instrumentarium. In *Proceedings of the international Computer Music Conference (ICMC 2009)*, pp. 411–414, Montreal, 2009. ICMC.

[27] R. Tünnermann and T. Hermann. Multi-touch interactions for model-based sonification. In M. Aramaki, R. Kronland-Martinet, S. Ystad, and K. Jensen, (eds.), *Proceedings of the 15th International Conference on Auditory Display (ICAD2009)*, Copenhagen, Denmark, 2009.

[28] R. Tünnermann, L. Kolbe, T. Bovermann, and T. Hermann. Surface interactions for interactive sonification. In M. Aramaki, R. Kronland-Martinet, S. Ystad, and K. Jensen, (eds.), *ICAD'09/CMMR'09 post proceedings edition*. Copenhagen, Denmark, 2009.

[29] B. Verplank, M. Mathews, and R. Shaw. Scanned synthesis. In *ICMC 2000 Proceedings*, Berlin, 2000. ICMC.

[30] J. Williamson, R. Murray-Smith, and S. Hughes. Shoogle: excitatory multimodal interaction on mobile devices. In *Proceedings of the SIGCHI conference on Human factors in computing systems*, pp. 121–124, San Jose, California, USA, 2007. ACM Press.

Part IV

Applications

Chapter 17

Auditory Display in Assistive Technology

Alistair D. N. Edwards

17.1 Introduction

This chapter is concerned with disabled people[1]. As soon as a label such as 'disabled' is applied, questions are raised as to its definition. For the purposes of this chapter, no formal definition is required, rather it should suffice to say that the people we are writing about have the same needs as everyone else, it is just that in some instances their needs are more intense and are sometimes harder to meet. If this book achieves anything, it should convince the reader that sound can be an immensely powerful medium of communication and the relevance of this chapter is that the full potential of the use of sounds can often be more completely realized when aimed at meeting the needs of people with disabilities.

The immediately obvious use of sounds is as a replacement for other forms of communication when they are not available. Specifically, blind people cannot access visual information. Much of this chapter will deal with this form of substitution, but it will also demonstrate the use of sounds in other applications.

It is a contention in this chapter that there is a great potential for the use of sound that has not yet been realized, but some progress has been made in the following areas which are reviewed in this chapter:

- computer access

[1] Language is powerful and sensitive. No other literature is more sensitive to the needs of being politically correct than that which deals with disability. It is recognized that inappropriate use of language can cause harm and offence, but at the same time perceptions of what is correct are constantly changing. For example, at the time of writing there are (sometimes fierce) arguments as to whether 'disabled people' or 'people with disabilities' is the better term. In this chapter we have attempted to be sensitive to all shades of opinion, and if we have failed and used any terminology felt to be inappropriate by any individual reader, then we can only apologize.

- mobility aids.

Then there are other potential uses and some of these are also discussed.

17.2 The Power of Sound

Of course, one of the most powerful (and the most used) form of auditory communication is speech. Even though the emphasis of this book is on non-speech sounds, the role of speech cannot be ignored and it will be discussed in this chapter, in the context of where speech has advantages over non-speech.

The potential power of non-speech sound is illustrated by the following extract, written by John Hull, who is blind.

> I hear the rain pattering on the roof above me, dripping down the walls to my left and right, splashing from the drainpipe at ground level on my left, while further over to the left there is a lighter patch as the rain falls almost inaudibly upon a large leafy shrub. On the right, it is drumming with a deeper, steadier sound, upon the lawn. I can even make out the contours of the lawn, which rises to the right in a little hill. The sound of the rain is different and shapes out the curvature for me. Still further to the right, I hear the rain sounding upon the fence which divides our property from that next door. In front, the contours of the path and the steps are marked out, right down to the garden gate. Here the rain is striking the concrete, here it is splashing into the shallow pools which have already formed. Here and there is a light cascade as it drips from step to step. The sound on the path is quite different from the sound of the rain drumming into the lawn on the right, and this is different again from the blanketed, heavy, sodden feel of the large bush on the left. Further out, the sounds are less detailed. I can hear the rain falling on the road, and the swish of the cars that pass up and down. I can hear the rushing of the water in the flooded gutter on the edge of the road. The whole scene is much more differentiated than I have been able to describe, because everywhere are little breaks in the patterns, obstructions, projections, where some slight interruption or difference of texture or of echo gives an additional detail or dimension to the scene. Over the whole thing, like light falling upon a landscape, is the gentle background patter gathered up into one continuous murmur of rain. [1, p. 26-27][2]

There are two important points to be taken from this extract. Firstly there is the immense amount of information that the writer was able to extract from sounds. Secondly, it has to be acknowledged that none of the attempts to use sounds in synthetic auditory displays has yet come close to conveying that amount of information. It can be done; we do not yet know how to do it. It has to be acknowledged that most of the devices and ideas described in this chapter are *not* embodied in commercially available, commonly-used products. For various reasons they are not sufficiently useful for widespread adoption, and yet the above extract clearly demonstrates the richness of information that can be usefully conveyed in non-speech sounds. Tony Stockman also describes how blind people can make use of environmental sounds, putting them in the context of attempts to supplement these with technology-generated

[2] *On Sight and Insight*, © John M. Hull, 1990, 1997. Reproduced by permission of Oneworld Publications.

sounds in [2].

Sight is a very powerful sense. By any measure, the amount of information that can be received visually is vast. Yet it is not simply the raw bandwidth of sight that makes it so powerful, it is the ability to (literally) focus on the information that is of relevance at any given time. Because of the amount of information available visually it is those people who do not have access to visual information who are the most obvious candidates to use auditory information as an alternative. There is a fundamental problem, though, in substituting for visual information. The capacity of the non-visual senses (including hearing) simply does not match that of sight. This is often referred to as the *bandwidth problem*.

Thus, the fundamental restriction is that sounds cannot be used to convey the same amount of parallel information as the visual sense can. There are two principal approaches that can be taken to address this problem:

1. Maximize the amount of information carried in the sounds;
2. Reduce the amount of information presented (i.e. filter it in some way).

Achieving (2) amounts to giving users a form of focus control corresponding to that of the visual sense. While (1) is the main topic of this chapter, it cannot be divorced from the necessity to provide the control implied in (2).

In this chapter a number of research projects are described in which non-speech sounds are used to convey information to blind people. In comparison to the example from John Hull, above, it will be evident that these attempts are quite crude. Nevertheless, this is surely a stage that has to be gone through in order to understand the nature of this style of communication, with the hope that eventually we will be able to create vast, rich and useable soundscapes.

17.3 Visually Disabled People

There are a large number of people with visual impairments. Although exact figures are hard to find, Tiresias [3] estimate that there are approximately six million people in Europe with a visual disability. Visual disabilities take a number of forms and the number of blind people - those with no useful sight - is relatively small (one million in Europe, according to Tiresias)[3].

Although the number with an impairment short of blindness (variously referred to as 'visually impaired' or 'partially sighted') is relatively large, the number of different forms of impairment make it difficult to meet their needs. (An impression of the effects of different forms of impairment can also be found on the Tiresias website, [3]). For instance, an adjustment that helps some people (such as text enlargement for people with cataracts) can even make vision worse for others (enlargement further reduces the material in view to someone with tunnel vision, perhaps due to glaucoma).

It might be suggested that any interaction that makes no use of vision - such as an auditory

[3]The figures are open to debate. For instance, [4] estimated the number of visually disabled Europeans as 2,000,000, while the proportion of people with visual disabilities has been estimated variously as 1.6% (in Europe, [3]) and 4.1% (in the USA, [5]). Also, one must beware that the population of Europe has changed since 1993 with the accession of new states.

interface designed for users who are completely blind - would be equally accessible to those with some vision. However, the fact is that those with some sight generally prefer to make as much use of that sight as possible. In other words, they do not need substitution of the visual information, but its enhancement to match their visual abilities.

Thus, this chapter really addresses the needs of those who must have an auditory substitute for visual forms of information, those who are blind - even though they are the minority of those with visual disabilities.

Visually disabled people have a variety of needs for non-visual information. This chapter looks at access to computers (through screen readers), electronic travel aids and other applications which make use of sounds.

17.4 Computer Access

Most human-computer interfaces are 'visually dominated' in that the principal channel for communication from the computer to its user is the monitor screen. For a blind user, all the information that is displayed on a computer screen has to be substituted by non-visual forms of communication, either tactual or auditory.

The dominant form of tactual communication is braille. Braille is mainly a translation of printable text. The greatest barrier to the use of braille, though, is the small number of (blind) people who have the skills to read it. Again accurate statistics are hard to compile, but Bruce at al. [6] suggest that in the UK the proportion of blind people who can read braille is as low as 2%. Computer braille displays are available [7]. These usually consist of 40 or 80 braille cells. They are electro-mechanical devices and are thus quite expensive and are also bulky and heavy.

While there is a significant community of enthusiastic braille users - including those who use braille for computer access - auditory interfaces have a lot of features which make them very attractive compared to braille, notably:

Ease-of-use: Unlike braille, sounds essentially require no training. Of course this is not strictly true of some of the more complex uses of sounds discussed in this book (e.g. chapters 8, 10, 12, 14), but the simplest sounds including speech can be used without training. Auditory interfaces are effectively accessible to 100% of blind people - as long as they do not also have a hearing impairment.

Cost: Sound cards are a standard component of all modern PCs, therefore the only additional cost is that of any special software.

Braille was originally designed for the presentation of literary text - that which can be expressed in the 26 letters of the alphabet plus 10 digits and a small number of punctuation marks. Its extension to other forms of communication (e.g., mathematics or music) is somewhat clumsy and labored. There is a similar problem with sound when applied to the complex information that can be displayed on a computer screen. On any computer screen there may be hundreds of different elements visible. The sighted user can cope with this large amount of information because they have the ability (literally) to focus on the item of interest at any time. Thus, the user can take in the information of importance and filter out that which is currently irrelevant. The non-visual senses (and here we are mainly concerned with hearing) do not have that ability.

In other words, if we were to take a simple-minded approach to the adaptation of a visual display for blind users we might try to associate a sound with each item on the screen. To glance at such a screen would imply having every one of those items make its sound. Clearly this would be a cacophony. Sounds would interact and mask each other and it would not be possible to spatially separate the sounds in the same way that vision can do.

Computer access for blind people is achieved by using a piece of software called a *screen reader*. Essentially this examines the contents of the screen and converts it into sounds[4].

Screen readers were first developed about the same time as the PC became available. The operating systems of the time (predominantly MS-DOS) were text-based. That is to say that the screen displayed text and commands were typed in on the keyboard. For instance, to display the contents of the current directory, the user would type `DIR`, or the contents of the file `foo.txt` could be displayed (*typed*) on the screen by entering `TYPE FOO.TXT`. It was relatively easy to render this kind of interaction (i.e., the text of the command line and the contents of the text file displayed in response to the command) in sounds by using a screen reader linked to a speech synthesizer. Some of these first-generation screen readers made some use of non-speech sounds. For example, the *Hal* screen reader [9] used beeps of different tones to guide the user between the different lines on the screen, but most of these screen readers relied mainly on speech.

The screen reader represented a major advance for blind people. The access to the computer that it gave, generated a degree of equality in job opportunities; jobs that had been inaccessible now became feasible for blind workers.

The next major development in the personal computer was the graphical user interface (GUI). This was firstly implemented commercially on the Apple Macintosh, but eventually was also found on 'IBM-compatible' PCs in the form of MicrosoftWindows. At first the GUI was seen as a real threat to blind people. The form of interaction was completely different and very much visually orientated. The mouse pointing device was added to the keyboard and screen. It was necessary to point at objects on the screen. The design and positions of those objects carried meaning. These properties and their meanings could not easily be translated into auditory forms. The emancipation that blind workers had experienced was in danger of being lost.

Edwards [10, 11] experimented with an auditory version of the GUI, *Soundtrack*. This was not a screen reader, but a word processor which retained most of the interactions of the GUI (windows, icons, scrollbars etc.) but represented them in an auditory form. The first level of interaction was based on tones of varying pitch, giving relative spatial information, but at any time the user could click the mouse and hear a spoken label. Double-clicking would activate the current object.

Soundtrack remains one of the few attempts to make mouse-based interaction with a GUI accessible in a non-visual form, but it was only a word processor, and not a generalized tool for making GUIs accessible. However, screen readers were eventually developed such that the modern GUI interface is about as accessible as the former text-based ones were. GUI screen readers obviate the need to use the mouse by taking over control of the cursor, which

[4]Most screen readers can also render the information on a braille display [7, 8], but that is outside the scope of this chapter.

is controlled through the keyboard. They also still tend to rely to a large extent on synthetic speech with minimal use of non-speech sounds.

Syntha-Voice's *WindowsBridge* was a screen reader which attempted to make the Windows operating system accessible through the mouse. Positional feedback on the cursor was given using musical tones, and mouse movements could be filtered so that only vertical and horizontal movements were detected (i.e. no diagonal movements). However, few users used this feature and, indeed, the product is no longer available.

Non-speech sounds were used more extensively in some experimental screen readers, notably *Mercator* and *Guib*. The contrasting approaches behind these different systems is written up in [12], but both tended to use the style of non-speech sound known as the *auditory icon* [13] (chapter 13). The Guib Project culminated in a commercial screenreader, *Windots*, but it did not make much use of the non-speech sounds developed in Guib.

Windots was never a great success commercially and is no longer available. In practice the most popular Windows screen reader is *Jaws for Windows*[5]. Jaws has quite extensive facilities for the use of non-speech sounds. A Speech and Sound Manager allows users to associate different utterances or sounds with screen objects. These include:

Control types These are widgets, such as buttons, scrollbars, check boxes.

Control state Widgets can be rendered differently depending on their state, a button that is pressed or a check box that is checked or not.

Attributes Different font attributes can be signaled.

Font name Changes in font can be signaled.

Color The color of the current item can be signaled.

Indentation An indication of the depth of indentation is presented.

HTML Different HTML elements (in webpages) can be signaled.

All of these properties can be rendered in different ways. Speech may be used (i.e., an explicit description of the attribute) or a change in the current voice, but there is also the option of playing a sound. Sounds are simply played from *.wav* files and a number of these are provided with the Jaws software. These include auditory-icon-style sounds such as recordings of door bolts being opened or closed (sample **S17.1**), a lamp being switched on (sample **S17.2**), the thump of a rubber mallet (sample **S17.3**) and the like. There are also musical sounds, such as a piano playing one or two notes (e.g., sample **S17.4**) that can be used in a more earcon-style of soundscape.

Different 'Speech and Sound' configurations can be created and stored. This means that users can load particular configurations for different purposes. For instance, they may wish to use one configuration when word processing and a different one when writing programs. Configurations can be stored in files. This means that they can easily be swapped and shared between users. A number of configurations are provided with the Jaws software and it is interesting that these make minimal use of the sounds option; they are (again) speech-driven.

Microsoft Windows is the operating system of choice for most blind users; it is best supported

[5]Freedom Scientific (`http://www.freedomscientific.com`).

by available screen readers. The Unix world has been slow to recognize the needs of blind users, but this changed with the advent of the Linux Gnopernicus Project which aimed to enable users with limited vision, or no vision, to use the Gnome 2 desktop and applications effectively. However, this project appears to have stalled due to lack of funding.

As mentioned above, the advent of the GUI was seen at the time as a serious blow to the emancipation of blind computer users. Apple Computers were responsible for the introduction of the GUI to the consumer market, with the release of the Macintosh. Although a screen reader (*OutSpoken*) was released for the Macintosh, it has never been used by many blind users (and is no longer marketed). However, in the release of Version 10.4 of its OS X operating system, Apple included *VoiceOver*, a built-in screen reading facility. Controversy exists regarding the efficacy of this screen reader [14, 15], but its use is growing as it is now part of the iPhone and the iPad. As with most screen readers, it is heavily speech-based, but does include the use of non-speech sounds.

A common theme in this book is that the true potential for the use of non-speech sounds has yet to be realized. This is clearly true in the application of computer access for blind individuals. Most screen readers have facilities for the use of non-speech sounds; however, few people use them. This implies that the kinds of sounds being used and the information they are providing is not perceived as valuable to the users.

17.5 Electronic Travel Aids

The need to access computers is growing, but is still a relative minority activity compared to moving around the world. There are two aspects to this for blind people: short-range obstacle avoidance and the broader-scale of navigating to desired destinations. Technologies (sometimes referred to as Electronic Travel Aids or ETAs) can be used in both of these applications.

By far the most popular technology for obstacle-avoidance is the guide cane (also known as the white cane). There are a number of reasons why this is so popular, which will be discussed in contrast to higher-technology approaches below. For a person walking through an environment, it is vital to know whether there are any obstacles in the path ahead. This is what the guide cane provides. Canes come in different lengths from approximately 60cm to 160cm. Shorter canes are symbolic-only, carried by the user (who is likely to have some vision) as a signal to others that they may need special assistance. It is only the longer ones that are used for obstacle avoidance.

The cane communicates information mainly through the haptic senses. In other words, the user detects forces on the cane handle as its tip collides with objects. However, it is important to be aware that there is an auditory component to the communication also. The sound that the cane makes as the tip is tapped on surfaces can communicate a lot of information. For instance the texture of the path (e.g., concrete versus grass) will be apparent from the sound the tip makes. Also the sounds generated will be modulated by the environment. A closed area surrounded by walls will generate echoes, whereas an open one does not. The amount of information available from such natural auditory sources should not be underestimated. John Hull describes [16] how he can recognize when he is walking by railings by the intermittent echo that they generate[6]. Snow is sometimes described as 'the blind person's fog' - because it dampens sounds rather as fog blocks sighted people's vision.

One of the major disadvantages of the traditional cane is that it operates only within a very narrow vertical range. That is to say that it will generally detect obstacles at ground level. While that is sufficient in many environments, there is clearly a danger from any obstacles up to head height. This is one advantage that high-technology sensors can have; they can scan the entire path ahead of the user.

There is then the question as to how to communicate the information to the user, in a non-visual form. Sound is the obvious medium to use [18]. There are, however, two particular problems with sound: auditory interference and the bandwidth problem.

A question arises as to how to present the auditory feedback from a guidance device. Headphones may seem the obvious choice. They can present the information privately. This is important because the information is not of any use to anyone else in the vicinity and is therefore likely to annoy them. More importantly, any audible sounds would draw attention to the person generating them and might be perceived as a label of their blindness. Headphones can also be used to present information spatially, either using simple binaural stereo or three-dimensional spatializations. Finally since the advent of the portable stereo player, it has become socially acceptable to wear headphones in public, so that their use is not a social faux pas.

However, headphones are in practice not necessarily appropriate. The main problem is that they are tend to interfere with environmental sounds. There are various headphones available designed to avoid this problem by not blocking external sounds [19], but their effectiveness is open to question.

Conspicuity and aesthetics are important factors, the importance of which are easy to underestimate. Most people do not like to stand out in the crowd, and this is just as true of people with visual disabilities as for sighted people. Modern white canes are usually foldable. That is to say that they can be dismantled and folded into a package around 20cm in length. This means that their visibility is under the user's control. As illustrated by the symbol cane, one of the features of the white cane is that it can be positively used as a signal to other people that the owner has a visual disability. However, on the other hand, the user can also choose to fold the cane away, removing that signal.

Some high-technology devices are not so discreet. For instance, the *Kaspa* [20] is a box worn on the forehead. While improvements in miniaturization will almost undoubtedly make it possible to conceal such devices better, most users will still not want to wear equipment that is too visible. Any such device makes the user stand out, clearly indicates that there is something different about them (it may not be obvious that the person has a visual disability) and may make them seem to be quite odd and freakish.

The importance of aesthetics should also not be underestimated; even a device which is very positive in the assistance and power that it gives the user, will be rejected by many if it is too ugly. (See also Chapter 7).

There has already been mention above of changing attitudes towards headphones, which

[6]Another interesting example is the experiments by McGrath et al. [17] in which it was found that blind people could locate and accurately describe objects (a sheet of aluminium, a sheet of aeroboard and a leather football) in a dark room using only the sound of their voice.

also suggests a fashion element. With the advent of the Sony walkman in the 1980s, it became acceptable (at least among young people) to wear headphones in public. For the most part they were small, discreet, and not very noticeable. Now the wearing of headphones is common. In fact, in some environments (such as on public transport) there may be as many people wearing headphones or earphones as not. Yet fashion has also moved on. There are those now who prefer not to wear barely visible earphones, but rather large, highly conspicuous headphones. They would, no doubt argue that their choice is based on acoustics, that the sound reproduction is so much better, but at the same time, the headphones are usually high quality ones - of sleek design and with the accompanying clear brand labels. In other words, there is an element of boasting in the wearing of these devices.

It would be ideal if the same kind of positive kudos could be attached to aids for visually disabled people. In other words, the device could become something 'cool' and not a label of deficiency. The headphones example illustrates, though, that aesthetics and fashion can involve complex interactions.

Yen [21] provides a comprehensive list of ETAs, some of which are explored in more detail in the following sections. While the emphasis is on the technical specification of these devices, it should be apparent that other factors - including aesthetics - are also important.

17.5.1 Obstacle Avoidance

There are a number of devices which operate as obstacle detectors. It is significant that the same approach to obstacle avoidance - a portable sensor generating auditory signals - has been tried many times. There is no point in trying to provide an exhaustive list of such experiments, but several devices are reviewed in [22], and [23] including:

- Russell Pathsounder,
- Nottingham Obstacle Detector,
- Laser Cane,
- Sonic Torch,
- Mowat Sensor,
- Sona,
- NavBelt.

Two exceptional examples are the *Bat 'K' Sonar-Cane*[7] and the *UltraCane*[8], exceptional in that they are commercially available products. The 'K' Sonar is a hand-held device resembling a flashlight or torch that can be clipped to a white cane. It has a cable connection to two miniature earpieces. The pitch of the echo sounds is proportional to distance: high-pitched sounds relate to distant objects and low-pitched sounds related to near objects. (Examples are provided, see Table 17.1).

One feature of many of these hand-held devices is that they are directional. That is to say that they generate feedback about obstacles only when they are within the (narrow) beam of the device. This means that spatial information is given directly by the device (i.e., through the

[7] `http://www.batforblind.co.nz`
[8] `http://www.ultracane.com`

Description	File name
Scanning a 4cm-diameter plastic pole at 1.5m from left to right and back.	**S17.5**
Walking towards a glass door from a distance of 5m to 1.5m and then retracing steps back to 5m.	**S17.6**
Person approaching from 5m to the halt position and then retracing his steps.	**S17.7**
Standing in front of a wooden fence with spacing between small panels and scanning the torch to the left and right to 'shine' the beam along the fence line.	**S17.8**
Scanning the torch down onto a grass lawn and up again.	**S17.9**
Standing in front of a 50cm wide tree. The tree has only a thin layer of bark and a thin (4cm) shoot growing out at the base of the tree; a clump of short flax on the other side. As the ultrasonic beam scans across it produces a strong 'warbling' sound from the trunk, a soft mushy sound from the flax, and a soft short whistle from the shoot.	**S17.10**
Standing in front of a concrete block wall with large thick well-developed ferns at the side of the standing position. The torch scanned across the ferns onto the wall and back again. The wall produces a tone sound. The ferns made a strong mushy sound.	**S17.11**

Table 17.1: Sample K-Sonar sounds.

kinesthetic information that the user has about the position of the hand grasping the device). There is no requirement to encode spatial (directional) information in the auditory signal. This makes the signal simpler - and hence generally easier to comprehend. In other words a hand-held device sends out a one-dimensional beam and scanning it horizontally adds a second dimension of information, whereas a representation of the entire scene includes all three dimensions. These might be represented directly by spatialization of the auditory representation (as in the *Kaspa*, [20]) or by applying some other modulation to the signal.

The UltraCane is also important in that it is commercially available. It avoids the problems of auditory output discussed above by presenting its information haptically. As such, it is outside the scope of this book, but it is interesting in that it possibly illustrates an attempt to sidestep some of the disadvantages of using sound. Sounds - and headphones - are not used, so there is no masking of the natural acoustic environment. The mapping of obstacles to vibrations of different buttons in the cane handle, with strength indicating separation, is a natural one.

As mentioned earlier, despite the advent of clever electronic aids, the white cane remains the most popular device. It is worthwhile looking at reasons for this. Differences between guide canes and electronic alternatives are summarized in Table 17.2.

The NavBelt is an experimental device of interest because of the ways it operates [23, 24, 25, 26, 27]. It takes the form of a belt worn around the user's waist. The belt contains an array of sonar devices. The sonars measure the distance to obstacles. NavBelt operates in two modes. In Guidance Mode it is designed to *actively guide* the user around obstacles towards a target, while in Image Mode it substitutes an auditory scene for the visual scene (more akin to the kinds of visual substitution systems explored in the next section).

The sonars detect obstacles from which the NavBelt calculates the *polar obstacle density*, a measure which combines the size of obstacles and distance to them. In Guidance Mode, the NavBelt calculates the area with the lowest polar obstacle density near the direction of travel and guides the user in that direction. In other words, Guidance Mode works best when the target is known. This might be achieved by integrating the NavBelt with a navigation aid. Bornstein [24] lists this as a potential future development, but there is no evidence of this having been subsequently implemented. In the absence of an absolute means of specifying the target, the device uses heuristic approaches to infer the intended direction of travel.

Cane	**Electronic obstacle detector**
Inexpensive. Losing one or accidentally swapping with another owner is not a major problem. It is feasible to own more than one in case of loss or damage. A standard guide cane costs of the order of €20 or $30.	Expensive. The 'K' Sonar costs of the order of €500 or $650, and the Ultracane is around €750 or $900.
Reliable.	Subject to faults and requiring maintenance.
Does not interfere with hearing.	Acoustic signals may block natural cues.
Senses only at ground level.	Can be designed to sense up to head height, but may not detect some important ground-level obstacles (e.g., kerbs).
Extensive training required - over 100 hours.	Estimates and claims as to the amount of training required vary.
Short-range - effectively the length of the cane (1 - 2 meters).	Can be designed to operate at longer ranges. Typical sonar devices can operate up to 10 meters. Video-based systems theoretically can operate up to the visual horizon.
Requires constant active exploration.	Requires constant active exploration.

Table 17.2: Comparison between the features of the traditional guide cane and electronic alternatives.

It is interesting that technological orientation devices have been under development for at least thirty years (e.g.. [28]). The 'K' Sonar had achieved sales of 1550 up to 2010[9] - but this is a tiny proportion of the market.

The guidance information is presented to the user as binaural sounds on headphones with interaural time difference to create the impression of directionality. In Guidance Mode the pitch and amplitude of the sounds are proportional to the recommended travel speed. The principle is that higher pitch and amplitude attract attention so that the user will instinctively slow down and concentrate on the direction of the signal. A special low-pitch signal (250 Hz, near to middle C) is generated when the direction of motion is approximately correct (i.e.,

[9] Personal communication

Sighted		1.30
Image Mode	Simulation	0.52
	Physical	0.40
Guidance Mode	Simulation	0.76
	Physical	0.45

Table 17.3: Average walking speeds (ms^{-1}) under different conditions. 'Sighted' refers to the speed of the average sighted walker. The other figures relate to the evaluation of the NavBelt in its two modes, both in simulations and in physical traversal of a laboratory. Note that the figure for Image Mode (Physical) was only attained after 'several hours' of training.

within $\pm 5°$). This provides simple positive feedback when the user is going in the correct direction. At the same time, using a low-frequency tone will have less of a masking effect on environmental sounds.

Image Mode is designed to invoke the impression of a virtual sound source sweeping across 180° in front of the user. The sweep is completed in 37 discrete steps separated by 5°. The sounds used are square waves modulated by amplitude, pitch and duration [26]. The duration of a signal varies between 20 and 40ms, where 20ms indicates the longest distance to an obstacle (5 meters) and 40ms indicates a very close object (0.5m). The amplitude varies inversely with the range reading from the corresponding sonar sector. Sixteen discrete amplitudes can be selected, where the lowest value (silence) represents no threat to the user from that direction, whereas the maximum value indicates a high risk. The intention is that 'the user's mind creates a mental picture of the environment that adequately describes the *obstacle density* around the user.' [24, p. 113].

Evaluations of the NavBelt have been based on simulations. In navigating randomly selected (simulated) maps the average travelling speed was $0.52ms^{-1}$ (compared to an average sighted person's walking pace of around $1.3ms^{-1}$). It was evident that the walking speed depends very much on the complexity of the environment. A more complex environment requires greater cognitive effort by the user and apparently leads to a slower walking speed. At the same time there appeared to be a learning effect, whereby experienced NavBelt users attained higher speeds. It was also noted, though, that users with 'reduced auditory perception capabilities travel slower than highly skilled people.' [26]. As well as the simulations, experiments were also carried out using the actual NavBelt in which blindfolded participants travelled from one side of the laboratory to the other. Walking speeds were slower than in the simulation because participants were more cautious. 'However, after a training period of several hours they traveled safely through the controlled environment of the laboratory with an average speed of $0.4ms^{-1}$.' (*ibid.*)

Simulation evaluations of Guidance Mode showed an average travel speed of $0.76ms^{-1}$ and an average deviation from the recommended direction of 7.7°. In similar experiments with the real prototype NavBelt whereby participants travelled 12 meters across the laboratory the average speed was $0.45ms^{-1}$. A more-realistic experiment was carried out in an office building corridor with which the participants were familiar. The length of the path was 25 meters and included several corners. No obstacles were positioned initially in the corridor, but passers-by did walk down the corridor. Participants were also able to avoid obstacles and attained an average walking speed of 0.6 ms^{-1}. The walking speeds attained under the various conditions are summarized in Table 17.3.

17.5.2 Visual Substitution

Developers of the NavBelt have experimented with two approaches to guidance. Its Image Mode is an example of the approach whereby the idea is to generate an auditory field representing the entire visual scene that a sighted person would see. The visual picture can be captured through video cameras and then translated into an auditory form.

The bandwidth problem was described above. The same problem applies in this application. Sensors such as video cameras can provide large amounts of (visual) information. The question is how much of that information to provide to the user. The more information, the greater the user's freedom to navigate, but the harder it becomes to interpret and understand. Obstacle detectors, such as those described above, can generate quite simple sounds, giving an indication of the location of the size and location of objects. At the other end of the scale there have been attempts to render the entire scene sonically.

One example is the *Voice* Project (see [29, 30] and the Voice web site[10]). This creates a representation of the visual picture pixel-by-pixel. The vertical positions of pixels are represented by pitch, horizontal positions (left-to-right) are represented by time, and brightness is represented by loudness. The sound effectively scans horizontally across the image so that a vertical column of pixels are all presented in a single, complex sound. The start of a scan is marked by a 'click' and the scanning repeatedly loops. An example of this sonification is shown in Figure 17.1.

This simple mapping is quite raw, implying minimal processing of the image. The system relies instead on brain plasticity. The intention is that with practice the user will learn to interpret the auditory scenes naturally. Some support for this approach is given in [31] which describes the examination of functional magnetic resonance images (fMRI) of the brains of blind and sighted participants performing sound localization tasks. They observe that blind people demonstrate a shift in activated brain areas towards more posterior areas - the areas that are involved in visual processing in sighted people[11].

González-Mora et al. [31, 37] have experimented with a prototype device which incorporates video cameras and headphones mounted on a pair of spectacles. Their sonification is described as follows:

> 'The basic idea of this prototype can be intuitively imagined as trying to emulate, using virtual reality techniques, the continuous stream of information flowing to the brain through the eyes, coming from the objects which define the surrounding space, and which is carried by the light which illuminates the environment. In

[10] `http://www.visualprosthesis.com/voice.htm`

[11] Modern brain imaging techniques such as fMRI have enabled researchers to shed new light on the idea that blind people's non-visual senses are in some ways heightened. Previously there was some skepticism about this apparent phenomenon (e.g., [32]) but now there is an increasing body of knowledge which suggests that the area of the brain usually referred to as the *visual* cortex is devoted largely to the processing of visual information simply because in sighted people that is the predominant source of stimulation. In people deprived of sight, the same area can be reassigned to the processing of non-visual information. Examples of this work include [33, 34, 35, 36].

Figure 17.1: Sample graphic which is sonified by the Voice system as illustrated in Sample **S17.12**. Note that the blurred style of the picture is deliberate, reflecting the pixel-by-pixel translation to sound.

> this scheme two slightly different images of the environment are formed on the retina, with the light reflected by surrounding objects and processed by the brain to generate its perception. The proposed analogy consists of simulating the sounds that all objects in the surrounding space would generate' [31, p. 371-372].

Of course in a real environment, inactive objects do not generate sounds, but in the prototype system a click sound is used:

> 'When a person is in front of a particular scene, he/she receives an acoustic input consisting of a set of a set of auralized[12] "clicks", with a randomized order of emission, corresponding to the calculated 3-D coordinates of the objects. This set of "clicks" is sent to the person in a time period of 153ms, after which the next acoustic image is sent. Depending on the number of coordinates that the objects occupy inside the perception field, there is a variable interclick interval, never less than 1 ms.' (*ibid.* p. 374).

Interestingly, in the context of the earlier quote from John Hull, the perceived effect is described as resembling 'a large number of rain drops striking the surface of a pane of glass'.

Spatial information is reproduced by spatialization of the sounds, using individualized head-related transfer functions (HRTFs)[13]. A field of 80° horizontally by 45° vertically is presented with a resolution of 17×9 and 8 levels of depth (although higher resolutions are being developed).

[12]The authors appear to use the word 'auralized' to mean 'spatialized'.

[13]Every individual is different in the way they perceive spatial sounds because of the shape of their ears and head. This can be modeled for artificially spatialized sounds by creating their HRTF. Best results are achieved using individual HRTFs, although an 'average' HRTF can be used, but it will be less effective.

Evaluations have yielded results which are claimed to be encouraging regarding blind people's ability to perceive the layout of a test room - although the evaluation is not described in detail. It is interesting that some participants reported experiencing apparent synaesthesic effects, whereby the sonification evoked a visual perception of 'luminous sparkles' coinciding with the spatial location of the sound sources. The system is very much a prototype and not yet a released product.

Sighted people rely on light reflecting off objects in the environment entering their eyes and forming an image on the retina. Babies learn to interpret these images through active exploration of their environment and hence learn to rely on them in interacting with the world. Visual substitution systems aim to create a similar representation using sounds. Most objects do not make sound, though, so there is no natural acoustic 'light', so instead a visual image is translated by technology into sound. The hope is that people can learn to interpret these soundscapes as well and as naturally as visual scenes. There is some hope for this approach in that the plasticity of the brain in interpreting acoustic input has been well demonstrated by the success of cochlear implants for deaf people. A cochlear implant generates artificial sensations in the auditory nerves. There is no reason to believe that the sensations thus generated resemble those generated by natural hearing, and yet - with practice - people become quite adept at interpreting those inputs as if they are (low fidelity) sounds [38].

It has to be stated that there is a dearth of formal evaluations of most of the systems described in this section. This is clearly a weakness in research terms, but furthermore there must always be a fear that the systems are ineffective and that to pursue them further would be a waste of time.

17.5.3 Navigation Systems

The obstacle avoidance and visual substitution systems described above are predominantly concerned with short-range mobility, mainly the avoidance of obstacles. A different problem is that of navigating to a chosen destination. For instance, when a person arrives by train in a strange city, they may need to know how to get to an office block which is known to be walking distance from the station. This is a problem for all travelers, but sighted people can rely on maps and similar aids to work out and follow the correct route. It is increasingly common now for car drivers faced with such navigation problems to rely on a SatNav global positioning device and the same option is available for blind pedestrians.

A number of systems have been developed. One feature which they all seem to share, though, is a reliance on the use of speech, and apparent minimal use of non-speech sounds.

Some of the systems developed are:

Trekker: Based on the Maestro, which is a PDA designed to be accessible to blind users, the Trekker is a talking GPS addition. Further details are available at `http://www.humanware.ca`.

BrailleNote GPS: BrailleNote is a portable braille PDA which also has an optional GPS attachment. It displays information in speech and braille. It is also available from Humanware (`http://www.humanware.ca`), and [39] is a (somewhat dated) comparative evaluation of Trekker and BrailleNote GPS.

Sendero: This company markets accessible GPS software for a number of devices, including the BrailleNote, the VoiceSense PDA, Windows Mobile and Symbian phones as well as non-mobile devices - the desktop PC for route planning (`http://www.senderogroup.com`). These rely almost entirely on voice output.

Mobic: Mobic was an experimental system developed with funding from the European Union. An evaluation of the system is documented in [40], but the output was entirely spoken.

Satellite navigation is a technology developed to assist in navigation tasks by providing the user's location in the world accurate to the nearest few meters. Originally this was for military personnel operating in unfamiliar territory (presumably because they had just invaded the territory!). However, it did not take long for developers to realize that the technology might be a valuable aid for those whose navigation problems arise from their not being able to see. The systems listed above are examples of this. Further developments and improvements will no doubt occur and it would seem to be an application in which there is potential for the use of non-speech sounds. The street is an environment in which naturally occurring sounds are invaluable to a blind pedestrian. Guidance information presented in a way which is complementary to the natural soundscape would be most valuable; it must be possible to improve on plain speech.

17.6 Other Systems

There are a variety of other attempts to translate graphical materials into sounds. They are all research projects which have not yet found their way into everyday use, but brief details are given here, along with links to further information.

17.6.1 Soundgraphs

A Cartesian graph is a simple but rich visual representation of information. See Figure 17.2, for instance. Many people have had the idea that the curve of such a graph could be represented by a soundgraph or auditory graph based on a sound, the pitch of which varies with the height of the curve, but one of the first published suggestions was [41]. A number of different groups have implemented the idea, including [42, 43, 44]. Walker & Mauney present guidelines on soundgraph design in [45]. (See also chapters 2, 6, 8).

A powerful visual facility is that of the *glance*. In other words, the viewer can look briefly at a visual representation and get an overall (but imprecise) impression of its meaning. Playing the waveform of the soundgraph curve, as in Sample **S17.13**, gives a similar overall impression of the curve's shape. To gain more precise information the user might interact with the sound. Thus a sound cursor can be moved left and right along the curve and by listening for the point of highest pitch, the user might locate the maximum in the curve in Figure 17.2. Having located the point, its coordinates could be found using speech.

The soundgraph implementation of Edwards and Stevens [43] facilitated the location of such turning points (maxima and minima) by allowing the user to hear the derivative of the curve. At a turning point the slope of the curve is zero and hence its derivative is a constant. Listening to the derivative, it should have constant pitch at such a point. Grond et al. [46] have taken this idea a step further by displaying the first m terms of the Taylor Series of a

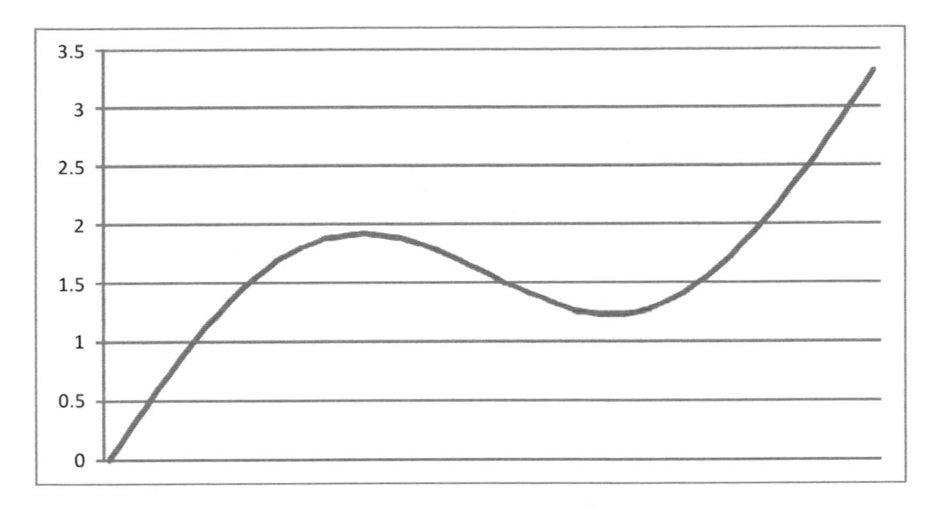

Figure 17.2: A graph of $y = \sin(x) + x/2$, a typical curve, which might be represented as a soundgraph. A sound representation of this graph can be found as Sample **S17.13**.

function. This work is in its early stages but the authors conclude that for suitable students (essentially those with experience of auditory media) 'the sonified functions are very supportive to grasp important characteristics of a mathematical function' [46, p. 20]. Examples of these sonifications are available at `http://www.techfak.uni-bielefeld.de/ags/ami/publications/GDH2010-SEW`.

A good soundgraph implementation should have the advantage of giving the user a feeling of very direct interaction with the *mathematics*. For instance, the user can move along the curve, sensing significant points (e.g., a maximum turning point) just by hearing the variation in the pitch (rising then falling). This contrasts with other representations such as algebra, where the requirement to manipulate the symbols can interfere with the appreciation of the mathematics that they represent. Yu, Ramloll, and Brewster [47] have gone a step further in facilitating direct interaction with soundgraphs by adding haptic interaction via the Phantom force-feedback device.

Soundgraphs have generally been used to represent curves on Cartesian graphs. Another form of graph is the scatter plot. Riedenklau et al. have developed a very novel non-visual representation of the scatter plot that uses sounds and Tangible Active Objects (TAOs) [48]. A TAO is a small plastic cube with on-board processing and wireless communication facilities. Placed on a Tangible Desk (*tDesk*) surface they can be tracked by camera. Scatter plots can be represented on the surface and the position of the TAO relative to clusters is fed back in an auditory form as a *sonogram* [49]. Sighted testers (who were blindfolded for the experiment) matched the TAO representation of different scatter plots to visual representations with a 77% success rate. Again this work is in its early stages and further developments - including testing with blind people - are proposed.

17.6.2 Audiograph

Audiograph [50, 51, 52] was a system originally designed to test how much information could be conveyed in non-speech sound, but it was soon realized that the most appropriate application would be as a means of presenting graphical information to blind people.

The following graphical information is communicated for each graphical object:

1. the position of each object;
2. the type, size and shape of each object;
3. the overall position of the objects using various scanning techniques.

All these used a similar metaphor - a coordinate point is described by using a musical mapping from distance to pitch (a higher note describing a larger coordinate value), and x and y co-ordinates are distinguished by timbre (Organ and Piano).

17.6.3 Smartsight

This is a simple form of translation from visual pixel information to non-speech sounds of different pitch, similar to soundgraphs [53, 54, 55]. An auditory cursor sweeps across the graphic horizontally. As it intersects a black pixel it makes a sound, the pitch of which represents the vertical height of the pixel. Figures 17.3 and 17.4 and their accompanying sound samples show how simple shapes are translated using this scheme.

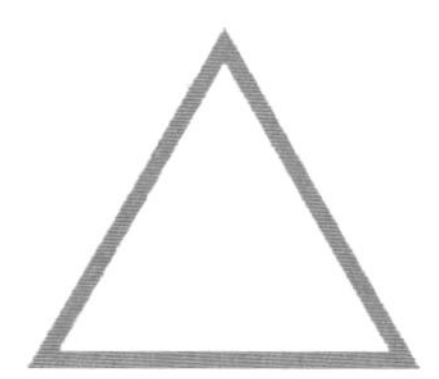

Figure 17.3: A triangle, which is rendered in sound by Smartsight as Sample **S17.14**. There is a constant (low) tone, representing the base of the triangle with rising and falling tones representing the other sides.

Simple graphics, such as those above can be perceived quite easily without training, but the developers claim that with training the same approach can be successfully used with much more complex layouts. Figure 17.5 is an example of a more complex, compound shape, but the developers claim that with training listeners can even interpret moving, animated graphics.

Smartsight originated in research in the University of Manchester, Institute of Science and Technology (now part of the University of Manchester) but was transferred to a spin-off company which has the objective of commercializing the idea. As yet, though, commercial success seems limited; during the writing of this chapter the company's web site disappeared.

Figure 17.4: A square, which is rendered by Smartsight as Sample **S17.15**. Notice that this starts with a sharp chord, representing the left-hand vertical edge, followed by a pair of notes representing the horizontal edges and finishes with another vertical edge.

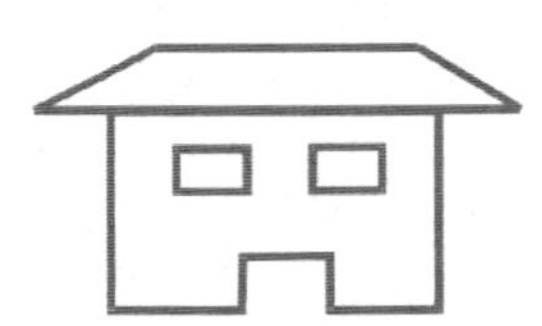

Figure 17.5: Sample graphic that would translate to Sample **S17.16** using the Smartsight system. The picture is a stylized, symmetrical house, with a trapezoid roof, two rectangular windows and a door.

17.7 Discussion

This chapter has been largely concerned with the use of hearing as a substitute for vision, but the two senses, and the stimuli which interact with them, are very different. The 'bandwidth problem' has been discussed earlier. Related to that is the fact that, in general, sight is passive. That is to say that, except in artificial conditions of darkness, visual information is always available. Thus, sighted people receive vast amounts of visual information constantly, and large parts of their brains are assigned to processing that information. Sound is also inherently temporal, while vision is more spatial. Of course sounds have a spatial origin and visual objects can move over time, but the emphasis is different in each case.

Sound, by contrast, is active, inasmuch as something must be moving to generate the sound. Most objects do not emit sounds and so to make them accessible to the auditory channel they must be made noisy. Sometimes objects can be embodied in sounds - as in screen readers which assign sounds to the elements of computer programs. Other systems work with the light analogy more directly. The Voice system operates on ambient light. It captures the visual scene, through video cameras, and converts them into sound, but in doing this it converts from the spatial domain to the temporal. In other words, the pixels are presented as an auditory raster scan, not in parallel. The problem is that the auditory sense is poorly equipped to interpret a scene thus presented.

Other devices work rather more like flashlights. The 'K' Sonar physically resembles a flashlight and it creates an auditory signal representing the portion of the environment captured in its (very limited) 'beam'. The signal presented is simple, but impoverished. With training, users can presumably interpret the sounds well (Table 17.1), but such perception is hardly comparable with vision.

The spatialized clicks of González-Mora's system are claimed to give a good picture. These are simple sounds. Serial presentation in a random order may overcome some of the problems of translation from the visual (spatial) to the auditory (spatial and temporal) domain, but the work is still experimental and yet to be proved.

17.8 Conclusion

When the sense of sight is missing other senses must be recruited. Sight is a very powerful sense and so it is very difficult (perhaps impossible) to completely substitute for it. Nevertheless, the auditory sense has great potential as an alternative. Much research effort has been expended into developing technologies which will do this, as described in this chapter. Yet it is significant that this chapter is almost solely concerned with the description of research projects; very few of the devices described are in commercial production and those which are tend to sell in small numbers (see also [56]). In other words, the great potential for the use of non-speech sounds as experienced in everyday life and highlighted by the passage from John Hull is not being realized.

This arrested development of auditory representation is a phenomenon which might be apparent in other chapters of this book; we authors and researchers know the potential for the use of auditory displays and are enthusiastic about promoting their adoption - yet the users are unconvinced. Within the context of this chapter specifically, one might expect that the users - those without sight - would in some senses be the easiest to convince, would be most willing to adopt an alternative for the sense which they lack, even if the alternative is less-than-perfect. Yet this is not the case.

Furthermore, while this chapter is concerned with disabled people, it has concentrated on those with visual disabilities. If auditory interaction has all the benefits and powers that we assert it has, then surely it could be a useful aid to those who have difficulties in interacting with technology due to other forms of impairment? Yet there seems to be almost no work that has demonstrated this to be the case.

It is not unusual for a researcher to conclude that what is required is more research, yet that is not always a cynical attempt at self-preservation. So it is in this case that there is a genuine need. We can develop auditory interfaces that are as good as the simple white cane, which give as much information as the cane (including the auditory feedback as it clicks on surfaces), and interfaces which can provide as much richness as rain falling on a garden - but as yet we do not know how.

Bibliography

[1] Hull, J., *On Sight and Insight: A Journey into the World of Blindness*. 1997, Oxford: Oneworld. 234.

[2] Stockman, T., *Listening to people, objects and interactions*, in *Proceedings of the 3rd Interactive Sonification Workshop (ISon 2010)*. 2010: Stockholm, Sweden. p. 3-8.

[3] Tiresias. *Users with disabilities - The numbers.* [Webpage] 2004 5 October 2004 [cited 2005 1 September]; Available from: http://www.tiresias.org/phoneability/telephones/user_numbers.htm.

[4] Carruthers, S., A. Humphreys, and J. Sandhu. The market for RT in Europe: A demographic study of need. in *Rehabilitation Technology: Strategies for the European Union (Proceedings of the First Tide Congress).* 1993. Brussels: IOS Press.

[5] Elkind, J., The incidence of disabilities in the United States. *Human Factors*, 1990. 32(4): p. 397-405.

[6] Bruce, I., A. McKennell, and E. Walker, *Blind and Partially Sighted Adults in Britain: The RNIB Survey.* 1991, London: HMSO.

[7] Weber, G., Reading and pointing - New interaction methods for braille displays, in *Extra-ordinary Human-Computer Interaction: Interfaces for Users with Disabilities*, A.D.N. Edwards, Editor. 1995, Cambridge University Press: New York. p. 183-200.

[8] Weber, G., Braille displays. *Information Technology and Disability*, 1994. 1(4).

[9] Blenkhorn, P., Requirements for screen access software using synthetic speech. *Journal of Microcomputer Applications*, 1993. 16(3): p. 243-248.

[10] Edwards, A.D.N., *Adapting user interfaces for visually disabled users.* 1987, Open University.

[11] Edwards, A.D.N., Soundtrack: An auditory interface for blind users. *Human-Computer Interaction*, 1989. 4(1): p. 45-66.

[12] Mynatt, E.D. and G. Weber. Nonvisual presentation of graphical user interfaces: Contrasting two approaches. in *Celebrating Interdependence: Proceedings of Chi '94.* 1994. Boston: New York: ACM Press.

[13] Gaver, W., Auditory Icons: Using sound in computer interfaces. *Human Computer Interaction*, 1986. 2(2): p. 167-177.

[14] Downie, A. *VoiceOver screen reader for Apple Macintosh Leopard.* 2008 April 2008 [cited 2011 2 February]; Available from: http://accessiblecli.wordpress.com/2009/07/28/voiceover-screen-reader-for-apple-macintosh-leopard/.

[15] Leventhal, J., Not what the doctor ordered: A review of Apple's VoiceOver screen reader. *AccessWorld*, 2005. 6(5).

[16] Hull, J., *Touching the Rock: An Experience of Blindness.* 1990, London: SPCK. 165.

[17] McGrath, R., T. Waldmann, and M. Fernström. Listening to rooms and objects. in *Proceedings of AES 16th International conference on Spatial Sound Reproduction.* 1999. Rovaniemi, Finland.

[18] Bradley, N.A. and M.D. Dunlop. Investigating context-aware clues to assist navigation for visually impaired people. in *Proceedings of Workshop on Building Bridges: Interdisciplinary Context-Sensitive Computing.* 2002. Glasgow: University of Glasgow.

[19] Walker, B.N. and J. Lindsay. Navigation performance in a virtual environment with bonephones. in *Proceedings of the 11th International Conference on Auditory Display (ICAD 2005).* 2005. Limerick, Ireland.

[20] Massof, R.W. Auditory assistive devices for the blind. in *Proceedings of ICAD 2003 (International Conference on Auditory Display).* 2003. Boston.

[21] Yen, D.H. *Currently Available Electronic Travel Aids for the Blind.* 2005 [cited 2006 3 April]; Available from: http://www.noogenesis.com/eta/current.html.

[22] Brabyn, J.A., New developments in mobility and orientation aids for the blind. *IEEE Trans. Biomedical Engineering*, 1982. BME-29(4): p. 285-289.

[23] Shoval, S., I. Ulrich, and J. Borenstein, Computerized obstacle avoidance systems for the blind and visually impaired, in *Intelligent Systems and Technologies in Rehabilitation Engineering*, H.-N.L. Teodorescu and L.C. Jain, Editors. 2001, CRC Press: Boca Raton. p. 413-447.

[24] Borenstein, J. The NavBelt - A Computerized Multi-Sensor Travel Aid for Active Guidance of the Blind. in *Proceedings of the Fifth Annual CSUN Conference on Technology and Persons With Disabilities.* 1990.

[25] Shoval, S., J. Borenstein, and Y. Koren. The Navbelt - A Computerized Travel Aid for the Blind. in *Proceedings of the RESNA '93 Conference.* 1993.

[26] Shoval, S., J. Borenstein, and Y. Koren. Mobile Robot Obstacle Avoidance in a Computerized Travel Aid for the Blind. in *Proceedings of the 1994 IEEE International Conference on Robotics and Automation*. 1994. San Diego.

[27] Shoval, S., J. Borenstein, and Y. Koren, Auditory Guidance With the NavBelt - A Computerized Travel Aid for the Blind. *IEEE Transactions on Systems, Man, and Cybernetics*, 1998. 28(3): p. 459-467.

[28] Kay, L., A sonar aid to enhance spatial perception of the blind: Engineering design and evaluation. *Radio and Electronic Engineer*, 1974. 44(11): p. 605-627.

[29] Meijer, P.B.L., An experimental system for auditory image representations. *IEEE Transactions on Biomedical Engineering*, 1992. 39(2): p. 112-121.

[30] Trivedi, B., Sensory hijack: rewiring brains to see with sound. *New Scientist*, 2010 (2773).

[31] González-Mora, J.L., et al., Seeing the world by hearing: Virtual acoustic space (VAS) a new space perception system for blind people, in *Touch Blindness and Neuroscience*, S. Ballesteros and M.A. Heller, Editors. 2004, UNED: Madrid, Spain. p. 371-383.

[32] Miller, L., Diderot reconsidered: Visual impairment and auditory compensation. *Journal of Visual Impairment and Blindness*, 1992. 86: p. 206-210.

[33] Rauschecker, J.P., Compensatory plasticity and sensory substitution in the cerebral cortex. *TINS*, 1995. 18(1): p. 36-43.

[34] Sadato, N., et al., Activation of the primary visual cortex by Braille reading in blind subjects. *Nature*, 1996. 380(6574): p. 526-528.

[35] Veraart, C., et al., Glucose utilisation in visual cortex is abnormally elevated in blindness of early onset but decreased in blindness of late onset. *Brain Research*, 1990. 510: p. 115-121.

[36] Zangaladze, A., et al., Involvement of visual cortex in tactile discrimination of orientation. *Nature*, 1999. 401: p. 587-590.

[37] Gonzalez-Mora, J.L., et al. Seeing the world by hearing: Virtual Acoustic Space (VAS) a new space perception system for blind people. in *Information and Communication Technologies, 2006. ICTTA '06. 2nd*. 2006.

[38] Tyler, R.S., ed. *Cochlear Implants: Audiological Foundations*. 1993, Singular: San Diego. 399.

[39] Denham, J., J. Leventhal, and H. McComas. *Getting from Point A to Point B: A Review of Two GPS Systems*. 2004 [cited 2006 24 July]; Volume 5 Number 6:[Available from: `http://www.afb.org/afbpress/pub.asp?DocID=aw050605`].

[40] Strothotte, T., et al. Development of dialogue systems for a mobility aid for blind people: Initial design and usability testing. in *Assets '96*. 1996. Vancouver: ACM.

[41] Mansur, D.L., M. Blattner, and K. Joy, Sound-Graphs: A numerical data analysis method for the blind. *Journal of Medical Systems*, 1985. 9: p. 163-174.

[42] Brown, L. and S. Brewster. Design guidelines for audio presentation of graphs and tables. in *Proceedings of the 9th International Conference on Auditory Display (ICAD 2003)*. 2003. Boston.

[43] Edwards, A.D.N. and R.D. Stevens, Mathematical representations: Graphs, curves and formulas, in *Non-Visual Human-Computer Interactions: Prospects for the visually handicapped*, D. Burger and J.-C. Sperandio, Editors. 1993, John Libbey Eurotext: Paris. p. 181-194.

[44] Sahyun, S.C. and J.A. Gardner. Audio and haptic access to Maths and Science: Audio Graphs, Triangle, the MathPlus Toolbox and the Tiger Printer. in *Computers and Assistive Technology, ICCHP '98: Proceedings of the XV IFIP World Computer Congress*. 1998. Vienna & Budapest.

[45] Walker, B.N. and L.M. Mauney, Universal Design of Auditory Graphs: A Comparison of Sonification Mappings for Visually Impaired and Sighted Listeners. *ACM Trans. Access. Comput.* 2(3): p. 1-16.

[46] Grond, F., T. Droßard, and T. Hermann, *SonicFunction: Experiments with a function browser for the visually impaired*, in *Proceedings of the The 16th International Conference on Auditory Display (ICAD 2010)*. 2010. p. 15-21.

[47] Yu, W., R. Ramloll, and S.A. Brewster, Haptic graphs for blind computer users, in *Haptic Human-Computer Interaction*, S.A. Brewster and R. Murray-Smith, Editors. 2001, Springer-Verlag: Heidelberg. p. 102-107.

[48] Riedenklau, E., T. Hermann, and H. Ritter, *Tangible active objects and interactive sonification as a scatter plot alternative for the visually impaired*, in *Proceedings of the The 16th International Conference on Auditory Display (ICAD 2010)*. 2010. p. 1-7.

[49] Hermann, T. and H. Ritter, Listen to your data: Model-based sonification for data analysis, in *Advances in intelligent computing and multimedia systems*, G.E. Lasker, Editor. 1999, International Institute for Advanced Studies in System Research and Cybernetics. p. 189-194.

[50] Rigas, D., *Guidelines for auditory interface design: An empirical investigation*, in *Department of Computer Science*. 1996, Loughborough University: Loughborough.

[51] Rigas, D.I. and J.L. Alty. The use of music in a graphical interface for the visually impaired. in *Proceedings of Interact '97, the International Conference on Human-Computer Interaction*. 1997. Sydney: Chapman and Hall.

[52] Alty, J.L. and D.I. Rigas. Communicating graphical information to blind users using music: The role of context. in *Making the Impossible Possible: Proceedings of CHI '98*. 1998: ACM Press.

[53] Cronly-Dillon, J.R., et al., Blind subjects analyse photo images of urban scenes encoded in musical form. *Investigative Ophthalmology & Visual Science*, 2000. 41.

[54] Cronly-Dillon, J.R., K.C. Persaud, and R.P.F. Gregory, The perception of visual images encoded in musical form: a study in cross-modality information transfer. *Proceedings Of The Royal Society Of London Series B-Biological Sciences*, 1999. 266: p. 2427-2433.

[55] Cronly-Dillon, J.R., K.C. Persaud, and R. Blore, Blind subjects construct conscious mental images of visual scenes encoded in musical form. *Proceedings of The Royal Society of London Series B-Biological Sciences*, 2000. 267: p. 2231-2238.

[56] Thierry, P., et al., Image and video processing for visually handicapped people. *J. Image Video Process.*, 2007. 2007(5): p. 1-12.

Chapter 18

Sonification for Process Monitoring

Paul Vickers

One of the tensions in auditory display is that sound is a temporal medium and so it is not always immediately apparent how one should approach the representation of data given that, prior to sonification, all data visualization used visual representations which rely on spatial organization. However, on the face of it, process monitoring would seem to be one of those tasks that was born to be sonified: at its heart is one or more series of temporally-related data. Monitoring entails the observation, supervision, and control of the process or system in question. To do this requires being attentive to changes in the state or behavior of the system over time so that appropriate interventions or other process-related activities may be carried out. Another feature of process monitoring is that it often has to be done as a background or secondary task or, perhaps, in parallel with one or more other primary tasks. This chapter looks at a range of auditory display and sonification applications that have tackled the problem of monitoring real-time data streams and concludes with some recommendations that further research should be informed by semiotic and aesthetic thinking and should explore the use of soundscapes, steganographic embedding, model-based sonification, and spatialization as profitable techniques for sonifying monitoring data.

18.1 Types of monitoring — basic categories

Because monitoring can be a primary or a secondary task we can classify monitoring activities into three categories: *direct*, *peripheral*, and *serendipitous-peripheral*. In a direct monitoring task we are directly engaged with the system being monitored and our attention is focused on the system as we take note of its state. In a peripheral monitoring task, our primary focus is elsewhere, our attention being diverted to the monitored system either on our own volition at intervals by scanning the system (what Jackson [66] would call a *state-vector inspection*) or through being interrupted by an exceptional event signaled by the system itself or through some monitor (such as an alarm).

In discussion of such peripheral information displays Maglio and Campbell [81, p. 241] identified that "the challenge is to create information displays that maximize information delivery while at the same time minimize intrusiveness or distraction." In Maglio's and Campbell's terms peripheral information is regarded as nonessential but nevertheless important. This, of course, is a question of context; if a network administrator is engaged in a primary task of, say, reconfiguring the firewall, then the monitoring of the network's behavior is peripheral and the information is not essential *to the task of firewall configuration* but it is, nevertheless essential to the overall goal of maintaining a healthy and stable network.

To distinguish between peripheral information that is important or essential to the overall goal but not to the immediate task in hand and that which is always only of secondary importance we can use Mynatt et al.'s [85] concept of *serendipitous* information. This is information that is useful and appreciated but not strictly required or vital either to the task in hand or the overall goal, hence the third class of process monitoring: *serendipitous-peripheral.*

Direct monitoring requires the user to be focused on the process under consideration, and so we can think of this as an information *pull* scenario: the user chooses to extract information from the interface. Peripheral monitoring is thus an information *push*: here the user is engaged in another task (or aspect of a process-related task) and it is up to the system to push significant information to the user. Serendipitous peripheral-monitoring, because the information is not vital and can be ignored by the user, is more of an information *nudge* situation: the system makes the information available and lets you know it's there, but doesn't press the matter. To summarize, we may think of three different modes of auditory process monitoring and their corresponding information push/pull type:

1. **Direct** (PULL) — the information to be monitored is the main focus of attention.
2. **Peripheral** (PUSH) — attention is focused on a primary task whilst required information relating to another task or goal is presented on a peripheral display and is monitored indirectly.
3. **Serendipitous-peripheral** (or just serendipitous for short) (PUSH/NUDGE) — attention is focused on a primary task whilst information that is useful but not required is presented on a peripheral display and is monitored indirectly.

Whilst visual displays are well suited to direct monitoring tasks (because our visual attention is focused on the display), they may not be so effective in peripheral or serendipitous-peripheral monitoring situations. It is in the monitoring of peripheral information that auditory displays come into their own, for the human auditory system does not need a directional fix on a sound source in order to perceive its presence. Tran and Mynatt [113] described sonic environmental monitors as "an extra set of eyes and ears" whilst Jenkins [69] discussed ways in which the auditory channel may be very useful for keeping users informed about background activities without being disruptive. Weiser and Brown [127] talk of 'calm computing' in which technology remains invisible in the background until needed [70].

In any case, sonification would appear to be a plausible part of the solution to the problem of monitoring processes as it allows visual attention to be focused elsewhere (perhaps on other tasks). Mountford and Gaver [84] described the difference between sound and vision in spatio-temporal terms: sound is a temporal medium with spatial characteristics whilst vision is primarily spatial in nature but with temporal features. The implication of this is that sound is good for communicating information that changes over time and can be heard over a range of spatial locations whereas vision is not so ephemeral but can only be observed at specific locations [111]. Audio is also good at orienting, or directing, a listener to key data [71] & [77, p. 57]. Schmandt and Vallejo [101] and Vallejo [114] argued that in peripheral monitoring activities audio is *more* useful than video as well as being less intrusive.[1] Mynatt et al. [85, p. 568] went as far as to say that non-speech audio is a natural medium for creating peripheral displays, and that speech requires more attention than non-speech audio.

Even in direct monitoring situations if the visual display is very complex or crowded, or many variables need to be monitored, an auditory display provides a useful complement to (and sometimes, a replacement for) a visual display. Several related variables can be mapped to the different parameters of a single sound source allowing information-rich monitors to be built with a minimum bandwidth. Consider Fitch and Kramer's [46] sonification of heart rate, respiratory rate, blood CO_2 & O_2 levels, pupilary reflex, atrio-ventricular dissociation, fibrillation, and temperature for the monitoring of patients during surgical operations. Through careful manipulation of pitch (frequency), timbre (spectral composition), and tempo, all these variables could be monitored using only three distinct timbres. Whilst a visual display could provide precise numeric values for these variables, the sonification allowed complex monitoring to take place without the need to focus visual attention on a single-display. Smith, Pickett, and Williams [103] discussed some of the common ways such multivariate monitoring is carried out together with a framework for empirical evaluation.

18.2 Modes of Listening

One factor that must be taken into account is the difference between hearing and listening. The distinction can be simply illustrated by thinking of hearing as a push activity (a sound is projected into our attention space) and listening as a pull activity (a listener deliberately attends to an audio stream in order to identify salient characteristics or extract information/meaning). In looking for easily understood mappings for his SonicFinder system, Gaver proposed a theory of *everyday listening* [53]. This says that in everyday situations people are more aware of the attributes of the source of a sound than the attributes of the sounds themselves: it is the size of the object making the sound, the type of the object, the material it is made of, etc. that interests the everyday listener. According to the theory we hear *big* lorries, *small* children, *plastic* cups being dropped, *glass* bottles breaking, and so on. Everyday listening is in contrast to what Gaver calls *musical listening* in which we are more interested in the sensations or attributes of the sounds themselves, their pitch, their intensity, and so on. Gaver summarizes thus:

> If the fundamental attributes of the sound wave itself are of concern, the experience will be one of musical listening. Another way to say this is that everyday listening involves attending to the *distal stimuli*, while musical listening involves attending to the *proximal stimuli*. [52, p. 4]

Note that these two types of listening are not categorical — they would seem more to be points along a continuum describing general listening approaches but sharing attributes. For example, in musical listening, though one may be primarily interested in the harmonic

[1] Here, intrusiveness refers to intrusion into the environment being monitored (a domestic setting, in this case) rather than intrusion into the environment of the monitor. We can think of it in terms of telephones: a regular audio-only phone is less intrusive than a video phone.

and temporal relationships of sounds, the individual timbres and their sources are also an important part of the experience. There are 'big' sounds and 'little' sounds even in music.[2]

Gaver's classification is in a similar vein to Pierre Schaeffer's [99] *reduced listening* and Michel Chion's [30] *causal* and *semantic* modes of listening.[3] Reduced listening is the opposite of Gaver's everyday listening. In reduced listening we listen "to the sound for its own sake, as a sound object by removing its real or supposed source and the meaning it may convey".[4] Causal listening is similar to Gaver's everyday listening in that the goal is to listen to a sound in order to infer information about its source. The difference lies in the modes' intentionalities: in causal listening we deliberately seek out source information whereas in everyday listening we are aware on a more subconscious level of the origin of a sound. In a sense, Chion's causal listening is truly a mode of *listening* whereas Gaver's everday listening is more of a response to *hearing*. The third mode, semantic listening, is one in which we are not interested in the properties of a sound *per se* but in the sound's code and, by extension, its semantic or linguistic meaning. For example, semantic listening enables us to interpret the message correctly even when given by two different speakers with different pronunciations.

Gaver's work (*v.i.*) is strongly motivated by a desire to maintain good acoustic ecology. R. Murray Schafer brought acoustic ecology to the fore with his *World Soundscape Project* [100]. Schafer saw the world around us as containing ecologies of sounds. Each soundscape possesses its own ecology and sounds from outside the soundscape are noticeable as not belonging to the ecology. In Schafer's worldview we are exhorted to treat the environments in which we find ourselves as musical compositions. By this we are transformed from being mere hearers of sound into active and analytic listeners — exactly the characteristic needed to benefit most from an auditory display. When the environment produces noises that result from data and events in the environment (or some system of interest) then we are able to monitor by listening rather than just viewing. This acoustic ecology viewpoint cuts across any possible 'everyday listening' vs. 'musical listening' dichotomy as Schafer sees the everyday world as a musical composition. In so doing he brings together into a single experience Gaver's distinct acts of everyday and musical listening — we attend to the attributes of the sounds and the attributes of the sound sources equally. This is facilitated by direct relationship between the physical properties of the source and the attributes of the sounds it produces. In a sonification there is always at least one level of mapping between the source data and the sound. Furthermore, the attributes of the sounds may be quite arbitrary and have no natural relationship to the data (as will be the case in metaphor-based sonification mappings) and so the question arises as to whether such equal attention is still possible.[5] In the natural world, two unfamiliar sounds emanating from two close-together sources might also lead to confusion in the listener over which attributes belong to what source: until the source and its range of sounds is learned there is room for ambiguity. The same, it could be argued, applies to sonification. The difference is merely that not all sonifications do build upon a lifetime of learned environmental sonic associations, but as long as the mappings are learnable, equal attention ought to be possible.

[2]For example, consider Phil Spector's *Wall of Sound* production technique which led to a very dense and rich sonic tapestry that sounded 'bigger' than other recordings of the time.

[3]Also see Chapter 7 which discusses the modes of listening within the context of sonification design and aesthetics.

[4]`http://www.ears.dmu.ac.uk`

[5]See also the discussion of mappings and metaphors in Chapter 7 section 7.2.2.

This way of approaching sonification has started to attract interest (for instance, influencing the discussions at the First International Symposium on Auditory Graphs (AGS2005) in Limerick, Ireland in July 2005.[6] Indeed, Hermann's work on model-based interactive sonification [60] has exploited the very characteristics of everyday and musical listening by mapping data to a physical sound model (e.g., a resonator) and allowing the user to perturb the model in order to infer meaning from the data (see also Chapters 11 and 16 for discussions of developments in this field). In model-based sonification the data become an instrument which the user plays in order to study its properties.

The sections that follow will discuss, first of all, some of the main examples of auditory process monitoring across a range of application domains. After that some of the higher-level issues that arise from auditory process monitoring such as intrusion, disturbance, annoyance, aesthetic considerations, and so forth are considered.

18.3 Environmental awareness (workspaces and living spaces)

18.3.1 'Industrial' monitoring

ARKola

Drawing lessons from acoustic ecology, Gaver, Smith, and O'Shea [54] built one of the first auditory process monitoring demonstrations, the ARKola system, an auditory monitoring system for a simulated soft drink bottling plant. The system was built using the SharedARK [102] virtual physics laboratory environment and comprised nine interconnected machines each carrying out a different function in the soft drink manufacturing and bottling process. Each machine communicated its state over time using auditory icons (see Chapter 13). The choice of auditory icons was analogic so as to create a complete ecology of bottling plant sounds. For example, heating machines made a sound like a blowtorch, bottle dispensers emitted sounds of clanking bottles, and so forth. Sounds were added to communicate additional information (such as a liquid splash sound to signal when materials were being wasted; the sound of breaking glass to signify bottles being lost). The ecology of the system was especially important given that as many as fourteen sounds could be played at once. Gaver et al. had to design the auditory icons carefully so as to avoid perceptual masking effects.

In a similar study Rauterberg and Styger [91] built a simulation of an assembly line of computer numeric control (CNC) robots. In one version of the system only visual feedback was provided whilst a second system was augmented with auditory feedback. Following Gaver et al.'s ARKola approach, Rauterberg and Styger created auditory icons and arranged their sonic spectra such that perceptual masking would not occur. However, whereas ARKola supported up to fourteen concurrent sounds the CNC simulator allowed for up to thirty-eight sounds at any one time. In an experiment it was found that users who had only the visual feedback needed to move to the system control station and request status reports much more frequently than those who had the benefit of the auditory icons. Like Gaver et al. [54] and Cohen [31, 32] Rauterberg and Styger concluded that for simulations of real-world situations the use of analogic (or even metaphoric) real-world sounds is appropriate and

[6] See `http://sonify.psych.gatech.edu/ags2005/`

reduces the possibility of annoyance that Gaver et al. warned could arise when using 'musical' messages.[7]

ShareMon

Acting on Buxton's observation that we monitor multiple real-world background activities by their associated sounds [26], Cohen constructed the ShareMon system [32] which notified users of file sharing activity on an Apple Macintosh network. Reactions to the system were varied and strong, and Cohen describes some users finding the band-filtered pink noise that was used to represent the percentage of CPU time consumed to be "obnoxious" [32, p. 514], despite them understanding the correlation between CPU load and the pitch of the noise. On the other hand, some users described the wave sounds used in the system as soothing. Cohen extended ShareMon with an ecoustic ecology of sounds (Cohen called it a *genre*) that comprised sounds from the original Star Trek television series [31]. The sounds were mapped to various system events and users were able to successfully monitor the system events.

18.3.2 Weather monitoring

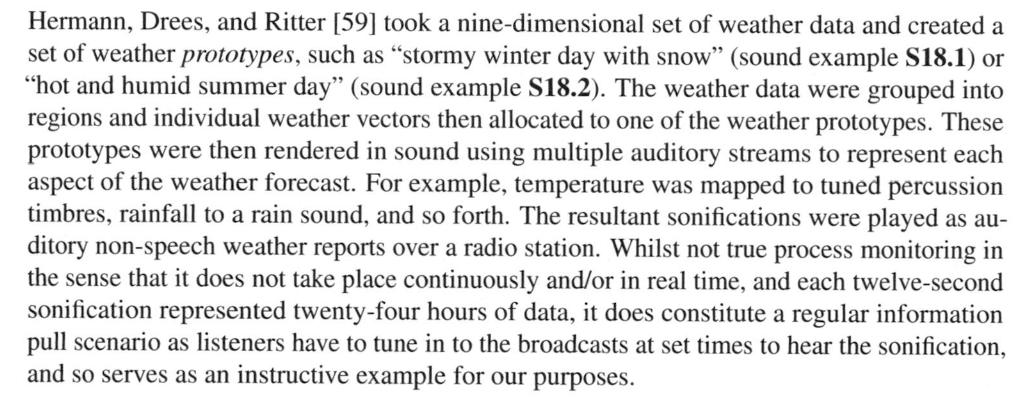

Hermann, Drees, and Ritter [59] took a nine-dimensional set of weather data and created a set of weather *prototypes*, such as "stormy winter day with snow" (sound example **S18.1**) or "hot and humid summer day" (sound example **S18.2**). The weather data were grouped into regions and individual weather vectors then allocated to one of the weather prototypes. These prototypes were then rendered in sound using multiple auditory streams to represent each aspect of the weather forecast. For example, temperature was mapped to tuned percussion timbres, rainfall to a rain sound, and so forth. The resultant sonifications were played as auditory non-speech weather reports over a radio station. Whilst not true process monitoring in the sense that it does not take place continuously and/or in real time, and each twelve-second sonification represented twenty-four hours of data, it does constitute a regular information pull scenario as listeners have to tune in to the broadcasts at set times to hear the sonification, and so serves as an instructive example for our purposes.

More recently, Bakker, van den Hoven, and Eggen [3] used auditory icons in a regular information-push scenario to play weather forecasts every thirty minutes in a shared work space. Participants in a study reported that they did not find the auditory icons distracting and after three weeks they found they noticed the auditory icons less than at the start. Although the system used a push design, participants who had no particular interest in the weather forecast did not find the sonifications annoying or distracting. The success of the system seems to have been due to a very careful sound design.

18.3.3 Home and shared work environments

ListenIN

Motivated by the desire to be able to monitor elderly relatives in their own homes remotely but still maintaining the relatives' privacy, Schmandt and Vallejo [101] built the ListenIN system. ListenIN has modules for a wide range of domestic situations, including detecting crying babies, and so is not restricted to the monitoring of older people. The ListenIN server

[7] Gaver's claim that auditory icons have less potential for annoyance than musical messages raises the question of role that aesthetics should play in sonification design. This is explored briefly at the end of this chapter (see section 18.7.2) and is discussed in more detail in Chapter 7 of this volume.

sits in the home being monitored and it determines what acoustic information is sent off the premises to the remote monitor clients. The principle of the system is that it identifies and classifies domestic noises and, if a sound matches one that the monitor has chosen to be reported, the server sends either a direct representation of the sound to the client (e.g., the sound of a baby crying for when a baby cry is detected) or a garbled version of the sound to protect privacy (e.g., garbled speech which is still recognizable as speech but in which no individual words, only the speed and intonation, can be made out). As yet, no formal evaluation has been reported. This is an example of peripheral monitoring as the sounds are not being produced at regular intervals and is akin to a sophisticated alarm.

Music Monitor

Tran and Mynatt [113] also built a system based on the user's own musical preferences for monitoring a domestic environment. Their Music Monitor used musical profiles to represent key information about the domestic activity being monitored. The intention was to provide an "extra set of eyes and ears" which could survey activities around the house and relay information about the activity states as real-time ambient music. Tran and Mynatt deliberately built the system to be what would here be classified as a serendipitous-peripheral monitoring application, the information it provides being interesting and useful but not of vital importance. In their thinking, vital information is left to alarms. In a similar approach to that of Barra et al.'s WebMelody [6] (*v.i.*), Music Monitor overlays the system information as a set of earcons onto a music track chosen by the user. Fishwick [43, 44] espouses such personalisation as an important principle of *aesthetic computing* [45]. In the realm of musical sonification it is, perhaps, even more important to cater for preference as individual musical tastes and cultural backgrounds vary a great deal. To allow the user to control the *style* of music is an important step forward in sonification practice. In a small experiment, participants were able to monitor activity state transitions displayed by Music Monitor, although situations in which there was a high level of background noise (such as cooking in the kitchen) required participants to stop what they were doing to attend to state changes played by the system. Other people in the environment who had not been told about the information display generated by Music Monitor reported only being aware of some pleasant background music — they were completely unaware of the information content in the signal.

WISP

Kilander and Lönnqvist [74, 75] sought to provide peripheral auditory monitoring through the construction of soundcapes. Their soundscapes were designed to be "weakly intrusive" to minimize the impact of the sonification on the listening environment. Kilander and Lönnqvist built two prototype systems, fuseONE and fuseTWO, which used their WISP (weakly intrusive ambient soundscape) [74] to communicate states and events in computational and physical environments through auditory cues.

RAVE

The Ravenscroft Audio Video Environment (RAVE) is described by Gaver et al. [50]. The motivation behind RAVE was to support collaborative working amongst colleagues dispersed throughout several rooms in a building. Each room had an audio-video node which comprised a camera, monitor, microphone, and speakers. All items in the node could be moved and turned on and off at will by the users. Users interacted with RAVE through GUI buttons. For example, pressing the background button would select a view from one of the public areas to be displayed on the audio-video node's monitor screen. The sweep button would cause a short sampling of each node's camera allowing users to find out who is present in the various offices. A specific location could be viewed by a three-second glance to its camera. Offices could also be connected via the *vphone* and *office share* functions which allowed creation of a larger virtual office. Auditory icons were introduced to allow greater levels of reasonable privacy. For example, when a glance connection was requested, an auditory warning was sounded at the target node's location three seconds prior to the glance being activated. When connections were broken other sounds, such as that of a door closing, were triggered. These auditory icons gave information about system state and user interactions.

OutToLunch

Cohen took the idea of monitoring background activity further with his OutToLunch system [33]. Cohen states:

> The liquid sound of keystrokes and mouse clicks generated by a number of computer users gives co-workers a sense of 'group awareness' — a feeling that other people are nearby and an impression of how busy they are. When my group moved from a building with open cubicles to one with closed offices, we could no longer hear this ambient sound. [p. 15]

OutToLunch was an attempt to restore this sense of group awareness through the use of auditory icons and an electronic sign board. Each time a user hit a key or clicked a mouse, a corresponding click sound would be played to the other users. Total activity over a thirty second period, rather than the exact timing of individual key strokes and mouse clicks, was recorded. Because the sounds did not convey much information most users soon found the system annoying. Unlike a real shared workspace, OutToLunch did not give directional clues to allow users to determine who was making the key clicks or mouse clicks. In a revised system each person in the group was represented by their own unique musical motif.

Workspace Zero

Eggen et al. [40] wanted to explore the interactions between people and their environment with a focus on understanding how a pervasive soundscape affects the behavior of an environment's inhabitants. With careful sound design they discovered that not only can sound be useful as an information carrier but it also has a "decorative value". Eggen et al. were greatly influenced by the concept of "calm technology" [127] which, in the context of auditory display, allows the user of a sonification to easily switch the sound between the perceptual foreground and background.

18.4 Monitoring program execution

Talk to engineers from the 1950s, '60s and '70s (and even home computer enthusiasts from the '80s) and many will tell stories about how they tuned AM radios to pick up the electrical noise given out by the processors of their computers. They learned to recognize the different patterns of sounds and matched them to program behavior during debugging activities [122] (see also section 9.12 in this volume). In fact, the use of audio as a tool for monitoring and debugging programs would seem to have a history almost as long as computer science itself. In his 2011 BCS/IET Turing Lecture Donald Knuth [76] made reference to the Manchester Mark 1 computer of 1949 (the successor to Baby, the world's first stored program computer) which had its circuits wired to an audio channel so that programs could be debugged by listening to them. Despite four decades of radio-assisted debugging it was not until the 1990s that research articles exploring the issue systematically began to appear. The case for using sound to aid programming was supported by Jackson and Francioni [64], although they felt that a visual resentation was also needed to provide a context or framework for the audio sound track. They argued that some types of programming error (such as those that can be spotted through pattern recognition) are more intuitively obvious to our ears than our eyes. Also, they pointed out that, unlike images, sound can be processed by the brain passively, that is, we can be aware of sounds without needing to listen to them. Francioni and Rover [49] found that sound allows a user to detect patterns of program behavior and also to detect anomalies with respect to expected patterns.

One of the first attempts at program sonification was described by Sonnenwald et al. [105] and was followed by DiGiano [37], DiGiano and Baecker [38], Brown and Hershberger [24], Jameson [68, 67], Bock [12, 13, 14], Mathur et al. [82, 10] and Vickers and Alty [121, 119, 120, 122, 118, 123]. These early systems all used complex tones in their auditory mappings but, like much other auditory display work, this was done without regard to the musicality of the representations (with the exception of Vickers and Alty). That is, simple mappings were often employed, such as quantizing the value of a data item to a chromatic pitch in the 128-tone range offered by MIDI-compatible tone generators. Furthermore, the pitches were typically atonal in their organization and were combined with sound effects (e.g., a machine sound to represent a function processing some data). Effort was largely invested in demonstrating that data could be mapped to sound with much less attention given to the aesthetics and usability of the auditory displays. Where aesthetic considerations are taken into account auditory displays become much easier to comprehend. Mayer-Kress, Bargar, and Choi [83] mapped chaotic attractor functions to musical structures in which the functions' similar but never-the-same regions could be clearly heard. The resultant music could be appreciated in its own right without needing to know how it was produced. Quinn's Seismic Sonata [90] likewise uses the aesthetics of musical form to sonify data from the 1994 Northridge, California earthquake. Alty [1] demonstrated how musical forms could be used to sonify the bubble-sort algorithm. When we turn from pure data sonification and look towards sonification techniques as a complement to existing visualization models we find that good progress has been made. Brown and Hershberger [24] coupled visual displays of a program during execution with a form of sonification. They suggested that sound will be a "powerful technique for communicating information about algorithms".

Brown and Hershberger offered some examples of successful uses of audio, for instance, applying sound to the bubble-sort algorithm. The main use of sound in this work was to reinforce visual displays, convey patterns and signal error conditions and was by no means the main focus of the work. Like most other visualization systems that employ audio, Brown and Hershberger's work used sound as a complement to visual representations. No formal evaluation of the approach was published. Early efforts at pure sonification were concerned with specific algorithms, often in the parallel programming domain. Examples include Francioni et al. [47, 48] who were interested in using sonifications to help debug distributed-memory parallel programs, and Jackson and Francioni [65, 64] who suggested features of parallel programs that would map well to sound. Those systems that are applied to more general sequential programming problems require a degree of expert knowledge to use, whether in terms of programming skill, musical knowledge, expertise in the use of sound generating hardware, or all three.

18.4.1 Systems for external auditory representations of programs

To date the following program sonification tools have been identified (as opposed to visualization systems that incorporate some sonification): InfoSound [105] by Sonnenwald et al, the LogoMedia system by DiGiano and Baecker [38], Jameson's Sonnet system [68, 67], Bock's Auditory Domain Specification Language (ADSL) [12, 14, 13] the LISTEN Specification Language, or LSL [82, 11, 10], Vickers and Alty's CAITLIN system [121, 119, 120, 122, 118, 123], Finlayson and Mellish's AudioView [41], Stefik et al. [109] and Berman and Gallagher [8, 7].[8]

Infosound

Sonnenwald et al.'s InfoSound [105] is an audio-interface tool kit that allows application developers to design and develop audio interfaces. It provides the facility to design musical sequences and everyday sounds, to store designed sounds, and to associate sounds with application events. InfoSound combines musical sequences with sound effects. A limitation is that the software developer is expected to compose the musical sequences himself. To the musically untrained (the majority) this militates against its general use. The system is used by application programs to indicate when an event has occurred during execution and was successfully used to locate errors in a program that was previously deemed to be correct. Users of the system were able to detect rapid, multiple event sequences that are hard to detect visually using text and graphics.

LogoMedia

DiGiano and Baecker's LogoMedia [38], an extension to LogoMotion [2] allows audio to be associated with program events. The programmer annotates the code with probes to track control and data flow. As execution of the program causes variables and machine state to change over time, the changes can be mapped to sounds to allow execution to be listened to. Like InfoSound, LogoMedia employs both music and sound effects. DiGiano and Baecker assert that comprehending "the course of execution of a program and how its data changes is essential to understanding why a program does or does not work. Auralisation expands the possible approaches to elucidating program behavior" [38].[9] The main limitation is that the sonifications have to be defined by the programmer for each expression that is required to be monitored during execution. In other words, after entering an expression, the programmer is prompted as to the desired mapping for that expression.

Sonnet

Jameson's Sonnet system [68, 67] is specifically aimed at the debugging process. Using the Sonnet visual programming language (SVPL) the code to be debugged is tagged with sonification agents that define how specific sections of code will sound. The sound example is of a bubble sort algorithm with an indexing error bug (sound example **S18.3**). Figure 18.1 shows a component to turn a note on and off connected to some source code. The "P" button on the component allows static properties such as pitch and amplitude to be altered. The component has two connections, one to turn on a note (via a MIDI note-on instruction) and one to silence the note (MIDI note-off). In this example the note-on connector is attached to the program just before the loop, and the note-off connector just after the close of the loop. Therefore, this sonification will cause the note to sound continuously for the duration of the loop. Thus, placement of the connectors defines the sonification.

[8] Stefik and Gellenbeck [108] have latterly reported their Sonified Omniscient Debugger in which speech is used as the auditory display.

[9] *Auralisation* was the term originally adopted by researchers working in the program sonification domain. It has fallen out of common usage with authors tending to use the more general (and well-known) *sonification*, so auralisation is now a (mostly) deprecated term.

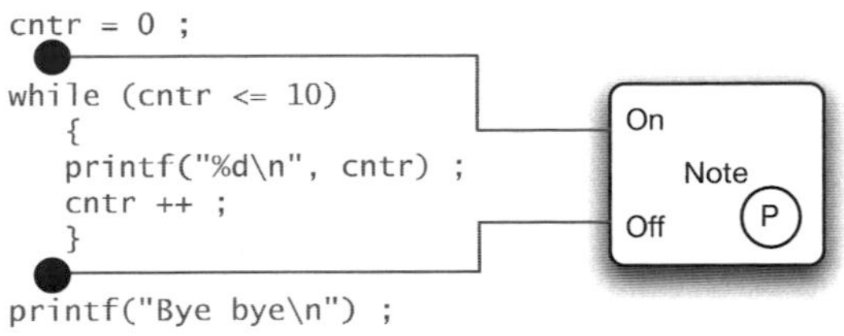

```
cntr = 0 ;
while (cntr <= 10)
   {
   printf("%d\n", cntr) ;
   cntr ++ ;
   }
printf("Bye bye\n") ;
```


Figure 18.1: A Sonnet VPL Component redrawn version of the original in Jameson [68]. The VPL component is the box on the right.

Other components allow the user to specify how many iterations of a loop to play. Program data could also be monitored by components that could be attached to identifiers within the code. The guiding principle is that the programmer, knowing what sounds have been associated with what parts of the program, can listen to the execution looking out for patterns that deviate from the expected path. When a deviation is heard, and assuming the expectation is not in error, the point at which the deviation occurred will indicate where in the program code to look for a bug.

Sonnet is an audio-enhanced debugger. This means that because it interfaces directly with the executing program it is not invasive and does not need to carry out any pre-processing to produce the sonifications. The visual programming language offers great flexibility to the programmer who wants to sonify his program. However, it does require a lot of work if an entire program is to be sonified even very simply.

```
Track_name=Loop
{
1 Track=Status('for'):Snd("for_sound");
2 Track=Status('while'):Snd("while_sound");
}
```

Figure 18.2: An ADSL track to monitor for and while loops redrawn from the original from Bock [12]

ADSL

Bock's Auditory Domain Specification Language (ADSL) [12, 14, 13] differs from the above three approaches in that it does not require sounds to be associated with specific lines of program code. Instead users define tracks using the ADSL meta-language to associate audio cues with program constructs and data. These tracks (see Figure 18.2) are then interpreted by a pre-processor so that the code has the sonifications added to it at compilation allowing the program to be listened to as it runs. The fragment of ADSL in Figure 18.2 specifies that `for` and `while` loops are to be signalled by playing the '`for_sound` and '`while_sound`' respectively. These two sounds have been previously defined and could be a MIDI note sequence, an auditory icon, or recorded speech. The sounds will be heard when the keywords `for` and `while` are encountered in the program. An advantage of this approach is that it is possible to define a general purpose sonification. That is, by specifying types of program construct to be sonified there is no requirement to tag individual lines of code with sonification specifications. When such tagging is required this can be done using the features of the

```
begin auralspec
specmodule call_auralize
var
  gear_change_pattern, oil_ check_ pattern, battery_ weak_pattern: pattern;
begin call_auralize
  gear_ change_pattern:="F2G2F2G2F2G2C1:qq'"+ "C1:";
  oil_ check_pattern:="F6G6:h"';
  battery_weak_pattern:="A2C2A2C2";
  notify all rule = function_ call "gear_change" using gear_ change_ pattern;
  notify all rule = function_ call "oil_ check" using oil_ check_ pattern;
  notify all rule = function_ call "battery_ weak" using battery_ weak_ pattern;
end call_auralize;
end auralspec.
```

Figure 18.3: An LSL ASPEC for an automobile controller redrawn version of the original in Mathur et al. [82]

specification language. Like Sonnet, ADSL is non-invasive as the sonifications are added to a copy of the source program during a pre-processing phase.

ADSL uses a mixture of digitized recordings, synthesized speech, and MIDI messages. The choice of sounds and mappings is at the discretion of the user, although a common reusable set of sonifications could be created and shared amongst several programmers. Tracks could be refined to allow probing of specific data items or selective sonification of loop iterations. The system is flexible, allowing reasonably straightforward whole-program sonification, or more refined probing (such as in Sonnet).

In a study [13] thirty post-graduate engineering students from Syracuse University all with some programming experience were required to locate a variety of bugs in three programs using only a pseudo-code representation of the program and the ADSL auditory output. On average, students identified 68% of the bugs in the three programs. However, no control group was used, so it is not possible to determine whether the sonifications assisted in the bug identification.

Listen/LSL & JListen

Mathur's Listen project [82, 10] (currently inactive) follows a similar approach to that used by ADSL. A program is sonified by writing an "auralisation specification" (ASPEC) in the Listen Specification Language (LSL). A pre-processing phase is used to parse and amend a copy of the source program prior to compilation. Again, the original source program is left unchanged. The accompanying sound file is LISTEN's sonification of a bubble sort algorithm (sound example **S18.4**). An ASPEC defines the mapping between program-domain events and auditory events, and an example for an automobile controller is shown in Figure 18.3. This example sonifies all calls to the program functions `gear_change`, `oil_check` and `weak_battery`. The ASPEC contains the sonification definitions and their usage instructions. For instance, in Figure 18.3 we see that a call to program function `gear_change` causes the sonification `gear_change_pattern` to be played. This sonification is defined earlier in the ASPEC as a sequence of MIDI note-on/off instructions.

As LSL is a meta-language it can, in theory, be used to define sonifications for programs written in any language; the practice requires an extended version of LSL for each target language. What is immediately apparent is that writing ASPECS in LSL is not a trivial task as it requires the programmer to learn the syntax of LSL. In addition, some musical knowledge is needed to know how to specify which pitches are used in the sonifications. The Listen project appeared dormant after 1996 but was reactivated in 2002 and applied to Java in the form of JListen [30]. Some informal studies and applications of JListen have been carried out (see Gopinath [56] and Prasath [89]) but, to date no formal experimentation or evaluation of the original or newer JListen systems has been published. The CAITLIN system

(a)

```
FOR counter := 1 TO 6 DO
   counter := counter + 1 ;
```

The Pascal code (a) results in the auralization (b). Bar 1 and beat 1 of bar 2 is the tune denoting entry to the loop. The notes remainder of bar 2 and bar 3 representsix iterations of the loop, the final iteration being supplemented by a sleighbell sound. Bars 4 and 5 denote exit from the loop.

(c)

```
a : = 10 ;
CASE a OF
   '1' : Writeln ('Found 1') ;
   '2' : Writeln ('Found 2') ;
   '3' : Writeln ('Found 3') ;
   '4' : Writeln ('Found 4') ;
   ELSE  Writeln ('Not found)
END ;
```

The Pascal code (c) results in the auralization (d). Bar 1 and beat 1 of bar 2 is the tune denoting entry to the CASE construct. A cowbell sound is played as each case instance is tested (bars 2, 3, and 4). The tune in bars4 to 6 signify exit from the construct. In this example, no match was found for the case selector, so the construct exits in a minor key. In the CAITLIN system the major mode was used to denote Boolean true and minor for Boolean false.

Figure 18.4: Example CAITLIN sonifications taken from Vickers [115]

CAITLIN

is a non-invasive pre-processor that allows the sonification of programs written in Turbo Pascal [121, 119, 120, 122, 118, 123]. Musical output is achieved by sending MIDI data to a multi-timbral synthesizer via the MIDI port on a sound card. The CAITLIN sonifications were designed around a tonal musical framework. Users were not required to design the sonification content as the system used unique pre-defined motifs (theme tunes) to represent the Pascal language constructs [120]. This meant that no musical knowledge was needed in order to produce musically-consistent output. Figure 18.4 shows example sonifications for two program fragments. Figure 18.4(d) can be heard in the accompanying sound file (sound example **S18.5**). Experiments with the CAITLIN system indicated that musical sonifications do provide information useful in bug-location and detection tasks [122, 123]. As the CAITLIN system was an experimental prototype it has some limitations. First, the sonifications were only applied to the language constructs (selections and iterations) which meant that other program features could not be inspected aurally. Secondly, constructs could not be individually marked for sonification meaning the entire program was sonified. In a real debugging situation programmers would not monitor an entire program but would choose candidate sections for close scrutiny. This whole-program approach meant that even short programs could take a long time to play back.

AudioView

Finlayson and Mellish [41] developed the AudioView to provide an auditory *glance* or overview of Java source code. Using non-speech-audio-only, speech-only, and combined non-speech & speech output modes, Finlayson and Mellish explored how good the auditory channel is at helping programmers to gain higher level understanding of their programs. A similar approach to Vickers and Alty was taken in the construction of the non-speech cues in that a structured hierarchy of earcons was built which corresponded to the hierarchy of Java programming constructs (see Figure 18.5).[10] Finlayson, Mellish, and Masthoff [42] extended the AudioView system to provide auditory fisheye views of the source code.

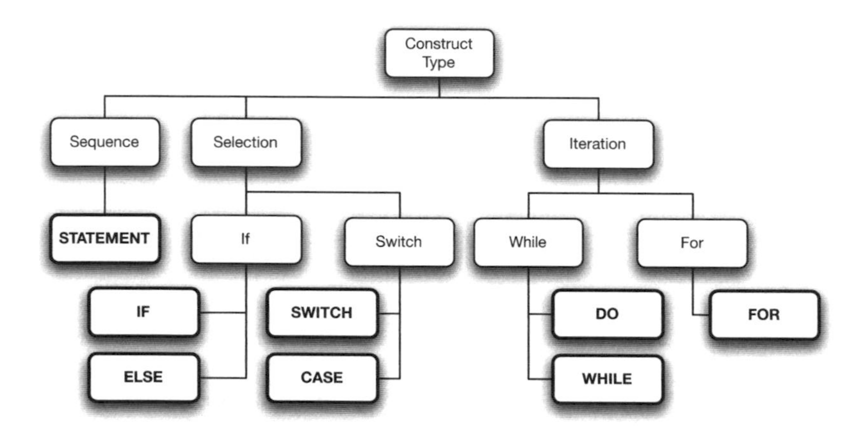

Figure 18.5: Basic Java construct hierarchy redrawn version of original from Finlayson and Mellish [41]

Program slices and low-level structures

Berman and Gallagher [7, 8] were also interested in higher-level or aggregated structural knowledge. To assist with debugging, rather than sonify the entire run-time behavior of a program they computed *slices* (sets of statements that are implicated in the values of some user-selected point of interest), (sound example **S18.6**). In a series of small studies Berman and Gallagher [7] found that programmers were able to use sonifications of the slices to infer knowledge such as the homgeneity of the slice, the amount of a particular function/method participating in a slice, etc. In follow-up work [8] they combined CSound (a music synthesis and signal processing system) with the Eclipse integrated development environment (IDE) to provide a framework for sonifying low-level software architecture.[11] The tool allowed programmers to sonify information about low-level and static program structures such as packages, classes, methods, and interfaces.

Sonification for program monitoring and debugging received the majority of its interest in the 1990s and the first decade of this century and examples of program sonification have been less common in recent years. This is not to say that software visualization has diminished *per se* (for example, see Romero et al. [96, 95, 94, 93]) but sonification work appears to have been focused more in other areas of endeavor. Perhaps this is related to the relative complexity of programming environments. In the 1980s a good deal of programming was done with

[10] Vickers and Alty did such a hierarchical analysis for the Pascal language [120].

[11] For information on CSound see `http://www.csounds.com/`. For Eclipse, see `http://www.eclipse.org/`.

line editors and command-line compilers on dumb mainframe terminals and the procedural programming paradigm was dominant. The late 1980s witnessed the start of the large-scale take up of the IBM PC-compatible machines and integrated development environments (IDE) began to get a foothold. Despite the more powerful programming environments graphical representations were still primitive and sonification was directed at communicating the run-time behavior of programs. Today, object-oriented programming is the prevailing paradigm, IDEs are much richer, code is concurrent and multithreaded, and programmers are looking for tools that offer insight into much more than run-time behavior. There are exceptions, of course. For example, Lapidot and Hazzan [78] describe a project that sets out to use music as a framework for helping computer science students to reflect upon the activity of debugging. Stefik and Gellenbeck [108] explored the use of speech for assisting the debugging process; although this is not, strictly, sonification, it is encouraging that the problem still continues to receive attention. What is particuarly interesting in Stefik and Gellenbeck's research is the reason for using speech rather than sonification techniques:

> While we find the idea of musical structures to represent code structures interesting, we have tested hundreds of combinations of musical structures in our laboratory and we generally find them unlikely to produce comprehension benefits in general-purpose integrated development environments. The reason for this may be that an obvious, intuitive, mapping between arbitrary program constructs and music does not exist. [110, p. 70]

Moving beyond the problem of run-time behavior Berman and Gallagher [7] and Finlayson et al. [41, 42] have provided insight into what future program monitoring environments will need to include for sonification to begin to get wider acceptance. Given the increased potential for external auditory representations of software it is somewhat suprising that more effort has not been engaged in such research. Hussein et al. [63] suggest that the main obstacle is that multi-disciplinary teams are needed with expertise in both music and software technology. However, this is arguably true for all sonification research (see Chapter 7) and so it is likely that the reasons are varied. Stefik and Gellenbeck [108] observe that understanding and debugging programs are intrinsically difficult activities and this might be one factor in the relative paucity of program sonification research. Another might be related to the IDEs that are used today. In the early days of program sonification work there were few IDEs and graphical programming environments were often the result of specialist research projects. Today there are several popular IDEs in use (e.g., Eclipse and NetBeans) and for sonification to make any real inroads it needs to be integrated with these environments. In the past researchers tended to build bespoke sonification environments, but this is no longer necessary. It is encouraging that the more recent projects have adopted the integrating-with-existing-IDEs approach and this would seem to be the way forward.

18.5 Monitoring interface tasks

SonicFinder

Gaver was one of the first to propose adding sounds to a general computer interface to allow users to monitor the progress and state of sundry activities and tasks. His SonicFinder system [53] used auditory icons [51] to enhance the existing Apple Macintosh Finder program. The visual feedback provided by the Finder was extended by adding sounds. The auditory icon for a progress bar representing a file copy operation was the sound of a jug being filled with water. An empty jug sounds hollow and the water pouring into it has a low pitch. As the file copy progressed, so the pitch of the water sound rose and the jug sounded increasingly full. The auditory icon did not tell the user anything that the visual display did not communicate, serving only to reinforce the feedback. However, unlike the visual version which showed the overall length of the progress bar and the amount completed at any point, the auditory icon gave no reference points for the start and end of the task. Thus, whilst the listener could hear the progress of the file copy, and could possibly infer that it was nearing its end, no absolute information to this effect was provided aurally. Gaver reports that the SonicFinder was appealing to users and that some were reluctant to give it up.[12] Unfortunately, no formal empirical or other evaluation was carried out so there is no firm evidence to indicate the efficacy of Gaver's system.

Lumsden et al. [80] identified three important principles for successful interface sonification widgets:

1. Minimize the annoyance
2. Simplify the mapping
3. Facilitate segregation of the sonification elements

Brewster has worked extensively with earcons to sonify interface activities [18, 17, 22, 19, 21, 23, 20]. Of particular interest here is Crease and Brewster's investigation of auditory progress bars [35, 36]. Parallel earcons were used to represent the initiation, progress, heartbeat, remainder, and completion information of a progress bar activity (these terms come from Conn's task properties for good affordance [34]). Experimentation with participants showed that users preferred the auditory progress bar to a visual one, that frustration experienced by participants fell in the auditory condition, and that annoyance levels were roughly the same for the auditory and visual progress bars. Furthermore, Crease and Brewster commented that the addition of sounds "allows users to monitor the state of the progress bar without using their visual focus . . . participants were aware of the state of the progress bar without having to remove the visual focus from their foreground task" [36].

18.5.1 Web server and internet sonification

Peep

Two good candidates for live monitoring applications are web servers and computer networks. Gilfix and Crouch [55] said of activity in complex networks that it is "both too important to ignore and too tedious too watch" [p. 109]. Their solution to this conundrum was to build the Peep network sonification tool. Peep employed a soundscape approach (an "'ecology' of natural sounds" in their words) in which network state information was gathered from multiple sources and was used to trigger the playback of pre-recorded natural sounds (sound example **S18.7**). This mapping between network events and natural sounds created a soundscape that represented the continuous and changing state of the network being monitored. Like most sonifications the mappings did not allow the communication of absolute values by the sounds but the person monitoring the network was able to infer rich knowledge about the network's state simply through hearing the relative differences

[12] One could also argue that the novelty factor of an auditory interface in 1989 would have added to the SonicFinder's appeal, but interface event sounds are now commonplace in modern desktop operating systems and so the novelty has worn off. However, it is interesting that despite Gaver's work back in 1989, auditory progress bars themselves still do not come as standard in operating system interfaces.

between and changes in the individual components of the soundscape. The power of this representational technique lay in the human auditory system's ability to recognise changes in continuous background sounds.

Ballora et al. [4] undertook a similar exercise to Gilfix and Crouch but instead of using pre-recorded natural sounds they used a technique more akin to model-based sonification in which the various channels of network traffic data being monitored drove the values of the parameters of a sound model in the SuperCollider synthesis program.[13] In their case they experimented with several approaches in which the the four octet values of an IP address controlled:

- The levels of the four main partials in a vibraphone model;
- The parameters of four separate instances of a water-like instrument;
- A set of formants in a vocal model.

They found that the mappings resulted in sonic backdrops that were neither annoying nor distracting.

WebMelody

Barra et al. [6, 5] sonified web server workload, severe server errors, and normal server behavior, by mapping the server data to musical soundtracks with their WebMelody system. Echoing Tran and Mynatt (see section 18.3.3) the motivation behind WebMelody was to allow long-term monitoring of the web server with minimum fatigue. A principal benefit of the auditory approach is that it is eyes-free. In an attempt to minimize fatigue and annoyance, Barra et al. moved beyond the use of simple audio alarms and tried "to offer an aesthetic connotation to sounds that border between music and background noises." [5, p.34]. In so doing it was claimed that WebMelody "let users listen for a long time without diverting them from their work". The approach taken was to allow the user to select any external music source of his or her choosing into which MIDI-based sonifications of the web server data were mixed. This allowed the administrator to hear the status of the server while listening to his or her preferred choice of music (sound example **S18.8**). One claimed benefit of WebMelody is that it allows real-time feedback to be provided, something that general log analyzers cannot do given their reliance on aggregated log files [6]. WebMelody uses an information push approach rather than the more typical pull needed to get information from a server's log file.[14] A potential drawback of real-time auditory information push is that the auditory stream could become tiring and/or annoying, as well as being environmentally intrusive (the stream is only of interest to those wishing to monitor the server activity). To mitigate these effects, WebMelody's designers made the system as configurable as possible.

Table 18.1 summarizes the characteristics of some of the systems discussed above showing their domain of application, their type (direct, peripheral, serendipitous-peripheral) and the primary sonification technique employed.

[13] SuperCollider is available from `http://supercollider.sourceforge.net/`.

[14] This is analogous in concept to using a data stream connection (WebMelody) instead of a state vector connection (log file analysis) in Jackson's JSD nomenclature [66].

System	Domain	Type (D,P,S-P)[1]	Sonification technique employed
Audio Aura (Mynatt et al)	Workplace monitoring	S-P	Sonic landscapes
OutToLunch (Cohen [33])	Workplace activities	P & S-P	Sound effects/auditory icons
ShareMon (Cohen [32])	Monitoring of file sharing	P & S-P	Sound effects/auditory icons
fuseONE, fuseTWO (Kilander and Lönnqvist [75])	Remote process monitoring	P & S-P	Ambient soundscapes
Music Monitor (Tran & Mynatt [113])	Monitoring domestic environments	S-P	Tonal music & ambient soundscapes
ListenIN (Schmandt & Vallejo [101])	Monitoring domestic environments	P & S-P	Auditory icons
LogoMedia (DiGiano & Baecker)	Program monitoring	D & P	Ad-hoc pitch mappings & sound effects
LISTEN/LSL (Mathur et al)	Program monitoring & debugging	D & P	Ad-hoc pitch mappings
ADSL (Bock [12, 14, 13])	Program monitoring & debugging	D & P	Ad-hoc pitch mappings
Sonnet (Jameson [68, 67])	Program monitoring & debugging	D & P	Ad-hoc pitch mappings
Zeus (Brown & Hershberger [24])	Program monitoring	D & P	Simple pitch mappings
Parallel programs (Jackson & Francioni [64])	Monitoring of parallel programs	D & P	Simple pitch mappings
Infosound (Sonnenwald et al. [105])	Monitoring of parallel programs	D& P	Ad-hoc pitch mappings
Nomadic Radio (Sawhney [98])	Contextual messaging in roaming environments	D & P	Speech & non-speech audio
SonicFinder (Gaver [53])	Interface monitoring	D & P	Auditory icons
ARKola (Gaver, Smith, & O'Shea [54])	Industrial plant monitoring	D	Auditory icons
CNC Robots (Rauterberg & Styger [91])	Industrial plant monitoring	D	Auditory icons
AudioView (Finlayson & Mellish [41])	Program monitoring & debugging	D & P	Earcons
Berman and Gallagher [8, 7]	Program architecture	D & P	Musical motifs
CAITLIN (Vickers & Alty [121, 119, 120, 122, 118, 123])	Program monitoring & debugging	D & P	Tonal music motifs/earcons

[1] (D)irect, (P)eripheral, (S-P)Serendipitous-peripheral

Table 18.1: Summary characteristics of a selection of monitoring systems.

18.6 Potential pitfalls

Previous sections showed how researchers have attempted to provide auditory displays for direct, peripheral, and serendipitous-peripheral auditory process monitoring. From this body of work we may identify a number of principal challenges that face designers of such sonifications:

1. The potential intrusion and distraction of sonifications;
2. Fatigue & annoyance induced by process sonification;
3. Aesthetic issues and acoustic ecology;
4. Comprehensibility and audibility.

18.6.1 Intrusion and distraction, fatigue and annoyance

There is a tension when designing auditory process monitors between the sonification being perceptible to its intended audience and being too intrusive or annoying. In their work on awareness support systems, Hudson and Smith [62] commented on the problem of intrusion in terms of awareness and privacy. They stated that this "dual tradeoff is between privacy and awareness, and between awareness and disturbance". The more information an auditory monitor provides the richer the sonification yet the greater the potential for disturbance, annoyance, and an upset in the balance of the acoustic ecology. Gutwin and Greenberg [57] claimed it is a tradeoff between being well informed and being distracted. Kilander and Lönnqvist [75] noted the effect of such sonifications on people sharing the workspace who are not part of the monitoring task:

> In a shared environment, one recipient may listen with interest while others find themselves exposed to an incomprehensible noise. [p. 4]

Indeed, commenting upon the design of their *nomadic radio* system, Sawhney and Schmandt [97, 98] cautioned that care must be taken to ensure that the auditory monitoring system intrudes minimally on the user's social and physical environment.[15]

In dealing with intrusiveness, Pedersen and Sokoler [88] framed the problem as a balance between putting a low demand on attention versus conveying enough information. In their Aroma system the goal was to communicate knowledge and awareness of the activities of people at remote sites, thus the privacy of the people being monitored by the system was of great importance. They studied this problem through an "ecology of awareness" showing awareness of the importance of the acoustic ecology of a sonification. Pedersen and Sokola made the auditory, visual, and haptic representations of the Aroma system highly abstract — abstraction would allow useful information to be communicated without divulging too many details that would violate privacy. It was hoped that abstract representations would be better at providing "peripheral non-attention demanding awareness" [88, p. 53]. It was also noted that such abstract representations lend themselves to being remapped to other media (what Somers [104] would call *semiotic transformation*), or, in turn, foster accommodating user preferences (an important aspect of aesthetic computing). Unfortunately, user studies showed that the

[15] In *nomadic radio* a mixture of ambient sound, recorded voice cues, and summaries of email and text messages is used to help mobile workers keep track of information and communication services.

abstraction led to users interpreting the representations in varied ways [87]. Furthermore, Kilander and Lönnqvist [75] warned that the "monitoring of mechanical activities such as network or server performance easily runs the risk of being monotonous" and Pedersen and Sokola reported that they soon grew tired of the highly abstract representations used in Aroma. It is interesting that they put some of the blame down to an impoverished aesthetic, feeling that involving expertise from the appropriate artistic communities would improve this aspect of their work.

Cohen [32] identified a general objection to using audio for process monitoring: people in shared office environments do not want more noise to distract them. Buxton [26] argued that audio is ubiquitous and would be less annoying if people had more control over it in their environments. Lessons from acoustic ecology would be helpful here. Cohen [32] defined an acoustic ecology as "a seamless and information-rich, yet unobtrusive, audio environment".

Kilander and Lönnqvist [75] tackled this problem in their fuseONE and fuseTWO environments with the notion of a *weakly intrusive ambient soundscape* (WISP). In this approach the sound cues for environmental and process data are subtle and minimally-intrusive.[16] Minimal- or weak-intrusion is achieved in Kilander and Lönnqvist's scheme by drawing upon the listener's expectation, anticipation, and perception; anticipated sounds, say Kilander and Lönnqvist, slip from our attention. For example, a ticking clock would be readily perceived and attended to when its sound is introduced into the environment (assuming it is not masked by another sound). However, as the steady-state of the ticking continues and the listener expects or anticipates its presence its perceived importance drops and the sound fades from our attention [75]). However, a change in the speed, timbre, or intensity of the clock tick would quickly bring it back to the attention of the listener. Intrusiveness can thus be kept to a necessary minimum by using and modulating sounds that fit well with the acoustic ecology of the process monitor's environment. The sonification is then able to be discriminated from other environmental sounds (either by deliberate attentiveness on the part of the listener, or by system changes to the sounds), yet is sufficiently subtle so as not to distract from other tasks that the listener (and others in the environment) may be carrying out. To increase the quality of the acoustic ecology further, Kilander and Lönnqvist used real-world sounds rather than synthesized noises and musical tones. They concluded that

> ... easily recognisable and natural sounds ... *[stand]* ... the greatest chance of being accepted as a part of the environment. In particular, a continuous background murmur is probably more easily ignored than a singular sound, and it also continuously reassures the listener that it is operative. [75]

Kilander and Lönnqvist's weakly intrusive ambient soundscapes would thus seem to be a suitable framework for the design of peripheral process monitoring sonifications in which monitoring is not the user's primary or sole task.

Schmandt and Vallejo [101] noted the perception-distraction dichotomy. Their ListenIN system for monitoring activity in a domestic environment attempted to provide continuous but minimally-distracting awareness. Unfortunately, no formal studies have been carried out with the system to test this aim. Mynatt et al. [85] aimed with their Audio Aura

[16]Kilander and Lönnqvist actually used the adjective 'non-intrusive' to describe their sonifications. One could argue that this term is misleading as any sonification needs to be intrusive to some extent in order to be heard. Their term 'weakly intrusive' is more helpful and more accurate.

scheme to provide environmental sonifications that enriched the physical world without being distracting. Tran and Myatt reduced the intrusiveness of their Music Monitor system [113] by overlaying earcons on top of music tracks chosen by the main user. The mixing of earcons with intentional music meant that other people in the environment would not be distracted as the encoded messages in the earcons would only be recognised by the main user: other people would just be aware of changes in the music. Of course, the fact that there is a music stream at all means that the system does still intrude into the environment. Rather, the attempt here was to minimize the negative distracting effects of that intrusion. Preliminary experimental results showed that the music was not distracting to those who were not monitoring the earcon messages embedded within it [113]. In a related project that combined auditory displays with tangible computing Bovermann et al. [16, 15] found that their Reim toolset could be used to provide peripheral monitoring without causing distraction or annoyance.

Fatigue is sometimes mentioned as a potential problem associated with auditory display but it is notable that hearing is "more resistant to fatigue than vision" [91, p. 42] and so it is not clear that auditory displays should cause more problems in this regard than visual representations.

18.6.2 Emotive associations

The degree and detail to which process data are sonified depends a great deal on the intended audience. Some may take a dispassionate view whilst others may attach emotional significance to the data. Cohen [31] found that it is difficult to construct "sounds which tell the right story and are also pleasant and emotionally neutral". For example, in their work on sonifying weather reports Hermann, Drees, and Ritter [59] noted that whilst meteorologists would be interested in exploring and analyzing long-term time-series weather data, the average public consumer of a weather forecast requires a much more abstract view in terms of what will the weather be this afternoon, tomorrow, or at the weekend. Choice of activity and clothing are dependent on the weather, so a simple forecast indicating likely temperature, wind, and precipitation levels is sufficient for most tasks. Furthermore, a weather forecast can trigger an emotional response in the listener. Hermann, Drees, and Ritter put it that rather than having the detailed quantitative interest of the expert meteorologist, the "listener is concerned with *[the weather's]* emotional value and contextual implications, which are not simply assessed from single ... attributes like temperature or humidity in isolation" [59]. For instance, to a northern European a temperature of 30°C would be very pleasant if the humidity is low, but quite unpleasant in overcast humid conditions. Thus, the sonification designer can take into account the fact that whilst the raw process data themselves are free of emotive content, their values and combinations can cause an inferential process of emotional coding on the part of the listener. Hermann, Drees, and Ritter [59] attempted to deal with this issue directly by deriving emotional relations from the high-dimensional 'weather vector'. This was accomplished by constructing an *Emo-Map* (see Figure 18.6). The Emo-Map is a two-dimensional plot of hourly weather vectors: each vector is displayed as a single point on the plot. From this plot were derived a number of prototype weather states (such as 'hot dry summer day', 'snowy winter day' and 'golden October day'). Each prototype had an associated emotive aspect (e.g., enervated and indifferent for the hot dry summer day, negative, calm, and indifferent for the snowy winter day, and positive emotional state for the

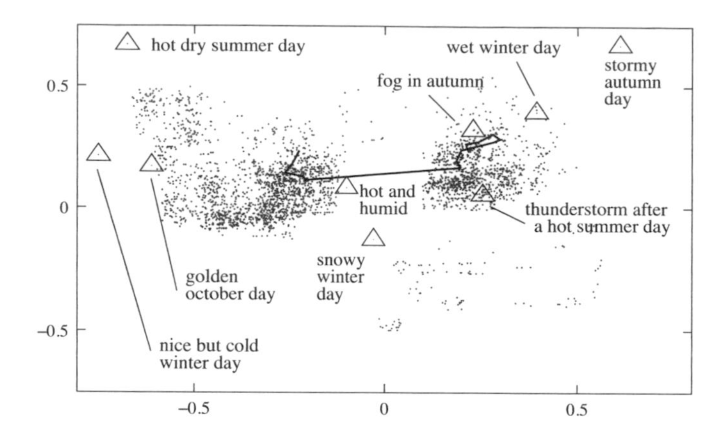

Figure 18.6: The 'Emo-Map' from Hermann et al. [59] showing some typical weather states and prototype markers.

golden October day). Sounds were chosen to correspond to the prototype-emotion classes (panting sound & cricket songs for the summer day, a shudder/shiver sound for the snowy winter day, and a rising 'organic sound' for the golden October day). A resultant sonification is shown schematically in Figure 18.7.

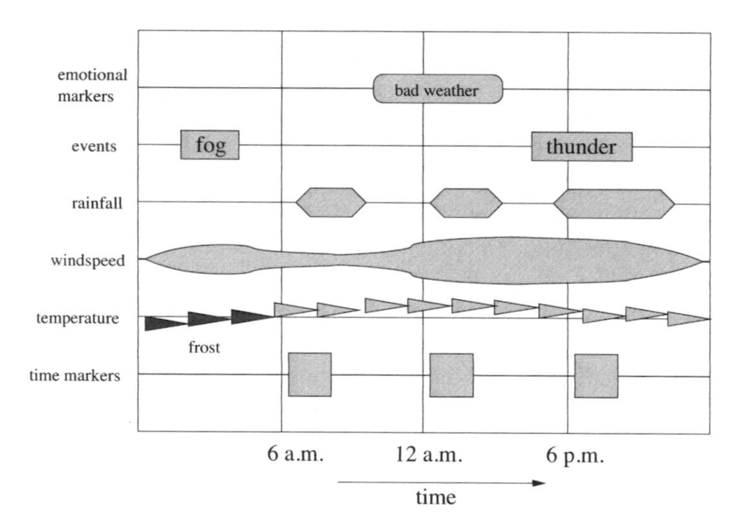

Figure 18.7: Graphical visualization of Hermann et al.'s weather sonification [59]. The y axis shows the six auditory streams with time shown along the x axis. As humidity and wind direction were not represented by their own streams but modulated existing streams, they are not shown on the figure.

18.6.3 Aesthetics and acoustic ecology

In some of the work discussed above, authors used terms such as *ecology* and *acoustic ecology* in reference to their sonification designs. In recent years there has been a growing realization of the important role to be played by aesthetics in the design of computer systems and artefacts. Fishwick [43, 44, 45] coined the term *aesthetic computing* to refer to application of art theory and practice to the design of computing systems. Fishwick [44] claims

> ...there is a tendency toward the mass-media approach of standardized design, rather than an approach toward a more cultural, personal, and customized set of aesthetics.

It is making use of cultural and personal differences that Fishwick claims will enlarge "the set of people who can use and understand computing". Whilst Fishwick has focused primarily on the aesthetics of visualizations and models, recent research in the auditory display community has begun to pay attention to aesthetic issues in sonification.[17]

Cohen used acoustic ecologies of sounds in his ShareMon system, only he called them collections of *genre* sounds [31]. A principle of aesthetic computing is that systems should be malleable according to the culture in which it is situated [43]. Cohen argued the strongest reason for *not* using 'genre' sounds is that they are less universal than everyday sounds (as used in auditory icons) or musical motifs (as used in earcons). He put the counter-argument thus: "everyday sounds vary for different cultures anyway, as do the ways of constructing musical motifs", so there is no reason in principle why these types of acoustic ecology cannot be successfully used in the right context. Cohen also identified the importance of allowing users to assign their own choice of sound sets to auditory monitoring applications: this catering for user preference is another principle of aesthetic computing. Indeed, Cohen suggested that users could "choose familiar genres based on aesthetic preference" [31].

For their WebMelody system (see section 18.5.1) Barra et al. [5] drew upon the ideas of futurist composer Luigi Russolo (1885-1947), principles of Pierre Schaeffer's *musique concrète*, and found inspiration in Edgard Varèse's *Poème Electronique* (1958) and John Cage's aleatoric compositions, to construct sonifications that were "neutral with respect to the usual and conventional musical themes".[18] In other words, they attempted to move away from the idioms of tonal and atonal music and towards the more abstract syntaxes found in the electroacoustic/*musique concrète*/organized sound traditions. *Musique concrète* approaches composition not by writing a tune which is then given to players to render in sound but instead by first recording existing or 'found' (concrete) sounds and assembling them into a musical piece.[19] It is not hard to see how auditory-icon-based approaches might fit into this mould given the auditory icon's origin in existing real-world sound. Vickers sounded a note of caution at such moves away from tonal music systems [116]:

> In the pursuit of aesthetic excellence we must be careful not to tip the balance too far in favour of artistic form. Much current art music would not be appropriate for a sonification system. The vernacular is popular music, the aesthetics of which

[17]The role of sonification in aesthetic computing has not gone unnoticed by Fishwick though as evidenced by the inclusion of a chapter by Vickers and Alty [124] on program sonification in the volume *Aesthetic Computing* [45]. The relationship between sonification design and aesthetics is discussed more fully in Chapter 7 of this volume.

[18]For example, *Music of Changes* (1951) and much of his music for prepared piano.

[19]This is in contrast to *musique abstraite* which is what traditional compositional techniques produce.

> are often far removed from the ideals of the music theorists and experimentalists. [p. 6]

This admonition was posited on the observations by Lucas [79] that the recognition accuracy of an auditory display was increased when users were made aware of the display's musical design principles. Furthermore, Watkins and Dyson [126] demonstrated that melodies following the rules of western tonal music are easier to learn, organize cognitively, and discriminate than control tone sequences of similar complexity. This, Vickers argued, suggested that the cognitive organizational overhead associated with atonal systems makes them less well suited as carriers of program information. However, this reasoning fails to account for the fact that electroacoustic music, whilst often lacking discernible melodies and harmonic structures, is still much easier to organize and decompose cognitively than atonal pieces [117]. The studies of Lucas [79] and Watkins and Dyson [126] were rooted in the tonal/atonal dichotomy and did not consider this other branch of music practice. Vickers's revised opinion [117] that the electroacoustic/*musique concrète*/organized sound traditions can, in fact, offer much to sonification design supports the position taken by Barra et al. [5] in their WebMelody web server sonification system. Indeed, Barra et al. avoided the use of harmonic tonal sequences and rhythmic references because they believed that such structures might distract users by drawing upon their individual "mnemonic and musical (personal) capabilities" [5]. Rather, they "let the sonification's timbre and duration represent the information and avoid recognizable musical patterns" [5, p. 35]. This, they said

> ...makes it possible to hear the music for a long time without inducing mental and musical fatigue that could result from repeated musical patterns that require a finite listening time and not, as in our case, a potentially infinite number of repetitions.

The system was evaluated in a dual-task experiment in which participants were given a primary text editing task and a secondary web server monitoring task. Here, the monitoring was peripheral because the information provided by the sonification was peripheral. The results indicated that the background music generated by WebMelody did not distract users from their primary task while at the same time alllowing meaningful information to be drawn from it. What is particularly encouraging about this study is that the sonification of web server data was far less distracting than similar visual displays. In a visual-only study, Maglio and Campbell [81] asked participants to perform a text-editing task whilst visually monitoring another process. Not suprisingly (given the need to keep switching visual focus) Maglio and Campbell observed a decrease in participants' performance on the text editing task. This suggests that, when designed well, a process monitoring sonification can provide a useful, minimally-distracting data display in situations where peripheral-monitoring is required.

When developing their Audio Aura system (a serendipitous-peripheral monitoring system to allow people to have background awareness of an office environment), Mynatt et al. [86, 85] created four separate sets of auditory cues which they called *ecologies*. The reason for the ecology label is that all the sounds within each set were designed to be compatible with the others not just in terms of frequency and intensity balance but in logical terms too. For example, their 'sound effects world' ecology was based around the noises to be heard at the beach: gull cries were mapped to quantities of incoming email with surf and wave noises representing the activity level of members of a particular group. Thus, each sound in a particular ecology would not sound out of place with the others. In all, four ecologies were constructed:

1. **Voice world** — vocal speech labels;
2. **Sound effects world** — beach noises: an auditory icon/soundscape set;
3. **Music world** — tonal musical motifs: a structured earcon set;
4. **Rich world** — a composite set of musical motifs, sound effects, and vocal messages.

Unfortunately, no formal studies have been published to discover how well the ecologies worked and which of the four was better received by users. In theory, this selection of different ecologies allows user preference to be catered for which is an important principle in aesthetic computing [43, 45]. Such principles can also be found in Tran and Mynatt's Music Monitor [113] which allowed the user to personalize the system by specifying their preferred music tracks upon which the main earcon messages were overlaid.

18.6.4 Comprehensibility and audibility

The audibility of sonifications is an important factor and is tightly coupled to the issue of intrusiveness (see section 18.6.1 above). The comprehensibility of sonifications depends on many factors including the production quality of the sounds, the quality of the playback system, and cultural and metaphoric associations. Many process data require metaphoric or analogic mappings for audio representation as they do not naturally possess their own sound. The choice of metaphor may determine how learnable and comprehensible the mapping is. For example, Kilander and Lönnqvist found that their sound of a golf ball dropping into a cup was difficult for listeners to recognize "except possibly for avid golfers" whilst the sound of a car engine was easy to identify [75]. This highlights the fact that when using real-world sounds it is important to assess the cultural attributes of those sounds. Investigating musical tones for the monitoring of background processes Søråsen [106] found that sudden onset or disappearance of a timbre is easier to detect than changes in the rhythm and melody of that timbre. He concluded "changes within one single instrument should be very carefully designed to represent non-binary changes in state or modus".

18.7 The road ahead

The above discussion has outlined several pitfalls that the researcher wishing to develop auditory display solutions for monitoring activities may face. The remaining sections discuss several strategies and techniques that look promising for avoiding these problems.

18.7.1 Representation and meaning making

Kramer portrayed auditory display as a representational continuum ranging from analogic to symbolic mappings [77, pp. 22–29]. Analogic representations are those which have an instrinsic correspondence to the data and a good example would be the field of audification (see Chapter 12). Symbolic representations are indirect, possibly involving abstractions or amalgamations of one or more data sets and much (if not most) sonification work lies towards this end of Kramer's continuum. Rauterberg and Styger [91] recommended the use of iconic sound mappings for real-time direct and peripheral process monitoring. They said that we should "look for everyday sounds that 'stand for themselves'". The question of representation is important and is richer than a simple anaolgic/symbolic continuum. Semiotics offers us the concept of *sign*. Signs are words, images, sounds, smells, objects, etc. that have no intrinsic meaning and which become signs when we attribute meaning to them [29]. Signs stand for or represent something beyond themselves. Modern semiotics is based upon the work of two principal thinkers, the linguist Ferdinand de Saussure and the philosopher and logician Charles Sanders Peirce. In Saussurean semiotics the sign is a dyadic relationship between the semiotics!signifiersignifier and the *signified*. The signifier represents the signified and both are psychological constructs, so Saussure's signifier has no real-world referent (though modern day interpretations have the signifier taking a more material form). The spoken word 'tree', for example, is the signifier for the concept of the thing we know as a tree. The sign thus formed is a link between the sound pattern and the concept. Peirce's semiotics is based upon a triadic relationship comprising:

- The *object*: a real-world referent, the thing to be represented (note, this need not have a material form);
- The *representamen*: the form the sign takes (word, image, sound, etc.,) and which represents the object);
- The *interpretant*: the sense we make of the sign.

Figure 18.8 shows two Peircean triads drawn as 'meaning triangles' (after Sowa [107]). It should be noted that the Saussurean signifier and signified correspond only approximately to Peirce's representamen and interpretant. Figure 8(a) shows the basic structure of a Peircean sign and Figure 8(b) shows the sign formed by the name Tom which represents a specific individual cat with that name.

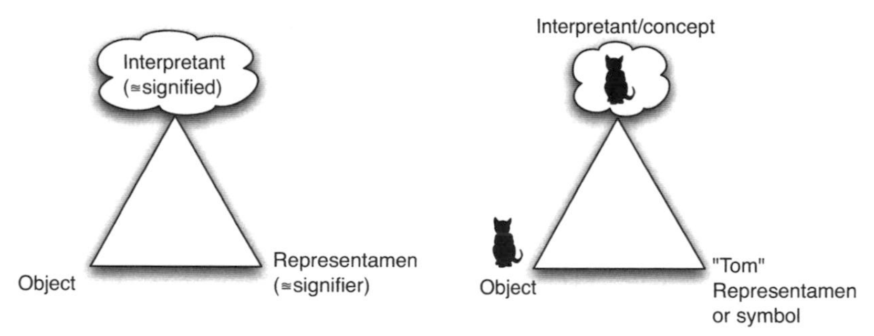

(a) A Peircean semiotic triad (after Sowa [107]). Approximations to de Saussure's semiotic terminology given in parentheses.

(b) Example: A real individual cat is the object. It is signified by the symbol "Tom" which brings into our mind the concept of Tom the cat.

Figure 18.8: Two semiotic 'meaning triangles'

To relate this to visualization consider Figure 18.9 which shows a semiotic relationship that exists in a spreadsheet application. The spreadsheet program takes a data set (in this case student grades) which has been collected from the real world of a cohort of students studying a course. These data are then presented to the user via the tabular visual representation we would be familiar with when launching our favourite spreadsheet program. It should be noted that this visual presentation is not the data set itself, but a particular representation of it. The tabular view, then, becomes the representamen of the data. The interpretant in this sign is the sense we make of the student marks by looking at the screen.

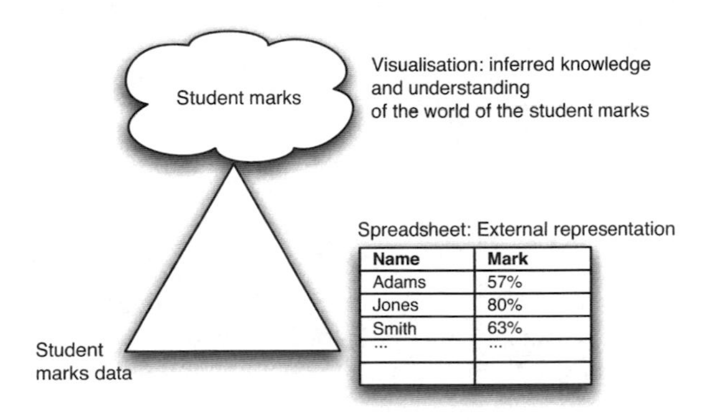

Figure 18.9: The common spreadsheet display is an external representation of the underlying data. The interpretant is the concept that is formed in our mind when we view the tabular representation.

Semiotics offers us three modes of representation: symbolic, iconic, and indexical. Symbolic signs are purely arbitrary (e.g., a "no entry" sign on a road). Iconic signs resemble the object in some way and this mode could include metaphors. Indexical signs are directly connected to the object (e.g., a rash as a sign of a particular disease). Thus Kramer's continuum distinguishes between indexical representations at its analogic end and symbolic representations at the other; iconic representations are not explicitly covered. It will be helpful in future work to consider representations from a semiotic perspective as it helps to clarify that central region of Kramer's continuum. For instance, earcons can be metaphoric but some are more symbolic than others. Consider Brewster's early example [18] which had a purely arbitary earcon hierarchy to represent various operating system events. Contrast this with the CAITLIN system [121] in which the sounds were metaphoric: programming language loops were represented by a repeating motif, selections by an up-and-down motif which resembles the intonation patterns of the human voice when asking and answering a question. The best example of an indexical auditory display would be audifications (e.g., Hayward's seismograms [58]) as the sounds are directly related to the data (the semiotic 'object'). In sonification we are primarily working with symbolic and iconic mappings. Rauterberg and Styger [91] suggested that iconic mappings are particularly well-suited to monitoring applications, but work in network and web server monitoring also indicates that symbolic mappings (e.g., packet size mapped to the sound of water) can also work well when combined in such a way as to provide a non-distracting soundscape. What is needed are studies that are designed specifically to explore the impact of iconic, symbolic, and indexical mappings in the three monitoring modes (direct, peripheral, and serendipitous-peripheral). Systematic investigation using these explicit classifications will provide a much clearer body of knowledge and evidence than is currently available.

18.7.2 Aesthetics

In their work on the ARKola system Gaver et al. [54] suggested that auditory icons had less potential for annoyance than musical messages because they can be designed to complement the auditory environment in which they will be used. This argument is influenced by principles found in acoustic ecology. Because they use real-world sounds it is said that auditory icons do not just sit in an acoustic environment but can also extend it in a complementary manner [54] unlike musical messages which are naturally alien to the acoustic ecology of the environment. However, a note of caution should be sounded at this point. Acoustic ecology as a field of study was first defined by R. Murray Schafer in his 1977 book "The Tuning of the World" [100] which came out of his earlier work on soundscapes. Schafer encouraged people to hear the acoustic environment as a musical composition for which the listener should own the responsibility of its composition (see Wrightson [128]). That Schafer sees ecologies of real-world sounds as musical compositions shows that the distinction between musical and non-musical sound is not as clear as proposed by Gaver et al. Indeed, Varèse referred to some of his electroacoustic music as *organized sound* (e.g., *Poème Electronique* (1965)); in the electroacoustic/organized sound/*musique concrète* schemes there is no requirement for the music to possess melodic components in the normal sense of tonal and atonal music schemes. Indeed, it could be argued that all sonification can be viewed (type cast) as music if the sonification is seen from the electroacoustic weltanschauung [125, 117]. What Gaver et al. are suggesting is that melodic motif-based sonifications are potentially more annoying than those using natural or real-world sounds. It is not clear from the evidence available that this is so. Indeed, Barra et al. [5] and Tran and Mynatt [113] used musical components to great effect in their peripheral and serendipitous-peripheral monitoring work. Tractinsky et al. [112] found that aesthetics plays a very strong role in user perception and there is a further growing body of evidence that systems which are designed with a conscious attention to the aesthetic dimension benefit from increase usability. In monitoring situations in which the auditory output needs to be continuous or frequent this dimension is especially important to ensure the avoidance of user annoyance and fatigue. The relationship between sonification and aesthetics and the art vs. science debate are discussed in more detail in Chapter 7.

Semiotics and aesthetics are two broad areas that provide the language for talking about representational schemata and which give us general theoretical foundations to inform the development of future sonification systems. In terms of the narrower problem of process monitoring and its attendant pitfalls several specific sound design techniques have emerged that offer potential as successful sonification strategies and which are worthy of further more focused research: soundscapes, steganography, model-based sonification, and spatialization. These are discussed briefly below.

18.7.3 Soundscapes

In the research discussed in the preceding sections some of the most successful results came from projects that delivered the sonifications through soundscapes or sound frameworks based upon soundscape principles. Many of these soundscapes are characterized in part by the use of natural real-world sounds (rather than musical instruments) containing large amounts of (modulated and filtered) broadband noise.

Cohen [32] claimed his acoustic ecology was unobtrusive meaning that problems with annoyance and distraction would be less than with other sonification designs and more recent research (e.g., Kainulainen et al. [73] and Jung [72]) seems to back up these early findings. At the same time new techniques have begun to emerge for studying how people interact with and understand soundscapes, and the stages of learning they pass through. For example, Droumeva and Wakkary [39] devised the concept of "aural fluency" which has three stages marking progression from familiarisation with the logic and syntax of a soundscape through to becoming fluent in the soundscape's language. With such tools it is now much easier to design detailed studies to investigate the issues involved in the production of soundscape-based auditory displays for monitoring tasks.

18.7.4 Steganographic embedding

Another technique that has led to increased user satisfaction and lower annoyance and distraction levels is the embedding of signal sounds inside some carrier sound. In the computer security and encryption worlds this would be called *steganography* (literally "concealed writing"). Tran and Mynatt's Music Monitor system [113] used just such an approach whereby the sonification signals were overlaid on a piece of user-selected music. Jung [72] describes a similar strategy for a user notification system. For tasks requiring user notification Butz and Jung [25] confirmed that auditory cues (in this case musical motifs) could be effectively embedded in an ambient soundscape. Such a steganographic approach allows monitoring to be carried out by those who need to do it without causing distraction to other people in the environment. As the Tran and Mynatt and Butz and Jung examples show, message embedding can be done with music or soundscapes alike and so this is another direction future research into sonification for process monitoring should explore. (Note, this sort of embedding is different from the piggy-backing of variables onto a single complex tone that Fitch and Kramer [46] describe.)

18.7.5 Physical modeling and model-based sonification

Going beyond direct parameter mapping, researchers have used physical models to allow more complex bindings between data and sound. Typically, a sound generating model (from a simple resonator to a multi-parameter model) is created in an appropriate audio environment or programming language (e.g., SuperCollider, Pure Data, Max/MSP, etc.) and then the data to be monitored is used to excite or perturb the model. For example, Ballora et al. used this technique successfully for sonifying network traffic [4]. We have seen that the majority of the soundscape approaches to process monitoring relied on pre-recorded audio files such that incoming data would trigger the playback of a discrete sound. Physical modeling offers the potential for increased expressivity because it allows more than just the amplitude or pitch of a sound to be controlled; slight changes in data would lead to changes in the continuous soundscape.

Model-based sonification takes this idea and turns it on its head so that the data set becomes the sound model (the resonator) and the user is left to manually excite or perturb the model to infer knowledge about the data. However, there is no reason why the excitation has to be manual (e.g., see Hermann and Ritter [61]). Indeed, Chafe et al. [27, 28] explored network sonification by setting the network state as the sound model and letting incoming ping messages 'play' the model. They offer some useful insight into how to create such sound models and so this is another avenue of research that should be further explored.

18.7.6 Spatialization and HRTFs

Spatialization of audio has also been shown to be important in user performance in real-time monitoring tasks. Roginska et al. [92] suggest the use of head-related transfer functions (HRTF) to ensure highest performance and Best et al. [9] found that spatial segregation of sound sources using HRTFs is advantageous when engaged in divided attention tasks.[20] In both pieces of research it was the spatialization that HRTFs afford that led to the performance increase. Of course, HRTFs are computationally expensive to produce and require headphones in use and so are not suitable for all monitoring scenarios (especially in shared environments where more than one person may be involved in the monitoring). However, where they are appropriate to use they work very well and will benefit from further research.

18.7.7 Closing remarks

Since work on auditory display and sonification began in earnest, the monitoring of environments and background processes has consistently attracted attention. Perhaps this is because audio seemingly provides a natural medium for this type of information. Process and environmental data are often supplementary to other primary task data and the affordances that sound offers through being able to occupy space in the perceptual background could be used to great effect. Despite the many different applications and approaches, common themes of dealing with intrusiveness, annoyance, and comprehensibility have risen to the fore. Furthermore, these themes are often linked to a common thread, the aesthetic design of the sonifications. Researchers who reported the most success also dealt directly with the issue of the aesthetics and acoustic ecology of their sonifications. It is suggested that the agenda for research in this field of sonification should be underpinned by a conscious attention to the role of semiotics and aesthetics and that these foundations should be used in conjunction with techniques involving soundscapes, steganographic embedding, model-based sonification, and spatialization to develop the next generation of real-time and process monitoring sonification applications. Chapter 7 in this volume presents a more detailed treatment of the relationship between sonification design and aesthetics.

Bibliography

[1] James L. Alty. Can we use music in computer-human communication? In M. A. R. Kirby, A. J. Dix, and J. E. Finlay, editors, *People and Computers X: Proceedings of HCI '95*, pages 409–423. Cambridge University Press, Cambridge, 1995.

[20] The HRTF uses mathematical transformations to produce realistic spatial audio by converting a recorded sound signal "to that which would have been heard by the listener if it had been played at the source location, with the listener's ear at the receiver location" (see `http://en.wikipedia.org/wiki/Head-related_transfer_function`). This results in highly accurate spatial reproduction of sound offering not just left-right discrimination of a sound's source in space but also front-back and up-down. Because every person's auditory system is unique (thanks, in part, to different shaped ears) to get an accurate spatial representation of a sound, the HRTF needs to be computed individually. Averaged HRTFs have been produced for general use, but their perceived accuracy depends on the closeness of match between the listener's auditory system and the HRTF.

[2] Ronald M. Baecker and John Buchanan. A programmer's interface: A visually enhanced and animated programming environment. In *Twenty-Third Annual Hawaii International Conference on Systems Sciences*, volume 11, pages 531–540. IEEE Computer Society Press, 1990.

[3] Saskia Bakker, Elise van den Hoven, and Berry Eggen. Exploring interactive systems using peripheral sounds. In Rolf Nordahl, Stefania Serafin, Federico Fontana, and Stephen Brewster, editors, *Haptic and Audio Interaction Design*, volume 6306 of *Lecture Notes in Computer Science*, pages 55–64. Springer Berlin / Heidelberg, 2010.

[4] Mark Ballora, Brian Panulla, Matthew Gourley, and David L. Hall. Preliminary steps in sonifying web log data. In Eoin Brazil, editor, *16th International Conference on Auditory Display*, pages 83–87, Washington, DC, 9–15 June 2010. ICAD.

[5] Maria Barra, Tania Cillo, Antonio De Santis, Umberto F. Petrillo, Alberto Negro, and Vittorio Scarano. Multimodal monitoring of web servers. *IEEE Multimedia*, 9(3):32–41, 2002.

[6] Maria Barra, Tania Cillo, Antonio De Santis, Umberto Ferraro Petrillo, Alberto Negro, and Vittorio Scarano. Personal WebMelody: Customized sonification of web servers. In Jarmo Hiipakka, Nick Zacharov, and Tapio Takala, editors, *Proceedings of the 2001 International Conference on Auditory Display*, pages 1–9, Espoo, Finland, 29 July–1 August 2001. ICAD.

[7] Lewis I. Berman and Keith B. Gallagher. Listening to program slices. In Tony Stockman, Louise Valgerður Nickerson, Christopher Frauenberger, Alistair D. N. Edwards, and Derek Brock, editors, *ICAD 2006 - The 12th Meeting of the International Conference on Auditory Display*, pages 172–175, London, UK, 20–23 June 2006. Department of Computer Science, Queen Mary, University of London, UK.

[8] Lewis I. Berman and Keith B. Gallagher. Using sound to understand software architecture. In *SIGDOC '09: Proceedings of the 27th ACM international conference on Design of communication*, pages 127–134, New York, NY, USA, 2009. ACM.

[9] Virginia Best, Antje Ihlefeld, and Barbara Shinn-Cunningham. The effect of auditory spatial layout in a divided attention task. In Eoin Brazil, editor, *ICAD 2005 - The 11th Meeting of the International Conference on Auditory Display*, pages 17–22, Limerick, Ireland, 6–9 July 2005.

[10] David B. Boardman, Geoffrey Greene, Vivek Khandelwal, and Aditya P. Mathur. LISTEN: A tool to investigate the use of sound for the analysis of program behaviour. In *19th International Computer Software and Applications Conference*, Dallas, TX, 1995. IEEE.

[11] David B. Boardman and Aditya P. Mathur. Preliminary report on design rationale, syntax, and semantics of LSL: A specification language for program auralization. Technical report, Dept. of Computer Sciences, Purdue University, 21 September 1993.

[12] Dale S. Bock. ADSL: An auditory domain specification language for program auralization. In Gregory Kramer and Stuart Smith, editors, *ICAD '94 Second International Conference on Auditory Display*, pages 251–256, Santa Fe, NM, 1994. Santa Fe Institute.

[13] Dale S. Bock. *Auditory Software Fault Diagnosis Using a Sound Domain Specification Language*. Ph.D. thesis, Syracuse University, 1995.

[14] Dale S. Bock. Sound enhanced visualization: A design approach based on natural paradigms. M.sc. dissertation, Syracuse University, 1995.

[15] Till Bovermann. *Tangible Auditory Interfaces: Combining Auditory Displays and Tangible Interfaces*. Ph.D. thesis, Bielefeld University, Germany, 15 December 2010.

[16] Till Bovermann, René Tünnermann, and Thomas Hermann. Auditory augmentation. *International Journal of Ambient Computing and Intelligence*, 2(2):27–41, 2010.

[17] Stephen Brewster. Navigating telephone-based interfaces with earcons. In H Thimbleby, B O'Conaill, and P Thomas, editors, *People and Computers XII*, pages 39–56. Springer, Berlin, 1997.

[18] Stephen Brewster. The design of sonically-enhanced widgets. *Interacting with Computers*, 11:211–235, 1998.

[19] Stephen A. Brewster. A sonically-enhanced interface toolkit. In Steven P. Frysinger and Gregory Kramer, editors, *ICAD '96 Third International Conference on Auditory Display*, pages 47–50, Palo Alto, 1996. Xerox PARC, Palo Alto, CA 94304.

[20] Stephen A. Brewster. Using earcons to improve the usability of tool palettes. In *CHI '98: CHI 98 conference summary on Human factors in computing systems*, pages 297–298, New York, NY, USA, 1998. ACM Press.

[21] Stephen A. Brewster. Overcoming the lack of screen space on mobile computers. Department of computing science technical report, Glasgow University, April 2001.

[22] Stephen A. Brewster, Veli-Pekka Raty, and Atte Kortekangas. Enhancing scanning input with non-speech sounds. In *ASSETS 96*, pages 10–14, Vancouver, BC, Canada, 1996. ACM.

[23] Stephen A. Brewster, Peter C. Wright, and Alistair D. N. Edwards. A detailed investigation into the effectiveness of earcons. In Gregory Kramer, editor, *Auditory Display*, volume XVIII of *Santa Fe Institute, Studies in the Sciences of Complexity Proceedings*, pages 471–498. Addison-Wesley, Reading, MA, 1994.

[24] Marc H. Brown and John Hershberger. Color and sound in algorithm animation. *Computer*, 25(12):52–63, 1992.

[25] Andreas Butz and Ralf Jung. Seamless user notification in ambient soundscapes. In *Proceedings of the 10th international conference on Intelligent user interfaces*, IUI '05, pages 320–322, New York, NY, USA, 2005. ACM.

[26] William Buxton. Introduction to this special issue on nonspeech audio. *Human-Computer Interaction*, 4:1–9, 1989.

[27] Chris Chafe and Randal Leistikow. Levels of temporal resolution in sonification of network performance. In Jarmo Hiipakka, Nick Zacharov, and Tapio Takala, editors, *ICAD 2001 7th International Conference on Auditory Display*, pages 50–55, Espoo, Finland, 29 July–1 August 2001. ICAD.

[28] Chris Chafe, Scott Wilson, and Daniel Walling. Physical model synthesis with application to internet acoustics. In *Proceedings of ICASSP02: IEEE International Conference on Acoustics, Speech and Signal Processing*, Orlando, Florida, 13–17 May 2002.

[29] Daniel Chandler. *Semiotics: The Basics*. Routledge, 2 edition, 2007.

[30] Michel Chion. *Audio-Vision: Sound on Screen*. Columbia University Press, NY, 1994.

[31] Jonathan Cohen. "Kirk Here": Using genre sounds to monitor background activity. In S. Ashlund, K. Mullet, A. Henderson, E. Hollnagel, and T. White, editors, *INTERCHI '93 Adjunct Proceedings*, pages 63–64. ACM, New York, 1993.

[32] Jonathan Cohen. Monitoring background activities. In Gregory Kramer, editor, *Auditory Display*, volume XVIII of *Santa Fe Institute, Studies in the Sciences of Complexity Proceedings*, pages 499–532. Addison-Wesley, Reading, MA, 1994.

[33] Jonathan Cohen. OutToLunch: Further adventures monitoring background activity. In Gregory Kramer and Stuart Smith, editors, *ICAD '94 Second International Conference on Auditory Display*, pages 15–20, Santa Fe, NM, 1994. Santa Fe Institute.

[34] Alex Paul Conn. Time affordances: The time factor in diagnostic usability heuristics. In *CHI '95: Proceedings of the SIGCHI conference on Human factors in computing systems*, pages 186–193, New York, NY, USA, 1995. ACM Press/Addison-Wesley Publishing Co.

[35] Murray Crease and Stephen A. Brewster. Scope for progress — monitoring background tasks with sound. In *Proc. INTERACT '99*, volume II, pages 19–20, Edinburgh, UK, 1999. British Computer Society.

[36] Murray G. Crease and Stephen A. Brewster. Making progress with sounds — the design and evaluation of an audio progress bar. In S. A. Brewster and A. D. N. Edwards, editors, *ICAD '98 Fifth International Conference on Auditory Display*, Electronic Workshops in Computing, Glasgow, 1998. British Computer Society.

[37] Christopher J. DiGiano. *Visualizing Program Behaviour Using Non-Speech Audio*. M.sc. dissertation, University of Toronto, 1992.

[38] Christopher J. DiGiano and Ronald M. Baecker. Program auralization: Sound enhancements to the programming environment. In *Graphics Interface '92*, pages 44–52, 1992.

[39] Milena Droumeva and Ron Wakkary. socio-ec(h)o: Focus, listening and collaboration in the experience of ambient intelligent environments. In Eoin Brazil, editor, *16th International Conference on Auditory Display*, pages 327–334, Washington, DC, 9–15 June 2010. ICAD.

[40] Berry Eggen, Koert van Mensvoort, David Menting, Emar Vegt, Wouter Widdsershoven, and Rob Zimmermann. Soundscapes at Workspace Zero — design explorations into the use of sound in a shared environment. In *Pervasive '08 Sixth International Conference on Pervasive Computing*, 19–22 May 2008.

[41] J. Louise Finlayson and Chris Mellish. The 'AudioView' — providing a glance at Java source code. In Eoin Brazil, editor, *ICAD 2005 - The 11th Meeting of the International Conference on Auditory Display*, pages 127–133, Limerick, Ireland, 6–9 July 2005.

[42] J. Louise Finlayson, Chris Mellish, and Judith Masthoff. Fisheye views of Java source code: An updated LOD algorithm. In Constantine Stephanidis, editor, *Universal Access in Human-Computer Interaction. Applications and Services*, volume 4556 of *Lecture Notes in Computer Science*, pages 289–298. Springer Berlin / Heidelberg, 2007.

[43] Paul Fishwick. Aesthetic programming: Crafting personalized software. *Leonardo*, 35(4):383–390, 2002.

[44] Paul A. Fishwick. Aesthetic computing manifesto. *Leonardo*, 36(4):255, September 2003.

[45] Paul A. Fishwick, editor. *Aesthetic Computing*. MIT Press, April 2006.

[46] W. Tecumseh Fitch and Gregory Kramer. Sonifying the body electric: Superiority of an auditory over a visual display in a complex, multivariate system. In Gregory Kramer, editor, *Auditory Display*, volume XVIII of *Santa Fe Institute, Studies in the Sciences of Complexity Proceedings*, pages 307–326. Addison-Wesley, Reading, MA, 1994.

[47] Joan M. Francioni, Larry Albright, and Jay Alan Jackson. Debugging parallel programs using sound. *SIGPLAN Notices*, 26(12):68–75, 1991.

[48] Joan M. Francioni, Jay Alan Jackson, and Larry Albright. The sounds of parallel programs. In *6th Distributed Memory Computing Conference*, pages 570–577, Portland, Oregon, 1991.

[49] Joan M. Francioni and Diane T. Rover. Visual-aural representations of performance for a scalable application program. In *High Performance Computing Conference*, pages 433–440, 1992.

[50] William Gaver, Thomas Moran, Allan MacLean, Lennart Lövstrand, Paul Dourish, Kathleen Carter, and William Buxton. Realizing a video environment: Europarc's rave system. In P. Bauersfeld, J. Bennett, and G. Lynch, editors, *CHI '92 Conference on Human Factors in Computing Systems*, pages 27–35, Monterey, CA, 1992. ACM Press/Addison-Wesley.

[51] William W. Gaver. Auditory icons: Using sound in computer interfaces. *Human Computer Interaction*, 2:167–177, 1986.

[52] William W. Gaver. *Everyday Listening and Auditory Icons*. Ph.D. thesis, University of California, San Diego, 1988.

[53] William W. Gaver. The SonicFinder: An interface that uses auditory icons. *Human Computer Interaction*, 4(1):67–94, 1989.

[54] William W. Gaver, Randall B. Smith, and Tim O'Shea. Effective sounds in complex systems: The arkola simulation. In S. Robertson, G. Olson, and J Olson, editors, *CHI '91 Conference on Human Factors in Computing Systems*, pages 85–90, New Orleans, 1991. ACM Press/Addison-Wesley.

[55] Michael Gilfix and Alva L. Couch. Peep (the network auralizer): Monitoring your network with sound. In *14th System Administration Conference (LISA 2000)*, pages 109–117, New Orleans, Louisiana, USA, 3–8 December 2000. The USENIX Association.

[56] M. C. Gopinath. Auralization of intrusion detection system using JListen. Master's thesis, Birla Institute of Technology and Science, Pilani (Rajasthan), India, May 2004.

[57] Carl Gutwin and Saul Greenberg. Support for group awareness in real time desktop conferences. In *Proceedings of The Second New Zealand Computer Science Research Students' Conference*, Hamilton, New Zealand, 18–21 April 1995. University of Waikato.

[58] Chris Hayward. Listening to the Earth sing. In Gregory Kramer, editor, *Auditory Display*, volume XVIII of *Santa Fe Institute, Studies in the Sciences of Complexity Proceedings*, pages 369–404. Addison-Wesley, Reading, MA, 1994.

[59] Thomas Hermann, Jan M. Drees, and Helge Ritter. Broadcasting auditory weather reports – a pilot project. In Eoin Brazil and Barbara Shinn-Cunningham, editors, *ICAD '03 9th International Conference on Auditory*

Display, pages 208–211, Boston, MA, 2003. ICAD.

[60] Thomas Hermann and Helge Ritter. Listen to your data: Model-based sonification for data analysis. In G. E. Lasker, editor, *Advances in intelligent computing and multimedia systems*, pages 189–194, Baden-Baden, Germany, August 1999. Int. Inst. for Advanced Studies in System research and cybernetics.

[61] Thomas Hermann and Helge Ritter. Crystallization sonification of high-dimensional datasets. *ACM Transactions on Applied Perception*, 2(4):550–558, October 2005.

[62] Scott E. Hudson and Ian Smith. Techniques for addressing fundamental privacy and disruption tradeoffs in awareness support systems. In *CSCW '96: Proceedings of the 1996 ACM conference on Computer supported cooperative work*, pages 248–257, New York, NY, USA, 1996. ACM Press.

[63] Khaled Hussein, Eli Tilevich, Ivica Ico Bukvic, and SooBeen Kim. Sonification design guidelines to enhance program comprehension. In *ICPC '09 IEEE 17th International Conference on Program Comprehension*, pages 120–129, 2009.

[64] Jay Alan Jackson and Joan M. Francioni. Aural signatures of parallel programs. In *Twenty-Fifth Hawaii International Conference on System Sciences*, pages 218–229, 1992.

[65] Jay Alan Jackson and Joan M. Francioni. Synchronization of visual and aural parallel program performance data. In Gregory Kramer, editor, *Auditory Display*, volume XVIII of *Santa Fe Institute, Studies in the Sciences of Complexity Proceedings*, pages 291–306. Addison-Wesley, Reading, MA, 1994.

[66] Michael A. Jackson. *System Development*. Prentice-Hall International, 1983.

[67] David H. Jameson. The run-time components of Sonnet. In Gregory Kramer and Stuart Smith, editors, *ICAD '94 Second International Conference on Auditory Display*, pages 241–250, Santa Fe, NM, 1994. Santa Fe Institute.

[68] David H. Jameson. Sonnet: Audio-enhanced monitoring and debugging. In Gregory Kramer, editor, *Auditory Display*, volume XVIII of *Santa Fe Institute, Studies in the Sciences of Complexity Proceedings*, pages 253–265. Addison-Wesley, Reading, MA, 1994.

[69] James J. Jenkins. Acoustic information for object, places, and events. In W. H. Warren, editor, *Persistence and Change: Proceedings of the First International Conference on Event Perception*, pages 115–138. Lawrence Erlbaum, Hillsdale, NJ, 1985.

[70] Martin Jonsson, Calle Jansson, Peter Lönnqvist, Patrik Werle, and Fredrik Kilander. Achieving non-intrusive environments for local collaboration. Technical Report FEEL Project Deliverable 2.2(1), Department of Computer and Systems Sciences, IT University, Kista, 2002.

[71] Abigail J. Joseph and Suresh K. Lodha. Musart: Musical audio transfer function real-time toolkit. In *ICAD '02 - 2002 International Conference on Auditory Display*, Kyoto, Japan, 2002.

[72] Ralf Jung. Ambience for auditory displays: Embedded musical instruments as peripheral audio cues. In Brian Katz, editor, *Proc. 14th Int. Conf. Auditory Display (ICAD 2008)*, Paris, France, 24–27 June 2008. ICAD.

[73] Anssi Kainulainen, Markku Turunen, and Jaakko Hakulinen. An architecture for presenting auditory awareness information in pervasive computing environments. In Tony Stockman, Louise Valgerður Nickerson, Christopher Frauenberger, Alistair D. N. Edwards, and Derek Brock, editors, *ICAD 2006 - The 12th Meeting of the International Conference on Auditory Display*, pages 121–128, London, UK, 20–23 June 2006.

[74] Frank Kilander and Peter Lönnqvist. A whisper in the woods – an ambient soundscape for peripheral awareness of remote processes. In *Continuity in Future Computing Systems Workshop I3*, Spring Days, Porto, Portugal, 2001.

[75] Frank Kilander and Peter Lönnqvist. A whisper in the woods – an ambient soundscape for peripheral awareness of remote processes. In *ICAD 2002 – International Conference on Auditory Display*, 2002.

[76] Donald E. Knuth. BCS/IET Turing lecture, February 2011.

[77] Gregory Kramer. An introduction to auditory display. In Gregory Kramer, editor, *Auditory Display*, volume XVIII of *Santa Fe Institute, Studies in the Sciences of Complexity Proceedings*, pages 1–78. Addison-Wesley, Reading, MA, 1994.

[78] Tami Lapidot and Orit Hazzan. Song debugging: Merging content and pedagogy in computer science

education. *SIGCSE Bull.*, 37:79–83, December 2005.

[79] Paul A. Lucas. An evaluation of the communicative ability of auditory icons and earcons. In *ICAD '94 Second International Conference on Auditory Display*, pages 121–128, Santa Fe, NM, 1994. Santa Fe Institute.

[80] Janet Lumsden, Stephen A. Brewster, Murray Crease, and Philip D. Gray. Guidelines for audio-enhancement of graphical user interface widgets. In *Proceedings of British HCI 2002*, volume II, pages 6–9. BCS, London, UK, September 2002.

[81] Paul P. Maglio and Christopher S. Campbell. Tradeoffs in displaying peripheral information. In *CHI '00: Proceedings of the SIGCHI conference on Human factors in computing systems*, pages 241–248, New York, NY, USA, 2000. ACM Press.

[82] Aditya P. Mathur, David B. Boardman, and Vivek Khandelwal. LSL: A specification language for program auralization. In Gregory Kramer and Stuart Smith, editors, *ICAD '94 Second International Conference on Auditory Display*, pages 257–264, Santa Fe, NM, 1994. Santa Fe Institute.

[83] Gottfried Mayer-Kress, Robin Bargar, and Insook Choi. Musical structures in data from chaotic attractors. In Gregory Kramer, editor, *Auditory Display*, volume XVIII of *Santa Fe Institute, Studies in the Sciences of Complexity Proceedings*, pages 341–368. Addison-Wesley, Reading, MA, 1994.

[84] S. Joy Mountford and William Gaver. Talking and listening to computers. In B. Laurel and S. Mountford, editors, *The Art of Human-Computer Interface Design*, pages 319–334. Addison-Wesley, Reading, MA, 1990.

[85] Elizabeth D. Mynatt, Maribeth Back, Roy Want, Michael Baer, and Jason B. Ellis. Designing Audio Aura. In *CHI '98: Proceedings of the SIGCHI conference on Human factors in computing systems*, pages 566–573, New York, NY, USA, 1998. ACM Press/Addison-Wesley Publishing Co.

[86] Elizabeth D. Mynatt, Maribeth Back, Roy Want, and Ron Frederick. Audio Aura: Light-weight audio augmented reality. In *UIST '97: Proceedings of the 10th annual ACM symposium on User interface software and technology*, pages 211–212, New York, NY, USA, 1997. ACM Press.

[87] Elin Rønby Pedersen. People presence or room activity supporting peripheral awareness over distance. In *CHI '98: CHI 98 conference summary on Human factors in computing systems*, pages 283–284, New York, NY, USA, 1998. ACM Press.

[88] Elin Rønby Pedersen and Tomas Sokoler. Aroma: Abstract representation of presence supporting mutual awareness. In *CHI '97: Proceedings of the SIGCHI conference on Human factors in computing systems*, pages 51–58, New York, NY, USA, 1997. ACM Press.

[89] R. Jagadish Prasath. Auralization of web server using JListen. Master's thesis, Birla Institute of Technology and Science, Pilani (Rajasthan), India, May 2004.

[90] Marty Quinn. Seismic Sonata: A musical replay of the 1994 Northridge, California earthquake, 2000.

[91] Matthias Rauterberg and Erich Styger. Positive effects of sound feedback during the operation of a plant simulator. In B. Blumenthal, J. Gornostaev, and C. Unger, editors, *Human Computer Interaction: 4th International Conference, EWHCI '94, St. Petersburg, Russia, August 2 - 5, 1994. Selected Papers*, volume 876 of *Lecture Notes in Computer Science*, pages 35–44. Springer, Berlin, 1994.

[92] Agnieszka Roginska, Edward Childs, and Micah K. Johnson. Monitoring real-time data: A sonification approach. In Tony Stockman, Louise Valgerður Nickerson, Christopher Frauenberger, Alistair D. N. Edwards, and Derek Brock, editors, *ICAD 2006 - The 12th Meeting of the International Conference on Auditory Display*, pages 176–181, London, UK, 20–23 June 2006.

[93] Pablo Romero, Richard Cox, Benedict du Boulay, and Rudi Lutz. A survey of external representations employed in object-oriented programming environments. *Journal of Visual Languages & Computing*, 14(5):387–419, 2003.

[94] Pablo Romero, Richard Cox, Benedict du Boulay, and Rudi K. Lutz. Visual attention and representation switching during Java program debugging: A study using the Restricted Focus Viewer. *Lecture Notes in Computer Science*, 2317:221–235, 2002.

[95] Pablo Romero, Benedict du Boulay, Richard Cox, Rudi Lutz, and Sallyann Bryant. Debugging strategies and tactics in a multi-representation software environment. *International Journal of Human-Computer Studies*,

65(12):992–1009, 2007.

[96] Pablo Romero, Rudi K. Lutz, Richard Cox, and Benedict du Boulay. Co-ordination of multiple external representations during java program debugging. In *IEEE 2002 Symposia on Human Centric Computing Languages and Environments (HCC'02)*, Arlington, Virginia, USA, 2002. IEEE Computer Society.

[97] Nitin Sawhney and Chris Schmandt. Nomadic radio: Scaleable and contextual notification for wearable audio messaging. In *Proceedings of the CHI 99 Conference on Human factors in computing systems: the CHI is the limit*, pages 96–103, New York, NY, 1999. ACM Press.

[98] Nitin Sawhney and Chris Schmandt. Nomadic radio: Speech and audio interaction for contextual messaging in nomadic environments. *ACM Transactions on Computer-Human Interaction (TOCHI)*, 7(3):353–383, 2000.

[99] Pierre Schaeffer. *Traité Des Objets Musicaux*. Seuil, Paris, rev. edition, 1967.

[100] R. Murray Schafer. *The Tuning of the World*. Random House, 1977.

[101] Chris Schmandt and Gerardo Vallejo. "ListenIN" to domestic environments from remote locations. In *ICAD '03 9th International Conference on Auditory Display*, pages 22–223, Boston, MA, USA, 6–9 July 2003.

[102] Randall B. Smith. A prototype futuristic technology for distance education. In *Proceedings of the NATO Advanced Workshop on New Directions in Educational Technology*, Cranfield, UK, 10–13 November 1988.

[103] Stuart Smith, Ronald M. Pickett, and Marian G. Williams. Environments for exploring auditory representations of multidimensional data. In Gregory Kramer, editor, *Auditory Display*, volume XVIII of *Santa Fe Institute, Studies in the Sciences of Complexity Proceedings*, pages 167–184. Addison-Wesley, Reading, MA, 1994.

[104] Eric Somers. A pedagogy of creative thinking based on sonification of visual structures and visualization of aural structures. In Stephen A. Brewster and Alistair D. N. Edwards, editors, *ICAD '98 Fifth International Conference on Auditory Display*, Electronic Workshops in Computing, Glasgow, 1998. British Computer Society.

[105] Diane H. Sonnenwald, B. Gopinath, Gary O. Haberman, William M. Keese, III, and John S. Myers. Infosound: An audio aid to program comprehension. In *Twenty-Third Hawaii International Conference on System Sciences*, volume 11, pages 541–546. IEEE Computer Society Press, 1990.

[106] Sigve Søråsen. Monitoring continuous activities with rhythmic music. In *Proceedings of the Student Interaction Design Research Conference*, Sønderborg, 27–28 January 2005. Mads Clausen Institute, University of Southern Denmark.

[107] John F. Sowa. *Knowledge Representation: Logical, Philosophical, and Computational Foundations*. Brooks/Cole, 2000.

[108] A. Stefik and E. Gellenbeck. Using spoken text to aid debugging: An empirical study. In *ICPC '09 IEEE 17th International Conference on Program Comprehension, 2009.*, pages 110–119, May 2009.

[109] Andreas Stefik, Kelly Fitz, and Roger Alexander. Layered program auralization: Using music to increase runtime program comprehension and debugging effectiveness. In *14th International Conference on Program Comprehension (ICPC 2006)*, pages 89–93, Athens, Greece, 2006. IEEE Computer Society.

[110] Andreas Stefik and Ed Gellenbeck. Empirical studies on programming language stimuli. *Software Quality Journal*, 19(1):65–99, 2011.

[111] Robert S. Tannen. Breaking the sound barrier: Designing auditory displays for global usability. In *4th Conference on Human Factors and the Web*, 5 September 1998.

[112] Noam Tractinsky, Adi Shoval-Katz, and Dror Ikar. What is beautiful is usable. *Interacting with Computers*, 13(2):127–145, 2000.

[113] Quan T. Tran and Elizabeth D. Mynatt. Music monitor: Ambient musical data for the home. In Andy Sloan and Felix van Rijn, editors, *Proceedings of the IFIP WG 9.3 International Conference on Home Oriented Informatics and Telematics (HOIT 2000)*, volume 173 of *IFIP Conference Proceedings*, pages 85–92. Kluwer, 2000.

[114] Gerardo Vallejo. ListenIN: Ambient auditory awareness at remote places. Master's thesis, MIT Media Lab, September 2003.

[115] Paul Vickers. *CAITLIN: Implementation of a Musical Program Auralisation System to Study the Effects on Debugging Tasks as Performed by Novice Pascal Programmers*. Ph.D. thesis, Loughborough University, Loughborough, Leicestershire, September 1999.

[116] Paul Vickers. External auditory representations of programs: Past, present, and future – an aesthetic perspective. In Stephen Barrass and Paul Vickers, editors, *ICAD 2004 – The Tenth Meeting of the International Conference on Auditory Display*, Sydney, 6–9 July 2004. ICAD.

[117] Paul Vickers. Ars informatica – ars electronica: Improving sonification aesthetics. In Luigina Ciolfi, Michael Cooke, Olav Bertelsen, and Liam Bannon, editors, *Understanding and Designing for Aesthetic Experience Workshop at HCI 2005 The 19th British HCI Group Annual Conference*, Edinburgh, Scotland, 2005.

[118] Paul Vickers. Program Auralization: Author's comments on vickers and alty, icad 2000. *ACM Trans. Appl. Percept.*, 2(4):490–494, 2005.

[119] Paul Vickers and James L. Alty. using music to communicate computing information. *Interacting with Computers*, 14(5):435–456, 2002.

[120] Paul Vickers and James L. Alty. musical program auralisation: A structured approach to motif design. *Interacting with Computers*, 14(5):457–485, 2002.

[121] Paul Vickers and James L. Alty. when bugs sing. *Interacting with Computers*, 14(6):793–819, 2002.

[122] Paul Vickers and James L. Alty. Siren songs and swan songs: Debugging with music. *Communications of the ACM*, 46(7):86–92, 2003.

[123] Paul Vickers and James L. Alty. Musical program auralization: Empirical studies. *ACM Trans. Appl. Percept.*, 2(4):477–489, 2005.

[124] Paul Vickers and James L. Alty. The well-tempered compiler: The aesthetics of program auralization. In Paul A. Fishwick, editor, *Aesthetic Computing*, chapter 17, pages 335–354. MIT Press, Boston, MA, 2006.

[125] Paul Vickers and Bennett Hogg. Sonification abstraite/sonification concrète: An 'æsthetic perspective space' for classifying auditory displays in the ars musica domain. In Tony Stockman, Louise Valgerður Nickerson, Christopher Frauenberger, Alistair D. N. Edwards, and Derek Brock, editors, *ICAD 2006 - The 12th Meeting of the International Conference on Auditory Display*, pages 210–216, London, UK, 20–23 June 2006.

[126] Anthony J. Watkins and Mary C. Dyson. On the perceptual organisation of tone sequences and melodies. In Peter Howell, Ian Cross, and Robert West, editors, *Musical Structure and Cognition*, pages 71–119. Academic Press, New York, 1985.

[127] Mark Weiser and John Seely Brown. The coming age of calm technology. 1996.

[128] Kendall Wrightson. An introduction to acoustic ecology. *Soundscape: The Journal of Acoustic Ecology*, 1(1):10–13, Spring 2000.

Chapter 19

Intelligent auditory alarms

Anne Guillaume

19.1 Introduction

When perceiving a sound-producing event, a person will try to find the meaning of the sound and locate where it comes from. Sound is used as a cue for identifying the behavior of surrounding sound producing objects, even if these objects are beyond the field of vision (McAdams, 1993). This spontaneous attribute probably corresponds to the most primitive role of auditory perception, which is to be warned of danger and prepare for counteraction. Because of this, for example, hikers can seek shelter as soon as they hear thunder, even if it is not yet raining. The sound of thunder plays the role of a natural alarm. This alerting function of sound signals is widely used in everyday life, and is also extensively used in the workplace, which is what we are interested in. In this case, the sound is no longer directly linked to the source of danger; the alarm is a synthetic sound, triggered to attract the operator's attention and result in a suitable reaction. This sound must distract the operator from the main task, and provide relevant information. Three kinds of information must be passed along:

- first, an indication of how serious the failure is, by helping the listener to perceive how urgent the situation is.
- The second type of information must provide clues about what triggered the alarm, using a customized sound iconography. This information must be transmitted while minimizing the attentional resources elicited by operators to manage alarms (Schreiber and Schreiber, 1989).
- A third type of information could be delivered concerning the location of the fault. For instance, in aeronautics and in road safety a rapid localization of the threat is of vital importance. Different studies effectively show that 3D sound enables a reduction in reaction times during search for a visual target (Begault, 1993; Bolia et al., 1999; Flanagan et al., 1998; Todd Nelson et al., 1998).

However, until the late 1980s, the detailed characteristics of sound alarms tended to be neglected. Alarms were installed here and there, without any real thought as to their acoustic features, or how to integrate them into a system of alarms. Operators started to complain about the mismatch between the properties of sound alarms and their purpose. This opened up a new field of investigation on how to design sound alarms, supported by an experimental approach.

In order to better understand the challenges and outcomes of research on this topic, the concept of the sound alarm will be described. Next, the perception of urgency will be addressed through considering the acoustic characteristics of sound sequences. A more cognitive approach to the problem helps in conceptualizing the many factors to be examined when designing an alarm.

Furthermore, an alarm cannot be designed in isolation, but as a component in a system of alarms customized to a specific environment. An ergonomic survey of the workstation is a prerequisite, prior to the development of any alarm system. Such a survey helps to prioritize emergencies. An intelligent alarm system is the final phase in the development, adapting its functionality whilst in use by the operator. The design of sound alarms is a complex task because of the many requirements imposed by their function, their context and operators' expectations.

19.2 The concept of auditory alarms

Hearing is a primary alert sense, and so sound alarms aim at alerting operators of any change in the state of the system they are interacting with. According to Schreiber and Schreiber (1989), a system of alarms must have five properties:

1. announcing any anomaly as quickly as possible, without its detection being hindered by false alarms or alarms of lesser importance,
2. making the localization and identification of new alarm messages as easy as possible,
3. minimizing any interference with other signals,
4. minimizing efforts devoted to its management, notably in critical moments, and
5. giving accurate information on the problem's cause.

Auditory alarms are an essential complement to visual alarms, which usually provide more information. This complementary effect of auditory alarms is due to the fact that they are effective in all directions in space, whatever the position of the operator's head and/or eyes: the operator can concentrate on the main task without the requirement to systematically scan a control panel. If this was the case, the visual system would soon be overloaded, since visual information is processed in sequence. Auditory alarms are also very useful when the operator is absent-minded or in a state of rest. In fact sound alarms have the advantage of increasing the probability of an operator reacting to emergency conditions and of reducing reaction time. Sound alarms are used to attract the operator's attention toward the relevant visual information during critical situations. In aeronautics, they have two additional advantages: they are economical in space compared to visual displays and they take advantage of the fact that audition offers a fairly good resistance to relative hypoxia[1] (in downgraded conditions) (Doll and Folds, 1986).

Sound alarms can be placed in one of two categories: speech or non-speech. The advantage of non-speech alarms is that they attract attention more effectively than speech alarms, which may be intertwined into the communication flow, and thus be unheard. Reaction time is shorter with non-speech alarms than with speech alarms (Simpson and Williams, 1980; Wheale, 1982). On the other hand, speech alarms provide more information, but the message delivered must be simple and concise. These two types of alarm may be associated in one system: non-speech to alert, speech to convey information, or even a possible solution. Under heavy workload, adding a non-speech signal to a speech alarm may be useful, because it helps to discriminate the speech alarm from the flow of speech-based information flooding the operator. However, depending on the context, speech-based alarms are not always suitable. For example, in intensive care units, delivering the message through speech might generate a considerable amount of stress for the patient. This was also the case, under specific operational conditions, for aircraft pilots during the Vietnam War (Doll and Folds, 1986). In those cases, it is essential to pay great attention to the sound design of non-speech alarms, to make sure they are well-suited to their application.

19.3 Problems linked to non-speech auditory alarm design

A number of specific organizations, such as intensive care units (ICU), are equipped with many alarms (examples of alarm sounds in operating room are **S19.1**, **S19.2**, **S19.3**, **S19.4**, **S19.5** and **S19.6**[2]). Their purpose is to help reduce the staff's workload during periods of intense activity. In reality, these alarms are often ill-adapted to this purpose: either too numerous, too loud or inaudible, or not adapted to the degree of urgency they are supposed to convey. Sometimes, in an ICU, up to twenty or thirty alarms are dedicated to monitoring a single patient, and from one patient to another relatively identical sound alarms can indicate very different problems. (Arnstein, 1997; Montahan et al., 1993; Stanford et al., 1985; Meredith and Edworthy, 1994).

ICUs are just one example. The problem of ill-adapted sound alarms can also be encountered in the monitoring systems of industrial facilities (Lazarus and Höge, 1986), or in aeronautics. In the latter environment, this problem is relayed perfectly by a pilot's report quoted by Patterson (1990). The pilot reports he was destabilized by several sound and visual alarms being activated simultaneously, making it impossible for him to react suitably, in this critical moment when he was supposed to analyze and manage the problem which triggered the alarms. This problem, noted by Wheale et al., (1979), Patterson et al., (1986), Sorkin et al., (1988) results from alarms being layered upon each other as the need arose, rather than having an all inclusive system designed in the first place. Sound levels are usually at maximum loudness, according to the "better safe than sorry" principle (Patterson, 1990). Very loud alarms are thus installed, to make sure they are perceived. However, the end result is to make

[1] Hypoxia: Hypoxia is defined as inadequate oxygen supply to the cells and tissues of the body. The major risk is brain hypoxia. In aeronautics, the main cause of hypoxia is altitude. Different technologies have been implemented to compensate for the altitude-related hypoxia, but it is important to warn the operators when a failure of these systems occurs.

[2] the alarm causes are explained on the website

them more harmful than helpful. They prevent any communication between team members, and disturb operators' cognitive activity.

19.4 Acoustic properties of non-speech sound alarms

In order to better meet operator's needs, many criteria come into play. These prerequisites are context-dependent, of course, but are quite identical in generic terms (James, 1996):

- sounds must be unique in the surrounding sound environment;
- sounds must be easily discriminated from one another;
- the sound warning must convey the right level of urgency, in relation to a degree of priority;
- the sound warning must be sufficiently audible to be detected, but should not be deafening, or prevent communication among team members.

19.4.1 Sound spectrum and intensity

A sound alarm must be designed while taking into account its surrounding sound environment. Taking into account the spectral content and noise level of ambient noise, the spectrum and sound level of alarms should be selected to interfere as little as possible with communications between crew members, yet to be sufficiently salient to be perceived reliably without being confusing or disturbing. An example from aeronautics, a very noisy environment, will help to clarify this design goal. Patterson (1982) developed a model in which the masking threshold was predicted for a large number of spectra. In this study, Patterson recorded the spectra of various helicopters flying at different speeds and altitudes, right at the position where the pilot's ear was located. He then obtained data indicating in which frequencies the greatest part of the cabin's sound energy was concentrated. The alarm's spectral content was then chosen to avoid being masked by the frequencies dominating cockpits. An alarm with at least four harmonic components scattered throughout the spectrum runs a lower chance of being masked by environmental noises than an alarm concentrating its entire energy on a single harmonic. Alarm intensity must be determined in relation to the threshold at which the different components of its spectrum are heard above the noise. To be sure the alarm is heard (100% detection), at least 4 of its spectral components must be 15dB above their specific audible threshold. Exceptions to this rule can be made, notably when background noise requires components to be above 85 dB.

19.4.2 Perceiving the urgency

Intensity is undoubtedly the most important factor to convey the sense of urgency (Loveless and Sanford, 1975). The louder the signal, the stronger the perception of urgency. This might be explained by the fact that the danger is perceived to be in the immediate proximity of the participant (Ho and Spence, 2009). However, in noisy environments (industry, aeronautics), or in critical environments (intensive care units, operating rooms), this parameter can only vary along a narrow scale: if the signal is too weak, it will go undetected. If it is too loud, it will become painful and distract the operator (Patterson et al., 1986). Therefore, even though intensity plays a major role, it must be systematically controlled. Given that the spectral content and intensity of alarms are set according to the noise level, the idea is to try, as much as possible, to define alarms with acoustic characteristics linked to the operator's perception of a given urgency.

Edworthy et al., (1991), Hellier et al., (1993), and Hellier and Edworthy (1999) studied the effects of sound parameters in non-speech alarms on the psychoacoustic perception of these alarms' urgency. Notably, these authors tried to identify the connection between spectral and/or temporal properties of acoustic signals (see section 3 for definitions) and the possibility of quantifying and predicting the urgency level perceived by listeners. Their starting point is Patterson's alarm design (1990). For Patterson, the alarm is designed with a structural hierarchy (see Fig. 19.1): the base unit is a 100 to 300 ms long pulse. This pulse is repeated several times, at different pitches and/or intensity, using different tempi. The resulting sound, made up of these consecutive components, is a sound burst. This burst is about 2s long, and is perceived as a rhythmic atonal melody. The combination of bursts makes up the entire alarm (hear sound examples **S19.7**, **S19.8**, **S19.9** and **S19.10**). The alarm provides for silences between bursts, to give the crew time to communicate and react adequately. Edworthy et al.'s study (1991) shows that the faster the rate, the higher the pitch and the more irregular the harmonics, the greater the perceived urgency. Authors have come to the conclusion that it is possible to design sound alarms with a predictable perceived urgency. This approach demonstrates the role of low level factors in determining the perception of urgency.

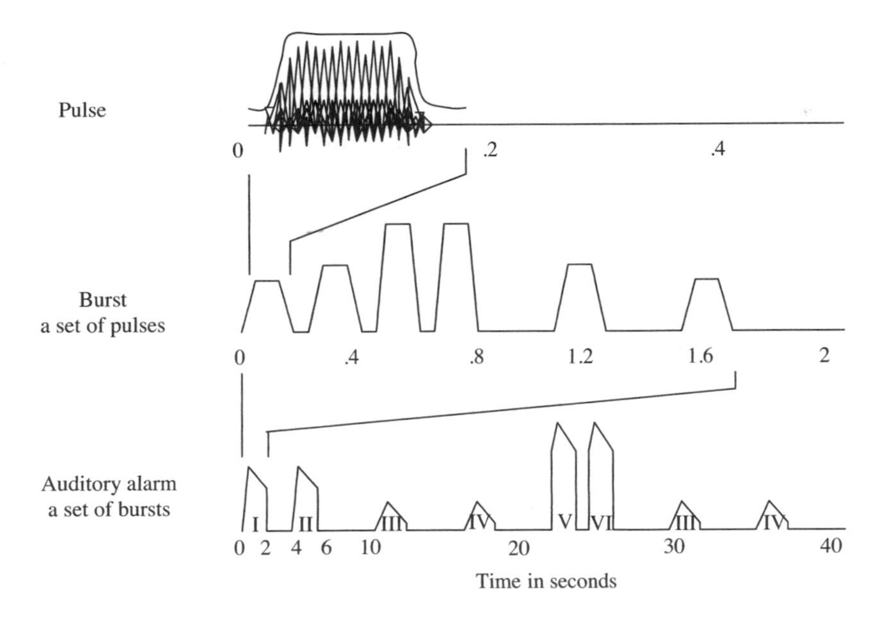

Figure 19.1: Design of alarms according to Patterson. The alarm consists of bursts that consist themselves in sets of pulses.

Alarm systems designed with these principles in mind were successfully tested in real-life conditions, either in noisy conditions (Haas and Casali, 1995), or under a moderate workload (Edworthy et al., 2000). However, the urgency of these alarms seems to no longer be perceived when the workload is more severe (Burt et al., 1995). Principles described by Edworthy appear to be relatively robust, but whether they still hold true under a significant workload is not clearly established. Bear in mind that sound alarms are usually triggered when situations become stressful.

19.4.3 Understanding the message of an alarm

Other authors recommend setting up a correspondence between the alarm's acoustic characteristics and the acoustic properties linked to the message: for example, imitating a heartbeat in order to monitor heart function (Fitch and Kramer, 1994). The message associated with the alarm is then easily identified and can help direct the operator's reaction. However, in some cases, the sound sequence is too close to the operator's everyday sound environment, and is no longer perceived as an alarm. Because it is a commonplace sound, it no longer conveys a sense of urgency (Stanton and Edworthy, 1999). This kind of approach can even be counterproductive, and disturb situational awareness. For example, the rotation of a helicopter rotor can be simulated by a sound sequence with varying intensity and tempo. If intensity or tempo decrease, it means that the rotor is slowing down, and that the helicopter is falling. But if these two parameters in the sequence change simultaneously, this means a decrease in the urgency level which goes totally against signaling a dangerous flight situation (Edworthy et al., 1995). Such an approach should therefore not be generalized but applied only on a case by case basis. It often applies to monitoring a critical physiological parameter through sound: for example, in the operating room, the conventional pulse oximeter 'beep' which gives heart rate through its rate and oxygenation level through its pitch, or respiratory sonification as described by Watson and Sanderson (2004). Sanderson et al., (2004) investigated the effectiveness of sonification to support patient monitoring. They showed that sonification triggered the fastest response, that visual displays resulted in the most accurate response and that sonification coupled with visual displays led to the slowest performance. Since Loeb and Fitch (2002) and Seagull et al., (2001) found no speed advantage when using sonification alone, the contribution of sonification with regards to its modalities requires further investigation. Sonification probably demands a learning phase. One of the main advantages of sonifying physiological signals in the operating room would be to provide the operator with information while performing surgery on the patient (and thus unable to access any external visual information). Sonification could provide the operator with a continuous stream of relevant information, to be consulted when the need arises, and interpreted according to context (Sanderson et al., 2005). However this approach should be very rigorous in order that these auditory displays add information rather than noise (Sanderson et al., 2009).

19.5 A cognitive approach to the problem

The alternative to the psychophysical approach could be to look for a link between the alarm's acoustic properties and the mental representation connected to the problem requiring action. The idea is to try and find whether the problem should be dealt with through the characterization of perceptive invariants conveying different degrees in the perception of urgency, or whether the approach should be broadened to consider that the notion of urgency is an abstract concept requiring the idea of a mental representation. This mental representation in turn may be modulated by the higher centers, according to a person's experience and surrounding context. Guillaume et al., (2003) carried out a series of parallel experiments (i) on signals designed according to the indications of Edworthy et al., (1991) to convey the perception of increasing urgency, or (ii) on real-life alarms recorded in military aircraft. When testing signals defined in (i), the results obtained validate Edworthy et al.'s findings. Sequences with increased pitch, fast tempo and irregular harmonics are perceived as having a greater urgency. However, the same results are not always obtained in the case of real alarms (ii). A number of sequences are classified as non-urgent, although their acoustic properties should have led listeners to perceive them as very urgent. These observations seem to demonstrate that different complex processes are involved in urgency perception. These processes seem to differ, depending on whether or not the sequence evokes a mental representation in the subject's mind. In cases where a mental representation may exist, hearing the sequence brings the mental representation to mind immediately. Judging as to whether or not this is an emergency depends on the association made by subjects between their mental association evoked by the sequence, and the emergency linked to it (Guillaume et al., 2003). According to Logan (1988), making this automatic depends on acquisitions stored in memory, and thus on representations which may be evoked when hearing various sequences. Automatic information processing (Schneider and Schiffrin, 1977) calling on a subject's personal experience probably comes into play when deciding whether a sound is perceived as conveying urgency or not, and allow for a fast and effective reaction. Furthermore, this activity is not attention-consuming, and may contribute to making the judgment on the urgency of a sound signal more robust, even under heavy workloads. On the other hand, in cases where no mental representation can be invoked, urgency is solely judged from the acoustic properties of the sound sequence, and context. This alternative is more attention-consuming, which would explain why the capacity to discriminate different emergency levels under heavy workloads tends to decrease. In such cases, listeners can determine the urgency of the alarm as theoretically planned, when their attention is focused on the task at hand. But if the hearer is busy doing another attention-consuming task, the acoustic characteristics of the sequence can no longer be properly analyzed, and the difference in urgency implied by the alarms presented is no longer perceived. Such an interpretation could help explain Burt et al.'s findings (1995) which show that the urgency of alarms is no longer perceived under high workload.

The cognitive approach involves searching for the mental representation of the cause for the alarm, and thus has a strong impact on the alarm's sound design. Designing the most relevant signal to evoke in as many minds as possible the same mental image requires implementing a very strict methodology to select the most representative sounds and to verify that choices made actually meet requirements.

Auditory icons (i.e., environmental sounds, see chapter 13) are good candidates for alarms which evoke mental representations. The challenge is then to find the more representative auditory icons for an event or a situation (McKeown et al., 2010). The association may be direct. The signal has a unique referent relation. For instance, if an aircraft is the target of a missile, the warning in the target aircraft could be rapid gunshots. More often the association is indirect. That means that the signal has more than one referent relation. In fact an indirect association may involve a real network of referent relations. For instance, the auditory icon associated with the lane departure warning in a car could be a horn, a car crashing, or knocking glasses.

A further improvement would be to use earcons (see chapter 14) dedicated to a specific threat. The design of these earcons could associate the cognitive and the psychophysical approaches. This would be done by slightly changing the acoustic characteristics of an environmental sound in order to render it more or less urgent. The challenge is that the modified sound should keep its referent in order to rapidly evoke the nature of the danger.

19.6 Spatialization of alarms

In order to improve the take up of information by operators in complex systems, new man-machine interfaces are presenting spatialized alarms. Information is presented in 3D sound, enabling operators to locate the virtual sound source rapidly and intuitively and direct their attention in this direction. This property of hearing is used to orient the direction of the gaze and/or the reaction of the operator. But, the act of localizing sources can also be favorable to segregating auditory streams. Thus, if two alarms go off at a short interval from one another, they are easier to segregate if they seem to be coming from different locations.

Determining the direction (given by its azimuth and elevation) of a sound by a subject depends on static and dynamic cues. Among the cues of static localization, the auditory system uses three types of cue:

(a) interaural intensity differences (IIDs) between the signals received by the two ears;

(b) interaural time differences (ITDs) – differences in phase and arrival time;

(c) spectral cues dependent on the shape of the pinna and of the head.

The first two cues are called binaural cues because they relate to the difference in information coming into the right and left ears. The third cue is “monaural” because it depends solely on information from one ear (Moore, 1997).

For binaural cues, auditory localization can be achieved by comparing the sound signals perceived by each ear. This comparison involves the differences in intensity, time and phase. The IIDs are due to the partial diffraction of sound waves in such a way that a signal reaching the ear opposite the source is weakened, and thus less intense compared to the signal coming into the ipsilateral ear. The ITDs correspond to both a difference in phase and a difference in time of the arrival of the signal between the two ears. The amplitude of interaural differences depends on the position of the source in relation to the listener. The interaural differences of phase and intensity vary according to source frequency. Differences in phase are only pertinent for low frequencies. On the contrary, IIDs are only a factor with high frequencies. They are linked to the head’s diffraction properties. This diffraction only occurs for signals with wavelengths smaller than the cranial diameter. Diffraction does not occur for low-frequency signals, but diffraction by the head becomes apparent at 1500 Hz, and then increases as the frequency grows. The third spatial cue is monaural. It is determined by the treatment of information coming from one ear, independent of the other ear. This information is extracted from the resonance and reflection properties of the pinna (Blauert et al., 1998). The pinna modifies the spectrum of incident sound, depending on the angle of the sound's incidence in relation to the head. It thus supplies useful information for discerning elevation and for improving front-back discrimination.

Thus, the head and the pinna together form a complex filter, dependent on the direction of sound. This role as a filter is often characterized by measuring the spectrum of the sound source and of the spectrum of the sound reaching the ear canal. The relationship between the two measurements (normally recorded in dB) gives the head-related transfer function (HRTF). The HRTF varies systematically with the direction of the sound source in relation to the head, and is unique for each direction of space (Searle et al., 1975). The spectral modulations produced by the head and the pinna can be used to discern the location of the source. The most pertinent information supplied by the pinna is obtained for sounds that have a large spectrum of frequencies. High frequencies, above 6000 Hz, are particularly important since only these high frequencies present wavelengths that are short enough to interact with the pinna.

Spatialized sound, or 3D sound, is a technology that aims to present an acoustic stimulation via headphones in such a way that the listener perceives it as coming from a precise point in space. It is a much more ecological[3] technique than classic stereophony, in which sound, although lateralized, seems to come from the inside of the head when listening to headphones.

The application of HRTFs to a sound presented via headphones reproduces the characteristics of the sound that would come to each ear from a sound source near the subject. The subject virtually perceives this source in a spatialized way.

The rendering is optimal on the condition that the HRTFs used by a subject are the HRTFs measured on that same subject (known as personalized HRTFs) (Middlebrooks, 1999). However, for cost reasons in terms of availability, team-expertise level, complexity of material, and infrastructure involved, supplying each individual with personalized HRTFs is not very realistic. In order to spread 3D sound technology to as many people as possible, the solution would be to use "nonindividualized" HRTFs, generally manufactured from a head dummy. However, when subjects achieve a localization task with HRTFs different from their own (which is like listening through someone else's ears), their performance is not as good as with their own personalized HRTFs (Middlebrooks, 1999 Wightman and Kistler, 2005).

It is possible to improve 3D sound perception by adding dynamic cues, which can be achieved by following a subject's head movements with an appropriate device. The benefits are substantial in terms of realism and precision, particularly in the front-back dimension, but the latency time must be short enough when sending signals during the time when the head is moving (Brungart et al., 2004). This requires sophisticated and expensive equipment, and can only be considered in certain workplace setups, such as the cockpit of a fighter aircraft (Bronkhorst et al., 1996; Nelson et al., 1998).

19.7 Contribution of learning

Low level acoustic properties influence the perception of urgency, as clearly demonstrated by Edworthy et al., (1991), then Hellier et al., (1993). However, other factors come into play and may influence the perception of urgency. The influence of these factors may be such that

[3] In the Gibsonian acceptance of the term

in some cases they may reverse the ranking of urgency that would have been expected by the analysis of the acoustic properties. (Guillaume et al., 2003).

It seems that the perception of urgency is in fact a judgment on the urgency of the situation, developed out of the mental representation evoked by the alarm and the context. This mental representation might result from two phenomena.

The first phenomenon is linked to learning. The mental representation evoked comes from professional experience or from acculturation. All subjects living in a given society have mental representations of alarms. These representations are acquired throughout life, through continuously associating these sounds to the notion of alarm, i.e., potential danger to others or to oneself (ambulance siren, fire brigade siren, fire alarm, anti-theft alarms etc.). For most people, these mental representations make up a database stored by the brain in memory. When activated by a sound, the judgment on perceived urgency is brought on by associating this sound with its emotional content. The cognitive processes implied range from identifying the source to judging the urgency associated with the mental representation evoked, taking context into account. In the work environment, a number of alarms are typically learnt, and this will supplement and/or reinforce the mental representations associated with the notion of alarm. A sound will be strongly associated to a specific cause, and to its urgency, allowing for faster and more appropriate motor reactions. Such acquired alarm sounds are abstract sequences, where the mental representation is built up through learning. This is the case most often found in the workplace, where operators connect the abstract sound they perceive to the origin of the alarm.

The second phenomenon relies on the fact that some sequences spontaneously evoke a mental representation, such as environmental sounds. Stephan et al., (2006) have shown that strong pre-existing associations between the signal and the referent facilitate learning and retention of auditory icon/referent pairings. This corresponds to Stanton and Edworthy's (1999) approach, mentioned earlier. These authors carried out a number of experiments on alarm design for an intensive care unit. They compared the alarm recognition performance of a well-practiced team with that of a team of employees new to the job. The alarms were either existing alarms on resuscitation equipment, or new alarms specifically designed to be more easily linked to the situation having triggered the alarm in the first place ("representational" sequences). Results show that for the experienced team, the old alarms are the most easily recognized, while the opposite is true for the freshman team, who recognize the new alarms better. However, both subject groups consider that the existing alarms are more suitable than the new ones ("representational" sequences). This result might be explained by the fact that even though the new alarms have a clearer connection to the failure they do not directly evoke emergency situations, because they are not connected to danger in everyday life.

Graham (1999) also carried out an interesting experiment. He compared the reaction time obtained to stop, using the brakes, in a driving simulation task. Operators hear either a horn, or tires screeching, or two more traditional sound alarms, i.e., a 600Hz tone, or a verbal alarm. The shortest reaction time is obtained with the horn and the screeching tires, the horn getting a slightly faster reaction time. The horn is an abstract alarm drivers are so familiar with that it is often considered as an environmental sound. The connection between horn and driving reaction is strongly established. As to the screeching tires, this is an environmental sound linked to a mental representation which evokes danger while driving.

Generally speaking, in the case of abstract alarms, connection with urgency is not direct, and only comes through learning. This connection is obtained when the subject knows what the alarm means and reacts accordingly. The confusion experienced by the different teams comes from the fact that the same alarm can be used for problems having different levels of seriousness.

In the case of "representational" alarms, learning is also highly important. Of course, making the connection with the cause of the alarm is easier, but the sound sequence in itself does not necessarily bring the notion of danger to mind, and consequently does not evoke the necessity of an urgent reaction in everyday life. Yet it is this notion of danger which helps the operator to react in the work environment. The motor reaction adapted to the sound sequence will be acquired by learning.

In both cases, the connection between the sequence and the cause of the sound is less direct than it would be in the everyday environment. Allocating a specific urgency to the alarm will thus gain more attention, notably from people who are not familiar with these alarms. In a second stage, learning will help reinforce the link between the alarm and the subject's appropriate reaction, coming from perceiving both the cause of the emergency and its urgency. Thus the difficulty in defining an alarm lies in finding a link, as direct as possible, between the alarm and its original cause, in order to minimize the attention allocated by the subject to "decode" the alarm.

19.8 Ergonomic approach to the problem

In order to evaluate the cause and the actual level of urgency of the alarms, a preliminary ergonomic approach is required, studying operator activity to pinpoint operators' real needs. Observing and questioning operators is the only way to obtain a realistic assessment of the degree of urgency associated with an alarm. Sanderson and Seagull (1997) carried out observations focused on variation in anesthetists' responses to alarm across different phases of surgery in varying kinds of surgical procedure. They observed that alarms do not function simply to warn of problems, but instead are used as tools with varying functions depending on type and phase procedure. They classified anesthetists' responses to alarms into four categories:

1. **correction or change** the alarm induced an action to correct an unexpected event;
2. **expected or intended**: the alarm indicated a state of affairs and no actions were required;
3. **ignore**: the alarm was an artifact and no action was needed;
4. **reminder**: the alarm was a reminder to initiate an expected action.

They pointed out that many more alarms were ignored than were the basis for corrective actions, and that the ignored alarms mainly took place during induction and emergence phases for which the context was quite different from the maintenance.

Similarly, Guillaume et al., (2005) carried out an ergonomics survey to assess the respective importance of the various alarms used in the operating room. Anesthesia procedures were observed in different operating rooms in order to classify the auditory signals. For each auditory signal, the team of anesthetists explained its meaning, the consequences for the patient or the monitoring systems, and whether they needed to interact with the patient, the monitoring equipment, or the warning signals.

The categories were represented in four sets:

1. Signals indicated a clinical problem. There was a vital risk for the patient. At least one physiological parameter was out the range of normal values;
2. Functioning signals reminded the anesthetist to act on the monitoring system;
3. Technical signals indicated a failure in the functioning of the monitoring equipment;
4. Interfering signals included auditory signals that originated from other parts of the operating room.

This classification aimed to allow a graded level of urgency to the alarms. As spectrum analyses were also performed on each warning signal, the authors pointed out that similar spectra were observed for alarms belonging to different sets (and thus with different levels of urgency). The use of a functional classification as described by Sanderson and Seagull (1997) or Guillaume et al., (2005) could help with the implementation of a realistic grading of the urgency level of the auditory alarms, that would result in a gradation of the acoustic properties of sound spectra. The conception of a well-designed alarm system requires an excellent knowledge of the application as well as practical experience that can be acquired by an ergonomic approach.

19.9 Intelligent alarm systems

Intelligent alarm systems generally use artificial intelligence for the automatic diagnosis of problems. These computerized systems collect information on system status from sensors and use artificial intelligence to organize this corpus of data into a data stream helping to diagnose the problem. The underlying assumption is that automatic diagnosis will reduce the time lag between the occurrence of a problem and its correction, by minimizing the time required to identify the problem (Westenskow et al., 1992). This approach to problem diagnosis also allows for the prioritization of problems. Being able to grade failures onto a scale is essential. A single failure may have side effects impacting system operation, and trigger off additional alarms. This, in turn, can result in several alarms going off simultaneously, confusing the operator, unable to decide which problem should be dealt with first (i.e., "cascading alarms") (Sorkin, 1988; Stanton, 1994). Under stress, the operator may choose to focus on a secondary problem and overlook the main failure, wasting precious time. Bliss and Gilson (1998) report the high-profile incident at the Three Mile Island nuclear power facility that underscored this problem. They cited Sheridan (1981) who noted that when detection of system failure is automatic, the sheer mass of display activity in the first few minutes of a major event can completely disturb the operators. For instance, at one loss-of-coolant incident at a nuclear reactor, more than 500 annunciators changed status within the first minute. Problem prioritization allows the system to trigger off only the one alarm corresponding to the main failure. The operator's attention is then directed to the real problem. Complex algorithms come into play to diagnose the problem and manage priorities within the system (Zhang and Tsien, 2001). This engineering approach helps to limit the number of alarms which may be triggered simultaneously and reduces false alarms.

It also reduces sound nuisance and the stress associated with it. It also decreases the operator's cognitive workload, by drawing attention solely to the main problem. This approach supplements the sound design approach, which requires in-depth work on the properties of acoustic signals to achieve the quickest information processing possible (see Figure 19.2). In this framework, the content of the sound signal may be supplemented by the signal's presentation mode, notably through the use of 3D sound. In aeronautics, 3D sound is being widely investigated (Begault, 1993; McKinley et al., 1994; Nelson et al.,1998; Bolia et al.,1999; Brungart et al., 2003) to save that extra few seconds which in turn will help to save the aircraft by immediately directing the pilot's attention to the threat at hand. To summarize, we could define the concept of an intelligent alarm as the alarm that takes all these approaches into account.

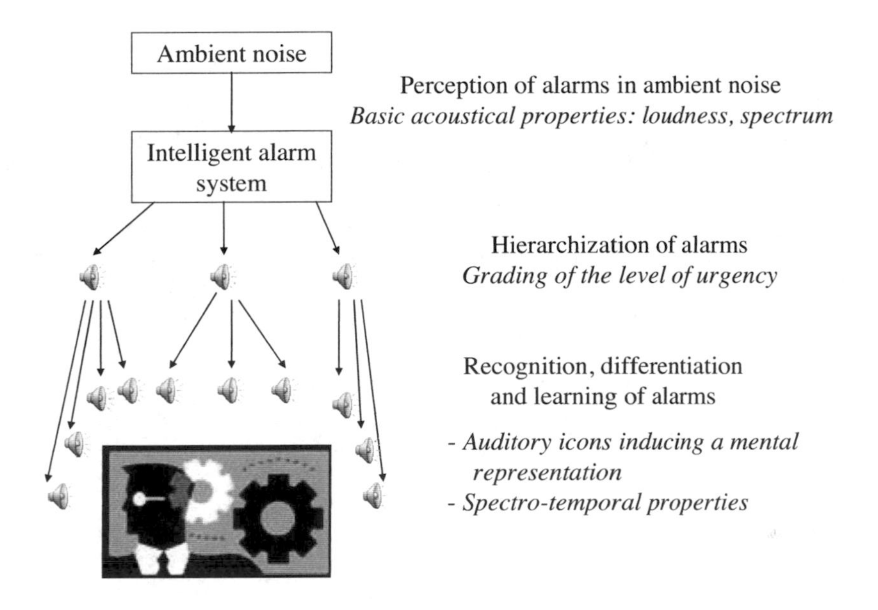

Figure 19.2: General scheme of an intelligent alarm system.

19.10 Conclusion

The complex issue of designing what can be called "intelligent sound alarms" requires bringing together multi-disciplinary teams, taking into account engineering, ergonomics and sound design aspects. The information-providing content of alarms has to convey the problem's degree of urgency and root cause. Psychophysical and cognitive approaches must be considered together, to evoke a mental representation while allowing the modulation of the degree of urgency perceived. Furthermore, reducing the attention consumed to manage alarm systems requires an ergonomic study of operators' real needs and the application of artificial intelligence in the management of the system of alarms.

Bibliography

[1] Arnstein F. Catalogue of human error. British Journal of Anaesthesia 1997; 79: 645–656.

[2] Begault, D. R. (1993). Head-up auditory displays for traffic collision avoidance system advisories: a preliminary investigation. *Human Factors 35*, 707–717.

[3] Blauert, J., Brueggen, M., Bronkhorst, A., Drullman, R., Reynaud, G., Pellieux, L., Krebber, W. and Sottek, R. (1998). The AUDIS catalog of human HRTFs. *J. Acoust. Soc. Am. 103, 3082.*

[4] Bliss, J. P., and Gilson R. D. (1998). Emergency signal failure: implications and recommendations. *Ergonomics 41*, 57–72.

[5] Bolia, R. S., D'Angelo, W. R., McKinley, R. L. (1999). Aurally aided visual search in three-dimensional space. *Human Factors 41*, 664–669.

[6] Bregman, A. S. (1990). *Auditory scene analysis: the perceptual organization of sound.* Cambridge, MIT Press.

[7] Bronkhorst, A.W., Veltman, J.A., Vreda, L.V.(1996). Application of a Three-Dimensional Auditory Display in a Flight Task. *Human Factors. 38*, .

[8] Brungart, D. S., Kordik, A. J., Simpson B. D., and McKinley, R. L. (2003). Auditory localization in the horizontal plane with single and double hearing protection. *Aviation, Space, and Environmental Medicine 74*, 937–946.

[9] Brungart, D.S., Simpson, B., McKinley, R.L., Kordik, A.J., Dallman, R.C., and Ovenshire D.A. (2004). The interaction between head-tracker latency, source duration, and response time in the localization of virtual sound sources. Proceedings of the 10th International Conference on Auditory Display, Australia.

[10] Burt, J. L., Bartolome, D. S., Burdette, D. W., and Comstock, J. R. (1995). A psychophysiological evaluation of the perceived urgency of auditory warning signals. *Ergonomics, 38*, 2327–2340.

[11] Doll, T. J., and Folds, D. J. (1986). Auditory signals in military aircraft: ergonomics principles versus practice. *Applied Ergonomics 17*, 257–264.

[12] Edworthy, J., Loxley, S., and Dennis, I. (1991). Improving auditory warning design: relationship between warning sound parameters and perceived urgency. *Human Factors 33*, 205–231.

[13] Edworthy, J., Hellier, E., and Hards, R. (1995). The semantic associations of acoustic parameters commonly used in the design of auditory information and warning signals. *Ergonomics 38*, 2341–2361.

[14] Edworthy, J., Hellier, E., Walters, K., Weedon, B., and Adams, A. (2000). The relationship between task performance, reaction time, and perceived urgency in nonverbal auditory warnings. In: *Proceedings of the IEA 2000/HFES 2000 Congress,* 3, 674–677.

[15] Fitch, W. T., and Kramer, G. (1994). Sonifying the body electric: superiority of an auditory over a visual display in a complex, multivariate system. In G. Kramer (Ed.), *Auditory Display: Sonification, Audification, and Auditory Interfaces* (pp. 307–325). Santa Fe, USA: Addison-Wesley Publishing Compagny.

[16] Flanagan, P., McAnally, K.I, Martin, R.L., Meehan, J.W., and Oldfield, S.R, (1998). Aurally and visually guided visual search in a virtual environment, *Human factors, 40*, 461–468.

[17] Graham, R. (1999). Use of auditory icons as emergency warnings: evaluation within a vehicle collision avoidance application, *Ergonomics 42*, 1233–1248.

[18] Guillaume, A., Pellieux, L., Chastres, V., and Drake, C. (2003). Judging the urgency of non-vocal auditory warning signals: perceptual and cognitive processes. *Journal of Experimental Psychology: Applied 9*, 196–212.

[19] Guillaume, A., Jacob, E., Bourgeon, L., Valot, C., Blancard, C., Pellieux, L., Chastres, V., Marie Rivenez, M., and Cazalaà, J.-B. (2005). Analysis of auditory warning signals in the operating room. In: *Proceedings of the International Conference on Auditory Displays*, Ireland, July 2005.

[20] Haas, E. C., and Casali, J. G. (1995). Perceived urgency of and response time to multi-tone and frequency-modulated warning signals in broadband noise. *Ergonomics 38*, 2313–2326.

[21] Hellier, E., Edworthy, J., and Dennis, I. (1993). Improving auditory warning design : quantifying and predicting the effects of different warning parameters on perceived urgency. *Human Factors 35*, 693–706.

[22] Hellier, E., and Edworthy, J. (1999). On using psychophysical techniques to achieve urgency mapping in

auditory warnings. *Applied Ergonomics 30*, 167–170.

[23] Ho, C., and Spence, C. (2009). Using peripersonal warning signals to orient a driver's gaze. *Human Factors, 51*, 539–556.

[24] James, S. H. (1996). Audio warnings for military aircraft. *AGARD Conference Proceedings,* 596, 7.1–7.17.

[25] Lazarus, H., and Höge, H. (1986). Industrial safety: Acoustic signals for danger situations in factories. *Applied Ergonomics 17*, 41–46.

[26] Loeb, R.G., and Fitch W.T. (2002). A laboratory evaluation of an auditory display designed to enhance intraoperative monitoring. *Anesthesia and Analgesia 94*, 362–368.

[27] Logan G. D. (1988). Automaticity, resources, and memory: theoretical controversies and practical implications. *Human Factors 30*, 583–598.

[28] Loveless, N. E. and Sanford A. J. (1975). The impact of warning signal intensity on reaction time and components of the contingent negative variation. *Biological Psychology 2*, 217–226.

[29] McKeown, D., Isherwood, S., and Conway, G. (2010). Auditory displays as occasion setters. *Human Factors, 52*, 54–62.

[30] McKinley, R. L., Erickson, M. A., D'Angelo W. R. (1994). 3-dimensional auditory displays: development, applications, and performance. *Aviation, Space, and Environmental Medicine 65*, A31–38.

[31] Meredith, C., and Edworthy, J. (1994). Sources of confusion in intensive therapy unit alarms. In N. Stanton (Ed.), *Human Factors in Alarm Design* (pp. 207–219). Basingstoke, England: Burgess Science Press.

[32] Middlebrooks, J.C. (1999). Individual differences in external-ear transfer functions reduced by scaling in frequency. *J Acoust Soc Am 106*, 1480–1492.

[33] Middlebrooks, J.C. (1999). Virtual localization improved by scaling nonindividualized external-ear transfer functions in frequency. *J Acoust Soc Am 106*, 1493–1510.

[34] Momtahan K, Hetu R, and Tansley B. (1993). Audibility and identification of auditory alarms in operating room and intensive care unit. *Ergonomics 36*, 1159–1176.

[35] Moore, B. (1997). An introduction to the psychology of hearing. London: Academic press 4th ed.

[36] Nelson, W.T., Hettinger, L. J., Cunningham, J. A., Brickman, B. J., Haas, M. W., and McKinley, R. L. (1998). Effects of localized auditory information on visual target detection performance using a helmet-mounted display. *Human Factors 40*, 452–460.

[37] Patterson, R. D. (1982). Guidelines for the auditory warnings systems on civil aircraft. London : *Cicil Aviation Authority*, paper 82017.

[38] Patterson, R. D., Edworthy, J., Shailer, M. J., Lower, M. C., and Wheeler, P. D. (1986). Alarm sounds for medical equipment in intensive care areas and operative theatres. *Report AC598.*

[39] Patterson, R. D. (1990). Auditory warning sounds in the work environment. *Phil. Trans. R. Soc. Lond., B 327*, 485–492.

[40] Sanderson, P. and Seagull F.J. (1997). Cognitive ergonomics of information technology in critical care: contexts and modalities for alarm interpretation. In: *Proceedings of the International Workplace Health and Safety Forum and 33rd Ergonomics Society of Australia Conference. Gold Cost 24–27 nov,* 43–52.

[41] Sanderson, P., Crawford, J., Savill, A., Watson, M., and Russell J. (2004). Visual and auditory attention in patient monitoring: a formative analysis. *Cognition, Technology and Work 6*, 172–185.

[42] Sanderson, P. M., Watson, M. O., and Russell, W. J. (2005). Advanced patient monitoring displays: Tools for continuous informing. *Anesthesia and Analgesia 101*, 161–168.

[43] Sanderson, P.M., Liu, D. Jenkins, S.A. (2009). Auditory displays in anaesthesiology. *Cur. Opin. Anaesthesiol., 22*, 788–795.

[44] Schneider, W. and Schiffrin, R. M. (1977). Controlled and automatic human information processing : detection search and attention. *Psychological Review 84*, 1–66.

[45] Seagull, F. J., Wickens, C.D., and Loeb, R.G. (2001). When is less more? Attention and workload in auditory, visual and redundant patient-monitoring conditions. In: Proceedings of the 45th Annual Meeting of the Human

Factors and Ergonomics Society, San Diego, CA, October 2001.

[46] Searle, C.L., Braida, L.D., Cuddy, D.R. and Davis MF (1975). Binaural pinna disparity: Another auditory localization cue. *J. Acoust. Soc. Am. 57*, 448–55.

[47] Sheridan, T. B. (1981). Understanding human error and aiding human diagnostic behaviour in nuclear power plants. In: Human detection and diagnosis of system failures, Rasmussen J. and Rouse W. B. (eds). Plenum Press; New York.

[48] Simpson, C. A., and Williams, D. H. (1980). Response time effects of alerting tone and semantic context for synthesized voice cockpit warnings. *Human Factors 22*, 319–330.

[49] Sorkin, R. D., Kantowitz, B. H., and Kantowitz, S. C. (1988). Likelihood alarm displays. *Human Factors 30*, 445–459.

[50] Sorkin, R. D. (1988). Why are people turning off our alarms? Journal of the Acoustical Society of America 84, 1107–1108.

[51] Stanford, L. M., McIntyre, J. W., and Hogan, J. T. (1985). Audible alarm signals for anaesthesia monitoring equipment. *Int J Clin Monit Comput 1*, 251–256.

[52] Stanton, N. A. (1994). Alarm initiated activities. In N. A. Stanton (Ed.), *Human factors in alarm design*. London: Taylor and Francis.

[53] Stanton, N. A., and Edworthy, J. (1998). Auditory affordances in the intensive treatment unit. *Applied Ergonomics 29*, 389–394.

[54] Stanton, N. A., and Edworthy, J. (1999). Auditory warning affordances. In N. A. Stanton and J. Edworthy (Eds.), *Human Factors in Auditory Warning* (pp. 113–127). Aldershot: Ashgate.

[55] Stephan, K.L., Smith, S.E., Martin, R.L., Parker, P.A., and McAnally, K.I. (2006). Learning and retention of associations between auditory icons and denotative referents: Implications for the design of auditory warnings. *Human Factors, 48*, 288–299.

[56] Todd Nelson, W., Hettinger, L.J., Cunningham, J.A., Brickman, B.J., Haas, M.W., and McKinley, R.L. (1998). Effects of localized auditory information on visual target detection performance using a helmet-mounted display, *Human factors, 40*, 452–60.

[57] Watson, M. and Sanderson P. (2004). Sonification helps eyes-free respiratory monitoring and task timesharing. *Human Factors 46*, 497–517.

[58] Westenskow, D.R., Orr, J.A., Simon, F.H., Bender, H.-J. and Frankenberger H. (1992). Intelligent alarms reduce anesthesiologist's response time to critical faults. *Anesthesiology 77*, 1074–1079.

[59] Wheale, J. (1982). Decrements in performance on a primary task associated with the reaction to voice warning messages. *British Journal of Audiology 16*, 265–272.

[60] Wheale, J., Goldino, D., and Doré C. (1979). Auditory warnings: signal variations and their effects on the operator. *30th annual conference of the Ergonomics Society*.

[61] Wightman, F., and Kistler, D. (2005). Measurement and validation of human HRTFs for use in hearing research. *Acta Acustica United with Acustica 91*, 425–439.

[62] Xiao, Y., and Seagull F.J. (1999). An analysis of problems with auditory alarms: defining the roles of alarms in process monitoring tasks. In: Proceedings of the Human Factors and Ergonomics Society 43rd Annual Meeting. Santa Monica, Calif. USA, 256–260.

[63] Zhang, Y., Tsien, C.L. (2001). Prospective trials of intelligent alarm algorithms for patient monitoring (abstract). In: Proceedings of AMIA Symposium, 1068.

Chapter 20

Navigation of Data

Eoin Brazil and Mikael Fernström

This chapter explores a range of topics concerning the navigation of data and auditory display. When we are navigating the world our minds continuously hunt for information about the place we are in and what things there are in our immediate environment. We pick up information that enables us to act in our current location. Our perception of the world can, with little effort, detect differences in structure, gradients and emerging patterns. Technology that aims at facilitating navigational behavior is often based on the idea of augmenting our human abilities.

One of the oldest technological examples of auditory display is the Geiger counter, invented by the German nuclear physicist Hans Johann Wilhelm Geiger in 1908. The original device detected the presence of alpha particles, and later developments resulted in devices that could detect different kinds of radioactive decay. One important characteristic of Geiger-Müller counters is the use of auditory display, i.e., each particle detected is converted to an electrical pulse that can be heard through headphones or loudspeakers as a click. The more radioactive particles per second, the higher the click-rate. Another characteristic of Geiger-Müller counters is that they are normally handheld and can be moved around by a mobile user to detect hot-spots and gradients of radiation, freeing up the user's visual modality, while navigating in the environment or making other observations. Several other kinds of instruments have been developed along the same principles, for example metal detectors and cable finders. An example of using this type of auditory display can be seen in this video. The geiger counter metaphor has been used in various domains including for the navigation of oil well exploration (video example **S20.1**).

For navigation, as a human activity, we first consider the concepts of the *navigation control loop* and *wayfinding*. Wayfinding is the meta level problem of how people build up an understanding of their environment over time.

20.1 Navigation Control Loop

To understand navigation the concept of what is called the *navigation control loop* is first outlined. Refining and navigating a view of a data space is complex and relates to spatial knowledge theory and the issue of feedback and interactivity with the data. A schematic illustration of the navigation control loop is shown in Figure 20.1, showing the human side as the cognitive and spatial understanding of a data space and the computer side representing a data space that can be updated and remapped to offer alternative perspectives on the data. To understand what navigation is and how the control loop relates to auditory displays, the next section explains the concept of way-finding.

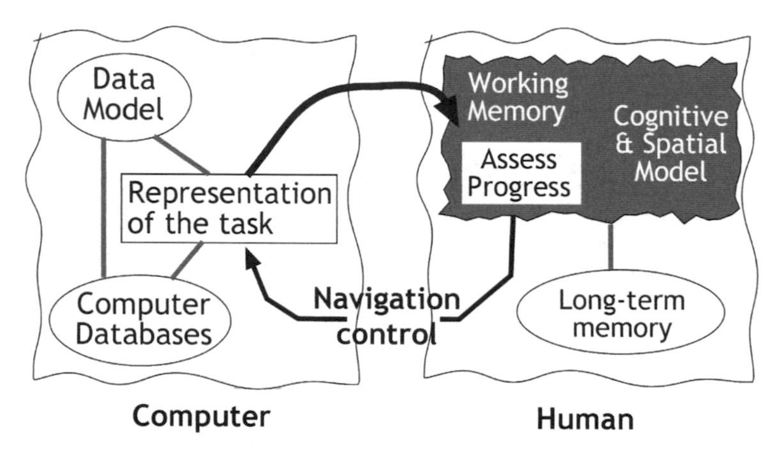

Figure 20.1: The basic navigation control loop

20.2 Wayfinding

Wayfinding is the method by which we create mental models of our environment. An example is how people use physical maps to support navigation. Peponis [27] defined wayfinding as "*how well people are able to find their way to a particular destination without delay or undue anxiety*" but this concept can also be interpreted as the understanding of a particular level or state within a computer environment. In the case of auditory display it can refer, for example, to a particular auditory beacon, with the sound representing a particular state or level. Wayfinding knowledge is built up using three types of knowledge [31]:

- Declarative or landmark knowledge allows for orientation within an environment as well as the ability to recognize destinations;
- Procedural or route knowledge allows routes to be followed to reach a destination;
- A cognitive spatial map supports the selection of the route most appropriate for the current context or task.

Each type of spatial knowledge is gathered from different properties within a given environment and via different experiences with that particular environment. The following sections look at each knowledge type in turn.

Declarative Knowledge

Landmarks are often the key details in an environment. In a visual scene it is the dominant item in the field of view. In an auditory scene it can be a dominant sound or for example an audio beacon [37]. A landmark is distinct and may provide directional information or something that has a personal meaning [23] and that is distinguishable by its acoustic attributes.

Route or Procedural Knowledge

This type of knowledge represents an egocentric viewpoint which considers the navigation process as a sequence of steps that are required in order to follow a route or set of routes. Where and when to turn, what actions are required along a route, and other critical points of information are part of this type of knowledge. It includes an implicit knowledge of distances in route segments, directions of turns, and the sequence of landmarks within a route. An auditory form of such knowledge can include recognition of sequences of, for example, earcons or auditory icons. The SWAN system [37] is an example of such a system[1] (video example **S20.2**). It used a selection of auditory icons and earcons as audio beacons that were spatialized around the user by means of a generalized Head-Related Transfer Function (HRTF). The audio beacons tested included a sonar pulse ("ping") sound, a pure sine wave of about 1000 Hz, and a burst of pink noise. These sounds were spatialized and the tempo or number of pings was changed depending on the distance from the next landmark.

Cognitive Spatial Map

This type of knowledge uses an exocentric (map-like) viewpoint and represents the navigation process within the context of a fixed coordinate system where objects are related by distances to other objects within this coordinate system. This type of knowledge allows for distance estimation between landmarks and for inferring new route possibilities.

20.3 Methods For Navigating Through Data

Using both static and dynamic representations can help us to develop a better understanding of a dataset. Mapping is the linking of numerical quantities to aural or visual representations. Three approaches of interest for designing auditory displays for data navigation are the auditory information-seeking principle, interactive sonification and virtual immersive environments.

20.3.1 Auditory Information-Seeking Principle

An important approach for navigating visual data was suggested by Shneiderman in his visual information seeking mantra [32]. The elements of visual information-seeking are "*overview first, zoom and filter, then details on demand*". Combined with direct manipulation

[1] http://sonify.psych.gatech.edu/research/SWAN/SWAN-audioVRdemo-movie.mov

and tight coupling, users experience an immediate response from the system for every action. Thus users can potentially attend to more information per unit time. The principle behind this approach is referred to as "*reducing the cost structure of information*" [7]. Auditory information-seeking [39] is a further development of the approach, in particular for auditory display. The four key areas in auditory information seeking are:

- to provide a gist of the data;
- to let the user navigate through the data space;
- to let the user filter the data space;
- to provide further details on an item or group of items on demand.

Zhao's [39] demonstration of auditory information-seeking for statistical data from the US government included the 'gist' concept, where two sets of sounds were used broken into spatialized piano pitches and non-spatialized string instrument pitches. These sounds represented five value categories using an ascending arpeggio in C major: C E G C E with the lower pitches representing lower values.

Gist

A gist is a short auditory message that conveys a sense of the overall trend or pattern of a data collection. Outliers or anomalies should be easily detected in the message and it should help in guiding exploration of the data space. As a data space may be large and a gist must be of a short duration, data aggregation can be used. A number of auditory displays have used  gists, for example in radio weather reports [16][2] (sound example **S20.3**). This example maps various weather forecast parameters to the sonification and shows how complex temperature, wind, and other climate data can be summarized and aggregated using an auditory display. It mapped the 9-dimensional weather vector including information on wind speed, wind direction, temperature, cloudiness, rainfall and humidity to a multi-stream auditory display on the basis of auditory icons. Figure 20.2 shows the range of mappings (including an emotional event set as shown in Table 20.1) and the mappings of auditory icons.

The Process of Navigation

Navigation is the activity when we find our way through a space, browsing or searching through data while listening to selective portions of the data set. This is an interactive cycle where we directly manipulate the data set, which results in the system giving feedback about our current focus of attention. Over time, we build up a mental map of the data space that allows us to recognize virtual objects or landmarks in the data space. This mental map allows for the easy location of a particular auditory stimuli representing a data item or group of items. In the navigation process, an auditory gist message of the current data item or group of interest can be a useful part of the system feedback to the user.

[2] http://www.techfak.uni-bielefeld.de/ags/ami/datason/demo/ICAD2003/weatherSonification.html

Emotional Event Sounds	
hot dry summer day	tired, forceless, exhausted, an indifferent emotion – panting sound, cricket songs
warm dry summer day	positive emotion, happy, optimistic – a bird sound associated with a walking occasion
hot and sweltry summer day	exhausting – a 'sigh' sound is played
nice cold winter day	positive emotion – e.g. uprising sound with shiver/strong vibrato
golden october day	positive emotion – e.g. uprising rising fifth with an pleasant organic sound
snowy winter day	negative, calm, indifferent – a shudder/shiver sound
fog on an autumn morning	mystic and curbed – distant reverberant scream
thunderstorm on a hot summer day	wild, anxious – maybe a kettledrum sound
wet winter day	depressing, pessimistic, negative – a downward tritone interval, crying or weeping sound
stormy rainy autumn day	depressing – a smooth diminished chord

Table 20.1: Mappings for emotional events used in the weather sonification [16].

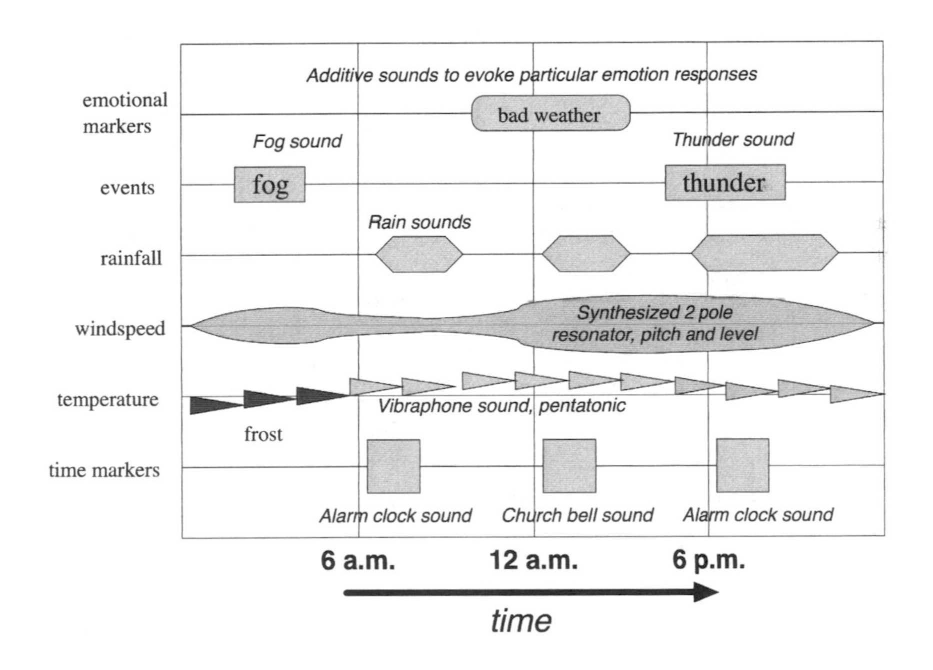

Figure 20.2: Auditory weather forecast with original figure taken from [16].

Filtering

The high dimensionality of complex data spaces necessitates mechanisms for reducing the complexity and size of the data being explored. Filtering is a mechanism that allows the overall cohesion and structure of the data set to be maintained while allowing certain aspects to be temporarily suppressed. This helps focus the presentation to only the items of interest within the data space.

Details On Demand

In a data space the user can select a particular item or group of items and get further details about the item or group. The information or detail can be represented using auditory icons, earcons, spearcons or speech. This depends on the particular item and the most appropriate mapping for its auditory display. In cases where the data or information is itself an audio resource, then this stage determines whether the resource is played back unaltered in a long or short form or whether it is representing using a sonification or audification.

20.3.2 Interactive Sonification

Interactive sonification has been defined by Hermann and Hunt [17] as "*the use of sound within a tightly closed human computer interface where the auditory signal provides information about data under analysis, or about the interaction itself, which is useful for refining the activity*". Interactive sonification is characterized by allowing the user control of the sonification process through tight coupling between the user's actions and auditory feedback from the system. Currently, there are two main approaches in interactive sonification: Parameter-mapping sonification as discussed in Chapter 15 and Model-based sonification discussed in Chapter 16. In a parameter-mapped sonification, the data *plays* the sonification (as an instrument or set of instruments or orchestra) and the user can modify the mappings to find items of interest and navigate through the data (almost like a DJ/turntablist uses scratching). In model-based sonification, the data is used for creating an *instrument* and the user interacts with the virtual instrument, i.e. the user's actions work as excitation of the model, to find or explore the structure of the virtual instrument, i.e. the information.

20.3.3 Navigating in Virtual Spaces

Virtual spaces can be either photorealistic (such as *Second Life* where entire cities [29] are being recreated), or abstract representations that can be experienced as 3D immersive virtual environments. Research by Lokki and Gröhn [22] found that the fastest and most accurate navigation is achieved with audiovisual cues. They suggested that 3D sound can be used to *highlight* objects of interest, which may be of use in complex data spaces where the user is surrounded by the data space and points of interest may be located outside the user's field of vision. A similar approach has been developed by Amatriain et al [1] with the *AlloSphere*, a large spherical audiovisual display system with both 3D visual and 3D auditory display that can accommodate up to 30 co-located users simultaneously.

20.4 Using Auditory Displays For Navigation Of Data

To highlight issues and possibilities of auditory displays being used for the navigation of data, this section gives a broad review of some existing auditory displays that support navigation in various forms. Auditory displays are beginning to see a wider use in many domains and by presenting this review this section gives a sampling of current use that may inspire others to expand these boundaries into new domains by highlighting the potential of using auditory displays for navigation.

20.4.1 Data Mining and Navigation

Data mining is one area where auditory display techniques such as interactive sonification and mechanisms for navigation of complex high dimensional data spaces are explored. Data mining has been defined as "the science of extracting useful information from large data sets or databases" [14]. Data mining can be regarded as having two stages: exploratory data analysis (EDA) followed by confirmatory data analysis (CDA). The focus of EDA is to detect patterns and computers can be used to augment the human ability to pick up patterns.Using an auditory display is one suitable method for data mining EDA as it allows listeners to interpret sonifications or audifications to improve their understanding of a data space as well as for pattern detection. Increasing digitization of scientific information has been a motivating factor in the development of auditory display systems in this domain. There is an explosive growth of sensor and data logging systems resulting in many scientific domains generating and recording vast quantities of data. Auditory displays excel at sifting rapidly through such large data sets.

20.4.2 Navigation Of Music and Sound Collections

Music and sound are interesting areas for auditory navigation as the resources themselves are aural. A number of systems have been developed to explore these areas allowing for the navigation of personal music collections to find related music or sounds based on a range of criteria. Many systems have been developed to assist with the navigation of music collections including *nepTune* [20] which provided a 2 1/2 D[3] visualization of a topographical map with song islands. The land masses were generated using audio analysis and corresponded to clusters of similar music. It was based on a self-organizing map (SOM), arranging a music collection as a 3D landscape. The vertical dimension represented the approximate density of items in a particular region of the terrain. The user heard the pieces of music closest to their virtual position as they navigated through the landscape, as shown in this video (video example **S20.4**).

Brazil and Fernström [6] developed an interactive system for navigating sound or music collections that was designed using the principles and techniques found in section 20.3. The Sonic Browser as shown in Figure 20.3 on page 516 provided three different 2D visualizations and an auditory display with multiple stream stereo-spatialized audio. The audio was activated by cursor/aura-over-icons that represented sound files, and these allowed the user to focus their attention by directly hearing sonifications of the objects under the

[3]This is a 2D visualization which includes the distance to the surface.

cursor/aura. The *aura* [13], in this context, is a function that defines the user's range of perception in a domain. All objects under the aura play simultaneously and their relative loudness and stereo panning is relative the centre of the aura[4] (sound example **S20.5**). It facilitated users switching their attention between different sounds in the auditory scene at will, by utilizing the "*cocktail party effect*" [9, 10], direct manipulation and tight coupling to navigate the sound collection. A similar system, *SoundTorch*, has been developed and investigated by Heise et al [15][5] (video example **S20.6**). Both of these applications use the navigation and details-on-demand approaches from auditory information seeking [39] to facilitate browsing audio resources.

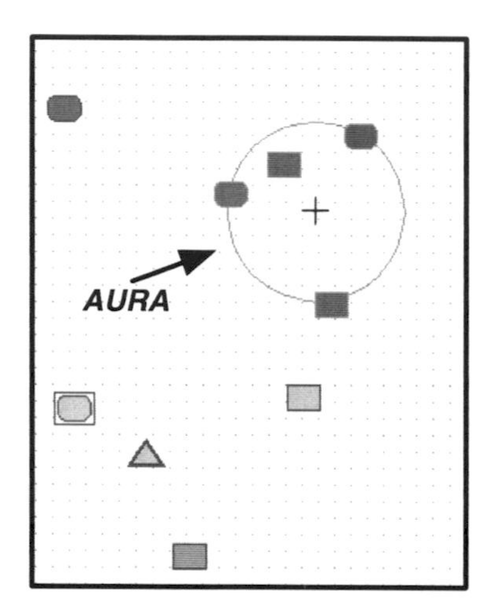

Figure 20.3: The Sonic Browser, an interactive sound collection interface by Brazil and Fernström et al. [6], with permission from authors.

Another consideration when navigating music or sound collections is to offer thumbnails of sound. A segmentation-based thumbnailing approach to help navigating music collections was developed by Tzanekis [35]. This presented a 2.5 D virtual space combining short slices of the sound with visualizations called *TimbreGrams* as a method of representing the audio files. Each TimbreGram had stripes corresponding to a short time-slice of sound (20 ms to 0.5 s), with time mapped from left to right within the symbol and the color of each stripe representing the spectral centroid of the slice. The similarity of sounds could be seen as color-pattern similarity between TimbreGrams. Chai [8] explored analysis of pop/rock music and suggested that thumbnails from musical section beginnings (incipits) are more effective than thumbnails from other parts of songs. Brazil [5] had similar results for browsing collections of sound files with recordings of everyday sounds.

20.4.3 Navigating Complex Geological Data

Barrass [3] developed two auditory displays for navigating complex geological data, PopRock and cOcktail. PopRock was designed for planning a mine-shaft where the sounds were used to

[4]http://www-staff.ichec.ie/~braz/sonicbrowser.mp3
[5]http://www.youtube.com/watch?v=eiwj7Td7Pec

present information about the amount of seismic activity that indicate faults or weaknesses in adjacent areas to the mine-shaft. Isolated events or differences in the activity of event groups were more easily distinguished when using audio to navigate the data space by presenting 90 days of seismic logs in approximately 3 minutes using 4 channels of the soprano sax playing the G above middle C with varying brightness indicating the size of the event. The application was used to display information about oxygen isotopes in 12 sea-bed drill-core locations. It allowed for the navigation of time-series data where the listener could switch attention between simultaneous time-series to compare locations[6] (sound example **S20.7**). The cOcktail application used speech segments of drinks orders, such as might be heard at a cocktail party, with three levels of spatial distance (near, medium and far) to represent the oxygen isotope value with the drink name representing a specific drill site. The mapping between sea-bed drill-core location and speech / drink name as well as oxygen isotope levels is shown in Table 20.2.

Mappings between drill sea-bed drill-core location and speech / cocktail names used	
drill-core site	**word, sex, location**
RC12-294	'blue-lagoon', male, 0 degrees
RC13-22	'shirley-temple', female, 8 degrees
RC24-16	'harvey-wallbanger', male, 16 degrees
V12-122	'bloody-mary', male, 24 degrees
V22-174	'martini', male, 32 degrees
V25-21	'gin-and-tonic', male, 40 degrees
V25-56	'margarita', female, 48 degrees
V25-59	'grasshopper', male, 56 degrees
V30-40	'golden-dream',female, 64 degrees
V30-97	'champagne-cocktail', female, 72 degrees
V30-49	'screwdriver',male, 80 degrees
RC11-120	'tequila-sunrise', male, 88 degrees
Mappings between oxygen isotope value and perceived distance to listener of sound	
O18 Isotope level	**ordinal variation**
low	far away
medium	medium distance away
high	close

Table 20.2: Mappings for the cOcktail application [3].

20.4.4 Navigation Of Biomedical Data

Pauletto and Hunt [26] have investigated the navigation of electromyographic (EMG) data through sonification. The auditory display was used to navigate information from 6 EMG sensors where each sensor mapped to the amplitude of a different sine oscillator. The frequencies of the oscillators were chosen to produce a single complex, but easily understandable sound. Therapists would normally spend several hours in a visual data mining task in order to interpret the patient's results for clinical diagnosis. Using an interactive sonification approach, therapists could focus on patients rather than visual screens.

Electroencephalogram (EEG) sonification for exploratory analysis [18] has been used to provide electrophysiological data about the brain's activity. Model-based sonification of EEG data has been developed to represent epileptic seizures [2]. EEG data normally shows

[6] `http://www.icad.org/websiteV2.0/Conferences/ICAD96/proc96/cocktail.au`

complex spatiotemporal patterns that indicate normal brain activity, but these turn into a globally ordered rhythmic pattern when an epileptic seizure [25] occurs. By sonifying this it is possible to explore the condition of epilepsy as well as providing a means for onset detection of an epileptic seizure. This type of sonification provides for an initial navigation of a data space to detect outliers as well as rhythmical or pitched patterns in the EEG data. This cursory exploration can be used by a medical specialist to help determine the areas of special interest in recordings from patients. A typical clinical session with a patient can consist of many hours of EEG recordings. Auditory display is particular well suited for the navigation of such EEG recordings[7] (sound example **S20.8**.)

Another navigation issue in the medical domain is for the placement or selection of surgical incisions, such as needle insertion points. There are many possible locations and directions and it may be difficult to navigate to the desired position within the patient's body. The work of Jovanov et al. [19] discusses the use of tactical audio and sonification to address this. Their design aimed at facilitating ultrasound-guided biopsy procedures by providing real-time navigation guidance using auditory feedback. The system maintained a 3D model of the patient using ultrasonic transducer and continuously compares this to a 3D anatomical model tracking the actual patient's orientation and actual position in context of this model. The biopsy needle was tracked continuously in 3D in relation to this model. A continuous audio signal allows the surgeon to be aware of the needle relative to a pre-planned trajectory and sonified using a polyphonic consonance/dissonance function. The testing and studies of this system were not included in the paper but the approach prompted others to continue in the track.

Similar research by Müller-Tomfelde [24] used auditory display combined with haptic feedback from a PHANTOM haptic force feedback device as a potential teaching aid for trainee surgeons to refine their surgical motor skills. The motor skills in this context where related to potential use by trainee medical surgeons for precise operations but this research did not be included any user studies of the interface. The positioning of the force feedback stylus used controls the difference in frequency of two sinusoids. When the force feedback stylus is close to a landmark this results in a single sounding sinusoid as there will be little frequency difference. When the force feedback stylus is far from a landmark, this results in a detuned sound with a high beating frequency. The distance to a landmark is also indicated through the use of reverberation, where the landmark functions as a loudspeaker and the positioning tool as a microphone to create a reverberant auditory space.

The application of auditory displays in surgery is particularly useful as Darzi et al. [11] have stated that a medical operation is 75% decision making and 25% dexterity, but is the dexterity comes from implicit knowledge which must be learned through practice rather than through studying texts or lectures. Interactive sonification can help to support the discovery of anatomical landmarks and enhance the performance of a sequence of distinct movements and also help to improve the "*naturalness*" of the trainees' movements [30].

[7] `http://www.techfak.uni-bielefeld.de/ags/ami/datason/demo/ICAD2006/EEGRhythms.html`

20.4.5 Navigation Of Geographical Sociological Data

Work by de Campo et al. [12] explored geographical sequences from sociology. The data space focused on the election results from the 2005 Styrian provincial parliament election. Styria is one of nine federal states in Austria, comprising 17 districts and 542 communities. In the 2005 parliamentary elections approximately 700,000 voters cast their ballot to elect their parliamentary representatives. This auditory display allowed the election results to be navigated by their geographical distribution and highlighted similarities in voting patterns between neighboring communities. The navigation metaphor was inspired by the idea of throwing a stone into a lake or pond and watching the ripples propagate from the point of impact out and across the water. A map of Styrian was presented using a 2D interface and when a point on the map was clicked, a ripple propagated the map sonifying each data point it encountered[8] (sound example **S20.9**). This is shown in Figure 20.4. Evaluation of the tool has found it to be useful for highlighting outliers in the data space where, for example, one party has 30% on average of the electorate, but in a particular community has 40% support, which will distinctly 'sound out' when exploring the sound space.

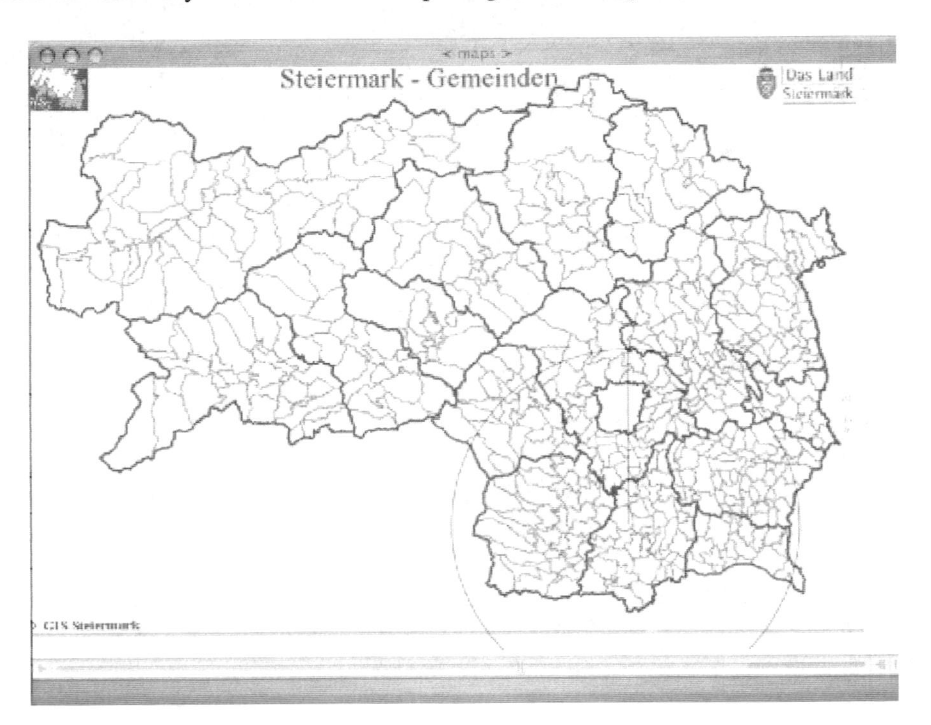

Figure 20.4: Original figure taken from de Campo et al. [12]: auditory display for navigating electoral results from Styria.

A similar navigation approach has been used to explore the interactive sonification of georeferenced data for visually impaired users [39]. Georeferenced data analysis and sonified choropleth maps were used to present geographical contextualised information. A choropleth

[8] http://www.sonenvir.at/data/wahlgesaenge

map has colored regions to show the differences between regions on the map. The geographically related auditory displays are an area that have been extensively used to highlight outliers in data trends and correlate it easily back to a spatial location or region.

20.4.6 Navigation Of Personal Information

Walker et al. [36] produced an auditory display for the navigation of calendar information on a mobile device or PDA. The application used an imaginary clock-face spatialized around the user's head as the spatial metaphor for the auditory display. The day's events in the user's calendar were heard with 9am/pm on the extreme left, 12am/pm being directly in front for the user, 3am/pm to the extreme right and 6am/pm heard as coming from behind the user. This is shown in Fig. 20.5. The auditory display used generalised HRTFs that for most individuals can provide a spatial resolution in the lateral plane smaller than 20°. The sound for the auditory display was a synthesised male voice (created using a text to speech application) speaking the appointment text.

Figure 20.5: Spatialized Auditory Display for navigating calendar information, original figure taken from [36].

20.4.7 Navigating the World Wide Web

The internet has become the single largest digital information source available in the whole of human history. Its size and complexity has resulted in the success of search engine portals such as Google or Yahoo. One use of navigation by auditory display for this resource is for blind or visually impaired users by providing mechanisms for navigating web pages and links between pages. Systems such as the *Web-based Interactive Radio Environment* (WIRE) [38] and *WebSound* [28] were designed to support navigation of the Web. The WIRE system was built as a non-visual browsing environment for the Web that rendered web pages using audio and with a keyboard-like interface for navigation within and between pages. The WebSound system was a 3D immersive virtual sound space using a joystick. The WIRE and WebSound systems provided alternatives to Braille or text-to-speech systems.

Other researchers investigated the issue of how to represent hyperlinks with auditory display. Hyperlinking has mostly focused on textual or graphical links but research by Braun et al. [4] and by Susini et al. [33] investigated the idea of sonifying hyperlinks. Braun's work concentrated on the use of annotating hyperlinks with sound that linked to a multimedia resource in the domain of interactive Web-based TV. Susini and colleagues [33] sonified hyperlinks by providing supplementary information about a collection of on-line radio programs. This supplementary information was conveyed by "*underlining*" a speech segment (a few words) with a non-speech signal, playing back a text-to-speech representation of the text with a pink noise sound. The underlining was achieved through the use of a pink noise sound which was investigated using three different auditory modifications:

- changing its spectral signature using a passband filter;
- changing the energy ratio between the pink noise and the target (weak, medium, or strong) as well as changing the onset of the underlining sound (pre-, simultaneous, or post-) with regard to the speech segment;
- modifying the attack time of the pink noise (gradual, rapid, abrupt) and the post-attack amplitude envelope over the duration of the underlining sound.

20.5 Considerations for the Design of Auditory Displays for the Navigation of Data

Designing displays for the effective presentation of visual information uses many generic techniques [32, 34]. Auditory displays that effectively present a navigable data space require more deliberate and considered mappings (parameter mappings) to ensure that mapping from data to sound also satisfies subjective and affective variables such as "value" or "beauty" [21] or the design of an interactive sonification model [17]. Methodologies such as the auditory information-seeking principle can provide a solid conceptual basis for the functionality that should be included in any auditory display for data navigation. Interactive sonification can help in constructing an interface which encourages exploration while allowing for complex and continuous interaction with the data. These techniques and methodologies can create successful designs without the inclusion of concepts from sound design or from auditory scene analysis but their inclusion makes for an easier and more informed design process.

The approach of visual information-seeking [32] with direct manipulation and interactive sonification providing overview first, zoom and filter, then details on demand using auditory or audio-visual means is the primary approach used for creating successful auditory displays for navigating data. The provision of a general overview or gist, a navigable data space, dynamic filtering and details on demand about items or groups has been shown to be successful.

Bibliography

[1] X. Amatriain, J. Kuchera-Morin, T. Hollerer, and S.T. Pope. The AlloSphere: Immersive multimedia for scientific discovery and artistic exploration. *Multimedia, IEEE*, 16(2):64 –75, april-june 2009.

[2] G. Baier, T. Hermann, S. Sahle, and U. Stephani. Sonified epileptic rhythms. In *International Conference on Auditory Display ICAD-06*, pages 148–151, 2006.

[3] S. Barrass. *Auditory Information Design*. PhD thesis, Australian National University, 1997.

[4] N. Braun and R. Dörner. Using sonic hyperlinks in web-tv. In *International Conference on Auditory Display ICAD-98*, pages 1–10, 1998.

[5] E. Brazil. Cue point processing: An introduction. In *Proceedings of the COST G-6 Conference on Digital*

Audio Effects (DAFX-01). Department of Computer Science and Information Systems, University of Limerick, 2001.

[6] E. Brazil and J. M. Fernström. Audio information browsing with the sonic browser. In *International Conference on Multiple Views in Exploratory Visualisation (CMV2003)*, pages 28–34, London, UK, 2003.

[7] S. K. Card, J. D. Mackinlay, and B. Shneiderman. *Information visualization: Using vision to think.* Morgan-Kaufmann, San Francisco, 1999.

[8] W. Chai. Semantic segmentation and summarization of music. *IEEE Signal Processing Magazine*, pages 124–132, March 2006.

[9] E. C. Cherry. Some experiments on the recognition of speech with one and two ears. *Journal of the Acoustical Society of America*, 25:975–979, 1953.

[10] E. C. Cherry and W. K. Taylor. Some further experiments on the recognition of speech with one and two ears. *Journal of the Acoustical Society of America*, 26:549–554, 1954.

[11] A. Darzi, S. Smith, and N. Taffinder. Assessing operative skill. *British Medical Journal*, (318):889–889, 1999.

[12] A. de Campo, C. Dayé, C. Frauenberger, K. Vogt, A. Wallisch, and G. Eckel. Sonification as an interdisciplinary working process. In *International Conference on Auditory Display ICAD-06*, pages 28–35, 2006.

[13] J. M. Fernström and C. McNamara. After direct manipulation - direct sonification. In *International Conference on Auditory Display ICAD-98*, Glasgow, Scotland, 1998. Springer-Verlag.

[14] D. J. Hand, H. Manila, and P. Smyth. *Principles of Data Mining: Adaptive Computation and Machine Learning*. MIT Press, Cambridge, 2001.

[15] S. Heise, M. Hlatky, and J. Loviscach. Aurally and visually enhanced audio search with soundtorch. In *Proceedings of the 27th international conference extended abstracts on Human factors in computing systems*, CHI '09, pages 3241–3246, New York, NY, USA, 2009. ACM.

[16] T. Hermann, J. M. Drees, and H. Ritter. Broadcasting auditory weather reports - a pilot project. In Eoin Brazil and Barbara Shinn-Cunningham, editors, *Proceedings of the International Conference on Auditory Display (ICAD 2003)*, pages 208–211, Boston, MA, USA, 07 2003. International Community for Auditory Display (ICAD), Boston University Publications Production Department.

[17] T. Hermann and A. Hunt. An introduction to interactive sonification. *IEEE Multimedia*, 12(2):20–24, 2005.

[18] T. Hermann, P. Meinicke, H. Bekel, H. Ritter, H. M. Müller, and S. Weiss. Sonifications for EEG Data Analysis. In *International Conference on Auditory Display ICAD-02*, pages 1–5, 2002.

[19] E. Jovanov, D. Starcevic, K. Wegner, D. Karron, and V. Radivojevic. Acoustic rendering as support for sustained attention during biomedical procedures. In *International Conference on Auditory Display ICAD-98*, pages 1–4, 1998.

[20] P. Knees, M. Schedl, T. Pohle, and G. Widmer. Exploring music collections in virtual landscapes. *IEEE Multimedia*, pages 46–54, July - September 2007.

[21] G. Kramer. *Auditory Display: Sonification, Audification and Auditory interfaces*, chapter Some Organizing Principles For Representing Data With Sound, pages 185–221. Santa Fe Institute Studies in the Sciences of Complexity. Addison-Wesley, 1994.

[22] T. Lokki and M. Gröhn. Navigation with auditory cues in a virtual environment. *IEEE Multimedia*, 12(2):80–86, 2005.

[23] K. Lynch. *The Image of the City*. MIT Press, 1960.

[24] C. Müller-Tomefelde. Interaction sound feedback in a haptic virtual environment to improve motor skill acquisition. In *International Conference on Auditory Display (ICAD-04)*, pages 1–4, 2004.

[25] E. Niedermeyer and A. Lopes da Silva. *Electroencephalography*. Lippincott Williams & Wilkins, 4 edition, 1999.

[26] S. Pauletto and A. Hunt. The Sonification of EMG data. In *International Conference on Auditory Display ICAD-06*, pages 152–157, 2006.

[27] J. Peponis, C. Zimring, and Y. K. Choi. Finding the building in wayfinding. *Environment and Behaviour*, 22(5):555–590, 1990.

[28] L. S. Petrucci, E. Harth, P. Roth, A. Assimacopoulos, and T. Pun. WebSound: A generic web sonification tool, and its application to an auditory web browser for blind and visually impaired users. In *International Conference on Auditory Display ICAD-00*, 2000.

[29] C. Poutine. Failte go d'ti dublin. http://virtualsuburbia.blogspot.com/2006/06/failte-go-dti-dublin.html, last checked: October 26 2011, June 2006.

[30] R. M. Satava, A. Cuschieri, and J. Hamdorf. Metrics for objective assessment. *Surgical Endoscopy*, (17):220–226, 2003.

[31] A. W. Seigel and S. H. White. *Advances in Child Development and Behaviour*, chapter The development of spatial representations of large scale environments, pages 9–55. Academic Press, 1975.

[32] B. Shneiderman. *Designing the User Interface: Strategies for Effective Human-Computer Interaction*. Addison-Wesley, Reading, MA, USA, 2 edition, 1992.

[33] P. Susini, S. Vieillard, E. Deruty, B. Smith, and C. Marin. Sound navigation: Sonified hyperlinks. In *International Conference on Auditory Display ICAD-02*, pages 1–4, 2002.

[34] E. Tufte. *Envisioning Information*. Graphics Press, 1990.

[35] G. Tzanetakis. *Manipulation, Analysis And Retrieval Systems For Audio Signals*. PhD thesis, Princeton University, 2002.

[36] A. Walker, S. A. Brewster, D. McGookin, and A. Ng. Diary in the sky: A spatial audio display for a mobile calendar. In *BCS IHM-HCI 2001*, pages 531–540, Lille, France, 2001. Springer.

[37] B. N. Walker and J. Lindsay. Navigation performance with a virtual auditory display: Effects of beacon sound, capture radius, and practice. *Human Factors*, 48(2):265–278, 2006.

[38] M. Wynblatt, D. Benson, and A. Hsu. Browsing the world wide web in a non-visual environment. In *International Conference on Auditory Display ICAD-97*, 1997.

[39] H. Zhao, C. Plaisant, B. Shneiderman, and R. Duraiswami. Sonification of geo-referenced data for auditory information seeking: Design principle and pilot study. In *International Conference on Auditory Display (ICAD-04)*, Sydney, Australia, 2004.

Chapter 21

Aiding Movement with Sonification in "Exercise, Play and Sport"

Edited by Oliver Höner

21.1 Multidisciplinary Applications of Sonification in the Field of "Exercise, Play and Sport"

Oliver Höner

This chapter deals with several applications of sonification in the field of sport and movement sciences. A selection of authors from various sciences (e.g., Electronics, Music, Technology and Sport Science) illustrate a wide scope of multidisciplinary applications in aiding human movement using interactive sound. These applications can be allocated to the comprehensive field of "Exercise, Play and Sport" i.e., physical activity in its widest meaning. For this field, we distinguish *health*-promoting exercises in rehabilitation programs (e.g., movements in physiotherapy, therapeutic games), *fun*-related movements in entertaining games (e.g., playing computer or sports games) and *performance*-related movements in competitive sport (e.g., diagnostics and training).

Figure 21.1 illustrates the framework for this chapter and presents an arrangement of the sections by assigning the applications to the three areas "Exercise, Play and Sport". Section 21.2 describes the use of sounds in measuring standardized movements in therapy and rehabilitation activities. Section 21.3 describes non-standardized movements based on open-skilled, situation-dependent and "fun-related" game activities within virtual spaces. Sections 21.4 and 21.5 take this further by linking sport with 'play' and 'exercise' respectively. Section 21.5 also examines the transfer of the enhancement processes in competitive sport to the field of motor rehabilitation.

In more detail, section 21.2 (by A. Hunt and S. Pauletto) demonstrates the multidisciplinary applications of sonification with the "*Use of sound for physiotherapy analysis and feedback*".

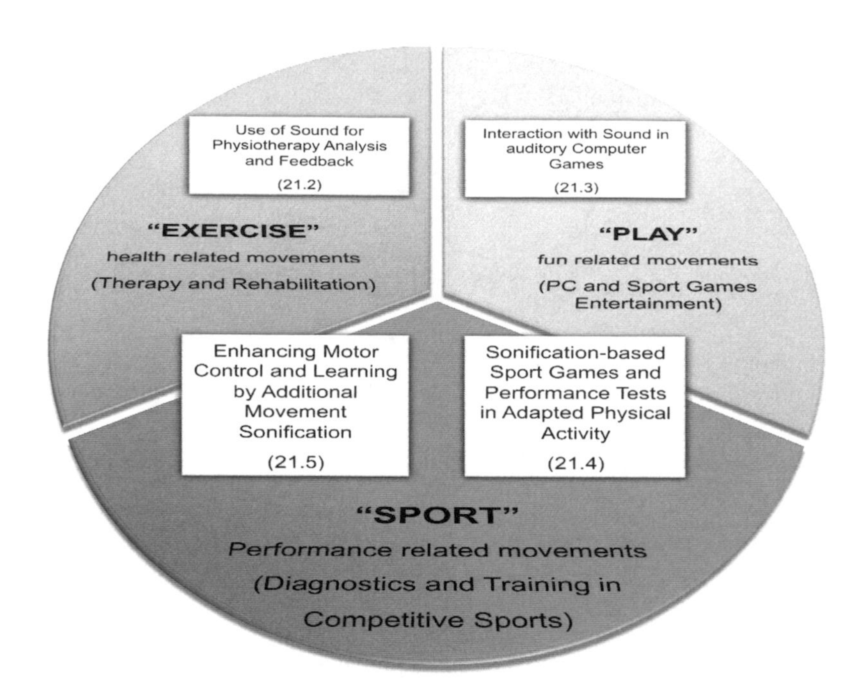

Figure 21.1: An arrangement of the multidisciplinary sections of chapter 21 "Aiding Movement".

The authors describe methods of sonification helping therapists to analyze the complex signals which originate from multiple EMG sensors on a patient's body. Converting electrical impulses from muscles into sound enables therapists to listen to the muscles contracting when physical activity is carried out. An adaptation of the standard hospital monitoring system allows this sound to be generated in real time. This has several advantages over visual displays, the most important of which is that medical staff and patients have an eyes-free display, which gives real-time feedback on the quality of muscle activity and is thus a new analytical component in therapy.

Section 21.3 *"Interaction with sound in auditory computer games"* (by N. Röber) extends the applications of the interactive use of sound to fun-related actions in playing computer games. Computer games are constantly increasing in popularity, but *audio-only* computer games still occupy only a very small niche market, mostly for people with visual impairment. This section reflects the reasons for this, presents an introduction into audio-only computer games and provides an overview on the current state of the art. Using the genres of narrative adventures and interactive augmented Audiogames, this section discusses the techniques which are necessary to interact with - and to explore - virtual auditory environments. It also provides a detailed look at the important methods for scene sonification and 3D sound rendering, and ends with a glimpse into future developments.

Just as in traditional (visual) computer games, visual information is the leading afferent information for players to regulate their actions in sports games. As a consequence, access to sports is particularly difficult for people with visual impairment. The field of *Adapted Physical Activity* explores new opportunities and enabling techniques to facilitate their participation. Section 21.4 introduces one approach to this task in "*Sonification-based sport games and performance tests in Adapted Physical Activity*" (by O. Höner and T. Hermann). This section relates to the previous one as it also reports on the use of a gesture-controlled audio system for playing non-visual, audio-only games. By contrast, the origins of these games are not computer games, but sports games such as badminton. Thus, the player is engaged in physically exhausting sporting activities and additionally gains audio-motor movement experience. Therefore, an Interactive Sonification System for Acoustic Motion Control ("AcouMotion") provides a link between body movements and auditory feedback through interactive sonification. As well as covering the development of new kinds of (adapted) sports, this section also covers new perspectives on performance diagnostics applied to traditional sports played by people with visual impairment.

This leads on to the field of performance-related movements. Section 21.5 focuses on the performance enhancement in competitive sports in "*Enhancing Motor Control and Learning by Additional Movement Sonification*" (by A. O. Effenberg). It explains the processes of motor control and learning which are based on perceptual functions and emergent motor representations. In contrast to Section 21.2 (where sonification is used to enhance the *knowledge* about the functional state of a muscle), Effenberg uses movement sonification to induce a *direct effect* on motor behavior. The author presents a theoretical framework and empirical evidence for the assertion that the auditory system can be involved in the processes of motor control and learning. This is done by providing additional movement acoustics ('movement sonification') resulting in more accurate motor perception and a better motor performance. Finally, the functionality of movement sonification for closed skills in competitive sports is discussed and moreover perspectives for motor rehabilitation are pointed out.

All sections of this chapter have a similar structure. Each section consists of at least five fundamental parts, i.e., the description of (i) general and core assumptions of the research approach, (ii) the main user / target groups (e.g., specific needs of these people, particular advantages of using sound for these groups), (iii) technical systems used for aiding movements (e.g., concerning gesture or movement analysis, type of sonified data, auditory display), (iv) empirical (case) studies and (v) future directions (e.g., concerning further user groups, further application in other areas of Figure 21.1, technical improvements, perspectives and expansions of the core applications).

21.2 Use of Sound for Physiotherapy Analysis and Feedback

Andy Hunt and Sandra Pauletto[1]

21.2.1 Introduction to EMG

EMG (electromyography) sensors detect the electrical activity associated with muscle movement. Electrodes on the skin's surface pick up electrical signals from the muscles below, and the signals are usually digitized for storage and analysis. Physiotherapists typically use various computer programs to capture the data, perform some basic statistics on it and display it in a graphical form. When working in a real-time situation the therapist is often distracted from contact with the patient because of having to operate the (visual) menu system and studying the (visual) results. EMG signals are believed to be full of information about the muscle activity and it is hypothesized that this visual and statistical analysis does not exploit the full information contained in the data.

21.2.2 Traditional analysis of the raw signal

The work described here is concerned with portraying as much of the raw signal as possible to the therapists, because it contains many clues about the health, motion and condition of the muscles.

> *"[The analyst] should monitor the raw signal, even though other signal processing may be used, so that artifacts can be detected and controlled as necessary"* [2].

Traditionally this monitoring work is carried out by visual inspection of a captured signal. The following section describes one example of how to use sound as a good alternative for monitoring the raw signal, and one that allows vital eye contact and focus with the patient to be maintained.

21.2.3 Designing sound to portray EMG signals

Initial experimentation was carried out using example data sets from patients at the Teesside Centre for Rehabilitation Sciences. A custom-designed *Interactive Sonification Toolkit* [3] was used to experiment with various methods of converting the EMG data into sound (sonification). This toolkit allowed researchers to take in multiple data sets, and try out a range of data-scaling and sonification techniques.

The design criteria for the sonification algorithm were:

1. It should portray an accurate *analogue* of the captured signal;
2. Sounds should be made in *real time*, in response to patient movement;

[1] The work described in this section is a collaboration between the University of York Electronics Dept., and Teesside Centre for Rehabilitation Sciences (a partnership between the University of Teesside's School of Health & Social Care and South Tees Hospitals NHS Trust). It was funded by EPSRC (Engineering & Physical Science Research Council), grant no. GR/s08886/0.

3. It should be *pleasant* to listen to (or at least not annoying);
4. Data should be audible when being analyzed at different speeds;
5. Signals from several EMG sensors can be listened to together.

The team's first experiments involved *audification* - the direct conversion of data samples into sound. However the EMG data sampling rate is rather slow compared to the data rate needed for sound, so when analyzing a signal slowly there was not fast enough change in the signal to make it audible (sound example **S21.1**). Also, when multiple sensors are used the resultant signal becomes very noisy.

Progress was made using MIDI notes to represent the values from more than one sensor. For example in the following sound (**S21.2**) two sensors are heard, panned left and right in the stereo field. Although this led to a useful form of comparison between two related sensors, this form of continually re-triggered sound (caused by the sensor value reaching a new MIDI note threshold) proved tiring for the clinicians to listen to and was overly quantized in pitch.

The final choice of sonification method involved amplitude modulation; each EMG sensor was mapped to the amplitude of a different sine wave oscillator. The frequencies of the different oscillators were set in a harmonic relationship with each other with the intention of making the sound pleasing (more instrumental than noise-like). This method also provides a tone whatever the speed of playback. It also allows the modulation of several sensors simultaneously, fusing their varying inputs into one complex, but easily understood, resultant sound (sound example **S21.3**).

21.2.4 Gathering and processing the clinical data

The EMG sensors are connected to the existing clinical *Biopac* [4] analogue-to-digital converter (which allows file storage and visual analysis), and also into a computer running the sound mapping software (written in PD [5]). Figure 21.2 shows this set-up, with a patient about to perform a leg extension (with resistance from the machine). This produces bursts of complex sound which can be heard by all in the room. A short video example (**S21.4**) shows the system in action and the resultant sound. The traditional visual analysis is also available for comparison.

21.2.5 Clinical testing of the sonification system

An experiment was conducted to verify the system's efficacy as an auditory display of the data. The sonification was found to be effective in displaying known characteristics of the data, comparing them to traditional analysis. Non-therapist listeners were able to gauge the condition of a client's muscles just from the sound.

A listening test was set up so that 21 subjects (average age 29, and all studying or teaching engineering with sound) could listen to 30 sonifications created from EMG data. Each sonification was then scored, on a scale from 1 to 5, for the following characteristics: Overall loudness; Speed of the sound's attack; Roughness; Presence of distinct pitches; Presence of structure in time. Loudness and attack speed are the variables that should vary with age and therefore they were clear candidates to test the validity of the sonification. The other

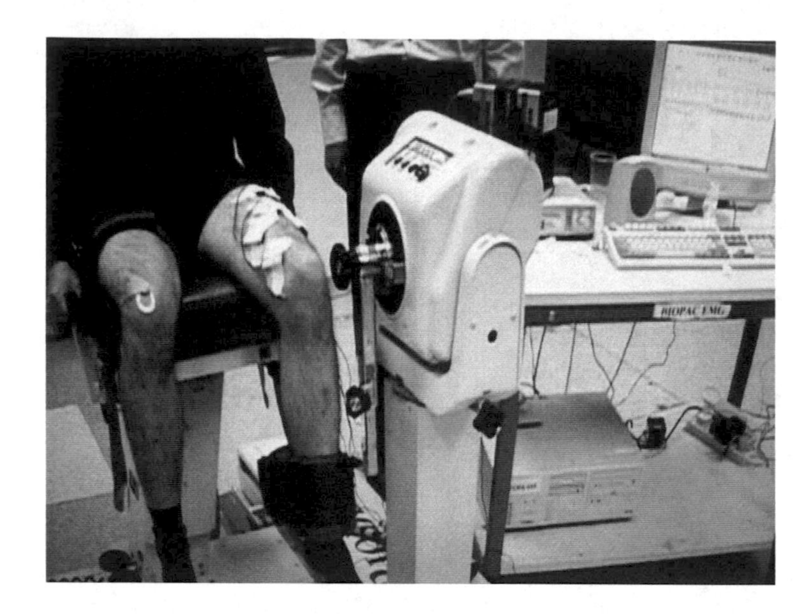

Figure 21.2: EMG Sensors on a client's leg with clinical equipment.

factors were included so that we could compare the effectiveness of the sonification with visual displays.

The data was gathered by Dr. John Dixon of Teesside University using the equipment in Figure 21.2, and from patients with a range of ages from 19 to 75. A testing interface was developed in Pure Data that was used to run the experiment and gather most of the experimental results automatically (see Figure 21.3).

Subjects were able to listen to each sonification as many times as they liked, and then scored the data according to the characteristics. Each subject received a different randomized order of sonifications so that any biases due to presentation order or layout were avoided.

21.2.6 Results

Though none of the experimental subjects knew anything about the ages of the participants they were listening to, the results showed a remarkable correlation between age and three of the scored parameters: attack speed, loudness and roughness (see Figure 21.4). For example loudness showed a significant negative rank correlation with age (non-parametric Spearman rank correlation factor $= -0.58$, significance test $p < 0.005$).

Thus a set of non-clinically trained listeners were able – by sound alone – to gain an insight into the age of participants and their muscle strength deterioration. Full details of the experiment are found in the proceedings of ICAD [6]. Further experiments have been carried out [38] which show that the same data is analyzed at least as well by sonification as by visualization as sonograms, and for some aspects (especially temporal changes) much better.

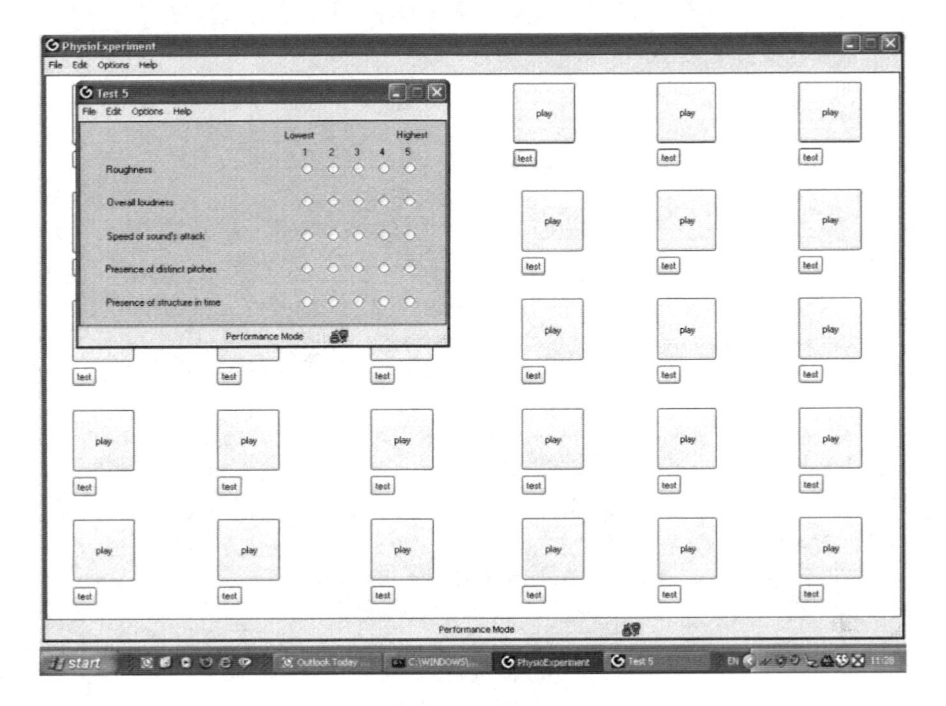

Figure 21.3: Pure Data testing interface.

21.2.7 Conclusions and further work

Muscle monitoring is a complex activity and currently involves therapists in many hours of visual data mining to interpret data for use in the clinical environment. The sonification of EMG data allows the health-care professional to observe the patient rather than the screen, using an auditory signal which may be better qualitatively understood than (and may provide additional information to) the more traditional visual displays. This is an innovative approach and has the potential to change clinical practice.

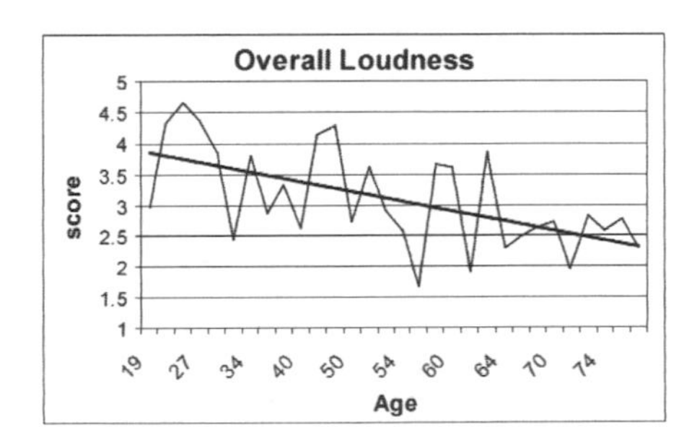

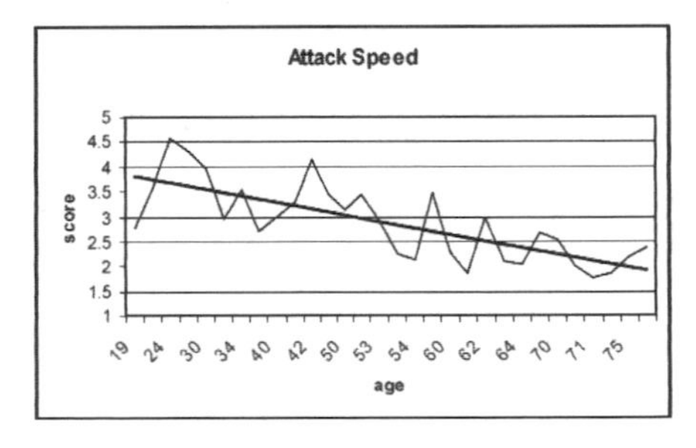

Figure 21.4: Correlation between age and (both) loudness and attack speed.

21.3 Interaction with Sound in auditory computer games

Niklas Röber

21.3.1 Sound in Computer Games

The first games to be played on a computer were designed in the early 1960s and 1970s. These games were very simple, having only primitive graphics and almost no sound. Since then, games have evolved tremendously and attract a huge range of followers today. Currently, games are one of the major industries in computer science and a huge driving force in research and computer development. Computer games have always been about fun, enjoyment and competition and are nowadays also employed in applied sciences in the area of health related computer games [7] (see also section 21.4 on sonification-based sports).

The first largely available computer games were played on custom consoles, then on the Commodore and Atari computers and later moved to the PC platform. Compared to its contemporaries (the Amiga and Atari systems), the PC of this time was very limited in its sound synthesis and playback capabilities. This changed in the late 1980s when the first add-on soundcards were introduced. Today's PC sound hardware is very well advanced, able to produce 3D sound and surround effects simulating virtual room acoustics, and in some cases even programmable through customized DSP[2] algorithms.

Sound is important for every game genre, and a bad acoustic environment can ruin an otherwise perfect game. 3D sound has proven to be advantageous, especially for very realistic games such as 3D FPS[3]. Here it assists the player to detect the opponent acoustically. The simulation of room acoustics plays an important role, as it intensifies the game's atmosphere and therefore the degree of immersion into the virtual game world.

Although sound hardware has not evolved as fast as graphics hardware, game audio has received a lot of attention in recent years, and the awareness of the capabilities of a good auditory design is present in both the developer's and the player's mind. Fast graphics hardware is employed for 3D sound rendering and a more accurate simulation of virtual room acoustics [8, 9].

21.3.2 Audiogames

Audiogames, also known as *audio-only* computer games, are played and perceived by auditory means alone. These games are often developed by, and for, visually impaired people. One of the first commercial audiogames developed was "Real Sound - Kaze no Regret" (1999), an audio adventure that was inspired by blind fans and available for Sega's Saturn and Dreamcast consoles. In the following years, several genres from the visual game domain have been adapted as audiogames, including adventures, action and racing games as well as simulations and role-playing games. The differences in game-play between a visual and an auditory implementation can be quite substantial. Audio is very well suited for presenting narrative content, but even action games that rely on precise listening and fast user reactions

[2]DSP - Digital Signal Processor

[3]FPS - First Person Shooting games

are available. The community of blind people is quite active in this area and many of the old text-adventures are still being played by people with visual impairment as they can be easily read out by speech synthesis software. An overview of the different genres and games can be found at the audiogames.net website [10].

A real advantage of audiogames is that they are able to provide greater stimulation to the player's imagination. This results in a higher immersion, similar to the experiences of radio listeners who often state that the "pictures *look* better on the radio". Other advantages include an accessible game-play and a simplified development cycle. Difficulties occur within the game itself in the estimation of distances and the mapping of sounds to specific events.

Although there is a potentially big market available, most audiogames are still rather simple and far less complex than their visual counterparts. But a current trend is moving towards more complex and challenging games as well as to the concept of augmented and real-world game play. The most important rule in designing audiogames is to immerse the player in a high quality virtual auditory world and to use techniques that support and enhance this sensation. Crucial here is the design of the user interface and its integration within the game. In some games, including visual ones, problems occur due to poorly designed interfaces and menus that break the illusion of being immersed in a virtual world.

Following is a discussion of two quite well designed audiogames:

1. *Terraformer*[4] is a so-called *hybrid* game. These are conventional audiovisual computer games that have been extended by certain sonification techniques to make them accessible for people with visual impairment. Hybrid games are quite common among audiogamers, as they are more attractive to a larger community, and sighted and blind people can play together. Terraformer is an action-adventure game and set on a foreign planet in a futuristic 3D world. The player's task is to fight against rebelling robots, find missing pieces of technology, and re-establish the terraforming process. Terraformer received a lot of attention at the time of its release due to some novel game sonification techniques. The acoustic orientation and navigation is supported by 3D sound and the user has a sonar-like technique for room exploration that provides a rough perception of distances, as well as identifies objects in front of the player. Other sonification tools include an auditory compass and GPS, as well as a voiced computer system that provides various feedback paths from the game.

2. *Seuss Crane: Detective for Hire*[5] is an audio adventure game in which the player is a detective who has to unveil a murder mystery. It is based on a radio play, in which the player chooses the locations to investigate, and after a while has to accuse someone for murder. The game has an interesting story and uses professional radio voices. Although the game does not rely on any visual information, the user interface is still in the form of a simple hypertext-like menu. It would have been nice to see this interface being represented using auditory means as well. Another drawback is that one has to follow a predefined sequence in order to get points and to solve the game.

The next section presents research on interactive auditory environments that extend and generalize the audiogames approach.

[4] `http://www.terraformers.nu/`
[5] `http://radio-play.com/`

21.3.3 Interactive auditory environments

Interactive auditory environments take audiogames one step further. They generalize the underlying ideas and combine the existing approach with common sonification and interaction techniques to form 3D auditory environments. Auditory environments can be thought of as being the acoustic analogue to a visual 3D game world. Applications exist not only in the areas of entertainment and edutainment, but also in the form of general auditory user interfaces and the development of tools to aid visually impaired people. It is imperative to have an intuitive and integrated design and the right balance between aesthetics and function.

The main components that characterize interactive auditory environments are:

- A 3D virtual scene/world described by a non-realistic auditory design.
- Intuitive sonification and interaction techniques to enable the user to explore, navigate, and interact with the environment.
- A narrative concept that focuses on an acoustic representation.

The following describes a research prototype that focuses on the implementation of such interactive auditory environments. The acoustic presentation that describes the scenery must have a non-realistic design, in which a realistic auditory representation is exaggerated and enhanced at certain points and also enriched with additional information. The auditory reality is not just augmented, but presented in a way that sounds' perception is more intuitive and clear. This can be done by exaggerating certain acoustic effects, such as the Doppler effect, or by making silent objects audible. The additional information can be conveyed through auditory textures, earcons, beacons and other sonification means. The quality of the sound rendering itself should thereby be as high as possible, especially the 3D sound spatialization and the simulation of environmental (room acoustic) effects, as they directly assist the player in orientation and navigation.

For experimenting with the various sonification and interaction techniques, a framework has been designed for interactive auditory environments, which also serves as a platform to prototype user interfaces and simple audiogames. Figure 21.5 shows an overview of the system. It is based on OpenScenegraph[6] to manage the various 3D scenes and uses OpenAL/EAX[7] for the sound rendering. The majority of sounds are spatialized using HRTFs[8] and the simulation of room acoustics is implemented using OpenALs EAX/EFX system.

Although the orientation and navigation within the 3D virtual auditory environments is challenging, it can be greatly improved by incorporating the user's (head-) orientation and movements. These motions are tracked using a Polhemus Fastrak® that is controlled by the VRPN library[9]. The modeling of the 3D environment takes place in 3DStudioMAX®, from which the data is exported and integrated into the system as an extended XML file [15].

The sonification and interaction techniques are closely related and depend on each other: Sonification is used to transfer information from the scene to the user, while interaction is required to input the user's changes into the virtual environment. Care has to be taken in

[6]OpenScenegraph - `http://www.opensg.org`
[7]OpenAL/EAX - `http://www.openal.org`
[8]HRTF - Head-Related Transfer Function
[9]VRPN - Virtual Reality Peripheral Network (`http://www.cs.unc.edu/Research/vrpn`)

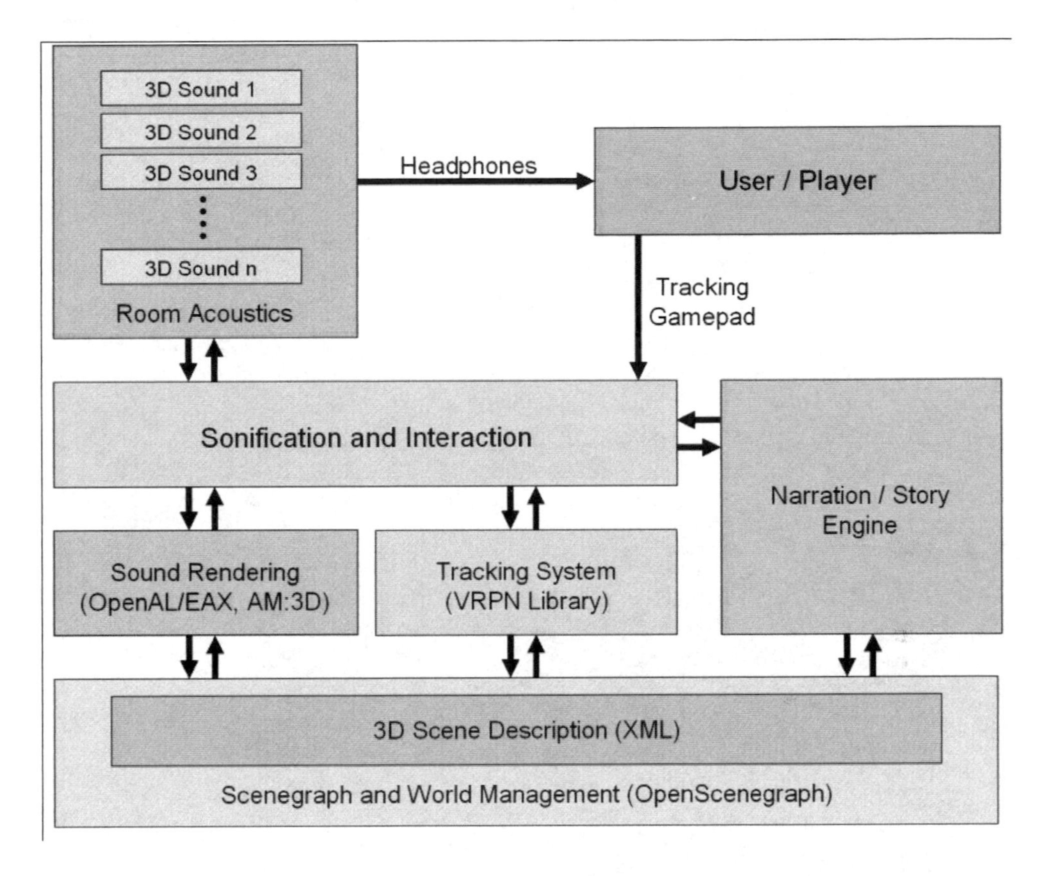

Figure 21.5: Overview of a System for interactive auditory environments.

the design of these methods, as they should integrate seamlessly. All techniques have to be implemented to perform in real time, otherwise the orientation and navigation would become extremely difficult. The tracking system allows the emulation of several listening and interaction behaviors, and therefore an easier and more intuitive orientation and navigation. It also permits an integration of basic gestures into the system, such as nodding or the drawing of symbols. A real advantage is the possibility of interacting within a spatial environment that allows the positioning of information and menus using a ring metaphor; sounds and interactable objects are thereby arranged in a ring 360° around the user. Additionally, a regular gamepad can also be used for user interaction.

The sonification techniques used can be divided into two different groups: the ones that are bound to interaction techniques, and the ones that solely sonify the scene without interaction. The first group consists of methods such as radar, sonar or an auditory cursor, in which the user probes and actively explores the environment and receives feedback through sonification. The second group describes the sonification of (non-interactive) scene objects and the auditory display of additional information using auditory textures, earcons, beacons and soundpipes [11, 12].

21.3.4 Designing interactive Audiogames

Example 1: Matrix Shot

One of the first experiments using this framework included spatial sonification and interaction techniques that led to the implementation of four simple action and adventure audiogames [11]. Figure 21.6 shows the principle and an action shot of the *Matrix* game, in which the player has to detect and avoid virtual acoustic bullets (see also the video example **S21.5**). We conducted several user tests to compare our implementations with other available audiogames, to investigate playability, usability as well as the quality of the sonification and interaction. Almost none of the participants had any prior experiences with audiogames, but everyone liked the idea and the simple concept of play. Initial difficulties occurred in the estimation of distances and the position of the virtual bullets. Very helpful was the later integration of 3D head-tracking, as it allowed the player an easier determination of the bullet's position simply by rotating the head.

Figure 21.6: *Matrix* Audiogame: a) Principle b) User Interaction.

Example 2: Interactive Audiobooks

A second project was called *Interactive Audiobooks* and aimed to unify the interactivity of computer games with the narration of books and radio plays [13, 14]. The first attempt consisted of an auditory adventure game to research the possibilities of storytelling within an auditory presentation. The story and the game were simple, linear and non-adaptive, leaving the user every freedom to explore the 3D environment. This caused several problems, as one could easily get lost in the virtual environment. The spatial representation with free user movement was replaced by a story engine that only allowed a certain movement depending on the development of the underlying storyline. The implementation is based on the previously introduced 3D audio framework, but also employs a storytelling engine that allows a non-linear game play with a varying degree of interactivity. Now it is possible to either play parts of it as an audiogame, or just listen to the complete story as a radio play. The presentation of the story has many similarities to common adventure games, but some differences exist, especially with the interaction and the design of the user interface (see video example **S21.6**)].

Example 3: Augmented Audiogames

For an implementation of so-called *Augmented Audiogames*, the system was made portable and extended by techniques to allow mobile head-tracking and user positioning [15]. Augmented auditory reality combines a real-world environment with an artificial auditory representation of this environment. The interaction and sonification is similar to other auditory displays, with the extension that the user can now freely walk around within the *virtual* scene. This narration in the real world largely increases the level of immersion, as more senses are addressed. The calculation of the player's position is important, but as the scene is described acoustically, an accuracy of about 1m has proven to be sufficient. For the positioning, GPS is used when outside, and several WiFi hotspots within buildings to track the position of the user. The use of WiFi emitters for position tracking is not easy, as the signal strengths decay differently, depending on the room size and objects therein. To increase the positioning accuracy a pre-sampled radio map with carefully selected WiFi emitter locations was used. Additionally, a digital compass was employed as simple head-tracking device to determine the player's orientation. A third challenge was the combination of the real sound environment with the artificial game world and the latency effects introduced by the tracking system. The application explored several game related possibilities along their potential for augmented audio edutainment, such as an augmented audio version of our audiogame *"The hidden Secret"* or an acoustic guiding and navigation system for the University's campus [15] (see example videos **S21.7**, **S21.8** and **S21.9**).

21.3.5 Rethinking Audiogames

Although the programming of simple audiogames is relatively easy, several guidelines should be observed to make the interface more intuitive and the audiogame more enjoyable. The most important goal is to immerse the player in a virtual auditory world and to use sonification and interaction methods to support and enhance this perception. The display must not be cluttered with too much information and should be designed in a way to balance function with auditory design. The quality of the sound and music used is of the utmost importance, as a poor sound design will otherwise ruin the game.

Because of the large design space for sonification and interaction techniques, a careful selection that concentrates on a clear presentation and an intuitive interaction will deliver a better performance. One of the most interesting genres for audiogames is adventure games, as they strongly focus on narration and storytelling. A rethinking of audiogames and their design will move them to the next level. Audiogames are not just acoustic adaptations of visual computer games; instead they present a new genre with different advantages and new possibilities.

21.4 Sonification-based Sport games and Performance Tests in Adapted Physical Activity

Oliver Höner and Thomas Hermann[10]

21.4.1 Core assumptions and objectives of the research approach

According to the framework presented in Fig. 21.1, this section deals with the sonification-based aiding of movements which are fun-related (movement as a part of *play*) and leads to the aiding of performance-related movements (in traditional and competitive *sports*). The core objective is to aim at joining principles and methods of the research program on interactive sonification (see chapter 11) with special requirements of the research field "Adapted Physical Activity" (APA). APA is a professional branch of kinesiology and physical education as well as sport and human movement sciences[11]. It is directed toward persons with physical disabilities or special needs who require adaptation for participation in the context of physical activity [1].

The use of sonification allows the investigation of two long-term topics relevant for APA:

1. Can motivating sport games be developed using non-visual, audio-only information for action control (which are specially suited for players with visual impairment)?
2. Can such games be used for testing and training abilities such as auditory-perception-based orientation in space?

The following section briefly discusses sport games and their suitability for people with visual impairment (section 21.4.2) then introduces a technical system called "AcouMotion" which is used to investigate the two topics mentioned above (section 21.4.3). Next are described some initial applications of the system: a new kind of audiomotor sport game adapted for people with visual impairment and an audiomotor performance test for paralympic goalball players (section 21.4.4). Finally, the future directions of the research are outlined (section 21.4.5).

21.4.2 Sport games for people with visual impairments

One of the main research objectives is to provide new opportunities for movement intensive games for people with visual impairments. This is a difficult task, as the leading (afferent) information in sports is obviously *visual*. This is especially true in ball games where players generally depend on their visual perception system to perceive information on the location and movement velocities of several objects, such as the ball, team mates or opponents. As a consequence, cognitive research on action control processes (such as anticipation and decision making) in sports is generally focused on visual perception. This is also true for the development of diagnostics for sensory-motor performance factors which usually investigate

[10] This project was conducted as an interdisciplinary research project between computer and sport scientists. It was funded by the Federal Institute of Sport Science ("Sonifikationsbasierter Leistungstest", VF 07 04 SP 00 69 05).

[11] See the definition by the International Federation of Adapted Physical Activity (IFAPA) on `http://www.ifapa.biz/?q=node/7`.

the visuomotoric competence of players, e.g., in using film-based stimuli for anticipation tests in games such as soccer, tennis, badminton and so forth [16].

We can draw at least *two conclusions* from the dominance of visual information for action control in sport games against the background of the APA perspective. One of the major tasks in the field of APA is to push the boundary of ordinary sport games in search of new opportunities or enabling techniques to facilitate the participation of people with visual impairment. Secondly, sensory-motor performance tests for blind-specific games cannot be adapted from ordinary sensory-motor performance tests which are in most cases visual-based (e.g., anticipation tests for goalkeepers). Therefore, there is a need to search for new ideas and new types of performance diagnostics in order to develop adequate, and in most cases blind-specific, performance tests. For both directions just outlined, the search begins with interactive sonification, i.e., the acoustic presentation of information in the presence of rapid feedback to users [17].

Sonification-based PC games

Interactive sonification presents information through non-speech sound and focuses on the development of systems integrating human actions into a tightly closed human-computer interaction loop. This interaction is enabled by providing immediate feedback on the user's actions with sound. Suitable interfaces for interacting with sonification systems can be conventional computer-based physical controllers, e.g., a computer mouse, but for sports applications these can also be tangible objects such as rackets.

To set the context there are promising examples in the field of *game entertainment*. One impressive example is described by Röber (see section 21.3). Further examples are given by Targett and Fernström [18], who present audio PC Games such as Mastermind or Tic Tac Toe. Using interactive sonification these games of strategy and dexterity can be played at a PC without using the visual display, but interacting solely with the auditory display. Further examples of audio PC games can be found with an internet search on "games for the blind" or "audiogames"[12]. These audiogames represent many of the genres found in traditional computer gaming (e.g., adventure games, but also sports such as SuperTennis). In all these examples the player receives no visual information and interacts with the computer by perceiving only auditory feedback following mouse or keyboard actions.

However, from the APA perspective there is one important disadvantage of the games mentioned above, i.e., their *lack of movement experience*. In contrast, "real" sports provide players with extensive movement experience and are expected to promote psychomotor development particularly for people with visual impairment [19].

Existing non-visual sport games

Several attempts to create new games for people with visual impairments are to be found in the field of (adapted) *physical education*, where new kinds of games result from modifications to traditional sports [20]. Further non-visual games are found in the competitive *Paralympics*. The game with the greatest tradition for people with visual impairment is the blind-specific

[12]See `http://www.gamesfortheblind.com` resp. `http://www.audiogames.net`

game *goalball*, played at the Paralympics ever since Toronto 1976. Goalball is played within the rectangular court of a gymnasium (9 × 18 m) by two opposing teams of three players. The aim of the game is to roll the sounding bell ball across the opponents' 9 meter wide goal line while the other team attempts to prevent this from happening. Since Athens 2004, there has been another non-visual sport at the Paralympics. The modified soccer game "Football 5-aside" is played with a sighted goalkeeper and with a guide behind the opponents' goal to direct the four non-seeing field players when they shoot. In both games players wear eyeshades so that it is impossible to perceive visual information during the games.

These and other blind-specific games are often based on certain *methodological principles.* These include the use of a rolling ball to generate a continuous sound and the use of tactile markers supporting players' spatial orientation in the playing area [21]. Another methodological principle is that these games use sound as the leading information. In particular for defensive movements the auditory information on the ball position and direction can be seen as the leading afferent information for motor control. As a consequence, *audiomotor abilities* are very relevant performance factors in competitive games of APA such as goalball.

A field inquiry was conducted at the Goalball European Championships in Belgium 2005 to investigate this proposition. Nearly all of the 22 questioned coaches of the international teams rated audiomotor abilities as "very important" for the performance of goalball. This inquiry also discovered that goalball-specific performance diagnostics to test audiomotor abilities do not exist [22]. Whereas goalball coaches can adapt performance diagnostics from ordinary competitive sports to test the players' physical condition (i.e., strength, movement velocity, general endurance, etc.), blind-specific tests for sensorimotor abilities have to be developed.

Conclusion and challenges

The examples mentioned above show impressively the adapted perceptual skills of players using non-visual information and prove the possibility of playing computer, sports and even ball games without any visual information. Is it possible to go beyond the existing audiogames by providing more intensive movement experiences, and beyond the existing games by using the sound in a more systematic way? The key question is whether it is possible to combine sport games, as practiced in APA, and sonification-based audio PC games to create motivating audiomotor games and blind-specific audiomotor performance tests. If so, it would be possible to initiate motivating sport games under educational and pedagogical perspectives as well as to conduct audiomotor performance tests for competitive sports. The next section describes the development of an Interactive Sonification System for Acoustic Motion Control ("AcouMotion").

21.4.3 The Interactive Sonification System for Acoustic Motion Control ("AcouMotion")

AcouMotion provides a link between body movements and auditory feedback through interactive sonification. The core idea behind AcouMotion is to employ sonification in order to create a new channel of proprioception aiding body movements in a virtual space whose properties and objects can be designed to support a wide range of different applications. The user's physical interactions with objects in the real world are mapped to manipulations of corresponding objects (e.g., a virtual racket) in a virtual model world. Reactions in the model world are displayed using sonification as the only feedback modality. From a technical perspective, AcouMotion connects *three system components* to implement this idea [23]: (*i*) a tangible sensor device providing movement-related information, (*ii*) a computer simulation model formalizing the coupling between body movements (reflected in the sensory data provided by the tangible device) and the object dynamics in the virtual space, and (*iii*) a sonification engine for the perceptual rendering of the joint dynamics of body and modeled object states.

Firstly, the *sensor device*in AcouMotion is provided by the Lukotronic motion analysis system[13]. The invention of AcouMotion pre-dated later controller developments such as the Nintendo Wii and Microsoft Kinect. Importantly, AcouMotion needs a proper localization of the sensor device in space, which many sensor-only devices rarely deliver. The Lukotronic system consists of a measuring beam and an active infrared marker set. Four infrared markers are fixed to a small handheld tangible device such as a table tennis racket or a self-made hemispherical wooden device (see Fig. 21.7). These markers can be used to assess the position and orientation of the racket in convenient accuracy (1-2mm) and frame rates (using here 100 Hz, which is sufficiently high to create the illusion of latency-free control in real-time interactions). Velocities and accelerations can be computed at high accuracy from successive frames. Secondly, a dynamic *computer simulation model* serves as the basis for

Figure 21.7: The hemispherical wooden device with fixed IR-Markers, used as a tangible device in the sports-related applications of AcouMotion.

representing processes and interactions in AcouMotion. The model represents the internal state of the AcouMotion system, and evolves according to its own "physical laws". For this, physical objects are modeled; for instance the tangible device (e.g., a racket), a ball and the walls limiting the virtual space. Physical parameters are also modeled, such as gravitation, damping of the ball or the general speed of the game. While real-world settings have to

[13] `http://www.lukotronic.com`

operate within existing physical laws, the computer simulation enables the control of any circumstance, for instance the viscosity of the air. This might cause a retardation of the ball due to increasing aerodynamic resistance, etc. The simulation needs to check at every point in time whether there are interactions with objects (e.g., ball and virtual racket), and respond accordingly with an update of the situation (e.g., an elastic impact).

Such event-based information is highly relevant for the auditory display created by the third component of AcouMotion, the *sonification engine*. The sonification serves as an interface between the computer simulation and the sensor device. It presents all information about the position of relevant virtual objects to the user via sound. Interactions such as the hit of a virtual ball update the ball's motion state in the simulation environment and thus the sonification. The user receives instantaneous auditory information to control and regulate his action. This real-time control and auditory feedback creates a closed interaction loop engaging the player in the game activities.

The basic elements of the auditory display are:

1. continuous sound streams which convey information by the change of acoustic attribute (an example is a pulsed sound whose pulse rate represents distance to the player),
2. discrete sound events, which are used to communicate discrete events (e.g., physical contact interactions in the model), and
3. ambient elements such as sound effects that influence the overall display.

The AcouMotion system connects the three components via OSC[14] (Open Sound Control) interfaces, allowing an easy exchange of sensors (e.g., webcam based sensor devices instead of the Lukotronic system), and distribution on different computers. Thus, AcouMotion integrates interactive sonification, movement experience and virtual game simulations and provides a technical basis offering new kinds of auditory sports that can be played with real motor activity by using non-visual sonification-based information alone. The following section presents the first applications of AcouMotion in the field of APA and illustrates the development of the sport game "Blindminton". Following that, it introduces a sonification-based performance test for the paralympic sport game goalball.

21.4.4 Applications of AcouMotion in the field of APA

"One-Player-Blindminton"

AcouMotion enables the creation of sport games operating within customizable "physical laws" (see 21.4.3). This means that the complexity and difficulty of the task can be controlled in detail to create a challenging game, even for novices. This is an advantage for APA in particular, because one of the most important barriers to participation in APA is that real sport games are too hard to learn for novices with impairments. Actually, this is often also true for "ordinary" sports, which tend to provide optimum excitement if the opponents reach a similar performance level when playing against each other. For example, badminton matches between two players at different performance levels are often boring for the more skilled player and overstrain the weaker player. From a motivational perspective, the matching of a game's challenges and players' skills and abilities is missing. But according to the *concept*

[14] `http://www.cnmat.berkeley.edu/OpenSoundControl/OSC-spec.html`

of flow-experience [24], this a necessary condition for high intrinsic motivation and therefore the aim is to provide this matching through computer simulation. Thus, the game concepts are built upon three ideas:

1. Game-relevant information needs to be transferred to an auditory display for people with visual impairment.
2. Audio sport games provide intensive movement experiences, and thus go beyond existing audio PC games.
3. A computer simulation enables the control and adaption of difficulty for each player in order to create exciting sport games independent of the performance level of each player.

Building on the above analysis, the first game concept created was "one-player-Blindminton". It is named Blindminton to denote that it is similar to the racket game badminton, but adapted to blind-specific needs. The name 'Blindminton' is used as a general term for one- or two-player ball games. The simulation engine can easily be modified to render the trajectory of a shuttlecock (as in Badminton) or a ball[15]. Although currently sound is considered as the main carrier of game-relevant information, certainly other modalities accessible to visually impaired players, such as haptic cues, can be added by integrating vibration motors into the racket.

In the pilot application shown in the Figure 21.8 the second player is replaced by a wall so that currently only one player is involved (in contrast to badminton). The player is expected to hit the ball against a wall (using a virtual racket) without bouncing it on the floor. The goal is to keep the (virtual) ball in the game as long as possible. The score increases with every contact and is also dependent on the speed of the ball when hitting the wall. This introduces an element which motivates the player to increase the amount of effort expended in order to obtain better scores.

All components of AcouMotion are required for implementing this game. In particular the AcouMotion *sensor device* is able to deliver position and orientation of the racket. Orientation is crucial since the ball reflects from the racket and this is an essential control for conducting the game. The *simulation model* creates a 3D model space with a limited number of objects represented by their coordinates, velocities and orientations. For instance, the objects in "one-player-Blindminton" are the racket, the ball, and a set of planes and walls to model the game field. The ball flies through the 3D space influenced by gravitational force and aerodynamic resistance.

The *sonification* is designed with a multilayered auditory display for the game concept. It consists of one multidimensional sound stream for the relative position of the player's racket, discrete sound events and verbal markers for game control. In detail, a 3-parameter stereo sound stream represents the three dimensional distance in space between the positions of the ball and the racket (see Fig. 21.8, sound and video example **S21.10**). The horizontal displacement of the racket, i.e., the x-distance between racket and ball, is presented by spatial sound cues such as stereo-panning: the sound for the ball is presented on the left speaker if the player has to direct his racket more to the left side and vice versa. The vertical displacement is presented using three levels of pitch: a high (or low) pitch directs the player to move the racket upwards (or downwards). The middle pitch indicates that the vertical

[15] In this case the game may perhaps better be called blind squash.

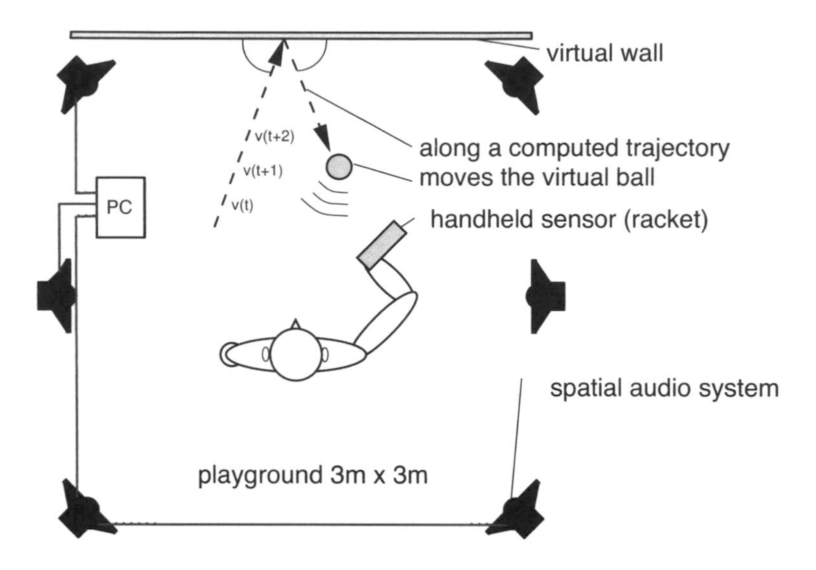

Figure 21.8: Illustration of the Blindminton game setting and the directions of racket's displacement (see also [23]).

position of the racket currently matches the ball's altitude. The third acoustic dimension is represented via a pulsing of the sound to represent the z-distance. A high the pulse rate indicates a large z-distance between racket and ball. Thus, the pulsing slows down as the ball approaches the player.

In games like Blindminton there are different types of *information-carrying variables*. In this system the 3D soundstream is continuous and thus accomplishes the "rolling ball" principle as known from goalball (see 21.4.2). It is extended by discrete sound events in the sonification design. To provide the player with a representation of the playing area, zone markers strike with sound as the player crosses them. Further discrete sound events represent certain ball collisions with the racket or the walls. Finally, the sonification design is completed by verbal markers to provide game control information about events such as the start of the game, ball out and so forth. The first application is illustrated in Figure 21.8 and can be seen on videos on the project homepage [25].

Goalball-specific performance test

AcouMotion's second application in APA aims at creating new kinds of performance tests, which are blind-specific and therefore very suitable supplements for the current inventory of diagnostic instruments in competitive sport games (such as goalball) for athletes with visual impairments. For a goalball-specific performance test AcouMotion is used as a virtual "*ball-throwing machine*". In the first pilot study with a female international goalball player, this machine throws (or rolls respectively) a virtual ball from a 4 meter distance to a 3 meter wide goal line. A test set consists of 15 trials with 15 different ball throws standardized by the computer simulation (see Fig. 21.9). The set was conducted seven times varying the ball velocities from 1 to 4 meters per second.

The sonification engine provides auditory information on the rolling of the ball by using an audio-setup with five speakers fixed on a segment of circle (see Fig. 21.9, and sound and video examples at **S21.11** and **S21.12**). Intensity panning and distance dependent level mapping was used to provide information about the ball's position. Further on, the player received verbal information on the start of the next trial as well as verbal feedback on the success of her movement behavior in defending her goal line (e.g. "0.3 meters left!" for 30 cm distance between the racket and the ball). Based on this auditory information it was the task of the goalball player to anticipate the ball as exactly as possible in order to defend her goal by moving her racket to the anticipated ball position on her goal line. The accuracy of her goal defending movements was measured by the sensor device held by the player.

In order to get a feedback from goalball experts, the international player and her coach were interviewed about the appropriateness of the performance test. Both gave positive feedback. The player stressed that - after a short adaptation phase which took about 15 minutes - she was able to perceive the start and final position as well as the course of the ball.

A quantitative analysis of the recorded data was in unison with the player's estimation. She was able to locate the position of the ball crossing the goal line with an absolute average error of 36 cm (median observed on all 105 throws). Additionally, the accuracy was clearly dependent on the ball velocity. The player reached her peak performance at ball velocities between 2 and 3 m/s (each median < 0.25 m). Balls with higher velocities and also lower – from a first glance surprisingly – led to reduced accuracy. On further inquiry, the player was

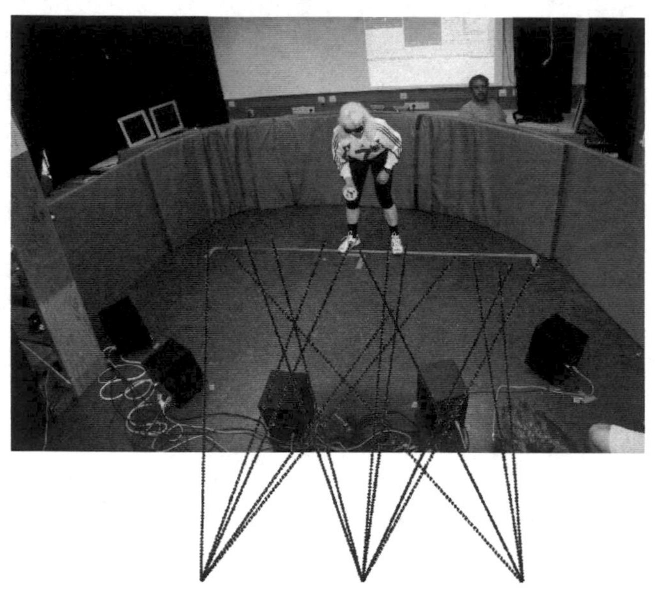

Figure 21.9: Laboratory setting for the goalball-specific performance test with the five-speaker audio-setup and the (not exactly to scale) illustration of the 15 standardized ball throws to the 3 meter wide goal line (here a test with Conny Dietz, the national flag bearer of the German paralympic team in Peking 2008 who played six paralympic goalball tournaments).

able to relate these results to her experience as she was unfamiliar with the slower balls from her goalball practice. She stated that she became insecure when waiting a long time for these balls. For a more demonstrative impression of this first goalball-specific performance test, videos are available showing parts of the test session [25].

21.4.5 Future directions

In addition to the first empirical case studies presented in this section, the two applications of AcouMotion in the field of APA shall receive further evaluation through empirical research. Current work includes the *evaluation of performance diagnostics* with the German national goalball teams (male and female) and collection of quantitative data on the performance level of each player as well as qualitative data from interviews with the goalball experts on the *validity of this test* [26, 27]. Concerning the audiomotor sport games it is an aim to develop experimental designs which scale up the speed of Blindminton and test the influence of matching a game's difficulty to a player's personal skills on the *flow-experience*. Further on, there are promising perspectives for the research on *audiomotor control and learning*. It is now possible to vary the parameters of the game in the computer simulation. This can be used to check whether standardized simplifications such as the enlargement of the racket size in games such as Blindminton lead to more effective learning processes. Thus, methodological learning principles like the simplification of an audiomotor task may be investigated in experimental settings designed with AcouMotion.

21.5 Enhancing Motor Control and Learning by Additional Movement Sonification

Alfred O. Effenberg[16]

21.5.1 Motor control and learning, vision and audition

When learning new closed skills[17] in sports or relearning basic skills in motor rehabilitation, the observation of the skill and the reproduction of it are key elements. These processes are dominated by visual perception; a well known theory in the field of motor learning is the theory of 'observational motor learning'. But vision is not the only sense providing information about movement patterns realized by trainers or therapists. Audition is another perceptual channel suitable for gathering information about movement patterns. You can hear the rhythm of a runner, even of a swimmer, and you can hear it more precisely than you can see it. Additional auditory information is achieved, because in some domains the ear is more precise than the eye, e.g., in temporal discrimination or in integrating sequenced sounds into a rhythm. Utilizing the ear in movement related perception can result in a broader spectrum and enhanced precision of perceived information supporting motor learning. Beside modality specific auditory benefits there are further perceptual effects achieved by multisensory integration [29, 30] and intersensory redundancy [31], if convergent visual information is available.

The main restriction for supporting auditory and multisensory information in motor control and learning seems to be the weak acoustical effectiveness of human movement, which is limited to short movement phases of 'getting in contact' - when the shoe hits the ground or the racket hits the ball. The movement itself is low-frequency and impossible to hear because it is below the human hearing range of about 20 - 20,000 Hz. Audification or Sonification of naturally silent movements can be helpful to motor learning and can result in a better performance, as shown in freestyle-performance (see, for example [32]). But the idea of creating additional movement sound is not completely new. There are different traditional forms of additional movement acoustics:

- Clapping the hands with the observed or aspired movement rhythm
- Simple forms of articulation or singing to emphasize duration and dynamic characteristics of selected movement phases
- Using simple musical instruments such as whistles or tambourines
- Using music as guiding rhythm and enhancing expression of movement (e.g., ice dancing, gymnastics)
- Using simple forms of body-instruments such as wrist- or ankle-bands with little bells etc.
- Using some technical sensors to detect discrete features of the movement and create an error-signal via electronic sound devices.

[16]The research was realized at the Institute of Sportscience and Sport at the University of Bonn, Prof. Dr. Heinz Mechling. It was funded by the 'German Research Foundation', grant no. ME 1526.

[17]Closed skills are skills "performed in a stable or predictable environment where the performer determines when to begin the action" [28]

21.5.2 Movement Sonification

Movement sonification, the sonification of human movement patterns is a new approach for creating 'authentic' acoustic movement sounds. This is achieved by transforming computed - kinematic as well as dynamic - movement parameters into sound. The 'MotionLab Sonify System' is a motion capture and sonification system, which is capable of real time sonification and computing force data by inverse dynamic algorithms [33].

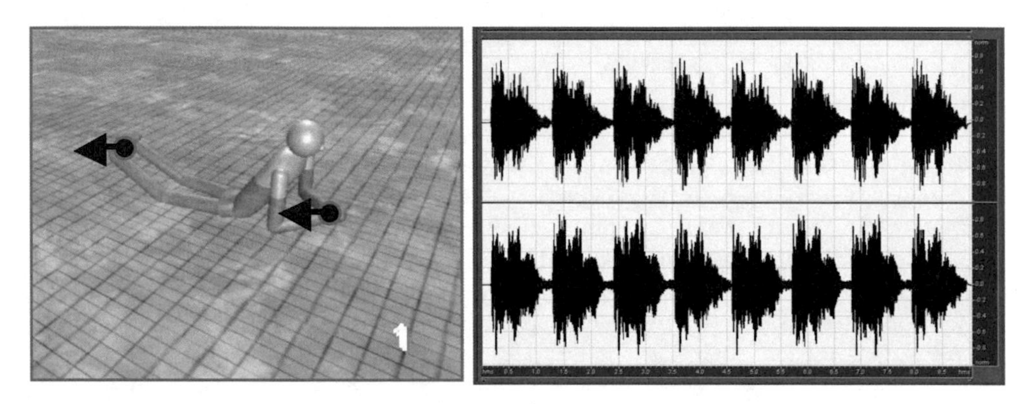

Figure 21.10: Breast stroke: Horizontal components of relative wrist and ankle motion - strokes only - are computed and used to modulate sound frequency and amplitude. 8 cycles in about 9 sec are shown, indicating a high stroke frequency.

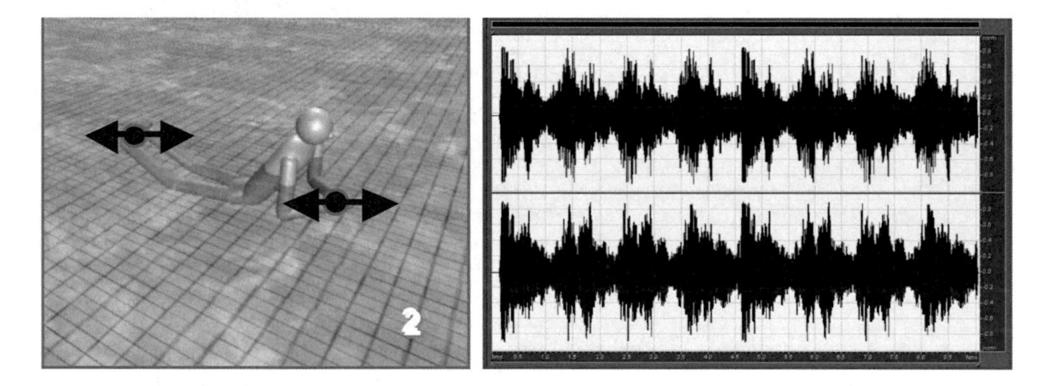

Figure 21.11: Horizontal components of relative wrist and ankle motion - complete cycles - modulating sound frequency and amplitude.

The system allows a direct acoustic transformation of movement parameters via MIDI[18]. MotionLab Sonify's architecture consists of a set of plug-ins for the MotionLab framework. The internal representation uses streams of motion data, which can be visualized as a skeletal

[18] MIDI ("Musical Instrument Digital Interface") is a serial control protocol customized for musical note event and control information.

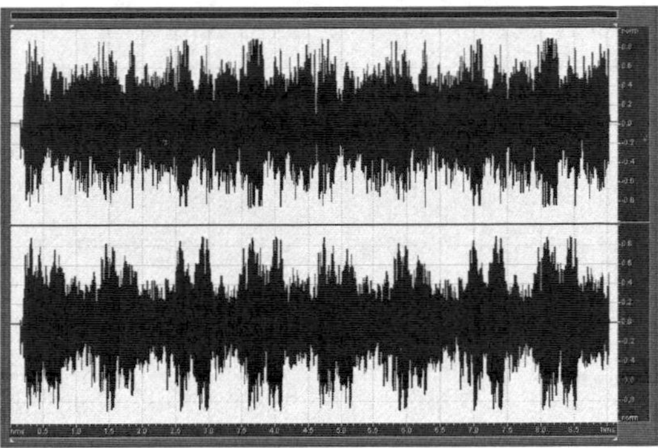

Figure 21.12: Horizontal components of relative wrist and ankle motion and vertical component of relative neck point motion.

representation. A parser for the AMC/ASF[19] file-format is implemented, which is available widely and is supported by a number of motion capture systems, such as the VICON[20] system. The MotionLab Sonify system can easily be adapted to handle motion data existing in different formats. Also the system can process streams of real-time motion capture data. Figures 21.10 - 21.12 represent a sonification of breast stroke based on kinematic parameters. Sonification was realized in three steps, and the example can be downloaded[21] or found in example videos **S21.13**, **S21.14** and **S21.15**.

Emerging sound patterns are typical for a sophisticated technique in breast stroke and contain concise temporal information about phase relations of arm cycle vs. leg cycle. Such kinds of movement sonification can be used to support the processes of motor learning.

21.5.3 Empirical data on effectiveness of movement sonification

Movement sonification of swimming breast stroke has not been used so far in an empirical study. But there is some information about the effectiveness of sonification on motor assessment and motor control related to another kind of movement, which is very common in sports; the counter-movement-jump (CMJ)[22]. Since the data has been reported elsewhere [34], here is just a brief summary. Investigations were realized in two distinct areas:

1. The precision of perception and judgment of sport movements was addressed. *Method*: Subjects sat in front of a video-/audio-projection watching videos of CMJs of different unknown heights. They were asked to judge the height-difference of two consecutive CMJs. Subjects were treated with visual, auditory and audiovisual stimuli comparatively. Audio consisted of a movement sonification based on the vertical component of the ground reaction force of a jump. The force parameter was measured with

[19] AMC/ASF: The ASF file holds the skeleton data while the AMC file holds the motion information.

[20] Vicon Peak is the new name for the combined businesses of Vicon Motion Systems and Peak Performance Inc..

[21] see `http://www.sportwiss.uni-hannover.de/alfred_effenberg.html` at "Mediadownload"

[22] A CMJ is a common athletic test of leg condition, and involves the subject standing straight, squatting, then leaping vertically off the ground and back into the standing position.

a KISTLER force plate 9287BA and acoustically transformed as a first order sonification: The force parameter modulates sound frequency and amplitude as shown with the breaststroke example above. *Results*: Judgment of jump-height differences was significantly more precise audio-visually compared to both unimodal conditions. Precision was enhanced by about 20 % without any learning experience.

2. Secondly the accuracy of perception and reproduction of sports movements was studied (motor control). *Method*: The method was nearly the same as reported above: Subjects observed a single CMJ as video- or audio/video-projection and were asked to reproduce or jump the same height as accurately as possible. Presented CMJ-heights ranged between 60% to 90% of subjects individual maximum jump height. *Results*: The precision of movement reproduction was significantly different between visual and audiovisual treatment. The absolute error under the audiovisual condition was reduced by about 20% compared to the visual condition.

Further research on the effectiveness of sonification on motor control had been realized by Chiari et al [35]. Also there are some references on the effectiveness of sonification on motor learning, in competitive sports [36] as well as in rehabilitation [37].

21.5.4 Discussion and Conclusion

This was the first investigation demonstrating that even a non-cyclic, non-rhythmic movement pattern (CMJ) can be perceived and reproduced more accurately with additional acoustic information created via movement sonification of dynamic parameters. The effectiveness of the movement sonification without convergent visual information was tested only for perception and judgment of movement patterns. Thereby no significant difference between accuracy achieved under auditory and visual treatment became evident, but absolute values had been more precise under the visual condition. Even though perceptual and motor control functions are also fundamental functions for motor learning, the effectiveness of additional movement sonification on motor learning has not been tested directly so far. That will be the next step of the empirical work described here. If effects on motor learning are detectable, there will be a broad application of movement sonification in the fields of competitive sports as well as in motor rehabilitation. Motor learning of closed skills in competitive sports could be shaped more effectively, supported by additional movement sonification, because audiovisual movement information is more precise and easier to keep in memory than visual or auditory information alone. Movement sonification could be used to enhance instruction by using audiovisual models as well as supporting feedback particularly when sonification is available in real time (see also section 21.4). Also mental training can be facilitated with simultaneous sonification. In motor rehabilitation, therapy could be started earlier for instance by addressing subliminal sensorimotor – audiovisuomotor – interconnections within the Central Nervous System (CNS) additionally by using movement sonification.

21.6 Concluding Remarks

Oliver Höner

The framework of this chapter integrated movements that are executed under distinguishable purposes and meanings. To help solve the problem of defining the common and distinctive elements of the terms "sport activities" and "movement activities", Figure 21.1 presents the applications to the field "Exercise, Play and Sport" in its widest meaning.

This figure could be divided into yet more sub-sections due to the distinction between more close and more open skills used to reach the specific action goal (e.g., promoting health for quality-of-life improvements, having fun and flow-experience in playing games, or enhancing performance due to yet unused resources and augmented information). A further dimension could be inserted with respect to a specific user group such as people with special needs (e.g., orthopedic patients, people with visual impairment).

Therefore, the applications of sonification described in this chapter could provide an important contribution for the cross-disciplinary research field of Adapted Physical Activity (APA). But in some sections, the target groups go beyond people with special needs. In particular, the contributions on the use of gestural audio systems for designing auditory computer games and for the enhancement of motor control and learning (sections 21.3 and 21.5) offer new perspectives for entertaining or performance enhancing activities that able-bodied people as well as top sport athletes may benefit from.

Owing to the multidisciplinary nature of these topics, this chapter has attempted to portray some examples of the manifold aspects of aiding physical activities. But such variety also implies difficulties in arranging and integrating these multidisciplinary applications into research programs such as gestural audio systems or interactive sonification, and this provides a challenge for the future. The common ground of the presented use of sonification systems in motor rehabilitation, game entertainment and competitive sports can be interpreted as a promising start for interdisciplinary approaches which provide useful systems for the future.

Bibliography

[1] G. Reid and H. Stanish, "Professional and disciplinary status of adapted physical activity," Adapted Physical Activity Quarterly, vol. 20, pp. 213–229, 2003.

[2] B. LeVeau & G. B. J. Andersson, "Output Forms: Data Analysis and Applications", in Selected Topics in Surface Electromyography for Use in the Occupational Setting: Expert Perspectives, Soderberg G. L., Ed.: U. S. Department of Health and Human Services, Public Health Service, Centers for Diseases Control, National Institute for Occupational Safety and Health, 1992, pp. 69–102.

[3] Pauletto, S. & Hunt, A., "A toolkit for interactive sonification" Proceedings of ICAD (the Int. Community of Auditory Display), Sydney, July. 2004. http://www.icad.org/websiteV2.0/Conferences/ICAD2004/papers/pauletto_hunt.pdf

[4] BIOPAC acquisition system: http://www.biopac.com, last seen: 2011-08-22

[5] PureData modular real-time programming environment: http://puredata.info, last seen: 2011-08-22

[6] Pauletto, S. & Hunt, A., The sonification of EMG data. *Proceedings of the International Conference on Auditory Display (ICAD)*, London, UK, 2006.

[7] D. R. Michael & S. L. Chen, "Serious Games. Games That Educate, Train, and Inform", Course Technology, 2005.

[8] N. Röber, U. Kaminski & M. Masuch, "Ray Acoustics using Computer Graphics Technology", *Proceedings of DAFx07*, Bordeaux, France, September, 2007.

[9] N. Röber, M. Spindler & M. Masuch, "Waveguide-based Room Acoustics through Graphics Hardware", *Proceedings of ICMC06*, New Orleans, USA, November, 2006.

[10] R. van Tol & S. Huiberts, "www.audiogames.net", 2006.

[11] N. Röber & M. Masuch, "Leaving the Screen: New Perspectives in Audio-only Gaming", Proceedings of 11th ICAD Conference, July 2005.

[12] N. Röber & M. Masuch, "Playing Audio-only Games: A compendium of interacting with virtual, auditory Worlds", *Proceedings of the 2nd DIGRA Gamesconference*, Vancouver, Canada, 2005.

[13] N. Röber, C. Huber, K. Hartmann, M. Feustel, & M. Masuch, "Interactive Audiobooks: Combining Narratives with Game Elements", *Proceedings of TIDSE 2006 Conference*, Darmstadt, Germany, 2006.

[14] C. Huber, N. Röber, K. Hartmann & M. Masuch, "Evolution of Interactive Audiobooks", *Proceedings of Audio Mostly Conference*, Ilmenau, Germany, September, 2007.

[15] N. Röber, E. C. Deutschmann & M. Masuch, "Authoring of 3D virtual auditory Environments", *Proceedings of Audio Mostly Conference*, Piteå, Sweden, 2006.

[16] A.M. Williams, C. Janelle & K. Davids, "Constraints on the search for visual information in sport," International Journal of Sport and Exercise Psychology, vol. 2 no 3, pp. 301–318, 2004.

[17] T. Hermann & A. Hunt, An introduction to interactive sonification, IEEE Multimedia, vol. 12 no. 2, 20–24, IEEE., 2005.

[18] S. Targett, M. Fernström. Audio Games: Fun for All? All for Fun?. E. Brazil, & B. Shinn-Cunningham (Eds.), Proc. Int. Conf. on Auditory Display (ICAD 2003), p. 216–219. Boston: Boston University Publications Production Department, 2003.

[19] H.-G. Scherer & H. Herwig, Wege zu Bewegung, Spiel und Sport für blinde und sehbehinderte Menschen, In V. Scheid (Hrsg.), Facetten des Sport behinderter Menschen. p. 116–154. Aachen: Meyer & Meyer, 2002.

[20] L.J. Liebermann, J.F. Cowart. Games for People With Sensory Impairments. Human Kinetics Europe, 1996.

[21] G. Friedrich, J. Schwier. Sportspiele für Blinde und Sehbehinderte. Motorik, 10, p. 101–110, 1987.

[22] O. Höner & T. Hermann. Entwicklung und Evaluation eines sonifikationsbasierten Gerätes zur Leistungsdiagnostik und Trainingssteuerung für den Sehgeschädigten-Leistungssport, [Final Report on the Research Project IIA1VF070404/05-06, Federal Institute of Sport Science], 2007.

[23] T. Hermann, O. Höner & H. Ritter. AcouMotion – An Interactive Sonification System for Acoustic Motion Contro. In S. Gibet, N. Courty & J.-F. Kamp (Eds.), Gesture in Human-Computer Interaction and Simulation [Lecture Notes in Artificial Intelligence, Vol. 3881] (pp. 312–323). Berlin, Heidelberg: Springer, 2005.

[24] M. Csikszentmihalyi, Beyond boredem and anxiety, Jossey-Bass, San Francisco, 1975.

[25] T. Hermann. AcouMotion: `http://www.sonification.de/AcouMotion.html`, last seen 04/2006.

[26] O. Höner. Development and Evaluation of a Goalball-specific Performance Test (4 pages). In J. Wittmannova (Ed.), Excellent research – Bridge between theory and practice. Proceedings of the European Conference on Adapted Physical Activity in Olomouc, Czech Republic, 2006 [online available as selected full paper under `http://www.ifapa.biz/imgs/uploads/PDF/EUFAPA2006/hoener1.pdf`].

[27] O. Höner & T. Hermann. Entwicklung und Evaluation eines sonifikationsbasierten Gerätes zur Leistungsdiagnostik und Trainingssteuerung für den Sehgeschädigten-Leistungssport. In Bundesinstitut für Sportwissenschaft (Hrsg.), BISp-Jahrbuch Forschungsförderung 2006/07 (S. 163–167). Bonn: Bundesinstitut für Sportwissenschaft, 2007, online available under `http://www.bisp.de/cln_090/nn_15924/SharedDocs/Downloads/Publikationen/Jahrbuch/Jb__200607__Artikel/Hoener__163,templateId=raw,property=publicationFile.pdf/Hoener_163.pdf`

[28] R. A. Magill, Motor Learning: Concepts and Applications. Boston, Massachusetts, p. 309, 1998.

[29] G. A. Calvert, C. Spence & B. E. Stein, "The handbook of multisensory processes". Cambridge, Massachusetts:

MIT Press, 2004.

[30] Scheef, L., Boecker, H., Daamen, M., Fehse, U., Landsberg, M. W., Granath, D. O., Mechling, H, & Effenberg, A. O. (2009). Multimodal motion processing in area V5/MT: Evidence from an artificial class of audio-visual events. Brain Research, 1252, 94–104.

[31] L. E. Bahrick & R. Lickliter, "Intersensory redundancy guides early perceptual and cognitive development", Advances in Child Development and Behavior, vol. 30, pp. 153–87, 2002.

[32] D. Chollet, N. Madani, & J. P. Nicallef, "Effects of two types of biomechanical bio-feedback on crawl performance", in Biomechanics and Medicine in Swimming. Swimming Science VI, D. Mac Laren, T. Reilly, and A. Lees, Eds. London: E & F Spon, pp. 57–62, 1992.

[33] A. O. Effenberg, J. Melzer, A. Weber & A. Zinke, "MotionLab Sonify: A Framework for the Sonification of Human Motion Data", The Ninth International Conference on Information Visualization (IV '05), London, UK, pp. 17–23, 2005.

[34] A. O. Effenberg, "Movement Sonification: Effects on Perception and Action", IEEE Multimedia, Special Issue on Interactive Sonification, vol. 12, pp. 53–59, 2005.

[35] Chiari, L., Dozza, M., Cappello, A., Horak, F. B., Macellari, V. & Giansanti, D. (2005). Audio-Biofeedback for Balance improvements: An Accelerometry-Based System. IEEE Eng Med Biol Mag, 52, No.12, 2108–2111.

[36] Schaffert, N., Mattes, K. & Effenberg, A. O. (2010): A Sound Design for Acoustic Feedback in Elite Sports. In: Ystad, S., Aramaki, M., Kronland-Martinet, R. & Jensen, K. (Eds.): Auditory Display. CMMR/ICAD 2009, Lecture Notes in Computer Sciences (LNCS) Vol. 5954. Springer-Verlag Berlin, pp. 143–165.

[37] Vogt, K., Pirro, D., Kobenz, I., Höldrich, R. & Eckel, G.. Physiosonic - Evaluated Movement Sonification as Auditory Feedback in Physiotherapy. In: Ystad, S., Aramaki, M., Kronland-Martinet, R. & Jensen, K. (Eds.): Auditory Display. CMMR/ICAD 2009, Lecture Notes in Computer Sciences (LNCS) Vol. 5954. Springer-Verlag Berlin, pp. 103–120.

[38] Pauletto, S. & Hunt, A.D., Interactive sonification of complex data (available online `http://dx.doi.org/10.1016/j.ijhcs.2009.05.006`). *International Journal of Human-Computer Studies; Special issue on Sonic Interaction*, Vol. 67, No. 11, pp. 923–933, November, 2009.

Index

3D sound, 493, 500, 501, 505

abstract data, 303
AcouMotion, 527, 538, 540
acoustic characteristics, 494, 497, 499, 500
acoustic ecology, 458, 473, 474, 479, 482
acoustic transformation, 548
action, 63
action-driven sound, 402
active sounds, 401
active use procedures, 137
Adams, Douglas, 147
Adapted Physical Activity, 527, 538
ADSL, 465, 466
aesthetic computing, 461, 473, 477
aesthetic experience, 155, 160
aesthetic oppositions, 163
aesthetic perspective space, 162
aesthetic premises, 163
aesthetic turn, 155
aesthetics, 145, 438, 439, 460, 463, 471, 477, 482, 484
    analytic, 159
    and art, 159
    and objective reality, 160
    and visualization, 156
    anti-sublime, 159
    at ICAD, 149–151
    beauty, 155, 156, 158, 160
    expressivity, 154
    first turn, 155
    for graph drawing, 158
    for sonification, 161, 162
    form and content, 156, 160
    function, 157, 165
    functionalism, 156
    in HCI, 163, 164
    in mathematics, 158
    in physics, 159
    interaction design, 160
    judgment, 157, 158
    metrics, 162
    of music, 162
    oppositions and premises, 163
    pattern recognition, 158
    perception, 157
    philosophy of art, 155
    pragmatist, 146, 159–161
    second turn, 156
    sensuous perception, 155, 158
    simplicity, 158
    skill, 164
    synaesthesia, relation to, 155
    talent, 164
    truth, 156, 158
    utility, 155
alarms, 13, 493
    in aeronautics, 493–496, 505
    spatialization, 500
alerts, 13
aliasing, 199
alpha inflation, 129
alpha level, 120
amateur radio, 279
ambient soundscape, 355
Ambisonics, 316
amusia, 118
anaesthetic, 155
analog to digital conversion, 199
analysis of variance (ANOVA), 127
analytic aesthetics, *see* aesthetics, analytic
analytical everyday listening, *see* listening, modes of
animal sounds, 304, 306
annnoyance, 471
annoyance, 159, 438, 473, 482
ARKola, 329, 459
Aroma, 159, 473
art, 15

art and science, 156, 166, 482
articulatory gestures, 80
ASPEC, 466
assistive technology, 431
astronomical data, 318
atonal music, 478, 482
attribute ratings, 122
audification, 17, 247, 288, 301, 547
    gestural control, 288
Audio Aura, 474, 478
audio games, 14, 526, 532
Audiograph, 448
audiomotor performance test, 538, 540
audiomotor sport game, 538, 546
AudioView, 468
audiovisual information, 550
audiovisual model, 550
auditory affordance, 326
auditory augmentation, 421
auditory cognition, 77
auditory computer games, 526, 532
auditory dimension, 64, 70
Auditory Domain Specification Language, *see* ADSL
auditory events, 78
Auditory Gestalt Formation, 422
auditory graphs, 14, 376, 446
auditory grouping, 51
auditory icons, 20, 23, 74, 325, 348, 350, 436, 459, 462, 469, 477, 499
auditory information, 547
auditory information-seeking, 512
auditory localization, 71, 72
auditory scene, 373
auditory scene analysis, 26, 28, 51, 75
auditory space, 71
auditory streaming, 51
auditory warnings, 326
auditory-visual interaction, 71, 72
aura, 516
aural fluency, 483
auralisation, 464, 466
automatic processes, 278
AWESOME, 151

Baby, 462
balanced interaction, 294
Ballancer, 332
bandlimited interpolation, 313
bandwidth, 433, 438, 443, 449
Baumgarten, Alexander Gottlieb, 155
beauty, *see* aesthetics, beauty
Ben Burtt, 326
between groups design, 119
binaural audio, 44
binaural localisation cues, 53
Blindminton, 542, 543
    one-player, 542
blood pressure monitoring, 345
body-instruments, 547
boxplot, 182, 188
braille, 434, 435, 445
BrailleNote, 446
brain plasticity, 443, 445
Bregman, Al, 75
    pond experiment, 402

Cage, John, 150, 306, 477
CAITLIN, 467
    semiotic aspects, 481
calibration, 252
calm technology, 462
car alarms, 500
cartoonification, 333
categorical sound dimension, 345
causal listening, *see* listening, modes of
causal uncertainty, 334
channel vocoder, 213
Chion, Michel, 146, 458
ChucK (computer language), 243
closed-loop
    control, 280
    interaction, 276
    system, 402
cochlea, 44
cochlear implant, 445
cocktail party problem, 41, 51
Cocteau, Jean, 306
cognition, 63
cognitive abilities, 29
cognitive approach, 494, 498, 499
command interface, 283
competitive sports, 527, 550
complexity, 422

computational complexity, 423
computer languages in sonification, 238
concatenative synthesis, 200
concurrent earcons, 347
cone of confusion, 54
confirmatory data analysis, 178
contact interactions, 419
content analysis, 122, 140
context, 26
continuous interactions, 419
control intimacy, 277
control loops, 277
convergent mapping, 371
Cordis-Anima, 424
correlation analyses, 126
Creative Commons, 156
Cronbach's alpha, 131
CSound, 242, 468
cumulative distribution function, 183

data, 239
    structure identification, 21
data analysis, 275
data crystallization sonification, 411
data mining, 419, 515
data parameter mapping, 345
data preparation, 367
data screening, 129
data sonograms, 404, 408
de Saussure, Ferdinand, 480
deafnesss, 445
demand characteristics, 113
DeMarini, Paul, 319
design, 145
    guidelines, 415
    practice, 146
detection, 28
Dewey, John Frederick, 160
diagnosis vs. interactive exploration, 276
digital audio, 198
digital to analog converter (DAC), 199
Dirac, Paul Adrien Maurice, 158
discrimination, 28
discrimination trials, 122
display duration, 114
display fidelity, 42
dissimilarity matrix, 124
dissimilarity rating, 123
dissipation, 405
distance perception, 55
distraction, 471, 473–475, 478
divergent mapping, 370
DSDM, 149, 154
dual task performance, 21
duplex theory of localisation, 54
dynamic model, 404

ear-witness accounts, 334
EarBenders, 334
earcons, 20, 23, 63, 74, 339, 436, 470, 475, 500
    concurrent, 347
    definition, 339
    design guidelines, 346
    semiotics, 481
EARCONSAMPLER, 151
earphones, 439
earthquakes, 463
Eclipse, 468, 469
EEG-data, 304, 317
effect size, 129
Einstein, Albert, 158
electroacoustic music, 145, 162, 477, 478, 482
Electronic Travel Aid (ETA), 434, 437, 440, 441
EMG, 526, 528
Emo-Map, 475
emotions, 92
entertainment, 14
environmental sounds, 81, 499, 502
epistemic things vs. technical things, 239
equal loudness contours, 47
equilibrium, 406
ergonomic approach, 503, 504
ergonomics, 422
ethical treatment of participants, 117
evaluation, 111
everyday listening, 400
evolutionary advantages, 49
evolutionary algorithms, 291
evolutionary objects, 331
experimental method, 237
experimenter bias, 113

expert appraisals, 140
exploratory data analysis, 175, 178, 399, 419
exploratory inspection, 21

false alarms, 494, 504
Fast Fourier Transform (FFT), 206
fatigue, 159, 471, 473, 475, 478
feedback, 550
filtering, 314, 514
finite state machine, 263
Fischinger, Oskar, 305
Fletcher-Munson curves, 47
flow, 278
fMRI, functional magnetic resonance imaging, 443
focus groups, 140
force-feedback, 356
formant, 216, 379
formant wave functions (FOFs), 218
Fourier analysis, 204
Fourier synthesis, 204
frequency modulation (FM) synthesis, 219
frequency place code, 45
frequency tuning, 45
friction, 405
fully formed objects, 331
functional artifacts, 334
fuseONE, 461, 474
fuseTWO, 461, 474

Geiger counter, 509
general acoustic data, 302
Gist, 512
Global Music, The World by Ear, 151
Gnopernicus, 437
goalball, 538, 540, 545
GPS, global positioning system, 445
granular synthesis, 221, 246
graphical user interface, 343, 352
grounded theory, 161
Growing Neural Gas sonification model, 413
GUI, graphical user interface, 284, 435, 437
guide cane (white cane), 437–441, 450

haptics, 437, 440, 447
harmonic series, 204
harmonics, 204
HCI, 163
  definition, 274
  history, 283
  sound in, 285
head-related transfer function (HRTF), 43, 444
hearing examination, 118
hearing range, 547
Hegel, Georg Wilhelm Friedrich, 155, 156
hierarchical cluster analysis, 132
histogram, 182
history of sonification, 303, 316
holistic encoding, 402
home notification system, 355
House of Cards, 154
HRTF, 484, 511
human interaction
  principles, 276

identification tasks, 120
immediate response, 403
inclined plane experiment, 267
indexicality, 156
information, 400
  richness, 403
information theory, 177
information visualization, 154, 156, 176
InfoSound, 464
initial state, 406
inner ear, 44
institutional review board (IRB), 117
instrusion, 473
intensive care units, 495
intentional active sounds, 401
interaction, 16, 275, 418
  auditory icons, 287
  earcons, 287
  ergonomics, 295
  guidelines, 282
  quality, 277
  with audification, 288
  with MBS, 291
  with PMSon, 289
  with sonification, 286
  with sound, 279

with tools, 278
Interaction Design, 87
interactions
continuous, 406
punctual, 406
interactive art, 99
interactive programming, 249
interactive sonification, 273, 514
definition, 274
guidelines, 293
sonic views, 293
interaural level difference (ILD), 53
interaural time difference (ITD), 53
interdisciplinarity, 2, 238
interfacing and external control, 249
intersensory redundancy, 547
interstimulus interval, 114
intrusion, 471, 473–475
intuitiveness, 423
invariance, 403, 420
IPA, 161
iPad, 437
iPhone, 437

Jaws, 436
JITLib, 249
just noticeable difference (JND), 255

'K' Sonar, 439, 441, 449
K-means clustering, 184
Kant, Immanuel, 155
beauty, 160
kernel density plot, 187
kernel smoothing, 183
kinetic energy, 405
Knuth, Donald, 462
Kubisch, Christina, 319

landmarks, 511
Laser Cane, 439
learning, 294, 422, 498, 501–503
linear predictive coding (LPC), 215
link-variables, 407
Linux, 437
listener characteristics, 408
ListenIN, 460, 474
listening
modes of, 146, 400, 457, 458
quatre écoutes, 147
analytical everyday, 400
causal, 146, 458
everyday, 457, 458
musical, 400, 457
reduced, 146, 458
semantic, 146, 458
skills, 403
Listening to the Mind Listening, 150
LogoMedia, 464
LogoMotion, 464
long-term memory, 114
loudness, 66
loudness perception, 46, 256

Müller, Wolfgang, 306
Macintosh, 435, 437
Many Ears, 166
Many Eyes, 156, 166
'many-to-one' mapping, 371
mapping, 18, 23, 70, 153, 458, 511
function, 367
in MBS, 153
in musical instruments, 281
in parameter mapping sonification, 153
learning time, 282
'many-to-one', 371
'one-to-many', 370
'one-to-one', 370
polarity, 24
scaling, 25
topology, 370
mass-spring systems, 404
mathematics, 158, 159, 434, 447
Max/MSP, 243, 483
MBS, 14, 17, 291, 459, 483
application, 419
definition of, 403
dynamics, 405
excitation, 405
generality, 421
guidelines, 415, 416
implementation shortcuts, 407
initial state, 406
interaction, 418
link-variables, 407

listener characteristics, 408
mappings in, 153
setup, 404
vs. Physical modeling, 424
vs. PMS, 423
when to use, 415
McGurk effect, 71
Meastro (PDA), 445
medical diagnosis, 276
mental representation, 78, 498, 499, 502, 505
Menus, 284
Mercator, 159
metaphor, 406, 417, 448
micro narratives, 335
Microsoft, 435, 436
middle ear, 44
MIDI, 464, 466
Milhaud, Darius, 306
minimum audible angle (MAA), 54
minimum audible field (MAF), 46
mirror neurons, 80
missing fundamental, 48
Mobic, 446
mobile computing devices, 354
mobile telephone, 340, 345
Modal Synthesis, 209
mode, 210
model dynamics, 405
model excitation, 405
model setup, 404
Model-Based Sonification, *see* MBS
modes of interaction, 16
modes of listening, *see* listening, modes of
modulation sensitivity, 49
Moholy-Nagy, László, 305
Mondrian, Piet, 305
monitoring, 19
Moore, George Edward, 159
Morris, William, 155
motif, 467, 481, 482
motion capture, 548
motive, 340
motor behavior, 527
motor control, 527, 547, 549, 550
motor learning, 527, 547, 549, 550
motor rehabilitation, 527, 547, 550
motor representation, 527
mouse, 435, 436
movement parameter, 548
movement pattern, 547
movement sonification, 525, 527, 547–550
moving sounds, 56
Mowat Sensor, 439
MS-DOS, 435
multidimensional scaling (MDS), 132
multimodal
displays, 294
feedback, 89
widget toolkit, 354
multimodal displays, 21
multisensory information, 547
multisensory integration, 547
multivariate, 185
Music Monitor, 461
steganographic sonification, 483
music notation, 369
Music of Changes, 150, 477
musical ability, 29, 119
musical instruments, 275, 280
characteristics, 281
musical listening, *see* listening, modes of
musical scales, 79
musical training, 65
MusicN languages, 242
Musicons, 352
musique abstraite, 477
musique concrète, 147, 150, 162, 477, 478, 482

National Science Foundation, *see* NSF
NavBelt, 439–443
navigation, 73, 509, 512
navigation control loop, 510
NetBeans, 469
neural phase locking, 46
nomadic radio, 473
non-linear synthesis, 229
non-parametric models, 197
Nottingham Obstacle Detector, 439
NSF, 148, 154
number of stimuli, 115

Nyquist frequency, 312

object sonore, 373
observational motor learning, 547
obstacle avoidance, 437–443, 445
Ondes Martenot, 306
open-ended questions, 122
OpenSoundControl (OSC), 244
operator based sonification, 266
optimization (of models), 418
order and sequence, 252
order effects, 123
order of stimuli, 115
outer filtering, 43
outlier analysis, 128
OutSpoken, 437
OutToLunch, 462

paco, 334
parameter mapping sonification, 16, 153, 245, 246, 289, 363
parametric models, 197
parametricity, 198
partial sight, 433
participant bias, 113
participant recruitment, 118
particle trajectory sonification, 412
passive sounds, 400
pattern (SuperCollider), 248
pattern classification, 184
PCM, 198, 200, 312
Peep, 470
Peirce, Charles Sanders, 351, 480
perception, 63, 255
  and action, 80
perceptual
  dimension, 68
  dimension, 68
  domain, 372
  mappings, 130
  organization, 75
perceptualization, 3, 153, 176, 251
phase relation, 549
phase vocoder, 206, 315
phoneme, 201
phonemic restoration effect, 77
physical data, 302
physical manipulation, 275
physical modeling, 483
physical modeling synthesis, 223
Physically Inspired Stochastic Event Modelling (PhISEM), 221
physiotherapy, 525, 528
pilot testing, 116
pilots, 73
pitch, 65, 340
pitch perception, 48
place code (pitch perception), 45
Poème Electronique, 150, 477, 482
polarity, 24, 65
post-hoc tests, 128
potential energy, 405
power analysis, 119
practical difference, 129
practice trials, 117
Pragmatist aesthetics, *see* aesthetics, pragmatist
preattentive, 176
presence, 42, 72
principal component analysis (PCA), 258
principal curve, 410
principal curve sonification, 410
probability, 180
problem prioritization, 504
process monitoring, 420, 455
  bottling plant, 459
  bubble sort, 463
  debugging, 462, 464, 467
  domestic activities, 461
  domestic activity, 460, 474
  emotive associations, 475
  file sharing, 460
  group awareness, 462
  interfaces, 469
  Internet, 470
  modes of, 471, 473, 482
    direct, 455
    peripheral, 455, 461
    semiotic aspects, 481
    serendipitous-peripheral, 455, 461
  networks, 470
  objections to, 474
  programming, 462, 469
  robots, 459
  weather, 460, 475

web servers, 470
with earcons, 470
workspace, 461
Processing, 156
product sound design, 94
product sounds, 94
programming environments, 242
proximal cues vs. distal cues, 251
proximity ratings, 123
pseudo-interactivity, 289
psychoacoustics, 41
pulse code modulation (PCM), 198
Pure Data, 483
PureData, 243

quantization, 200
quatre écoutes, *see* listening, modes of

RAVE, 461
Ravenscroft Audio Video Environment, *see* RAVE
reaction time, 120
reading data files for sonification, 249
real time motion capture, 549
reduced listening, *see* listening, modes of
redundancy mapping, 70
register, 340
regression analysis, 136
reified theories, 240
Reim, 475
reliability analysis, 131
Renaissance music, 267
resampling, 313
residue pitch, 48
resonant filter, 212
reusability, 423
rhythm, 73
Rilke, Rainer Maria, 305
Russell Pathsounder, 439
Russolo, Luigi, 150, 477

sample size, 119
sampling, 200
sampling theorem, 313
SatNav, 445
scaling, 25
scanned synthesis, 424
scatterplot, 189
Schaeffer, Pierre, 146, 150, 306, 458, 477
Schafer, R. Murray, 458, 482
Schelling, Friedrich Wilhelm Joseph, 155
Schiller, Friedrich, 155
screen reader, 434–437, 449
Hal, 435
Jaws, 436
VoiceOver, 437
Windots, 436
WindowsBridge, 436
seismological data, 305, 317–319
semantic differential ratings, 122
semantic listening, *see* listening, modes of
semantics
neglecting of, 261
of sound, 401
semiotic triad, 480
semiotics, 22, 157, 349, 473, 480, 482, 484
interpretant, 480
meaning triangles, 480
modes of representation, 481
iconic, 481
indexical, 481
symbolic, 481
object, 480
Peircean, 480
representamen, 480, 481
Saussurean, 480
sign, 480
signified, 480
sensory memory, 114
SharedARK, 329, 459
ShareMon, 460, 477
shock wave, 404, 406
Shoogle, 292, 330, 415
SID, 87
sign, 350
signified, 350
signifier, 350
similarity ratings/scaling, 334
sketching, 98
Smartsight, 448, 449
Snow, C. P., 152
soft-buttons, 332
Sona, 439
sonar, 440–442

Sonic Interaction Design, 87
sonic loop, 402
sonic maps, 334
Sonic Torch, 439
sonic variability, 403
SonicFinder, 159, 286, 325, 469
sonification, 547
    and the arts, 305, 318
    definition of, 149, 152, 153
        by Hermann, 274
        by Kramer, 274
    laboratory, 237
    musical, 461, 464, 467, 469, 477
    networks, 470, 471
    of algorithms, 264
    of causal and logical relations, 264
    of graphs, 261
    of interfaces, 469
    of program slices, 468
    of programs, 462–469
    of trees, 263
    of vector spaces, 258
    realm of, 146
    recommender system, 291
    software, 320
    steganographic, 483
    toolkits, 241
    variables, 267
sonification design space map, 251
sonification models, 399, 403, 408
    macroscopic, 408
    microscopic, 408
sonification operator, 369
sonification time vs. domain time, 240, 267
Sonnet, 464, 465
sonograms, 404
sorting tasks, 124
sound design, 162
sound graph (auditory graph), 446, 447
sound localisation, 53
sound object, 373
sound pressure level (SPL), 46
sound recording data, 302
sound synthesis, 197
Soundgraph (word processor), 435
soundscape, 14, 15, 20, 462, 470, 474, 482
SoundShark, 329
Soundtrack, 435
space perception, 52
spatial cueing, 73
spatialization, 438
spearcons, 24, 74, 349, 351
SPL, 46
sport, 525, 527, 538, 539, 542
sport games, 527, 538
Star Wars, 326
statistically significant difference, 129
statistics, 175
steganography, 483
stereo, 438
stethoscope, 304
stiffness of strings and bars, 226
stock market data, 70
subject characteristics, 118
subject reliability, 115
subliminal interconnections, 550
subtractive synthesis, 213
SuperCollider, 237, 243, 471, 483
surveys, 138
SWAN, 511
Symbian, 446
synaesthesia, 155, 445
synth definition (SynthDef, SuperCollider), 245
system of alarms, 494, 505

t-test, 127
Tactons, 349, 357
TaDa, 148, 334
tangible data scanning, 409
tangible interaction, 296
Task (SuperCollider), 246
task time, 116
tasks, 19, 20
templates, 417
tempo, 74
temporal fine structure, 49
testing conditions, 116
The Tuning of the World, 482
Theory, 9
Theremin, 306

threshold shift, 48
timbre, 67, 340, 448
time, 239
tonotopicity, 45
top-down processing, 77
training, 30
Trautonium, 306
Trekker, 445
trend identification, 21
tuning, 374
Two Cultures, The, 152, 155
Two-pole filter, 212

ubiquity of sound, 402
UltraCane, 439, 440
unit generator (SuperCollider), 245
urgency level, 497
usability testing, 137

Varèse, Edgard, 150, 306, 477, 482
VBAP, 316
ventriloquist effect, 71
verbal protocol, 139
virtual auditory space, 44
virtual environment, 72
visual impairment, 30, 356, 433
visualization, 176
vocal, 379
vocoder, 206, 304, 315
Voice Project, 443, 444
VoiceOver, 437
VoiceSense, 446
vowels, 379

waveguide synthesis, 224
waveshaping synthesis, 219
wavetable synthesis, 200
wayfinding, 509, 510
WebMelody, 461, 471, 477, 478
Whitehead, Alfred North, 158
Wilde, Oscar, 145, 152, 155
Windots, 436
Windows Mobile, 446
Windows, Microsoft, 435, 436
WindowsBridge, 436
WISP, 461, 474
within groups design, 119
working memory, 114
Workspace Zero, 462